Experimental Thermodynamics
Volume I

Calorimetry of Non-reacting Systems

Experimental Thermodynamics
Volume I
Calorimetry of Non-reacting Systems

Prepared under the sponsorship of the
INTERNATIONAL UNION OF
PURE AND APPLIED CHEMISTRY
COMMISSION ON THERMODYNAMICS AND
THERMOCHEMISTRY

Editors

JOHN P. McCULLOUGH

Mobil Research and Development Corporation, Princeton,
New Jersey, U.S.A.

and

DONALD W. SCOTT

Bartlesville Petroleum Research Center, Bureau of Mines, Bartlesville,
Oklahoma, U.S.A.

LONDON

BUTTERWORTHS

Distributed in the United States by
CRANE, RUSSAK & COMPANY, INC.
52 Vanderbilt Avenue
New York, New York 10017

ENGLAND:	BUTTERWORTH & CO. (PUBLISHERS) LTD. LONDON: 88 Kingsway, W.C.2
AUSTRALIA:	BUTTERWORTH & CO. (AUSTRALIA) LTD. SYDNEY: 20 Loftus Street MELBOURNE: 343 Little Collins Street BRISBANE: 240 Queen Street
CANADA:	BUTTERWORTH & CO. (CANADA) LTD. TORONTO: 14 Curity Avenue, 16
NEW ZEALAND:	BUTTERWORTH & CO. (NEW ZEALAND) LTD. WELLINGTON: 49/51 Ballance Street AUCKLAND: 35 High Street
SOUTH AFRICA:	BUTTERWORTH & CO. (SOUTH AFRICA) LTD. DURBAN: 33/35 Beach Grove

IUPAC Publications

Chairman, Editorial Advisory Board: H. W. THOMPSON
Scientific Editor: B. C. L. WEEDON
Assistant Scientific Editor: C. F. CULLIS
Assistant Editor: P. D. GUJRAL

©
International Union of Pure and Applied Chemistry
1968

Suggested U.D.C. number: 536.62 : 541.11

Printed in Great Britain by Page Bros (Norwich) Ltd., Norwich

Foreword

The IUPAC Commission on Thermodynamics and Thermochemistry is entrusted with the task of defining and maintaining standards in the fields of study covered by its title. The establishing and surveillance of the international scales of temperature and pressure, the choice of reference standards for the calibration and testing of calorimeters of all types, the recommendation of proper procedures to be followed in making measurements of thermodynamic properties, and in the presentation of results, are in the forefront of Commission business. During recent years, the Commission has broadened its activities by the promotion of International Symposia on Thermodynamics and Thermochemistry in Europe and in North America, and by the publication annually of the *Bulletin of Thermodynamics and Thermochemistry*.

In pursuance of these broader aims, the Commission sponsored the preparation of two books on *Experimental Thermochemistry*, the first of which was published in 1956 (Interscience Publishers, Inc., New York: Frederick D. Rossini, Editor) and the second of which appeared in 1962 (Interscience-Wiley, New York–London: H. A. Skinner, Editor). Both these volumes aimed to present to investigators and would-be investigators in thermochemistry authoritative knowledge on the techniques then available for the accurate measurement of the heats of chemical reactions. In 1961 at Montreal, the Commission proposed that two further books to cover *Experimental Thermodynamics* should be prepared, and called upon Dr. John P. McCullough and Dr. Donald W. Scott to act as Editors of the first volume which you now have before you.

The twenty-four contributors, each distinguished as experimentalists, have freely given of their experience in the pages to follow. Thanks are due to each of them for their willing co-operation in this venture, and no less so to the Editors, faced with the task (common to Editors of all co-operative efforts) of synthesis of a book from the several independent parts. I am sure that readers interested in the art of measurement of thermodynamic properties will benefit from this volume, and many will find in it the means to improve their future contributions to experimental thermodynamics.

University of Manchester H. A. Skinner
Manchester 13, England *President*
October 1967 *Commission on Thermodynamics and Thermochemistry*
International Union of Pure and Applied Chemistry

Preface

In early 1961, the Subcommission on Experimental Thermodynamics of the IUPAC Commission on Chemical Thermodynamics considered the desirability of preparing books that would place before the scientific and technical world the best knowledge available relative to experimental thermodynamics. The books would serve the same purpose for that field as was served for the related field of experimental thermochemistry by the books on *Experimental Thermochemistry* prepared under the Subcommission on Experimental Thermochemistry, of which Volume I, edited by F. D. Rossini, had been published in 1956, and Volume II, edited by H. A. Skinner and published in 1962, then was in preparation. At its meeting in Montreal, Canada, 2–4 August 1961, the Subcommission on Experimental Thermodynamics embarked on the preparation of the first of such books, to be entitled '*Experimental Thermodynamics, Volume I*. It was to cover broadly the determination of heat capacity and enthalpies of phase change. To distinguish the subject matter from the calorimetry of chemical reactions treated in the two volumes on *Experimental Thermochemistry*, the subtitle "Calorimetry of Non-reacting Systems" was chosen. This book is the culmination of the effort embarked on at that time.

At the Montreal meeting, the Commission was renamed the Commission on Thermodynamics and Thermochemistry, and the two Subcommissions were abolished. This book was completed under the auspices of the renamed Commission. A second volume of this series, tentatively subtitled "Thermodynamics of Non-reacting Fluids," is in preparation under the editorship of Professor B. Vodar, Centre National de la Recherche Scientifique, Bellevue, France.

This book was planned to cover heat capacity determinations for chemical substances in the solid, liquid, solution, and vapor states, at temperatures ranging from near the absolute zero to the highest at which calorimetry now is feasible, and measurements of enthalpy changes for all kinds of phase transformations. The first four chapters are intended to supply background information and general principles applicable to all types of calorimetry of non-reacting systems. The remaining ten chapters deal with specific types of calorimetry. Most of the types of calorimetry treated have been developed over a considerable period and brought to a relatively sophisticated state. For such calorimetry, the approach adopted was to give detailed accounts of a few examples of apparatus and techniques representative of the best current practice in the field. For the few types of calorimetry still in the exploratory or pioneering stage, a general review of the field was considered more appropriate.

In planning and editing this book, actual duplication was avoided carefully, but some overlap was considered not only inevitable but desirable. Certain topics are relevant to two or more of the types of calorimetry covered in the different chapters. Because of their backgrounds in different types of calorimetry the authors treat such topics with different emphasis and from

PREFACE

different points of view. The different treatments should produce desirable "cross-fertilization" of ideas between the different areas of calorimetry.

The editors of this book wish to express their thanks to all the authors for their enthusiastic cooperation, and to acknowledge the continued encouragement given by Prof. G. M. Schwab and Sir Harry Melville as presidents of the Physical Chemistry Division of IUPAC, Lord Todd, F.R.S., and Prof. W. Klemm as presidents of IUPAC, and Dr. R. Morf as Secretary General of IUPAC. Special thanks are due the present and former presidents of the Commission and Subcommission, Dr. H. A. Skinner (1965–), Professor K. Schafer (1961–1965), and Professor D. M. Newitt (1959–1961); the present and former secretaries, Dr. S. Sunner (1965–) and Dr. Guy Waddington (1961–1965); and, in another capacity, Dr. Skinner, editor of the Commission (1961–1965).

Princeton, New Jersey   J. P. McCullough
Bartlesville, Oklahoma   D. W. Scott
October 1967

Contributors to this Volume

Th. Ackermann (*Germany*)
C. W. Beckett (*U.S.A.*)
A. Cezairliyan (*U.S.A.*)
H. Chihara (*Japan*)
J. R. Clement (*U.S.A.*)
A. J. B. Cruickshank (*U.K.*)
T. B. Douglas (*U.S.A.*)
G. T. Furukawa (*U.S.A.*)
P. A. Giguère (*Canada*)
D. C. Ginnings (*U.S.A.*)
L. Hartshorn (*U.K.*)
R. W. Hill (*U.K.*)

E. G. King (*U.S.A.*)
D. R. Lovejoy (*Canada*)
D. L. Martin (*Canada*)
J. P. McCullough (*U.S.A.*)
A. G. McNish (*U.S.A.*)
J. A. Morrison (*Canada*)
D. W. Osborne (*U.S.A.*)
H. F. Stimson (*U.S.A.*)
J. W. Stout (*U.S.A.*)
G. Waddington (*U.S.A.*)
E. D. West (*U.S.A.*)
E. F. Westrum, Jr. (*U.S.A.*)

Contents

Foreword v
Preface vii
Contributors to this Volume ix
Summary of Notation xvii

1. Introduction 1
DEFOE C. GINNINGS
    I. General Principles and Terminology 1
   II. Definitions and Symbols 4
  III. Heat Units and Constants 7
  IV. Standard Reference Substances 8
   V. Choice of Calorimetric Method 9
  VI. Summary 13
 VII. Bibliography and References 13

2. Temperature Scales and Temperature Measurement 15
H. F. STIMSON, D. R. LOVEJOY and J. R. CLEMENT
    I. Introduction 15
   II. Temperature Scales 16
      1. The Kelvin Temperature Scale 16
      2. Practical Scales 23
  III. Temperature Measurement 30
      1. Liquid-in-Glass Thermometry 30
      2. Electrical Thermometry 32
      3. Optical Pyrometry 50
      4. Vapor Pressure Thermometry 54
      5. Magnetic Thermometry 55
  IV. References 56

3. Energy Measurement and Standardization 59
L. HARTSHORN and A. G. MCNISH
    I. The Problem 59
   II. The Basis for Energy Measurement 60
  III. The Derived Mechanical Units 63
  IV. Working Standards for Dynamics 64
   V. The Electrical Units 65
  VI. Absolute Electrical Standards 66
 VII. Standard Inductors 68
VIII. Absolute Determination of the Ampere 70
  IX. Absolute Determination of the Ohm 72
   X. Working Standards 75
      1. Standard Resistors 76
      2. Standard Cells 77
  XI. Application to Thermodynamic Measurements 78

4. Principles of Calorimetric Design 85
D. C. GINNINGS and E. D. WEST
    I. Introduction 85
   II. Chemical, Mechanical, and Electrical Considerations 86
      1. Chemical 86
      2. Mechanical 87

CONTENTS

        3. Electrical .. 91
   III. Heat Flow Considerations .. 99
        1. General .. 99
        2. Heat Transfer by Radiation .. 100
        3. Heat Transfer by Convection .. 103
        4. Heat Transfer by Thermal Acoustical Oscillation .. 108
        5. Heat Transfer by Mechanical Vibration .. 108
        6. Heat Transfer by Conduction .. 108
   IV. Applications to Calorimeter Design .. 122
        1. Temperature in a Calorimeter Wall at Constant Heating Rate .. 122
        2. Uncertainty in Heat Leak Due to Temperature Gradients .. 124
        3. Methods of Minimizing Heat Leak Due to Temperature Gradients .. 124
        4. The Calorimeter Heater Lead Problem .. 127
   V. References .. 130

5. Adiabatic Low-temperature Calorimetry .. 133
Edgar F. Westrum, Jr., George T. Furukawa and John P. McCullough

   I. Introduction and Historical Survey .. 135
   II. Apparatus .. 137
        1. The Cryostat Environment .. 137
        2. The Calorimeter Vessel .. 150
        3. Measurement Techniques and Adjuvant Circuitry .. 156
   III. Procedures .. 173
        1. Preliminary Procedures .. 173
        2. Sequence of Experiments .. 173
        3. Crystallization Procedures .. 175
        4. Observational Procedures .. 176
        5. Melting Point and Purity Determinations .. 180
        6. Studies of Polymorphic Substances .. 181
        7. Vapor Pressure and Enthalpy of Vaporization Measurements .. 184
   IV. Data Reductions .. 184
        1. Treatment of Experimental Data .. 184
        2. Reduction of Observed Data to Molal Basis .. 186
        3. Smoothing the Experimental Data .. 195
        4. Extrapolation of Heat Capacity below about 10°K .. 203
        5. Calculation of Thermodynamic Properties .. 206
   V. General Discussion .. 207
        1. Other Modes of Operation .. 207
        2. Precision and Accuracy .. 208
        3. Calorimetry Conference Standards .. 208
        4. Comparison of Adiabatic Calorimetry with other Methods of Calorimetry 209
        5. Future Developments .. 210
   VI. Acknowledgement .. 211
   VII. References .. 211

6. Low-temperature Calorimetry with Isothermal Shield and Evaluated Heat leak .. 215
J. W. Stout

   I. Introduction .. 216
   II. Description of Typical Calorimetric Apparatus .. 217
        1. General Description of a Typical Cryostat .. 217
        2. A Cryostat for Calorimetry of Condensed Gases .. 221
        3. Calorimeters, Thermometers, and Heaters .. 224
        4. Equipment for Measurement of Resistance, Energy, and Time .. 229

CONTENTS

	5. Experimental Procedures	230
III.	Calculation of Heat Capacity Data	232
	1. Corrections for Heat Leak	232
	2. Electrical Energy Measurements and Calculations	251
	3. Correction for Heat Capacity of Empty Calorimeter, Solder, and Helium	255
	4. Temperature Scales	255
	5. Calculation of $C_P°$ and other Thermodynamic Properties	256
	6. Precision and Accuracy	259
IV.	Acknowledgements	259
V.	References	260

7. Calorimetry Below 20°K — 263
R. W. Hill, Douglas L. Martin and Darrell W. Osborne

I.	Introduction	263
II.	Cryostats for Calorimetry below 20°K	264
	1. Cryostats for Use between 4° and 20°K (with possible extension in each direction)	265
	2. Cryostats for Use in the Liquid Helium Range	266
	3. Cryostats for Measurements below 1°K	267
III.	Thermal Isolation of Calorimeter	272
IV.	Methods of Cooling Calorimeters	273
	1. With Helium Exchange Gas	273
	2. With a Condensing Pot	273
	3. By Waiting	274
	4. With Mechanical Heat Switches	274
	5. With Superconducting Heat Switches	278
	6. Comparison of Heat Switches for Use below 1°K	279
V.	Thermal Contact with Solid Samples	279
VI.	Thermometry	282
	1. Germanium Thermometers	282
	2. Measuring Circuits for Germanium or Carbon Resistance Thermometers	283
	3. Magnetic Thermometry	284
VII.	Isothermal versus Adiabatic Calorimetry	287
VIII.	Special Techniques below 1°K	288
IX.	References	290

8. High-temperature Drop Calorimetry — 293
Thomas B. Douglas and Edward G. King

I.	Introduction	293
	1. General Requirements	294
	2. Advantages and Disadvantages of the Drop Method	295
II.	The Furnace	296
	1. Furnace Design and Operation	296
	2. Temperature Measurement	304
	3. The Sample in the Furnace	306
III.	The Calorimeter	310
	1. The Isothermal Calorimeter	310
	2. The Isothermal-Jacket, Block-Type Calorimeter	318
IV.	Treatment of the Data	322
	1. Correcting to Standard Conditions	322
	2. Smoothing and Representing Enthalpy Values	325
	3. Derivation of Other Thermodynamic Properties from Relative Enthalpy	328
	4. Precision and Accuracy	329
V.	References	330

CONTENTS

9. Adiabatic Calorimetry from 300 to 800°K 333
E. D. West and Edgar F. Westrum, Jr. ..
   I. Introduction 333
        1. Methods of Calorimetry above 300°K 333
        2. Comparison with Low-temperature Adiabatic Calorimetry 335
   II. Heat Exchange in Adiabatic Calorimetry 335
   III. A High-precision Adiabatic Calorimeter Utilizing Intermittent heating over the Range 300 to 800°K 338
        1. The Calorimeter (Sample Container) 339
        2. The Calorimeter Heater 341
        3. The Adiabatic Shield 342
        4. Adiabatic Shield Temperature Control 345
        5. Environmental Control 347
        6. Measurement of Electrical Power 348
        7. Experimental Procedures 349
        8. Determination of Enthalpies of Phase Transitions 350
        9. Determination of Transition Temperatures 352
   IV. An Adiabatic Calorimeter for Thermal Measurements from 250 to 600°K 353
        1. Intermediate-Temperature Thermostat 353
        2. The Silver Calorimeter 356
        3. Operational Method 358
        4. Calibration of Calorimeter 359
   V. General Discussion 360
   VI. Survey of Adiabatic Calorimeters in the Intermediate and Higher Temperature Ranges 361
   VII. Acknowledgement 365
   VIII. References 365

10. Vapor-flow Calorimetry 369
John P. McCullough and Guy Waddington
   I. Introduction 370
        1. Brief Survey 370
        2. Constant Flow Methods 370
   II. Non-adiabatic Calorimeters 372
        1. Apparatus 373
        2. Operating Procedures 379
        3. Calculation of Experimental Results 383
        4. Derived Results 388
   III. Discussion 390
        1. Accuracy of Results 390
        2. Reference Substances 392
   IV. References 393

11. Calorimetry of Saturated Fluids Including Determination of Enthalpies of Vaporization 395
D. C. Ginnings and H. F. Stimson
   I. Introduction 395
   II. Theory of the Method 396
   III. General Calorimetric Design and Procedures 399
   IV. Calorimeters for Measuring Properties of Water 403
        1. Calorimeter for 0–270°C 403
        2. Calorimeter for 100–374°C 404
        3. Calorimeter for 0–100°C 407
        4. Comparison of results with the three calorimeters 413

CONTENTS

V.	Calorimeter for Hydrocarbons	413
VI.	Other Calorimetric Methods for Measuring Enthalpy of Vaporization	418
VII.	Summary	417
VIII.	References	419

12. Heat Capacity of Liquids and Solutions near Room Temperature .. 421
A. J. B. Cruickshank, Th. Ackermann and P. A. Giguère

I.	Survey of Experimental Methods	423
	1. General Considerations	423
	2. Methods of Limited Precision	425
II.	Adiabatic Heating Methods	435
	1. Theoretical Introduction	435
	2. Calorimeter Vessels	449
	3. Adiabatic Shields and Regulators	477
	4. Control Thermocouples for Adiabatic Shields	500
	5. Experimental Procedure	515
III.	Isothermal Drop Calorimetry	521
	1. Introduction	521
	2. Isothermal Drop Calorimeter Receivers	523
	3. Drop Calorimeter Furnaces and Vessels	529
	4. Procedure and Calculations	531
IV.	References	533

13. Calorimetric Studies of Some Physical Phenomena at Low Temperatures 537
H. Chihara and J. A. Morrison

I.	Introduction	537
II.	Thermodynamic Properties of Films Adsorbed on Solids	538
	1. Measurable Quantities	538
	2. Adsorption Calorimeters of Moderate Precision	538
	3. High Precision Adiabatic Calorimeters	540
	4. Measurements at Liquid Helium Temperatures	545
III.	Particle Size Effects	545
	1. Heat Conduction between Solid Particles	545
	2. Calorimetric Measurements	546
IV.	Stored Energy Experiments	547
	1. General Remarks	547
	2. Radiation Damage at Low Temperatures	547
	3. Clustering in Alloys	548

14. High-speed Thermodynamic Measurements and Related Techniques 552
Charles W. Beckett and Ared Cezairliyan

I.	Introduction	552
II.	Generation of Heat	554
	1. General Considerations	554
	2 Heating by Electrical Pulses	554
	3. Heating by Pulsed Radiation	556
	4. Electrical Circuitry Associated with Single Pulse Generators	557
III.	Measurement of Heat	558
	1. General Considerations	558
	2. Current Measurements	559
	3. Voltage Measurements	559
	4. Conclusions Regarding Measurement of Heat	561
IV.	Measurement of Temperature	561
	1. General Considerations	561

CONTENTS

	2. Photoelectric Temperature Measurement Techniques	564
	3. Photographic Temperature Measurement Techniques	565
	4. Other Temperature Measurement Techniques	567
V.	Other Measurements	568
	1. High-speed Photographic Techniques	568
	2. Dynamic Pressure Measurement Techniques	569
	3. X-ray and Interferometric Techniques for Density and Velocity Measurements	571
VI.	Applications of High-speed Measurements	572
	1. General Considerations	572
	2. Measurement of Properties	572
	3. Other Applications	577
VII.	Appendices	579
VIII.	References	581
Author Index		587
Subject Index		599

Summary of Notation

A	Area (also a); Helmholtz energy ($E - TS$)
A/E	Property of the thermal resistance within a calorimeter
a	Area (also A); ratio of total heat capacity to that of the empty vessel; temperature coefficient of resistance
B	Second virial coefficient
B_N	Newton's law coefficient of heat exchange (also k)
C	Capacitance; Curie constant; heat capacity (extensive or molal), often with appropriate subscripts, e.g.:
	$\quad C_P$ Heat capacity at constant pressure
	$\quad C_S$ Heat capacity at saturation pressure
	$\quad C_V$ Heat capacity at constant volume
C_1	First radiation constant
C_2	Second radiation constant
c	Molal heat capacity related to solutions, e.g.:
	$\quad \bar{c}$ Partial molal heat capacity
	$\quad \phi_C$ Apparent molal heat capacity
c	Specific heat (heat capacity per unit mass); speed of light
D	Debye function; diameter (also r)
d_M	Density of mercury
ds	Element of circuit
E	Einstein function; electrical energy; electrical potential, electromotive force, or voltage (also ε); thermoelectric power or thermocouple sensitivity; emittance (also ϵ); error term; internal energy (occasionally U)
F	Axial force; fraction of sample melted; mass rate of flow
F_{12}	Emissivity factor
f	Function
G	Galvanometer sensitivity, Gibbs energy ($H - TS$)
g	Acceleration of gravity
H	Enthalpy ($E + PV$); magnetic field strength
h	Heat transfer coefficient, interval of numerical integration; Planck's constant
I	Electrical current
i	Thermometer current (also I)
J	Intensity of radiation; mechanical equivalent of heat, Planck's function
J_T	Radiated power at temperature T
K	Calibration factor of a calorimeter; thermal conductivity (also k and λ)
k	Boltzmann's constant; calibration factor; distribution coefficient; Newton's law coefficient of heat exchange (also B_N); ratio of heating rates; temperature drift of shield; thermal conductivity (also K and λ); time constant (also m and τ)
L	Length (also l); specific enthalpy of vaporization (also Δhv)

SUMMARY OF NOTATION

l	Height, length (also L); width
M	Mass (also m and w); molecular weight or formula mass (also W); mutual inductance
m	Mass (also M and w); molal concentration; time constant (also k and τ)
N	Neumann's integral; mole fraction (also x)
N_{Gr}	Grashof number
$\mathbf{N}$	Avagadro's number
n	Number of moles; number of thermocouple junctions; number per unit time
P	Power; pressure; probable error
p	Power per unit length, area, etc.; vapor pressure
Q	Heat (usually q)
Q'	Integral heat (enthalpy) of absorption
q	Heat (occasionally Q)
q_{ST}	Isosteric (differential) heat (enthalpy) of absorption
R	Electrical resistance (occasionally r); gas constant; thermal resistance
r	Diameter (also D); distance; electrical resistance (usually R); radial distance; radius
S	Entropy
s	Standard deviation of the mean
T	Kelvin temperature; torque
T^*	Magnetic temperature
T_{TP}	Triple point temperature
t	Celsius temperature; time
t_W	Relaxation time (also α^{-1})
U	Energy, internal energy (usually E)
u	Speed of sound (also W)
V	Volume
V_λ	Visual relative luminous efficiency factor
v	Specific volume; speed
W	Molecular weight or gram formula mass (also M); ratio of thermometer resistances at two temperatures; speed of sound (also u)
w	Mass (also m and M); thickness (also x)
x	Mole fraction (also N); observation in general; thickness (also w)
Z	Impedance; vertical distance
z	Distance along axis of a cylinder
α	Coefficient of cubical expansion (also β); coefficient of linear expansion; defined property with dimensions of enthalpy; intrinsic temperature drift of a calorimeter; thermal diffusivity (also κ)
α^{-1}	Relaxation time (also t_W)
β	Coefficient of cubical expansion (also α); compressibility; heat exchange modulus; heating rate; vapor correction term with dimensions of enthalpy
γ	Correction factor; defined property with dimensions of enthalpy; ratio of heat capacities, C_P/C_V
Δ	Increment; Weiss constant or effective Weiss constant
$\Delta C_P m$	Premelting correction to heat capacity
$\Delta H m$	Enthalpy of fusion (melting)

SUMMARY OF NOTATION

ΔHv	Enthalpy of vaporization
Δhv	Specific enthalpy of vaporization (also L)
δ	Increment; inexact differential
ε	Electrical potential, electromotive force, or voltage (also E)
ϵ	Emittance (also E)
ζ	Thermal signal
η	Efficiency of thermocouple junctions; viscosity
Θ	Amplified thermocouple voltage for shield control; characteristic temperature
θ	Angle; temperature difference
κ	Thermal diffusivity (also a)
λ	Thermal conductivity (also K and k), heat-leak coefficient
ν	Frequency
ξ	Heat-leak coefficient (also λ)
ρ	Density
σ	Stefan-Boltzmann constant
τ	Duration of heating interval (2τ used also); time constant (also k and m)
υ	Shield heating (cooling) rate
ϕ	Regulator backlash
χ	Magnetic susceptibility; time delay or lag
ψ	Thermal response coefficient
ω	Angular frequency; electrical resistivity

CHAPTER 1

Introduction

DEFOE C. GINNINGS

National Bureau of Standards, Washington, D.C., U.S.A.

Contents

I.	General Principles and Terminology	1
II.	Definitions and Symbols	4
III.	Heat Units and Constants	7
IV.	Standard Reference Substances	8
V.	Choice of Calorimetric Method	9
VI.	Summary	13
VII.	Bibliography and References	13

I. General Principles and Terminology

The word *calorimeter* is derived from the Latin word for *heat* and the Greek word for *measure*. A calorimeter is accepted today to mean an instrument to measure the heat involved in a known change in state of a material. This can involve a change in phase, temperature, pressure, volume, chemical composition, or any other property of the material which is associated with the change in heat. Thus, it is possible that a calorimeter may really be used to measure several quantities besides heat. Ordinarily, however, in non-reacting systems, only one or two other quantities are significant. For example, in measuring the heat capacity of a solid at moderate temperatures and pressures, the temperature change in the solid is usually the only significant quantity, although the volume of the solid and the pressure on it have some slight effect.

A calorimeter can measure the heat involved either directly by comparison with a known electrical energy or indirectly by comparison with materials with known properties. Direct measurement usually involves the addition of a known amount of electrical energy in the form of heat, because with modern techniques, electrical energy can be measured more accurately and conveniently than other forms of energy. Frequently, one can avoid the use of electrical energy measurements by utilizing some thermal property which has already been measured electrically. An example of this is the "calibration" of a calorimeter, with a "standard" material such as aluminum oxide (sapphire), in heat capacity measurements. Whether direct or indirect measurements of heat are used, the heat involved in the change in state usually includes some heat (called *heat leak*) which is due to the non-ideality of the calorimeter and its application. This heat leak is usually significant for accurate heat measurements. The importance of heat leak in accurate calorimetry was pointed out in 1928 by W. P. White in his book *The Modern Calorimeter*[20]. He says "There is a difference of opinion as to whether thermal leakage is necessarily the chief source of error in calorimetry, but it is undoubtedly responsible for most of the experimental features and devices in accurate work". With increased emphasis on accuracy this is more true today than it was in 1928!

The term *calorimeter* is commonly used with two different meanings. In the broad sense, it is used to describe an entire calorimetric apparatus. Although this meaning is very useful in a general description, it is more useful in much of this book to limit it to a particular part of the calorimetric apparatus. Specifically, when the word calorimeter is used in this limited sense, it is defined as that part of the apparatus in which the change in energy is accounted for. Sometimes, such terms as *calorimeter proper* and *calorimeter vessel* have been used to have the above meaning.

The calorimeter as defined above in the limited sense can consist of either few or many parts, depending on the type of calorimeter, accuracy desired, and many other factors. The exact geometric location of the boundary between the calorimeter (used in this limited sense) and its surroundings usually has little significance because of the method of using the calorimeter, as described later in this book.

Calorimeters may be classified in several ways. One method of classification is according to which physical variables are kept constant. Thus *isothermal calorimeters*, as the name implies, are those in which there is ideally no temperature change during the experiment. The general usage of the term isothermal here implies constancy of temperature with time of any part of the calorimeter, rather than the uniformity of temperature over the calorimeter at any one time. Many isothermal calorimeters utilize a phase change to maintain constancy of temperature. The solid–liquid phase change is used in the ice calorimeter[5], the resulting volume change being used as a measure of either the heat added or removed. The liquid–vapor phase change also has been used[7, 18]. Other methods that have been used to keep constancy of temperature during the experiment include use of electric heating to balance removal of heat and electric cooling (Peltier effect) to balance the addition of heat[9, 14]. Sometimes a heat flow is provided to balance out a heat source. Such a calorimeter may be called a *heat flow isothermal calorimeter* or simply a *heat flow calorimeter*. In a heat flow calorimeter, the heat may be removed either by fluid flow, such as in a *flow calorimeter*, or by using heat leak by radiation[21], convection or conduction. Twin *"conduction" calorimeters* have been used by Calvet in the microcalorimetric study of slow thermal processes.

A second class of calorimeters are referred to as *adiabatic*. An adiabatic calorimeter is one which ideally has no heat transfer across its boundary. Strictly speaking, the term adiabatic is used in a vague sense because no practical calorimeter is truly adiabatic. A calorimeter can be made more adiabatic by reducing the heat exchange between the calorimeter and its surroundings. This can be done by (*i*) minimizing the temperature difference between calorimeter and surroundings, (*ii*) minimizing the heat transfer coefficient (sometimes called leakage modulus), or (*iii*) minimizing the time for the heat exchange. It is true that common usage implies mainly the

INTRODUCTION

minimizing of temperature difference when describing an adiabatic calorimeter, even though the other factors may be just as important. There is also a tendency in modern calorimetry to restrict the term adiabatic calorimeter to one which both changes in temperature and minimizes heat leak by control of the temperature of the surrounding surface of the shield to the same temperature as the calorimeter. The term *adiabatic shield* (or jacket) is used to describe the above shield. Such a restricted definition of adiabatic calorimeter would not include an ice calorimeter, even though its shield temperature is the same as that of the calorimeter. Possibly the greatest use of the term adiabatic calorimeter is to differentiate it from the calorimeter whose surrounding shield is kept at a constant temperature, rather than at the calorimeter temperature (*See* Chapter 6). The latter calorimeter could be named appropriately an *isothermal shield calorimeter* or *isothermal jacket calorimeter*. Since these names are somewhat cumbersome to use, this calorimeter has sometimes loosely and erroneously been called *isothermal* calorimeter. As an alternative name for this calorimeter, the term *isoperibol* has been suggested by Kubaschewski and Hultgren[15].

The use of the term adiabatic calorimeter has been questioned when electric energy passes through the calorimeter boundary. In this regard, it should be noted that the basic definition of adiabatic refers to heat, not any other form of energy. The accepted meaning of an adiabatic calorimeter permits measured electric energy to enter the calorimeter and be transformed to heat, as in a calorimeter heater. By carrying this principle even further, it would be possible in an adiabatic calorimeter to absorb γ-radiation from the surroundings, the resulting temperature change being a measure of the radiation energy[13]. Perhaps the dividing line in the case of radiation would depend on its frequency; that is, the radiation frequency must not be in the range produced by the temperatures involved in the calorimeter and shield.

Sometimes, a type of calorimeter is named after the scientist who first used it extensively. Examples are the *Bunsen ice calorimeter* (Chapter 8), which is an isothermal calorimeter using the ice–water phase change, the *Nernst calorimeter* (Chapter 6), which is the isothermal shield type (isoperibol) used frequently in combustion calorimetry and in low-temperature calorimetry, and the *Joule twin calorimeter* system, which uses two calorimeters differentially to minimize errors due to heat leak. Frequently, the latter calorimeters are called simply *twin* calorimeters or *differential* calorimeters. Of historical interest is the *Dewar low-temperature calorimeter* using evacuation and surfaces with low radiation emittance for insulation.

The *Calvet calorimeter*[3] is a twin microcalorimeter characterized by use of a metal block for isothermal surroundings, with evaluation of heat in one twin either by compensation (Peltier cooling or Joule heating) or by time integration of the temperature difference between the twins. The *Callendar radiation balance*[2], sometimes called *Callendar radio balance*, consists of twin microcalorimeters used to measure heats of radioactivity or solar radiation. Peltier cooling in a thermocouple is used to balance out the unknown heat.

The names of several calorimeters describe a characteristic property or function of the calorimeter. A *microcalorimeter* refers to a calorimeter used to measure small amounts of heat. It is not necessarily small in size[17] although in modern microcalorimetry it usually is. A *flow calorimeter* is one used with fluid samples flowing through the calorimeter. Of historic value are the names *continuous* or *continuous electric calorimeters* used to describe early flow calorimeters[1]. Flow calorimeters have been used both to measure the thermal properties of a fluid in terms of electric energy (*See* Chapter 10) and to measure a steady heat source, by using the known thermal properties of the fluid. An *aneroid calorimeter* is usually interpreted as one which depends on solid conduction rather than liquids for distributing heat. Thus the usual combustion bomb calorimeter, with its reaction vessel immersed in a stirred liquid bath, is not an aneroid calorimeter. One aneroid calorimeter is the *copper block calorimeter* which has been used frequently in the drop method for measuring enthalpies at high temperatures. Sometimes, a calorimeter which is used in the drop method is called a *receiving calorimeter*. The term *radiation calorimeter* has sometimes been used to describe an isothermal calorimeter in which heat radiated to the surroundings is used to balance a heat evolution in the calorimeter plus a measured electric heat. In the field of chemical reactions such names as combustion, bomb, reaction, flame, and solution calorimeters are used to describe the various types of reactions studied[15].

The *method of mixtures* uses a mixing process to measure the thermal properties of one material in terms of the known thermal properties of another material. For example, when a known mass of sample at a known temperature is "mixed" with a known mass of water at another known temperature, the mixture will come to thermal equilibrium at a third known temperature. By using the known heat capacity of water, the average heat capacity of the sample over its change in temperature can be calculated. Sometimes the method of mixtures is considered perhaps loosely to include the case of the "standard" material changing in phase, rather than temperature, such as in the ice calorimeter. In this case, the amount of the phase change, and its corresponding heat change, must be known.

As pointed out earlier, a large part of the effort in accurate calorimetry is usually devoted to either minimizing and/or measuring the heat transfer between the calorimeter and its surroundings. The terminology for the immediate surroundings varies considerably among different laboratories. Among the various names for the surroundings are *shield, jacket, guard, envelope, thermostat,* etc. The reader must depend upon the author of a particular chapter to define his terms for the calorimeter surroundings. The term *shield* will be used in this introduction to describe the calorimeter surroundings.

II. Definitions and Symbols

In calorimetry there are a few thermodynamic properties which relate directly to the experimentally measured heat. The simplest property is the *internal energy*, E, of a system, which depends only on the state of the system. Thus, the difference in internal energy, ΔE, between two states is independent of the "path" the system follows in going from one state to the other.

An example of a direct calorimetric measurement of E as a function of temperature is that of a gas at a constant volume. In this case, because no

INTRODUCTION

work is done by the gas on its surroundings, the heat input, q, corresponds directly to the change in internal energy, E, of the gas, so that $q = (E_2 - E_1)$.

Most calorimetry, however, is not carried out with the sample at constant volume. This is especially true with solids and liquids for which this procedure would be very difficult. Most calorimetry is carried out at approximately constant pressure so that the sample does work on the surroundings and the heat input does not correspond to the change in internal energy E. In this case, the heat measured, q, corresponds to the change in the thermodynamic quantity enthalpy, H, which is defined by $H = E + PV$, in which P and V represent the pressure and volume of the sample. In other words, at constant pressure P, $q = H_2 - H_1 = \Delta H = (E_2 - E_1) + P(V_2 - V_1)$. The quantity enthalpy was called *heat content* (or *total heat*) in earlier calorimetry. These older names are sometimes misleading because they describe a quantity whose change does not always correspond to the heat input. Such a case is the heat input, q, to a system in which the pressure changes so that $q \neq \Delta H$.

A system with changing pressure is frequently encountered in calorimetric measurements on liquids having significant vapor pressure (*see* Chapter 11). If the liquid sample is in equilibrium with its vapor in a calorimeter, the pressure in the calorimeter will vary with temperature.

The previous discussion covers measurements of heats corresponding to the three "paths", (*i*) constant volume, (*ii*) constant pressure, and (*iii*) "saturation" pressure (defined as equilibrium vapor pressure). In all these measurements if no transition or phase change is involved, a heat input q results in a temperature change ΔT. The ratio $q/\Delta T$ defines the term *average heat capacity* of the system over the temperature interval ΔT. The limit of this ratio as q is made indefinitely small is called the *heat capacity*, C, (sometimes *true heat capacity* or *thermal capacity*). Thus the dimensions of heat capacity are (energy) (temperature)$^{-1}$ and it is an extensive property. The three types of heat capacity defined by the three "paths" previously given are $C_V [= (\partial E/\partial T)_V]$, heat capacity at constant volume, $C_P [= (\partial H/\partial T)_P]$ heat capacity at constant pressure, and C_S, heat capacity at saturation pressure. The term *molal heat capacity* (or *molar heat capacity*) refers to the heat capacity of one gram-formula-weight of a material, and is an intensive property.

The term *specific* (*heat* or *specific heat capacity*) is frequently used interchangeably with heat capacity. Specific heat is generally interpreted as the *heat capacity per unit mass* of the sample, so that it is an intensive property. From these definitions, it is seen that the term heat capacity applies to *any* system, such as a calorimeter with many separate components. When the system is restricted to one mole of a compound, the term molal (or molar) heat capacity is used. In a strict sense, molal heat capacity is a specific heat, but the term specific heat is usually limited to other units of mass than one mole, such as one gram, etc. Although the distinction between the quantities heat capacity and specific heat is not always sharply defined, there should be no confusion if the authors give the units of the quantity. In most general discussions, they can be used interchangeably.

In the case of solutions involving two or more components, the terms *partial molal heat capacity* and *apparent molal heat capacity* are used to describe the behaviour of single components (*See* Chapter 12). In a similar way, *partial molal*, *apparent molal*, and *relative molal* are applied to other thermodynamic properties of solutions.

The term *water equivalent* has been used as the heat capacity of the calorimeter (sometimes called *energy equivalent*). In modern calorimetry, use of the term water equivalent has little meaning because heat units are not defined in terms of the heat capacity of water.

The terms *heat of vaporization* and *latent heat of vaporization* are both used to describe the heat required to transform unit mass of liquid to vapor, so they have dimensions of (energy) (mass)$^{-1}$. The older term latent heat is used very rarely in modern calorimetry. While in principle the term heat of vaporization can be applied to any process in which liquid is vaporized, it is generally interpreted to be limited to vaporization at a constant temperature and pressure. In this case, the term heat of vaporization, ΔH_v, is synonymous with *enthalpy of vaporization*. Since most heat of vaporization experiments correspond to enthalpies of vaporization, it is recommended that the term enthalpy of vaporization be used if it applies.

The terms *heat of fusion* and *latent heat of fusion* are analogous to the case of vaporization except they apply to the solid–liquid instead of the liquid–vapor phase change. Here again, the term latent heat is used rarely in modern calorimetry. The term *enthalpy of fusion* is synonymous with heat of fusion for a constant-pressure melting. Similarly, the term *enthalpy of sublimation* applies to a *heat of sublimation* at constant pressure.

The term *entropy*, S, applied to calorimetric experiments, is defined simply by the relation $dS = \delta q/T$, in which dS is the change in entropy in a *reversible* process and δq the heat involved at the temperature T. The term *free energy* has been used for many years to include the *Gibbs energy*, defined as $H - TS$ or *Helmholtz energy*, defined as $E - TS$. Of these two quantities, the Gibbs energy has been used more frequently. There has not been international agreement on symbols for these two quantities. European textbooks have used the symbol F and the name *free energy* or *Helmholtz energy* for the function, $E - TS$. These books also use the symbol G and names such as *Gibbs function*, *Gibbs energy*, or *free enthalpy* for the function, $H - TS$. On the other hand, in America, the quantity, $H - TS$, has been represented by the symbol F, and the quantity, $E - TS$, by the symbol A. As a result of the international confusion on these symbols, I.U.P.A.C. has recommended that the function, $H - TS$, be named the *Gibbs energy*, represented by the symbol G, and the function $E - TS$ be named the *Helmholtz energy*, represented by the symbol A. This notation will be used throughout in this book. A summary of notation used is given in the Appendix.

It seems appropriate to mention here the terms precision and accuracy for those who may not realize their difference in usage in modern science. The term *precision* is used to measure the *reproducibility* of experiments and may not be a measure of the accuracy of experiments. The precision may be so good that this source of error is negligible in comparison with the systematic error. On the other hand, the term *accuracy* is used to describe how close the experimental quantities are to the *true* quantities. In other words, systematic errors as well as random errors should be considered in estimating accuracy. In most calorimetric measurements it is believed that high precision is necessary in order to obtain high accuracy. In giving either precision or accuracy, the number given is necessarily statistical, so that it is imperative that the statistical basis be given in order that the number be meaningful. In early calorimetry, the term *average deviation from the mean* was used frequently as a measure of precision of a set of experiments. This term, defined as the average of differences between each measurement and the average for the set, is not nearly so useful and meaningful as the term *standard deviation of the mean*, which is commonly used now[14] as an indication of the precision of a set of experiments. Standard deviation of the mean, s, is defined:

$$s = \sqrt{\frac{\Sigma(x_i - \bar{x})^2}{n(n-1)}}$$

in which $\bar{x}$ is the average of the observations, x_i, and n is the number of observations. The term *standard deviation of the mean* should not be confused with the term *standard deviation of an individual experiment*, which is defined as $\sqrt{n}$ times the standard deviation of the mean. Since there has been some question as to which of the above two meanings the term *standard deviation* applies to, the term *standard error* has been suggested to be synonymous with standard deviation of the mean. Most of the literature uses σ as the symbol for the standard deviation. However, statisticians use the symbol s when derived from a *finite* number of experiments, so that $s \rightarrow \sigma$ when $n \rightarrow \infty$. *Probable error* has also been used as a measure of precision. Representing an error which has a 50 per cent chance of being exceeded, the probable error, P, is related to the standard deviation of the mean by $P = 0.6745\,s$, if there are a large number of observations. Another term is the *variance* which is the square of the standard deviation. This term is rarely used because it is not convenient to compare with the original measurements. The term *confidence limit* is used sometimes as a measure of the reproducibility of results. If an infinite number of observations have been made, the relation between confidence limits and standard deviation of the mean is fixed by the normal distribution curve. For 50 per cent confidence limit, error $= 0.6745\,s$; for 90 per cent confidence limit, error $= 1.64\,s$; for 95 per cent confidence limit, error $= 1.96\,s$; for 99 per cent confidence limit, error $= 2.58\,s$. A more complete discussion of assignment of calorimetric uncertainties has been given previously[17].

III. Heat Units and Constants

In early work in calorimetry in which most measurements were made near room temperature, it was convenient to both measure and express the heat in terms of a unit called the calorie which was defined as the heat required to raise the temperature of one gram of liquid water one degree C. As calorimetric measurements became more accurate, it was realized that the magnitude of the calorie so defined depends on the temperature of the water. Consequently, a variety of "calories" originated, such as the 4°-calorie, the 15°-calorie, the 20°-calorie, the mean (0–100°) calorie, etc.

Although the comparative heat measurements could be made with relative ease, it was recognized very early that it was necessary to determine the quantity of energy corresponding to a given calorie. The energy corresponding to a specific calorie was called J, the *mechanical equivalent of heat*[10].

Investigations of this quantity were carried out over most of a century with either mechanical or electrical sources of energy.

With the development of accurate electrical standards and instrumentation, it became possible to measure electrical energy more precisely than heat could be measured in terms of the heat capacity of water. After this time, the calorie and the mechanical equivalent of heat had little significance except as measurements of the heat capacity of water. Out of habit, however, the calorie continued to be used as a heat unit, so that it became desirable in accurate calorimetry to separate the definition of calorie from the heat capacity of water. As a result, there are two "dry" or defined calories used today[11]. In engineering steam tables, the *I.T. calorie* (International Table calorie) is used and defined as 4·1868 joules. In the field of calorimetry as applied by physicists and chemists, if the term calorie is used, it usually refers to the *thermochemical calorie*, which is defined as 4·1840 joules. The joule referred to here is by international agreement the absolute joule which was accepted in 1948 and is discussed in Chapter 3.

The name *mechanical equivalent of heat* is now obsolete, because it refers merely to a conversion factor to change from one to another unit of heat. Although the term calorie has lost its original utility, its use has persisted in many fields. No doubt it will be used for a long time, but logically it is destined ultimately for obsolescence along with many other superfluous units. It should be noted that in a Resolution which was adopted in 1948 by the 9th International General Conference on Weights and Measures[16], the recommendation on the unit of heat was the joule, and that it was requested that the results of calorimetric experiments be expressed in joules when possible. If the calorimetric results are expressed in units other than joules, the unit should be defined in terms of the joule.

IV. Standard Reference Substances

In evaluating the results of calorimetric research, uncertainties in the final results may arise from two sources, namely from errors in the calorimetry itself (energy, temperature, mass, etc.), or from uncertainties in the purity and state of the material whose properties are measured. In order to minimize the errors in the calorimetry, it may sometimes be a considerable advantage to check the reliability of the calorimetric measurements with a "reference" material whose state is reproducible and known. If the thermal properties of this reference material are known, this check yields evidence on the reliability of the calorimetric measurements. Of course, this requires that someone must have made accurate measurements on this reference material. When moderate calorimetric accuracy is adequate, it may not be difficult to find a reference material whose thermal properties have been measured with higher accuracy. When high calorimetric accuracy is required, this becomes more difficult because the accuracy of other calorimetric measurements may not be significantly better. However, even in this case, a reference material is useful. By having samples of any one material taken from one source of high purity, one has a means of *comparing* calorimetric measurements made in different laboratories under different experimental conditions. It was with this in mind that in 1949 the Fourth Calorimetry Conference[4] recommended three substances to be used as "standard" reference

INTRODUCTION

materials for heat capacity measurements. These three substances were benzoic acid, n-heptane, and aluminum oxide (corundum). Benzoic acid and n-heptane were intended for the moderate to low temperature range, whereas aluminum oxide was intended primarily for the high temperature range, although it is also useful at lower temperatures.

For calorimetric measurements near room temperature, water has always been used as a reference material, partly because of its universal availability in a pure state, and partly because of the numerous measurements on its heat capacity. In the range 0–100°C, the specific heat of liquid water is believed to be known to within 0·01–0·02 per cent[12].

When using standard reference materials, some experimenters regard their use as a "calibration" of their calorimeters, with the conclusion that when using their calorimeters as a comparison device, the accuracy on the unknown sample is comparable to the accuracy of the standard reference materials. The experimenter should be warned against this conclusion. Although it is true that a comparison method does reduce certain errors, the degree of reduction depends upon the experimental conditions. The comparison method is no "cure-all" for experimental defects! The comparison method is most effective in reducing errors when the physical properties and amount of the material being investigated are identical with those of the standard reference material. Departure from this ideal results in errors whose magnitudes depend upon the particular calorimetric apparatus. Even when the material has identical physical properties, there is always the possibility of the variation of the performance of a calorimetric apparatus from day-to-day.

V. Choice of Calorimetric Method

The first question raised after embarking on a calorimetric project is the choice of method. There are numerous factors which should be considered in choosing a calorimetric method. Some of these factors will be discussed now.

One of the most important factors to be considered is the accuracy required. The effort required increases exponentially as higher accuracy is desired. Many calorimetric measurements near room temperature can be made with 1 per cent accuracy with an ordinary household Dewar flask and very simple instrumentation[15]. Measurements to 0·1 per cent accuracy, if possible, usually require at least one and possibly two or three orders of magnitude increased effort, depending upon circumstances. Measurements accurate to 0·01 per cent are usually impractical with most calorimeters, although there are special examples near room temperature in which this accuracy has been approached with great effort[12]. At extreme temperatures or other conditions, the difficulties in attaining 1 per cent may be as great as attaining 0·01 per cent near room temperature.

Another important factor related to the accuracy required is whether to make absolute or relative calorimetric measurements. If the accuracy required is not as high as the accuracy obtained in published calorimetric measurements on a certain material, then it may be advantageous to use this reference material to "calibrate" a less complicated calorimeter. In this regard, one must keep in mind that this "calibration" usually only partially compensates for calorimeter errors, depending upon a number of factors. If possible, a reference material should be chosen which has thermal properties similar to those of the material being investigated.

The physical properties of the sample and its thermal property to be measured usually have considerable influence on the choice of method and calorimetric design. A flow method may be used with advantage to measure the heat capacity of a gas which requires large volumes to give the desired mass. The temperature and pressure range involved in the measurements are important, as well as the amount of sample available. If only a few milliliters of solid or liquid sample are available and if a large number of samples are to be measured, the use of the drop method offers simplicity in changing samples. The drop method is most useful for deriving heat capacity values if the heat capacity is not changing rapidly with temperature.

If the measurements involve very small heats, it may be advantageous to use twin calorimeters to minimize errors introduced by uncertainties in heat leak. In microcalorimetry, twin calorimeters are usually used for this reason. In measurements for which the time of experiment is necessarily very long, it may be necessary to choose a method having small heat leak (power) uncertainty.

As a practical matter, the choice of method is frequently determined by available apparatus and personnel. If a minimum of electrical measuring equipment is available, an isothermal calorimeter (such as the Bunsen ice calorimeter) might be considered because it does not require electrical instrumentation. Choice of a method using a platinum resistance thermometer for temperature measurement might not be possible if an expensive resistance bridge (or potentiometer) is not available. As a general rule, an adiabatic calorimeter would require more instrumentation (or personnel) than either an isothermal calorimeter or an isothermal shield calorimeter. The quality of personnel available may be a determining factor in choice of a method and apparatus. It is of little value to use a very elaborate and complicated apparatus if qualified personnel are not available for its operation.

It seems appropriate here to consider some of the advantages and disadvantages of various calorimetric methods. There is no "best" method for a general calorimetric problem. Even for a very specific problem, there may be no significant over-all advantage of one method over another. Consider first the *isothermal shield calorimeter* (*isoperibol calorimeter*) which has been used so extensively. Here, the calorimeter may change temperature (such as in heat capacity experiments) while the shield is kept at a constant temperature. One obvious advantage of this method is the simplicity of the operation of the shield, controlled either manually or automatically. Another advantage may be the relative independence from variable thermal contact in the shield, since it is at constant temperature. On the other side of the ledger, the corrections for "heat leak" from the calorimeter may be relatively large in some experiments, with the requirement that the heat leak coefficient (leakage modulus) be both constant and accurately known. In this regard, it should be pointed out that it is only the uncertainty in the heat leak correction which affects the results, not the magnitude of the correction. This is no doubt why in combustion bomb calorimetry, the isothermal shield

INTRODUCTION

has been used almost universally. The rapid change in the calorimeter immediately following the combustion has made it very difficult to control the shield temperature to that of the calorimeter. Even though the actual heat leak corrections could no doubt be reduced by changing the temperature of the jacket, the uncertainties in the heat leak might be larger than if the jacket is kept at a constant temperature. With modern instrumentation and advanced jacket design this situation may not be the same! Isothermal shield calorimeters for measuring heat capacities are still used for accurate measurements in low temperature calorimetry for which heat leak becomes less important owing to the decrease in radiative heat transfer. The use of isothermal shield calorimeters at higher temperatures depends upon successfully keeping heat leaks down to a reasonable value. In a very special type of calorimetric experiment, it is possible to use an isothermal shield at room temperature and still measure heat capacities at high temperatures with little heat leak error[19]. However, this is accomplished by making the time of the experiment so short that the total energy transferred from the sample (wire) is still small compared to the energy used to heat the wire.

The *adiabatic calorimeter* has increased in use along with the availability of modern electronic instrumentation for temperature control and recording. However, the design of the calorimeter and its auxiliary parts is at least as important as the external instrumentation in obtaining good temperature control. For example, large thermal "lags" in parts of the apparatus can make it difficult if not impossible to control an adiabatic calorimeter properly. Also, the calorimeter becomes more difficult to use in going to higher temperatures, because of increased heat leak coefficients. At 500°C, it is very difficult to obtain 0·1 per cent accuracy and at higher temperatures difficulties are much greater. Consequently, most of the heat capacity measurements in the range above 500°C have been made by using the "drop" method which keeps the heat leak small by allowing unwanted heat transfer for only a very short time during the drop.

The *drop method* is an example of the method of mixtures in which the receiving calorimeter measures the change in heat in the sample between its initial and final temperatures. The principal application of the drop method has been in measuring enthalpies at high temperatures. With the drop method, the initial temperature of the sample (in a furnace) is varied, but the final temperature is kept essentially constant, usually near room temperature. This restriction of varying only one temperature places an important limitation on the application of the drop method. For this method to be valid requires that the sample be at a thermodynamically reproducible state at both temperatures. If the sample (or its container) has a solid–solid transition, it is possible that the transition will not be complete in the time of drop so that some energy of transition will be "frozen in"[6]. Consequently, whenever the drop method is used with materials having such transitions, the experimenter must be aware of this possibility of error. Another disadvantage of the drop method is that it obtains values of heat from differences in heats which may be quite large by comparison. For example, it may be necessary to measure over-all heats (relative enthalpies) to 0·01 per cent to obtain 0·1 per cent accuracy on the enthalpy derivative, heat capacity, if this heat capacity changes significantly with temperature. This disadvantage is partly overcome in the drop method because the calorimeter is usually operated in a favorable temperature range (room temperature) so that measurements can be made very precisely. The drop method has an important advantage in its ease of changing from one sample to another. Sometimes, the drop method has been applied to enthalpy measurements at lower temperatures than the "receiving" calorimeter[8]. The principal value of this is its use when a receiving calorimeter already exists for measurements at higher temperatures. The drop method usually finds little application at low temperatures because (*i*) the heat capacity frequently changes more rapidly than at high temperatures, and (*ii*) the quantities of heat are smaller.

The *isothermal calorimeter* has an advantage that its heat capacity does not affect the results since its temperature does not change. Also the problem of minimizing uncertainty in heat leak is frequently made easier. In the ice calorimeter, for example, the heat in the inside ideally goes to melting ice and the outer part of the calorimeter is unaffected. In the history of calorimetry, this advantage has been considerable; however, with modern instrumentation, it is not so great. Even today, however, the isothermal calorimeter offers excellent heat leak control without elaborate instrumentation. With experiments of long duration, this control may be necessary. The isothermal calorimeter using vaporization (liquid–vapor equilibrium) may provide high sensitivity. If liquid hydrogen or helium is used, for example, a very small quantity of heat results in the evolution of a large volume of gas to be measured. This is partly because the heats of vaporization of hydrogen and helium are relatively low whereas the specific volumes of gas are large.

Recent developments in thermoelectricity have provided calorimetry with a useful tool with the availability of semi-conductor materials for refrigeration. With available thermocouples, it is now possible to make an isothermal calorimeter to measure heat quantities without using a change in phase of a material. The heat in a calorimeter can be balanced out with thermoelectric cooling to provide isothermal conditions, and the thermoelectric device can be calibrated in place with an electric heater. Even before semi-conductor thermocouples were available, metallic thermocouples were used in microcalorimetry for heat compensation[9, 14].

The use of *twin calorimeters* may be a real advantage if the necessary control of heat leak is difficult. In microcalorimetry, in which only very small heats are measured, the use of twin calorimeters minimizes the effect of changes in temperature of the environment. To be effective, the twin calorimeters should be as nearly alike as possible and located so that they are affected the same by their environment. One disadvantage of the twin calorimeters is the effort required to build an extra calorimeter and in many cases to prove that the two calorimeters have identical thermal characteristics. Another advantage of twin calorimeters is that differential thermo-elements can be used to best advantage for measuring the temperature difference between twin calorimeters. In most calorimetry involving large amounts of heat, twin calorimeters are not used.

Flow calorimeters have been used with both gases and liquids. When used to measure heat capacities, the flow calorimeter offers an advantage that the heat capacity of the calorimeter is not involved since ideally the

INTRODUCTION

calorimeter does not change temperature during an experiment. Although the flow calorimeter was used in early calorimetry on liquids, its principal use in accurate calorimetry today is with gases. Here, the flow method provides a relatively large volume of sample per experiment, which would be impractical with non-flow calorimeters.

VI. Summary

The principal fundamental physical quantities measured in calorimetry are temperature, energy, and mass. Whereas the first two are covered in Chapters 2 and 3, the techniques for measuring mass are so generally known or available, that no treatment is given here. In calorimetry it is frequently necessary to measure other quantities such as pressure and volume. Since measurements of these two quantities are usually secondary in calorimetric measurements, a discussion of them will be deferred to Volume II, which will cover P–V–T and transport properties.

For the student in calorimetry, a brief bibliography is given. The field of reacting systems has recently been covered[14, 15] and some of the techniques are applicable to non-reacting systems. The specialized field of microcalorimetry is covered in two books in the following bibliography.

VII. Bibliography and References

Bibliography

White, W. P., *The Modern Calorimeter*, Chemical Catalog Co., New York (1928).
Swietoslawski, W., *Microcalorimetry*, Reinhold, New York (1964).
Calvet, E., and H. Prat, *Microcalorimétrie*, Masson et Cie, Paris (1956).
Roth, W. A., and F. Becker, *Kalorimetrische Methoden zur Bestimmung chemischer Reaktionswärmen*, F. Vieweg, Braunschweig (1956).
Rossini, F. D. (Ed.), *Experimental Thermochemistry*, Vol. I, Interscience, New York (1956).
Weissberger, A. (Ed.), "Calorimetry" in *Technique of Organic Chemistry—Vol. I. Physical Methods of Organic Chemistry*, Chap. X, Interscience, New York (1959).
Skinner, H. A. (Ed.), *Experimental Thermochemistry*, Vol. II, Interscience, London (1962).

References

[1] Callendar, H. L., *Phil. Trans. Roy. Soc. (London)*, **A199**, 55 (1902).
[2] Callendar, H. L., *Proc. Phys. Soc. (London)*, **23**, 1 (1911).
[3] Calvet, E., and H. Prat, *Microcalorimétrie*, Masson et Cie, Paris (1956).
[4] *Chem. Eng. News*, **27**, 2772 (1949).
[5] Ginnings, D. C., T. B. Douglas, and A. F. Ball, *J. Research Natl. Bur. Standards*, **45**, 23 (1950).
[6] Ginnings, D. C., *J. Phys. Chem.*, **67**, 1917 (1963).
[7] Kraus, C. A., and J. A. Ridderhof, *J. Am. Chem. Soc.*, **56**, 79 (1934).
[8] Lucks, C. F. and H. W. Deem, *Special Publication No. 227, Am. Soc. Testing Materials* (1958).
[9] Mann, W. B., *J. Research Natl. Bir. Standards*, **52**, 177 (1954).
[10] Mueller, E. F., *Mech. Eng.*, **52**, 139 (1930).
[11] Mueller, E. F., and F. D. Rossini, *Am. J. Phys.*, **12**, 1 (1944).
[12] Osborne, N. S., H. F. Stimson, and D. C. Ginnings, *J. Research Natl. Bur. Standards*, **23**, 197 (1939).
[13] Pruitt, J. S., and S. R. Domen, *J. Research Natl. Bur. Standards*, **60A**, 371 (1962).
[14] Rossini, F. D. (Ed.), *Experimental Thermochemistry*, Vol. I, pp. 239, 297, Interscience, New York, (1956).
[15] Skinner, H. A. (Ed.), *Experimental Thermochemistry*, Vol. II, pp. 343, 189, Interscience, London (1962).
[16] Stimson, H. F., *Am. J. Phys.*, **2**, 617 (1955).
[17] Swietoslawski, W., *Microcalorimetry*, Reinhold, New York (1946).
[18] Tong, L. K. J., and W. O. Kenyon, *J. Am. Chem. Soc.*, **67**, 1278 (1945).
[19] Wallace, D. C., P. H. Sidles, and G. C. Danielson, *J. Appl. Phys.*, **31**, 168 (1960).
[20] White, W. P., *The Modern Calorimeter*, p. 17, Chemical Catalog Co., New York (1928).
[21] Wittig, V. F. E., and W. Schilling, *Z. Elektrochem.*, **65**, 70 (1961).

CHAPTER 2

Temperature Scales and Temperature Measurement

H. F. STIMSON

National Bureau of Standards, Washington, D.C., U.S.A. (Retired)

D. R. LOVEJOY

National Research Council, Ottawa, Canada

and

J. R. CLEMENT

U.S. Naval Research Laboratory, Washington, D.C., U.S.A.

Contents

I.	Introduction	15
II.	Temperature Scales	16
	1. The Kelvin Temperature Scale	16
	A. Realization of Thermodynamic Scale by Gas Thermometry	19
	B. Realization with Speed of Sound	22
	2. Practical Scales	23
	A. International Temperature Scales	23
	B. Provisional Temperature Scale 10°K to 90°K	28
	C. 1958 He[4] Scale of Temperature	29
	D. 1962 He[3] Scale	30
III.	Temperature Measurement	30
	1. Liquid in Glass Thermometry	30
	2. Electrical Thermometry	32
	A. Resistance Thermometry	32
	(1) Platinum Resistance Thermometers	32
	(2) Semiconductors	34
	(a) Thermistors	35
	(b) Germanium Thermometers	35
	(c) Carbon Thermometers	35
	(3) Effects of Magnetic and Radio Frequency Fields	37
	B. Thermoelectric Thermometry	37
	C. Electrical Measurements in Thermometry	40
	(1) Bridge Methods	41
	(a) Mueller Bridge	42
	(b) Smith Bridge	44
	(c) General Remarks	45
	(d) Bridges for Semiconductors	46
	(2) Potentiometric Methods	47
	3. Optical Pyrometry	50
	4. Vapor Pressure Thermometry	54
	5. Magnetic Thermometry	55
IV.	References	56

I. Introduction

Temperature may be regarded as the degree of hotness. When the temperature of a body is measured, a value is obtained to assign to its hotness.

In order to measure temperatures so that they can be reproduced accurately at some other time or place, there must be a precisely defined scale on which to assign values to these hotnesses. For experimental thermodynamics it would be ideal if all temperatures could be expressed on the Kelvin scale, because it is an absolute scale based on the laws of thermodynamics, and is independent of the properties of any substance. It is inherently difficult, however, to measure temperatures on this scale; hence practical scales have to be defined so that the measurements can be precise and reproducible. In general, these practical scales are defined to correspond to the Kelvin scale as closely as can be agreed upon by first assigning values to certain reproducible temperatures (fixed points) and then specifying the means for interpolating values of other temperatures.

Temperature measurements are ordinarily made with instruments called thermometers, although some other instruments not normally called thermometers are also used, e.g. thermocouples. All are instruments for making observations on some property whose magnitude changes reproducibly with change in temperature. There are thus many types of thermometric instruments, each using the characteristics of some selected property to measure temperature. Measurements made with these instruments can be stated on a defined scale by comparing them directly or indirectly with measurements on the defined scale. On the defined practical scales, the interpolation is usually specified, over limited temperature ranges, by stating the temperature relationship of the property used and often also by stating specifications for the instrument.

II. Temperature Scales

1. *The Kelvin temperature scale*

From the second law of thermodynamics it can be shown that, if q_1 and q_2 are the quantities of heat flowing to a reversible heat engine operating on a Carnot cycle between two heat reservoirs at the respective temperatures t_1 and t_2 then

$$q_1/q_2 = -\mathrm{f}(t_1)/\mathrm{f}(t_2) \tag{1}$$

in which $\mathrm{f}(t)$ is any function of temperature. The scale of temperature obtained by putting $\mathrm{f}(t) = T$ so that

$$T_1/T_2 = -q_1/q_2 \tag{2}$$

is known as the absolute thermodynamic temperature scale. The zero of this scale is not arbitrary but is fixed by the second law of thermodynamics. This scale, now known as the Kelvin scale, was proposed by William Thomson (later Lord Kelvin) in 1854 from a consideration of Carnot's theorem, which is a consequence of the second law of thermodynamics.

When Kelvin proposed this scale he said that, in order to fix on a unit or degree for the numerical measurement of temperature, either some number may be assigned to a definite temperature, such as the ice point, or

TEMPERATURE SCALES AND TEMPERATURE MEASUREMENT

the number of degrees in some temperature interval may be chosen, such as a hundred degrees from the ice to the steam point. He then said "The latter assumption is the only one that can be made conveniently in the present state of science, on account of the necessity of retaining a connection with practical thermometry as hitherto practised; but the former is preferable in the abstract, and must be adopted ultimately." He went on to say that when the value for the ice-point temperature became known to a tenth of a degree or so, the foundation on which the degree is defined may be altered. He said the value was not known at that time to better than two or three tenths of a degree but thought the value would probably be found to be about 273·7.

A hundred years later, in 1954, the Tenth General Conference on Weights and Measures, on the recommendation of the International Committee on Weights and Measures, adopted a resolution redefining the size of the Kelvin degree in the way which Kelvin had said "must be adopted ultimately", and assigned the value 273·16°K to the triple point of water[20]. Six years earlier the Ninth General Conference had adopted a resolution "that the zero of the thermodynamic Celsius (centigrade) scale should be defined as being the temperature 0·0100° below that of the triple point of pure water[18]. This definition now makes 0°C = 273·15°K.

The thermodynamic temperature scale can be established from experimental measurements of quantities that appear in any second law equation. At present the gas thermometer, using the principle of an ideal or perfect gas, is regarded as the primary instrument for realizing the thermodynamic scale in the range over which this thermometer can be used. It can be shown that if there were a gas which obeyed both Boyle's law and Joule's law, its equation of state would be that of a perfect gas,

$$PV/n = RT. \tag{3}$$

In this equation, P is the pressure of the gas, V is the volume of n moles of gas, R is the gas constant, and T is the absolute thermodynamic or Kelvin temperature. There are no perfect gases, but for a few real gases, at pressures ordinarily used in gas thermometry, sufficiently high accuracy may be attained if a simple equation of state such as

$$PV/n = RT(1 + BP) \tag{4}$$

is used, in which B, called a second virial coefficient, is a function of temperature. With this equation, it can be seen that for n constant, at two states 1 and 2, the relation for the ratio of two thermodynamic temperatures is

$$T_1/T_2 = (P_1V_1/P_2V_2)[(1 + B_2P_2)/(1 + B_1P_1)], \tag{5}$$

in which the factor in brackets is the correction factor for the imperfection of the gas. The evaluation of this correction factor is important in gas thermometry.

It may be seen from equation (5) that the gas thermometer provides the means for determining the ratios of Kelvin temperatures. Starting with the value assigned to the triple point of water, these ratios enable us to find experimental values of higher and lower temperatures within the range of gas thermometry. Gas thermometer determinations have been made at temperatures ranging from above the gold point at 1063°C down into the liquid helium region, to about 1·3°K.

At high temperatures, the Planck radiation formula can be used to determine the values of temperature on the thermodynamic scale. This formula, which is derivable from quantum statistical mechanics, contains a term which can be shown to be identical to the temperature as defined by Kelvin on the basis of the second law of thermodynamics. The formula is

$$J_T = C_1 / \left[\lambda^5 \left(\exp \frac{C_2}{\lambda T} - 1 \right) \right] \tag{6}$$

in which J_T is the radiated power per unit solid angle per unit area per unit wavelength interval, at wavelength λ, from a perfectly absorbing surface, or blackbody, at the thermodynamic temperature T. Such a blackbody source may be approximated by a very small hole into an enclosure at a uniform temperature. The constants C_1 and C_2 are known as the first and second radiation constants respectively.

Since neither absolute measurements of radiated power, nor relative measurements with the triple point of water as a reference temperature can be made with sufficient accuracy, it is customary to make relative measurements with the gold point as a reference. In this case, the thermodynamic temperature defined by Planck's formula will be on the Kelvin scale provided that the correct values of the gold point and of C_2 are used. C_2 is given by

$$C_2 = hc/k \tag{7}$$

in which c is the speed of light, h is Planck's constant and k is Boltzmann's constant.

At low temperatures, measurements of the magnetic susceptibility of a paramagnetic salt are used to determine temperatures. For an ideal paramagnetic material the relation is given by Curie's law

$$\chi = C/T \tag{8}$$

in which χ is the magnetic susceptibility, C is the Curie constant, and T is the thermodynamic temperature. As no ideal paramagnetic material exists, a more useful equation for real materials, one somewhat analogous to the virial equation for a real gas, is the Curie–Weiss law

$$\chi = C/(T - \Delta) \tag{9}$$

in which the second constant, Δ, is the Weiss constant.

The two constants, C and Δ, cannot be determined accurately enough at the triple point of water, and so the procedure usually employed is to determine these constants at temperatures already known in the liquid helium or hydrogen region. The magnetic susceptibility measurements should be made on a single crystal in the form of a sphere, or corrections to this form should be made[29]. Values of temperature derived by means of the Curie–Weiss law are on the thermodynamic scale if the values of the temperature used for calibration are thermodynamic and if the Curie–Weiss law is valid for the material used.

For every paramagnetic material the Curie–Weiss law becomes inaccurate at some low temperature. Below that temperature values derived directly by means of the law are called magnetic temperatures and are denoted by T^*. The relation between T and T^* is obtained from the second law of thermodynamics, which may be written as

$$T = \delta q/dS. \tag{10}$$

Here δq is the quantity of heat absorbed in a reversible process and dS is the corresponding entropy change. In practice, one method[48] is to measure the heat capacity on the magnetic temperature scale, $C^* = \delta q/dT^*$, at various values of T^*. Also adiabatic demagnetization experiments are made to obtain a curve of S versus T^* at zero magnetic field. Then the absolute temperature is obtained from the equation

$$T = (\delta q/dT^*)/(dS/dT^*). \tag{11}$$

Data on $T^* - T$ are available[29] for the paramagnetic salts commonly used as thermometers.

At some still lower temperature, the susceptibility reaches a maximum, and below that temperature the procedures described above are not useful. In this region, however, the paramagnetic material may be used for thermometry, but the useful parameters are either the imaginary component of the a.c. susceptibility or the remanent magnetization[29].

In the above discussion, the significant phenomenon was the electron spin system. At temperatures of the order of 10^{-5}°K a useful thermometric parameter is, analogously, the susceptibility of the nuclear spin system.

A. Realization of Thermodynamic Scale by Gas Thermometry

It has been seen that the gas thermometer provides a means for measuring the ratio of an unknown temperature to the defined temperature, 273·16°K. In principle, measurements with an idealized gas thermometer filled with a perfect gas give this ratio directly. In practice, however, the deviations of the real from the idealized instrument and of the real from a perfect gas require that several troublesome corrections be applied to the experimental observations.

Four different methods of gas thermometry may be used: the constant volume method, the constant pressure method, the constant bulb temperature method, and the PV isotherm method. In all four methods, the accuracy with which an unknown temperature can be determined is limited by the accuracy with which the pressure–volume products can be measured. The discrimination of small differences in temperature is exacting and becomes more exacting at higher temperatures; for instance, at the steam point, a change of a thousandth of a degree changes the pressure–volume product of the gas by only one part in about 373 000.

In the idealized constant volume method (constant density method), n moles of a perfect gas are contained in a bulb of invariant volume, V. A pressure indicator of negligible volume is connected with a tube of negligible volume to the thermometer bulb. When the bulb is first at the unknown temperature, T, and then at the reference temperature $T_0 = 273 \cdot 16°K$, the corresponding pressures, P and P_0, are measured. The ratio of the Kelvin temperatures of the gas in the bulb is given by the equation

$$T/T_0 = P/P_0. \tag{12}$$

In the idealized constant pressure method, a bulb of invariant volume, V, is connected by a tube of negligible volume to a pressure indicator of negligible volume and a pipette of variable volume, which is thermostated at the reference temperature, $T_0 = 273 \cdot 16°K$. The gas pressure is the same in the bulb and the pipette. The bulb contains all n moles of the gas at the lower temperature. When the temperature is changed to a higher temperature, part of the gas is transferred into a volume ΔV in the pipette in order to keep the pressure constant in the system. One of the two temperatures is T_0. The accounting for mass in the system reduces to the equations

$$V/T = (V - \Delta V)/T_0, \quad \text{when} \quad T > T_0, \tag{13}$$

and

$$V/T = (V + \Delta V)/T_0, \quad \text{when} \quad T < T_0. \tag{14}$$

In the idealized constant bulb temperature method, the bulb is connected to the pressure indicator and pipette, as in the previous method, but the bulb temperature is not changed during a set of measurements. The pressure, P_0, is first measured when all the gas is in the bulb. Some of the gas is then transferred into the pipette until the pressure, P, is reduced to half, $P_0/2$. At this pressure, the accounting for mass in the system reduces to the equation

$$V/T = \Delta V/T_0. \tag{15}$$

In the idealized PV isotherm method, the gas constant, R, is assumed to be known exactly. Measuring the pressure, P, the volume, V, and the number of moles, n, in the bulb gives the Kelvin temperature by means of the equation

$$T = PV/nR. \tag{16}$$

The value of R, however, must have been determined by similar measurements for $T_0 = 273 \cdot 16°K$; hence the ratio of the unknown to the defined temperature has been determined indirectly. When this method is used for a real gas, the PV isotherms are measured for different values of n and extrapolated to zero pressure, or zero density.

The experimental gas thermometer necessarily differs from the idealized instrument in many details. For high accuracy, several corrections must be made for the departures from ideality. Fortunately most of the corrections are small, but some are difficult to evaluate with certainty. Some can be minimized by the adoption of certain refinements in the design of the apparatus, but few can be entirely eliminated. It is important to make adequate corrections for the accumulated experimental error to be sufficiently small. Many of the corrections have been discussed in detail elsewhere[5]. Some of the departures from ideality are mentioned in the following paragraphs.

The volume of the gas thermometer bulb is not invariant. All materials change their dimensions with change of temperature, and some do not change equally in all directions. It is necessary, therefore, to know the thermal dilation of the bulb. The bulb volume is also changed by elastic deformations caused by pressure changes. The effects of pressure have been minimized in some instruments by enclosing the bulb within another which contains the same kind of gas at the same pressure as that within the thermometer bulb.

Some of the thermometric gas is outside the thermometer bulb and not at the bulb temperature. To connect to some device for measuring pressure, and to a pipette if it is used, a capillary tube extends from the bulb to a thermostated enclosure outside the gas thermometer. The tube traverses a temperature gradient which changes the gas density along the tube, so integrated corrections have to be made for this effect. The diameter of the tube can be made small, but the smaller the tube the greater is the correction for the thermomolecular pressure through the temperature gradient.

If the capillary tube extends all the way to a mercury manometer there is a "dead space" (volume) above the mercury meniscus. To minimize the capillary depression error, the meniscus should be large, but a larger meniscus increases the uncertainty of the meniscus covolume. Compromises have to be made for these effects.

One method of making these corrections relatively small is to use a thermometer bulb of large volume so that the ratio of the volume in the dead space to that in the bulb is small. The larger bulb, however, increases the difficulty in making the bulb temperature uniform.

A second method to minimize the dead space correction is to end the capillary at a small diaphragm cell which transmits the pressure without significant loss in sensitivity. This cell can be thermostated near room temperature and its volume computed from its dimensions. Beyond this cell there can be a manometer with large menisci to make the manometer measurements sensitive and reproducible.

A third method is to end the capillary at a small sharp circular edge, say 3 mm in diameter, which dips into mercury. The sharp edge captures a mercury meniscus whose curvature depends on the pressure balance. A reproducible pressure balance is achieved when the meniscus has the right curvature to focus light from a lamp into a microscope. This method was used in a gas thermometer at the Physikalish–Technische Bundesanstalt (PTB)[46] with the constant bulb temperature method.

By the constant bulb temperature method, the sensitivity at the gold point, dP/dT, is only about one fourth that by the constant volume method; however a set of measurements can be made in a short time. It was reasoned that errors from gas adsorption or desorption would be less by this method because of its greater speed. A set of measurements has been made with the PTB apparatus in minutes—it takes hours by other methods. Regardless of the validity of this reasoning, sorption can be a source of obscure errors in gas thermometry, and must be given serious consideration in these gas thermometers.

One obvious departure from ideality in the gas thermometer is the fact that there are no perfect gases. Methods for correcting for the imperfections of the thermometric gas have been the subject of many papers since the time scientists began to refer temperatures to the thermodynamic scale. During the first part of this century, it was thought that the results of Joule–Thomson experiments offered the best means for correcting for gas imperfections[9, 52]. A little later the results of compressibility measurements (pressure–volume–temperature measurements) were preferred[23]. These measurements yield values of virial coefficients from observations over a wide range of pressures. At temperatures above 400°C, however, it is difficult to make compressibility measurements, so values of these virial coefficients become uncertain because they must be derived either from extrapolations or from theoretical equations of dubious accuracy.

The second virial coefficients are functions of temperature and appear multiplied by pressure in equation (5). Making the measurements at lower initial gas pressures would reduce the errors of these products. Such measurements, however, increase the demands for higher precision in the pressure measurements.

The present trend in correcting for the imperfections of the gas is to make the gas thermometer measurements at different initial pressures (different initial fillings) and to extrapolate the results to zero pressure where gases are assumed to become perfect. For the gases commonly selected for gas thermometry, viz. helium, neon, argon, nitrogen, etc. the extrapolations are not large and are nearly linear. Therefore, highly accurate measurements, at low pressures, seem to be a necessity for accurate gas thermometry.

B. Realization with Speed of Sound

Another method for determining Kelvin temperatures with a thermometer by use of gas has been developed recently, particularly for temperatures below 20°K[12]. This is the speed of sound method which depends on the equation

$$W^2 = \left(\frac{C_P}{C_V}\right)_{P=0} \frac{RT}{M} (1 + aP + \beta P^2 + \ldots) \qquad (17)$$

in which W is the speed of sound at temperature T, C_P/C_V is the ratio of the specific heats, R is the gas constant, M is the molecular weight of the gas, P is the pressure, and a and β are constants into which second and higher virial coefficients and their derivatives enter. The thermometer, now called an ultrasonic thermometer, is an adaptation of an acoustic interferometer which is maintained at the temperature T while measurements are being made. The speed of sound is obtained from the measurement of the displacement of the interferometer piston for a whole number of half wavelengths of sound of a known frequency (of the order of a megacycle). The constants a and β are eliminated by making measurements at different pressures and extrapolating the results to zero pressure.

This method requires precise length measurements in place of precise measurements of pressure and volume. The method has the advantage that it avoids the troublesome corrections which the gas thermometer requires for dead space, thermomolecular pressures, and sorption, but there may be some new disadvantages.

2. *Practical scales*

In the decades following Kelvin's proposal for his thermodynamic scale, there was an apparent need for a practical scale which would be sufficiently close to the thermodynamic scale for international scientific purposes. Gas thermometers using air had been used, but it was known that the composition of air varied somewhat and that other gases were more nearly perfect. These considerations prompted the International Committee on Weights and Measures to adopt the Normal Hydrogen Scale in October 1887[50]. Two years later, in September 1889, the First General Conference on Weights and Measures sanctioned the use of the Normal Hydrogen Scale[16]. This scale was defined by assigning the values 0°C and 100°C to the ice and steam points, and one meter of mercury pressure (1000/760 atm) for the initial pressure of hydrogen in the gas thermometer (at 0°C).

In 1907, after helium had been discovered and some of its properties determined, the laboratory at the University of Leiden began using the normal helium scale instead of the normal hydrogen scale. Helium has the advantage that it is a gas at lower temperatures than hydrogen and is somewhat closer to a perfect gas.

Between 1907 and 1912, Day and Sosman, at the Geophysical Laboratory of the Carnegie Institution of Washington, made gas thermometer measurements, in the range 200°–1600°C, to obtain values for a number of fixed reference points[28]. Many of these values remained unchallenged for over four decades. From these data the Geophysical Laboratory formulated a practical thermocouple scale for measuring temperatures of minerals. The definition of this scale, known as the "Geophysical Laboratory Scale", was published in 1919[1]. That laboratory still uses this scale unchanged—temperature values reported in papers from there represent the same temperatures as they did 50 years ago.

A. International Temperature Scales

In order to establish a practical scale for international use on which temperature could be conveniently and accurately measured, the directors of the national laboratories of Germany, Great Britain, and the United

States agreed in 1911 to undertake the unification of the temperature scales in use in their respective countries. In 1927, when a practical scale was finally agreed upon, it was recommended to the Seventh General Conference on Weights and Measures, which adopted the scale and the name "International Temperature Scale"[10, 17].

The International Temperature Scale was designed to be reproducible and to represent the thermodynamic scale as closely as possible. This scale was regarded as susceptible to revision and amendment as more accurate methods of measurement were evolved. It was based on assigned values for six reproducible equilibrium temperatures (fixed points) and on formulas for the relation between temperature and the indications of instruments calibrated by means of the fixed points. The six basic fixed points were the ice point, 0°C, the normal boiling points of oxygen, −182·97°C, water, 100°C, and sulfur, 444·60°C, and the freezing points of silver, 960·5°C, and gold, 1063°C, all at atmospheric pressure.

The means available for interpolation led to the division of the scale into four parts. From 0°–660°C, the temperature was to be deduced by means of the Callendar formula in the form

$$R_t = R_0(1 + At + Bt^2) \tag{18}$$

in which R_t is the resistance at temperature t of the platinum resistor of a standard resistance thermometer, and R_0 is the resistance at 0°C. The constants R_0, A, and B were to be determined from calibrations at the ice, steam, and sulfur points. The standard resistance thermometer was to be made of platinum wire of such purity that R_t/R_0 should not be less than 1·390 for $t = 100$°C.

From −190° to –0°C, the temperature was to be deduced by means of the Callendar–Van Dusen formula in the form:

$$R_t = R_0(1 + At + Bt^2 + C(t - 100)\,t^3) \tag{19}$$

in which the additional constant C was to be determined from an additional calibration at the oxygen point.

From 660°C to the gold point, the temperature was to be deduced by means of the formula:

$$E = a + bt + ct^2 \tag{20}$$

in which E was the electromotive force of a standard platinum vs. platinum–rhodium thermocouple, one junction of which was at 0°C and the other at temperature t. The constants a, b, and c were to be determined from calibrations at the freezing point of antimony, and at the silver and gold points.

Above the gold point, the temperature was to be determined by means of the ratio of the intensity J_2 of monochromatic visible radiation of wavelength λ cm, emitted by a blackbody at the temperature t_2, to the intensity J_1 of radiation of the same wavelength emitted by a blackbody at the gold point, by means of the Wien formula:

$$\log_e \frac{J_2}{J_1} = \frac{C_2}{\lambda}\left[\frac{1}{1336} - \frac{1}{(t+273)}\right] \qquad (21)$$

The constant C_2 was taken as 1·432 cm deg. The equation was valid if $\lambda(t+273)$ was less than 0·3 cm deg. (For $\lambda = 0·65 \cdot 10^{-4}$ cm this requirement made the upper limit of temperature about 4342°C.)

After the definition of the scale, the recommended procedures were given. The methods then used for realizing the temperature at each of the basic fixed points were described in varying detail. The recommendations for the standard resistance thermometer had statements about its wire size, its mounting, its leads and its annealing. The recommendation for the standard thermocouple specified the purity of the platinum element to be such that the initial value of R_t/R_0 was not less than 1·390 for $t = 100°C$ and specified that the alloy element was to consist of 90 per cent Pt with 10 per cent Rh. It gave the allowable limits for the electromotive force of the thermocouple at the gold point, and the usual range for the wire diameter.

The procedure for calibration at the freezing point of antimony was given but it was stated that the value of this temperature was to be determined with a standard resistance thermometer. After the recommended procedure was a list of 13 secondary points which might be used for the calibration of secondary temperature measuring instruments.

The Seventh General Conference on Weights and Measures recommended that international thermometric conferences be called to revise the temperature scale as occasion required, and the Eighth General Conference in 1933 made official provision for these conferences. The International Committee then approved an Advisory Committee on Thermometry which met in 1939 and again in 1948, proposing a revision of the scale. The International Committee then recommended this proposal to the Ninth General Conference, where it was adopted in October 1948 with the name "The International Temperature Scale of 1948"[19, 55].

The changes made by this revision can best be summarized by quoting from the Introduction in the text of this scale.

> The experimental procedures by which the scale is to be realized are substantially unchanged. Only two of the revisions in the definition of the scale result in appreciable changes in the numerical values assigned to measured temperatures. The change in the value for the silver point from 960·5°C to 960·8°C changes temperatures measured with the standard thermocouple. The adoption of a different value for the radiation constant, C_2, changes all temperatures above the gold point, while the use of the Planck radiation formula instead of the Wien formula affects the very high temperatures. The Planck formula is consistent with the thermodynamic scale and consequently removes the upper limit which was imposed by Wien's law in the 1927 scale.
>
> Other important modifications, which cause little or no change in the numerical values of temperatures, but serve to make the scale more definite and reproducible are (a) the termination of one part of the scale at the oxygen point instead of at −190°C; (b) the division of the scale at the freezing point of antimony (about 630°C) instead of at 660°C; (c) the requirement for higher purity of the platinum of the standard resistance thermometer and standard thermocouple, and for smaller permissible limits for the electromotive force of the standard thermocouple at the gold point.

The value for the silver point was changed by 0·3° not because the new value was known to make the value closer to that on the thermodynamic scale, but merely because it made the part of the scale defined by the standard thermocouple join more smoothly with the part defined by the standard resistance thermometer. The new value for the radiation constant C_2 was rounded from the value derived at that time from atomic constants, and the change to Planck's formula made the scale consistent with the thermodynamic scale. At 4000°C these changes amounted to about 43°, most of which was due to the change in the radiation constant (*Figure 1*).

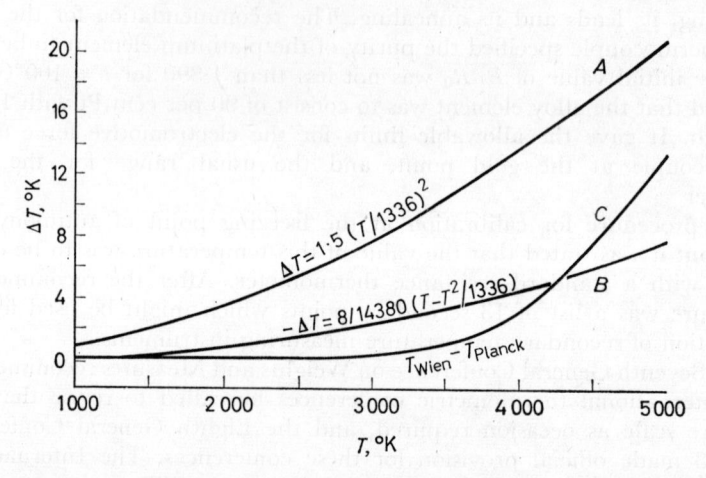

Figure 1. Effect of changes in the definition of the temperature scale above 1063°C. *A*: correction to be applied if the gold point is redefined to be 1064·5°C instead of 1063·0°C; *B*: correction to be applied if the second radiation constant, C_2, is redefined to be 0·014388 m-deg. instead of 0·014380 m-deg.; *C*: correction to be subtracted if Wien's law has been used instead of Planck's law

The scale was terminated at the oxygen point because gas thermometry subsequent to 1927 had shown that values using the Callendar–Van Dusen formula differed increasingly from those on the thermodynamic scale at temperatures farther below the oxygen point. The platinum wires of both the standard resistance thermometer and the standard thermocouple were required by definition to be of such purity that R_{100}/R_0 was greater than 1·3910.

Before 1948 it had been shown that the triple point of water would make a more precise thermometric reference point than the ice point. The triple point was not adopted as a fixed point of the scale in 1948, but its use was recommended for work of the highest precision. A resolution was adopted at the Ninth General Conference, however, stating that "the zero of the thermodynamic Centesimal scale should be defined as being the temperature 0·0100° below the triple point of pure water"[18]. During the General Conference an impromptu suggestion led to the adoption of the name Celsius to replace the names Centigrade and Centesimal. The name Celsius had already been used in many countries, so its use in the remaining countries would do much to make the nomenclature of temperature uniform in all countries[58].

The redefinition of the size of the Kelvin degree by the Tenth General Conference in 1954 made it necessary to change the introduction in the text of the International Temperature Scale of 1948. It became evident that the entire text of the scale would profit by a revision which would bring definitions and practices up to date. In the revision, care was taken that the definitions in the new text would keep all values of temperature the same as they were on the 1948 scale within the experimental error of measurement. The tentative text was discussed in detail at the meeting of the Advisory Committee on Thermometry in 1958 and many suggested changes were agreed upon. The text was then proposed to the International Committee, which changed the name to "International Practical Temperature Scale of 1948, Text Revision of 1960". The text was adopted at the Eleventh General Conference in October 1960[22, 57]. Some of the changes made in the text[59] will now be mentioned.

In the new text, section 2 is the definition of the scale. The six fixed points are now called "Defining fixed points," and their values are exact by definition. In place of the ice point there is the triple point of water with the value 0·01°C. This value makes one defining fixed point common to both the international and the thermodynamic scales.

The normal boiling points of oxygen, water, and sulfur are still retained as defining fixed points. The accurate realization of normal boiling points, however, requires a somewhat elaborate apparatus for measuring pressure, particularly at the sulfur point where 0·0001° corresponds to about 1·1 μ of mercury pressure. By contrast, the freezing point of pure zinc can be realized with high reproducibility, yet it requires only a rough determination of pressure. In the table of defining fixed points there is a footnote after the value of the sulfur point which states: "In place of the sulfur point, it is recommended to use the temperature of equilibrium between solid zinc and liquid zinc (zinc point) with the value 419·505°C (Int. 1948). The zinc point is more reproducible than the sulfur point and the value which is assigned to it has been so chosen that its use leads to the same value of temperature on the International Practical Temperature Scale." This footnote not only promotes the zinc point to a status comparable to that of the defining fixed points of the scale but also states that its use is preferable to that of sulfur.

To ensure greater purity of the platinum in the standard resistance thermometers and the standard thermocouples, the lower limit of the ratio R_{100}/R_0 has been raised to 1·3920.

In the part of the scale above the gold point, the former restriction that λ be a wavelength of the visible spectrum has been removed in order to permit the use of photocells at wavelengths outside the visible region.

Section 3 has the recommendations. No substantial changes have been made in this section, but many new pieces of information are included which were not available in 1948.

In section 4, which has supplementary information, there is a table showing the relations of four different temperature scales. These four scales are the international Celsius and Kelvin scales, and the thermodynamic

Celsius and Kelvin scales. The Celsius scales have their zeros at 0·01° below the triple point of water, and the Kelvin scales have their zeros at the absolute zero.

Values for Kelvin temperatures are commonly obtained merely by adding the values on the international Celsius scale, $t°C$ (Int. 1948), to the value 273·15°K, which is derived from definitions. These values, however, are not exactly on the thermodynamic Kelvin scale but are on the international scale. The thermodynamic scales are ideal and are difficult to realize accurately, but they supply useful concepts and are accepted as the scales to which all temperature measurements should ultimately be referable.

B. Provisional Temperature Scale 10°K to 90°K

The International Practical Temperature Scale is undefined below the oxygen point. However, in 1939 Hoge and Brickwedde at the U.S. National Bureau of Standards set up a provisional temperature scale from 10°K to 90°K. They calibrated a group of helium-filled capsule-type platinum resistance thermometers against a gas thermometer, using the oxygen point, taken as 90·19°K, as the reference temperature. The results were published in the form of a table giving $W_T = R_T/R_{273}$ against T, for one of the six thermometers in the group, in which R_T was the resistance at $T°K$ and R_{273} the resistance at the ice point. An accuracy of $\pm 0·02°$ with respect to the thermodynamic scale was claimed, and the scale was propagated to other resistance thermometers by intercomparison at several points in the range. The redefinition of the size of the Kelvin degree in 1954, by assigning the value 273·16°K to the triple point of water, made the value of the oxygen point, at −182·97°C, become 90·18°K. In 1955, to take account of this value, and to favor the value then preferred for the normal boiling point of hydrogen, all thermometers on the current provisional scale, known as NBS 1955, were defined to have values 0·01° lower than they were on the first provisional scale set up in 1939.

Since 1939, several other laboratories have calibrated groups of resistance thermometers against gas thermometers. Results of such calibrations have been reported from the Physical-technical and Radio-technical Measurements Institute in Moscow, the National Physical Laboratory in Teddington, and the Pennsylvania State University. These resistance thermometer calibrations have been intercompared, and Barber[4] has shown that if the scales are adjusted for a common reference temperature, for common virial coefficients and for common values of the thermal expansion of copper, the agreement of the four scales with the mean is about 0·01° over the range 20–90°K.

The present situation is unsatisfactory in the following respects.

(i) There are now three practical scales being propagated by various national laboratories. Although the agreement among them is good enough for most purposes it is still one or two orders of magnitude worse than the reproducibility of the interpolating instrument.

(ii) The dependence of the scale on the arbitrary group of resistance thermometers, which may undergo systematic changes with time, is clearly unsatisfactory. The scale, NBS 1955, for example, is now being maintained in part by thermometers which are three "generations" removed from the group on which the scale was originally based.

(*iii*) For calorimetry it is necessary that adjacent scales join smoothly. It is difficult to satisfy this condition at the oxygen point, where the various 10–90°K scales join the International Practical Temperature Scale, because the resistance–temperature curve for platinum has an inflection point at about 87°K. It seems unlikely that any simple algebraic function, such as the Callendar–Van Dusen equation above the oxygen point, can be joined in the required way to a scale determined by gas thermometry below the oxygen point. The solution for these difficulties may become clearer when more accurate gas thermometry above the oxygen point has been done. This gas thermometry is now being undertaken in at least two national laboratories.

At about 17°K the sensitivity, $(1/R)\,\mathrm{d}R/\mathrm{d}T$, of platinum is at a maximum, but its resistivity is very low. At lower temperatures the sensitivity falls off so rapidly that platinum thermometers are seldom used below about 10°K. In the temperature range from 10°K down to 5·2°K there are no accepted practical scales. In calorimetry, simple helium gas thermometers with the thermometer bulb built into the calorimeter have been used, also carbon resistance thermometers. More recently, germanium resistance thermometers have been made whose reproducibility is satisfactory. From 5·2°K down to 1°K the vapor pressure of He^4 has been used widely for defining a practical scale, and the vapor pressure of He^3 can also be used down to about 0·3°K.

C. 1958 He^4 Scale of Temperature

In 1948 a small group of low temperature physicists, meeting informally in Amsterdam, agreed to use and recommend the vapor pressure table for He^4 which was then used in Leiden for temperature measurements between 1°K and 5·2°K. Later it was shown that this table did not represent a smooth function of temperature and that it differed significantly from the thermodynamic scale. In 1955 a He^4 vapor pressure table, known as the 55E scale, was published by the Naval Research Laboratory, and in the same year a similar table, known as the L55 scale, was published by the Leiden Laboratory. Below 4·2°K the maximum difference between these scales was about 4 millidegrees.

At a conference in Leiden in June 1958, a compromise scale was finally agreed upon. About a week later the Advisory Committee on Thermometry, having recognized the need for establishing a single temperature scale for general use in the low temperature range, and having the agreement of specialists in this field, proposed a resolution recommending the 1958 He^4 vapor pressure scale, as defined by an annexed table. The values of temperature on this scale were to be designated by the symbol T_{58}. In October 1958 the International Committee on Weights and Measures approved this resolution, and it was adopted, along with the text revision of the International Practical Temperature Scale of 1948, at the Eleventh General Conference on Weights and Measures in October 1960[8, 21].

The "1958 He⁴ Scale of Temperature" gives the value 4·2150°K for the normal boiling point of He⁴, and the value 2·1720°K for the λ-point at a pressure of 37·80 mm of mercury. The critical point is at 5·1994°K if the critical pressure is taken as 1712 mm of mercury pressure.

Tables of the 1958 He⁴ Scale are given in NBS Monograph 10. Table I of the monograph gives the computed values of the vapor pressure in microns of mercury for every millidegree from 0·500°K to 5·22°K to at least five significant figures. Such high precision is merely for keeping the table smooth, though the accuracy is estimated to be only about two millidegrees. Tables II and III give values of temperature as a function of the vapor pressure. Table IV gives the temperature derivatives, dp/dT. Table V is for use in making hydrostatic head corrections and Table VI shows deviations of earlier scales from the 1958 He⁴ scale.

D. 1962 He³ Scale

At a meeting in September 1962, a temperature scale based on the vapor pressure of He³ was proposed to the Advisory Committee on Thermometry by the American Los Alamos Scientific Laboratory. The committee considered that this scale should also be recommended for general use, with the designation $T_{62}{}^{60}$. Either of the helium vapor scales, T_{58} or T_{62}, may be used in the ranges in which they are valid. The committee warned, however, that when the time comes for adoption of this new scale as part of the International Practical Temperature Scale, care should be taken to avoid any ambiguity in the range where the two scales now overlap.

III. Temperature Measurement

1. *Liquid-in-glass thermometry*

Liquid-in-glass thermometers are the most familiar and among the least expensive. They are still very useful laboratory instruments when an accuracy of only a few tenths of a degree is sufficient. At moderate temperatures the best of these thermometers may realize an accuracy approaching a thousandth of a degree when suitable precautions are taken and the relevant corrections applied. Individual thermometers made for this high accuracy cover only a small range of temperature because they need to have an elongated scale. They must also have a short auxiliary scale for a reference, usually at the ice point.

Calorimetric and Beckmann (metastatic) thermometers are intended for accurate measurements of temperature differences rather than absolute temperatures, and they are not provided with an auxiliary scale for reference. In the Beckmann thermometer the mercury column can be separated in such a way as to make possible the selection of the limited range of temperature (5 or 6 degrees) over which measurements are to be made.

The following corrections, discussed in more detail elsewhere[11, 61], must be made if the mercury-in-glass thermometer is to be used for high accuracy. The accuracy of the location of the graduations on the thermometer stem must be determined by calibration. Such a calibration is either supplied by the manufacturer or may be done by a standardizing laboratory. The reference temperature check (usually at the ice point) is necessary if accuracy is important. The bulb of a Celsius thermometer has a volume equivalent to about 6000° of scale, depending on the type of glass. Small changes in bulb volume with time are due either to slow annealing or to straining of the glass, and cause a shift of the ice-point reference reading. The corrections obtained from this reading must be applied to the whole scale. It usually will not exceed one or two degrees after several hundred hours of use at temperatures up to 400°C. With a good thermometer, used mostly near room temperature, it should not exceed 0·2° over the lifetime of the thermometer. The use of fused silica instead of glass means that much better stability is possible.

Thermometers are normally calibrated either for "total immersion", i.e. to the top of the liquid column in the stem, or for "partial immersion", i.e. to a fiducial mark. Total immersion thermometers are to be preferred in circumstances in which their use is practical. When a total immersion thermometer is used with only partial immersion, or when a partial immersion thermometer is used under conditions such that the temperature of the exposed stem differs from that which is obtained during calibration, an emergent stem correction must be made. This correction may be deduced from a determination of the emergent stem temperature—which may itself require a special auxiliary capillary thermometer—and a knowledge of the differential volume expansion of mercury and glass. This expansion is about 0·00016 per degree C, but depends on the kind of glass.

Corrections for pressure effects due to the change of effective volume of the bulb resulting from hydrostatic pressures may be required. These effects may arise externally from the depth of immersion, or internally from changing the thermometer from vertical to horizontal operation, or as a result of temperature changes in the filling gas. The pressure coefficient is usually about 0·1°C/atm and is 10 per cent greater for inside pressure changes because the mercury is compressed too.

Hysteresis effects in the glass require correcting. After a thermometer has been heated to some moderate temperature, say 100°C, the bulb reaches its new equilibrium volume within a few minutes. An ice point taken immediately after such a reading is too low and takes about three days to reach equilibrium. After readings at much higher temperatures the ice-point readings will also be low, but the recovery is unpredictable, and often is not significant after months or years at room temperature. After at least three days rest at room temperature, calibrations are done at the ice point and then successively for increasing temperatures. If readings are not taken in order of increasing temperature, they must be corrected for hysteresis by immediately taking an ice point and comparing this non-equilibrium ice point with one taken after three days. Such corrections may be as high as 0·1°C for 100°C rise.

If the above corrections are made, and if due care is taken to allow adequate response time (usually a few seconds) and to avoid parallax errors, the best thermometers may still be limited in accuracy by sticking of the mercury. This limitation is caused by changes in meniscus shape which affect the capillary pressure and hence the volume of the bulb. The effect is proportional to the pressure coefficient and inversely proportional to the bore diameter; for some recent fused silica thermometers with thick-walled bulbs, it is less than 0·001°C. To some extent the "sticking" error may be reduced by taking readings when the temperature is rising and tapping the thermometer before reading.

The range of use of mercury-in-glass thermometers is from −38°C to as high as 600°C depending on the kind of glass used. Some recent experimental thermometers have used gallium in fused silica at temperatures ranging up to 1100°C. By use of a mercury–thallium eutectic the lower limit of the range may be pushed to −56°C. For less accurate measurements, alcohol or toluene fillings give −80°C, and isopentane fillings −196°C.

2. Electrical thermometry

A. Resistance Thermometry

(1) Platinum Resistance Thermometers

The International Practical Temperature Scale of 1948 is defined between −182·97°C and 630·5°C by means of a standard resistance thermometer[57]. The standard resistance thermometer is now defined to be of platinum wire which is annealed and of such purity that R_{100}/R_0 is not less than 1·3920. It is recommended that it should be constructed so that the wire of the platinum resistor is as nearly strain-free as practicable and will remain so during continued use. The completed resistor should be annealed in air at a temperature higher than the highest temperature at which it is to be used, but in no case below 450°C. It is recommended, for better stability, that the tube protecting the completed resistor be filled with gas containing some oxygen.

The purity of platinum can be judged, in part, by the ratio R_{100}/R_0—usually a higher ratio indicates purer platinum. It is reasonable to suppose that, for properly annealed thermometers made of perfectly pure platinum, the thermometer constants, A, B, and C would be the same for all thermometers. This ideal has not yet been reached, so calibrations are still necessary, especially for the constant, A. The nearer one approaches to this ideal, the more uniform are the characteristics of different thermometers. It is particularly desirable for thermometers to be close to this ideal if they are to be used at temperatures in the range below the oxygen point. Although the allowable lower limit for the ratio in a standard resistance thermometer is defined as $R_{100}/R_0 = 1·3920$, wire is now readily available with the ratio higher than 1·3925, and the ratio 1·3927 has been reached in a few thermometers. The ultimate limit for this ratio may be about 1·3929.

The annealing of the wire in the completed thermometer is important. A useful criterion for the adequacy of the annealing and the reliability of the thermometer is the constancy of its resistance at some reference temperature. For example, the resistance of a thermometer at the triple point of water should change by less than the equivalent of 0·001° when the thermometer is subjected to temperature cycles such as are necessary for its calibration.

TEMPERATURE SCALES AND TEMPERATURE MEASUREMENT

Thermometers have been made in wire diameters ranging from 0·05–0·5 mm, depending on the purpose for which the thermometer is to be used, and with R_0 resistances ranging over an even greater factor. For temperatures in the present resistance thermometer range of the International Practical Temperature Scale, a popular resistance is about 25·5 ohms. This choice of resistance has no virtue in itself except that it makes the room temperature sensitivity approximately the round number of 0·1 ohm per degree. In recent years thermometer resistors have been made smaller; it is now common to find them confined within a length of 3–5 cm, a diameter of 4–7 mm, and with wire diameters from 0·07–0·1 mm.

Except for low temperature thermometers, these resistors are usually enclosed near the sealed end of a protecting tube, of Pyrex glass, fused silica, or platinum, up to 50 cm long. Four leads extend inside the tube from the resistor to a head, then through a hermetic seal to flexible copper leads extending to the measuring instrument. The resistors are mounted in several ways, all of which are intended to leave the platinum wire strain-free. They are mounted as a helix of coiled wire supported on a mica cross[45], as a single layer helix mounted on a mica cross[56], as a helix encased in an inner U-shaped tube[2], or as a helix on opposite sides of a twisted ribbon of fused silica inside a fused silica protecting tube[7]. The four leads from the resistor are usually of gold, but platinum leads have also sometimes been used. The object of these long leads is to permit the resistor to be inserted far enough into the constant temperature region to prevent heat conduction along the lead wires from significantly influencing the temperature of the thermometer resistor.

The measurement of the thermometer resistance requires a small electric current which is usually a milliampere or more. Although small, this current produces enough heating to raise the temperature of the resistor a significant amount above the temperature of the material being measured, often several millidegrees. Measuring the resistance at two different currents permits an extrapolation to zero power input to get the resistance at the temperature being measured[56]. The amount of heating depends not only on the current used but also on the thermal resistance through which the heat has to flow. A large part of this thermal resistance is through the gas inside the thermometer-protecting tube. By constructing the thermometer with the platinum wire nearer the wall of the protecting tube, the thermal resistance is smaller and so, consequently, is the heating[56]. Similarly, the closer the leads are to the wall of the protecting tube the better their thermal connection to the wall and the smaller the immsersion needed.

Platinum resistance thermometers are also used in the range of the provisional scales from 10°K to 90°K[3, 7]. These thermometers usually have the resistors enclosed in sealed capsules containing gas, part of which is helium because of its thermal conductivity. Helium is especially necessary at the lower end of this temperature range at which the vapor pressure of all other gases approaches zero. The capsules are only a little longer than the resistors themselves so that they can be installed in calorimetric apparatus. The capsules should be in good thermal contact with the calorimetric apparatus.

In the relatively near future the resistance thermometer will probably replace the standard thermocouple for defining the scale in the range of the International Practical Temperature Scale from 630·5°C to 1063°C. This replacement will result in greater accuracy of temperature measurements in that range. At higher temperatures of this range, however, there are increasing difficulties due to electrical insulation, contamination of the platinum, and drifting of the resistance with continued use. Some of these difficulties may be overcome, but it seems unreasonable to expect the thermometers to have the same stability as they have at lower temperatures.

To reduce the effects of some of these difficulties, larger wire diameters are indicated. Larger wire sizes result in lower resistances which make accurate measurements more difficult. Because our more common insulators become conducting to electricity at higher temperatures, sapphire seems to be one of the best choices, though it is none too good. The drifting of the resistance can be the result of: changes in the dimensions, such as from evaporation of the platinum; changes in the physical properties, such as may result from strains, recrystallization, or changes in the lattice structure of the crystals; and changes in the chemical properties, such as may be the result of contaminations from supports and the protecting tubes[34].

Platinum appears to be the best of the metals for resistance thermometers which are to cover a wide range of temperature. It is a noble metal which does not oxidize easily and can be made very pure. It is solid up to 1769°C and is free from transitions which cause sudden changes in the slope along the resistance–temperature curve. A simple quadratic equation can be used to define a temperature scale remarkably close to the thermodynamic scale all the way from the ice point to the gold point. However, the sensitivity, $(1/R)\,dR/dT$, for platinum is not high and it decreases with increase in temperature.

Copper has a sensitivity, at room temperature, slightly greater than that of platinum and it does not fall off as rapidly with increase in temperature[25]. However, copper has a low resistivity so it is not always convenient for thermometers. Also, it is not a noble metal and oxidizes easily. Nickel, on the other hand, has a sensitivity about twice that of platinum and a resistivity slightly higher. However, it has a transition near 350°C which limits its useful range. Both copper and nickel have been used for simple resistance thermometers in calorimeters for which precise adherence to the adopted international scale is not demanded. In some calorimeters the same resistors have served both as thermometers and as heaters. Tables of resistance ratios are used to translate resistances into values of temperature or temperature difference[37].

(2) Semiconductors

In recent years semiconductors have come to be used as resistance thermometers for measuring temperature[35]. Semiconductors have resistivities intermediate between those for metals and for insulators, and temperature coefficients of resistance of opposite sign to that for metals. Over a restricted range their resistance–temperature relations may be approximated by:

$$R_T = R_\infty\, e^{B/T} \tag{22}$$

in which R_T is the resistance at Kelvin temperature T, R_∞ is the limiting resistance when $1/T$ approaches zero, and B is a constant number of degrees.

(a) *Thermistors.* One class of semiconductor thermometer is known as the thermistor. Thermistors are usually made of precise mixtures of metallic oxides fired in a carefully controlled atmosphere[30]. Their temperature coefficients of resistance at room temperature are over 10 times as great as those for platinum, but negative. A resistance–temperature relation which fits some resistors better than the equation above is:

$$R_T = R_\infty \, e^{B/(T+\theta)} \tag{23}$$

in which θ is another constant number of degrees. A typical value for B is around 3500° for one type of thermistor.

Because of its large temperature coefficient, the convenient range of any one thermistor is necessarily limited, but a series of different thermistors can be made to cover a large temperature range. Thermistors usually have resistances so high that their lead resistances are relatively insignificant. The reproducibility of thermistors, after they have been subjected to large temperature cycles, has not yet been improved to the state in which it approaches that of standard (platinum) resistance thermometers. Thermistors are useful in calorimetry where very small temperature changes are to be measured, especially if these changes are near room temperature so that large temperature cycling of these instruments can be avoided.

(b) *Germanium Thermometers.* Another kind of semiconductor thermometer is the germanium thermometer, which can be used for temperatures from about 1°K to 35°K. Thermometers made of single crystals of germanium doped with a few parts per million of arsenic or gallium have been intensively investigated during the past several years. Like other semiconductors, germanium thermometers have a large negative temperature coefficient. This property makes them desirable for thermometers if they can be made sufficiently reproducible. Thermometers have been constructed with four leads mounted strain-free and encased in helium-filled capsules less than 0·5 cm in diameter and less than 1·5 cm long. Reproducibilities of better than a millidegree at 4·2°K, even after many cyclings between liquid helium and room temperatures, have been reported[32, 40]. These results suggest that semiconducting germanium thermometers, such as are available from Minneapolis-Honeywell, Texas Instrument Company, or Radiation Research Corporation may shortly be accepted as standard instruments. It is also convenient that the upper range of these thermometers overlaps the lower range of platinum resistance thermometers.

(c) *Carbon Thermometers.* The resistance of carbon, in the physical form in which it is used for electrical resistance elements, varies with temperature in a manner resembling that of a semiconductor. Certain commercial radio resistors (notably those manufactured by the Allen–Bradley Company, Speer Resistor Division of Speer Carbon Company, and Radio Resistor Company Ltd.) are especially useful in the temperature range below 20°K. For these resistors, the useful range is usually limited at a few tenths of a degree K by the resistance becoming so high (around 1 megohm) that it is inconvenient to measure.

A relatively exhaustive examination of the Allen–Bradley resistors[13] has shown that, under certain conditions, they are reproducible to the equivalent of at least 0·1 per cent of the temperature at liquid helium temperature even after several cycles up to room temperature. An even higher reproducibility was indicated in some calorimetric experiments in which they were used to measure the transition temperatures of indium and lead from the superconducting to the normal state[14]. They have small magneto-resistances and are only slightly sensitive to measuring power, the slight sensitivity being due apparently to thermal gradients caused by the measuring power. Several schemes have been proposed for converting measured resistances to values of temperature. Clement and Quinnell proposed a three-constant equation:

$$\log R + K/(\log R) = A + B/T \qquad (24)$$

and Clement suggested a two-constant version of the same equation

$$(\log R/T)^{\frac{1}{2}} = a + b \log R. \qquad (25)$$

The first of these equations, with a temperature difference curve, ($T_{meas} - T_{calc}$) vs. T_{calc}, serves adequately for calibration from the lowest He^4 temperatures up to 20°K. The second with a similar temperature difference curve is easier to use and works as well. A preferable scheme is to determine b from a best fit of all data to the second equation and then construct a curve of a vs. $\log R$ directly from calibration data. Any of these procedures provides a relatively simple means of spanning the gap from 4°K to 12°K if calibration points in both He^4 and H$_2$ are available.

The Allen–Bradley resistors have also been shown to fit a single normalized function to ± 1 per cent in temperature[15]. Two normalizing factors are required, one for resistance and one for temperature. This normalized function at present represents the only known single relation covering the entire range from 0·3°K to 100°K. Although this function is less accurate than might be desired, and it cannot be realized with a single thermometer, it does have the advantage that a single thermometer, calibrated at two points, gives a usable temperature scale over any decade of temperature in the range 0·3°K to 100°K.

The properties of the Speer resistors have been less thoroughly investigated. Their special usefulness stems from their lower sensitivity and their consequent application to temperatures lower than those possible with the Allen–Bradley resistors.

The LAB resistors, manufactured by the Radio Resistor Company Ltd., have been much used in the U.K. Their characteristics are also similar to, but not identical with, those of the Allen–Bradley resistors.

Neither the Speer nor the LAB resistors fit so well to equation (24) or (25) as do the Allen–Bradley resistors. However, other simple calibration schemes have been devised[43] for these.

(3) *Effects of Magnetic and Radio Frequency Fields*

Magnetic fields affect the electrical resistance of all conductors. In general, the effect is greater the greater the field and the lower the temperature but is insignificant at ordinary temperatures and fields. With high magnetic fields at temperatures below say 20°K, the effect can be troublesome in resistance thermometry. The effect is much greater for some conductors than others and may depend on their size and shape. The effect may also depend on crystal anisotropy and on the direction of the field, whether it is longitudinal $(H \parallel I)$ or transverse $(H \perp I)$ to the current. It is warned, therefore, that whenever resistance thermometers may be affected significantly by magnetic fields, they should be calibrated under conditions that are identical to those under which they are to be used.

In a somewhat analogous way the Joule heating produced by radio frequency fields has been found to increase the temperature of carbon thermometers and probably also of germanium thermometers. Adequate shielding of the thermometer appears to be the only way to avoid this trouble. Two points should be borne in mind: (*i*) adequate shielding is difficult because radio frequency fields penetrate apparatus in astonishing ways, and (*ii*) thermal contacts become poorer as the temperature decreases below 1°K so the Joule heating increases.

B. Thermoelectric Thermometry

Thermocouples are very important and widely used instruments for temperature measurement. This fact is evident from some 300 pages devoted to thermoelectric thermometry in Volume 3 of *Temperature, Its Measurement and Control in Science and Industry*. Thermocouples of one type or another can be used over the entire range of temperature from 1°K up to temperatures above 2000°C. In spite of their limitations, they do have certain advantages over other kinds of thermometric instruments because they do not require power at their junctions. They are especially valuable at low temperatures because they can be made small and at high temperatures because they are free from the uncertainties in surface emittance inherent in radiation thermometry.

Some definitions and descriptions will be useful in the following discussion. A thermocouple is a pair of dissimilar electrical conductors so joined that an electromotive force (e.m.f.) is developed by the thermoelectric effects when the two junctions are at different temperatures. A thermocouple is thus inherently an instrument for determining temperature differences. One of the thermocouple junctions may be called the measuring junction and the other the reference junction. When the reference junction is at a reference temperature, e.g. the ice point, the e.m.f. produced in the thermocouple can be used to determine the temperature at the measuring junction. A thermocouple used in this way is effectively a thermoelectric thermometer.

The thermal e.m.f. of any electrical conductor versus platinum can be represented as a function of the temperature of the measuring junction, by means of a graph or table. For any temperature, all these electrical conductors can be listed in a series with each entry more thermoelectrically negative than its predecessor. The relative order in these entries, however, may be different at other temperatures depending on the e.m.f.–temperature relationships of the conductors.

Any pair of conductors in the series could presumably be used for a thermocouple but none is ideal, and practical considerations limit thermocouples to certain recognized types. Some of the characteristics to be consdered in the choice of wire conductors for a desirable thermocouple are: the ductility of the wire, its temperature range, freedom from oxidation, thermal and electrical conductivity, e.m.f.–temperature relationship and sensitivity, the homogeneity of the wire, and its stability and reproducibility.

The conductor combinations most commonly used for thermocouples are now designated by the Instrument Society of America as types S, R, J, T, K, and E. Tables are available for these types of thermocouples[53]. For designating any thermocouple, the nomenclature used here gives the thermoelectrically positive conductor first and, when possible, the nominal weight per cent of the chemical elements of alloy conductors. The designation 90Pt–10Rh versus Pt thus represents the platinum–rhodium versus platinum thermocouple which is specified for interpolation in part of the International Practical Temperature Scale. This thermocouple is type S. Most of its characteristics are desirable except that its sensitivity is relatively low, about 10 μV/deg.

Type R, which is 87Pt–13Rh versus Pt, has characteristics almost identical to those of type S. These two types are stable for many hours up to 1300–1400°C. They can also be used above these temperatures nearly up to the melting point of platinum at 1769°C, but only with increasing danger of changing their calibrations.

For still higher temperatures there are no thermocouples in this list of types given above but it may be appropriate here to mention some which can be used. The 70Pt–30Rh versus 94Pt–6Rh thermocouple has stability superior to type S at high temperatures and has a practical limit of about 1800°C. Tables are available for the 40Ir–60Rh versus Ir thermocouple[6] but the similar 50Ir–50Rh versus Ir thermocouple is preferred for temperatures up to 2100°C. These iridium-based thermocouples have a sensitivity of around 6 μV/deg. and are not as homogeneous as the platinum-based thermocouples. For temperatures up to 2300°C and possibly as high as 3000°C, thermocouples such as W versus 74W–26Re or 97W–3Re versus 75W–25Re, with a sensitivity of about 15 μV/deg., appear to be promising[42]. The problem of sheathing and insulation, however, becomes more difficult at the higher temperatures. Alumina cannot be used above about 1900°C and, at present, beryllia and iron-free thoria appear to be the most promising materials.

Type J is the iron versus constantan thermocouple. It is the most widely used thermocouple in industrial thermometry, partly because of its low cost and its serviceability in oxidizing atmospheres. Its sensitivity increases from about 26 μV/deg. at −190°C to 63 μV/deg. at 800°C. Special iron for thermocouples, however, contains small amounts of several other elements which affect its thermoelectric properties. It is possible now to secure iron which is especially selected and controlled so it matches the first iron used for tables of this type of thermocouple. For precise measurements, inhomogeneities in the wire are a handicap because they produce spurious e.m.f.s along the temperature gradients and hence cause errors in the determination of temperature differences. The thermoelectrical properties of thermocouple-grade constantan may also not be sufficiently reproducible. Constantan is approximately 57Cu–43Ni but has traces of other elements.

Type T is the copper versus constantan thermocouple. Its sensitivity increases from about 15 μV/deg. at −200°C to 60 μV/deg. at 350°C. It is more reliable than type J for precise measurements because oxygen-free electrolytic copper is usually used, which is very homogeneous and reproducible. More care needs to be taken in the selection of the constantan because it is less homogeneous. Copper, however, is limited to a temperature of about 350°C on account of oxidation, and its high thermal conductivity is sometimes a disadvantage.

Type K now designates any thermocouple having, within limits, the thermal e.m.f. relationship of Chromel versus Alumel as given in the Chromel versus Alumel tables in NBS Circular No. 561 over the range of temperatures from −200°C to 1371°C. Chromel P and Alumel are registered trade marks of the Hoskins Manufacturing Company. Chromel P is an alloy of 90Ni–10Cr which has a high positive e.m.f. against Pt, and Alumel is an alloy of 94Ni–3Mn–2Al–1Si which has a negative e.m.f. against Pt. Both alloys stand up well at high temperatures in oxidizing atmospheres. The sensitivity is about 15 μV/deg. at −200°C, 30 μV/deg. at −100°C, 40 μV/deg. at 0°C, and does not fall below 35 μV/deg. until the temperature is above 1300°C. At temperatures below 500°C the greater sensitivity often makes this type more advantageous than types S or R. Another thermocouple, called "Platinel", has been developed to have a longer life at high temperatures yet match the characteristics of Chromel versus Alumel above 400°C so closely that no change in instrumentation is needed. This thermocouple is 3Au–83Pd–14Pt versus 65Au–35Pd.

Type E is the Chromel versus constantan thermocouple. Its sensitivity increases from about 68 μV/deg. at 100°C to 81 μV/deg. at 500°C, and then falls to about 77 μV/deg. at 900°C. Its stability is good over this entire range. The low conductivity of the Chromel is usually desirable. Its high sensitivity makes this type of thermocouple useful for differential temperature measurements.

It has been seen that thermocouples are primarily instruments for determining temperature differences. This property is very useful in calorimetry in which temperatures should be made equal, such as in adiabatic calorimeters. High sensitivity thermocouples are desirable for this purpose. More sensitivity can be gained by using several thermocouples in series.

As an example, assume that a precision of a thousandth of a degree is desired for the temperature equality between a calorimeter and its shield. As a rule one may take one microvolt as the practical limit of accuracy of small e.m.f. measurements. Greater accuracy is possible but only at the expense of meticulous vigilance to avoid obscure spurious e.m.f.s. This rule requires a combined sensitivity of 1000 μV/deg. If Chromel versus constantan thermocouples of 67 μV/deg. are used, 15 thermocouples in series are needed. The 15 reference junctions can be distributed for integration over the surface of the shield and the 15 measuring junctions over the calorimeter surface. The wires can be in the nearly isothermal space between the calorimeter and its shield so there will be no significant temperature gradients to produce spurious e.m.f. values from inhomogeneities in these wires. The lead wires out to the measuring instrument, however, should be of homogeneous wire.

For such differential thermocouples, to be used where silver-soldered junctions of type K thermocouples had failed frequently, a thermocouple of 90Pt–10Rh versus 60Au–40Pd was found useful[63]. This thermocouple had a sensitivity of about 40 μV/deg. at 100°C and over 50 μV/deg. at temperatures in the range 300–1000°C.

It should be emphasized that the thermojunctions of the thermocouples, like other measuring instruments, must be at the temperatures being measured. This warning is especially pertinent if large temperature gradients must be encountered along thermocouple wires. Thermocouples must usually be electrically insulated from the body whose temperatures they are measuring but they should be in good thermal contact with that body. These two requirements are somewhat incompatible. A wire traversing a temperature gradient carries a heat flow which is proportional to the gradient, to the wire cross-section, and to its thermal conductivity. When this heat flows across the thermal resistance from the junction to the body being measured, there is a temperature drop which is a temperature error. This heat flow can sometimes be minimized by choosing small wires of low thermal conductivity. Before reaching the junction, the heat flow should be shunted to the body, by one means or another, so that the junction will be sufficiently close to the temperature being measured.

The sensitivity of all thermocouples approaches zero at 0°K. Thermocouples using an alloy of gold with 2·11 atomic per cent cobalt versus copper or "normal" silver have a considerably greater sensitivity than other types of thermocouples at low temperatures. Tables of these thermocouples and of types T, K, and J are available from 1°K to 280°K[49]. "Normal" silver is an alloy of silver with 0·37 atomic per cent gold which has about the same e.m.f. as copper but is often preferable at low temperatures because it has much smaller thermal conductivity. Inhomogeneities in the gold–cobalt wires have been found to be sources of error to about 1/500 of the e.m.f. Types K and J are not as suitable for low temperatures because of their low sensitivities and relatively high inhomogeneities. Both types T and gold–cobalt thermocouples are useful down to 4°K but the greater inhomogeneities in the gold–cobalt alloy may offset the advantage of its higher sensitivity. For measuring small temperature differences, however, the errors from inhomogeneities are small and the gold–cobalt thermocouples are superior.

C. Electrical Measurements in Thermometry

Two types of electrical measurements are made for thermometry, those of resistance and those of electromotive force (e.m.f.). The range of magnitudes of these properties is large. The resistances of platinum or other metal thermometers seldom exceed a few hundred ohms at their highest temperatures but may be as low as one or two hundredths of an ohm at the lowest temperatures. The resistances of semiconductor thermometers, on the other hand, are generally larger and increase exponentially as the temperature is lowered, therefore resistances as large as a megohm have to be measured. Specialized types of Wheatstone bridges are commonly used for measuring resistances, but it is sometimes preferable to use potentiometric methods which compare unknown resistances with standards. Thermoelectric thermometry almost necessarily requires potentiometric methods.

(1) Bridge Methods

The bridge, designed by Callendar and Griffiths about 1890, was an equal-arm slide-wire bridge for thermometers having about 1 ohm resistance. The thermometer leads were compensated for by dummy leads connected into the adjustable resistor arm, sometimes called the rheostat arm. In place of Callendar's slide wire, the bridge designed by Waidner in about 1902 used shunted decades, now known as Waidner–Wolff elements. These elements reduce the effect of the variability of dial contact resistances by large factors. The shunts add resistance to the adjustable arm which is compensated for by an equal resistance in the thermometer arm.

In about 1912 Meuller used a platinum thermometer which had four leads so it could be used with a Kelvin double bridge. This thermometer was connected into the Waidner bridge as a three-lead thermometer with the current introduced at a branch point at one end of the thermometer resistor, *Figure 2(a)*. He then reversed the thermometer to use the branch point at the other end of the resistor, *Figure 2(b)*.

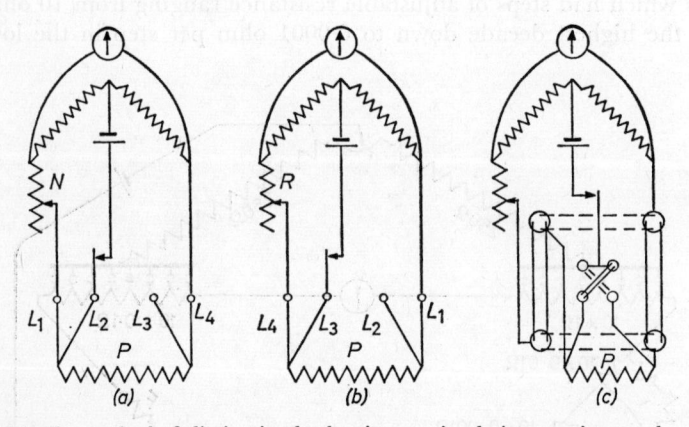

Figure 2. Mueller method of eliminating lead resistances in platinum resistance thermometry

The equations for bridge balance are:

$$N + L_1 = P + L_4 \tag{26}$$

and

$$R + L_4 = P + L_1 \tag{27}$$

in which N and R are the resistances in the adjustable arm for the normal and reversed connections, L_1, L_2, L_3, and L_4 are the lead resistances, and P is the resistance of the platinum resistor. Summing up the equations gives:

$$P = (N + R)/2. \tag{28}$$

This development led to his design of the mercury contact commutator, *Figure 2(c)*, which could be lifted and rotated 90° to perform the reversals quickly and reproducibly. Essentially this same method of reversal was also proposed by Smith in 1912[54].

Temperature measurements with platinum thermometers can, of course, be no more accurate than the resistance measurements. Even with thermometers having 25 ohms at the ice-point, the resistance change is only about a tenth of an ohm per degree. A ten-thousandth of an ohm thus corresponds to a thousandth of a degree, and greater precision is often sought. The variations in contact resistances at the various places in equal-arm bridges, therefore, are of prime importance in bridge design.

On the other hand, some thermal or intrusive e.m.f.s are almost inevitable in the thermometer circuit when its electric conductors traverse temperature gradients. In well made thermometers and bridges, these e.m.f.s are generally small, but they can be detected by the deflection of the galvanometer when it is connected without the battery. If the galvanometer is kept connected, however, the bridge balancing can proceed by current reversals as if the thermal e.m.f.s were not present.

(*a*) *Mueller Bridge*. In 1913 Mueller designed a bridge (*Figure 3*) with six decades which had steps of adjustable resistance ranging from 10 ohms per step in the highest decade down to 0·0001 ohm per step in the lowest[47].

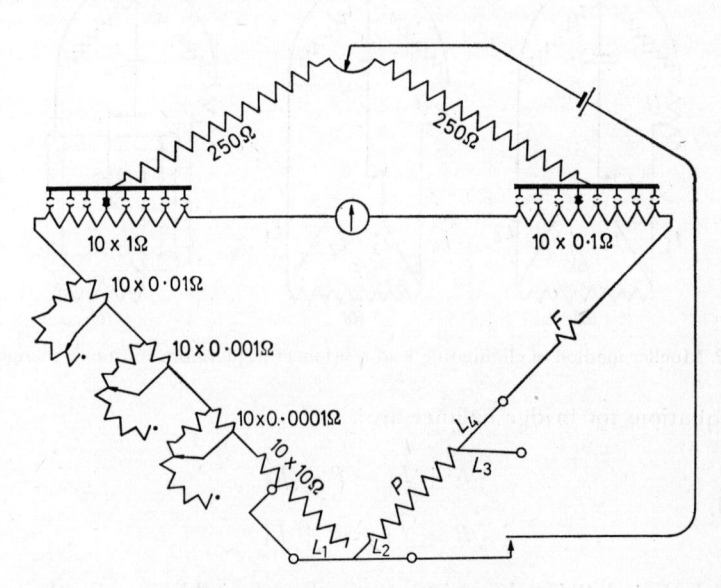

Figure 3. Mueller bridge, 1913 design

In describing this bridge he stated that the probable variations of resistance at contacts when kept in good condition was estimated to be about 0·00002 ohm for binding posts, 0·0001 ohm for plugs, and 0·0002 ohm for switches. (For mercury links the variation can be made less than 1 microhm.) He put the decade with 1 ohm steps at the end of the adjustable arm and the decade with 0·1 ohm steps at the end of the thermometer arm, with the galvanometer between the ends of these decades. The ratio arms were connected to these two decades with plugs at the appropriate positions for bridge balance. The ratio arms had 250 ohms resistance so that the effect of variability of plug contacts to them (0·0001 ohm) would be reduced tenfold when 25 ohm were being measured. The 10 ohm step decade had binding posts to one of the leads from the thermometer. This contact resistance, however, would remain constant if undisturbed. The three lower range decades were Waidner–Wolff elements with shunts ranging from 1·12 to 0·112 ohm. The sum of the three shunts was compensated for by the 0·1 ohm decade plus a fixed resistor, F. The shunting resistances on the three elements were alike and ranged from 11·5 ohms to ∞. The largest error from variation of contact resistance at the dial switches was at the zero position of the 0·01 ohm step decade and even here the effect of variation was reduced by a factor of over 100. The adjustable resistors were of manganin and thermostated in an oil bath. Provisions were made for adjusting the ratio coils to equality and for measuring the bridge zero.

During the next 30 years Mueller continued to add refinements to his designs which increased the accuracy of the bridge till it was 1–2 microhms. The latest design had manganin coils thermostated in an aluminum block a few degrees above room temperature where the temperature coefficient of manganin is nearly zero. In order to be able to use leaf contact dial switches for the 1 ohm and 0·1 ohm step decades, the resistances of the ratio coils were increased to 3000 ohms. The 10 ohm decade had mercury contact dial switches. Another decade was added with 0·00001 ohm steps at the lower end of the range and three extra coils added at the upper. Because it is more accurate and a great convenience not to have to make large changes in dial settings after reversals, a provision was made to adjust the resistance in one of the thermometer lead lines such that the normal and reverse balances would be very nearly equal. An extra pair of connections, operating with the mercury commutator, reversed the ratio arms as well and practically eliminated the error from inequality of the ratio arms. This method was proposed by Hoge at the National Bureau of Standards, and later essentially the same method was found to have been described by Smith in 1912[54]. A provision was made to use small snap switches to reverse the current when balancing the bridge.

When seeking the precision this bridge is capable of, care must be taken to keep the thermometer current steady. Sufficient current to discriminate between small increments of temperature heats the thermometer resistor above its surroundings by an amount many times as great as the precision of measurement. As mentioned earlier, the resistance at the temperature being measured can be found by measuring with two currents and extrapolating to zero power input. For correct measurements, however, the resistor temperature must be sufficiently near its asymptotic limit. Interruptions of the current stop this heating and the resistance starts to fall rapidly below its steady current value. The snap switches for reversing the current help to make these interruptions sufficiently brief. There is another interruption when the connections are reversed at the commutator. To minimize the time of interruption a recent practice is to use microswitches for the battery connections on the commutator. When lifting the commutator the microswitches stop the current before breaking the connections with the mercury links. As soon as the commutator is lifted, other microswitches connect the thermometer to a dummy bridge without a galvanometer. The same current then flows in the thermometer until the commutator has been rotated and the links start down to their new positions.

"An Improved Resistance Thermometer Bridge" of the Mueller type, described by Evans[33], is now in use at the National Bureau of Standards. In the design of this bridge the aim was to reduce the error in resistance measurements to a few tenths of a microhm. This aim was attained by (*i*) using mercury-wetted contacts of large area for the 1, 0·1, and 0·01 decades in place of the leaf switches, (*ii*) using totally enclosed wafer-type switches for the lower decades, (*iii*) adding a decade with 1 microhm steps, (*iv*) electrically guarding and shielding the measuring circuit, and (*v*) providing for digital read-out of the bridge dial settings.

(*b*) *Smith Bridge*. In 1912, Smith described four different "Bridge Methods for Resistance Measurements of Precision in Platinum Thermometry"[54]. A Smith bridge Type III was first made commercially in 1922 for use at the National Physical Laboratory. This bridge is a form of Kelvin double bridge which avoids some of the errors from variation of contact resistances by using a ratio of 100:1 instead of 1:1. After certain preliminary adjustments have been made, the measurement of resistance with this bridge requires only one observation, i.e. one current reversal for indicating the bridge balance. This feature is especially valuable when temperatures are changing. Another valuable feature is that small changes in lead resistances have much less effect on the measurements than they would with the Mueller bridge.

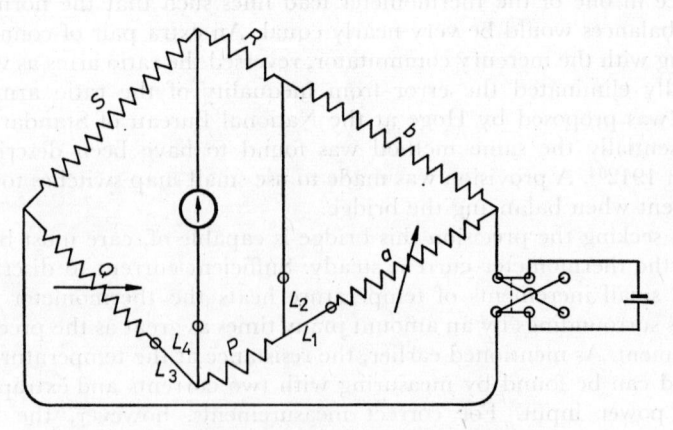

Figure 4. Smith bridge, Type III

The circuit of the bridge is shown in *Figure 4* where P is the platinum resistor, and L_1, L_2, L_3, and L_4 are the lead resistances. The ratio arms are S and R, usually of 1000 and 10 ohms respectively. Q has the adjustable resistors and a and b are shunt resistances. The ratio of the shunt resistance $(a + L_1)$ to b, however, is not the same as the ratio of $(Q + L_3)$ to S, but differs from it by an amount depending on the ratio of R to S. The equation of balance is:

$$P = QR/S + RL_3/S + \frac{bL_2}{(a + b + L_1 + L_2)} \left(\frac{Q + L_3}{S} - \frac{a + L_1}{b} \right). \quad (29)$$

If the lead resistances are equal and small with respect to S, Q, b, and a, it can be shown that if $a = b(R + Q)/(S - R)$ the equation reduces to:

$$P = QR/S. \quad (30)$$

To keep the relation, $a = b(R + Q)/(S - R)$, however, requires that the shunt a be varied with Q. Smith proposed to make $b = S$. This condition requires that the resistor coils in a differ from the corresponding coils in Q by a factor of $1/0.99$, and the first bridge at the National Physical Laboratory and subsequent bridges up to 1953 were made that way. Gautier[36] showed that if b were made 990 ohms instead of 1000 ohms the resistances of the coils in a should correspond exactly to those in Q. He also showed that a very small second order correction would be made even smaller by this change.

If the small second order correction is neglected, the equation of balance reduces to:

$$P = R/S(Q + L_3 - L_2). \quad (31)$$

This equation shows that, for the best measurements, L_3 and L_2 should be made equal. One way of achieving this is by a preliminary adjustment of the lead resistances till the balance is unaffected by changing the order of the leads from 1, 2, 3, 4 to 4, 3, 2, 1. Other methods have been described.

The bridge is complicated by the fact that the need for a change of the shunt a to correspond to the change of Q requires a double set of resistance coils with switches to go with them. These resistance coils, furthermore, are 100 times the resistances being measured, and coils of such large resistance are often difficult to stabilize.

(c) *General Remarks*. Temperatures are interpolated on the International Practical Temperature Scale with standard resistance thermometers by means of either a quadratic or a quartic formula relating the ratios of resistances to Celsius temperatures. The measurement of temperature, therefore, requires a determination of the ratio of the resistance at the temperature t to that at 0°C. It is not necessary, however, that the bridge coils used for these measurements be calibrated in absolute ohms but only necessary that they be calibrated in "bridge ohms". For temperature measurement a thermometer should have its resistance determined at some reference temperature, such as the triple point of water. If this determination is carried out and the bridge is self-consistent (autocalibrated), then measurements at some other temperature with the same bridge can yield the correct temperatures regardless of the actual magnitude of its bridge unit. It is to be recommended, therefore, that the user of a bridge calibrate it himself as often as is necessary to ensure its dependability and also that he check his thermometer with that bridge at his reference point as many times as he needs to ensure the accuracy he demands.

(d) Bridges for Semiconductors. Thermistor temperatures can be measured by use of still another specialized type of bridge called the "ratio set"[30]. The general form of resistance–temperature relation for thermistors in equation (22) can be written:

$$\ln (R_T/R) = B/T. \tag{32}$$

The characteristics of the ratio set correspond to the first term of a series development:

$$\ln (R_T/R) = 2 \left[\frac{R_T/R - 1}{R_T/R + 1} + \cdots \right]. \tag{33}$$

The principle of the bridge is represented in *Figures 5(a)* and *(b)*.

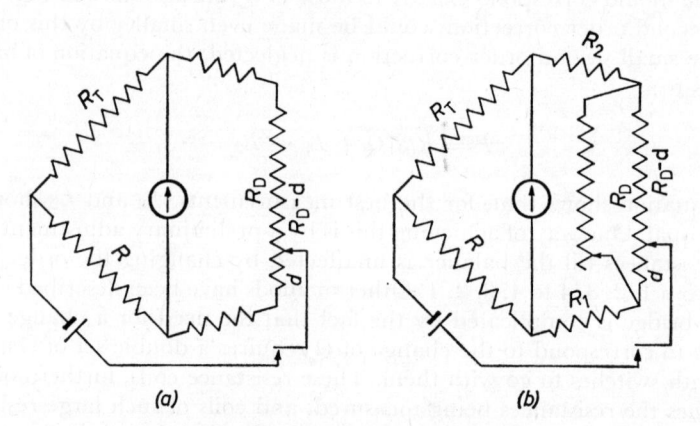

Figure 5. "Ratio set" bridge for thermistors

In these diagrams R_T is the thermistor resistance, R is a balance resistor, R_D is the resistance in the right branch of which d is one part and $R_D - d$ is the other part. The range of R_T/R can be limited between 1/3 and 3/1 by adding the resistances R_1 and R_2 at the ends of R_D. This arrangement permits a range of about 50 deg. One type of thermistor, for example, had a resistance of 2000 ohms at 25°C, so R was made 2000 ohms to obtain a setting of d in the center of R_D which was 2000 ohms with R_1 and R_2 1000 ohms each. For a different range of the same thermistor a different value could be taken for R.

To accommodate different values of the constant B of equation (32) an adjustable shunt is provided across the resistance R_D. The dial movement on R_D in this bridge varies nearly linearly with the temperature of the thermistor. Refinements can be applied to this type of bridge to give all the precision that may be needed for temperature measurements with semiconductors.

(2) *Potentiometric Methods*

Four leads on a resistance thermometer suggest that its resistance should be measured by potentiometric methods because they avoid lead resistance problems. Ordinary lead resistance problems have been fairly well solved in the Mueller bridge by means of the commutator, and in the Smith bridge by means of the high resistance shunts. Toward the extremes of temperature, however, the lead resistances are relatively large and more variable so are less perfectly eliminated in measurements with bridges. Potentiometric methods have always been used in some laboratories for resistance thermometry, but now new developments in potentiometers and circuitry make these methods even more desirable than formerly.

Potentiometric methods depend on the determination of the ratio of the e.m.f. across the unknown resistor to that across a standard or reference resistor when the same current is flowing in both. For the temperature discrimination now sought in present-day resistance thermometry, e.m.f. changes nearly as small as 0·001 µV need to be measured when thermometer currents of 1 mA are used. Thermal e.m.f.s of many times this magnitude are almost inevitable, so they have to be eliminated by reversing the current in both the unknown and the standard resistor, thus making four observations for each determination. These measurements also require that there be no drift of current during the period of measurements or change of current on the reversals.

The more common method for determination of the ratio of resistances is illustrated in *Figure 6(a)*[27]. The potentiometer, *P*, is balanced successively

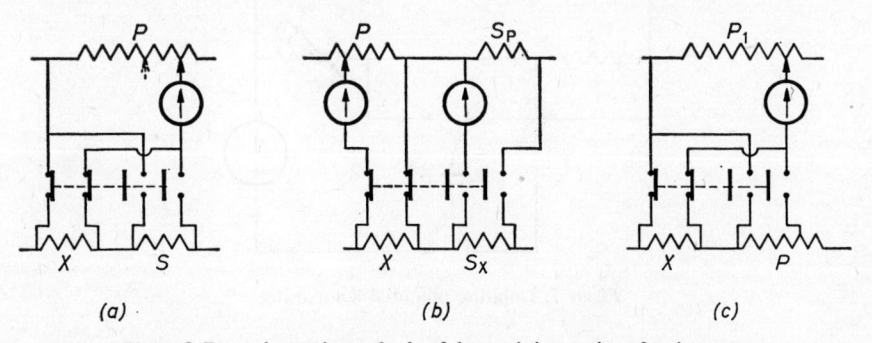

Figure 6. Potentiometric methods of determining ratios of resistances

to the e.m.f.s across the resistances *X* and *S* when a constant direct current is flowing through them both, and also when the current is reversed. The unknown resistance, *X*, is found by the equation:

$$X = (P_X/P_S) S, \tag{34}$$

in which S is the standard resistance and P_X and P_S are the potentiometer resistances for balance. The values of P_X and P_S are seldom alike, so two quite different potentiometer settings must be used. A second method is illustrated in *Figure 6(b)*, which has two standard resistors, S_X and S_P. With a constant current through X and S_X, the current in S_P and P is adjusted until the e.m.f.s across S_P and S_X are balanced. The potentiometer is then balanced to X, and the unknown resistance is found by the equation:

$$X = P(S_X/S_P). \tag{35}$$

Although this method appears more complicated, it may be more convenient because the potentiometer setting is only for the e.m.f. across X. A third method, illustrated in *Figure 6(c)*, does not appear to have been used commonly but has advantages in some instances. Here the current in P_1 is adjusted to give a balance between X and P_1. P is then balanced to P_1 so:

$$X = P. \tag{36}$$

In the second and third methods, the resistance in X is determined in the units of P.

An adaptation of the third method was described by Dauphinee[24] in a device called an "isolating potential comparator". In *Figure 6(c)* the e.m.f. across X is referred to the e.m.f. across P_1 and then measured with P. In the isolating potential comparator (*Figure 7*), the e.m.f. produced by a current

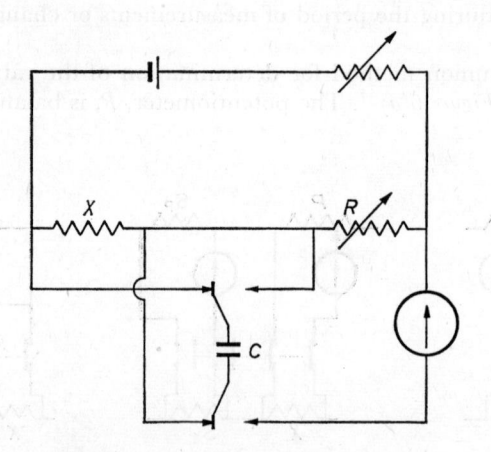

Figure 7. Isolating potential comparator

across X is stored in a high quality capacitor, C, of say 100 µf, and compared a fraction of a second later with the e.m.f. produced by the same current across a resistance, R, which is adjustable to make $R = X$. The capacitor C is charged with a double-pole double-throw chopper which operates at some selected frequency between 20 and 80 c/s. The balance can be made irrespective of any reasonable resistance in the leads. Thermal e.m.f.s are easily eliminated by reversing the battery current and preferably reversing the capacitor at the same time. Thus, in this comparator, the four observations for a determination are essentially reduced to one reversal.

The chopper is somewhat more efficient if it is a synchronous double or triple unit with appropriate phase differences. Pickup from 60 cycle a.c. power lines can be avoided satisfactorily by using frequencies of 37·5 or 51 c/s. The adjustable resistors in a Mueller bridge can sometimes be used for the adjustable resistance, R, in *Figure 7*.

The isolating potential comparator was also used for a "direct reading resistance thermometer bridge"[26] to read directly in degrees Celsius at temperatures ranging from $-50°C$ to $700°C$ with a precision of $0·001°$. With this bridge any nominal 25·5 ohm (R_0) standard resistance thermometer can be used with the equation $R = R_0(1 + At + Bt^2)$. The bridge consists of a loop with three adjustable resistors, one for R_0, one for the constant A, and a third for the ratio of the constants B/A of the equation. Six decades in one of the three loop resistors adjust a resistance to compare directly with the thermometer resistance by means of the comparator. This bridge also has a provision for recording.

In thermoelectric thermometry the range of e.m.f.s to be measured is generally within the range mentioned above for potentiometric methods in resistance thermometry. The highest e.m.f. for a thermocouple, listed in thermocouple tables, is scarcely more than 50 mV, which corresponds to the e.m.f. across a 25 ohm thermometer when 2 mA are flowing through it. At the lower end of the range there is more difficulty with the thermocouple than the resistor. The small e.m.f.s caused by currents flowing across variable contact resistances have been overcome in bridges by various methods, and similar methods have been used in the design of potentiometers. The thermocouple, however, is also plagued by certain extraneous e.m.f.s, caused by temperature gradients in dissimilar conductors in the circuit, which have been known in different laboratories as thermal, stray, spurious, unwanted, unwelcome, intruding, or vagabond e.m.f.s. In resistance thermometry these e.m.f.s can usually be eliminated from the measurements by reversing the current, but in thermoelectric thermometry they have to be avoided as far as practicable. They can sometimes be avoided by using homogeneous conductors such as copper and by thermal shielding to avoid temperature gradients in the dissimilar conductors. For small e.m.f.s, accuracies within 1 μV are easy, those of 0·1 μV require care and those of 0·01 μV require meticulous care, but claims of 0·001 μV accuracy are evidences of narcissism.

The choice of instrument for e.m.f. measurements with thermocouples depends on the magnitude of the e.m.f.s and the purpose for which the thermocouples are to be used. Fundamentally, thermocouples are for measuring temperature differences which may be either large or small, whereas resistance thermometers are for measuring temperature ratios (e.g. the ratio of the temperature t to the temperature of the triple point). If thermocouples are to be used for indicating temperature equalities in calorimeters, multiple junction thermocouples (thermopiles) can be used with a sensitive indicating instrument such as a galvanometer, but in that situation the unit of the magnitude of departure from balance may be of secondary importance. Galvanometer deflections are measures of current and are proportional to the e.m.f.s in different thermocouples, provided the resistances in the circuits are the same.

Where two-figure accuracy is ample, a Lindeck potentiometer is a convenient instrument which may easily be made in the laboratory. *Figure 8*

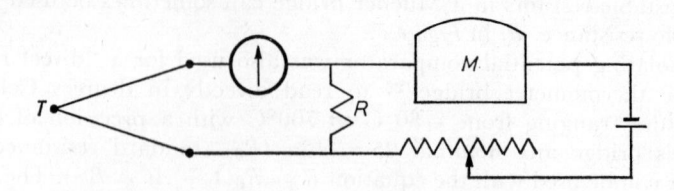

Figure 8. Principle of Lindeck potentiometer

is a diagram to illustrate its principle. Assume, for example, that R has a resistance of 0·001 ohm. When an e.m.f. in the thermocouple, T, is balanced by the e.m.f. across R, as indicated on the galvanometer, then the e.m.f. is equal (in microvolts) to the number of milliamperes indicated on the milliammeter, M.

If greater than two-figure accuracy is necessary, more elaborate instruments have to be used. White[65] described the development of potentiometers having up to five decades, which had been designed by Diesselhorst, White, and Wenner. He claimed that the success of potentiometers for thermoelectric work was not due as much "to unprecedented refinement in the construction of switches" as it was "to devices or arrangements which put the switches where they could not do harm."

Dauphinee[27] described various potentiometers made by such manufacturers as Leeds and Northrup, Tinsley, Rubicon, and Cambridge Instrument Company and diagrammed the circuits of many of them. Design development has now gone to potentiometers with as many as seven decades. Some of these potentiometers can be used with the isolating potential comparator mentioned above.

3. *Optical pyrometry*

Temperature measurement by means of the optical pyrometer takes advantage of Planck's law at one wavelength. As discussed above, Planck's law makes the optical pyrometer of fundamental importance in the realization of the International Practical Temperature Scale above 1063°C. However, optical pyrometers (either visual or photoelectric) are often less useful than thermocouples in the range where thermocouples can be used (up to 1300°C or possibly 2000°C). They are less useful because the optical pyrometer needs blackbody conditions or well-defined emittance, it needs a sighting tube, it lacks convenient recording, and it is not well adapted to measuring small differences or changes of temperature. Fortunately some of these disadvantages do not apply to the photoelectric pyrometers now being developed.

Figure 9 is a diagram of a visual optical pyrometer. It consists of an objective lens, O, which images the source, S, in the plane of the filament of a reference lamp, L. The source image and reference lamp filament are viewed together through the field lens, F, and eyepiece, H, which together give a magnification of about 20 times. The red filter, R, reduces the spectral band-width and makes it possible to define the effective wavelength to be used in Planck's law, more accurately. This wavelength is

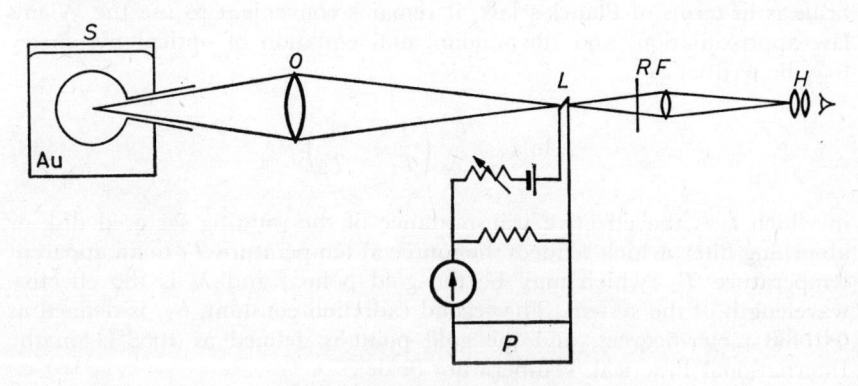

Figure 9. Visual optical pyrometer

usually chosen to be about $0.65\ \mu$, and is known to an accuracy of about $1\ m\mu$. The reference lamp circuit includes a battery and rheostat to control the lamp current, a standard resistor, and a potentiometer, P, to measure the current.

A primary calibration of an optical pyrometer begins with the calibration of the pyrometer at the gold point. This calibration is done by matching the brightness of the reference lamp filament to that of a blackbody at the gold point, the match being determined by the "disappearances" of the filament into the background of the source. Higher source temperatures are determined by reducing the apparent brightness to that of the gold point by means of rotating sectored disks of accurately known transmission. This transmission gives the intensity ratio, J_t/J_{Au}, to be used in the formula

$$\frac{J_t}{J_{Au}} = \frac{\exp\left[\dfrac{C_2}{\lambda(t_{Au} + T_0)}\right] - 1}{\exp\left[\dfrac{C_2}{\lambda(t + T_0)}\right] - 1} \qquad (37)$$

in which $T_0 = 273.15°$. This equation defines the international scale above 1063°C, and must be solved for t.

By proceeding along these lines, it is possible to build up a current–temperature calibration of the reference lamp, which is normally a vacuum lamp with a flat horizontal filament about 0·05 mm wide and about 1·5 cm long. Such a lamp can be safely calibrated from about 750°C to 1350°C without risk of being over-run. The rotating sectored disks can also be used to calibrate absorbing filters. These filters are not only more convenient to use than the sectored disks, but they also can be used in combination to make possible the reduction of sources at indefinitely high temperatures to apparent temperatures within the calibration range of the reference lamp.

An optical pyrometer calibrated in this way can be used to calibrate secondary standard tungsten strip or ribbon filament lamps, which can in turn be used in the routine calibration of other optical pyrometers.

Although the definition of the International Practical Temperature Scale is in terms of Planck's law, it remains convenient to use the Wien's law approximation, and the fundamental equation of optical pyrometry may be written as

$$\ln t_e = \frac{C_2}{\lambda_e}\left(\frac{1}{T_1} - \frac{1}{T_2}\right) \tag{38}$$

in which t_e is the effective transmittance of the rotating sectored disk or absorbing filter, which reduces the source at temperature T_1 to an apparent temperature T_2 (which may be the gold point), and λ_e is the effective wavelength of the system. The second radiation constant, C_2, is defined as 0·01438 meter-degrees, and the gold point is defined as 1063°C on the International Practical Temperature Scale.

The calibration of effective wavelength is done through the relation

$$t_e = \frac{J(\lambda_e, T_2)}{J(\lambda_e, T_1)} = \frac{\int_0^\infty J(\lambda, T_2)\, t_\lambda V_\lambda\, d\lambda}{\int_0^\infty J(\lambda, T_1)\, t_\lambda V_\lambda\, d\lambda} = \frac{\int_0^\infty J(\lambda, T_1)\, t'_\lambda t_\lambda V_\lambda\, d\lambda}{\int_0^\infty J(\lambda, T_1)\, t_\lambda V_\lambda\, d\lambda} \tag{39}$$

in which $J(\lambda, T)$ is Planck's function at wavelength λ and temperature T (but Wien's function is usually a sufficient approximation), t_λ is the spectral transmission factor for the red filter at wavelength λ, t'_λ is the spectral transmission factor for the absorbing filter or sectored disk at wavelength λ, V_λ is the visual relative luminous efficiency factor at wavelength λ (the Comité International Éclairage standard observer values are normally used for this function), and λ_e is a function of T_1 and T_2. Relationships exist which greatly simplify the computations involved in integrating equation (39).

For absorbing filters, the effective transmittance, t_e, is in general a function of source temperature. However, by choosing an absorbing filter of the right spectral transmission characteristic, it is possible to get a filter for which

$$\frac{\lambda_e}{C_2} \ln t_e = -A = \frac{1}{T_1} - \frac{1}{T_2} \tag{40}$$

in which A is a constant of the filter known as the pyrometric absorption, which is usually given in 10^{-6} deg^{-1} or mireds. The use of such filters greatly facilitates high temperature measurements, since they result in a perfect color match (not merely a luminance match) and the absorption of such a filter is independent of source temperature. For a fuller discussion of the theory of optical pyrometry see Kostkowski[39] and Lovejoy[44].

Optical pyrometers can be calibrated with an accuracy uncertainty of about 1 deg. near the gold point which increases to about 2° around 2200°C. To realize this accuracy with a commercial pyrometer, however, it is necessary to use a lamp current–temperature calibration of the pyrometer and a pyrometric absorption calibration of the filters rather than to rely on the manufacturer's graduated scale calibration. An external current source, rheostat, standard resistor, and potentiometer should be used for the measurement. Again, for the most accurate work, a temperature measurement should be made by averaging, say, four readings (disappearances), preferably by each of two observers. In this way a temperature discrimination of better than 1 deg. should be possible near the gold point.

In using an optical pyrometer, it is important to see that the optical aperture or stop really is the front objective lens of the pyrometer and not, for example, an undersized hole in a radiation shield. Also there must be no obscuring of the field. For example, a single thickness of glass, as in an observation port, reduces the transmission by about 8 per cent (owing to reflection), equivalent to about 8 deg. near the gold point. Furthermore, the transmission loss may increase with time owing to the deposition of a film on the port. This is a serious problem in using a vacuum furnace and is usually handled by means of a shutter, which is left closed over the port except when observations are in progress. The shutter, which of course is inside the vacuum, may be actuated magnetically or by means of a sealed shaft going into the vacuum.

It is of great importance to have blackbody conditions for the source. A cylindrical hole, the walls of which are at a uniform temperature, closed at the far end and ten diameters in length, will usually have an emittance of better than 0·99 (unless the inner walls are highly polished). Published emittance data should be used with great caution since, for a given material, emittances vary with surface finish (rough or polished), temperature, and wavelength (although published data are, in most cases, for the usual optical pyrometer wavelength of 0·65 μ). In addition reflection from neighboring hot surfaces often increases the apparent temperature of a non-black body.

If the emittance of a non-blackbody is known, the following equation enables the true temperature, T, to be calculated:

$$\frac{1}{T} = \frac{1}{T_B} + \frac{\lambda_e}{C_2} \ln E_{\lambda_e, T} \tag{41}$$

In equation (41), λ_e is the effective wavelength (between measured brightness temperature T_B and color temperature T_C) and $E_{\lambda_e, T}$ is the emittance at wavelength λ_e and temperature T.

An error of 10 per cent in the emittance corresponds to an error, at a wavelength of 0·65 μ, of around 10 deg. near the gold point. If the unknown emittance of the source were the same at two wavelengths (graybody condition), then it would be possible, by determining the ratio of the emitted intensities at the two wavelengths, to determine the true temperature. In general this condition is not fulfilled, however, and the two wavelength or color pyrometer finds little application in the laboratory.

4. *Vapor pressure thermometry*

The vapor pressure of a saturated vapor in equilibrium with its liquid is, in many respects, a nearly ideal thermometric property because it is sensitive, reproducible and requires but a single measurement, and also because the experimental precautions required are relatively minor compared to those for gas thermometry. Furthermore, over the usable range of temperature for a saturated fluid, the vapor pressure furnishes a continuous series of temperature values. For measuring temperatures below 20°K the substance most frequently used is He^4, which covers the range, 1–5°K. Next is He^3, which covers the range, 0·3–3°K, and then H$_2$, which covers the range, 10–20°K.

In many cryogenic experiments, the thermal environment is provided by a cryostat containing one of these three liquids boiling at the appropriate pressure. Except when maximum accuracy is required, it is often possible to determine the temperature adequately, in such cryostats, merely by measuring the pressure over the bath. It should be well noted, however, that this practice may lead to gross error unless the bath is carefully cooled to the desired temperature. Only in He II (He^4 below the λ-point at 2·18°K) is it possible to realize a uniform temperature by warming. In cryostats at temperatures below 14°K, using H$_2$ in the solid state, it is not so reliable to judge the temperature by the vapor pressure because of the lack of mixing.

To obtain the maximum precision of which the vapor pressure method of temperature measurement is capable, it is necessary to adopt more refined procedures than for cryostats. A little of the thermometric fluid should be condensed into a small bulb which is in intimate thermal contact with the object whose temperature is to be determined. The pressure exerted by this liquid should be measured by sufficiently refined manometric methods. Several precautions are necessary. With the possible exception of measurements involving He II it is generally preferable to use a vacuum-shielded line to transmit the pressure in the vapor pressure bulb to the manometer because of the danger of having a spot colder than the bulb along the line. The diameter of the pressure-transmitting tube should be large enough to avoid the thermomolecular pressure differences which arise because of the large difference in temperature at the two ends of the tube. Alternatively, of course, corrections for this effect can be made to the measured pressure[51, 62]. The liquid in the vapor pressure bulb should be thoroughly shielded from radiation or other heat flow down the pressure-transmitting tube, else there is likely to be a temperature difference between the liquid in the bulb and the object whose temperature is sought.

Special problems arise for both He^4 and H$_2$. For He^4, the extraordinary changes in properties of the liquid at the λ-point and the existence of the Rollin film at temperatures below this point can complicate matters. "λ-leaks", i.e. leaks large enough to pass He II but no other liquid or gas, must be eliminated from the system. The effect of the film on, and in, the pressure-transmitting line is still not a settled question. A brief discussion of this point appears in NBS Monograph 10.

For H_2, the special problem arises from the existence of the ortho and para arrangements of the H_2 molecule, the ortho form being stable at high temperatures and the para form stable at low temperatures. The vapor pressures of these two forms are not the same. Hydrogen gas which has been at room temperature for a long time has approximately 25 per cent para–H_2 and 75 per cent ortho–H_2. This mixture is commonly referred to as "normal" hydrogen. Liquid hydrogen which has been at its normal boiling point for a long time, or has been converted rapidly by the use of a catalyst, will consist of nearly pure para–H_2, only about 0·2 per cent of the molecules remaining in the ortho form. In "clean" apparatus the transformation from the ortho to para form proceeds fairly slowly at the normal boiling point; the vapor pressure of freshly condensed normal hydrogen changes at the rate of about 0·23 mm Hg per hour[31, 66]. The experimenter thus has a choice of two possible modes of operation. He may condense normal hydrogen into "clean" apparatus and make all his measurements as quickly as possible, say within an hour or so, or he may put some catalyst such as neodymium oxide, hydrous ferric oxide, or chromic anhydride, into the vapor pressure bulb and wait long enough after condensation for a complete conversion to be made. If liquid hydrogen is stored in containers containing a catalyst, a plentiful supply of para–H_2 is always available.

Which of these procedures to adopt is, of course, a matter for the individual experimenter to decide. However, if the first approach is adopted, it should be well understood that "clean" is a rather stringent requirement. For example, 0·01 per cent O_2 in the condensed hydrogen will raise the rate of change of the vapor pressure at the normal boiling point from 0·23 mm Hg per hour up to about 0·7 mm Hg per hour; 0·7 mm Hg change of pressure corresponds to about 0·003 deg. change in temperature at the normal boiling point. On the other hand, if the second method is chosen, the temperature at some fixed pressure (most conveniently atmospheric) can be monitored after liquid has been condensed into the bulb, and measurements begun only after this temperature has approached satisfactorily close to its asymptotic value.

5. *Magnetic thermometry*

It was noted earlier that the magnetic susceptibility of a paramagnetic material (salt) is a thermometric parameter useful for establishing the thermodynamic scale at very low temperatures. In addition, such paramagnetic salts are commonly used for practical temperature measurements.

The quantity usually measured is the self or mutual inductance of a coil system surrounding a sample of paramagnetic salt. This inductance, θ, is related to the temperature by the formula:

$$\theta = \theta_0 + C/(T + \Delta). \tag{42}$$

In this equation, θ_0 is the inductance of the coil system when the salt is at a temperature sufficiently high for its susceptibility to be negligible, C is the effective Curie constant, and Δ the effective Weiss constant which may include a geometric factor not related to the true Weiss constant. The values of these three constants are usually found by fitting the equation to numerous calibration points where the temperatures are determined with the vapor pressure of He^4. Calibration points obtained with liquid He^3, H_2, or even N_2 may also be useful.

When the magnetic thermometer is used at temperatures so low that values of temperature derived from equation (42) are inaccurate, these values are called magnetic temperatures and denoted by T^*, as mentioned earlier. These can be corrected to thermodynamic temperature values by means of a table or graph of T versus T^*, such as already exists for a number of salts. Among these salts are iron ammonium alum, $FeNH_4(SO_4)_2 \cdot 12 H_2O$, chromium potassium alum, $CrK(SO_4)_2 \cdot 12 H_2O$, manganese ammonium alum, $Mn(NH_4)_2(SO_4)_2 \cdot 6 H_2O$, cerium magnesium nitrate, $Ce_2Mg_3(NO_3)_{12} \cdot 24 H_2O$, and chromium methylamine alum, $Cr(NH_3CH_3)(SO_4)_2 \cdot 12 H_2O$[64].

Although none of these materials can be considered an accepted standard, chromium methylamine alum is often preferred because of the good agreement between the various existing measurements of its T versus T^* relation[29].

The susceptibility of the nuclear spin system may also be used for practical thermometry as well as for establishing the thermodynamic scale. The effective Curie constant for this system has so far been obtained only by theoretical calculation[41].

IV. References

[1] Adams, L. H., *Bull. Am. Inst. Mining Met. Eng.*, **1919**, 2111.
[2] Barber, C. R., *J. Sci. Instr.*, **27**, 47 (1950).
[3] Barber, C. R., *J. Sci. Instr.*, **32**, 416 (1955).
[4] Barber, C. R. In *Temperature**, I I I–1, p. 345.
[5] Beattie, J. A., D. D. Jacobus, J. M. Gaines, Jr., M. Benedict, and B. E. Blaisdell, *Proc. Am. Acad. Arts Sci.*, **74**, 327 (1941).
[6] Blackburn, G. F., and F. R. Caldwell. In *Temperature**, I I I–2, p. 161.
[7] Borovick-Romanov, A. C., P. G. Strelkov, M. P. Orlova, and D. N. Astrov. In *Temperature**, I I I–1, p. 113. See also F. Z. Alieva, *Tr. Vses. Nauchn. Issled. Inst. Metrol.*, **51**, 49 (1961).
[8] Brickwedde, F. G., H. van Dijk, M. Durieux, J. R. Clement, and J. K. Logan, *Natl. Bur. Std. (U.S.), Monograph, No.* 10, 1960; *J. Res. Natl. Bur. Std.*, **64A**, 1 (1960).
[9] Buckingham, E., *Bureau Standards, Bull.*, **3**, 237 (1907).
[10] Burgess, G. K., *Bur. Std. J. Res.*, **1**, 635 (1928).
[11] Busse, J. In *Temperature**, I, p. 228.
[12] Catalano, G., M. Edlow, and H. H. Plumb. In *Temperature**, I I I–1, p. 129.
[13] Clement, J. R., and E. H. Quinnell, *Rev. Sci. Instr.*, **23**, 213 (1952).
[14] Clement, J. R., and E. H. Quinnell, *Phys. Rev.*, **85**, 502 (1952); **92**, 258 (1953).
[15] Clement, J. R., R. L. Dolecek, and J. K. Logan, *Advances in Cryogenic Engineering* (Ed. K. D. Timmerhaus), Vol. II, p. 104, Plenum Press, Inc., New York (1960).
[16] *Comptes Rendus, Première Conf. Gén. des Poids et Mesures*, p. 38, Paris, Gautheir-Villars (1889).
[17] *Comptes Rendus, Septième Conf. Gén. des Poids et Mesures*, pp. 58, 94, Paris (1927).
[18] *Comptes Rendus, Neuvième Conf. Gén. des Poids et Mesures*, p. 55, Paris (1948).
[19] *Comptes Rendus, Neuvième Conf. Gén. des Poids et Mesures*, pp. 57, 89, Paris (1948).
[20] *Comptes Rendus, Dixième Conf. Gén. des Poids et Mesures*, p. 79, Paris (1954).
[21] *Comptes Rendus, Onzième Conf. Gén. des Poids et Mesures*, p. 64, Paris (1960).
[22] *Comptes Rendus, Onzième Conf. Gén. des Poids et Mesures*, p. 124, Paris (1960).

* *Temperature, Its Measurement and Control in Science and Industry*, Vol. I, American Institute of Physics, Reinhold, New York (1941); Vol. I I (Ed. H. C. Wolfe), Reinhold, New York (1955); Vol. I I I (Editor-in-Chief C. M. Herzfeld), Part 1 (Ed. F. G. Brickwedde), Part 2 (Ed. A. I. Dahl), Reinhold, New York (1962). In the references *Temperature* I I I–1, for example, would refer to Vol. I I I, Part 1.

23 Cragoe, C. S. In *Temperature**, I, p. 89.
24 Dauphinee, T. M., *Can. J. Phys.*, **31**, 577 (1953).
25 Dauphinee, T. M., and H. Preston-Thomas, *Rev. Sci. Instr.*, **25**, 884 (1954).
26 Dauphinee, T. M., and H. Preston-Thomas, *Rev. Sci. Instr.*, **31**, 253 (1960).
27 Dauphinee, T. M. In *Temperature**, I I I–1, p. 269.
28 Day, A. L., and R. B. Sosman, *Am. J. Sci.*, **29**, 93 (1910); **33**, 517 (1912).
29 de Klerk. In *Temperature**, I I, p. 251; *Handbuch der Physik* (Ed. S. Flügge), Vol. XV, pp. 38–209, Springer-Verlag, Berlin (1956).
30 Droms, C. R. In *Temperature**, I I I–2, p. 339.
31 Durieux, M. Thesis, Leiden (1960).
32 Edlow, M. H., and H. H. Plumb. In *Temperature**, I I I–1, p. 407.
33 Evans, J. P. In *Temperature**, I I I–1, p. 285.
34 Evans, J. P., and G. W. Burns. In *Temperature**, I I I–1, p. 313.
35 Friedberg, S. A. In *Temperature**, I I, p. 359.
36 Gautier, M., *J. Sci. Instr.*, **30**, 381 (1953).
37 Grant, D. A., and W. F. Hickes. In *Temperature**, I I I–2, p. 305.
38 Hall, J. A., and V. M. Leaver. In *Temperature**, I I I–1, p. 231.
39 Kostkowski, H. J., and R. D. Lee. In *Temperature**, I I I–1, p. 449.
40 Kunzler, J. E., T. H. Geballe, and G. W. Hull, Jr. In *Temperature**, I I I–1, p. 391.
41 Kurti, N., F. N. H. Robinson, F. Simon, and D. A. Spohr, *Nature*, **178**, 450 (1956).
42 Lachman, J. C., and J. A. McGurty. In *Temperature**, I I I–2, p. 177.
43 Lindenfeld, P. In *Temperature**, I I I–1, p. 399.
44 Lovejoy, D. R. In *Temperature**, I I I–1, p. 487.
45 Meyers, C. H., *Bur. Std. J. Res.*, **9**, 807 (1932).
46 Moser, H., J. Otto, and W. Thomas, *Z. Physik*, **147**, 59, 76 (1957); Moser, H. In *Temperature**, I I I–1, p. 167.
47 Mueller, E. F., *Bur. Standards, Bull.*, **13**, 547 (1916).
48 Partington, J. R., *An Advanced Treatise on Physical Chemistry*, Vol. I, p. 536, Wiley, New York (1949).
49 Powell, R. L., L. P. Caywood, Jr., and M. D. Bunch. In *Temperature**, I I I–2, p. 65.
50 *Procès-Verbaux, Comité International des Poids et Mesures*, p. 86 (1887).
51 Roberts, T. R., and S. G. Sydoriak, *Phys. Rev.*, **102**, 304 (1956).
52 Roebuck, J. R., and T. A. Murrell. In *Temperature**, I, p. 60.
53 Shenker, H., J. I. Lauritzen, Jr., R. J. Corruccini, and S. T. Lonberger, *Natl. Bur. Std. (U.S.), Circ. No.* 561 (1955).
54 Smith, F. D., *Phil. Mag.*, **24**, 541 (1912).
55 Stimson, H. F., *J. Res. Natl. Bur. Std.*, **42**, 209 (1949).
56 Stimson, H. F. In *Temperature**, I I, p. 141.
57 Stimson, H. F., *J. Res. Natl. Bur. Std.*, **65A**, 139 (1961).
58 Stimson, H. F., *Science*, **136**, 254 (1962).
59 Stimson, H. F. In *Temperature**, I I I–1, p. 59.
60 Sydoriak, S. G., T. R. Roberts, R. H. Sherman, and F. G. Brickwedde, "Proposition pour une Échelle Internationale de Température a Tension de Vapeur de He^3", *Rapport* du Comité Consultatif de Thermométrie au Comité Internationale des Poids et Mesures, 6e Session (1962).
61 Thompson, R. D. In *Temperature**, I I I–1, p. 201.
62 Weber, S., *Leiden Communications Suppl.*, **71b** (1932).
63 West, E. D., *Rev. Sci. Instr.*, **31**, 896 (1960).
64 White, G. K., *Experimental Techniques in Low-Temperature Physics*, p. 235, Clarendon, Oxford (1959).
65 White, W. P. In *Temperature**, I, p. 265.
66 Woolley, H. W., R. B. Scott, and F. G. Brickwedde, *J. Res. Natl. Bur. Std.*, **41**, 379, 455 (1948).

* *Temperature, Its Measurement and Control in Science and Industry*, Vol. I, American Institute of Physics, Reinhold, New York (1941); Vol. I I (Ed. H. C. Wolfe), Reinhold, New York (1955); Vol. I I I (Editor-in-Chief C. M. Herzfeld), Part 1 (Ed. F. G. Brickwedde), Part 2 (Ed. A. I. Dahl), Reinhold, New York (1962). In the references *Temperature* I I I–1, for example, would refer to Vol. I I I, Part 1.

CHAPTER 3

Energy Measurement and Standardization

L. HARTSHORN

National Physical Laboratory, Teddington, U.K. (Retired)

A. G. MCNISH

National Bureau of Standards, Washington, D.C., U.S.A.

Contents

I.	The Problem	59
II.	The Basis for Energy Measurement	60
III.	The Derived Mechanical Units	63
IV.	Working Standards for Dynamics	64
V.	The Electrical Units	65
VI.	Absolute Electrical Standards	66
VII.	Standard Inductors	68
VIII.	Absolute Determination of the Ampere	70
IX.	Absolute Determination of the Ohm	72
X.	Working Standards	75
	1. Standard Resistors	76
	2. Standard Cells	77
XI.	Application to Thermodynamic Measurements	78

I. The Problem

Thermodynamics begins with the concept of energy as an accurately measurable physical quantity, and the primary requirement for the attainment of the present level of the science as a cooperative human enterprise is that quantitative reports about experiments concerning energy and related quantities shall have precisely the same meaning for all competent workers in this field in all parts of the world. This degree of standardization, to use that term for the moment in its broadest sense, involves two more detailed requirements: (*i*) a universally agreed unit of energy, that is to say, a precisely defined particular quantity of energy, by comparison with which any other quantity of energy can be expressed and (*ii*) some means by which it is practicable for any competent worker to determine the magnitude of an unknown quantity of energy in terms of the unit.

The unit of energy is necessarily an abstract quantity, since it must be equally available to every observer if it is to serve its purpose. The unit can indeed be considered as a purely theoretical quantity. However, the second requirement includes apparatus and practical techniques of a somewhat complex character, and the main purpose of the present chapter is to outline those techniques which have been developed to satisfy this basic requirement. These techniques are not the ones most employed in calorimetric practice, even that of the highest precision. Such techniques will appear in other chapters. The concern at present is with the links that have been established between the abstract unit and the methods and apparatus of actual practice. These links are, in an important sense, the foundations of the science as practiced today, and every worker should appreciate their significance, though he will, no doubt, be glad to leave the details to the standardizing laboratories.

The complex character of the links arises from the fact that energy, notwithstanding its fundamental role in physical science, is not directly measurable in the way that other quantities, notably length, mass, and time, are measurable, i.e. independently of one another and of any other physical quantity. Other quantities, in fact, are often used as a matter of convenience in the basic measurements of length, mass, and time, but they are not essential to such measurements, which depend ultimately on the facts that length, mass, and time-interval are additive, and that comparators have been devised by which the equality of lengths, masses, and time-intervals can be compared with great precision. Energy, though also additive, must be measured indirectly in terms of other quantities which can be measured directly. Length, mass, and time are the quantities chosen by common consent as the ultimate basis of measurement not only of energy but also of all physical quantities. The justification for this choice is simply the practical one that it has been generally accepted and found to provide the required coordination of scientific work over a very wide field, features of great importance in thermodynamics. The term "absolute measurement" commonly applied to measurements made on this basis, should not be taken to imply any superiority in principle to possible alternatives.

On theoretical consideration one might be tempted to hold that the natural unit of energy is the Planck quantum at a standard frequency, and recent research has shown that a standard frequency can be realized with, say, the cesium atomic resonator, with a precision that has not been surpassed in any other measurement. If the energy comparator should become a reality so that energy could be measured by direct comparison with a quantum standard, our present indirect measurements based on standards of length, mass and time would probably no longer be described as "absolute". However, this system, as now established throughout the world by the cooperative work of the national standardizing laboratories and international organizations, has been of immense service in the development of science. It has, among other things, provided workers in thermodynamics with practicable scales of reference for all thermodynamic quantities. The means by which these scales have been established and their accuracy steadily improved will be outlined in this chapter and the present position noted.

II. The Basis for Energy Measurement

It is a matter of common experience among practical workers in thermodynamics that the most convenient methods of measuring power and energy are the electrical methods. The apparatus is now so readily available and the technique so convenient, flexible, and accurate that the ordinary electrical methods can, for our present purpose, be taken for granted as *the* methods of modern practice. In establishing a comprehensive system of physical measurement it has been found expedient to begin with length, mass, and time. It follows that the basis of the subject will include a sequence of definitions of units beginning with length, mass, and time and including energy and the various electrical units employed in practice. These definitions will be given in the form commonly used by the international organizations, such as the General Conference of Weights and Measures and its advisory committees and by the national standardizing laboratories. The scientific laws underlying the definitions are of course taken for granted, since the units are merely defined for use in conjunction with the network of scientific laws that is already accepted as the framework of the science by users of the definitions. The definitions do, however, depend mainly on the laws for their meaning, and the exact form of the verbal statement is not important provided it is adequate in conjunction with the relevant laws to indicate the magnitude of the quantity adopted as the unit.

It will be well at this stage to note a distinction that is usually made at the standardizing laboratories between units and standards. *The unit* of a physical quantity is a quantity assigned the magnitude corresponding to the numerical value of one exactly in the scale of measurement adopted. *A standard* of a physical quantity is a material object or a physical system which establishes the size of the unit of that quantity. Its value is determined in conformity with the definition of the unit. Thus, the unit is abstract whereas the corresponding standard is concrete. The unit is said to be "realized" by the standard. A unit can be defined by constructing a special prototype standard and by general agreement assigning some convenient value to it. Alternatively, the unit can be defined theoretically and standards then constructed in such a way that their numerical values can be determined by reference to the definition and the relevant scientific laws, with the highest precision obtainable by the known experimental techniques. A unit must be defined for every quantity, but a standard for every quantity is not at all necessary. In view of the difficulties encountered in constructing standards and ensuring that they do not vary in value, the number of standards employed is kept to the minimum. A single standard, together with the techniques of measurement associated with it, serves to realize multiples and sub-multiples of a unit as well as the unit itself; it is merely a question of a change of numerical value. The standard and associated techniques can be described as a realization of the complete scale of reference for the quantity concerned.

From the standpoint of science all legal considerations can be ignored since they are only of civil interest. Only the units, standards, and laws recognized by common consent throughout the scientific world will be considered here.

The unit of length and distance is the *meter* (m). It was originally defined by a prototype standard in the form of a specially constructed bar of platinum–iridium designed to secure the highest attainable degree of constancy and rigidity. The meter was defined as the distance between two graduation lines engraved on the neutral plane which was exposed near each end of the bar. The prototype standard and the associated comparator were set up and maintained at the International Bureau of Weights and Measures (BIPM) near Paris, and the unit was made available in other parts of the world by distributing copies of the standard which had been compared with the original and given their appropriate values in terms of the meter as defined.

It was subsequently established that by using the optical interferometer as a distance comparator and the wavelength of a precisely specified radiation as standard, the accuracy of length measurement could be improved and a more permanent standard obtained. The new standard was adopted by the General Conference on Weights and Measures in 1960 in the place of the original prototype, but its numerical value was adjusted as a result of comparative measurements so that the unit was unchanged. The meter is therefore now defined as equal to 1 650 763·73 times the wavelength in vacuo of the orange radiation corresponding to the $^2P_{10}$–$^5D_5$ transition of the isotope Kr^{86}. The unit has remained sensibly unchanged, but it is now realized by a reproducible standard, which can be set up at any time in any standardizing laboratory. The radiating atom is so much simpler in structure, from the standpoint of modern science, than the solid bar incorporating a closely packed crowd of atoms with a somewhat indefinite boundary that the new standard is accepted without question as the more nearly constant.

The unit of mass is the *kilogram* (kg). It is defined by a prototype standard in the form of a solid cylinder, of a diameter equal to its height, made of the same alloy as was used for the meter standard. This standard has remained at the International Bureau since 1889, and the unit is made available in all parts of the world by distributing copies of the standard, calibrated at the Bureau so as to conform to the unit. Any such standard provides the user with the complete reference scale of mass, including the gram, milligram, etc; the process of calibrating laboratory standard "weights" in terms of any single one is well-known to depend solely for its validity on the law that mass is additive.

The unit of time is the *second* (sec). It was originally defined in terms of the rotating earth as the prototype standard, the mean solar day having by definition the value 86 400 sec. As astronomical observations became more and more precise, it was established that the observed periods of the earth's rotation, the day and the year as deduced from stellar and solar observations, were all subject to detectable variation. It therefore became a more and more complicated matter to assign to the earth, as the prototype standard, a numerical value that would serve in practice to determine a constant second. It is unnecessary for the purposes of thermodynamics to pursue the matter in detail as the variations are for the most part too small to be significant. It is sufficient to state that the latest astronomical definition to secure acceptance internationally is that the second is the fraction 1/31 556 925·9747 of the tropical year at 1900·0. Fortunately for the purposes of physical science, a better standard than the rotating earth has now been developed. It depends on the law that the transition between two energy states U_1 and U_2 of an atom is accompanied by the emission or absorption of radiation of a frequency v, or periodic time $1/v$, given by $hv = U_1 - U_2$ where h is Planck's constant. Thus, any atomic system which is capable of generating radiation corresponding to a definite spectral line can be used to realize a definite time-interval. A precisely defined line of this kind has now been obtained in the range of frequencies covered by laboratory oscillators and clocks by observations on cesium atoms which are undergoing the transition F,m(4,0) ⇌ F,m(3,0), and reproducible standard clocks of this kind now serve to maintain the second and the complete reference scale of time-interval at the principal standardizing laboratories. A movement is now underway to redefine the second in terms of the cesium or some other atomic transition. For practical purposes the second is taken to be 9 192 631 770 times the period of the cesium radiation mentioned above, a value derived from a long series of careful observations to relate the frequency to the second defined by the tropical year at 1900·0. This unit is made available throughout the world by means of time-signals broadcast by the national laboratories and other agencies.

III. The Derived Mechanical Units

Having at this stage established independent scales of measurement for the quantities length, mass and time, all the mechanical quantities, including energy, can be determined because each of them can be expressed as a definite function of measured values of length, mass, and time by universally recognized scientific laws. It is only necessary to make the appropriate measurements of length, mass, and time and to calculate the corresponding value of the function to obtain the numerical value of any mechanical quantity. The unit of the quantity is evidently determined by the function employed, and all such functions necessarily include a proportionality factor, which provides for any desired changes of unit. By a generally accepted convention, the unit adopted is that which corresponds to a proportionality factor of unity. A term and symbol for the unit are then obtained by combining the units of length, mass, and time employed, in the appropriate functional arrangement, a procedure which is an elaboration of the traditional convention followed when naming the unit of velocity as, say the meter per second, m/sec. These self-explanatory names are, for convenience, supplemented by shorter ones, such as the joule, in special cases, but it is a great advantage to be able to give a self-explanatory name and symbol for any unit of any quantity when necessary The following units of mechanical quantities are defined on these principles:

> Unit velocity, meter per second, m/sec—a rate of change of distance from a fixed point of 1 m/sec.
> Unit acceleration, meter per second squared, m/sec^2—a rate of change of velocity of 1 m/sec^2.
> Unit force, the newton, N, kg/m/sec^2—the force which gives to a mass of 1 kg an acceleration of 1 m/sec^2.
> Unit of energy or work, the joule, J—the work done when the point of application of a force of 1 N is displaced a distance of 1 m in the direction of the force.
> Unit of power, the watt, W—the power which gives rise to the production of energy at the rate of 1 J/sec.

The above-mentioned mechanical units are selected from the comprehensive system adopted for general use in science by the General Conference of Weights and Measures of 1960 and named by them the International System of Units (S.I.). The electrical units of this system will be given later. The units derived on the same principles from the centimeter, gram, and second are also widely acceptable since they are based on the same standards and are merely sub-multiples of the above units, with names and symbols that are for the most part self-explanatory when the same convention is followed. The special names, dyne for the unit of force, and erg for the unit of energy, in this system should, however, be noted:

The dyne or g cm/sec^2 = 10^{-5} kg m/sec^2 = 10^{-5} newton
The erg or g cm^2/sec^2 = 10^{-7} kg m^2/sec^2 = 10^{-7} joule.

IV. Working Standards for Dynamics

The mechanical units defined above are abstract and theoretical, and it has still to be shown that they can be realized in laboratory practice with satisfactory precision. Although one does not need prototype standards to define the units, the system only becomes practicable for the purposes of scientific investigation when it can be demonstrated that mechanical quantities can be measured with precision in terms of these units, with laboratory equipment that is reproducible and therefore in principle available to every worker. Such equipment is the means of realizing the unit and will for convenience be referred to as the working standard for the unit.

The physical system that is most prominent in our experience of dynamics is the earth with its gravitational field. Thus it is natural to find the working standards for dynamical quantities in the equipment for measuring the acceleration of gravity, g. Two types of equipment have been widely used for such measurements: reversible pendulums of Kater's type used for determining g in terms of pendulum length and time of vibration, and free-fall apparatus used for determining g in terms of distance fallen and time of fall. This second method has become feasible in recent years because of the development of electronic timing techniques.

Determinations of g performed by both methods at the principal metrology laboratories of the world agree within a few parts per million after allowing for local differences in g. These local differences can be established with high precision by comparison pendulums and gravimeters in accord with the metrological principle that like quantities can be compared with each other more precisely than either can be determined in terms of the elemental quantities of which it is composed. Collation of the results for these variations leads to the belief that an adjusted average of them is accurate to within about one part in a million and that by comparison measurements a value of g may be assigned at any station with high confidence that it is not in error by any more than this amount.

Laboratory balances are widely used for comparative measurements of mass with an accuracy of 1 in 10^7 or even better. Thus, in experiments in which the technique of weighing can be applied to the measurement of force, the accuracy now obtainable is not significantly worse than that with which g has been determined at the place of the experiment, and in favorable circumstances may be from 1 to 4 parts per million. It will be seen later that this accuracy has an important bearing on the realization of the electrical units used in thermodynamic practice.

The joule has not been realized in the form of mechanical work with comparable precision. Joule's classical experiments on the mechanical equivalent of heat constitute one such determination, and in more recent times proposals have been made to use experimental equipment incorporating calibrated weights falling repeatedly through a measured distance inside a rotating calorimeter, but such developments do not appear to have attained an accuracy comparable with that of electrical methods and therefore for the present purpose need not be considered further.

V. The Electrical Units

Between 1908 and 1940 the electrical units actually employed in practice were, by international agreement, based on material standards made to definite specifications in the form of a silver voltameter defining the International ampere, and a mercury resistor defining the International ohm. It was always recognized that theoretical units defined by reference to the meter, kilogram, and second by means of the universally accepted laws of electrodynamics should form the ultimate basis of electrical work. The empirical standards were adopted because the theoretical units had not been realized with the precision often obtained in many comparative electrical measurements and therefore indispensable in the working standards. However, by 1940 it had become clear that the theoretical units had been realized in various national laboratories with a precision at least equal to that of the empirical standards, which were therefore abandoned. The theoretical units only will be considered here. They are derived from the laws of electrodynamics on the same principle as was applied in the derivation of the mechanical units. The definitions of relevant units in the form used by the International Committee of Weights and Measures are given below. This form was adopted merely for verbal simplicity and as it does not in every case clearly indicate the pre-supposed law, this has been noted in parenthesis, where necessary, together with unit symbols, which are commonly used, following the convention mentioned above.

> Unit of electric current, the ampere, A—the constant current, which, if maintained in two straight parallel conductors of infinite length, of negligible circular sections, and placed 1 m apart in a vacuum, will produce between these conductors a force equal to 2×10^{-7} newton per meter of length (i.e. 2×10^{-7} N/m, Ampere's Law).
>
> Unit of p.d. (potential difference) and e.m.f. (electromotive force), the volt, V—the difference of electric potential between two points of a conducting wire carrying a constant current of 1A, when the power dissipated between these points is equal to 1 W [Joule's Law, $V = W/A = J/(A. \sec)$].
>
> Unit of electric resistance, the ohm, Ω—electric resistance between two points of a conductor, when a constant p.d. of 1 V applied between these points produces in this conductor a current of 1 A, this conductor not being the seat of any e.m.f. (Ohm's Law, $\Omega = V/A$).

Unit of quantity of electricity, the coulomb, C—the quantity of electricity transported in 1 sec by a current of 1 A (C = A sec).

Unit of electric capacitance, the farad, F—the capacitance of a capacitor, between the plates of which there appears a p.d. of 1 V when it is charged by a quantity of electricity equal to 1 C (Faraday's Law, F = C/V).

Unit of electric inductance, the henry, H—the inductance of a closed circuit, in which an e.m.f. of 1 V is produced when the current in the circuit varies uniformly at the rate of 1 A/sec (Henry's Law, H = V sec/A).

It will be noted that the ampere is defined by reference to the mechanical units, and is in that sense a derived unit, but for some purposes and particularly for that of obtaining a set of unambiguous unit symbols, covering all the electrical and mechanical quantities, on the convention previously mentioned, it is preferable to regard the ampere as a fourth independent unit similar in status to the m, kg and sec. The whole system then becomes coherent and relatively free of ambiguity, the proportionality factor in all the defining equations being unity. The definition of the ampere is sufficient to show that this would not be the case if the ampere were treated as a derived unit. It is in practice a great advantage to be able to name any unit of the system unambiguously without burdening the memory with an excessive number of independent names, and this consideration provides ample justification for the convention of four independent units serving the complete system of electrical and mechanical quantities. The corresponding system of units, often called the MKSA system, is accepted internationally and forms part of the International System mentioned earlier.

It is to be noted that the volt depends directly on the law that electrical energy, mechanical work, and heat are different forms of the same physical quantity, energy, which is conserved in transformations from one form to another. Heat is therefore measured in joules and can in principle be determined in terms of the ampere, ohm, volt, and second as well as the newton and meter, that is to say, by measurements of the electrical quantities current, voltage, and resistance in addition to time-interval, as well as by measurements of force and distance. The ampere, volt, and ohm have, however, so far only been defined as theoretical units. Their realization in actual laboratory standards must be considered next and shown to be practicable with the accuracy likely to be required in thermodynamic practice.

VI. Absolute Electrical Standards

The standards by which electrical units are realized by reference to the primary mechanical units and the theoretical definitions and laws are commonly called absolute standards, and the term is retained here for its convenience, its limitations already having been emphasized sufficiently. To establish such standards, apparatus is constructed, on which it is feasible to observe some electrical quantity with the highest precision, and also to make such measurements of those mechanical quantities as are theoretically necessary to determine the value of that quantity. Thus, it is necessary to satisfy in the same apparatus the conditions for precise mechanical measurements, for precise electrical measurements, and those postulated in the theoretical equations. These are very severe conditions. For mechanical precision it is necessary to use only materials that are sensibly rigid and of a low temperature coefficient of expansion. For electrical precision the materials are limited to perfect conductors and perfect insulators, and theoretical requirements include the best possible approximation to linear conductors surrounded by material that is perfectly non-conducting and non-magnetic. Small wonder that for many years the authorities at the various national standardizing laboratories could not believe that a precision adequate for the practical purposes of international standardization was within reach in this direction, and therefore adopted an elaborate specification for a silver voltameter to define a practical unit, the International ampere, and an even more elaborate specification for a mercury resistor to define the International ohm. The specifications were drafted so as to make these practical units, as nearly as was possible at the time, equal to the theoretical units, but from 1908–1948 all international standardizing work was actually expressed in terms of the empirical units. However, before the end of that period it had become evident that the accuracy of the absolute standard that had been constructed was probably better than the reproducibility of the silver voltameters and mercury resistors made to the agreed specifications. Thus the International Units became merely an undesirable complication and were abandoned from 1 January 1948. They are now, therefore, of purely historical interest and will not be considered further except to note that in making the change of units the International Committee agreed that the changes were as follows:

1 International ohm = 1·00049 ohm (absolute)
1 International volt = 1·00034 volt (absolute).

It was also agreed that the word "absolute" should be omitted from the terms for the units then adopted, as soon as the International Units had disappeared from use, so that there was no longer any risk of ambiguity.

Returning to absolute standards, it should be noted that at least two such standards are needed to establish the measurements of electrical quantities at any one laboratory, and to make the system effective internationally, comparable standards must also have been set up in other countries and some means found for comparing the units realized by these standards. The agreement obtained in such comparisons provides direct evidence for the success of the whole system as a means of coordinating scientific work involving electrical measurements on a world-wide front.

The absolute standards themselves are not usually transportable without loss of accuracy, and therefore, for international comparisons of units and for conveying the units realized at the national laboratories to workers in general, yet another group of standards is required, portability with constancy of value over a long period being their main characteristics. Portable standards of at least two electrical quantities are necessary, and it is a matter of common experience that the two in most general use are the standard cell, which, in conjunction with a standard potentiometer as comparator, serves to realize the complete scale of reference for voltage, and the standard resistor, with the precision Wheatstone bridge as comparator, which serves to realize the ohm and the whole scale of values of resistance. The combination of standard cell, resistor, and potentiometer serves as the working standard of current, since it makes possible the realization of a precisely determined current, namely that current which when passing through the resistor gives rise to a p.d. which just balances the e.m.f. of the cell. This combination, therefore, comprises the working standard for both voltage and current, and consequently also for their product, power. Further, when used in conjunction with some form of standard clock it becomes the working standard of energy, the energy realized being that liberated as heat in a particular resistor when the current and terminal p.d. are maintained at definite values throughout a definite time-interval.

The forms adopted for the absolute electrical standards have not been mentioned so far. As stated earlier, they are links between mechanical and electrical quantities. Those that have actually been used at the various national laboratories fall for the most part into three groups: *current balances* or dynamometers relating current to mechanical force and linear dimensions, *inductors* providing inductance in terms of linear dimensions, and *generators*, which, like the well-known Lorenz apparatus, generate an e.m.f. determined by a current, the speed and a calculable geometrical factor, and therefore serve to determine the quotient of an e.m.f. by a current i.e., a resistance, in terms of speed and measured lengths. Various forms of induction balance also give resistance in terms of lengths and rotational speed or frequency and therefore belong to this last class. A discussion of the details of all these methods is beyond the scope of the present work, but a few examples will serve to illustrate the nature of the work involved.

It may be noted that mutual inductance is a common factor in all the absolute standards mentioned above, and it follows that the accuracy of the whole system of measurements turns very largely on the accuracy with which inductors of a form and size suitable for both mechanical and electrical measurements of the highest precision can be constructed and measured. A typical inductor embodies two current circuits, 1 and 2, approximately linear, and the value of their mutual inductance is calculated from their measured dimensions by the use of some formula that is mathematically equivalent to Neumann's integral

$$N = \oint_1 \oint_2 ds_1.ds_2.\cos\theta/r \tag{1}$$

(in which ds_1 and ds_2 are elements of the circuits separated by a distance r and oriented at an angle θ with respect to each other) with a proportionality factor corresponding to the definitions of the electrical units. The "straight parallel conductors of infinite length" of the definitions of the ampere are an idealized highly special case of such circuits, serving merely as a means of indicating the proportionality factor adopted.

VII. Standard Inductors

The standard of mutual inductance at the National Physical Laboratory, Teddington, U.K., may be taken as an illustration of actual practice in this field. The evolution of the type of construction now usually adopted began with the first model of this standard constructed by Albert Campbell in 1907. The circuit whose linear dimensions were required with the highest accuracy took the form of a uniform helix of thin bare copper wire wound under tension on a marble cylinder. The tension in the wire ensured that the geometrical form of the circuit would be maintained nearly constant by the supporting marble, in spite of inevitable fluctuations in temperature. Marble was chosen for its dimensional stability, low coefficient of expansion (about $5 \times 10^{-6}/°C$) and tolerably good insulating and machining properties. Accuracy in both radial and axial dimensions are essential, and thus every turn must be precisely located. This requirement was achieved by winding the wire in a helical groove cut into the surface of the cylinder, which had previously been turned to uniformity of diameter. The highest possible degree of uniformity in the finishing winding is essential, not only because the theoretical formula can be derived only for circuits of regular form, but also because only in such cases can the actual geometrical form of the circuit be determined by a finite number of measurements of length. Thus the helical groove must be uniform in pitch, diameter and cross-section, and the wire itself must be of uniform and circular cross-section and precisely located by line-contact along each of the two sides of the V-groove.

The construction of such a coil demands every available refinement of workshop and laboratory practice, but length measurement is always somewhat more precise than any operation of mechanical shaping, so some detectable departures from uniformity are inevitable, and it becomes necessary to measure and tabulate all such deviations for a large number of points distributed throughout the length of the circuit, and to make allowance for them by calculating the appropriate "correction" that must be applied to the theoretical value calculated on the assumption of uniformity.

Other deviations from the theoretical conditions are inevitable, and in every instance the corresponding "correction" must be calculated and applied to the theoretical value. Thus the ideal conductor "of negligible circular section" and, by implication, of negligible resistance, can never be realized, and as a first approximation the current is regarded as concentrated in the axis of the wire, but this approximation is not close enough for modern requirements, and to obtain a better one it is necessary to know the distribution of current over the cross-section. If the wire is regarded as a bundle of parallel filaments, allowance must be made for the fact that the outer will be longer than the inner ones and therefore of higher resistance if the resistivity is uniform. Thus the outer filaments must be credited with slightly less current than the inner ones and the equivalent single filament will not be in the axis of the wire. However, the assumption of uniform resistivity is itself an oversimplification; the wire being under tension, the inner filaments will be compressed by the pressure of the support, whereas the outer ones will be stretched, and the differences of mechanical stress are accompanied by corresponding differences of resistivity and therefore of current distribution. This complication was ignored in the original NPL standard, but shown to be of practical significance in later developments at the National Bureau of Standards in Washington D.C., U.S.A. The estimation of the corresponding corrections need not be pursued here. It is sufficient to note the gap between theory and practice.

The method of construction outlined above, single-layer coils of bare copper wire located in grooves cut very slowly by means of a lathe in large marble cylinders (up to 30 cm diameter), was used at the NPL in a number of investigations between 1907 and 1936 with results which suggested that the unit of inductance had been realized with an uncertainty of about $\pm$ 10 parts per million; e.g., three different standards made in 1921, 1926 and 1935 were found by electrical comparisons to show a total spread of less than 10 parts per million in the unit realized, though the extreme range of uncertainty obtained by summing the estimated limits of error on the measurements of the various quantities involved was about 30 parts per million. The accuracy was mainly limited by the unavoidable irregularities in the diameter and pitch of the successive turns of the coils, and no significant improvement in this accuracy was within sight until a new method of construction was developed at the NBS in Washington D.C. Cylinders of Pyrex glass (a borosilicate) were used instead of marble, and they were shaped to uniformity of diameter and subsequently grooved for the location of the wire by a very laborious process of grinding followed by lapping. It is unnecessary to go into details, but it will be appreciated that with such a process the observed irregularities can be reduced stage by stage until the required degree of uniformity is reached. The borosilicate glass is moreover greatly superior to marble in insulating properties and in dimensional stability. In order to obtain the desired uniformity and circularity of section of the wire, the final drawing of the wire and winding of the coil were carried out in one operation, the wire passing through the die straight on to the cylinder, the die itself being rotated about its axis during the drawing. This type of construction has also been used with a cylinder of fuzed quartz at the Canadian Standards Laboratory and in the most recent version of the Campbell Standard at the NPL. It has become an important feature of modern absolute standards.

The theoretical definitions refer primarily to circuits or current carrying conductors in a vacuum. Thus all the material supporting or surrounding the circuits of an absolute standard must be magnetically equivalent to a vacuum. Materials like marble, glass, and fuzed quartz are mostly diamagnetic, having a permeability about 10 parts per million less than that of a vacuum. The permeability of all the materials actually used in the construction of an absolute standard must be measured and the appropriate correction estimated. By appropriate design it can usually be kept less than 3 parts per million but cannot be assumed to be negligible.

These structural details have been considered at some length since they apply to many absolute standards and are characteristic of much of the practice of the modern electrical standards laboratory. The necessary measurements of axial and radial lengths are matters of standard practice in metrology and do not call for special consideration here, any more than do the computations involved in calculating the required functions of the measured values.

VIII. Absolute Determination of the Ampere

Consistency with the definition requires that the ampere shall be realized by measuring the mechanical force between two circuits of precisely determined geometrical form produced by the current flowing through them. Two types of standard have proved to be capable of the precision required today, the current balance and the Pellat type of electrodynamometer. In both, the two circuits of precisely determined geometrical form are realized by constructing coils by the techniques discussed in the preceding section. In the electrodynamometer the coils are mounted with their axes at right angles, one coil being fixed and the other capable of rotation through a small angle, θ. The mechanical forces between the two circuits produced by the current flowing through them result in a torque, T, tending to displace the movable coil and given by Ampere's Law according to the equation

$$T = I^2 \, (\partial M/\partial \theta) = 10^{-7} \, I^2 \, (\partial N/\partial \theta) \qquad (2)$$

in which M denotes the mutual inductance of the coils and N is Neumann's integral as in the previous section. Equilibrium will be maintained when the current is switched on only if an equal but oppositely directed mechanical torque is applied, e.g. by weights applied to a horizontal arm.

In the current balance, the two coils are coaxial with the common axis vertical, and one coil is fixed, whereas the other is suspended from the beam of a sensitive balance so as to be capable of vertical displacement. The axial force, F, produced by the current can in this case be expressed by Ampere's Law in the form

$$F = I^2 \, (\partial M/\partial Z) = 10^{-7} \, I^2 \, (\partial N/\partial Z) \qquad (3)$$

in which Z denotes vertical distance. The mechanical force can be measured by the well-known technique of precision weighing, and the condition of balance can be expressed as

$$mg = 10^{-7} \, I^2 \, (\partial N/\partial Z) \qquad (4)$$

to determine I in terms of the mechanical quantities.

The purely geometrical quantities $\partial N/\partial Z$ and $\partial N/\partial \theta$ can, for suitably constructed coils, be calculated from radial and axial dimensions with a precision comparable to that for N; indeed the coils are in each case a special form of standard mutual inductor, and the considerations of the previous section must be applied.

The experimental technique cannot be pursued in detail here. Its general character will be sufficiently obvious; the current I passing through the standard is in either case evaluated in terms of a geometrical factor calculated from the measured dimensions, an observed mass required to preserve the initial state of equilibrium when the current is switched on, and of g a mechanical quantity which is determined by the methods mentioned earlier. The complete equipment therefore constitutes an absolute standard of current. The current measured by the standard also passes through a working standard in the form of the combination of standard cell and standard resistor mentioned earlier, and the final result is recorded by assigning a numerical value to this combination. These electrical measurements are repeated at intervals and serve to show if changes of any sort in either the coils of the balance or the resistor and standard cell have occurred. A new absolute determination, including the remeasurement of all the mechanical quantities as well as the electrical observations and involving months of work, is made only when there is unmistakable evidence of change. The agreement obtained between determinations of current made with standards established in various national laboratories constitutes a major part of the evidence for the validity of present day electrical measurements.

IX. Absolute Determination of the Ohm

Several methods of high precision have played a part in realizing the ohm in use today, and nearly all depend on an absolute standard of inductance constructed as described in Section VII. Here the type of experimental technique which has been used to derive a resistance from an inductance will be considered.

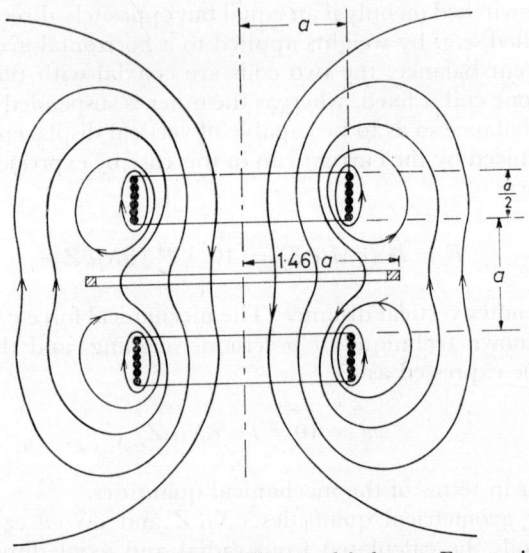

Figure 1. Schematic cross-section of Campbell mutual inductor. For a primary winding in two sections of the proportions shown, the magnetic field is almost zero in the annular space occupied by the secondary coil. Hence an error in the measured dimensions or location of the secondary coil will have little effect on the calculated mutual inductance.

One of the most successful is the a.c. bridge network in which an alternating current I of angular frequency, ω is passed through the resistor, R to be measured, and also through the primary coil of a standard mutual inductor, M (*Figure 1*). By the law of electromagnetic induction an alternating e.m.f., $M\omega I$, is generated in the secondary coil of the inductor and can be observed as a p.d. between its terminals. The object of the method is to balance this p.d. against the p.d. across the terminals of the resistor, namely RI. If equality could be precisely established, the relation $R = M\omega$ would hold and would serve to determine R in terms of a frequency and of an inductance calculated from lengths. The two p.d.'s cannot be directly balanced simply by connecting them in opposition and using a null-detector, as in the potentiometer because they are out of step, having a phase difference of 90 degrees. However, by an arrangement of two inductors in cascade (M_1 and M_2) and an additional resistor, a further change of phase is produced and adjusted until the final terminal p.d. is in step with RI

and the two can be balanced by means of a null-detector. A condition of balance can be expressed

$$R_1 R_2 = M_1 M_2 \omega^2 \tag{5}$$

giving the product of two resistors in terms of the frequency and two inductances which can be determined by simple electrical comparison with an absolute standard. The ratio of the two resistances can easily be measured by a simple bridge measurement, and thus both known, and the whole equipment becomes in effect an absolute standard of resistance. This is Albert Campbell's method, which has been developed at the NPL to a high

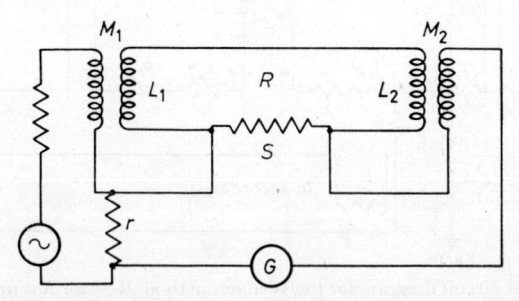

Figure 2. Circuit for Campbell method for the absolute measurement of resistance. Resistors r and S have resistances R_1 and R_2, respectively. The values of the two mutual inductors are derived by comparison with a computable standard inductor. The secondary e.m.f. of M_2 is nearly equal in magnitude and opposite in phase to the p.d. of the primary current in r.

degree of precision (*Figure 2*). The details of the a.c. bridge technique are beyond the scope of this chapter. Other methods involving a similar technique have been developed at other national laboratories.

There is, however, space for only one more example of standard methods and the Wenner method is chosen because it is outstanding in combining high precision with extreme simplicity in principle. The method is a d.c. one (*Figure 3*). The resistor R, to be calibrated, is supplied with constant current I from a battery B_1. A commutator C_p rotating at a constant speed is included in the circuit and so constructed that in two portions occupying corresponding positions in the successive halves of each rotation the same current I also passes through the primary coil of the standard inductor M, first in one direction and then the opposite, the direction reversing in every half-rotation. Thus, there is a total change of current in the primary coil of $2I$ and $-2I$ alternately in successive half-rotations and corresponding pulses of e.m.f. are induced in the secondary coil of the inductor. The secondary p.d. is rectified by a second commutator C_s, driven synchronously with C_p and, after smoothing, the mean secondary p.d. is balanced against that across R, a d.c. galvanometer serving as null-detector. If n denotes the number of reversals per second, the mean secondary p.d. is $2nMI$ and the condition of balance is therefore

$$R = 2nM \tag{6}$$

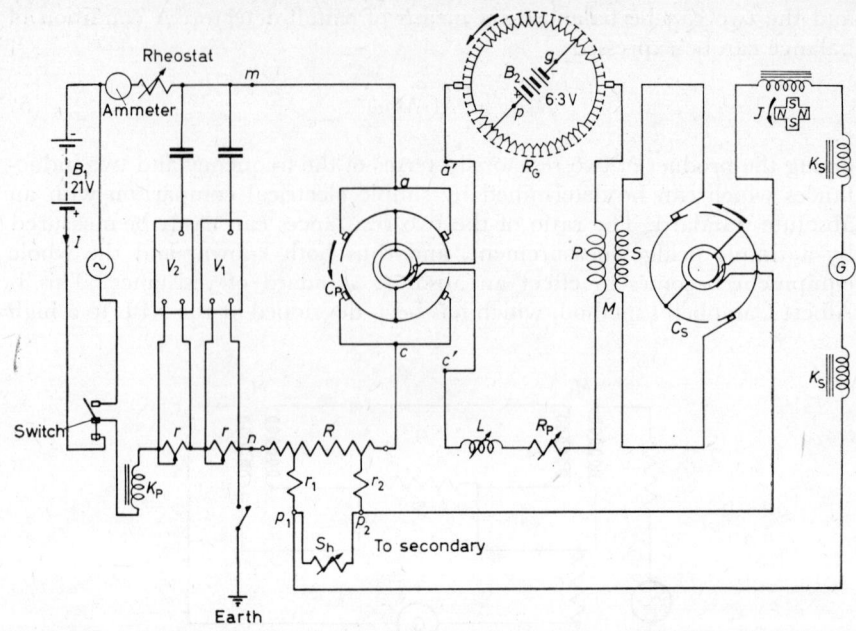

Figure 3. Schematic circuit diagram for the Wenner method. Resistor R is measured in terms of mutual inductor M, and frequency of reversal of primary and secondary commutators C_p and C_s. Electronic amplifiers V_1 and V_2 hold the current in R constant, while that in P reverses cyclically. Details of the circuit not discussed here or in text are practical complications not essential to an understanding of the method in principle.

so that the resistance is determined from a frequency and an inductance as before. The method is extremely simple in principle and convenient in operation, once the necessary experimental conditions have been realized, but the necessity of commutation and the existence of a very large ripple in the induced e.m.f. call for considerable practical complications. Thus chokes and electronic equipment are used to maintain the current I through the resistor constant; a special generator R_G, in the form of a battery and rotary potential divider controls the current in the inductor during the change-over and therefore the form of the a.c. ripple, which must in effect be eliminated from the detector circuit; an auxiliary inductor-alternator J, which is matched to R_G compensates the main part of this ripple, leaving only a small residue which is prevented from reaching the galvanometer by chokes on either side of it.

The agreement obtained between the results of such widely different electrical techniques provides convincing evidence for the validity of the whole scheme.

An entirely independent and highly accurate method of realizing the ohm has been developed by Thompson and Lampard. It uses a capacitor whose value can be calculated with great accuracy by application of a newly discovered theorem in electrostatics. A capacitor, constructed to satisfy the conditions of this theorem, consists of four cylindrical conductors with parallel axes which intercept a perpendicular plane at the apices of a square, the conductors themselves being close to, but insulated from each other.

The theorem states that the average of the capacitance per unit length between diametrically opposite pairs of cylinders, the other pair being grounded, is constant, subject to corrections involving the difference between the capacitances of the two pairs. Symmetrical arrangement makes this difference very small. The theorem itself is more general.

The average capacitance, for length L, is given in MKSA units by

$$\bar{C} = \epsilon_0 L \,(\ln 2/\pi)\, [1 + 0{\cdot}087\, (\Delta C/\bar{C})^2]. \tag{7}$$

All other terms are fourth order or higher in $\Delta C/\bar{C}$. The constant ϵ_0 is given by

$$\epsilon_0 = 1/(4\pi \times 10^{-7})\, c^2 \tag{8}$$

in which c is the speed of light.

The great advantage of this capacitor for establishing a standard for capacitance is that its capacitance is proportional to the length of the cylinders only, a quantity which can be measured with great accuracy. It requires no measurement of radii or calculation of areas as do ordinary spherical or cylindrical condensers. Since the impedance of a capacitor is given by $Z_c = 1/C\omega$, and of an inductor by $Z_L = L\omega$, these may be compared if expressed in MKSA units, Z_c and Z_L being giving in ohms, and both may be compared with the impedance of a pure resistor which is independent of frequency. The value of the standard resistors maintained by NBS has recently been redetermined by this method, the results agreeing to within a few parts per million with those obtained by the inductor method at the NBS and other laboratories.

X. Working Standards

In the day-to-day practice of the research laboratory, as distinct from the standardizing laboratory, measurements of resistance are made as straightforward comparative measurements by some form of bridge circuit, a carefully preserved standard resistor or group of resistors serving as the ultimate standard to which all such measurements made in that laboratory are referred. Similarly all the measurements of voltage are comparative measurements made with some form of potentiometer circuit by reference to carefully preserved standard cells, and measurements of current and power are made in terms of resistance and p.d. by reference to the same standards and employing the same techniques in a manner that has already been outlined. The standard resistors and standard cells, called for convenience the working standards, determine the accuracy of all the measurements made, for the accuracy of the comparator equipment can always be checked by processes of substitution and addition, such as are commonly used for establishing the validity of any equipment that has been calibrated for comparative measurements of an additive quantity, whereas such checks can give no information about the ultimate standard of reference.

The standard resistors and standard cells are readily obtained from specialist manufacturers of electrical standards, and their construction need not therefore be considered in detail here. It is sufficient to note that there is no difficulty in obtaining bridge and potentiometer equipment that will provide a sensitivity of one part per million in comparative measurements of resistance and p.d. Users of such equipment naturally record their observations in favorable circumstances to six significant figures, but when they do, they should understand that the ohm, ampere, volt, watt and joule of their measured values are not the internationally agreed units, but approximations to them peculiar to their own laboratory at the time of the experiment, and the results will not necessarily be significant within the same limits to a worker in any other laboratory, or even to a worker in the same laboratory at a later date. This circumstance arises partly because absolute measurements have not yet been developed to the point at which the uncertainty in the realization of any of the electrical units is as little as ± 1 part per million, but mainly because no working standard has been found to remain constant within these limits over a period of years. Indeed it is not uncommon for a standard to change by a considerably greater amount in the course of a few weeks or even days. For this reason standard resistors and standard cells are studied in great detail at the standardizing laboratories, and some account of this work must be given since it forms an essential link between the absolute measurements and ordinary laboratory practice.

1. *Standard resistors*

The working standard resistor usually consists of a coil of wire having a nominal value of 1 ohm made of one of the "resistance alloys" having a very small temperature-coefficient of resistance, and provided with four termainals, two "current terminals" through which the measuring current I is led in and out of the resistor, and two "potential terminals", which provide connections to precisely located points or equipotentials on the resistor and thereby identify the precise voltage E which defines the value R of the resistance in the resistor in accordance with the law $E = RI$. Coils of manganin wire, annealed after winding, and hermetically sealed in metal containers (*Figure 4*) have been found at the NBS and NPL to make the most satisfactory standards, the chief criterion being constancy of value over several years.

In work of the highest precision, it is impracticable to use a single coil as the reference standard, as it is impossible to predict that any coil will have the necessary constancy of resistance over a period of years. At the standardizing laboratories a group of several such coils serves as the reference standard. The evidence for their constancy is ultimately obtained from the absolute measurements, but these measurements are long-term investigations and subject to much greater uncertainties than the mere comparisons of resistance, so that their evidence only becomes conclusive after many years. The purely electrical observations of the Wenner or Campbell methods can be repeated at short intervals and may be reproducible to one part per million, but when deviations are observed they are as likely to be in the absolute standard as in the resistor and the uncertainty remains until the whole of the absolute determination can be repeated. Thus, in practice, the chief evidence of constancy from month to month is obtained from comparative measurements of the resistances of the several coils in the

Figure 4. Double-walled (Thomas type) standard resistor. The manganin wire shown at the extreme right is annealed in vacuum. It is tied down on the inner form shown next to it and then sealed between the inner and outer brass cylinders.

group chosen as the working standard. At the NPL a group of five such coils, specially selected for relative constancy and compared at intervals over 20 years, showed an overall deviation of only 0·8 part per million. The mean value of the group is assumed to have remained constant, and values are assigned to the separate coils on this basis. Similar procedures are followed at the NBS, Washington, and at the BIPM (International Bureau of Weights and Measures) near Paris, where the reference group is formed from working standards received from the various national laboratories, each with its value recorded in terms of the ohm as realized at the laboratory of origin. The comparisons made at the International Bureau serve to realize a "mean international ohm", i.e., a mean value for the ohm as realized in various countries and conveyed to the Bureau in the form of a working standard. By using values based on these comparisons, workers throughout the world can make measurements that are comparable with a greater precision than can be claimed for any absolute determination, though the values should not, strictly speaking, be labelled "ohms" but, say, "mean international ohm, 1960". The relative constancy of the groups of coils at the NBS and BIPM is found to be of the same order as that at NPL, but comparisons between the three groups showed deviations in mean value amounting to 2×10^{-6} in 10 years, which fact shows that the mean values that were assumed to have remained constant at each laboratory are in fact drifting relative to one another by amounts not markedly less than the deviations from the mean within each group. Absolute measurements, therefore, repeated at long intervals, provide the only reliable evidence for the constancy of the ohm employed in practice, and the highest accuracy will only be available in laboratories where the reference standard is checked at suitable intervals at a standardizing laboratory.

2. *Standard cells*

The working standard of p.d. is the standard cell, and each of the national standardizing laboratories employs a selected group of cells as the reference standard. The cells now used are all of the Weston type, the electrodes being pure mercury and a pure cadmium amalgam, and the electrolyte a solution of cadmium sulfate (*Figure 5*). In the construction of reference standards, stability is the main criterion, and experience has shown that the highest stability is obtained with cells in which the electrolyte contains excess cadmium sulfate crystals, and the mercury electrode is covered with a depolarizing layer which consists of a paste of finely divided mercurous sulfate, cadmium sulfate and mercury. The e.m.f. is always about 1·0183 V, and the precise value is of little importance so long as it is known and constant in the conditions of potentiometric work. It is obtained ultimately from the absolute determinations of the ohm and ampere, but since the uncertainty of such determinations is at least 10 times as great as that of the comparison of cells with a potentiometer, a working unit is realized within closer limits at each national laboratory by the assumption of the constancy of the mean e.m.f. of a group of specially selected cells.

It is not to be expected that a chemically active system like a standard cell would show a stability comparable with that of an inert object like a resistance coil, and indeed when a batch of cells has been made from the

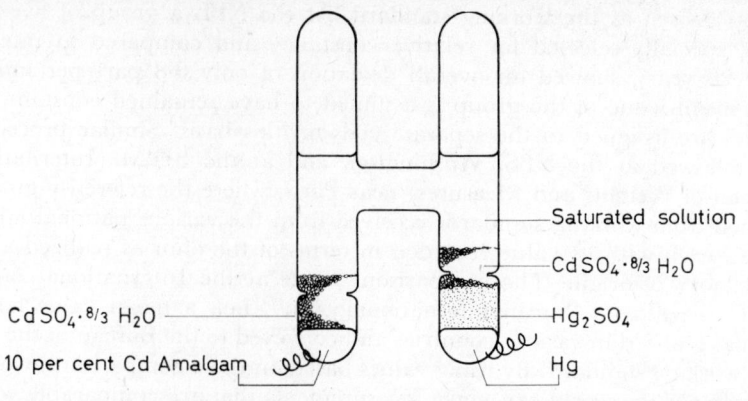

Figure 5. Cross-section of a saturated cadmium standard cell. The primary reaction in such a cell is:
$$\text{Cd (2-phase amalgam)} + \text{Hg}_2\text{SO}_4(s) + (8/3)/(m\text{-}8/3)\ \text{CdSO}_4.m\text{H}_2\text{O}(l) = m/(m\text{-}8/3)\ \text{CdSO}_4.(8/3)\ \text{H}_2\text{O}(s) + 2\ \text{Hg}(l).$$

same supply of specially purified materials, it may be necessary to discard one here and there because it shows unmistakable variations relative to the rest. As in the case of resistors, relative stability from day to day within the group is the only evidence available, but the cell has a more limited life and is discarded when it shows an increasing instability, which may be in 10 or 20 years. Thus the cells in a reference group are changed as the years go by, and the group includes cells of different ages. The unit is preserved by assuming that the mean value of the group remains constant so long as the deviations of members of the group relative to the mean remain within tolerable limits. Experiments alone, with such cells as are available, can fix these limits.

The national standardizing laboratories send to the International Bureau from time to time, standard cells that have been calibrated by comparison with their own reference group, and a "mean international volt" becomes available as a result of systematic intercomparisons similar to those mentioned above for resistors. Such comparisons, made with selected groups of cells in specially favorable conditions, e.g., the cells remain undisturbed at a constant temperature for several weeks before any measurement is made, show a drift in mean value of about 2×10^{-6} in 10–20 years. International consistency in p.d. measurements is therefore obtainable at any particular time with an uncertainty little greater than that for resistance measurement, but there can be no certainty that the unit realized by either the national or international reference groups remains constant from year to year, and none of these units can, strictly speaking, be identified with the volt of the theoretical definition.

XI. Application to Thermodynamic Measurements

Electrical measurements have two important functions in thermodynamic measurements because of the great convenience of electric measurements and the high precision which is achievable with them. One of these functions is the interpolation between fixed points on the practical scale of temperature and the other is determination of the energy involved in an exothermic or endothermic process. The accuracy requirements for the electric standards are different in the two applications. It is useful to distinguish between these two cases.

In realizing the practical temperature scale, electrical measurements are involved in determining the resistance of a platinum thermometer in ohms or the electromotive force of a thermocouple in volts. In neither of these cases is it necessary that the electric standards used embody either the idealized ohm or the idealized volt, or even that the volt and ohm as embodied by these standards satisfy the power relationship, $P = E^2/R$, where P is given in watts. It is necessary only that the multiples and submultiples of each of these units be consistent with the respective unit, i.e. 1 000 milliohms = 1 ohm, etc. Thus, imperfection in the knowledge of the value of an electrical standard in terms of the standards of length, mass, and time in no way impairs the accuracy with which the practical temperature scale can be realized.

The reason for this is clear. In calibrating a resistance thermometer, its resistance is measured at several fixed points and coefficients calculated for an interpolation equation by means of which the resistance at various temperatures between these fixed points can be obtained. If the resistance standards used are in error, compensating errors will be present in the coefficients, and temperatures indicated by the interpolation equation will be correct, *provided* the same resistance standards are employed in measuring temperatures with the thermometer as are employed in calibrating it. Similar considerations apply for thermocouples. In most practical situations it is of little concern that *identical* standards for resistance and electromotive force be used in both the calibrating and the measuring processes. The art of calibrating electrical standards is so highly developed that standards for resistance and electromotive force are disseminated from the leading national standardizing laboratories with an uncertainty of only one or two parts per million in terms of the unit as maintained by those laboratories. The important consideration is that it does not matter, as far as temperature measurements are concerned, if the units maintained by those laboratories differ appreciably from the idealized value. It is otherwise in heat measurements.

In heat measurements, one is concerned with determining the thermal energy, in joules or ergs, involved in certain processes. The results of such determinations are used in calculations involving various physical constants such as the gas constant R_0, Boltzmann's constant k, Avogadro's number $\mathbf{N}$, and Planck's constant h. R_0 is ordinarily expressed in joules or ergs per mole-degree. Its value is obtained by mechanical measurements of pressure, volume, and temperature and is thus independent of the electrical standards. $\mathbf{N}$ and h are determined from a complicated chain of measurements, and $k = R_0/\mathbf{N}$. In order that measurements and calculations be consistent, it is necessary that the energy units involved in R_0, h, and k be consistent with energy units in the equation, $q = EIt$, expressing the heat evolved when a current I flows through a resistor for a time t, E being the potential drop across the resistor. The extent of the required consistency is set by the accuracy with which the various quantities are known.

Calorimetry by electrical methods is performed by supplying heat to the calorimeter by a coil or other heating element. It is necessary to measure the current flowing, the duration of flow, and the potential drop across the heating element. The heat supplied electrically is then given by the preceding equation, $q = EIt$. Adequate accuracy is not achieved by measuring the resistance, R_h, of the heating element and calculating the heating by the equation $q = I^2 R_h t$ because the resistance of the heating element may not be constant during the procedure.

With great care, the electrical measurements in calorimetry may be conducted in such a manner that the full precision of the electrical units of voltage and resistance, as maintained by the national standardizing laboratories, may be approached (*Figure 6*). This would involve use of a heating

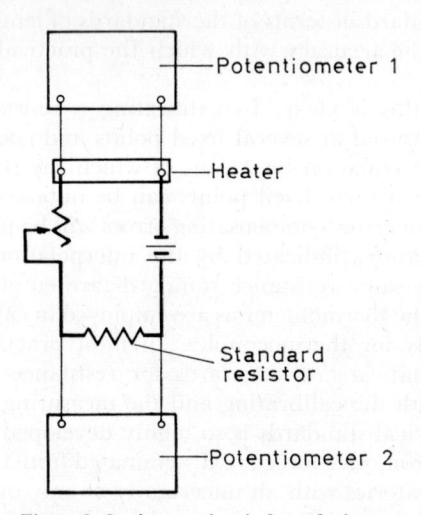

Figure 6. Optimum circuit for calorimetry

element connected in series with a resistor of known value. The voltage drop across the resistor would be measured by a potentiometer, giving the current in the heating element by the relationship $I_h = E_s/R_s$. A similar potentiometer connected across the heating element will measure the potential drop E_h across the heating element. The heat evolved is then $q = E_h I_h t$, where t is the time the current flows, which can be measured by modern electronic methods with an accuracy far beyond the accuracy of any of the other measurements involved. It must be assumed that all of the heat evolved between the points where the potentiometer is connected to the heating element, and no other heat, is delivered to the calorimeter. Optimum results in the electrical measurements are achieved when the heater and standard resistor have resistance of about one ohm and the current is about one ampere. The power input to the calorimeter will then be about one watt and the voltages about one volt. Under such conditions the heat developed in the heater element can be known with a standard error between one and two parts per million in terms of the electrical units *as maintained* by the national laboratories.

The potentiometer (*Figure 7*) enables a very precise comparison of an unknown voltage with that of a standard. In commercial forms the instrument consists of a system of resistors and their controls mounted on a panel. Terminals are supplied for connecting the standard cell, auxiliary battery, and galvanometers. This arrangement permits the standard cell to be used while in a temperature-controlled oil bath. Since the current drawn from a standard cell must be very small and intermittent, if the cell is not to be impaired, an auxiliary battery is employed. This battery is placed in series with a calibrated fixed resistor and a variable resistor. The current through

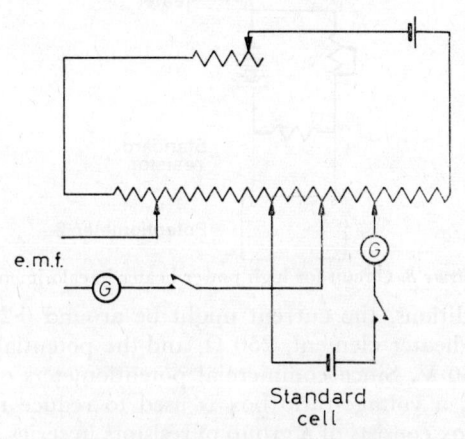

Figure 7. Typical potentiometer circuit

this circuit is adjusted by the variable resistor until the voltage drop across a part of the calibrated resistor bridged by the standard cell equals the voltage of the cell. This condition is indicated when no reading is given on a galvanometer in series with the cell when the galvanometer key is depressed. In actual use there are protective resistors in series with the galvanometer to prevent too large a current from flowing when the circuit is unbalanced. The resistors may be removed progressively as balance is approached, to allow more sensitive adjustment.

The voltage to be measured is bridged across a part of the calibrated resistor in series with a galvanometer. The portion of the resistor bridged is varied until a null reading is given by the galvanometer as before. If I_a is the current flowing through the calibrated resistor then $E_s = I_a R_s$ where E_s is the standard cell voltage and R_s the resistance of the segment bridged by it. Similarly the unknown voltage is given by $E_u = I_a R_u$ where R_u is the resistance of the segment bridged by it; then $E_u/E_s = R_u/R_s$. It is noteworthy that it is not necessary to know the values of the resistance segments in ohms, but only their values relative to each other. In commercial potentiometers the controls are so arranged that the unknown may be read directly in volts from the setting of the controls.

The conditions which optimize the accuracy of the electrical measurements may not be those which optimize the accuracy of the entire calorimeter experiment (*Figure 8*). In practice, a power input of 10 W may be used.

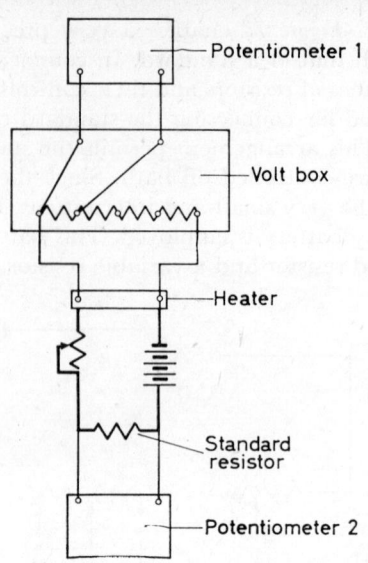

Figure 8. Circuit for high power heater in calorimetry

Under such conditions, the current might be around 0·2 A, the standard resistor 5 Ω, the heater element, 250 Ω, and the potential drop across the heater element 50 V. Since commercial potentiometers ordinarily operate up to 1·5–3·0 V, a voltage-ratio box is used to reduce the voltage to be measured. The box consists of a group of resistors in series and the potentiometer is connected to taps across one of these resistors. The total resistance may be 50 000 Ω and the potentiometer used to measure the drop across a 1 000 Ω section. The standard resistor now carries both the heater current and the current through the box. The current through the heater is now $I_h = I_{total} - I_{box}$. The current through the voltage box can be obtained by the relationship $I_{box} = E_p/R_p$ where E_p is the voltage measured by the potentiometer across the segment in the box and R_p the resistance of that segment; in the example given this current would be 0·001 A. The power being supplied to the heater is then $P_h = E_h I_h = K E_p I_h = K E_p (E_s/R_s - E_p/R_p)$ in which K is the multiplication factor for the voltage-ratio box.

This method does not lead to as accurate a determination of the power dissipated in the heater as the first method for several reasons. The resistors are not as close in value to unit resistance and consequently their values are not as well-known, there are more unknowns to be measured and calculated, and the leads to the voltage-ratio box are current-carrying elements and must be figured into the calculation. Unless the leads from the heater to the voltage-ratio box are of negligible resistance, corrections must be made for them in the multiplying factor K. The leads are likely to change in resistance with temperature changes, and this change must be taken into account. However, when proper precautions are observed, it appears that the power can be determined to about 10 parts per million in terms of the electrical units. However, this power is that dissipated in the heater; the total heat flow into the calorimeter may differ from this amount owing to heating in the various leads.

ENERGY MEASUREMENT AND STANDARDIZATION

To evaluate the accuracy with which the determination of the heat generated can be expressed in terms of the mechanical units of length, mass, and time requires an assessment of the accuracy with which the electrical units are maintained by the national standardizing laboratories. There are differences between the units maintained by the standards of the various national laboratories, amounting to not more than two or three parts per million, which are well-known through international comparisons of the standards. Corrections may be made for these differences if calorimetric measurements made in one country are to be related with high precision to those made in another. Absolute determinations indicate that the units maintained by the various laboratories differ from the theoretical values of the units by more than this amount. Recent redeterminations at NBS of the value of the ampere as realized by the NBS electricity standards show, after correcting for what is believed to be the best value for gravity

$$1 \text{ NBS ampere} = 1 \cdot 000\ 010 \text{ ampere}$$

and

$$1 \text{ NBS ampere} = 1 \cdot 000\ 015 \text{ ampere}.$$

A recent determination at NPL, corrected for the difference between the two laboratories indicates

$$1 \text{ NBS ampere} = 1 \cdot 000\ 005 \text{ ampere}.$$

Estimates of the errors from various sources in experiments of this sort suggest that, on the average, each value has an uncertainty associated with it, corresponding to a standard error, of about 6 parts per million. This uncertainty is consistent with the discrepancy between individual values. This leads to the belief that the ampere as realized by the standards at NBS is greater than the theoretical ampere by about 10 parts per million, and that the uncertainty associated with this correction is about 4 parts per million. Setting an outer limit of error as three times this value, one concludes that it is highly unlikely that this correction is in error by as much as 12 parts per million. This estimate applies to the ampere as realized by the NBS standards, but it also applies to the ampere realized by the standards of other national laboratories when interlaboratory differences are corrected for in accordance with international comparison of standards.

Estimates of the uncertainty in individual determinations of the ohm average somewhat smaller and amount to about 4 parts per million, which is likewise consistent with the discrepancies between individual values. Recent redeterminations of the ohm as maintained by the NBS indicate

$$1 \text{ NBS ohm} = 0 \cdot 999\ 997 \text{ ohm (NBS, 1938–49)}$$

$$1 \text{ NBS ohm} = 1 \cdot 000\ 004 \text{ ohms (NPL, 1951)}$$

$$1 \text{ NBS ohm} = 0 \cdot 999\ 996 \text{ ohm (NRC,* 1957)}$$

$$1 \text{ NBS ohm} = 1 \cdot 000\ 002 \text{ ohms (NBS, 1961)}.$$

* National Research Council, Canada.

From these redeterminations one concludes that the NBS ohm is equal to the theoretical ohm to within 1 part per million and the uncertainty in this value is about 2 parts per million.

Since the watt is related to the electrical quantities by $P = I^2R$, the uncertainty in the watt as realized by the electricity standards is

$$\epsilon_W = 2\,\epsilon_A \pm \epsilon_\Omega$$

and is 6–10 parts per million. The outer limit of error would be 18–30 parts per million.

This is the uncertainty associated with determination of the heat evolved using optimum methods in electrical calorimetry, provided all necessary corrections to the units maintained by the national laboratories are applied. If less precise calorimetric methods are employed, the estimated error in the method should be added to the estimated errors in the electrical standards. When it is recognized that the standard error of the gas constant is estimated as about 40 parts per million, it is clear that uncertainty in the theoretical value of the electrical units does not make an important contribution to the uncertainty in thermodynamic measurements when they are to be used in calculation of other thermodynamic quantities.

CHAPTER 4

Principles of Calorimetric Design

D. C. GINNINGS and E. D. WEST

National Bureau of Standards, Washington, D.C., U.S.A.

Contents

I.	Introduction	85
II.	Chemical, Mechanical, and Electrical Considerations	86
	1. Chemical	86
	2. Mechanical	87
	3. Electrical	91
	A. Thermocouples	91
	B. Calorimeter Heater	94
	C. Heaters for Temperature Control	97
	D. Galvanometers and Electronic Amplifiers	97
III.	Heat Flow Considerations	99
	1. General	99
	2. Heat Transfer by Radiation	100
	3. Heat Transfer by Convection	103
	4. Heat Transfer by Thermal Acoustical Oscillation	108
	5. Heat Transfer by Mechanical Vibration	108
	6. Heat Transfer by Conduction	108
	A. Steady-State Heat Flow	111
	(1) Rod	111
	(2) Wire Carrying Current	111
	(3) Cylindrical Shell	112
	(4) Radial Heat Flow in a Hollow Cylinder and Sphere	112
	(5) Thin Disk with Uniform Heating and Cooling	113
	(6) Tempering of Electrical Leads and Other Conductors	113
	(7) Continuous Tempering	113
	(8) Tempering of Wires Carrying Current	115
	(9) Step Tempering	115
	(10) Tempering of Fluid in a Tube	116
	B. Unsteady-State Heat Flow	117
	(1) The Principle of Superposition	119
	(2) Transients in One-Dimensional Heat Conduction	121
IV.	Applications to Calorimeter Design	122
	1. Temperature in a Calorimeter Wall at Constant Heating Rate	122
	2. Uncertainty in Heat Leak Due to Temperature Gradients	124
	3. Methods of Minimizing Heat Leak Due to Temperature Gradients	124
	4. The Calorimeter Heater Lead Problem	127
V.	References	130

I. Introduction

The design of calorimetric apparatus involves consideration of many factors. The first factor is obviously the choice of the calorimetric method, as discussed in Chapter 1. This choice depends upon temperature range, accuracy desired, properties of the sample, amount of heat involved, duration of experiment, available apparatus and personnel, cost of the apparatus, and many other factors. After the choice of method comes the detailed design of the calorimetric apparatus needed. If the scientific literature describes calorimeters suitable for the investigation, it sometimes is possible to obtain constructional details either from the published papers or directly from the authors. More often, however, either this is impractical or certain variations in design are needed for the particular investigation. In any case, the construction and operation of a calorimetric apparatus is probably facilitated if certain fundamentals of design are known. It is the purpose of this chapter to review some general principles which underlie the design and operation of accurate calorimeters. For details of design and construction, the reader is referred to the various chapters describing apparatus. To illustrate some of the principles, simple examples will be given together with corresponding calculations of significant quantities.

In the references at the end of this Chapter are listed some of the handbooks and other sources of values of properties of materials useful in calorimetric design. Such information may be found also in *Mechanical Engineers Handbook*, McGraw-Hill Book Co., Inc., New York, 1958 and *Handbook of Chemistry and Physics*, Chemical Rubber Publishing Co., Cleveland, Ohio, published annually.

The specific design of calorimetric apparatus involves mechanical, chemical, electrical, thermal, and perhaps other properties of construction materials. Usually, the actual design is a compromise with at least some of these properties. For example, it may be necessary to sacrifice mechanical rigidity to obtain low thermal conductance in certain parts, or in selecting the size of the calorimeter heater current leads, it is necessary to compromise between electrical heat developed in the leads and thermal conductance of the leads. Some discussion will be given of the mechanical, chemical, and electrical design factors, but the major part of this chapter will be devoted to consideration of heat flow because it is believed that lack of understanding of heat flow is the source of most errors in accurate calorimetry.

II. Chemical, Mechanical, and Electrical Considerations

1. *Chemical*

For most calorimetry at moderate and low temperatures, the problem of chemical properties of materials is not difficult to solve. In this temperature range, it is usually possible to select inexpensive metals or alloys which are sufficiently unreactive either to the sample material or to the environment. Sometimes the use of noble metals such as platinum and gold is warranted. Glass is usually considered to be unreactive to most materials in this temperature range, and in earlier calorimetry it was frequently used. However, its use in modern calorimetry is limited because of its other properties. If good thermal conductance is desired, glass is undesirable because of its low thermal conductivity. If poor thermal conductance is desired, it frequently turns out that for mechanical reasons it is necessary to use a greater cross sectional area than would be required with metal, so there may be no significant gain. For example, in a glass tube, it might be necessary to use a wall thickness many times that required for a metal tube. Although the use of metals for most calorimetry is most convenient, at higher temperatures they become more reactive, both with the sample and with air when they are in contact with it. Although there has been some progress in the development of materials inert at high temperatures, the chemical problem here still may be the major one. Further discussion of the chemical problem will not be given here, partly because it is not difficult in most calorimetry, and partly because any problems involved are usually characteristic of the particular experiment.

2. *Mechanical*

In all calorimetric apparatus, it is necessary to contain the sample, and to position and support the various parts. The mechanical design is necessarily a compromise between mechanical and other considerations, such as thermal and chemical. First of all, in order to contain the sample, the calorimeter must be strong enough to withstand any pressure and temperature encountered. The term "strong" here means not only avoiding rupture, but also keeping down to a reasonable tolerance any change in dimension due to pressure. In accurate calorimetry, for example, if there is any significant change in dimension due to either pressure or temperature, the dimensions should be a reproducible function of these variables. An example of irreproducibility is exceeding the mechanical elastic limit of the material. The *elastic limit* (which has values essentially the same as the *proportional limit*) of a material may be defined as the greatest stress to which a material may be subjected without developing a permanent set. The calculation of the pressure effect will be given for simple cases.

If the mechanical properties of the calorimeter are likely to be a serious limitation, the shape of the calorimeter should be designed to minimize strain. In this respect, a spherical shape might be ideal mechanically and also offers the smallest area for heat transfer to the outside. Since a sphere is not always convenient to fabricate, a cylindrical section having hemispherical ends frequently is used. In both cases the use of supporting internal struts gives added strength with relatively little additional material. The internal struts can serve also to distribute heat to reduce temperature gradients.

In considering the strengths of calorimeter sample containers, one must consider the effect of both external and internal pressure. Usually, the external pressure is limited to atmospheric pressure so that the sample container does not have to withstand more than one atmosphere differential pressure. Therefore, a small amount of metal can be used in order to minimize the heat capacity of the empty calorimeter. This is especially important at low temperatures where the heat capacity of the metal sample container may be a large fraction of the total. The reader is referred to Chapters 5, 6, 7, and 13 for details of low-temperature sample containers, including those designed for one atmosphere external pressure.

For sample containers with *internal* pressure, calculations are now given for the sphere and cylinder, which approximate most cases. Assume that the calorimeter is a spherical shell having a diameter D and thickness w, and containing an internal pressure P. The force operating to stretch the wall is $P(\pi D^2/4)$ and is exerted on a wall cross-sectional area $(\pi D w)$, so that the stress on the material (force/area) is $PD/4w$. If $P = 100$ lb/in^2, $D = 4$ in, and $w = 0{\cdot}02$ in, the stress is then 5000 lb/in^2. If the material is cold-worked copper, we find in tables of mechanical properties† that the "proportional limit" of this copper at room temperature is about 15 000 lb/in^2. This means that there is a safety factor on pressure of about three before the pressure will cause permanent set. In actual design, this safety factor seems too small, so that either the material or some other factor probably would be changed. Now calculate the change in the diameter D due to the pressure P. The material property which determines this change is called *modulus of elasticity*. The *fractional* increase in length of a specimen under tension is the ratio of stress to modulus of elasticity. For the copper, with a modulus of elasticity of about 16×10^6 lb/in^2, the pressure of 100 lb/in^2, which results in a stress of 5000 lb/in^2, increases the diameter D by $[(5 \times 10^3)/(16 \times 10^6)] D = 0{\cdot}00125$ in.

Using a length (l) of the same material in a cylinder of the same thickness (w) gives a force (PDl) operating on an area of $2wl$, so that the stress is $PDl/2wl = PD/2w = 10\,000$ lb/in^2. The increase in diameter is then $[10^4/(16 \times 10^6)]\,4 = 0{\cdot}0025$ in.

There is another way in which modulus of elasticity enters into calorimetric design, namely in calculating stress in a material caused by increase in temperature in the material which is not allowed to expand significantly in one dimension. Assume in a simplified example that a specimen disk is held flat between two parallel fixed plates so that there is perfect contact at temperature T_1. Now heat the specimen to a temperature T_2. If the disk were free to expand, its thickness would increase by the factor $a(T_2 - T_1)$ in which a is the coefficient of linear thermal expansion. Since we keep the disk from expanding axially, a stress is developed which can be calculated from the modulus of elasticity by using the relation, stress $= [a(T_2 - T_1)]$ $\times$ [Modulus of Elas.]. Consider as an example a silver disk heated from 0° to 200°C. Taking an average value of $a = 18 \times 10^{-6}$ and modulus of elasticity $= 10^7$ lb/in^2, one obtains stress $= 36\,000$ lb/in^2. However, this figure is considerably greater than the elastic limit of silver which is about 1000 lb/in^2, so that actually the silver would flow radially. When the specimen is cooled to room temperature, it will have a permanent set, its thickness being less than its original thickness. If the specimen is steel with an average coefficient of expansion of $10^{-5}/°$C, a modulus of elasticity of 3×10^7 lb/in^2, and an elastic limit of 40 000 lb/in^2, then the calculated stress is 60 000 lb/in^2, which is also above the elastic limit, so that there would be some permanent set.

One application of the above principles to calorimetry is in the use of gasket materials for sealing two parts together, as in a "union". The above examples indicate that if a union is to be used over a wide temperature range, two precautions should be considered. First, the union body material might be made of a material having about the same coefficient of expansion, thereby minimizing the change in stress due to differential expansion. Second, the union might be designed to provide some flexure, so that large

---

† English units are used here because of their general use in engineering tables of properties of materials.

changes in stress in the gasket are avoided. This spring quality could be obtained either in a separate spring, or possibly in the configuration of the union body, if a material with high elastic limit is used.

The effect of thermal expansion in calorimetry over a wide temperature range is important also in choosing materials which are in contact, such as in soldered joints. As a general rule, the materials should be chosen so that the change in termperature results in a force tending to tighten the joint (compression) rather than loosen it (tension). For example, in a low temperature calorimeter, if there is a differential thermal expansion in a joint, the outer part preferably should have the higher coefficient of thermal expansion in the low temperature range.

A universal mechanical problem in calorimetry is the use of soldered joints, both in giving a "tight" joint and in providing strength. A variety of available solders[17] provides a melting temperature over a wide range of temperature. This is frequently useful in assembly of various calorimetric components. However, the lower melting solders are not generally as strong as the higher melting solders, especially at temperatures approaching melting. For example, in the design and preliminary testing of the 0–100° calorimeter described in Chapter 11, it was found that a lead-tin soldered joint holding the two hemispherical shells together failed at 100°C. When the lead-tin solder was replaced with a tin-antimony solder having both higher melting point and better mechanical properties at 100°C, the soldered joint was successful. Soldered joints should be subjected to the least possible stress and should depend as little as possible on the strength of the solder, which is usually considerably less than the strength of the metals being joined. In practice this means avoiding thick layers. The design of soldered joints is especially important in calorimeters used over large temperature ranges where differential expansion may cause large stresses in the joints.

In calorimeters operating over a large temperature range, account should be taken of the change in mechanical properties with temperature. Most materials are weaker at high temperatures so that their properties at the highest operational temperature must be used in the design. In the discussion earlier on modulus of elasticity, the modulus under tension or compression was used. Under special circumstances, the modulus in shear might be more appropriate.

A warning should be given on the use of certain "cast" metals or alloys if vacuum tightness is required. As an extreme example, when silver is cast in an air atmosphere, considerable oxygen in solution is expelled on solidification, thereby giving a porous structure. In this case a vacuum casting should be used to give a non-porous solid. The presence of oxygen in copper may lead to less desirable mechanical properties. For example, ordinary copper work hardens much faster than oxygen-free copper, which is now available commercially.

The design of the calorimeter supports, together with the choice of materials, is always a compromise among various requirements, which include mechanical, chemical, and heat flow. Discussion of the heat flow problems will be given later in this chapter. The shape of the mechanical support is a large factor in obtaining maximum support strength for a given thermal conductance. The optimum shape is determined partly by the direction in which strength is needed. For example, with certain calorimeters which have no connecting tubes to the outside, vertical strength may be needed primarily for suspending the calorimeter, with little strength needed horizontally. In this circumstance it might be possible to use one or more small wires or "threads" under tension. If more horizontal strength is needed, the mechanical supports may consist of thin "struts" which may be tubes. For a given cross section of material (given thermal conductance) a tube gives greater mechanical rigidity than a rod. In one apparatus described in Chapter 11 the shield was suspended by three short thin mica strips spaced 120° apart so that they could be considered as narrow sections of a large "mica tube". This arrangement gave considerable strength horizontally, as well as vertically with the mica under tension. In general, if a minimum of material is desired, it is more stable under tension than compression. Also, because the deflection of "beams" is usually directly proportional to the cube of the length of the beam, it is better to use a short support piece, and keep to a minimum its dimension in a direction perpendicular to the force.

In choosing materials for mechanical supports, the thermal diffusivity of the material, together with dimensions, determines the time lag in the material as its ends are heated. Because ceramic materials usually have lower thermal diffusivities than metals, their time lags should be considered before using them in calorimeter design.

The above brief consideration of mechanical design is intended only to point out a few of the simple problems encountered in calorimetry. For detailed computations, the reader is referred to texts and handbooks covering the strength of materials[16] and to the various chapters in this book for specific calorimetric applications.

A question which arises repeatedly in calorimetric design is the thermal conduction between two surfaces in mechanical contact. The mechanism of heat transfer is by gas conduction, radiation, and by direct conduction at solid–solid contacts. Under pressure, the solid–solid contacts increase in number and in area and it is possible to gain considerably over gas conduction and radiation. The small amount of information available is for metallic surfaces.

For rather rough laminated steel surfaces at temperatures slightly above room temperature, aluminiun foil placed between surfaces increases the thermal contact conductance in air by a factor of about two with moderate contact pressures. An organic material in the space usually improves conduction considerably. For solid steel blocks, the largest contact conduction is for lapped surfaces. They show only slight dependence on contact pressure in air. Rougher surfaces show increased conduction by about a factor of two for pressures of 300 lb/in^2 (ref. 3).

Data for smooth aluminum alloys, steel, and bronze show considerable variation for different assemblies[20]. The conduction of aluminum joints is considerably better than steel at each pressure measured. Furthermore, copper plating one of the surfaces improves the conduction of the joints. Oil in the joint increases conduction, especially at low pressure, owing to the higher conductivity of oil compared to that of air.

For polished gold, silver and copper surfaces in vacuum at 25°C, the conduction of a metallic joint is negligible at zero contact pressure and increases linearly for copper to 0·3 W cm^{-2} deg^{-1} at 4 kg/cm^2. For gold and silver, the conductance is 0·22 W cm^{-2} deg^{-1} at a constant pressure of 1 kg/cm^2, increasing to 0·4 at 4 kg/cm^2 (ref. 11).

Qualitatively, conductance through metallic contact is increased by using softer metallic surfaces, smoother surfaces, and higher pressure. If such joints must be used, a large number of bolts or screws should be used to obtain a high pressure. If thermal conductance is needed in a contact in a vacuum, precaution should be taken either to apply sufficient force or to use a softer material at the contact. Some measurements of thermal contact resistance in vacuum at low temperatures have been made[2] on a few metals and non-metals.

3. *Electrical*

A. Thermocouples

Most of the factors involving the electrical system in the calorimetric apparatus also involve heat flow, so that both must be considered in the design. Consider first the sizes of electrical leads of thermocouples. Of course, the minimum size is that which can be handled safely during assembly and disassembly. In deciding how much larger to make the leads, one must compromise between electrical resistance and thermal conductance. If heat flow is important, one uses the smallest thermocouple wire compatible with the two requirements of strength for convenient handling and of tolerable resistance of the thermocouple circuit. This point depends upon the electrical resistance of the circuit external to the thermocouple, which frequently consists of a galvanometer and potentiometer. Consider the schematic circuit in *Figure 1*, in which E_1 and R_1 represent the e.m.f. and resistance of the

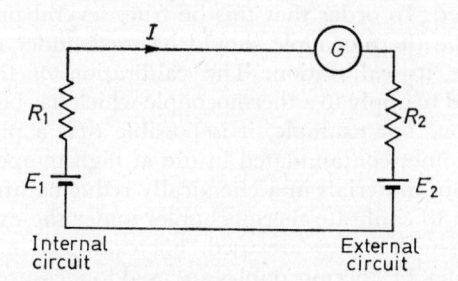

Figure 1. Thermocouple measuring circuit.

thermocouple and E_2 and R_2 represent the potentiometer bucking e.m.f. and the total resistance of the external circuit, considered to be a potentiometer and a galvanometer, G, for the simple case. At ideal balance, the current, I, is zero and $E_2 = E_1$. The actual balance occurs at some finite current, I_1, representing the smallest current which the galvanometer can detect. In this actual case, this error signal is $(E_2 - E_1) = I_1 (R_1 + R_2)$. If $R_2 \gg R_1$, then an increase in R_1, the thermocouple resistance, does not change significantly the error signal $I_1 (R_1 + R_2)$. However, if $R_1 \gg R_2$, this error signal increases in proportion to the resistance R_1. With a given measuring circuit, the resistance of the thermocouples may determine the practical limit of the number of thermocouples which can be used in series (thermopile) to increase the sensitivity. If $R_1 \gg R_2$, then increasing the number of thermocouples increases the resistance $(R_1 + R_2)$ in proportion to the increase in E_1 so that the *relative* value of the error signal (precision of measurement) to the total e.m.f. (E_1) remains the same.

A typical resistance of a galvanometer–potentiometer circuit is perhaps 50 Ω, so that as a practical matter the resistance of any combination of thermocouples used should not be much larger than this resistance. Although higher resistance galvanometers can be used with thermocouple circuits having higher resistances, it is usually more practical to use electronic amplifiers because of their flexibility of input impedance. Even with low-resistance circuits, the use of electronic amplifiers has increased greatly in recent years. A comparison of the galvanometer with the electronic amplifier will be given later in this section (II–3–D).

In designing a temperature measuring system involving thermocouples, it is necessary to decide on thermocouple materials. Since this decision depends greatly upon the temperature range involved, only a brief discussion will be given here. In Chapter 2, the general subject of thermocouples has already been covered. In most accurate calorimetry, thermocouples have their main utility in measuring temperature differences in different parts of the apparatus. For this purpose, frequently it is convenient to use thermocouple materials having high thermoelectric powers, low thermal conductivities, and good mechanical properties. Reference should be made to the various chapters for the authors' choices of materials.

In the use of thermocouples in accurate calorimetry, it is necessary that the thermocouple e.m.f. be a true measure of difference in temperature between the components to which the principal and reference junctions of the thermocouple are attached. In order that this be true, several precautions should be taken. First, the thermocouple should be used under conditions which do not invalidate its calibration. The calibration of the thermocouple cannot be expected to apply to a thermocouple which has been contaminated after its calibration. For example, it is possible that a platinum–rhodium thermocouple becomes contaminated in use at high temperatures if it is in contact with certain materials in a chemically reducing atmosphere. Sometimes it is possible to calibrate thermocouples under the exact conditions of the experiment in which they are used. For example, in the calorimeters described in Chapter 11, thermocouples are used to measure the temperature difference between the calorimeter and a "reference block" containing platinum resistance thermometers. The thermoelectric coefficients of these thermocouples can be determined *in place* at any time during the experiments by comparing with the platinum resistance thermometer. This "in place" calibration also has the advantage that certain thermocouple errors are also accounted for. Another example of "in place" calibration of thermocouples is with differential thermocouples between the calorimeter and the surrounding shield. In this use, the calibration of the e.m.f. in terms of temperature difference is unnecessary. Instead, the e.m.f. is calibrated

PRINCIPLES OF CALORIMETRIC DESIGN

directly in terms of heat flow by observing the resultant change in temperature of the calorimeter whose heat capacity is known with adequate accuracy.

A second precaution in the use of thermocouples concerns spurious e.m.f.s due, for example, to inhomogeneities in the thermocouple material. If an inhomogeneity is located in a region of temperature gradient in the thermocouple wire, there may be a spurious e.m.f. which will be measured along with the "true" e.m.f. of the thermocouple. Frequently, the thermocouple wires are tested for inhomogeneities before their use by moving a sharp temperature gradient along the wire and observing any resulting e.m.f. If the thermocouples are used only to measure small temperature differences, most of the effect of the inhomogeneities can be avoided by keeping the thermocouple material in relatively isothermal regions and using pure metal electrical leads which are relatively free from inhomogeneities. An example of this technique is described in Chapter 11 where many thermocouples measure temperatures relative to a "reference block" whose temperature is very close to that of the calorimeter.

A third precaution in using thermocouples is to make certain that their junctions are really at the temperatures of the components whose temperatures they measure. This problem usually is not serious if the thermojunction is soldered or welded to the component. However, it is not always possible or convenient to make this attachment. Frequently, it is necessary to insulate electrically the thermojunction from the component, so that thermocouples can be connected in series either for differential measurements, for increasing sensitivity, or for integration of temperature gradients. Electrical insulation of the thermojunction usually incurs some thermal insulation, so that if there is a heat flow from the thermojunction, its temperature necessarily will be different from the temperature of the component to which it is attached. Much of the heat flow from the thermojunction is along wires connected to it. It may be necessary, therefore, in accurate calorimetry to make sure that there is no significant heat flow along these wires. This can be accomplished by a technique sometimes called *"tempering"* of the thermocouple wires. Tempering of these wires may be defined as the process of bringing them to the temperature of the component *before* the wires reach the thermojunction. The tempering of wires will be discussed later in this chapter. A common method of tempering wires at moderate to low temperatures is to fasten the insulated wires in thermal contact with the component by using an organic cementing material to attach them, thereby providing both thermal contact and electrical insulation. At somewhat higher temperatures[5] in the presence of gas, bare wires may be placed in ceramic tubes or between strips of mica to be held in contact with the component.

Sometimes the choice of thermocouple material depends not only on temperature range but also on the environment. For example, at high temperatures, some materials should not be used in oxidizing atmospheres whereas other materials should not be used in reducing atmospheres. For an appropriate discussion of thermocouples, the readers are referred to books[4] covering the subject.

In addition to the requirement in accurate calorimetry that the thermocouple e.m.f. be a true measure of temperature difference, there is the problem of proper location of the junctions to evaluate properly the effective temperatures so that proper accounting can be made for both heat leak and temperature change, even if temperature gradients exist. Further discussion will be given later in this chapter on the problem of location of thermocouples.

B. Calorimeter Heater

The design of the calorimeter heater and its leads is important to accurate calorimetry in two different ways. The first way relates to the location of the potential leads in order to evaluate that part of the electrical heat in the leads which belongs to the calorimeter. Although this problem is discussed later in Section IV, it may be mentioned here that the calorimeter heater resistance should be large compared to its lead resistance, and that the thermal contacts between the ends of the heater current lead segment and the calorimeter and shield should be either good or equal and preferably both.

The second way design is important as it relates to changes in resistance in the heater and leads due to their temperature change. These changes in resistance make the accurate measurement of power more difficult. In the calorimeter heater, the change in temperature may be considered in two parts, the first due to the *initial* thermal transient as the heater wire attains its normal temperature excess, and the second due to the slower change in temperature as the calorimeter temperature increases†. When the current is first switched to the heater, its temperature is very near that of the calorimeter. Before it can deliver power to the calorimeter, its temperature must rise to provide a thermal head for heat flow. Since the heater itself has some heat capacity, usually an appreciable time is required for the initial rise to take place. In general, the resistance of the heater wire will increase during this time and the effect on the d.c. power measurements must be considered.

The thermal transient may be considered a simple exponential with time constant (time for 63 per cent of change) equal to the product of the heat capacity of the heater and the thermal resistance to the calorimeter. Since the total effect is small, only the order of magnitude is important and this "lumped constant" approximation is adequate. For small heater wires varnished to the calorimeter surface, the time constant is probably so small that the effect is negligible. More massive heaters and poor thermal contact depending on gas conduction or radiation across a mechanical space make the time constant larger and warrant considering it.

A typical case is a heater with heat capacity of 2 J/deg and thermal resistance to the calorimeter of 10 deg/W. This heater has a time constant of 20 sec, so that the effect would not be accounted for by "conventional" measurements of current and potential for which the first measurement is usually after 20 sec. A plot of resistance against time after turning on the heater is given in *Figure 2* for a linear resistance-temperature dependence

---

† The *electrical* transients in a calorimeter heater circuit are always over in such a short time that it is usually not necessary to consider them.

PRINCIPLES OF CALORIMETRIC DESIGN

and constant heating rate. This heater resistance curve approaches exponentially the straight line which would be obtained by extrapolating from the current and potential measurements made later during the experiment. If a constant voltage is imposed on the heater, the electrical power is $P = E^2/R$, and there is a small error because the extrapolated resistance (dashed line)

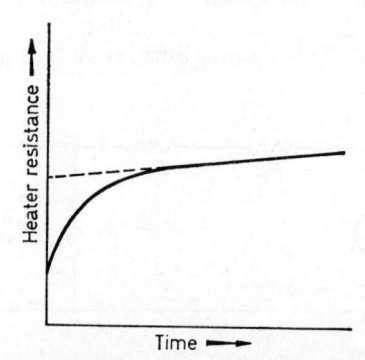

Figure 2. Change of heater resistance during experiment.

between start of the heating and the first measurement is different from the actual resistance. If the eventual excess temperature of the heater is ΔT, and R is a linear function of the temperature, the resistance at any time t can be written

$$R = R_0 \{1 + a [\Delta T(1 - e^{-t/\tau}) + \beta t]\}, \tag{1}$$

in which a is the temperature coefficient of resistance, β is the heating rate of the calorimeter, τ the heater time constant, and R_0 is the resistance at the beginning of the experiment. The difference between the extrapolated resistance and the true resistance is due to the exponential term in Equation (1). The difference ΔR between the true average value and observed average value is obtained by integrating over the heating period of length t_f. The fractional error when $t_f \gg \tau$ is then given approximately by the equation

$$\frac{\Delta R}{R_0} = a \Delta T \frac{\tau}{t_f}. \tag{2}$$

For our typical case $a = 4 \times 10^{-4}$ deg, $\Delta T = 30$ deg for 3 watts of heater power, so that $(\Delta R/R_0) = (0.24/t_f)$. In a 600 sec experiment, the fractional error is four parts in 10 000. If the heating rate is doubled, ΔT is doubled and t_f is halved so that the fractional error is 4 times as great. If the calorimetric measurements have sufficient sensitivity, the error may be detected by changing the heating rate. In addition to the starting transient the slower change of resistance must be considered, but usually this is not bothersome except in very accurate calorimetry (see Chapter 5).

Several schemes have been developed to maintain constant power input when the calorimeter heater resistance changes owing either to the starting transient in the heater or to the gradual change of heater resistance with the temperature of the calorimeter. In one scheme, approximately constant power with a moderate change in heater resistance can be obtained from a *constant current source* supplying the calorimeter heater shunted by an external resistance which is equal to the average calorimeter heater resistance[18]. In another scheme using a *constant voltage source*, the equal resistance is placed in series with the calorimeter heater[10]. Sometimes (see Chapter 5) the calorimeter heater leads are important in that they also change resistance during an experiment. This case is illustrated by *Figure 3* in which R_s is

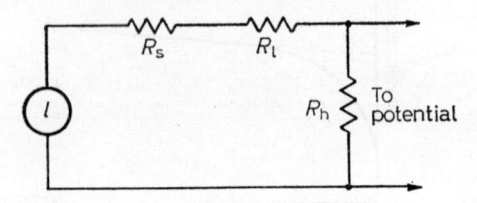

Figure 3. Calorimeter heater equivalent circuit.

constant whereas R_l and R_h are functions of the temperature of the calorimeter. For this circuit the power in the calorimeter is given by the equation

$$P = E^2 \frac{R_h}{(R_s + R_l + R_h)^2}. \tag{3}$$

To find the condition for no change in the power with a change in the heater and lead resistance we take the total differential of Equation (3) and set the result equal to zero. The condition for constant power is then given by the equation

$$\frac{R_s + R_l - R_h}{R_l} \frac{dR_h}{R_h} = \frac{2dR_l}{R_l}. \tag{4}$$

If the fractional changes in the heater and lead resistance are related by some proportionality constant m, $(mdR_h/R_h) = (dR_l/R_l)$, then the choice of R_s is determined from the following equation

$$R_s - R_h = (2m - 1) R_l. \tag{5}$$

A more convenient method is to make the change in the lead resistance dR_l negligible so that the condition for constant power is simply $R_s + R_l = R_h$.

The obvious solution to the problems resulting from change in resistance of the calorimeter heater and its leads is (*i*) to minimize the starting transient by keeping the heater in good thermal contact with the calorimeter, (*ii*) to construct the heater with a material having a small temperature coefficient of resistance, and (*iii*) to either make the heater leads of low electrical resistance or use a material with small temperature coefficient. Although the use of such a material is not always practical, it is available for calorimetry at low to moderate temperatures, as described in Chapter 5.

PRINCIPLES OF CALORIMETRIC DESIGN

C. Heaters for Temperature Control

In the design of other heaters in calorimetric apparatus, it is usually also advantageous to have good thermal contact and low heat capacity in order to avoid lags. In addition, there are two other important factors. First, the resistance of a heater should be made to be compatible with an available power supply. With the extensive use of automatic controls in the past, the use of moderately high resistance heaters facilitated the use of electron tubes as power sources when the power required was not excessive. With the recent development of the transistor which has a lower impedance output than the electron tube, it may be more advantageous to use low-resistance heaters because of the high current available with the transistor. The availability of low-cost transistors controlling over 100 watts of power ensures their increasing use in calorimetric measurements.

In apparatus requiring larger power, it is frequently convenient to use magnetic amplifiers (saturable reactors) for control. The magnetic amplifier is usually used with a 60 c/s power supply and has a time constant longer than that of a transistor. In addition the output of the magnetic amplifier is either a.c. or d.c. with an a.c. component. As pointed out in the next section, this a.c. may cause trouble with pickup in other circuits if high sensitivity electronic amplifiers are used. Of course the rectified output of the magnetic amplifier can be filtered to eliminate the a.c. component at the expense of lengthening the time constant to perhaps a second. It should be pointed out that most magnetic amplifiers have the disadvantage in many control applications that their output powers cannot be reduced to zero.

Two other devices used for control of a.c. power are the thyratron and its modern solid-state equivalent, the SCR (silicon controlled rectifier) whose cost is decreasing steadily. The output of the SCR is pulsed so that it has a large a.c. component.

D. Galvanometers and Electronic Amplifiers

The use of d.c. electronic amplifiers in place of galvanometers has increased in recent years, so that it is believed appropriate here to compare their application to calorimetric measurements. The galvanometer has two fundamental disadvantages due to its mechanical nature. Most sensitive galvanometers are affected by mechanical vibration, so that considerable effort may be necessary to avoid motion of the moving coil due to this vibration. In addition, if a relatively large voltage is impressed on the galvanometer, the moving coil system may be damaged, possibly so that a new suspension wire is required. The electronic amplifier is essentially free from mechanical difficulties. The electronic amplifier also has the advantage of flexibility of power output and input impedance. The flexibility in the magnitude and impedance of the output of the electronic amplifier facilitates its use in recording and automatic control. In most calorimetric measurements the signal source is less than 100 ohms, so that a typical measuring circuit using a sensitive galvanometer is reasonably matched to the signal source; in other words, the power developed in the galvanometer approaches the maximum possible from the signal source. If the signal source has a high resistance, a low-resistance galvanometer would be relatively insensitive to a voltage source. The electronic amplifier can easily be made with either high or low impedance input.

Most modern electronic d.c. amplifiers "chop" the d.c. signal before amplifying, so that an a.c. amplifier can be used. The purpose of this is two-fold. First, by chopping, any drift with time of the amplifier characteristics only changes its sensitivity so that its use as a null device is satisfactory†. Second, by use of a step-up transformer after the chopper, the signal voltage from a low-impedance source can be amplified before the first electron tube (or transistor). The practical limit of this transformer gain is determined by the ratio of impedances of the signal circuit to that of the grid circuit of the first amplifier tube. By using a phase detector (synchronized chopper on the amplifier signal), the amplifier gives a d.c. output.

Although the electronic amplifier can be designed for use with signal sources having high impedances, the reader should be aware of possible difficulties resulting from the high impedance. When using any detector, the theoretical minimum of the electrical noise in the input circuit is that due to "thermal" noise, which in a resistive circuit is directly proportional to the square root of the resistance. With circuits having resistance of about 100 Ω and with detectors having time constants about one second, the thermal noise equivalent (root mean square voltage) in the signal source at room temperature is about 0·001 μV. More often, the high impedance signal source gives trouble because of increased "pickup" of both transients and power line frequency (60 c/s). Consequently, when it is necessary to have very high sensitivity, high impedance signal sources should be avoided if possible.

When a typical galvanometer is used with a low-impedance signal source, not only is the "pickup" relatively small, but the inertia of the galvanometer serves as a high frequency filter which attenuates 60 c/s pickup. In most sensitive electronic amplifiers (d.c.), the effects of 60 c/s pickup are attenuated by an electrical filter in the early stages of the amplifier. To be completely free from first stage amplifier "saturation", the filter should precede the amplifier. However, this location of the filter requires it to have a low impedance when used with a low impedance signal source in order to avoid increased 60 c/s pickup in the signal circuit. Most commercial electronic amplifiers do not have this low-impedance input filter so that they are most susceptible to large 60 c/s pickup than galvanometers having the same time constant.

It follows from the previous discussion that electronic d.c. amplifiers have several advantages over galvanometers in many calorimetric measurements. For many applications, galvanometers are frequently less expensive. When very sensitive electronic amplifiers are used with signal sources susceptible to power line frequency pickup, a low-impedance signal source should be used and suitable steps should be taken to avoid using a.c. (60 c/s) close to any component leading to the input to the amplifier and to follow the usual precautions of shielding, twisted leads, etc.

---

† Negative feed-back is used sometimes to minimize change in sensitivity, to obtain a constant gain, and to make the amplifier have a high input impedance.

III. Heat Flow Considerations

1. *General*

For accurate calorimetry, the essential quantities to be measured are mass, temperature, and energy. Sometimes, additional quantities are measured, such as pressure and volume. Except in extreme ranges, it is possible to measure most of these quantities with all the accuracy that is needed. In most experiments, the major part of the energy is either derived from or compared with electric power which, with precautions, can be measured very accurately. However, the energy also includes heat from other sources, such as that resulting from heat leak. It is believed that the uncertainty in the evaluation of this heat leak is the principal source of error in a large fraction of measurements for accurate calorimetry. Of historical interest is the comment of W. P. White[23] on the first page of his book *The Modern Calorimeter* [1928]. He says "There is a difference of opinion as to whether thermal leakage is necessarily the chief source of error in calorimetry, but it is undoubtedly responsible for most of the experimental features and devices in accurate work." This is more true today than it was in 1928!

It should be pointed out that in calorimetric measurements, the absolute uncertainty in the evaluation of heat leak is not necessarily proportional to the value of the heat leak. For example, in Chapter 1, it has been pointed out that in calorimeters using an isothermal shield (surroundings), the absolute heat leak is usually considerably larger than with calorimeters with adiabatic shields. However, it must be emphasized that it is the absolute uncertainty in the heat leak that results in error, rather than the magnitude of the heat leak.

The problem of reducing heat leak is usually much more difficult than the problem of reducing electrical leakage. This is more obvious when one compares heat and electrical transport properties at room temperature. In electrical measurements, there are solid insulating materials available which have electrical resistivities about a factor of 10^{25} greater than the best electrical conductors. This means usually that it is not difficult, at least at moderate temperatures, to reduce electrical leakage to a very small amount. However, the best solid thermal insulators at room temperatures have thermal resistivities only a factor of about 10^4 greater than the best thermal conductor. Therefore, one cannot expect to minimize heat leakage to the extent that one can usually minimize electric current leakage. Furthermore, in thermal measurements, even when all material connections are removed, heat transfer by radiation may still persist. As shown later, at low temperatures, the radiative heat transfer may be relatively unimportant because radiation is proportional to T^4. On the other hand, at very high temperatures, radiation may be the predominant means of heat transfer.

In the ideal adiabatic calorimeter, for example, there is no heat transfer between the calorimeter and its surroundings. This ideal condition can be approximated if the heat transfer coefficient and the temperature difference between the calorimeter and surroundings are both very small. In the real adiabatic calorimeter, there are five principal steps which can be taken to reduce errors due to heat leak: (*i*) design the calorimeter to have a large thermal resistance to its surrounding shield; (*ii*) design both calorimeter and shield to minimize temperature gradients in them; (*iii*) design the calorimeter and shield to have adequate temperature sensing devices at suitable locations to measure the *effective* temperature differences between the calorimeter and its surroundings, even with temperature gradients; (*iv*) design the calorimeter so that the experimental method or procedure can be made to compensate for at least some of the unknown heat leaks, such as by making two series of experiments; (*v*) make corrections for any measurable heat leak in an effort to account for it. Of the above five steps, all but the last involve calorimetric design requiring heat flow calculations. Many of these calculations can be made easily by assuming simple boundary conditions which are adequate for design purposes. It seems desirable to review briefly the fundamentals of heat flow before applying them to calorimeter design.

Heat flow is by radiation, convection, and conduction in most calorimetric operations. Although the separation of these three modes of heat flow is not always sharply defined, in most calorimetry at moderate temperatures, these modes can be treated independently for design purposes.

2. *Heat transfer by radiation*

In calorimetry we are concerned with heat radiated from one body at temperature T_1 to a cooler body at temperature T_2. The rate of heat (P_1) transferred to body (2) from unit area of body (1) is given by the relation

$$P_1 = \sigma A_2 F_{12} [T_1^4 - T_2^4] \qquad (6)$$

in which P is in watts cm^{-2}, σ is the Stefan–Boltzman constant ($5 \cdot 67 \times 10^{-12}$ watts cm^{-2} deg K^{-4}), T_1 and T_2 are absolute temperatures (°K), A_2 is the area on body 2, and F_{12} is a dimensionless factor determined by the radiation emittances and geometric configuration of the surfaces. If T_1 and T_2 are nearly the same at about temperature T, the differential form of Equation (6) applies, i.e.,

$$\frac{dP}{dT} = 4\sigma A_2 F_{12} T^3. \qquad (7)$$

The degree of approximation of Equation (7) can be shown by taking the case of $T = 300°$K and $\Delta T/2 = 5$ deg K. In this case the use of the value of dP/dT at 300°K for the temperature difference between 295 to 305°K gives the same average heat transfer coefficient within less than 1 part in 5000 as that calculated from Equation (6). Nevertheless, the experimenter must be aware that the radiative heat transfer coefficient is greatly dependent upon absolute temperature, so that it cannot be considered constant over as large a temperature interval as heat transfer by conduction or convection. Between 295 and 305°K the increase in the radiative heat transfer coefficient due to temperature increase is about 7 per cent. Any experiment which depends on *constancy* of this coefficient should be examined to determine how great an error will result.

PRINCIPLES OF CALORIMETRIC DESIGN

A discussion of the evaluation of the emissivity factor F_{12} is given in various books[13]. For the simplest case of a completely enclosed body, *small* compared to the enclosing body, the factor is given by $F_{12} = \epsilon_1$ in which ϵ_1 is the emittance of the enclosed body. For a completely enclosed body comparable in size and shape to the enclosure, the case becomes essentially the same as two infinite parallel planes for which

$$F_{12} = \frac{1}{\dfrac{1}{\epsilon_1} + \dfrac{1}{\epsilon_2} - 1} \tag{8}$$

in which ϵ_1 and ϵ_2 are the emittances of the two surfaces. In using these or other approximations, it is usually relatively easy to calculate radiative heat transfer to the accuracy needed in most design calculations, if values of ϵ are known. The difference in the usage of terms emissivity, effective emissivity, and emittance should be pointed out here. As used in this chapter, the above relation $F_{12} = \epsilon_1$ really defines ϵ_1 as either the *effective emissivity* or the *emittance* of the small enclosed body, such as a sphere. The outer surface A_1 of the sphere is considered in this case to be the ideal surface area $4\pi r^2$, even though the actual surface is rough and several times this area. If this surface is actually perfectly smooth such as with a highly polished metal, the value of its emittance is the same as what is called *emissivity*. In other words, the emissivity of a metal is for a smooth polished surface (no concave surfaces) but its emittance is for any surface. In most cases we do not have very smooth surfaces, so that the term emittance is more appropriate. As used in this chapter the term emittance is dimensionless, so that it should not be confused with a second meaning of emittance (radiant emittance) which has the dimensions of power.

In considering the effect of radiation on calorimetric design, simple calculations show how radiative heat transfer usually becomes relatively small at low temperatures. On the other hand, at room temperature with two infinite parallel planes each having an emittance of 0·6, and with a 1-cm still air space between them, the radiative heat transfer is about the same as the gaseous heat transfer (see *Figure 4*). At higher temperatures, the radiative heat transfer soon becomes the predominant heat transfer. In calorimetric design for accurate work, radiation shields can be quite useful even below room temperature, and at high temperatures extreme precautions may have to be taken.

Radiative heat transfer between two surfaces can be decreased by inserting one or more "floating" shields between them. One type of shield is a thin metallic surface completely surrounding the calorimeter. As soon as the steady state of heat flow has been established to a shield, the same amount of power, P_t, flows across each space. (The transient effect for a single shield is discussed in the next section.) This power can be written for the ith space as

$$P_t = \sigma A_{i+1} F_{i, i+1} (T_i^4 - T_{i+1}^4)$$

A series of equations can be written for n spaces and summed

$$P_t \frac{1}{\sigma A_1 F_{12}} = T_1^4 - T_2^4$$

$$P_t \frac{1}{\sigma A_2 F_{23}} = T_2^4 - T_3^4$$

$$\ldots\ldots = \ldots\ldots$$

$$P_t \frac{1}{\sigma A_n F_{n,\,n+1}} = T_n^4 - T_{n+1}^4$$

$$P_t \left(\sum_{i=1}^{i=n} \frac{1}{\sigma A_i F_{i,\,i+1}} \right) = T_1^4 - T_{n+1}^4$$

whence P_t can be found by

$$P_t = \frac{T_1^4 - T_{n+1}^4}{\sum_{i=1}^{i=n} \frac{1}{\sigma A_i F_{i,\,i+1}}}$$

The overall heat flow is still proportional to the difference in the fourth powers of the temperature, but is smaller than in the unshielded case. If the shields have equal areas and emittances, as is approximately true in many calorimeters the $A_i F_{ij}$ factors are equal and the heat flow is inversely proportional to $m + 1$, in which m is the number of "floating" shields. A simple case shows the advantage of the radiation shield. If a bright shield (such as one gold plated and polished to give an emittance of 0·02) is inserted between two black ($\epsilon = 1$) parallel planes, the radiative heat transfer is reduced by a factor of 100. Even a black shield would reduce the radiative heat transfer by a factor of 2. *Figure 4* given under Section III–3, illustrates the effectiveness of shields, both for radiation and convection.

In shielding a calorimeter or furnace, the area of the shield increases with the distance from the center and the effectiveness of successive shields decreases proportionately. It follows that the spacing of multiple shields should be the smallest compatible with mechanical and gas conduction considerations. At high temperatures the shields may be in the form of a fine powder which effectively inserts many shields in a narrow space. Because of surface roughness and higher intrinsic emissivities, a single powder surface is less effective than a metallic surface by perhaps a factor of 10, but ten particles of 1000 mesh powders at ~50 per cent bulk density occupy only 0·25 mm. Powder used for radiation shielding in vacuum should be placed in hot parts of the apparatus where it will outgas readily. Another consideration in selecting a powder is that a particle should not be transparent to the radiation at the highest temperature of operation.

A combination of this sort of radiation shielding with spaces shorter than the mean free path has resulted in an air-filled silica gel insulator being better than still air[14]. Although a powder is a useful insulating material, especially in furnaces, it is not generally used directly next to the calorimeter where energy must be accounted for. This is because of the heat capacity of the powder and the uncertainty of the calorimeter boundary, within which energy must be accounted for.

In making computations of heat transfer by radiation, it is sometimes difficult to select the proper value of the emittance for a particular problem. This is not so much because the data lack precision but because they may not apply to the particular surfaces under consideration. For example, the emittance of an aluminum surface changes by a factor of about 12 as the oxide layer increases from 0·25 μ to 7 μ. Another uncertainty in radiation computations is in the geometry of the calorimeter. The effective emittances of indentations and corners can be quite different from the emittances of the plane surface. The best approach is perhaps to choose the most unfavorable value of the emittance to make the calculation. More detailed treatment of radiant heat transfer is given by Jakob[13].

3. *Heat transfer by convection*

Heat transfer by convection is the result of mass transfer of a fluid (gas or liquid) from one region to another at another temperature. Convection is classified into two general types, (1) *forced convection* and (2) *free convection*. In forced convection the fluid is forced to move, such as by a stirrer in a liquid bath. In free convection the motion of the fluid is due to gravitational forces acting because of different densities of the fluid at different temperatures. The laws governing the behavior of convective heat transfer are the same for both liquids and gases, although values of their physical properties vary greatly. If these physical properties are known, calculations of both forced and free convection can be made by using engineering formulae[12]. Consideration will be given first to free convection.

Where the calorimeter is surrounded by a gas such as air, heat transfer by *free convection* (as well as by conduction) must be considered in designing and using the calorimeter. In precise calorimetry at room temperature and below, evacuation can be used advantageously to eliminate both gaseous convection and conduction between calorimeter and shield. At higher temperatures, at which radiation becomes more important, the use of evacuation depends on the particular design, which must weigh the advantages against the disadvantages of evacuation. Sometimes, partial evacuation is used to avoid convection while leaving gaseous conduction. In some apparatus, the gaseous conduction may be desirable to provide thermal contact between two components in poor mechanical contact. As a general rule, atmospheric gas pressure can be reduced by several orders of magnitude without much change in thermal conductivity while the effect of convection is made negligible.

In the case of convection with a gas between two parallel vertical surfaces at different temperatures, the denser gas near the cold surface moves down to displace the gas near the bottom of the warm surface which has been warmed by conduction. Warm gas at the top of the space is moved into contact with the top part of the cold surface, where it is cooled. Heat is thus transported by motion of the air as well as by direct gas conduction across the space. For horizontal surfaces with the hot surface above the cold surface, the low density gas is in stable equilibrium and ideally no convection occurs. When the hot surface is below, convection takes place as a local phenomenon, with upward and downward streams close to one another, sometimes resulting in considerable turbulence.

Early workers in calorimetry, recognizing the requirements for the linear relationship between heat transfer and temperature difference, performed experiments on the gaseous heat transfer in their apparatus. As long as heat transfer is *only* by gaseous conduction, this heat transfer is nearly proportional to the temperature difference (ΔT) because the thermal conductivity of a gas does not change very much with temperature. Of historical note, when this proportionality is valid, *Newton's Law of Cooling* is said to apply, i.e. the rate of cooling of a hot body is directly proportional to the temperature difference from its surroundings. As the temperature difference is made larger, convection begins to be significant. Since gas convection is not proportional to ΔT, the total heat transfer is no longer proportional to ΔT when the convection is significant.

In the calculation of the magnitude of free convection, the dimensionless quantity Grashof Number (N_{Gr}) is needed. This quantity is

$$N_{Gr} = \frac{\beta g \rho^2}{\eta^2} l^3 (T_1 - T_2) \qquad (9)$$

in which β is the cubical expansion coefficient (deg K^{-1}), g is the acceleration due to gravity (cm sec^{-2}), ρ is the density (g/cm^3), l is the width of the gas space (cm), η is the dynamic viscosity (poise), and T is temperature (°K). Jakob[12] gives dimensionless plots of the equivalent conductivity (defined as conduction plus convection) against the Grashof Number, for cases of air layers bounded by horizontal planes, by vertical planes, and by coaxial cylinders. As a general rule, the convection is almost negligible for Grashof Numbers less than 1000. For horizontal planes and a Grashof Number of about two thousand, the ratio of equivalent to true conductivity is about 1·05; in other words, for this case the contribution of the convection is about 5 per cent of the thermal conductivity. The other cases yield similar results. For air near room temperature and one atmosphere pressure one can write that $l^3(T_1 - T_2)$ must be less than 11 cm^3 deg K to be reasonably sure that heat transfer is directly proportional to temperature difference. This is in reasonable agreement with the work of White[24] and Barry[1].

From the formula for the Grashof Number it is evident that the most useful experimental ways of decreasing this number are to work with small temperature differences, to reduce the spacing (at the expense of increasing the gas conduction) or to reduce the density. Of course, the ultimate step in reducing the effect of convection is to remove all of the gas, reducing the density to zero. In cases for which it is desirable to reduce convection but to retain gas conduction in order to maintain the thermal contact of heaters or temperature measuring devices with the calorimeter, it is possible to reduce the density of the gas substantially without impairing the thermal conduction of the gas. Until the pressure is reduced to the point at which the mean free path of the gas molecule is comparable to the dimensions of the gas space, the thermal conductivity of a gas is substantially independent of its pressure. At 0·1 atmosphere, the Grashof number is reduced by a factor of 100, but the mean free path in air is about 9×10^{-4} cm, so that gas conduction in spaces around thermometers and heaters is only slightly less than at 1 atm.

When calorimeters are used with gas at atmospheric pressure surrounding the shield (and perhaps the calorimeter also), the problems of radiation and convection frequently can be solved simultaneously by the use of thin metal heaterless shields which act *both* to reduce radiative heat transfer (as noted in Section III–2) and to reduce convective heat transfer by reducing the thickness of the gas gap. *Figure 4* illustrates the effectiveness of

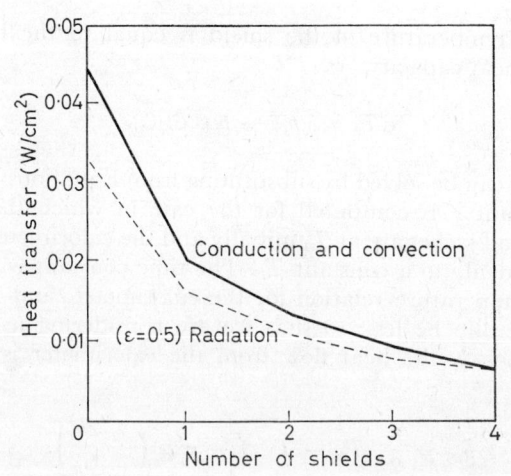

Figure 4. Effect of number of shields on heat transfer.

these radiation–convection shields in reducing the heat transfer coefficient between two parallel horizontal planes, spaced 5 cm apart with the lower plane at 400°K and the upper at 300°K with air between them. The emittance (ϵ) of both plane and shield surfaces is assumed to be 0·5. The figure shows that even with the large temperature difference of 100°, the use of shields 1 cm apart reduces the convection to a relatively small value. The figure also illustrates that the shields reduce the radiative heat transfer by the factor $1/(m + 1)$ in which m is the number of shields, if all surfaces are assumed to have the same emittance. In using convection shields, precautions should be taken to avoid "chimney" type convection which might occur if the shields allow gas to leak through.

One or more heaterless or "floating" shields can appreciably reduce both convection and radiation between the calorimeter and its surroundings, but only at the expense of introducing an additional time lag in the calorimeter temperature whenever either the calorimeter or surrounding temperature is changed. This lag may be calculated if the geometric and heat transfer data are known. The calculation will be illustrated for one floating shield between a calorimeter and its constant temperature surroundings, a water jacket, for example. Usually, the surfaces involved are of comparable areas, and it is an adequate approximation to treat them as parallel planes with heat flow perpendicular to the planes. Let p be the power (heat flow) per unit area, h the heat transfer coefficient, C the heat capacity per unit area of the floating shield, T the temperature, t the time, and the subscripts c, s and j refer to calorimeter, floating shield and jacket, respectively. For small temperature differences, heat flow is proportional to the temperature difference

$$p_{cs} = h_{cs}(T_c - T_s) \tag{9a}$$

$$p_{sj} = h_{sj}(T_s - T_j). \tag{9b}$$

The change of temperature of the shield is equal to the heat absorbed divided by the heat capacity

$$dT_s = (p_{cs} - p_{sj})\,dt/C_s. \tag{9c}$$

These equations can be solved by substituting from Equations (9a) and (9b). The time constant τ is computed for the case in which the calorimeter, floating shield and jacket are at T_j initially and the calorimeter temperature is increased suddenly to a constant T_c. The time constant will be the same for any time–temperature relation for the calorimeter, and the amplitude factor will generally be less, so that we are considering a "worse" case. When $h_{cs} = h_{sj} \equiv h$, the heat flow from the calorimeter is given by the equation

$$p_{cs} = \frac{h}{2}(T_c - T_j)\,1 + \exp\left(-\frac{2ht}{C_s}\right). \tag{9d}$$

To reduce the time constant, $C_s/2h$, the only practical step is to reduce the heat capacity of the shield, because increasing the heat transfer coefficient increases the uncertainty in heat loss from the calorimeter.

It is instructive to consider a particular case. For an aluminum shield spaced 1 cm from aluminum-surfaced calorimeter and shield in air at room temperature, the heat transfer coefficient is about $2 \cdot 5 \times 10^{-4}$ W cm^{-2} deg^{-1}. The heat capacity is $2 \cdot 4$ J cm^{-3} deg^{-1} and the time constant as a function of the thickness d of the shield is

$$\tau = 4 \cdot 8 \times 10^3\,d$$

in which τ is in sec if d is in cm. For ordinary aluminum foil, $d = 2 \cdot 5 \times 10^{-3}$ cm, and $\tau = 12$ sec. For p_{cs} to reach within 1/1000 of its ultimate value, $6 \cdot 9\,\tau$ or 83 sec are required. If aluminum sheet having $d = 0 \cdot 025$ cm is used for its greater mechanical strength, $6 \cdot 9\,\tau = 830$. If the aluminum floating shield is used in an evacuated space, the time constants for the 0·0025-cm and 0·025-cm thick shields are about 200 and 2000 sec, respectively, so that the total times for the power to reach within 1/1000 of its ultimate value are 1380 and 13 800 sec respectively. These figures show that in calorimetry using floating shields in evacuated spaces near room temperature, serious consideration should be given to errors resulting from the time lags.

In experiments using furnaces, frequently we need to know the power required to keep the inner core of the furnace at some given temperature and the corresponding temperature of the outside surface of the furnace. The calculation of the power required to maintain a given temperature difference in the furnace can be made simply by methods described later.

However, the temperature on the outside of the furnace depends upon the coefficient of heat transfer away from the outside surfaces. In this case, it is usually considered desirable to have a large enough heat transfer coefficient to hold the outside to a reasonable temperature. Radiative heat loss can be calculated as described earlier. The heat loss by free convection may be calculated with sufficient accuracy for furnace design, using methods described by Jakob[12]. If the heat flow is written proportional to the temperature difference ($P = h\Delta T$), the constant of proportionality, h, which we term the heat transfer coefficient, is a function of ΔT and may be calculated approximately for typical conditions by the following relations

horizontal surface — heat lost up
$$h = 2\cdot 2 \times 10^{-4} (\Delta T)^{\frac{1}{4}} \qquad (10)$$

horizontal surface — heat lost down
$$h = 1\cdot 1 \times 10^{-4} (\Delta T)^{\frac{1}{4}} \qquad (11)$$

vertical surface — (height > 30 cm)
$$h = 1\cdot 5 \times 10^{-4} (\Delta T)^{\frac{1}{4}} \qquad (12)$$

in which ΔT is the difference between room temperature and the surface temperature in °K (or °C) and h is in W cm^{-2} deg^{-1}. For vertical surfaces less than 30 cm high, h may be somewhat larger, perhaps by as much as a factor of 2 or 3.

In addition to free convection which may or may not be desired in calorimetry, we sometimes use forced convection to obtain good thermal contact between a fluid and its surrounding tube. In forced convection, there are two distinct types of heat transfer, depending upon whether the fluid flow is *laminar* (streamline) or *turbulent*. For fluid flow in smooth tubes, the transition between laminar and turbulent flow occurs at a value of the dimensionless Reynolds Number ($vD\rho/\eta$) of approximately 2000. In this expression, v is the average velocity of the fluid (cm/sec), D is the diameter of the tube (cm), ρ is the density of the fluid (g/cm^3), and η is the dynamic viscosity of the fluid (poise). Turbulent flow is sometimes used to obtain a maximum heat transfer between the fluid and a solid surface. For laminar flow in a tube, the flow of gas at pressures for which the molecular mean free path is much less than the diameter of the tube may be calculated from Poiseuille's equation

$$\hat{v} = \frac{\pi D^4}{128\eta} \frac{(\Delta P)}{L} \qquad (13)$$

in which $\hat{v}$ is the rate of volume flow (cm^3/sec), ΔP the pressure drop between the ends of the tube (in dynes/cm^2; note that 1 atm $\sim 10^6$ dynes/cm^2), L and D the length and diameter of the tube (cm), and η the dynamic viscosity of the fluid (poise). For laminar flow, for calculating the heat transfer coefficient between a tube and a fluid inside it, it is convenient for design purposes to use a simple approximation to estimate a value of heat transfer coefficient which will be slightly less than the true value. This approximation is that the heat transfer coefficient (h) per unit length of tube per unit temperature difference is $5\pi\lambda$ in which λ is the thermal conductivity of the fluid in the tube. If λ is in W cm^{-1} deg^{-1}, then h will be in watts per cm length of tube per degree C difference in temperature between tube and the fluid. For turbulent flow in a tube, the relation

$$h = 0.1\lambda \left(\frac{FC}{D\lambda}\right)^{0.786} \tag{14}$$

will be adequate for most design purposes. In this relation, h is the heat transfer coefficient (W cm^{-1} deg^{-1}) for unit length of tube and unit temperature difference, λ is thermal conductivity (W cm^{-1} deg^{-1}), F the rate of mass flow of fluid (g/sec), C the heat capacity (J g^{-1} deg^{-1}), and D the inner diameter of the tube (cm).

4. Heat transfer by thermal acoustical oscillation

In apparatuses which have gas contained in a tube having a temperature gradient, there may be another mechanism of heat transfer which is unknown in most calorimetry. This mechanism, which may be called "thermal acoustical oscillation" may be important in low temperature calorimetry, especially in the liquid helium range. Under certain conditions depending on the tube dimensions, gas present, temperature gradient, etc., this gaseous oscillation may result in large pressure variations, together with a relatively large heat transfer which may make calorimetric measurements difficult or impossible. While a change in design of the apparatus may avoid this oscillation, a recent study[8] has developed a method to minimize the oscillation by using a "damping" device attached *externally* to the apparatus. With this device, consisting basically of a Helmholtz resonator, it is not necessary to redesign the apparatus.

5. Heat transfer by mechanical vibration

Another mechanism of energy transfer which may result in heat developed in a calorimeter is by mechanical vibration. This is a problem which sometimes arises where measurements are made of very small heats. In localities with large mechanical vibrations, detectable mechanical power has sometimes been transmitted to the calorimeter where it has been transformed into heat. This "anomalous" heat has been a problem in some calorimeters operating at liquid helium temperatures where a small amount of heat can be detected. One solution here is to mount the apparatus on a solid pier which is free from vibration. In case this is not practical, a mounting can be designed to attenuate the vibration before it reaches the calorimeter.

6. Heat transfer by conduction

In design calculation in most calorimetry except at high temperatures, heat transfer by solid conduction is probably more important than other modes of heat transfer. Frequently, conduction plays the predominant role in determining temperature gradients in the calorimeter and on its surface.

PRINCIPLES OF CALORIMETRIC DESIGN

At low to moderate temperatures, solid conduction usually is the main source of heat transfer between calorimeter and shield. In many cases, dependence on thermal conduction is necessary in the evaluation of temperature, etc. As a consequence of the variety of calorimetric design problems in which conduction is important, this heat transfer mode will be considered in more detail, and some examples of calculations involving conduction will be given.

In most calorimetric work, thermal conductivity changes more slowly as a function of temperature than coefficients for heat transfer by radiation and convection. With most materials the changes with temperature of their thermal conductivities are sufficiently small over 10°C so that for design purposes, the heat conduction-heat transfer coefficient can be considered constant in an average experiment. One noteworthy exception is where a phase change takes place in the material. Consequently, materials of construction preferably should have no phase transitions in the temperature range of calorimeter operation. In all the following examples of heat flow calculations, it will be assumed that the thermal conductivity is constant and that there are no transitions in the material.

Heat flow calculations are usually classed into two types, *steady-state* and *unsteady-state*. In steady-state heat flow, the temperature at any point in the material does not change with time, whereas in unsteady-state heat flow, the temperature is a function of time.

The quantity thermal conductivity (λ) is the proportionality constant *defined* by the steady-state relation $P_x = -\lambda A \, (\partial T/\partial x)$ in which P_x is the power (rate of heat flow at x) in the x direction through an isotropic material having a cross sectional area A, and $\partial T/\partial x$ is the temperature gradient at x in the x direction. The minus sign in the defining equation is used to indicate that heat flow (power) is positive in the direction of increasing x if the temperature decreases with increasing x ($\partial T/\partial x$ is negative). Let us consider now a case in which ($\partial T/\partial x$) is a function of x. If at x the power is $-\lambda A(\partial T/\partial x)$, then at $x + \mathrm{d}x$ the power is

$$-\lambda A \frac{\partial}{\partial x} \left(T + \frac{\partial T}{\partial x} \mathrm{d}x \right) = -\lambda A \frac{\partial T}{\partial x} - \lambda A \frac{\partial^2 T}{\partial x^2} \mathrm{d}x.$$

Thus, the *difference* in power at x and at $(x + \mathrm{d}x)$ is

$$P_x - P_{(x+\mathrm{d}x)} = \left[\lambda A \frac{\partial^2 T}{\partial x^2} \mathrm{d}x \right]. \tag{15}$$

The magnitude of this difference in power is determined by the boundary conditions for the particular heat flow problem. In solving any heat flow problem, the starting point is to set up the "heat balance" equation expressing the conditions of the particular problem. In principle, the solution of any such equation can be obtained if the boundary conditions are all known. The heat balance differential equation for a homogeneous isotropic solid has been derived for the general *three-dimensional* case in which temperature is changing with time and there is a power source (or sink) in the material. This general equation can be written as

$$\lambda \left[\frac{\partial^2 T}{\partial x^2} + \frac{\partial^2 T}{\partial y^2} + \frac{\partial^2 T}{\partial z^2} \right] - \rho c \frac{\partial T}{\partial t} = -F(x, y, z, t) \tag{16}$$

in which ρ and c are density and specific heat of the material, $\partial T/\partial t$ is the rate of change of temperature at a point (x, y, z) with time, t, and $F(x, y, z, t)$ is the power source per unit volume. The above equation is written conventionally as

$$\nabla^2 T - \left(\frac{\rho c}{\lambda}\right)\frac{\partial T}{\partial t} = \frac{-F(x, y, z, t)}{\lambda} \tag{17}$$

in which $(\lambda/\rho c)$ is a quantity known usually as thermal diffusivity, a. In the case of steady-state heat flow, $(\partial T/\partial t) = 0$, so that we have Poisson's equation

$$\nabla^2 T = \frac{-F(x, y, z, t)}{\lambda}. \tag{18}$$

If no heat is supplied to the body, this equation reduces to Laplace's equation $\nabla^2 T = 0$.

If instead of rectangular coordinates, cylindrical coordinates are used, the heat balance differential equation for the case of no heat generated in the cylinder is

$$\lambda \left[\frac{\partial^2 T}{\partial r^2} + \frac{1}{r}\frac{\partial T}{\partial r} + \frac{1}{r^2}\frac{\partial^2 T}{\partial \theta^2} + \frac{\partial^2 T}{\partial z^2}\right] = \rho c \frac{\partial T}{\partial t} \tag{19}$$

in which r is the radial distance from the cylindrical axis, θ the azimuth around the axis and z the distance along the axis. Many cylindrical heat flow problems of interest in calorimetry can be approximated by assuming T to be independent both of θ (symmetrical heat flow around axis) and of z (effectively infinitely long cylinders). In this case the cylindrical heat flow equation reduces to

$$\lambda \left[\frac{\partial^2 T}{\partial r^2} + \frac{1}{r}\frac{\partial T}{\partial r}\right] = \rho c \frac{\partial T}{\partial t}$$

or

$$\frac{\partial^2 T}{\partial r^2} + \frac{1}{r}\frac{\partial T}{\partial r} = \frac{1}{a}\frac{\partial T}{\partial t} \tag{20}$$

in which r is the radial distance from the axis of the cylinder.

There is a corresponding general heat flow differential equation for spherical polar coordinates[6]. Most applications in calorimetric design use the approximation that isothermal surfaces are concentric spheres. In this case the general equation reduces to

$$\frac{\partial^2 T}{\partial r^2} + \frac{2}{r}\frac{\partial T}{\partial r} = \frac{1}{a}\frac{\partial T}{\partial t} \tag{21}$$

in which r is the radial distance from the center of the sphere.

Steady-state heat flow will now be discussed because it is usually simpler and more frequently encountered in calorimetric design.

PRINCIPLES OF CALORIMETRIC DESIGN

A. Steady-State Heat Flow

(1) *Rod.* The simplest steady-state heat flow case is for linear (unidirectional) heat flow in a rod having constant cross sectional area A in the direction of heat flow. We have linear heat flow if there is no lateral heat transfer. Here, the *defining* equation of thermal conductivity (λ) applies, so the power $P = -\lambda A \, dT/dx$ and $dT/dx = -P/(\lambda A)$. Integrating this equation between x_1 and x_2, corresponding to temperatures T_1 and T_2, gives

$$(T_2 - T_1) = \int_{x_1}^{x_2} -\frac{P}{\lambda A} \, dx = \frac{P}{\lambda A}(x_1 - x_2). \tag{22}$$

In calorimetry, this simple equation may be used to calculate heat flow along a conductor (such as a wire or tube) corresponding to a known or estimated temperature difference across a known length.

The linear temperature–distance relation is common, but by no means general. For example, if the cross sectional area is not constant, as in a cone for which $A = Mx^2$, integration gives

$$(T_2 - T_1) = \int_{x_1}^{x_2} \frac{-P}{\lambda A} \, dx = \frac{-P}{\lambda M} \int_{x_1}^{x_2} \frac{dx}{x^2} = \frac{P}{\lambda M}\left[\frac{1}{x_2} - \frac{1}{x_1}\right]. \tag{23}$$

A more practical example is the heat flow along a tube of constant cross section for which the thermal conductivity changes drastically over the temperature variation on the tube. For example, take a stainless steel tube with constant cross section A, length l, and variable thermal conductivity λ. If this tube is used between a calorimeter at $4°K$ and a liquid nitrogen bath at $90°K$, its thermal conductivity changes by about a factor of thirty over the temperature range. As a first approximation, adequate for most calorimetric design, $\lambda = 0.0009\,T$ in this temperature range, with λ in watt cm^{-1} deg^{-1} and T in deg K. Integrating, we get $(0.00045)\,T^2 = -(P/A)\,x + C$. From the boundary condition that $T = 4°$ when $x = 0$, the integration constant is 0.0072. Using the boundary condition that $T = 90°$ when $x = 10$ cm, and with $A = 0.05$ cm^2, we get $P = 0.018$ watt. The temperature distribution along this tube is $T = \sqrt{(808.4\,x + 16)}$.

(2) *Wire Carrying Current.* An important example of a non-linear temperature–distance relationship arises when power is developed in the body, as in a heater lead. In calorimetry it is important that the heater lead wire not be excessively hot. A heater lead can be treated as a small rod of length l, cross sectional area A, and thermal conductivity λ, with power p developed per *unit length* of the wire. To simplify the problem, both ends of the wire are taken to be at temperature T_0 and losses from the surface of the wire by radiation and conduction are assumed negligible. The power developed in a differential length dx of the wire is simply $p\,dx$. As shown earlier in Equation (15), this power can also be expressed in terms of the second differential

$\partial^2 T/\partial x^2$, so that the heat balance equation is

$$\lambda A \frac{\partial^2 T}{\partial x^2} \, \mathrm{d}x = -p \, \mathrm{d}x,$$

and by integrating,

$$T = \frac{-p}{2\lambda A} x^2 + C_1 x + C_2 \tag{24}$$

in which C_1 and C_2 are integration constants to be determined by the boundary conditions. Then

$$C_2 = T_0 \quad \text{and} \quad C_1 = \frac{pl}{2\lambda A},$$

so that

$$T = \frac{p}{2\lambda A} (lx - x^2) + T_0. \tag{25}$$

The above equation shows that the maximum temperature for this case is at $x = l/2$. As an example, take a 5 cm length of AWG#26 copper wire with a current of 0·5 amp. The excess temperature at the middle of the wire is calculated to be about 0·6°C.

(3) *Cylindrical Shell.* Consider the gradient in the axial (z) direction in a thin cylindrical shell (length l) due to constant power supplied per unit length of cylinder and removed at one end of the cylindrical shell ($z = l$). The heat balance equation is

$$-\lambda A \left(\frac{\partial T}{\partial z}\right)_{z=0} = pz$$

so that

$$(T_{z=0} - T_{z=l}) = \frac{p}{2\lambda A} l^2 \tag{26}$$

(4) *Radial Heat Flow in a Hollow Cylinder and Sphere.* A frequent problem in calorimetric design is steady-state radial heat flow in a cylinder for which the temperature is a function only of the radius r. Applying the defining equation for thermal conductivity (λ) to plane polar coordinates, and assuming a *constant* power P flowing radially for *unit length* of the cylinder, then one finds at any radius r, $P = -\lambda(2\pi r) \, \mathrm{d}T/\mathrm{d}r$. Integrating this between r_1 and r_2 gives the corresponding temperature difference

$$(T_2 - T_1) = \frac{-P}{2\pi\lambda} \ln\left(\frac{r_2}{r_1}\right) \tag{27}$$

PRINCIPLES OF CALORIMETRIC DESIGN

A similar treatment for steady-state heat flow in a sphere gives a heat balance equation $P = -\lambda(4\pi r^2)\,dT/dr$ in which P is the total power flow. Integrating this between r_1 and r_2 gives

$$(T_2 - T_1) = \frac{-P}{4\pi\lambda}\left(\frac{1}{r_2} - \frac{1}{r_1}\right) \tag{28}$$

(5) *Thin Disk with Uniform Heating and Cooling.* A simple heat conduction case involving radial heat flow (such as in the top or bottom of a calorimeter sample container) is the calculation of the temperature in a thin disk in which there is introduced a constant power p per unit area which is dissipated at its circumference. The power crossing the cylindrical surface of radius r is $p(\pi r^2)$ so that $p(\pi r^2) = -\lambda(2\pi rw)\,dT/dr$ in which w and λ are the thickness and thermal conductivity respectively. The solution here for the temperature (T_r) at radius r is

$$T_r = -\left(\frac{p}{4\lambda w}\right)r^2 + C_1.$$

If we impose the boundary conditions that $T = T_0$ at the circumference ($r = a$, with a the radius of the disk), then

$$C_1 = T_0 + \frac{p}{4\lambda w}a^2$$

and

$$T_r = \frac{p}{4\lambda w}(a^2 - r^2) + T_0. \tag{29}$$

This solution also holds (except for sign) for the reverse case for which a constant power p per unit area is *removed* from the disk.

(6) *Tempering of Electrical Leads and Other Conductors.* Consider now a slightly more complicated case which is basic to most calorimetry. This is the problem of "tempering" of electrical leads, defined as bringing the temperature of leads to the temperature of the body by thermal contact. This problem is exceedingly important in accurate calorimetry if the true temperature of lead wires must be known in order to minimize and measure heat leak. Tempering of wires may be accomplished either *continuously* over a length of wire or in *steps*. These two methods will be considered separately.

(7) *Continuous Tempering.* For continuous tempering, consider a small wire having a thermal conductivity λ and cross sectional area A. Let the temperature of a wire be T_1 at $x = 0$, where the wire first comes in thermal contact with a body at a temperature T_0. Let h represent the *constant* heat transfer (power) from wire to body per unit length of the wire and per unit temperature difference. The heat transfer (power) from dx length of wire to the

body is therefore $h(T - T_0)\,dx$ in which T is the temperature of the wire. From previous consideration Equation (15), this power must be

$$\lambda A \frac{d^2 T}{dx^2}\,dx.$$

Hence the heat balance equation is

$$\lambda A \frac{\partial^2 T}{\partial x^2}\,dx = h\,(T - T_0)\,dx \qquad (30)$$

or

$$\frac{\partial^2 T}{\partial x^2} = \left(\frac{h}{\lambda A}\right)(T - T_0). \qquad (31)$$

The general solution to this linear differential equation is

$$(T - T_0) = C_1 \exp\left[-x\sqrt{(h/\lambda A)}\right] + C_2 \exp\left[x\sqrt{(h/\lambda A)}\right]. \qquad (32)$$

Imposing the boundary conditions that $T = T_1$ at $x = 0$ and $T = T_0$ at $x = \infty$ we get $C_2 = 0$ and $C_1 = (T_1 - T_0)$, so that

$$(T - T_0) = (T_1 - T_0) \exp\left[-x\sqrt{(h/\lambda A)}\right]. \qquad (33)$$

The physical meaning of this equation is that when $x\sqrt{(h/\lambda A)} = 1$, the excess temperature of the wire has been reduced from $T_1 - T_0$ to $(T_1 - T_0)/e$. Solving for this distance x_e, we get $x_e = \sqrt{(\lambda A/h)}$. To reduce the temperature excess of the wire to $(T_1 - T_0)/10$ requires $x_{10} = (2\cdot 303\,x_e) = 2\cdot 303\,\sqrt{(\lambda A/h)}$ length of wire. If it is desired to bring the temperature of the wire to $(T_1 - T_0)/1000$, we must have $x_{1000} = 3(2\cdot 303)\sqrt{(\lambda A/h)}$ length of wire. This type of tempering problem has application in calorimetry in a number of problems in which the heat flow can be approximated in the above manner. The heat transfer coefficient h (W cm^{-1} deg^{-1}) must be evaluated for the particular problem. In the case of a wire insulated from a tube, the cylindrical heat flow formula (Equation 27) is used to evaluate it. In the case of radiative heat transfer frequently it is sufficiently constant for design purposes so that a "radiative" h can be calculated. In case the calorimeter is used over a temperature range, sometimes the "worst case" is assumed.

There is a similar example for a *finite* length (l) of wire with the boundary conditions that $T = T_1$ at both $x = 0$ and $x = l$. In this case, the values of C_1 and C_2 satisfying the boundary conditions are

$$C_1 = \frac{(T_1 - T_0)\,[\exp(ml) - 1]}{[\exp(-ml) - \exp(ml)]} \quad \text{and} \quad C_2 = \frac{(T_1 - T_0)\,[\exp(-ml) - 1]}{[\exp(-ml) - \exp(ml)]}$$

in which $m = \sqrt{(h/\lambda A)}$.

A problem that sometimes arises is to estimate the extent of tempering of a wire which cannot be treated as having an infinite length in contact

PRINCIPLES OF CALORIMETRIC DESIGN

with the body. Such a case arises when a lead is brought through a body at a temperature between that of the surroundings (e.g., room temperature) and the temperature of the calorimeter. This body will supply some of the heat flowing along the wire and reduce the amount which must be supplied from the direction of the calorimeter. Equation (32) still applies, but with different boundary conditions. Neglecting heat flow from the direction of the calorimeter (a less favorable case) gives $dT/dx = 0$ at $x = 0$. The other boundary condition is obtained by equating heat flow along the external part of the wire to the total heat flow from the body to the wire, with the result that

$$C_1 = C_2 = \frac{(T_s - T_0)\sqrt{(h/\lambda A)}}{[1 - L\sqrt{(h/\lambda A)}]\{\exp[l\sqrt{(h/\lambda A)}] - \exp[-l\sqrt{(h/\lambda A)}]\}}$$

in which T_s is the temperature of the surroundings, T_0 is the temperature of the body, and L is the length of the wire between the body and the surroundings.

(8) *Tempering of Wires Carrying Current.* The problem of tempering a wire carrying current is similar to the problem of a wire with no current. If p is the power developed per unit length of wire, the heat balance equation is

$$\lambda A \frac{\partial^2 T}{dx^2} dx + p\,dx = h(T - T_0)dx$$

or

$$\frac{\partial^2 T}{\partial x^2} = \left(\frac{h}{\lambda A}\right)(T - T_0) - \left(\frac{p}{\lambda A}\right). \tag{34}$$

Solving this linear differential equation gives

$$(T - T_0) = C_1 \exp[-x\sqrt{(h/\lambda A)}] + C_2 \exp[x\sqrt{(h/\lambda A)}] + (p/h)$$

for the same boundary conditions that $T = T_1$ at $x = 0$ and $x = l$, and with

$$C_1 = \frac{[T_1 - T_0 - (p/h)][\exp(ml) - 1]}{[\exp(ml)] - [\exp(-ml)]}$$

and

$$C_2 = \frac{[T_1 - T_0 - (p/h)][1 - \exp(-ml)]}{[\exp(ml)] - [\exp(-ml)]},$$

in which $m = \sqrt{(h/\lambda A)}$.

(9) *Step Tempering.* Another method of tempering leads is by steps in contrast to continuous tempering. This is done by "thermal tiedowns", as used extensively in calorimeters described in *Figure 2* in Chapter 11. A thermal tiedown, as used here, is defined as a thermal connection between

a point on the wire and a small region on the body whose temperature is to be approached. For a given electrical insulation between wire and the body, this method may have an advantage in that it can give better tempering by effectively increasing the length of path of heat flow along the wire.

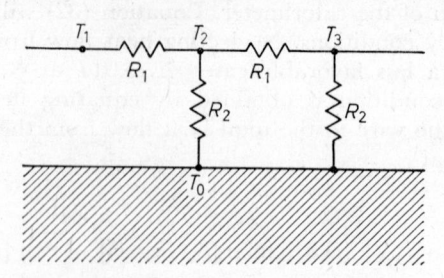

Figure 5. Principle of thermal tiedowns.

This method may also have certain other advantages, such as ease of disassembly, etc. The principle of this step tempering is shown in *Figure 5*. In this figure, T_1, T_2, and T_3 represent the temperatures along a wire (referred to the temperature T_0 of the body) in the tempering network, R_1 represents the thermal resistance of a chosen length of the wire, whereas R_2 represents the thermal resistance of the thermal tiedown. If $R_1 = 100\ R_2$, then $(T_2 - T_0) \simeq 0.01\ (T_1 - T_0)$ and $(T_3 - T_0) \simeq 0.01\ (T_2 - T_0)$, so that $(T_3 - T_0) \simeq 0.0001\ (T_1 - T_0)$. In one calorimeter described in Chapter 11, the thermal resistance R_1 was that of about 10 cm of small wire whereas R_2 (thermal tiedown) was *equivalent* in thermal resistance to about 0.1 cm length of the same wire, thus giving an attenuation factor of 0.01 for each tiedown. Details of the thermal tiedown which has proved effective even in vacuum are given in Chapter 11.

(10) *Tempering of Fluid in a Tube.* In some calorimetry, such as flow calorimetry, a fluid (liquid or gas) is brought close to the temperature (T_0) of a bath by flowing through a "tempering" tube in the bath. In designing this tempering tube, it may be advantageous to know the minimum length required to bring the temperature of the fluid sufficiently close to the bath temperature. If F is the mass rate of flow of fluid (g/sec), c is the specific heat of the fluid (J g^{-1}deg^{-1}), T_1 is the temperature of the fluid at $x = 0$, and h is the heat transfer coefficient (Wcm^{-1}deg^{-1}) between tube and fluid[12] per cm length of tube, then the heat balance equation is

$$h(T - T_0) = -cF\frac{dT}{dx} \tag{35}$$

or

$$\frac{dT}{dx} + \left(\frac{h}{cF}\right)T = \left(\frac{h}{cF}\right)T_0 \tag{36}$$

which is a linear differential equation whose solution is

$$T = T_0 + (T_1 - T_0)\exp{-(h/cF)x}. \tag{37}$$

The above solution assumes that the radial gradient in the fluid is small by comparison with the difference in temperature between fluid and tube. This type of approximation is usually adequate for design purposes. The length of tube necessary for the fluid to come to the fraction $1/e$ (63 per cent) of its starting temperature difference ($T_1 - T_0$) is cF/h.

These examples of steady-state heat flow are only a few of the many calorimetric problems that may arise. Solutions to many more complicated problems are given in Carslaw and Jaeger[6]. In most calorimetric problems, the heat balance differential equation is a *linear* differential equation so that its solution can be treated as the combination of several solutions. This is known as the Principle of Superposition, which is discussed in the following section (B).

B. Unsteady-State Heat Flow

An unsteady-state heat flow problem is one in which temperature changes with time. Although perhaps the majority of calorimetric design problems can be solved using steady-state heat flow, there are some important unsteady-state problems which arise. In general, the *complete* solution of a heat flow problem, in which temperature is a function of both distance and time, is considerably more difficult than that for the steady state. The complete solution usually consists of a *"transient"* term, which may be short-lived, and a "steady" term, which changes only owing to either change in the material physical properties or time variable boundary conditions. If the transient terms decay before the final temperature is observed, West[21] has shown that in an adiabatic calorimeter heated intermittently, heat leaks due to initial and final transients are equal and opposite in sign. It is usually advantageous, therefore, to design the calorimeter and shield with fast transients, i.e. with small "time constants". In unsteady state heat flow, in which the transient may be approximated by a single exponential term, the term "time constant" frequently is used in a manner similar to the electrical time constant RC. If a capacitor is charged to some potential difference ($E_1 - E_0$), and then connected to the resistance, the potential difference across the capacitor will decay exponentially through the relation $\Delta E = (E_1 - E_0) \exp[-(t/RC)]$. When $t = RC$, $\Delta E = (E_1 - E_0)/e$, so that RC is called the electrical time constant (τ) defined as that time in a single exponential decay for the potential difference to come to $1/e$ of its initial value. In the analogous heat case, the capacitor corresponds to a body with a heat capacity c, which is heated to an initial temperature T_1 and then put in thermal contact through thermal resistance R with another body held at temperature $T = T_0$. This is mathematically similar to the discussion of tempering a wire in which there is a distance to bring the wire to $1/e$ of its initial temperature difference. In both cases we are considering a decay through a single exponential term. The time required to come to $1/10$ the initial temperature difference is simply $2 \cdot 303 \, (\tau)$, the time to come to $1/100$ is $2 \, (2 \cdot 303 \, \tau)$, etc. Sometimes it is desired to estimate this time constant τ from cooling experiments. In principle, this can be done from any two measurements of temperature at different times. As a simple short cut, it is sometimes convenient to measure the initial cooling rate and divide the initial temperature difference by this initial cooling rate to get the time constant (τ). For example, if a body with a heat capacity of 1000 joules/deg C, has an initial cooling rate of 10 deg C/sec, for an initial temperature difference of 1000 deg C, then the time constant (τ) = 100 sec. If the thermal resistance is known, the time constant is merely the product of the heat capacity and thermal resistance. For example, the above case corresponds to a thermal resistance of 0·1 deg C/watt.

Although in some calorimetric design the use of a single exponential may be adequate for design purposes, the actual transient is usually not so simple. Fortunately, a variety of unsteady state heat problems have already been solved[6], and in some important cases the solution is shown graphically to facilitate its use.

Consider the case of a thin cylindrical shell similar to that treated earlier in a steady-state case. Suppose this shell has a specific heat c, thermal conductivity λ, density ρ, and length l, and the cylinder is initially at temperature $T = T_0$. Assume also that the thermal properties are independent of temperature. Now at time $t = 0$, start to heat one end ($z = 0$) at a constant rate β, so that $T_{z=0} = \beta t$. Putting the initial and boundary conditions into the general heat balance differential equation gives

$$\lambda \frac{\partial^2 T}{\partial z^2} = \rho c \frac{\partial T}{\partial t}. \tag{38}$$

The complete solution is given as

$$(T - T_0) = \beta t + \frac{\beta z^2}{2a} + [\text{transient term}] \tag{39}$$

in which a is thermal diffusivity ($\lambda/\rho c$). If the transient term is short-lived as it is frequently in good calorimetric design, the "steady" temperature gradient on the cylindrical shell is of primary interest. This temperature difference (after transient has died out) over the length l is $\beta l^2/2a$. Note that the average temperature is not at $z = l/2$.

Consider now a case of radial instead of linear heat flow. In terms of cylindrical coordinates, the general differential heat balance equation for a cylinder infinitely long for which the temperature is dependent only on radius r and time t, is Equation (20)

$$\frac{\partial^2 T}{\partial r^2} + \frac{1}{r}\frac{\partial T}{\partial r} = \left(\frac{1}{a}\right)\frac{\partial T}{\partial t}. \tag{40}$$

Now suppose that we have a circular disk of radius a which initially is at temperature $T = T_0$, but at time $t = 0$, the circumferential area of the disk is heated at a constant rate β so that $T_{r=a} = T_0 + \beta t$. Upon assuming that there is no heat transfer at the plane surfaces of the disk, this problem becomes the same as for a cylinder of infinite length, whose solution is

$$T - T_0 = \beta\left[t - \frac{(a^2 - r^2)}{4a}\right] + [\text{transient}]. \tag{41}$$

Although the transient in Equation (41) can be calculated[6], it is usually short-lived so that the "steady" temperature difference between $r = 0$ and $r = a$ is most important. This temperature difference is simply $\beta a^2/4\alpha$.

(1) *The Principle of Superposition.* Before consideration of thermal transients in calorimetry, the Principle of Superposition should be mentioned. This principle, which can be applied to *linear differential equations*, states that solutions to linear differential equations with linear boundary conditions may be summed to give other solutions; a complex solution can be obtained by adding the solutions to simpler problems. This method not only makes numerical solutions easier but also provides physical insight into the heat flow process in calorimetry. A concise statement of superposition theorems is given by Korn and Korn[15]. In brief, the temperature for some particular problem may be considered as a sum of terms all satisfying the heat flow equation, but separately taking account of (*i*) heat generated in the calorimeter, (*ii*) heat exchange between calorimeter and surroundings, and (*iii*) an initial temperature distribution in the calorimeter.

A problem involving two thermocouple errors will be used to illustrate the use of the principle of superposition. Suppose that the shield control in an adiabatic calorimeter is subject to errors due (*i*) to an inhomogeneity in a segment of thermocouple wire between the calorimeter and the shield and (*ii*) to a constant zero bias, which might be caused by an inhomogeneity in a region outside of the shield or simply to a zero offset in an electronic control device. Since the shield temperature is not matched to the calorimeter temperature because of these two spurious e.m.f.s, there will be a temperature gradient in the segment of thermocouple wire and an additional offset due to the effect of this gradient on the inhomogeneity†. When the calorimeter is heated, the heat flowing into the wire segment from the calorimeter and shield will have an additional effect on the inhomogeneity. The problem is to account for the two effects in the measurement process.

To set up the equations, the segment of wire is taken to be of length l, the constant heating rate $dT/dt = \beta$ deg/sec, and the constant control offset ΔT_c. Assuming negligible current in the thermocouple, the heat balance equation is

$$\alpha \frac{\partial^2 T}{\partial z^2} = \frac{\partial T}{\partial t} \qquad (42)$$

which is just Equation (17) with the power term zero. Since the solution for constant heating rate is desired, the substitution $dT/dt = \beta$ can be made at once. Equation (42) is linear either before or after the substitution, since the temperature and its derivatives appear only in the first power. (Linear equations also cannot have cross products such as $T(\partial T/\partial z)$). The boundary condition at the calorimeter end of the wire is that the temperature is rising at a constant rate and that its temperature would have been T_0 at $t = 0$

$$T = T_0 + \beta t \quad \text{at } z = 0.$$

† In adiabatic calorimetry, this secondary effect is probably negligible; however, similar reasoning applies in isothermal shield calorimetry for which the effect on the spurious e.m.f. of the temperature difference between calorimeter and shield may be appreciable.

The temperature at the shield end of the wire differs from that on the calorimeter by the constant offset ΔT_c and the effect of the inhomogeneity

$$T = T_0 + \beta t + \Delta T_c + b(T_1 - T_2) \quad \text{at } z = l$$

in which T_1 and T_2 represent the temperatures at the ends of the inhomogeneity and b is the ratio of the thermoelectric power of the inhomogeneity to that of the thermocouple.

The problem can be written in terms of $T - T_0 = U + V$ [noting $\partial T/\partial t = \partial (T - T_0)/\partial t$] in which

$$\frac{\partial^2 U}{\mathrm{d}z^2} = \frac{\beta}{a}$$

$$U = \beta t \text{ at } z = 0$$

$$U = b(U_1 - U_2) + \beta t \text{ at } z = l \tag{43}$$

and

$$\lambda \frac{\partial^2 V}{\partial z^2} = 0$$

$$V = 0 \text{ at } z = 0$$

$$V = \Delta T_c + b(V_1 - V_2) \text{ at } z = l. \tag{44}$$

Evidently summing Equations (43) and (44) for U and V will result in the equation for $T - T_0$. Inspection of these two sets of equations for U and V shows that the contribution V to the temperature due to the control offset has been separated from the effect of heating the calorimeter, even with respect to the spurious e.m.f. This effect is constant and independent of the heating rate, so that it can be evaluated in separate experiments, as in fore and after rating periods, assuming only that ΔT_c is the same in both cases. The effect on U of a change in the heating rate is now easily deduced. If the rate is changed by a constant ratio to $k\beta$, then it is easily verified by substitution that kU satisfies the equations for U, including the effect of the spurious e.m.f. Any error associated with U, such as the effect of the spurious e.m.f., is directly proportional to the heating rate. It does not follow that heat leak due to U or similar functions can be detected by measuring the heat capacity of a calorimeter at various heating rates. The time required to heat through the temperature rise for which comparison is made is inversely proportional to the heating rate. Heat transfer between the calorimeter and its surroundings is proportional to the product of the time and functions like U, but the heating rate constant cancels from the product.

The more general problem of the effect of heating rates on heat exchange in adiabatic calorimeters is discussed in Chapter 9, based on detailed mathematical arguments given in the literature[21].

By contrast, the inclusion of a simple variation of the thermal conductivity with temperature, such as $\lambda = \lambda_0 (1 + aT)$, results in the following *non-linear* equation

$$\lambda_0 \frac{\partial}{\partial x} (1 + aT) \frac{\partial T}{\partial x} = \beta c, \qquad (45)$$

and the substitution $T = U + V$ results in cross terms in U, V, and their derivatives. Mathematical tractability is gone, but more important is the loss of validity of the correction for the control offset. In general, of course, the thermal properties of materials are functions of temperature, so that the linear equations are approximations. The designer must keep in mind that temperature differences in the experiment should be small enough that the linear differential equation provide an adequate description. The tolerable temperature differences are greater when the temperature variations of thermal properties are small and when the total heat transfer between the calorimeter and its surroundings is small. Obviously, if the heat leak is zero, it cannot produce a temperature distribution to interact with that due to heating.

In testing an apparatus after assembly, it seems worth while to investigate the linearity by deliberately exaggerating the magnitudes of the various contributions to the temperature distribution. In the simple case of the terms U and V discussed above, different rates of rapid heating and different offset of controls might set an upper limit on permissible rates and offsets. Variable heating rates commonly used in adiabatic heat capacity calorimeters rarely reveal a corresponding variation in the observed heat capacity. One obvious interpretation is that it is not very difficult to obtain an apparatus which is adequately described by linear partial differential equations.

(2) *Transients in One-Dimensional Heat Conduction.* Transients may be important to the design of calorimeters. Calorimeters must come to equilibrium in a reasonable time so that the uncertainty in the heat leak correction remains tolerable. Transient problems are difficult, but most cases of interest for the approximate design calculations in calorimetry have been worked out[6]. A transient problem will be illustrated with a calculation of the flow in one direction (one-dimensional) with no lateral losses. This problem treats approximately the temperature distribution in a lead wire, with one end fastened to the constant temperature bath and the other heated at the rate β, with the initial temperature that of the bath. Although the problem seems restricted in scope, by the principle of superposition we can apply it also to the case for which both ends are heated and the wire has an initial temperature distribution which might be the steady state between a calorimeter at one temperature and a bath at another temperature.

The solution to this problem is given by Churchill[7]

$$T - T_0 = \beta \left[\frac{1}{6} \left(\frac{x^3}{l^3} - \frac{x}{l} \right) + \frac{xt}{l} + \frac{2}{\pi^3} \sum_{n=1}^{\infty} \frac{(-1)^{n-1}}{n^3} \exp[(-\lambda n^2 \pi^2 t / \rho c l^2)] [\sin(n\pi x/l)] \right]. \qquad (46)$$

The exponential series does not converge rapidly for small values of the argument, but for $\lambda t/\rho c l^2$ greater than 0·1 only the term for $n = 1$ is significant in the time required for the calorimeter to come to equilibrium. Qualitatively, in order to get rapid equilibrium in the calorimeter, we would like to have the thermal conductivity large and the length, which appears to the second power, small. For a copper wire 10 cm long at room temperature, the time constant for the first term is equal to 10 sec (time to decrease to $(1/e)$ of the initial value). For an insulated copper wire we can use the approximation that heat is conducted along the copper to the insulating material which contributes to the heat capacity but not to conduction. For a copper wire of 0·02 cm diam and 0·0025 cm thickness of insulation of approximately the same heat capacity per unit volume, the time constant is increased to 14 sec. Doubling the length increases the time constant for the insulated wire to 56 sec. The time constant is greatly increased for alloy wires. If we consider a constantan wire instead of the insulated copper wire above, the time constant is about 20 times longer or 220 sec. If such a wire has a spurious e.m.f. or sufficiently large heat capacity so that we must wait until it is within one per cent of its final value then the time required is about 2(2·3) times the time constant or 1000 seconds. If it is necessary to install long wires in the apparatus, their temperatures should be kept constant during an experiment.

If supports for a calorimeter are chosen only on the basis of low thermal conductivity, harmful lags may be introduced. A calorimeter supported on glass rods provides an example. Equation (46) applies with the thermal diffusivity $\lambda/\rho c \sim 5\cdot 7 \times 10^{-3}$ for a glass near room temperature, and the time constant for the first term is 176 sec for a 1 cm length, and 704 sec for a 2 cm length. Of course, if the calorimeter merely rests on the glass rods, the additional thermal resistance of the contact will give a still longer time constant.

Although Equation (46) applies to the case of turning on power in a calorimeter, the same arguments apply when the power is turned off. To obtain the solution for this case we merely subtract the right hand side of Equation (46) in which we have substituted $(t - t_f)$ for t, in which t_f is the time at which the power is turned off. In the general case there will be two sets of decaying exponentials, but those in $t - t_f$ will be predominant and the equilibrium time depends on these.

When some portion of the apparatus is subjected to automatic control it is important that the transient response be rapid. The thermal conductivity should be as large as possible, the distances between heaters should be kept small and heat capacities should be kept small. These problems are discussed in detail in other chapters. The previous example of transients is for a linear heat flow. Many other problems, including transients in radial heat flow, are solved in Carslaw and Jaeger's book[6].

IV. Applications to Calorimeter Design

1. *Temperature in a calorimeter wall at constant heating rate*

One of the most important problems in accurate calorimetry results from temperature gradients in the calorimeter. In most calorimeters the heater is not distributed in such a manner as to avoid gradients, and even if it were for one amount of filling in the calorimeter, the same condition probably would not hold for another filling.

Consider as a simple example the temperature difference in a cylindrical calorimeter wall which receives heat only at one end ($z = 0$). Assume that the wall has a length l, thickness w, radius r, thermal conductivity λ, density ρ, and heat capacity c, and that all these quantities are constant during an experiment so that each part is heating at a constant rate β. The heat balance differential equation here is $\lambda(2\pi rw)\, \partial T/\partial z = p(l - z)$ in which p is the power per unit length of cylinder necessary to heat it at a rate β. The power p is simply $p = (2\pi rw)\rho c\beta$, so that

$$\lambda(2\pi rw)\frac{\partial T}{\partial z} = (2\pi rw)\rho c\beta\,(l - z) \tag{47}$$

or

$$\frac{\partial T}{\partial z} = \beta\left(\frac{\rho c}{\lambda}\right)(l - z) = \frac{\beta}{a}(l - z) \tag{48}$$

in which a is thermal diffusivity, $\lambda/\rho c$.

Integrating between $z = 0$ and $z = l$ gives

$$(T_l - T_0) = -\frac{\beta l^2}{2a} \tag{49}$$

in which T_l and T_0 are both increasing at the rate β. This relation is the same as that given by Equation (39) after the transient. It is apparent that the largest temperature gradient occurs near the heat source, so that the assumption of linear temperature change on the surface is invalid. The average temperature (relative to T_0) over the surface can be calculated from the equation

$$(T - T_0)_{av} = \frac{\beta \int_0^l [(z^2/2) - lz]\,dz}{a \int_0^l dz} = \frac{\beta}{a}\frac{l^2}{3} \tag{50}$$

However, $(T_{l/2} - T_0) = -\beta/a (3/8)l^2$, so that the difference between the average temperature and $T_{l/2}$ is about $(0\cdot042\,\beta/a\,l^2)$.

The example given above is for a cylindrical calorimeter wall heated at one end. In a similar manner there may be obtained the temperatures in a shell heated equally from both ends. The solution here using boundary conditions that $T = T_0$ at $z = 0$ and $z = l$ is

$$(T_{l/2} - T_0) = -\frac{\beta}{2a}(lz - z^2) \tag{51}$$

The minimum temperature is at $z = l/2$, and its value is $(T_{l/2} - T_0) = -(\beta l^2/8a)$, so that the use of two heaters reduces the maximum gradient by a factor of four.

To avoid temperature differences on the surface, various techniques are used. Stirred liquids are used to distribute the heat in many calorimeters and in others metal vanes run from the heat source to the surface. In large calorimeters, the stirred liquid technique is usually preferable. Conduction is less attractive for large dimensions because the temperature difference increases rapidly with the distance. In the simple case described by Equation (49), the temperature increases with the square of the distance. In small calorimeters, good metallic conductors serve well and avoid the heat and mechanical problems of stirring.

2. Uncertainty in heat leak due to temperature gradients

It has been illustrated that for the different methods of heating the calorimeter wall the temperature gradients in the wall may be different. When the heater is located at one end of the cylindrical wall, the *average* temperature of the cylindrical surface occurs at a distance of about $(0.43l)$ in which l is the length of cylinder. If the heat leak is entirely by radiation from the wall surface (assume constant emittance), then a single thermocouple should be located at $0.43l$ from the end for proper accounting of radiation heat leak. If the thermocouple had been located at the middle of the cylindrical wall (at $l/2$), then the thermocouple temperature would differ from the average surface temperature. Let us examine an actual case to see the magnitude of this difference. As an unfavorable case, take a calorimeter wall of steel having length $l = 10$ cm, thermal diffusivity $a = 0.1$, and heated at a rate of 0.01 deg/sec. The temperature at its middle $(l/2)$ referred to the heated end is less by $\beta/a \, [(3/8 \, l^2)] = 3.75$ deg. However, the average temperature of the surface is $\beta/a \, (l^2/3) = 3.33$ deg. Therefore, the location of the thermocouple at $l/2$ gives a heat leak error corresponding to $0.42°C$ over the surface. If copper were used instead of steel, the temperature difference would be about 0.04 deg.

3. Methods of minimizing heat leak due to temperature gradients

The previous example indicates that temperature gradients in a calorimeter or shield wall can result in errors in heat leak measurement if the thermocouple (or couples) used are not located properly to account for the temperature gradients. Where possible, one obvious solution to the problem is to design the calorimeter and shield to minimize temperature gradients. This may be done by stirring or by use of materials with high conductivity or proper distribution of heater so that heat has to flow only a short distance. This latter solution may be complicated by different relative power requirements at different times. For example, in heating experiments with most of the power going to changing the temperature, the power distribution needed in the shield will be approximately according to mass. However, at steady temperatures with the power going only to heat leak, the power distribution needed probably will be entirely different. One partial solution to this problem has been to use two separate heaters, one distributed according to mass and one according to heat leak.

PRINCIPLES OF CALORIMETRIC DESIGN

Another method of minimizing temperature gradients in a shield is to minimize the power required to keep it at a given temperature. This may be accomplished with better insulation from its cooler surroundings. Generally, it is better to provide insulation with a minimum of heat capacity because of effects of time lag in the insulation. One such insulation with a minimum of lag is a number of concentric thin shells, such as discussed later in this chapter. Another method of reducing the power required in a shield, thereby reducing temperature gradients, is to provide external concentric shells, with the temperature of each independently controlled. In this manner, the power in the shield (inner shell) at a constant temperature can be negligible. An example of this is given in Chapter 11. However, use of such a system does not avoid temperature gradients on the shield while heating.

Another solution of the temperature gradient problem is to use a sufficient number of thermocouples to provide adequate information on the temperature distribution. One calorimeter (Chapter 11) used eight thermocouples on both the calorimeter and shield wall surfaces to evaluate average temperatures. In the case of heat capacity experiments, it is possible in principle to reduce errors in heat leak due to temperature gradients by using two series of experiments, a tare and a gross series. If the gradients on both calorimeter and shield are the *same* (considering each separately) in both series, then ideally the absolute heat leak errors are the same. They then cancel out when taking the difference between the tare and gross series. The effectiveness of this method for reducing heat leak errors depends upon the extent that the temperature gradients are the same. The shield gradients do not have to be the same as the calorimeter gradients. This method is used commonly with the assumption that the gradients are nearly the same in the two series. The fallacy in this assumption is that the temperature gradients in the calorimeter are necessarily changed by the change in the amount of sample, the magnitude of the change being a function of the design.

The obvious solution to this problem is to provide a calorimeter surface whose temperature is *independent* of the presence of the sample. This ideal condition can be approached in many calorimeters by a technique applied in Chapters 9 and 11. The principle of this technique is illustrated simply in *Figure 6* which shows a portion of a sample container with a heaterless

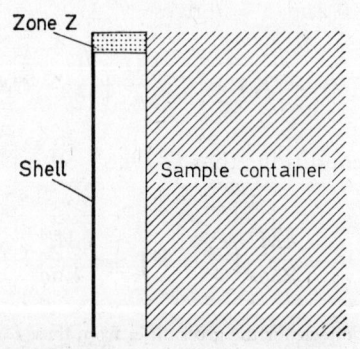

Figure 6. Method of reducing heat leak errors due to temperature gradients on sample container.

thin metal shell S thermally attached *only* at the zone Z. In heating experiments, in which it is assumed for simplicity that the calorimeter is solid metal, the heater produces a temperature gradient. After the starting transient has diminished, *all* of the calorimeter, including the ring at Z and the heaterless shell, will be heating at the rate β. This, of course, assumes constancy of λ, ρ, c, and no heat transfer except by solid conduction at zone Z. Under this condition, the heaterless shell will also have gradients. Since the gradients will be *independent* of the gradients on the calorimeter, any heat leak will also be independent of the gradients in the calorimeter. In this case, in principle, the use of the tare and gross series of experiments avoids heat leak errors *if* the temperature at the zone Z follows the same time-temperature function in the two series of experiments[†]. The success of this technique depends on the extent of thermal isolation of the heaterless shield S from the sample container C over the cylindrical surface. In one adiabatic calorimeter near room temperature with a single heaterless shell on the sample container, the thermal isolation was obtained by evacuation and use of polished gold surfaces. The effectiveness of this single shell can be demonstrated by considering the effect of the vertical gradient in the sample container (*Figure 6*) on the temperature of the shell. The principle of superposition can be used to separate the vertical gradient in the shell into two components, one due to heat flow through the zone Z to heat the shell at the rate β, and the other due to heat flow from the sample container directly across to the shell by radiation. For simplicity, consider only the cylindrical portions. Let h = heat transfer coefficient between the two surfaces, λ, w, r, and l the thermal conductivity, thickness, radius, and length of the heaterless cylindrical shell. Also assume there is a gradient on the sample container so that its surface temperature (referred to zone Z) is Mz^2 in which $z = 0$ at the zone Z. Also as a first approximation, assume that the resulting gradient in the shell is much smaller than the gradient on the sample container. The heat balance equation here is

$$\lambda(2\pi rw) \frac{\partial^2 T}{\partial z^2} \, dz = - h \, (2\pi r dz) \, (Mz^2) \tag{52}$$

in which h is the heat transfer coefficient between sample container and the shell, so that $\partial^2 T/\partial z^2 = - (hM/\lambda w) z^2$. Integrating this equation between the limits $z = 0$ and $z = l$ gives

$$\Delta T = (T_l - T_0) = - \frac{hM}{\lambda w} \left(\frac{l^4}{12} + C_1 l \right). \tag{53}$$

Using the boundary condition that $\partial T/\partial z = 0$ at $z = l$, gives $C_1 = - hMl^3/3\lambda w$ so that

$$\Delta T = - \frac{hM}{3\lambda w} \left(\frac{l^4}{4} - l^4 \right) = \frac{hMl^4}{4\lambda w}. \tag{54}$$

[†] In general, the transient effects for the gross differ from those for the tare. The statement is valid, but depends on a more sophisticated argument which does not even require the same heating rate[21].

For an example, use $l = 5$ cm, $w = 0.025$ cm, $\lambda = 1.0$ W cm^{-1} deg^{-1}, and $M = 1$, giving a 25 deg temperature difference on the sample container. If the space is evacuated and the surfaces are polished gold plated, the radiative heat transfer coefficient is about 10^{-5} W cm^{-2}deg^{-1}. Substituting these values gives $\Delta T = 0.0625$ deg C. This contrasts with the 25°C temperature difference on the sample container. The conclusion is that in this case, one heaterless shell has attenuated the effect of the sample container gradient by a factor of approximately 400. It is interesting to note the magnitude of the temperature difference in this shell due to heating the shell at a constant rate β. This temperature difference after the transient has vanished can be calculated using Equation (49), giving $(T_l - T_0) = -\beta l^2/2a$. Using $l = 5$ cm, a heating rate of 0.01 deg/sec, and $a = 1$ cm^2/sec (for copper) we get $\Delta T = 0.125$ deg C.

With this technique of providing a calorimeter surface whose temperature is independent of temperature differences in the calorimeter, it is no longer necessary to position thermocouples on the surface of the shell in order to obtain an average surface temperature. The tare and gross experiment will have the same heat loss, independent of the location of the thermocouple. It usually is convenient to locate one or more thermocouples at the zone [22].

The use of the heaterless shell is equally valuable in experiments such as heat of vaporization for which the calorimeter temperature is not changing with time. In this case, the shell attached to the calorimeter at one zone is isothermal in spite of temperature gradients in the calorimeter due to vaporization. A similar shell attached to the inside of the shield at one zone also is isothermal, in spite of temperature gradients in the shield due to heat loss to the surroundings.

It is not always possible to obtain as low a heat transfer coefficient as there was in the evacuated space bounded by polished gold surfaces. In a high-temperature adiabatic calorimeter[22] in which the space was not evacuated, the effectiveness of an unheated shield was not so great, so that it was necessary to use several shells.

4. The calorimetric heater lead problem

A more complicated case in steady-state heat flow is the problem of the calorimeter heater leads between calorimeter and shield. This problem has been considered in great detail in a recent publication[9], so that it will be discussed here only briefly.

The calorimeter heater current leads must be considered differently than the other calorimeter leads because the electric current in them develops heat which must be properly accounted for and apportioned to calorimeter and shield. The other leads merely serve as heat conductors to contribute to heat leak coefficients. The calorimetric error resulting from any uncertainty in the apportionment of the heat in the current lead segment between calorimeter and shield may be made small by using current lead segments of low electrical resistance relative to the calorimeter heater resistance. The problem here is that electrical and thermal resistances are usually proportional under ordinary conditions with metal wires, so that the smaller electrical resistance gives smaller thermal resistance. In turn, this results in a larger heat transfer coefficient which increases the heat leak uncertainty. The optimum size of lead depends on the particular calorimeter.

There is, however, a method by which any error in current lead power apportionment can be made smaller without this increased heat leak uncertainty. This method is simply to increase the relative resistance of the calorimeter heater. Accordingly, in most calorimeters at moderate temperatures, a calorimeter heater resistance of at least 100 Ω is considered desirable if accuracies approaching 0·01 per cent are desired. Sometimes at low temperatures, heaters are used which have resistances of thousands of ohms.

In accounting for the power in the current lead segments, the convention assumes that the heat developed in a current lead segment divides equally between calorimeter and shield, so that the potential lead is attached to the center of the current lead segment. It has been pointed out earlier (Equation 25) that this assumption is strictly true only when the ends of a current lead segment are at the same temperature. This condition for validity is satisfied if the calorimeter and shield are at the same temperatures and the thermal connections to the calorimeter and shield at the ends of a current lead are also the same. It has also been pointed out that even if these thermal connections are the same, the use of potential leads at mid points is valid with the calorimeter and shield at *different* temperatures only *if* a heat leak correction is made for the "apparent" heat flow along the current lead segments *assuming* their ends are at the temperatures of calorimeter and shield. This procedure is approximated in accurate calorimetry by merely measuring an overall heat transfer coefficient between calorimeter and shield when there is no current in the calorimeter heater and using this coefficient to calculate heat leak corrections, including periods when there is current through the heater lead.

The general problem of calorimeter current leads is when (*i*) the calorimeter and shield are at different temperatures (T_c and T_s), and (*ii*) the thermal connections at the ends of the current lead segments are different.

Consider calorimeter current leads, each having cross section A and length l between calorimeter and shield, and having a thermal conductivity λ. Now suppose that there is electric heat developed at a time rate p for unit length of the wire, and that all heat lost from the current lead segment is by solid conduction along the wire. Also suppose that each current lead is thermally and electrically insulated from the calorimeter or shield and that these thermal resistances per unit length between the wire and the calorimeter or shield are R_c and R_s, respectively. With these conditions a solution[9] of the heat balance equations gives the rate of heat flow P_c into the calorimeter by conduction along the current lead segment.

$$P_c = \frac{p}{2}\left[\frac{l^2 + 2l\sqrt{(\lambda A R_s)} + 2\lambda A(R_s - R_c)}{l + \sqrt{(\lambda A R_s)} + \sqrt{(\lambda A R_c)}}\right] + \left[\frac{\lambda A(T_s - T_c)}{l + \sqrt{(\lambda A R_s)} + \sqrt{(\lambda A R_c)}}\right]. \tag{55}$$

The solution is shown in the two terms to demonstrate its physical significance. The *second* term in the solution is independent of power developed in the wire and is the heat flow due to the temperature difference ($T_s - T_c$)

between shield and calorimeter. It is important to note that this second term (involving $T_s - T_c$) is usually accounted for by conventional calorimetric techniques, which experimentally measure and correct for the *overall* heat transfer coefficient between shield and calorimeter when there is no current in the leads. The first term in Equation (55) is proportional to the power developed in the current lead so that it accounts for that part of the heat flow caused by the current in the lead. The resulting overall calorimetric error resulting from this first term has been calculated[9] to be

$$\text{Calorimetric Error (\%)} = 100 \, r \left[\frac{2(L_s^2 - L_c^2) + (L_s - L_c)}{1 + L_s + L_c} \right] \quad (56)$$

in which r is the ratio of electrical resistance of length l of the current lead segment to the total electrical resistance of the calorimeter heater, including leads out to the potential terminal, and

$$L_s = \frac{\sqrt{(\lambda A R_s)}}{l}, \text{ and } L_c = \frac{\sqrt{(\lambda A R_c)}}{l}.$$

The relation between these dimensionless quantities L_c and L_s and $[(1/r) \cdot \text{Calorimetric error}]$ is illustrated in *Figure 7*.

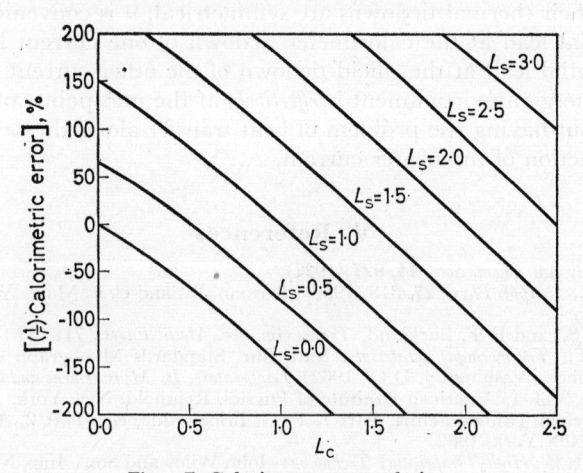

Figure 7. Calorimeter error relation.

In order to judge the practical importance of the error described above, it is necessary to calculate values of L_s and L_c and r for specific examples. Suppose that the current lead is circular, that the length l of the current lead segment is 3 cm, and that the lead is copper with o.d. = 0·02 cm Also suppose for thermal contact in the calorimeter this lead is electrically insulated by a cylindrical solid surrounding the wire to an o.d. of 0·2 cm and having a thermal conductivity of 0·002 W cm^{-1} deg^{-1} corresponding to a typical organic material. In this case, $R_c = 184$ degW^{-1} cm^{-1} and $L_c = 0.166$. Now suppose that in the shield there is an air gap of 0·005 cm between the wire and the insulation (o.d. of 0·2 cm). The resulting value of $R_s = 411$ deg W^{-1} cm^{-1} so that $L_s = 0.238$ and $[(1/r) \cdot$ (calorimetric error)] is 9·3 per cent. If the calorimeter heater has a resistance of 10 Ω, $r = 0.0016$ so that the calorimetric error in the power measurements due to using a potential lead at the mid point of the current lead is 0·0016 (9·3) =0·015 per cent. This error is only significant in the most accurate calorimetry. On the other hand, it is possible to have considerably larger error in a poor design. Take a case in which R_c is reasonably small but R_s is perhaps 2000 degW^{-1} cm^{-1} due, for example, to an evacuated powder insulation. In this case, the calorimeteric error is 0·083 per cent.

In the previous discussion of calorimeter heater leads, it has been assumed that the potential leads were attached to the current lead segments at their mid points and that heat flow along these leads is negligible. If this heat flow is not negligible, consideration should be given to the location of their attachments. The mid point attachment has the disadvantage that the temperature of the mid point is a function of the current in the calorimeter heater, but the resulting heat conduction along the potential lead is not accounted for in the first term in Equation (55). Although probably an analysis for the heat flow could be obtained, since the effect is usually a small one, it is easier to change the design to avoid the error. One method which has been used[22] is to attach thermal tiedowns on the current leads close to the calorimeter and shield mechanical boundaries. If the two current leads and their thermal tiedowns are symmetrical, it is convenient to attach one potential lead at the calorimeter tiedown of one current lead and the other potential lead at the shield tiedown of the other current lead. In this way, the energy apportionment is *effectively* at the mid points of the current leads without having the problem of heat transfer along the potential leads being a function of the heater current.

V. References

[1] Barry, F., *J. Am. Chem. Soc.*, **44**, 927 (1922).
[2] Berman, R., *J. Appl. Phys.*, **27**, 318 (1956); Berman, R., and C. F. Mate, *Nature*, **182**, 1661 (1958).
[3] Brunot, A. S., and F. F. Buckland, *Trans. Am. Soc. Mech. Engrs.*, **71**, 253 (1949).
[4] Caldwell, F., *Thermocouple Materials*, Natl. Bur. Standards Monograph 40, U.S. Govt. Printing Office, Washington, D.C., 1962; *Temperature, Its Measurement and Control in Science and Industry*, Vol. I, American Institute of Physics, Reinhold, New York, 1941; Vol. III, C. M. Herzfeld, editor-in-chief, Part 1, F. G. Brickwedde, ed., Part 2, A. I. Dahl, ed., Reinhold, New York, 1962.
[5] Campbell, I. E., *High-Temperature Technology*, John Wiley and Sons, Inc., New York, 1956.
[6] Carslaw, H. S., and J. C. Jaeger, *Conduction of Heat in Solids*, Oxford University Press, London, 1959.
[7] Churchill, R. V., *Operational Mathematics*, McGraw-Hill Book Co., Inc., New York, 1958.
[8] Ditmars, D. A., and G. T. Furakawa, *J. Res. Natl. Bur. Std.*, **69C**, 35 (1964).
[9] Ginnings, D. C., and E. D. West, *Rev. Sci. Instr.*, **35**, 965 (1964).
[10] Hoge, H. J., *Rev. Sci. Instr.*, **20**, 59 (1949).
[11] Jacobs, R. B., and C. Starr, *Rev. Sci. Instr.*, **10**, 140 (1939).
[12] Jakob, M., *Heat Transfer*, Vol. I, John Wiley and Sons, Inc., New York, 1949.
[13] Jakob, M., *Heat Transfer*, Vol. II, John Wiley and Sons, Inc., New York, 1957.
[14] Kistler, S. S., and A. G. Caldwell, *Ind. Eng. Chem.*, **26**, 658 (1934).
[15] Korn, G. A., and T. M. Korn, *Mathematical Handbook for Scientists and Engineers*, McGraw-Hill Book Co., Inc., New York, 1961.
[16] Lyman, T., ed., *Metals Handbook*, 8th ed., American Society for Metals, Novelty, Ohio, 1961; *American Institute of Physics Handbook*, McGraw-Hill Book Co., Inc., New York, 1957; *International Critical Tables*, McGraw-Hill Book Co., Inc., New York, 1928.

[17] Mebs, R. W., and W. F. Roeser, *Solders and Soldering*, Natl. Bur. Standards Circular 492, U.S. Govt. Printing Office, Washington, D.C., 1950.
[18] Petree, B., and G. Ward, *Natl. Bur. Standards Technical Note* **163** (1962), U.S. Govt. Printing Office, Washington, D.C.
[19] Scott, R. B., *Cryogenic Engineering*, D. Van Nostrand and Co., Inc., New York (1959).
[20] Weills, N. D., and E. A. Rider, *Trans. Am. Soc. Mech. Engrs.*, **71**, 259 (1949).
[21] West, E. D., *J. Res. Natl. Bur. Std.*, **67A**, 331 (1963).
[22] West, E. D., and D. C. Ginnings, *J. Res. Natl. Bur. Std.*, **60**, 309 (1958).
[23] White, W. P., *The Modern Calorimeter*, The Chemical Catalog Co., Inc., New York, 1928.
[24] White, W. P., *J. Am. Chem. Soc.*, **40**, 382 (1918).

CHAPTER 5

Adiabatic Low-temperature Calorimetry

EDGAR F. WESTRUM, JR.
University of Michigan, Ann Arbor, Michigan, U.S.A.
GEORGE T. FURUKAWA
National Bureau of Standards, Washington D.C., U.S.A.
JOHN P. MCCULLOUGH[†]
Bartlesville Petroleum Research Center, Bureau of Mines, Bartlesville, Oklahoma, U.S.A.

Contents

I.	Introduction and Historical Survey		135
II.	Apparatus		137
	1. The Cryostat Environment		137
	A. Aneroid-Type Adiabatic Cryostat		137
	B. Immersion-Type Cryostat		140
	C. Combination Immersion-Aneroid-Type Cryostat		140
	D. Filling-Tube-Type Cryostat		142
	E. Automatic Adiabatic Calorimetry		143
	F. Control of Heat Exchange to Exterior		144
	G. Control of Heat Exchange with the Calorimeter		146
	(1) The Adiabatic Shield		146
	(2) The Tempering Ring		148
	(3) The Heater Lead Wires		148
	(4) Thermal Switching		149
	H. Cryogenic Techniques		150
	2. The Calorimeter Vessel		150
	A. General Requirements		150
	(1) Thermal Diffusivity and Heat Distribution		150
	(2) Heat Capacity of the Empty Calorimeter Vessel (Tare)		151
	B. Typical Designs		152
	(1) For Solid Samples		152
	(2) For Liquids and Condensable Gases		156
	3. Measurement Techniques and Adjuvant Circuitry		156
	A. Cryogenic Thermometry		157
	(1) Thermometers and Temperature Scales		157
	(2) Measurement of Thermometer Resistance		160
	B. Energy Input and Measurement		162
	(1) Calorimeter Heaters		162
	(2) Electrical Energy Measurements		163
	(3) Electrical Standards		167
	(4) Time Standards		167
	(5) Determination of Duration of Heating		167
	C. Shield Control		168
	(1) Automatic Adiabatic Shield Control		168
III.	Procedures		173
	1. Preliminary Procedures		173
	A. Calorimeter Filling		173
	B. Preparation of Cryostats		173

[†] Present address: Central Research Division Laboratory, Mobil Oil Corporation, Princeton, New Jersey, U.S.A.

2. Sequence of Experiments 173
   A. Preparations for Measurements 174
   B. Order of Measurements 174
3. Crystallization Procedures 175
4. Observational Procedures 176
   A. Heat Capacity Measurements 176
      (1) Determination of Temperature and Energy Increments 177
      (2) Effect of Non-Adiabaticity 177
      (3) Equilibration Rates 178
      (4) Heat Capacity of the Calorimeter 179
   B. Enthalpy of Phase Changes 180
5. Melting Point and Purity Determinations 180
6. Studies of Polymorphic Substances 181
   A. Enantiotropic Substances 181
      (1) Detection of Transitions 182
      (2) Completing Transformations 182
      (3) Equilibration 182
      (4) Observational Procedures 183
      (5) Hysteresis Effects 183
   B. Monotropic Substances 183
7. Vapor Pressure and Enthalpy of Vaporization Measurements 184

IV. Data Reductions 184
1. Treatment of Experimental Data 184
   A. Description of Primary Observed Data 184
   B. Adjustment for Heat Leak 185
2. Reduction of Observed Data to Molal Basis 186
   A. Heat Capacity of the Calorimeter Vessel 187
   B. Curvature Corrections 187
   C. Premelting Corrections 189
      (1) Theoretical Formulation 189
      (2) Practical Applications 190
   D. Vaporization Corrections 191
   E. Other Corrections 194
      (1) Heat Developed in Leads 194
      (2) Helium Gas Correction 194
      (3) Mass Corrections 194
3. Smoothing the Experimental Data 195
   A. Smoothing Heat Capacity Data 195
      (1) Graphical Methods 195
      (2) Analytical Methods 196
         (a) Short Temperature Range Method 196
         (b) Representation of the Data by Equations .. 197
   B. Smoothing Enthalpy Increments 198
   C. Application of High Speed Digital Computers to Calorimetric Data 198
      (1) Energy Calculation for a Heat Capacity Determination 199
      (2) Energy Calculation for an Enthalpy Determination 199
      (3) Temperature 201
      (4) Digital Computer Smoothing of the Experimental Data 203
4. Extrapolation of Heat Capacity below about 10°K 203
5. Calculation of Thermodynamic Properties 206

V. General Discussion 207
1. Other Modes of Operation 207
2. Precision and Accuracy 208
3. Calorimetry Conference Standards 208
4. Comparison of Adiabatic Calorimetry with other Methods of Calorimetry 209
   A. Comparison with Isothermal Low-temperature Calorimetry 209
   B. Comparison with Intermediate Temperature (>250°K) Methods 209
5. Future Developments 210

VI. Acknowledgement 211
VII. References 211

I. Introduction and Historical Survey

Low-temperature heat-capacity data are of basic significance in physics, chemistry, and technological thermodynamics. Because of the difficulty of interpreting and analyzing the contributions of several degrees of freedom to thermal properties, heat-capacity studies at temperatures above about 10 or 20°K are of considerably less interest to the physicist than are those at lower temperatures. On the other hand, the entire range is of significance to the chemical thermodynamicist concerned with thermodynamic properties at 300°K or higher. In most instances, the major contribution to these properties originates above 10°K. But since, in many instances, significant and important contributions come from lower temperatures or are associated with zero-point entropy, the special techniques needed for optimum precision in calorimetry below 20°K (which are treated in Chapter 7) also are relevant to the chemical thermodynamicist. This chapter, however, is concerned primarily with measurements at temperatures above 10°K with occasional reference to the lower limit of 4°K, depending upon the refrigerant used to achieve the lowest temperature.

The desiderata of the physical chemist (or the chemical thermodynamicist) are definitive measurements of the highest practicable accuracy possible and even higher precision. In the final utilization of these data for the evaluation of the driving force of chemical reactions, differences in the Gibbs energy increments or entropy increments of (or even between) chemical reactions are important. Since these increments tend to be small compared with the measured quantities, the importance of reproducibility is so great that most (but not all) chemical thermodynamicists engaged in thermal measurements in the United States, for example, use similar resistance thermometers calibrated against the International Temperature Scale above 90°K and against a national standard scale below 90°K[63, 121]. They also compare experimental determinations on standard substances[33] for the elimination of systematic errors. Although the importance of the temperature scale is less for the derived thermodynamic properties, the accuracy of heat-capacity data is probably limited at the present time by uncertainties in the temperature scale. An experimental precision of the order of a few hundredths of a per cent is in principle possible and has been claimed by several laboratories.

The concept of evaluating the heat capacity of condensed phases by determining the temperature rise accompanying the input of a measured quantity of electrical energy was first applied by Gaede[46]. In about 1910, the concept was developed by Nernst[91] and his collaborator Eucken[36], leading to the precursor of the modern, adiabatic, vacuum calorimeter. A simple schematic representation of such a calorimeter is presented in *Figure 1*. Here the crystalline substance, 6, whose heat capacity is to be determined is placed in the sample vessel (calorimeter), 5, equipped with a heater, 3, and a thermometer, 7, and suspended within an evacuated enclosure (or submarine), 8. This in turn is immersed in a suitable refrigerant, 2, contained in a Dewar vessel, 1. The interposition of a controlled-temperature shield, 4, between calorimeter and submarine allows operation at temperatures considerably higher than those of the refrigerant bath.

In most calorimetric measurements, determination of the energy lost by heat leak is the factor which most severely limits the accuracy; therefore, a decision usually must be made between retaining a significant—but calculable—heat leak or endeavoring to eliminate the heat leak to the

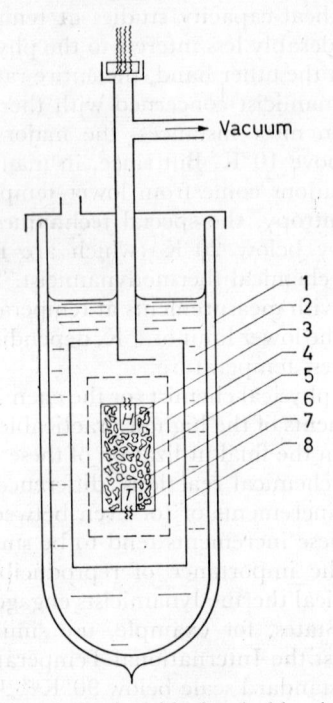

Figure 1. Schematic cryogenic calorimeter with electrical heating.
1, Dewar vessel
2, Liquified gas refrigerant
3, Electrical heater
4, Adiabatic jacket
5, Calorimeter vessel
6, Crystals
7, Thermometer
8, Evacuated submarine

point at which it is insignificant—but *not* calculable. If the controlled-temperature shield, 4, in *Figure 1* is designed with enough heat capacity to retain an essentially constant temperature during an experiment, the technique is described as one involving an essentially *isothermal shield* (cf. Chapter 1). The technique of operation with an isothermal shield was that first used by Nernst and his collaborators. Its current cryogenic counterpart (described in Chapter 6) still finds many applications today.

If, on the other hand, the temperature of the shield is so regulated as to be as nearly identical as possible with that of the calorimeter, the technique may be spoken of as involving an adiabatic shield (cf. Chapter 1). In practice, since exact equality is difficult to achieve, and since some minor deviation occurs, the usual *adiabatic* technique may best be described as *quasi-adiabatic*. The introduction of the adiabatic method at low temperatures is properly credited to Southard and Andrews[116].

This chapter is especially concerned with the application of the adiabatic shield technique (or, as it is usually referred to—adiabatic calorimetry) over the temperature range between 10 and 350°K. The basic requirements which are involved in such calorimetry are (*i*) to bring the sample to the desired temperature, (*ii*) to isolate it from its surroundings, (*iii*) to add heat or energy to the calorimeter and sample, (*iv*) to maintain adiabatic conditions, and (*v*) to provide accurate thermometry. Detailed information on the techniques of achieving these ends will be given in subsequent sections of this chapter. It is apparent that many alternative solutions have been devised, for example, those by Lange[74], Southard and Brickwedde[117], Aston and Eidinoff[2], Ruehrwein and Huffman[65, 108], Yost, *et al.*[146], Johnston, *et al.*[67], and one of a somewhat different type of Hill[57]. In addition, a low-temperature calorimeter that can be operated either adiabatically or isothermally has been described by Busey[15]. Stull has also described a pre-programmed, automatic, adiabatic calorimeter for use from 5 to 330°K[122]. In this calorimeter, all factors necessary for the calculation of data are controlled and recorded automatically as functions of time, with only a slight sacrifice in accuracy. Many calorimeters are supplied with automatic temperature control for adiabatic shields, not only as a labor-saving device, but also to permit more definitive study of transitions involving relatively more thermal hysteresis and long equilibration periods. The system developed by Furukawa[41] has been employed in several laboratories with some elaboration and adaptation. Other schemes for automatic adiabatic shield control have been described by Todd, *et al.*[131], Zabetakis, *et al.*[147], Stull[122], Dole, *et al.*[31], and West and Ginnings[137]. Some of the special calorimetric features introduced in the various laboratories are considered in subsequent sections.

II. Apparatus

1. *The cryostat environment*

Several contemporary, high-precision cryostats suitable for the determination of heat capacities from about 4 to 350°K will be described briefly. Considerations common to the design of all of them and specific features of the several types of cryostats will be discussed.

A. Aneroid-Type Adiabatic Cryostat

The first cryostat to be described, one of essentially completely metallic construction, is depicted in *Figure 2*[139]. The entire cryostat assembly is suspended from the cover plate so that a hoist may be used to remove it from the vacuum tank for loading of the sample. The primary function of this assembly is to maintain the calorimetric sample vessel at any desired temperature between 4 and 350°K in a state of thermal isolation, such that virtually no heat is lost or gained by the calorimeter vessel except when introduced by an electrical heater during measurement. Two chromium-plated, copper, refrigerant tanks for holding liquid nitrogen and helium

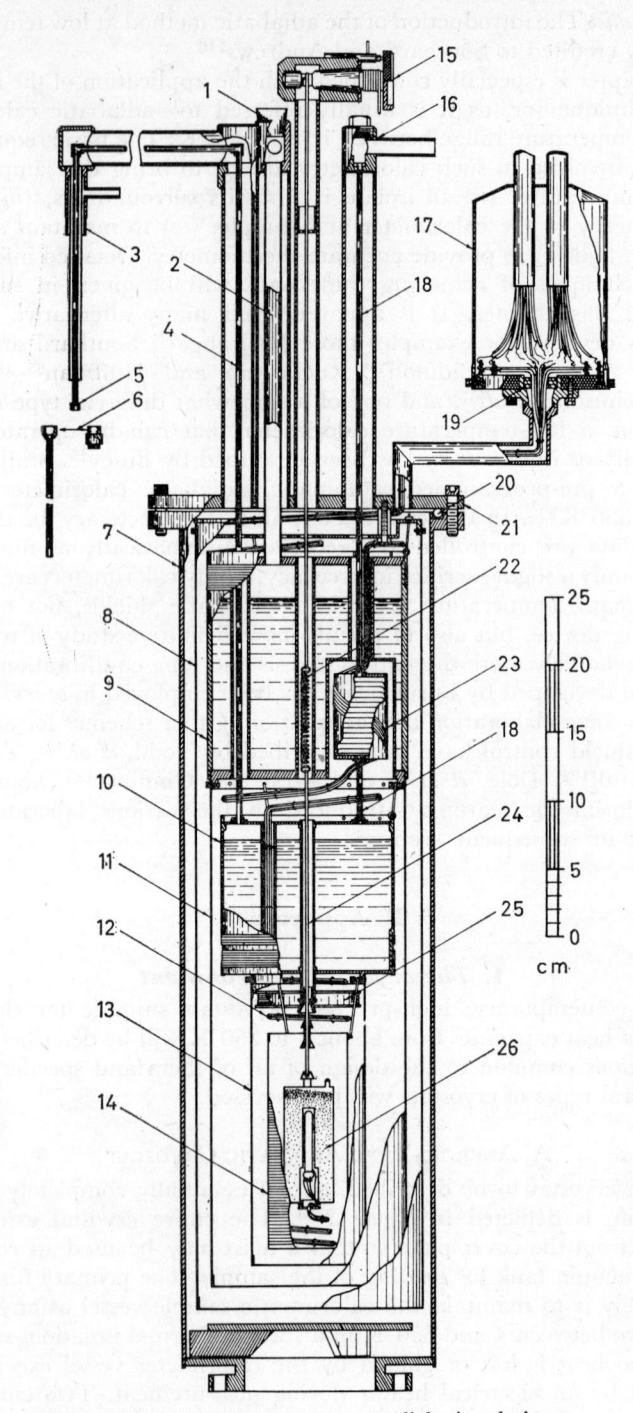

Figure 2. Aneroid-type cryostat for low-temperature adiabatic calorimetry.

Figure 2 legends (cont'd.)
1, Liquid nitrogen inlet and outlet connector
2, Liquid nitrogen filling tube
3, Sleeve fitting to liquid helium transport Dewar
4, Liquid helium transfer tube
5, Screw fitting at inlet of the liquid helium transfer tube
6, Liquid helium transfer tube extender and cap
7, Brass vacuum jacket
8, Outer "floating" radiation shield
9, Liquid nitrogen tank
10, Liquid helium tank
11, Nitrogen radiation shield
12, Bundle of lead wire
13, Helium radiation shield
14, Adiabatic shield
15, Windlass
16, Helium exit connector
17, Copper shield for terminal block
18, Helium exit tube
19, Vacuum seal and terminal plate for leads
20, O-ring gasket
21, Cover plate
22, Coil spring
23, "Economizer" (effluent helium vapor exchanger)
24, Supporting braided silk line
25, "Floating" ring
26, Calorimeter assembly[139]

provide low-temperature heat sinks. The calorimeter (i.e., the sample container) is suspended from a windlass by a braided silk line, and the adiabatic shield surrounding it is suspended from the helium tank by three fixed silk cords. The windlass is used to bring coaxial cones on the calorimeter, adiabatic shield, and lower tank into direct thermal contact, thereby cooling the calorimeter and shield. At the desired operating temperature, thermal contact is broken by lowering the calorimeter and the adiabatic shield, and adiabatic conditions are established in preparation for measurements.

Adiabatic conditions are established at any operating temperature by careful control of all factors that would lead to heat exchange between the calorimeter and its environment. Gas conduction is eliminated by maintaining a high vacuum (10^{-6} torr) within the vacuum tank. Chromium-plated copper radiation shields serve not only to conserve refrigerant but also to set up zones of uniform and progressively lower temperatures. Suspensions of low thermal conductivity (such as the stainless steel wire supporting the upper tank and the silk cord supporting the lower tank, adiabatic shield, and calorimeter) minimize thermal conduction. In a calorimeter of this design, it is also important to take special precautions to minimize the thermal conduction along the electrical leads. They enter the cryostat through a perforated Lucite disk covered with Apiezon-W wax. These leads are thermally anchored to the liquid nitrogen tank and to an "economizer", which serves as a heat exchanger utilizing the heat capacity of the cold, effluent, helium gas to absorb most of the heat conducted down the leads, thus conserving liquid helium. The leads are then anchored to the lower tank, from which they pass to a "floating ring", which can be maintained at the same temperature as the adiabatic shield and calorimeter by an electrical heater. Between each of the stages enumerated, the bundle of leads makes loops in the vacuum space to lengthen the conduction path.

The final tempering of the leads before they reach the calorimeter is accomplished by placing them in a helical groove beneath the heater winding on the cylindrical portion of the adiabatic shield. Upon establishment o adiabatic conditions, the equilibrium temperature of the sample is measured by a platinum-encapsulated, platinum resistance thermometer mounted in the calorimeter vessel. Electrical energy is then supplied to the calorimeter and sample by the heating element installed in the vessel.

The principal advantages of this design of calorimeter are its economy of liquid helium, ruggedness, and ease of reloading with a different sample. The only apparent disadvantage is a possibly higher initial construction cost.

B. Immersion-Type Cryostat

This type of cryostat is particularly simple—both in respect to minimal metal fabrication involved and in the facility with which it may be constructed. In many instances it may consist of little more than an evacuated container (or submarine) in which the calorimeter and/or specimen is suspended. The container is then immersed in a suitable refrigerant, e.g., liquid nitrogen, liquid hydrogen, or liquid helium, for operation over corresponding temperature ranges. *Figure 3* represents a calorimeter devised by Furukawa and coworkers[33, 42], which has been used extensively for measurements on both inorganic and organic substances. The construction, as is readily apparent from the figure, consists of dual submarines, the inner one containing a tempering ring, an adiabatic shield, and the calorimeter vessel itself. This submarine is immersed in the refrigerant (capable of producing the lowest temperatures desired), which is contained in a glass or metal Dewar. This Dewar, in turn, is entirely surrounded by a second submarine to facilitate pumping upon the refrigerant and thus achieving lower temperatures. This second submarine is contained within the liquid nitrogen bath of an outer Dewar. In order to cool a calorimeter and its contents and to calibrate thermometers (a process which also requires good thermal contact with the bath) some form of thermal switching is required. In an immersion cryostat, this is usually effected by means of helium exchange gas in the inner submarine. In order to obtain the best possible thermal insulation for the measurements, it is important to obtain a sufficiently good vacuum after removal of the exchange gas and cooling to the lowest temperatures to be employed. Because helium gas is strongly adsorbed on most surfaces at liquid-helium temperatures, considerable time and high-speed pumping may be required to obtain a sufficient degree of degassing. Moreover, as in the previously described aneroid cryostat, it is possible to use a mechanical heat switch between the calorimeter, adiabatic shield, and tempering ring (cf. Section II–1–G–4). The calorimeter shown in *Figure 3* has not actually been used with liquid helium as refrigerant for reasons of economy. The lower temperature limit of the calorimeter was designed to be 10°K, obtained by using liquid hydrogen.

C. Combination Immersion-Aneroid-Type Cryostat

A combination of the two preceding designs has been constructed by Sterrett, *et al.*[120] especially for measurements on materials in the vitreous

ADIABATIC LOW-TEMPERATURE CALORIMETRY

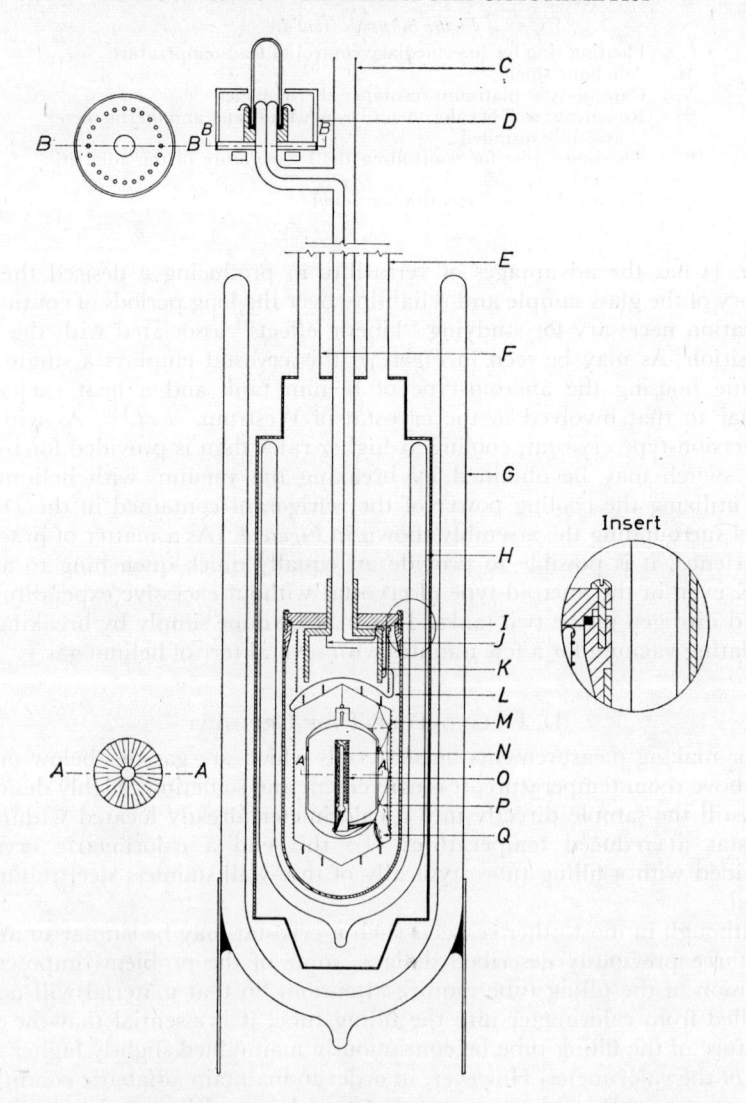

Figure 3. **Immersed-in-Dewar** type of adiabatic cryostat for solids or liquids of low volatility[42, 111].

- A–A, Cross-section of calorimeter sample vessel—vertically and radially arranged vanes are shown
- B–B, Cross-section of terminal box
- C, Tube to high vacuum pump
- D, Terminal box for leads
- E, Tube for filling the inner Dewar with liquid hydrogen
- F, Tube to vacuum pump to reduce the pressure of hydrogen (liquid or solid)
- G, Outer Dewar for liquid nitrogen
- H, Inner Dewar for liquid hydrogen
- I, Gasket seal with screws for the vacuum can—gold wire gasket is shown
- J, Shield for any radiation down high vacuum tube C
- K, Ring for cooling the leads to bath temperature

141

Figure 3 legends (cont'd.)

L, Floating ring for intermediary control of lead temperature
M, Adiabatic shield
N, Capsule-type platinum resistance thermometer
O, Re-entrant well of calorimeter vessel with heater and thermometer assembly installed
P, Thermocouples for controlling the temperature of the adiabatic shield
Q, Electrical leads to calorimeter vessel

state. It has the advantages of versatility in producing a desired thermal history of the glass sample and reliability over the long periods of continuous operation necessary for studying "kinetic effects" associated with the glass transition. As may be seen in *Figure 4*, this cryostat employs a single submarine housing the aneroid-type of helium tank and a heat exchanger similar to that involved in the cryostat of Westrum, et al.[140]. As with the mmersion-type cryostat, cooling at higher rates than is provided for by the heat switch may be obtained by breaking the vacuum with helium gas and utilizing the cooling power of the refrigerant contained in the Dewar vessel surrounding the assembly shown in *Figure 4*. (As a matter of practical experience, it is possible to provide an equally quick quenching to about 70°K even in the aneroid-type of cryostat without excessive expenditure of liquid nitrogen in the two tanks. This may be done simply by breaking the insulating vacuum for a few minutes with several torr of helium gas.)

D. Filling-Tube-Type Cryostat

For making measurements on materials which are gaseous below or not ar above room temperature, it is convenient and sometimes highly desirable to distil the sample directly into a calorimeter already located within the cryostat at reduced temperatures. To this end a calorimetric cryostat provided with a filling tube (typically of thin-wall stainless steel tubing) is useful.

Although in most other respects such a cryostat may be similar to any of the three previously described designs, some of the problems imposed by inclusion of the filling tube require attention. So that material will not be distilled from calorimeter into the filling tube, it is essential that the temperature of the filling tube be continuously maintained slightly higher than that of the calorimeter. However, in order to maintain adiabatic conditions, it is necessary that this excess temperature be carefully regulated. To this end, it is customary to provide a heater winding about the tube and, in addition, suitably-placed thermocouples to survey the temperature of the tube during measurements (and during loading). Another useful feature of this design is the possibility of distilling material directly from the calorimeter into a suitable receiver and thereby determining directly the enthalpy of vaporization of the sample. The technique here is rather similar to that of the corresponding operation in a calorimeter with isothermal shield. The design of a successful adiabatic filling-tube-type cryostat[44, 111] is shown in *Figure 5*. *Figure 6* shows a slightly more complex calorimeter used by Furukawa[40] especially for heat of vaporization measurements.

The provision of a throttling valve to control the rate of vaporization is a

ADIABATIC LOW-TEMPERATURE CALORIMETRY

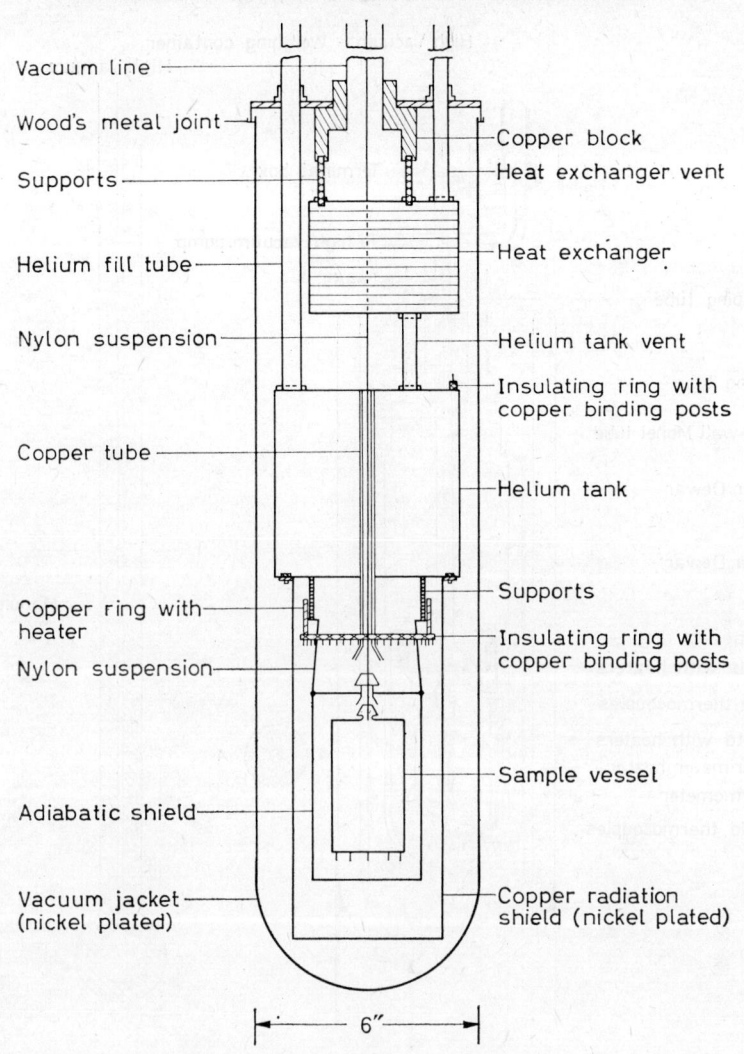

Figure 4. Submarine of an immersion-aneroid-type cryostat. (The entire submarine shown here is immersed in a Dewar of liquid nitrogen)[120]

valuable asset in determining the enthalpy of vaporization of substances. In some respects, it is also a convenience because it minimizes the required exactness of the temperature control of the filling tube, since the valve may be closed during the course of ordinary heat capacity and enthalpy of transition measurements.

E. Automatic Adiabatic Calorimetry

Although, in principle, fully automated calorimetry is possible with the type of apparatus described previously in this section, the automatic adiabatic-type calorimeter using discontinuous heating recently described

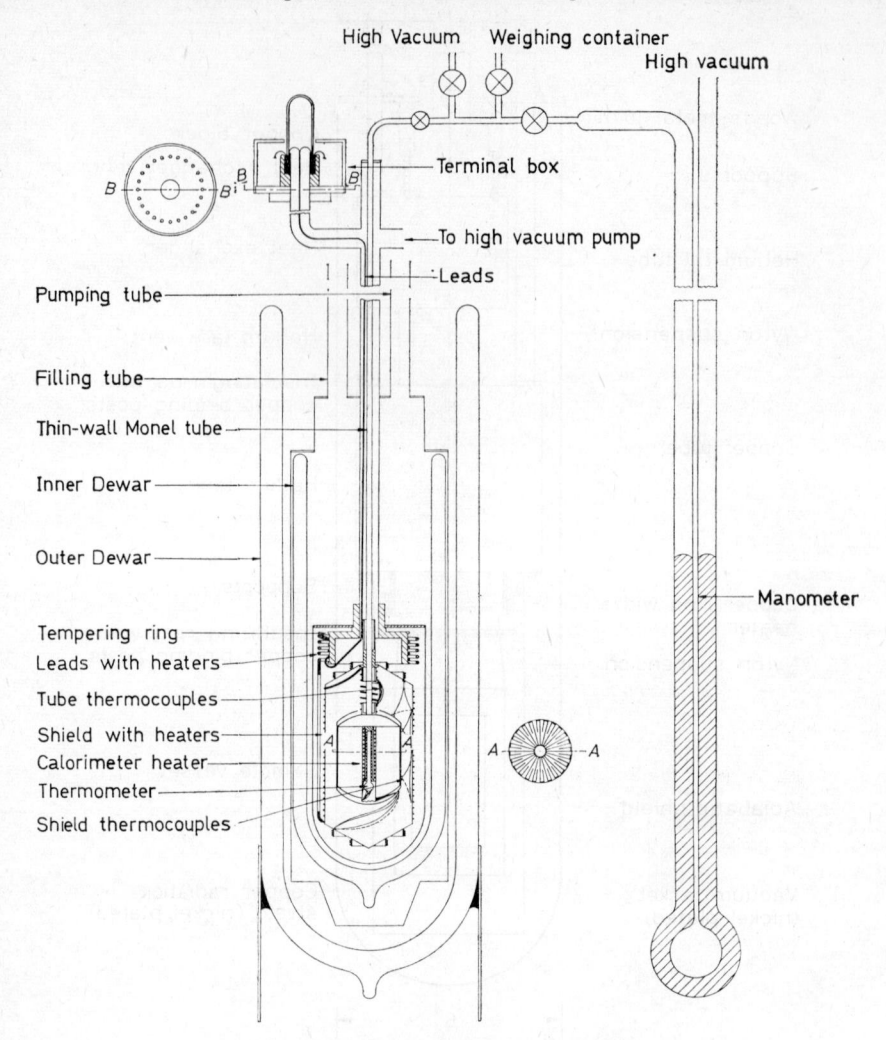

Figure 5. Filling tube-type adiabatic cryostat[44, 111]

by Stull[122] uses a different and rather more complex arrangement of shields. This device achieves an accuracy of about 0·3 per cent, and thus represents somewhat less than is possible by manual operation, although it does not reflect on any basic inferiority of ultimate automatic operation.

F. Control of Heat Exchange to Exterior

All the previously described cryostats have used liquid refrigerants for the achievement of low temperatures and minimization of the convective, conductive, and radiative heat exchange with the exterior of the cryostat. Vacua of the order of 10^{-6} torr or better have proven adequate in reducing to negligible amounts the convective transfer of heat. The primary conduc-

ADIABATIC LOW-TEMPERATURE CALORIMETRY

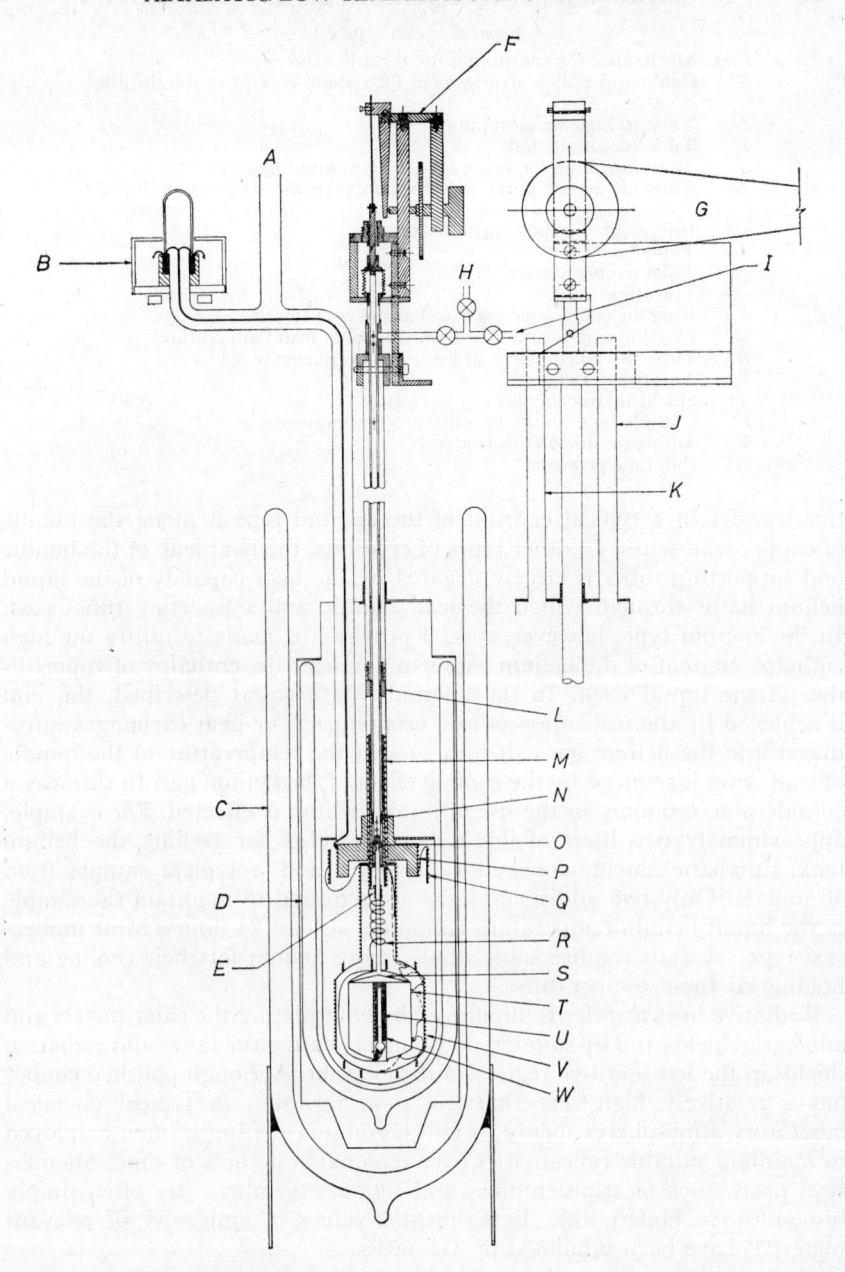

Figure 6. Filling tube-type adiabatic cryostat with throttling valve for enthalpy of vaporization measurements.

    A,   Tube to high vacuum pump
    B,   Terminal box for leads
    C,   Outer Dewar for liquid nitrogen
    D,   Inner Dewar for liquid hydrogen
    E,   Tube thermometer for sensing the temperature of vapors being removed

Figure 6 legends (cont'd).

- F, Mechanism for controlling the throttle valve
- G, Cable and pulley arrangement for remote control of the throttle valve
- H, Tube to high vacuum pump
- I, Tube to sample flask
- J, Tube for filling the inner Dewar with liquid hydrogen
- K, Tube to vacuum pump to reduce the pressure of hydrogen (liquid or solid)
- L, Intermediate tube control heater
- M, Tube heater
- N, Tube temper sleeves (4)
- O, Throttling valve
- P, Ring for cooling the leads to bath temperature
- Q, Floating ring for intermediary control of lead temperatures
- R, Tube thermocouples and leads to calorimeter vessel
- S, Anti-droplet gauge
- T, Shield thermocouple
- U, Capsule-type platinum resistance thermometer
- V, Adiabatic shield with heaters
- W, Calorimeter heater

tive transfer in a typical cryostat of the aneroid type is along the bundle of copper lead wires. In other types of cryostats, the heat leak of the bundle and supporting tubes is largely negated by the heat capacity of the liquid helium baths through which the lead bundle and supporting tubes pass. In the aneroid type, however, special provision is made to utilize the high enthalpy content of the helium vapor to augment the enthalpy of vaporization of the liquid itself. In the aneroid-type cryostat described, this end is achieved by the utilization of an "economizer" or heat exchanger introduced into the helium gas exit line, so that the temperature of the bundle of lead wires is reduced by the cooling effect of the helium gas. In this way a considerable economy in the use of liquid helium is effected. For example, approximately two liters of liquid helium suffice for cooling the helium tank, adiabatic shield, enclosed calorimeter and a typical sample from 50 to 4°K. Only two additional liters are required to maintain the sample in the liquid helium temperature region for at least 24 hours. Most immersion-type cryostats require substantially more helium for their cooling and holding at these temperatures.

Radiative heat transfer is minimized by gold-plating the calorimeters and adiabatic shields and by suitably polishing or plating the tanks and radiation shields in the less sensitive regions of the cryostat. Although polished copper has a relatively high reflectivity, it soon tarnishes in typical chemical laboratory atmospheres; hence, bright chromium-plating is often employed to maintain suitable reflectivities over reasonable periods of time. Stainless steel parts, such as transfer tubes and support members, are often simply brought to a high polish. Experimental values of emissivity of relevant materials have been tabulated by White[143].

G. Control of Heat Exchange with the Calorimeter

(1) The Adiabatic Shield. The calorimeter vessel (defined as that part of the apparatus in which heat input and corresponding temperature changes are accurately measured) and the adiabatic shield are designed to have their relative heat-leak coefficients as small as possible. This allows reductions

both in the uniformity of temperature over the parts of the shield and in the precision of control required, without excessive total heat transfer. As in other regions of the cryostat, the convective heat-leak coefficient in the calorimetric system is eliminated by maintaining a vacuum; that for radiative heat transfer is kept small by gold-plating the inner surface of the radiation shield and the outer surface of the calorimeter vessel to achieve a high reflectivity. To minimize the heat-transfer coefficient by conduction, the conductive cross-sections of the lead wires and the filling tube are kept as small as practicable to minimize their thermal conductivity.

Two temperature-controlling functions are incorporated in adiabatic shields. The first is the control of the radiation shield temperature to approximate that of the calorimeter vessel which it encloses. The second is the provision for controlling the temperature of the electrical leads to the calorimeter vessel to minimize heat conduction along the leads. As previously noted, a third temperature-controlling function must be provided for the filling-tube calorimeter to equalize the temperature of the calorimeter vessel and that of the segment of the filling tube between the adiabatic shield and the calorimeter.

The relative temperature between the adiabatic shield and the calorimeter vessel is determined by differential thermocouples. These thermocouples are selected to give a high sensitivity (i.e., large dE/dT) and to have a low thermal conductivity for minimizing heat transfer. Copper *vs.* Constantan, Chromel-P *vs.* Constantan, Chromel-P *vs.* gold-cobalt (2·1 atomic per cent Co), or other thermocouples may be used. Electrical resistance heaters, bifilarly wound on the adiabatic shield, control the shield temperature as indicated by the thermocouple signals. (Section II–3–C may be consulted for a brief description of automatic shield control.) Both a.c. and d.c. currents have been used to energize the heaters. If manual control of the adiabatic shield is to be employed, provision should be made for both exact (fine) setting of current values (e.g., with suitable multiple-turn potentiometers) as well as rapid (coarse) addition of increments of energy with reproducible settings (e.g., a step-switch bank of resistors with suitable detents).

The location of the thermocouples on the calorimeter vessel radiation shield is important in order to attain as nearly as possible the same temperature or the same temperature distribution as that of the calorimeter vessel. West and Ginnings[137] describe a calorimeter design in which corresponding girth rings of the radiation shield and calorimeter vessel are controlled at the same temperature. By employing multiple shells, the temperature distributions on the innermost shell of the radiation shield and on the outermost shell of the calorimeter vessel are made as nearly reproducible as possible for the measurements with and without sample. The electrical leads to the calorimeter are attached in good thermal contact with the girth rings of both the shield and the calorimeter vessel; they are thus controlled simultaneously with the radiation shield. In the filling-tube calorimeter described by Scott, *et al.*[111], the thermocouples are distributed over the surfaces of the radiation shield and the calorimeter vessel; the apparent average temperatures of the two surfaces are detected and controlled.

The location of heaters on the radiation shield is equally as important as that of thermocouples. During the equilibration period, the outer surface of the calorimeter vessel is nearly isothermal; thus the radiation shield must also be isothermal at the temperature of the vessel. During the heating period, the calorimeter vessel has, however, a temperature gradient; the shield should be controlled to have exactly the same gradient and temperatures. In the design of most low-temperature calorimeters, the heater wire is wound either evenly over the outer surface of the shield in order to introduce electrical heat uniformly or with resistance in proportion to the mass of that part of shield. The control of electrical current in the upper, middle, and lower sections of the radiation shield is often done separately, in conjunction with differential thermocouples placed at the corresponding sections of the shield and the calorimeter vessel, in order to approximate the temperature gradients on the calorimeter vessel. Since ideal temperature control of the radiation shield cannot be achieved, the calorimeter is designed to have the smallest heat-transfer coefficient possible.

The control of the electrical lead wires is somewhat simpler. The leads are wound in good thermal contact with the shield at one end and with the calorimeter vessel at the other. Differential thermocouples are properly located to indicate the temperature of the two ends, and the shield heater is adjusted to maintain the thermocouple indication at zero. Alternatively, the temperature of the leads may be controlled either coupled, by placing them in good thermal contact with the adiabatic shield, or separately, by winding the leads in good thermal contact with, for example, the temperature-controlled segment of the tube of the filling-tube calorimeter (cf. *Figures 5 and 6*), or by winding them on a "reference block" maintained at the same temperature as that of the calorimeter vessel[40].

(2) The Tempering Ring. In order to facilitate the exact control of the adiabatic shield, it has been found convenient to provide a ring (cf. *Figure 2*) to temper the gradient in the lead bundle preceding its junction with the adiabatic shield. A differential thermocouple between the ring and the adiabatic shield and a suitable manually- or electronically-controlled heater is also provided on the ring.

(3) The Heater Lead Wires. An important heat leak between calorimeter and adiabatic shield involves the conduction along the lead bundle. Although diameters of most copper potential and thermocouple leads may be minimized to practicable mechanical limits, that of the heater current leads requires consideration in the design, for the ohmic heat generated in these leads between the adiabatic shield and calorimeter vessel must be accounted for, even though it is often less than 0·1 per cent of the total. When the same temperature and thermal contact for both ends of the current leads (at the adiabatic shield and at the calorimeter vessel) obtain, the heat generated in the leads is dissipated equally to the shield and to the calorimeter vessel[51]. Therefore, by attaching one potential lead at the shield end of one current lead and the other potential lead at the calorimeter vessel end of the second current lead, the heat generated in the current leads will be correctly accounted for in the power measurements. The potential leads also may be attached at the mid-points of each current lead. To

maintain symmetry in the heat flow, one potential lead is brought out directly to the shield and the other, first to the calorimeter vessel and then to the shield. The lengths of the potential leads from the mid-point of the current leads to the shield and to the calorimeter vessel must be equal.

If the potential leads are attached in any other configuration, corrections must be applied, since the actual amount of heat introduced to the calorimeter vessel is not measured. For example, if both potential leads are joined to the current leads at the calorimeter vessel end, the heat generated in one of the current leads is added to the measured heat. Or, if the two potential leads are connected at the shield ends of the current leads, the heat generated in one of the leads is subtracted from the measured heat.

The best method for actually measuring the heat generated in the leads is to attach potential leads at the ends of the current leads and observe the voltage across the current leads, the current being already determined. The temperature excess (between the leads and the shield and calorimeter vessel) in the current leads can be significantly high, depending upon the length and size of the wire and the current. Therefore, much information on the temperature and current dependence of the lead resistance must first be accumulated before reliable adjustment can be made by simple calculation for the heat generated in the current leads. If these data are available to sufficient accuracy, the heat generated is $\int I^2 R \, dt$. Fortunately, the heat generated in the current leads can often be minimized with respect to that generated in the heater by selecting a suitable resistance ratio.

If the potential leads are attached to the current leads at the mid-points (or any other positions between the shield and calorimeter vessel) and brought directly out to the shield, the unsymmetrical heat flow generated in the current leads must be calculated from the thermal conductivity of the wires involved. By using relatively fine wire, thermal resistance of the potential leads can be made high and symmetrical heat flow maintained.

(4) Thermal Switching. Although helium exchange gas provides a convenient and effective means of cooling, as has been noted, the tendency for helium adsorption on surfaces at liquid helium temperatures usually requires provision of a properly designed pumping system and adequately long times for degassing. As operating temperatures approach those attainable only with pumped liquid helium (below 4°K), there is an increased incentive for the complete elimination of an exchange gas by a mechanical thermal switch, e.g., by aneroid thermal contact between the calorimeter and some portion of the apparatus in contact with the refrigerant bath.

Keesom and Kok[69] obtained thermal contact by lowering the specimen onto a projection from the base of the surrounding vacuum can, lifting the sample free of the base with a simple windlass when making measurements. Considerable frictional heating apparently occurs at low temperatures in using this arrangement, according to Kostryukova and Strelkov[71].

The device already described by Westrum, Hatcher and Osborne[140] (cf. Section II–1–A) has been used effectively between room temperature and 4°K. Little frictional heating seems to result during the process of isolation at 4°K. If the surfaces are kept scrupulously clean, cooling from room temperature to 4°K may be obtained—even on a sample of large heat capacity—within a period of 4 to 8 h. Should very rapid cooling be desirable, it is possible even in this apparatus to break the vacuum with a small amount of helium gas and cool directly to about 80°K within a period of 20 to 30 min. Other simple heat switches have been described by Ramanathan and Srinivasan[100], Rayne[101] and Webb and Wilks[135]. Because of the complication introduced by the presence of the adiabatic shield, each of these devices would require some modification to render them applicable to adiabatic calorimetry.

H. Cryogenic Techniques

The general techniques of design and operation at cryogenic temperatures are described in standard monographs, such as those of White[143], Hoare, Jackson, and Kurti[60], Scott[110], Vance[134], Rose-Innes[102], Din and Cockett[30] and the chapters by Hill[58] and Timmerhaus[130]. These techniques will not be detailed further here except as required for special emphasis or reference.

2. The calorimeter vessel

A. General Requirements

As in other types of calorimetry, the calorimeter vessel provides not only a container for the sample itself, but also a well-defined part of the apparatus where heat input and the corresponding temperature changes are accurately measured. Perhaps the simplest type of calorimeter is a vacuum-tight vessel provided with a suitable thermometer and heater, some means of introducing specimens into the vessel, and perhaps apparatus for adding exchange gas to facilitate thermal equilibration. The general requirements of this calorimeter vessel are usually that it be strong enough to withstand a differential pressure of at least 1 atm, that it not react chemically with the calorimetric sample, that it provide good thermal diffusivity, and that it also provide good thermal contact throughout the heater-thermometer-calorimeter vessel-sample system. In addition, it is desirable that the calorimeter (or tare) heat capacity be relatively small compared to that of the sample. Further, provision must also be made for convenience in filling and emptying the calorimeter of sample and in providing a vacuum-tight seal over the temperature range to be employed. Simultaneous satisfaction of all these requirements usually requires a certain amount of compromise in the design and fabrication. Although the requirements are somewhat overlapping, they will be discussed in terms primarily of the thermal diffusivity and the tare heat capacity.

(1) Thermal Diffusivity and Heat Distribution. The importance of using metals of high thermal diffusivity will be immediately evident because the main problems are in providing a proper mounting for the platinum resistance thermometer and heater and in putting these in good thermal contact with the sample. Gold, copper, and silver—and to a lesser extent, other pure metals, e.g., platinum—have high thermal conductivity and are most frequently used in the construction of calorimetric vessels for the low-temperature range. These substances, especially gold and platinum, are useful also because they have high resistance to chemical corrosion and

reaction. If copper or silver is used in the general construction, it is still desirable to electroplate gold or platinum onto the interior surfaces to reduce the extent of chemical corrosion and tarnishing. Because gold-plating tends to be very porous, copper calorimeters are typically first silver-plated to provide a more impermeable noble-metal surface and to minimize the diffusion of gold, especially into seams which have been joined with lead–tin solder.

Even when using materials of high thermal diffusivity, it is still necessary to position the thermometer and heater centrally within the calorimeter to minimize temperature gradients. Although it is possible to provide a heater on the exterior, typically-cylindrical surface of the calorimeter, this arrangement is relatively inconvenient for adiabatic calorimetry since it usually decreases the reflectivity of the boundary and tends to introduce awkward temperature gradients. If this type of heater is employed, it should be appreciated that certain corrections are mandatory, as discussed in Chapter 6. Provision of a suitable thermal conduction medium between the thermometer-heater-calorimeter assembly is also pertinent, as well as a means for bringing the lead wires to the temperature of the calorimeter surface prior to their departure in the bundle.

The use of exchange gas to provide thermal conductivity between the sample and the calorimeter vessel creates another problem, for although the amount and the heat capacity of the exchange gas in the free state may be well-known and readily corrected for, it may indeed be expected that the gas will be adsorbed to some extent on the sample, particularly if this is a finely divided material. The consequent desorption of the gas as the temperature rises during the heat capacity measurement will be part of the apparent thermal effect. This effect, of course, becomes more serious as the temperature is lowered nearer the boiling point of the exchange gas used. Although in some instances the desorption effect can be seen in the heat capacity (or at least strongly suspected), it is difficult to apply an appropriate correction for this adsorption; attempts to minimize it should, however, be made whenever possible. Perhaps the most effective minimization is obtained by using (when practicable) samples with a minimum specific surface; this provides an additional advantage also in that macroscopic crystals are better defined thermodynamic reference states than fine powders. The adsorption may be further minimized by using the smallest pressure of exchange gas that will provide adequate thermal conductivity. Since it is possible to calculate what error would result if the entire quantity of gas became desorbed in the course of one measurement, the maximum error may be estimated upon the assumption of a reasonable value for the enthalpy of adsorption.

Although normal helium (isotopic mass four) is typically used as an exchange gas, the considerably-lower boiling temperature and consequently higher vapor pressure of helium of isotopic mass three also help to minimize this error[138]. The small quantities of gas required and the recent reductions in the price of this isotope make such a procedure feasible.

(2) *Heat Capacity of the Empty Calorimeter Vessel (Tare)*. A sample of relatively high Debye characteristic temperature may typically have a heat capacity

near 300°K comparable to that of the calorimeter vessel (tare), but at 10°K its heat capacity will be only a small fraction of that of the vessel. Therefore, in order to maintain a high relative accuracy in the net heat capacity, measurements of higher accuracy must be obtained. The situation is improved with substances of low Debye characteristic temperature, such as lead or organic compounds, because the heat capacities of these substances are relatively high at 10°K. Most metals of high thermal diffusivity (from which calorimeter vessels are preferably constructed) change in heat capacity by a factor of about 100 from 10°K to 400°K. For example, aluminum, copper, and magnesium change about 500-fold, platinum and silver, 100-fold, and gold, about 50-fold. On the other hand, substances such as Al_2O_3 and MgO change almost 10^4-fold. Lead changes only ten-fold and n-heptane, about 100-fold. The heat capacity at 10°K of one gram of lead is equivalent to that of about 150 g Al_2O_3 and 120 g MgO. Therefore, the amount of lead–tin solder used in the tare measurements must be accurately known relative to that used in the sample-plus-tare measurements. Also, a minimum amount of lead–tin should be used in the fabrication of the calorimeter vessel. Because of its lower Debye characteristic temperature and low mechanical strength, gold is one of the least desirable noble metals for construction of the calorimeter vessel intended for use in investigating substances of high Debye characteristic temperatures. If the mass of the calorimeter vessel is slightly different for the nearly-full and tare measurements, the mass of materials which differ between the two sets of measurements must be accurately known. This is particularly true for materials of low Debye characteristic temperatures.

B. Typical Designs

(1) For Solid Samples. Typical samples of small crystals of ionic or molecular solids, or even samples of zone-melted or flame-fused crystalline substances, may be sealed within a calorimeter, such as that shown in *Figure 7*, with a small amount (10^{-2} torr pressure) of gaseous helium added in the sample for thermal equilibration. Most of the details of the calorimeter may be seen by reference to the figure legend. The heater and thermometer are inserted within an entrant well, and the bundle of leads passes from them around a spool to insure that the leads will attain the temperature of the calorimeter surface prior to departure in the bundle. In this calorimeter, Apiezon-T is used to obtain thermal contact between the heater-thermometer-assembly. It will be noted that the heater is wound bifilarly on a grooved copper mandrel which is closely fitted within the reamed well of the calorimeter. The thermometer fits tightly to the copper heater sleeve. The differential thermocouple (or thermopile) used in the adiabatic shield control system is mounted with the same grease for thermal contact in a sleeve (or sleeves) attached to the side of the vessel. The vertical vanes of thin copper foil aid in the distribution of heat from the heater and in the subseqeunt reestablishment of thermal equilibrium.

Occasionally the specimen is obtained in the form of a suitably-shaped (machined, cast, or grown single crystal) block or cylinder. It is possible to use a similar style of calorimeter if the heater-thermometer well is displaced from the cylindrical axis of the calorimeter and the sample encased in a

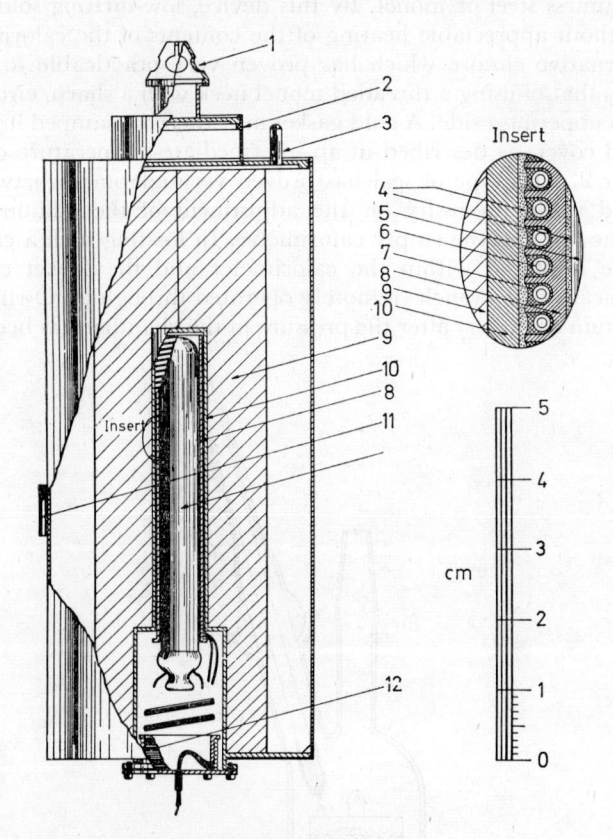

Figure 7. Cross-sectional diagram of calorimeter for solids.

    1, Thermal contact cone
    2, Monel cupola
    3, Solder-capped monel tube for helium seal-off
    4, Grease for thermal contact
    5, Capsule-type platinum resistance thermometer
    6, Fiberglas-insulated No. 40 Advance wire
    7, Formvar enamel
    8, Gold-plated copper heater core
    9, Copper vanes
    10, Gold-plated copper heater sleeve
    11, Differential thermocouple sleeve
    12, Spool to thermally equilibrate leads with calorimeter[139]

suitable vane of foil attached to the entrant well, and, if necessary, the entire top of the calorimeter may be made removable in order to insert the specimen. Low-melting solder may then be required to seal the sample suitably in the calorimeter.

Because certain samples readily decompose in the inevitable temperature rise of the calorimeter during the soldering operation, it is desirable to provide a cover or cap such as the one to which the thermal contact cone is attached in *Figure 7* with a small neck or cupola of a poor conducting material, such as stainless steel or monel. By this device, low-melting solder may be affixed without appreciable heating of the contents of the calorimeter.

An alternative closure which has proven very practicable to 1·5 cm in diameter is that of using a threaded monel neck with a sharp, circular knife-edge on its uppermost side. A gold gasket may then be clamped into place by a threaded cover, as described in an intermediate-temperature calorimeter in Chapter 9. This type of seal has proven very adaptable between 1 and 800°K and avoids necessity for the adjustment of the amount of solder between the full and the empty calorimeter. In loading such a calorimeter, the sample is placed within the calorimeter and the gasket clamped in place by means of a wrench (remotely operated through an O-ring seal in a metal vacuum chamber) after the pressure in the chamber has been adjusted

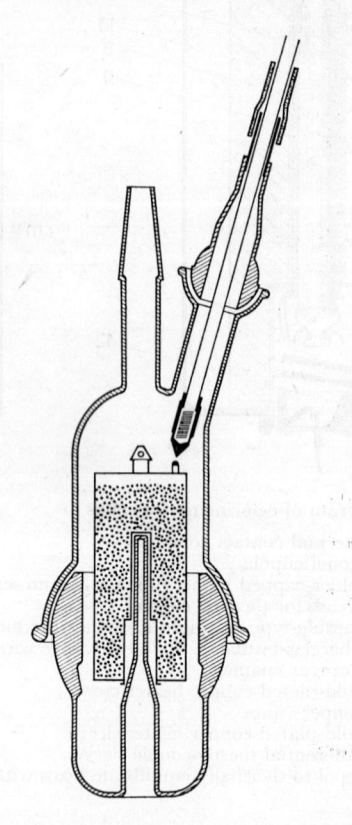

Figure 8. Device for evacuation of calorimeter and soldering tip of helium seal-off tube.

to that desired within the sample space. If it is desired to test the seal, a small solder-capped monel or stainless steel tube may be provided, as shown in the calorimeter in *Figure 7*. After the seal is tested, the calorimeter is evacuated in the bulb shown in *Figure 8* and, after adjustment of the helium pressure, sealed by means of a vacuum-encased soldering iron operated from without the apparatus.

For samples of which suitably-shaped cylinders or blocks of material may be obtained, the calorimeter vessel may be largely dispensed with, as in the well-known method of Keesom and his collaborators[70], who used a heater-thermometer assembly which screws directly into the specimen. For ionic crystals, Webb and Wilks[135] have devised an arrangement in which heater and thermometer are separately mounted on rings clamped on opposite

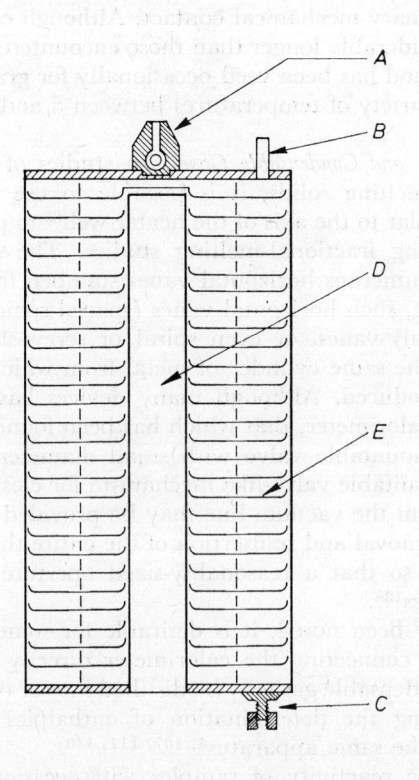

Figure 9. Calorimeter vessel.
    A, Thermal-contact cone
    B, Tube for filling and emptying
    C, Screw and nut to attach thermocouple
    D, Re-entrant well for thermometer-heater unit
    E, Perforated disks and spacers (somewhat schematic).

ends of the specimen. Corak, et al.[21] have favored an arrangement in which the heater and thermometer are pressed as a unit into a tapered hole in the specimen. These methods are obviously of very limited applicability, especially at higher temperatures, since the importance of radiation makes more stringent the need for real adiabaticity. In another instance, a cylindrical section and the ends of an essentially cylindrical calorimetric "vessel" were mounted over the machined cylindrical sample of rare-earth metal to provide suitably reflective coatings in fair contact with the sample.

Another technique which has been employed on occasion is the disc

155

technique, in which a specimen of regular shape and small heat capacity has been mounted on an open disc or dish and brought in thermal contact therewith by the use of a little grease. Because the tray, unlike a cylindrical calorimeter, is not subject to appreciable mechanical forces, its heat capacity can be minimized. This technique has been employed by Martin[79]. In yet another method, a very lightweight open bucket has been used for the sample container, with no provision for heat exchange apart from that provided by ordinary mechanical contact. Although equilibrium times will normally be considerably longer than those encountered when exchange gas is used, this method has been used occasionally for granular and powdered specimens at a variety of temperatures between 5 and 300°K[13, 29, 135].

(2) For Liquids and Condensable Gases. In studies of condensable vapors, liquids, or low-melting solids, it is possible to use circular (horizontal) vanes perpendicular to the axis of the heater well[65] to prevent settling of the solid phase during fractional-melting studies. The calorimeter depicted in *Figure 9* has numerous horizontal vanes stamped from thin noble-metal foil. Alternatively, such horizontal vanes (spaced somewhat further apart), or vertical (radial) vanes, or even spiral or screw-shaped vanes may be machined from the same cylinder of metal from which the heater-thermometer well is produced. Although many devices have been used to seal liquids into the calorimeter, that which has been found to be the most convenient is a demountable valve with small diameter aperture and gold-gasketed seal. A suitable valve-like mechanism for closing this demountable valve directly from the vacuum line may be provided with a suitable snap ring to permit removal and reinsertion of the entire threaded plug carrying the gold gasket, so that a reasonably-sized aperture is available for the distillation process[138].

As has already been noted, it is desirable for some purposes to have a thin-walled tube connecting the calorimeter directly to the vacuum line, from which a condensable gas may be distilled into or out of the calorimeter, thereby permitting the determination of enthalpies of vaporization (or sublimation) in the same apparatus[4, 107, 111, 146].

Frequently, the reactivity of samples will occasion the use of special metals for calorimeter construction, as in the use of nickel calorimeters for transition-metal hexafluorides[138]. A glass-lined copper calorimeter was employed for perchloric acid[132]. Staveley[119] and others[140] have described microcalorimetry vessels suitable for studies of samples smaller than customarily used. By these means the thermal properties of prohibitively-expensive, explosively-dangerous, or difficultly-prepared materials may be determined.

3. *Measurement techniques and adjuvant circuitry*

Adiabatic calorimetry involves a number of electrical measurement and control features. In addition to regulating the electrical power input to the calorimeter and determining the energy input and the resultant temperature increments under adiabatic or quasi-adiabatic conditions by resistance thermometry, the temperatures of the adiabatic shield surfaces and of the electrical lead bundle relative to the temperature of the surface of the calorimeter are controlled by heaters which operate in conjunction with differential thermocouples.

In the *continuous heating* method, the instantaneous derivative of the temperature with respect to time is determined at the time (or temperature) at which the power input is measured and the heat capacity, C_{T_i}, is calculated from the relation

$$C_{T_i} = (dq/dt)_{T_i}/(dT/dt)_{T_i}. \qquad (1)$$

In practice, periodic measurements of temperature and power input are made at prescribed times and the values of $(dT/dt)_{T_i}$ computed. Ideally, the continuous heating method requires instantaneous distribution of heat to the calorimeter vessel and its contents. In endeavoring to fulfill this requirement, the calorimeter vessel is typically designed for small samples of high thermal conductivity or is provided with means of achieving rapid thermal equilibration. Both heater and thermometer are placed in good thermal contact with the calorimeter so that the sample can be heated uniformly and both temperature and power input determined simultaneously. The power input is made small enough to insure that the deviation in temperature, δT, of any part of the calorimetric system from the observed temperature is small enough that $\delta T (dC/dT)_{T_i}$ is negligible[25, 80].

Most laboratories conducting low-temperature calorimetry use the *interrupted heating* method, in which measurements are made of the power and duration of the energy input and the corresponding temperatures before and after heating. The electrical energy supplied to the calorimeter heater is the time integral of the power: $q = \int_0^\tau EI\,dt$, in which E and I are the instantaneous voltage across and current through the heater, and the duration of the heating interval is denoted by τ. In this method, the equilibrium or steady state temperatures of the calorimeter are determined before and after the electrical energy input. Heat interchange corrections computed from the observations of temperatures at known times are applied if significant. (see Section IV-1-B on heat-leak corrections.)

It is essential to remember that the process of determining the thermometer resistance introduces electrical energy to the calorimeter vessel. A 2 mA current flowing through a thermometer with 25 Ω resistance introduces 0.1 mW of power, which corresponds to a change in temperature of 1 $\mu°$C/sec for a calorimetric system of 100 j/°K heat capacity. If temperature observations are made continuously over 20 min, for example, a significant temperature increase from this source of energy should be detectable.

A. Cryogenic Thermometry

(1) Thermometers and Temperature Scales. Although in principle any temperature-dependent, stable, easily-measured property may be used for thermometric purposes, most adiabatic calorimeters use resistance thermometers as basic temperature sensing devices because of their high sensitivity and suitability for temperature-interpolation. Corrections for the energy input of their operation can be made easily in conjunction with the heat leak by observing the temperature drift. (see Section IV-1-B on heat leak corrections.) For the temperature range under discussion in this

chapter, the platinum-resistance thermometer is most practical, although lead (Pb) wire thermometers have been used by Clusius and Vaughen[19] and others[73], and other metallic resistance wires have been proposed or used[26, 47, 105, 144]. Germanium-resistance thermometers[72], for example, may be used up to temperatures of 20°K or higher, depending upon the carrier impurity.

Standard platinum-resistance thermometers[8, 85, 114, 118] may be sent to national standardizing laboratories for calibration on both the IPT Scale and provisional temperature scales. (For discussions on the temperature scales see Chapter 2.) For temperatures below 10°K, individual laboratories may calibrate their resistance thermometers against the helium vapor-pressure scale[11] and augment the range by gas thermometry (relative to temperatures in the hydrogen vapor-pressure range) between 5·22°K, the upper limit of the helium vapor-pressure scale, and 14°K, the lower limit of the hydrogen vapor-pressure scale[7, 14]. The temperature scale of the particular resistance thermometer, calibrated on the above two vapor-pressure scales, may also be used in conjunction with a suitable interpolation equation[93].

A helium-gas thermometer may be used to calibrate the resistance thermometer over the complete range of use. For calibration of a working resistance thermometer, the helium-gas thermometer scale may be maintained and preserved in terms of the e.m.f. of a thermocouple[6, 105] or a resistance thermometer. Alternatively, a platinum-resistance thermometer calibrated at a national standardizing laboratory may be used to calibrate the working thermometers. The calibrated, encapsulated, platinum-resistance thermometer may be placed directly in the calorimeter vessel or used to calibrate another thermometer which is, or will be, incorporated in the calorimeter vessel. A convenient method of installing the encapsulated thermometer is illustrated in *Figure 10*. The thermometer is shown with the heater coil (discussed in Section II–3–B). The thermometer may be installed in the well with solder or grease such as Apiezon-T. The solder or grease should preferably not have any phase transitions in the temperature range of use. The well should fit closely with the thermometer case; therefore, a vent should be provided to avoid entrapment of air during installation or the installation should be done in vacuum. To avoid possible damage to the thermometer (e.g., if the thermometer requires frequent removal from the calorimeter) it can be mounted in a secondary case, such as shown in *Figure 10*, which can be inserted in a close-fitting, preferably slightly tapered, well of the calorimeter. A thin film of grease should be used and the outer surface of the secondary case should be grooved along its length to provide a vent. Alternatively, the thermometer may be made an integral part of the calorimeter vessel. The platinum thermometer coil may be sealed directly into the re-entrant well of the calorimeter vessel to improve the thermal contact between the calorimeter vessel and the thermometer. Westrum, Hatcher, and Osborne[140] have designed a semi-microcalorimeter vessel in which the platinum thermometer wire is wound on a mica form in a narrow outer annular space filled with helium gas. Dauphinee, MacDonald, and Preston-Thomas[25] and Martin[80] have used copper wire thermometers wound on cylindrical copper calorimeters of about 8 cm^3 capacity. The temperature scale and calibration of such thermometers have been described by Dauphinee and Preston-Thomas[26]. A semi-micro calorimeter in use at the National Bureau of Standards has a calibrated thermometer in the adiabatic shield system so that thermometers attached to the calorimeter vessel are calibrated during the course of the calorimetric measurements[40].

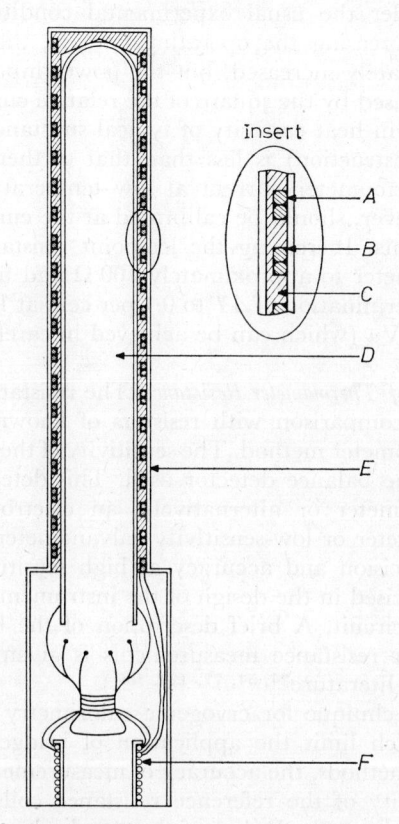

Figure 10. Thermometer-heater assembly.
- A, Heater wire
- B, Solder
- C, Copper mandrel with helical grooves to accommodate heater wire
- D, Capsule-type platinum resistance thermometer
- E, Re-entrant well of calorimeter vessel or secondary case designed to be inserted into the re-entrant well in good thermal contact
- F, Ring for tempering the leads

At 10°K, the dR/dT of a standard platinum-resistance thermometer is about 3 per cent of that at the ice-point temperature. This decrease in sensitivity arises principally from more than a thousand-fold decrease in the resistance, because the temperature coefficient of the logarithm of the resistance, $(dR/dT)/R$, of platinum is actually larger at 10°K than at 273°K (0·12/°K at 10°K and 0·004/°K at 273°K). Commercially-available standard, platinum-resistance thermometers have an ice-point resistance, $R_{0°C}$, of about 25·5 Ω, and a dR/dT of 0·1 Ω/°K at 273°K and 0·0027 Ω/°K at 10°K. The accuracy with which ΔT is determined is related directly to the accuracy with which ΔR is measured. For example, in order to determine a ΔT of $1 \pm 0.001°$ at $10°K$, a ΔR of $0.0027\ \Omega$ must be determined within $2.7\ \mu\Omega$, corresponding to about $0.005\ \mu V$ for a 2 mA thermometer current. At present the best galvanometer systems with optical and/or the best electronic amplification barely detect $0.005\ \mu V$ imbalance with certainty under the usual experimental conditions associated with calorimetry. By increasing the operating current, the available potential may be proportionately increased, but the power input to the calorimeter vessel will be increased by the square of the relative current ratio. Since the fractional decrease in heat capacity of typical substances (and of materials for calorimeter construction) is less than that of thermometer sensitivity, an increase of thermometer current at low temperatures is feasible. The thermometer, however, should be calibrated at the current to be employed in the measurements. Increasing the ice-point resistance of the platinum resistance thermometer to approximately $100\ \Omega$ and using a 2 mA current will permit the determination of ΔT to 0.1 per cent at $10°K$ upon the detection of about $0.02\ V\mu$ (which can be achieved in careful work).

(2) Measurement of Thermometer Resistance. The resistance of a thermometer is determined by comparison with resistors of known values by either a bridge or a potentiometer method. The sensitivity of the method is ultimately dependent upon the balance detector used. This detector may be a high-sensitivity galvanometer or alternatively, an electronic or photoelectric amplifier with a meter or low-sensitivity galvanometer for output observation. Both the precision and accuracy at high sensitivities are dependent upon the skill exercised in the design of the instrument and arrangement of the measurement circuit. A brief description of the bridge and potentiometric methods for resistance measurements is given in Chapter 2. (For details, consult the literature[24, 54, 77, 129, 136].)

The choice of technique for cryogenic calorimetry depends on the following factors which limit the application of bridge and potentiometric methods. In both methods, the accuracy of measurement is dependent upon the absolute stability of the reference resistance coils in the instrument; in the potentiometric method, that of the standard resistor also is relevant. In precision instruments of both types, the stability of the decade resistors is comparable to those of primary standard resistors. As long as the above stability requirements are met, internally consistent calibration alone is adequate for the determination of temperature of a platinum-resistance thermometer or of any other stable temperature-sensitive resistance device.

The two instruments differ with respect to the constancy of the measuring current required. In the bridge method, variations in the current too small to change the heating effect of the thermometers may be tolerated. In the potentiometric method, however, both the working current and the measuring current must remain constant during all measurements, or the variation of the currents with time must be determined accurately. In principle, for both methods, the *IR* voltage in the resistors is compared by means of a voltage sensitive detector. Therefore the "load" conditions during measurement should be made as small as practical and should be carefully assessed when comparing the absolute accuracy of the resistances measured. The measurement procedure should include steps to eliminate or cancel non-IR voltages. The IR voltage in the measurement circuit must also be independent of the current direction. Therefore, all connections should be carefully made with good metal-to-metal contact to avoid rectification effects.

Unless the overall current constancy can be maintained higher than one part in 10^6, the potentiometer will not yield accurate measurements beyond the six-digit range. In a series of papers, Dauphinee and colleagues[23, 27] described an isolating potential comparator (see Chapter 2) for potentiometric comparison of a four-terminal variable resistor and the thermometer. The IR voltages are compared by rapidly switching high grade capacitors between them while adjusting the "potentiometer resistors" to balance. In this method, the balance is independent of the current as in the case of the Mueller bridge method[89].

During precision measurements of resistance in a four-lead thermometer, any non-IR voltage in the measurement circuit is a source of error in the potentiometer method. Current reversal techniques will eliminate this source of error[24]. In the bridge method, any change in resistance of the potential leads between the "normal" and "reverse" readings causes errors. This latter effect cannot be detected without additional measurements.

When the same ultimate detector sensitivity is available, and the current-stability requirement is met in the potentiometric method, the number of decades or significant digits covered by the potentiometer and bridge can be the same. For example, with a 2 mA current in the thermometer and an ultimate detector system sensitivity of $0 \cdot 02$ μV, the corresponding detectable δR is 10 $\mu \Omega$. A 6-decade potentiometer with lowest steps of $0 \cdot 01$ μV can be used to determine the resistance of almost 5 Ω to 6 digits. The G-3 Mueller bridge[89] can be used to make the same measurement. Although the two instruments yield the same number of significant digits under the above conditions of thermometer current and detector sensitivity, the bridge method is at a disadvantage because the leads are in the bridge circuit and during the measurements the resistance of the potential leads must be very stable. Although the resistance of the platinum thermometer is about $25 \cdot 5$ Ω at 273°K, it is only about $0 \cdot 02$ Ω at 10°K, and that of the leads is often at least 2 Ω. With both instruments, the number of significant digits can be increased at extremely low resistances by increasing the thermometer current (to a limited extent) and thereby causing greater voltage change for a given δR.

The potentiometer can conveniently be used at resistances above the range of a typical Mueller bridge since the IR voltages can be adapted to the range of the potentiometer by adjusting the current and using a detector for circuits of high resistance. In order to obtain maximum accuracy in the measurement, the standard resistor should be of comparable resistance to the unknown resistor. However, a Mueller bridge may be adapted for operation over an extended range upon suitable circuit modification.

The Mueller bridge and other bridge methods are superior in the resistance range in which the relative stability of the resistors exceeds that of the current. The bridge method is limited by the detector sensitivity corresponding to

δR which, as mentioned earlier, is dependent upon the characteristics of the detector and on the load (total resistance of the circuit).

One of the advantages of using the potentiometer in calorimetry involving electrical calibration is that the instrument can also be used for measurement of power. Since, for temperature determination, the potentiometer should permit measurement of lowest practical voltage, a range selector for measuring the highest voltage to be encountered in power measurement is a desideratum. In most cases this requirement is not met, so a voltage divider is employed. Any non-IR voltages in the potential leads between heater and voltage divider should, however, be a negligible fraction of the whole, and potential leads from voltage divider to potentiometer should be free of non-IR voltages.

B. Energy Input and Measurement

(1) Calorimeter Heaters. The heater assembly is designed to distribute the generated heat as rapidly as possible to the calorimeter vessel, which is, in turn, designed to distribute the heat to the sample. The heater resistance wire should preferably have a small temperature coefficient of resistance, as do, e.g., Manganin, Constantan, Evanohm, and Karma[45]. The wire is wound bifilarly so that reactance effects are minimized during the transient periods at the beginning and end of a heating period.

Figure 10 shows a thermometer-heater-assembly design in which the heater wire, insulated with Fiberglass, is wound tightly in helical grooves on a copper mandrel. This assembly is soldered in the closely-fitting central well of the calorimeter vessel or within the removable secondary case as described in Section II–3–A. The platinum-resistance thermometer is soldered in the mandrel. The four thermometer leads and the two heater leads are wound on the tempering ring as shown for equilibration with the exterior temperature of the calorimeter.

Another design uses the same mica cross for the heater wire and the platinum-resistance thermometer[65]. This assembly is encapsulated with a small amount of helium gas in a platinum tube and sealed with soft glass in a manner similar to that described for platinum thermometers. This capsule is soldered within the close-fitting tube of the calorimeter vessel. Westrum, Hatcher, and Osborne[140] designed a heater which was wound on mica forms and placed in an outer, annular space filled with helium gas. Alternatively, copper wire wound on the cylindrical copper vessels as described by Dauphinee, *et al.*[25] and by Martin[80] serves simultaneously as thermometer and heater.

The size and resistance of the heater should be selected to meet maximum power requirements. Most measurements made in low-temperature calorimetry involve samples of 100 cm^3 or less, and the upper limit of power required is about 5 watts. At this power level, the heat capacity of the calorimeter system could be as large as 500 j/°K at a heating rate of 1 deg/min. Higher rates tend to require an increased increment in power for the control of the adiabatic shield between drift and heating periods. Furthermore, the possibility of increase in "oscillation" or "hunting" in the control action will be greater.

The resistance of the heater should also be large enough to insure that the power developed in the leads is a small fraction of that developed in the heater resistor. The upper limit of resistance is dictated by the power requirement and the voltage available from the power supply. Because it is necessary to achieve optimum thermal contact between the leads and metallic surfaces (which serve as intermediaries in controlling the temperature of the leads), electrical insulation of acceptable thermal conductivity is not suitable for very high voltage. The relative voltage between the leads and the metallic surface should not exceed about 100 V. When the power supply is "floating", the insulation between the leads limits the voltage. High resistance and consequent high voltage in the heater require a voltage divider of high resistance in order to maintain a relatively low current flow in the voltage divider. This may increase the resistance of the volt-box tap and hence increase the resistance of the potentiometer circuit and therefore decrease the sensitivity of the voltage measurements. A heater resistor of 100 to 300 Ω is generally used in low-temperature calorimetry. A voltage divider of 10^4 to 10^5 Ω may be used in conjunction with potentiometers having upper voltage limits of approximately 0·2 or 2 V. (For details regarding voltage dividers, cf. refs. 34 and 113.)

(2) *Electrical Energy Measurements.* A general circuit arrangement for the introduction and measurement of electrical power to the calorimeter is shown in *Figure 11*, and a simplified calorimeter heater circuit during electrical energy input is shown in *Figure 12*. When switch SW_1 is in position 2, the current from the battery, BA, made variable by means of a potential divider, is circulated through the dummy heater R_D (of resistance equal to that of the calorimeter heater) in order to maintain a constant drain on the battery. When both switches SW_1 and SW_2 of *Figure 11* are in position 1, the battery current flows through the calorimeter heater.

The resistance of Constantan wire increases by about 5 per cent over the range 10 to 100°K and an additional one per cent on going to 400°K. Manganin wire increases in resistance by over 9 per cent from 10 to 200°K but by less than one per cent on going to 400°K. However, Karma and Evanohm wires change in resistance only about 0·9 per cent from 10 to 400°K[45]. Since the battery potential decreases with time during discharge, the potential difference across the heater changes relatively less than the current through the heater. In heat-capacity measurements, if the temperature interval of heating is at most 10° and the time of heating is from 10–20 min, the voltage variation across the calorimeter heater is virtually linear with time.

Standard precision potentiometers cannot measure directly the potential difference across the heater. The simplified circuit given in *Figure 12* shows the application of a potential divider or volt box consisting of two resistors, R_E and R_2. Over the range 10 to 400°K, the potential difference across the heater, R_H, may be changed as much as 20-fold. To utilize the maximum precision of the available potentiometer, suitable alternate resistors for R_S, R_E, and R_2 should be provided for different temperature ranges so that all of the decades of the potentiometer are employed whenever possible. Obviously, selection of the resistors should be compatible with achievement of maximum sensitivity of the galvanometer or other balance detector.

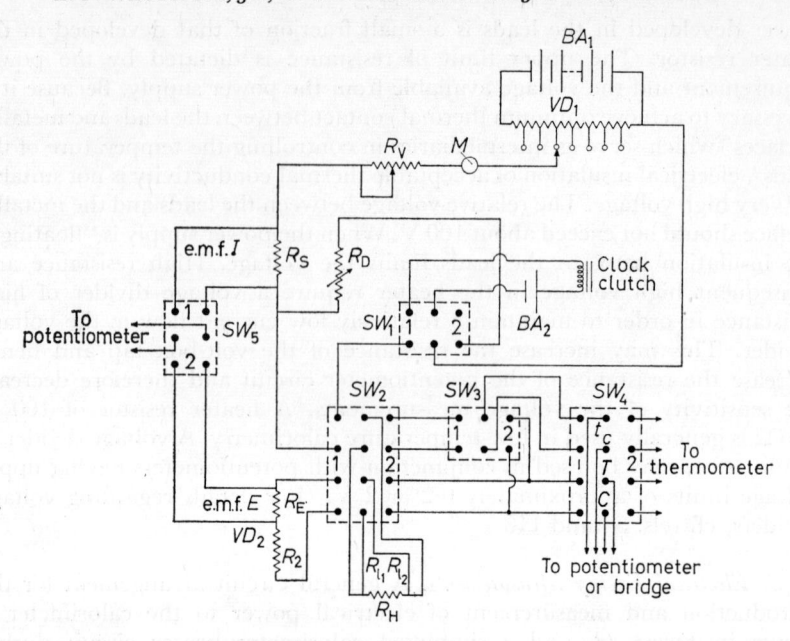

Figure 11. General circuit arrangement for measurement of electrical power to calorimeter:

BA_1, Batteries
VD_1, Voltage divider
M, Ammeter
R_V, Variable resistor for adjusting current
R_D, Dummy heater
R_S, Standard resistor for determining the calorimeter heater current
BA_2, d.c. supply for controlling the clock clutch
R_H, Calorimeter heater
R_{l_1} and R_{l_2}, Potential leads to calorimeter heater R_H
VD_2, Volt box with the divider sections R_E and R_2
SW_1, Switch for controlling the clock and the current to the calorimeter heater
SW_2, When this switch is in position 1 the current from SW_1 flows through the calorimeter heater. Position 2 is used in conjunction with switches SW_3 and SW_4 to determine the resistances R_H and $R_H + R_{l_1} + R_{l_2}$. With SW_4 in position 1, position 1 of SW_3 yields R_H and position 2 of SW_3 yields $R_H + R_{l_1} + R_{l_2}$. All leads connecting VD_2, SW_2, SW_3 and SW_4 have negligible resistance.
SW_5, Switch for selecting e.m.f. I and e.m.f. E

At 10°K, the heater current may be 10 mA or less; therefore, for a heater resistance of about 100 Ω, the potential difference can be measured directly. For convenience and maximum accuracy, the standard resistor, R_S, should also be 100 Ω at this temperature. The power requirement for a reasonable heating rate (i.e., one deg/min), however, increases rapidly with the temperature below *ca.* 100°K for most substances. Thus, in the region in which the heat capacity changes rapidly, a compromise must be made between the highest voltage measurement possible and the number of sets of resistors needed for power measurement. A switching circuit arrangement permitting selection of different sets of R_S and $R_E - R_2$ combinations

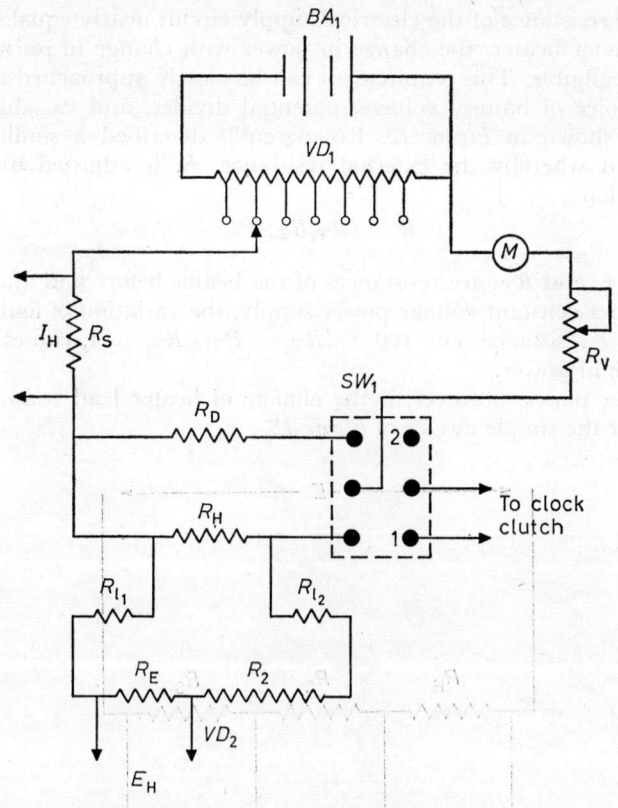

Figure 12. Simplified calorimeter heater circuit during electrical energy input. The figure symbols have the same significance as those of *Figure 11*.

corresponding to different current levels for a given heater resistance is convenient.

When a potential divider is used, the following considerations must be applied in the determination of the actual potential and current associated with the heater from the measurements of the potential difference across R_S and R_E (*Figure 12*). In addition to the resistors R_E and R_2, the resistances of the potential leads, R_{l_1} and R_{l_2}, from the heater are a part of the potential divider. The resistance of the leads, $R_l = (R_{l_1} + R_{l_2})$, which is dependent upon the length and size of the copper wire used, is determined as a function of temperature over the measurement range. Depending upon the method used for calculating the power, the resistance of the heater as a function of the temperature may also have to be determined.

Transients in the heater current are not entirely eliminated by use of the dummy resistor. When the switch to the calorimeter heater is closed, there may be initial rapid changes in the current and in the potential difference across the heater because its resistance is changed by the sharp rise in temperature. During this period accurate, rapid measurements of the current and potential difference cannot be made. Hoge[62] has shown that by making the resistance of the electrical supply circuit nearly equal to that of the calorimeter heater, the change in power with change in resistance can be made negligible. This requirement can be closely approached through a suitable choice of battery voltage, potential divider, and variable resistor located as shown in *Figure 12*. Rosengren[103] described a similar circuit arrangement whereby the external resistance, R, is adjusted to conform to the relation

$$R = (R_{T_1}R_{T_2})^{1/2}, \qquad (2)$$

in which R_{T_1} and R_{T_2} are resistances of the heater before and after energy input. With a constant voltage power supply, the variation of four per cent in the heater resistance, i.e., $100 \times (R_{T_2} - R_{T_1})/R_{T_1} = 4$, causes 0·01 per cent change in power.

In neither paper, however, is the change of heater lead resistance considered. For the simple circuit of *Figure 13*,

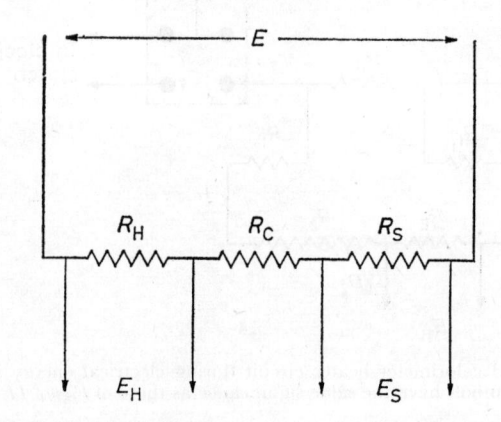

Figure 13. Schematic heater circuit.
E, d.c. power supply voltage
R_H, Calorimeter heater resistance with corresponding voltage drop E_H
R_C, Current lead resistance
R_S, Standard resistor with corresponding voltage drop E_S

$$I = E/(R_H + R_C + R_S); \qquad (3)$$

$$P = R_H E^2/(R_H + R_C + R_S)^2; \qquad (4)$$

$$\frac{dP}{P} = [1 - 2R_H/(R_H + R_C + R_S)](dR_H/R_H) - 2\,dR_C/(R_H + R_C + R_S); \qquad (5)$$

in which E = constant voltage supply, R_H = heater resistance, R_C = current lead resistance, R_S = standard resistor or other resistor, and P = power. Obviously, if $R_C = 0$ (consequently $dR_C = 0$) and $2R_H = R_H + R_S$, Hoge's[62] condition is met. Rosengren's[103] description is a variation of this. In low-temperature calorimetry, R_C is changed when the adiabatic shield temperature is changed. By selecting $2R_H$ to closely approach $(R_H + R_C + R_S)$ and having R_H and R_S sufficiently large, dP/P can be made small.

Constant current power supplies now commercially available have better than 0·001 per cent stability over more than the usual heating interval and are ideally suited for application with resistance heater wires of low temperature coefficient, such as Karma, Evanohm, and other Ni(75)–Cr(20) type alloys[45]. Most of the effort for determining the heater power can be concentrated on measuring the voltage across the heater with only occasional measurements across the standard resistor for determination of current. At a heating rate of one °K/min, the voltage change across the heaters constructed from the above wires averages about 0·002 to 0·003 per cent/min, which can be conveniently followed. Most models of constant current power supplies have transient response times that are well below 10^{-3} sec. Therefore, with properly designed heaters of the above alloy wires and for heating intervals of the usual length, the effect of the initial thermal transient just after the heater switch is turned on is negligible[152].

(3) Electrical Standards. The electrical measurements are based on the standard resistors and standard cells calibrated in terms of the national standards (see Chapter 3 for details).

(4) Time Standards. The Twelfth General Conference of Weights and Measures authorized in 1964 the atomic definition of the second: "The standard to be employed is the transition between the two hyperfine levels $F = 4$, $M_F = 0$ and $F = 3$, $M_F = 0$ of the fundamental $^2S_{1/2}$ of the atom of cesium 133 undisturbed by external fields and the value 9 192 631 770 Hz is assigned"[151]. (see Chapter 3.) Standard radio frequencies broadcast in the United States by the National Bureau of Standards over stations WWV, WWVB, and WWVL (Fort Collins, Colorado) and WWVH (Maui, Hawaii), are maintained by means of quartz oscillators relative to the cesium frequency. (For information on time standards consult the refs. 9, 32, 78, 149–151.)

For measurement of time, an accurate clock or a highly stable oscillator which can provide known reference frequencies is compared with the standard radio broadcasts. If no precision oscillator is available and the much less desirable expedient of using the local 60-Hertz power for running the synchronous clock is used, the variation from standard time may be obtained from the local power company. Any adjustment that is needed by the local power company is usually done at odd hours. The interval times also may be checked directly with standard radio broadcasts.

(5) Determination of Duration of Heating. The duration of a heating period may be determined by a variety of methods. In the DPDT switch, SW_1, shown in *Figure 11*, one of the poles closes or opens the circuit to the calorimeter heater and the other pole closes or opens a relay circuit of a clock clutch which in turn engages or disengages a synchronous motor and the movement of a clock. The clock should permit reading of the time interval to the nearest 0·01 sec. The switch, SW_1, may be replaced by a fast action latching relay or activated by electric pulses. Second "ticks" or electric pulses spaced by accurately known time intervals may be used with the above relay. The calorimeter circuit is closed by pulsing the latching coil

of the relay, and, after a lapse of a desired number of pulses, a pulse is directed to the unlatching coil of the relay. The heating period will then be an integral number of accurately known pulse intervals. An electronic counter of high input impedance may be connected directly across the calorimeter heater. When the calorimeter heater circuit is closed, the potential developed across the heater initiates the counting. The counter is operated in conjunction with an oscillator of precisely known frequency. When the calorimeter switch is opened, the drop in the potential stops the counter. However, at low currents the potential developed across the calorimeter heater may not be adequate to initiate the counter. In that case, the potential developed in a larger external resistor in series with the calorimeter heater may be used instead to drive the counter.

C. Shield Control

The circuitry used for sensing the temperature differential between the calorimeter and adiabatic shield and for controlling the electrical energy to the shield will vary considerably from one cryostat to another. Although manual control by a skilled operator is feasible and often employed, the full advantage of the adiabatic method can be conveniently realized only with automatic, electronic control so that a true equilibrium state may be approached even over a period of many hours without undue strain on the operators. As indicated in Section II–1–G, thermocouples are commonly used to detect temperature differences, and the main requirements for manual control are those of adequate sensitivity and stability with relatively small thermal lag for the thermocouples and shield heaters. To minimize the heat leak to the calorimeter, it is often convenient to refer temperature increments of various parts of the shield and the ring to that of the most important control point on the shield rather than to the calorimeter. Either a.c. or d.c. is employed for shield control, and if rectifiers are employed to attain the latter choice, similar control circuits may be employed. Typically, parallel channels consisting of one or more autotransformers (to select the power level appropriate to a given temperature range), a potential divider step switch (to provide surges of power), and a fine continuously-variable, multiple-turn resistor (for fine adjustment) are convenient. Bifilar winding is, of course, appropriate for the heaters.

(*1*) *Automatic Adiabatic Shield Control.* Automatic shield controls have been employed by several authors and notably by Zabetakis, *et al.*[147], who developed an ingenious optico-mechanical device, and by Furukawa[41] who adapted commercial electronic circuits. Much of the present development in the United States stems from discussions at early meetings of the Calorimetry Conference.

An automatic adiabatic shield control system must be capable of rapid response in supplying a broad range of electric power to the shield system in order to maintain adiabatic conditions. In principle, the voltage of the thermocouples (sensed as temperature difference) must be suitably amplified and the corresponding electric energy applied to the shield heaters to minimize the temperature difference. When the operating temperature of the calorimeter is close to that of the cryostat, essentially zero power is needed in the adiabatic shield. With increasing operating temperature of the calorimeter, the power requirement increases. In addition to the power needed to offset the heat leak from the adiabatic shield to the cryostat, the control system must quickly increase the power necessary to follow the temperature rise of the calorimeter vessel during the heating interval, reduce the power when the heating is terminated, and provide power at an increased level to offset the increased heat leak at the new higher equilibration temperature. The demand on the control system is most severe during the short periods immediately following the beginning and the end of a heating period, when the largest and most rapid changes in the power level are required. Thermal lags in the thermocouples and in the shield heaters add to the problems and demands made on the shield control system.

A typical automatic control system that has been applied successfully in a number of laboratories engaged in low-temperature calorimetry is shown schematically in *Figure 14*. The thermocouple shown is usually a

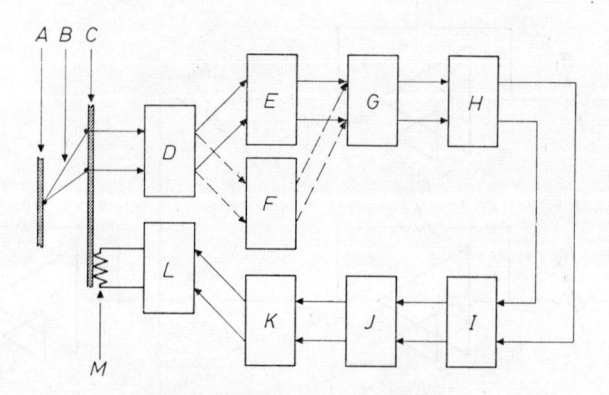

Figure 14. Block diagram of adiabatic shield control system.

　　　　　　　　　　A, Calorimeter vessel surface
　　　　　　　　　　B, Thermocouple
　　　　　　　　　　C, Adiabatic shield
　　　　　　　　　　D, d.c. pre-amplifier
　　　　　　　　　　E, Strip-chart recorder
　　　　　　　　　　F, Intermediate amplifier
　　　　　　　　　　G, Three action control (control)
　　　　　　　　　　H, d.c. to a.c. converter
　　　　　　　　　　I, Transformer
　　　　　　　　　　J, Audio range "Hi Fi" amplifier
　　　　　　　　　　K, Isolating and step-up transformer
　　　　　　　　　　L, Rectifier and a.c. filter
　　　　　　　　　　M, Shield heater

multiple-junction thermopile. Moreover, several such channels are used for independent control of the various portions of the adiabatic shield.

Under the steady-state conditions of an equilibration period, the input power to the adiabatic shield system may be adjusted to balance exactly the heat leak from the shield to the cryostat to obtain a zero temperature imbalance. After heating to a higher temperature, a new input power level to the shield must be found to balance the new heat leak condition. Thus, an automatic reset action is needed to meet this mode of shield control.

With automatic reset action, the input power level to the shield heaters will change continuously as long as a temperature difference exists and will settle to a constant level only when zero temperature imbalance is achieved. Reset action is an "integral action" because the reset effect depends on the time integral of the temperature imbalance. Thus, this action may be obtained by any control system whose output is a time integral of the temperature deviation.

There are many methods by which the reset (integrating) action can be achieved. One method, which is most instructive, may be illustrated by the use of an electronic integrating circuit with an operational amplifier as shown schematically in *Figure 15*. Details may be found in any text book on analog computers[66, 88].

The "integrating factor" or the "ramp" for a constant input voltage

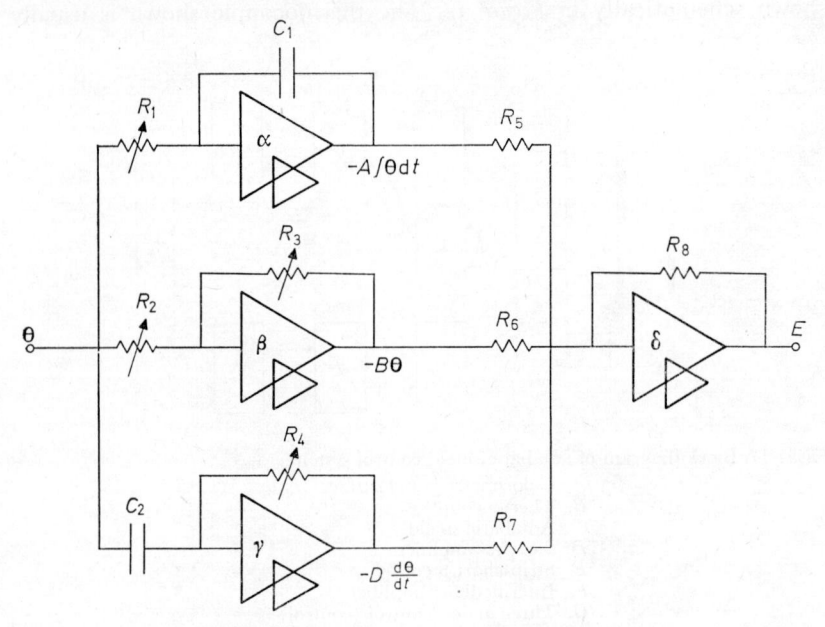

Figure 15. Idealized circuit for shield control, showing combined reset, proportional, and rate control actions.
$A = 1/R_1C_1$, $B = R_3/R_2$, $D = R_4C_2$, and $R_5 = R_6 = R_7 = R_8$. Amplifiers α, β, γ, and δ are very high gain; δ is a non-inverting amplifier.

(temperature imbalance) corresponds to the reset rate and is controlled by setting the input resistance to the desired value.

With proportional control action, the input power to the shield can be made proportional to the temperature imbalance. An operational amplifier with suitable resistors can be used to illustrate this action[66, 88] (cf. *Figure 15*). The proportionality "factor" or the proportional "band", which corresponds to the relation between temperature imbalance and maximum input power, is controlled by setting the input or feedback resistance to the desired value.

In proportional control, however, there is one input power level corresponding to each value of temperature imbalance. Therefore, when the input power requirement changes, e.g., for a change to a higher equilibrium temperature, an offset from zero temperature imbalance will occur with proportional control action alone. Reset action, described earlier, reinforces the proportional action by removing this offset.

Proportional action reacts immediately to the temperature imbalance, whereas reset action reacts with the time integral of the imbalance. The effect of these control actions is dependent only on the magnitude of the existing temperature imbalance. When rapid changes in the temperature of the calorimeter vessel occur immediately after the beginning and end of a heating period, rate control action corresponding to the time derivative of the temperature imbalance helps to anticipate the desired input power level in the adiabatic shield. The idealized differentiating circuit illustrated in *Figure 15* with the operational amplifier would provide this control action[66,88], which is effective only when changes in the temperature imbalance occur. When the rate action is too great, it approaches on-off type control and causes "hunting". Therefore, the rate factor (RC) should be suitably set to avoid this effect.

The combined reset, proportional, and rate control actions shown in *Figure 15* may be represented mathematically as

$$E = -A \int \Theta \, dt - B\Theta - D(d\Theta/dt), \qquad (6)$$

in which t is time and E is the voltage corresponding to the combined effects of the three control actions or the output from the three operational amplifiers appropriately wired with resistors and capacitors to yield the desired control action. The constants A, B, and D are dependent upon the values of the resistors and capacitors selected. The thermocouple voltage is amplified by a highly stable and sensitive d.c. pre-amplifier whose output Θ is appropriately directed to the control units. The effects of the three control actions may be summed according to Equation (6) by means of a summing amplifier[66,88]. For the purposes of illustration, the input and feedback resistors of the summing amplifier have been made equal. The output of the summing amplifier must be appropriately amplified to a power level sufficient to heat the adiabatic shields.

When the voltage E is zero, the output from the power amplifier must be zero. The simplest method for meeting this requirement is to use an a.c. power amplifier whose output is subsequently either rectified and smoothed to provide d.c. power for the shield heaters, or whose a.c. power is used directly. The voltage, E, is converted to a.c. by means of a "chopper" (60-Hertz) and directed to the a.c. power amplifier. A transformer between the chopper and the a.c. power amplifier could improve the general performance of the system. Since the a.c. is in the low audio-frequency range, the usual audio amplifier with good fidelity in the low frequency range is suitable for the application. The output voltage of such an amplifier may be increased before rectification by a suitable step-up transformer and, at the same time (if not previously), "isolated" from the input.

Several useful d.c. pre-amplifiers of high stability and sensitivity are now commercially available for amplifying the thermocouple voltage. Since they are operated at zero or very close to zero voltage (zero temperature imbalance), high linearity of the output is not a serious requirement. Because of the high sensitivity requirement, the pre-amplifier system must attenuate the extraneous voltages that may occur in the input.

The output from the pre-amplifier may be connected to a strip-chart recorder to record the quality of control and, if necessary, to make heat leak corrections (see Section IV–1–B). Often the deviations from the desired control conditions, which occur particularly during the critical period immediately after the beginning and end of a heating period, balance out or differ only negligibly from equality.

The three-action control (henceforth to be referred to in this section as the control) may be connected to the pre-amplifier in parallel with the recorder with an intermediate stage of amplification, if needed.

Many controls somewhat more condensed than that illustrated in *Figure 15* are now commercially available; these controls are designed to operate in conjunction with an auxiliary transmitting slide wire coupled to the action of the recorder. (In a chosen commercial control, the corresponding output voltage E should be zero when the input Θ is zero.) The slide wire is a part of a bridge, and the bridge imbalance corresponds to the temperature imbalance sensed by the thermocouples. The power for the bridge is generally supplied by the control. The recorder serves, therefore, as an electro-mechanical linkage between the pre-amplifier and the control and is provided with the accessories needed to offset the control point from the apparent zero temperature imbalance. If the control is connected directly to the pre-amplifier without the recorder linkage, a source of bias voltage is required to provide any necessary offset. Particulars regarding any control components should be available from the manufacturers.

The voltage of the control thermocouples is directly related to the power requirements needed to maintain adiabatic conditions. Since the current or the voltage impressed on the adiabatic shield heaters by the control system described above is directly related to the thermocouple voltage, the input power would, therefore, be related to the square of the thermocouple voltage. Although the square root of the voltage E (Equation 6) may be obtained by an appropriate circuit, the electronic control action is so rapid and the thermal lags in the calorimeter are (relatively) so large, that experience in actual application has shown no need for the "square root" circuit.

With such circuits, adiabatic regulation over the entire low temperature range is accurate to within a few thousandths of a degree over the entire operational sequence, except for a few seconds during the initiation and end of the energy input during which deviations (largely compensating) occur.

Cruickshank[22] has presented an analysis of an approximate method of computing the 'average' behavior of two-valued response regulators applied both to externally-compensated adiabatic calorimeters and to facsimile compensated adiabatic calorimeters. These predictions have been compared with experimental results from an externally compensated adiabatic calorimeter and from various thermostats.

III. Procedures

The procedures used for adiabatic low-temperature calorimetry vary in detail from laboratory to laboratory because the basic requirements of accurate calorimetry often can be met in alternative ways. Apart from those due chiefly to personal preference, some variations in method are necessary because of the nature of the substances being investigated in a particular laboratory. The following general procedures are composites of practices developed especially in the last 30 years. Important variations dictated by the characteristics of materials studied are also described.

1. *Preliminary procedures*

A. CALORIMETER FILLING

The calorimeter containing the sample to be investigated must be filled in such a way that contamination of the sample is avoided and accurate determination of the sample mass is possible. When the calorimeter is attached to a fixed filling tube within the cryostat, samples are distilled into it from an external reservoir, as described by Scott et al.[111] and by Aston and Eidinoff[2]. The amount of material in the calorimeter is determined from the loss in weight of the reservoir. This method can be used only if the material studied is volatile enough. Demountable calorimeters which can be filled with a sample and sealed outside the cryostat system are more convenient for solids and for liquids of low volatility.

B. PREPARATION OF CRYOSTATS

One advantage of systems with calorimeters attached to fixed filling tubes (*Figures 5 and 6*) is that the cryostat is not disturbed between investigations and therefore requires little preparation. Systems with demountable calorimeters (*Figures 2, 3 and 4*) must be partially disassembled to install or remove one. Before the calorimeter is installed, the heat-transfer cones must be inspected and cleaned, if necessary, because perfectly matching, clean surfaces are essential for rapid cooling at low temperatures. The thermometer and heater leads are connected by soldering. Care must be taken during measurement to duplicate the amount of solder used during calibration. Before the radiation shields are assembled, the electrical circuits are checked, and the position of the calorimeter with respect to the windlass indicator is noted. When assembly of the cryostat is completed, the system is evacuated. Because the interior parts of the cryostat are exposed to air, the cryostat should be outgassed by heating the calorimeter and adiabatic shield to the highest expected operating temperature. After outgassing (at the highest expected operating temperature) dry helium gas only is used as exchange gas with the Dewar-type calorimeter. The system is then ready for use.

2. *Sequence of experiments*

In the interrupted heating method most commonly used, all measurements involve determination of an enthalpy increment and corresponding temperature increment. One measurement thus consists of three observations: initial steady state temperature, electrical energy input, and final steady state temperature. In a sequence of measurements, the observations of the final temperature of one determination will define the initial temperature of the following determination, and so on. (See Section III–4–A for details.)

A. Preparations for Measurements

The calorimeter is usually brought to the temperature desired for the start of a series of observations by cooling. A variety of refrigerant combinations, such as liquid helium, liquid and solid hydrogen, liquid and solid nitrogen, solid carbon dioxide, ice, and water[111], may be used to achieve the desired operating temperature range. However, two refrigerants are enough for operating in the range 4–400°K in the system similar to that in *Figure 2*. Between about 4 and 50°K, solid nitrogen is used in the upper refrigerant tank, and either liquid helium or solid and liquid hydrogen is used in the lower tank. Between about 50 and 90°K, solid nitrogen is used in the lower tank and either liquid or solid nitrogen is used in the upper tank. From about 90 to 300°K, liquid nitrogen is used in both tanks, although the lower tank may be left empty for measurements just below 300°K. At higher temperatures, liquid nitrogen is used in the upper tank only or, especially for prolonged measurements at high temperatures, no refrigerant is used. A stream of dry gas may be blown through the lower tank to maintain it at a steady temperature and thereby facilitate control of the adiabatic shield. With the calorimeter of the design shown in *Figure 3*, a suitable refrigerant is placed in the Dewar immediately surrounding the calorimeter. The double Dewar system shown is used when liquid hydrogen is the refrigerant. Liquid hydrogen is placed in the inner Dewar after the system is cooled to the temperature of liquid nitrogen in the outer Dewar. When liquid nitrogen is used to cool the double Dewar system, the inner Dewar system need be just stoppered. The small amount of air pressure is adequate to cool the calorimeter.

B. Order of Measurements

The order of measurements is not important for a substance existing as a single phase in the temperature range of interest. For substances that undergo phase changes, however, the order of measurements may be quite important. The general sequence to be followed will be illustrated with reference to measurements on a substance that is liquid at 300°K.

To avoid wasting time on studies of an impure material, a purity determination should be made as early as possible. The sample is crystallized and the completeness of crystallization checked as described in Section III–3. The enthalpy of fusion and the melting point curve are then determined; from these data the sample purity is calculated. After recrystallization, the heat capacity curve of the solid is determined. Measurements on the solid may be made by first cooling to the lowest temperature, or perhaps only within the liquid nitrogen range, and then making a continuous series of observations up to the melting point. However, if the substance undergoes a solid–solid phase change, this procedure may not allow the transformation to take place completely, and invalid results for partly transposed crystals may be obtained before the transition is discovered. For this reason, the practice of "backing down" the heat capacity curve is often followed. The solid is first cooled only to about 50° below the melting point, for example, and a series of measurements is made to a temperature just below the melting point. The next series is made from an initial temperature perhaps 50° below that of the first series to a final temperature somewhat higher than the starting temperature of the first series, and so on until the entire temperature range is covered. In this way, the heat capacity *vs.* temperature curve is defined in overlapping segments, each successive segment starting at a progressively lower temperature. The temperature range covered by any one segment is a function of the temperature intervals used in the measurements, the rate at which the sample equilibrates after a heating period, and, of course, the length of time the investigator chooses to spend in a continuous series of measurements. This procedure allows a transition to be detected early in the experiments, regardless of how sluggish it may be, and allows the investigator to make certain that transformation to a low-temperature phase is complete before taking more heat capacity data.

Because some solid–solid phase changes do occur fairly quickly, it may be convenient to make a preliminary search for transitions before making heat capacity measurements as described in the preceding paragraph. One method involves observation of the temperature of the calorimeter as it is cooled. A marked diminution in cooling rate indicates that the sample is evolving energy as it transposes to a polymorph that is stable at low temperatures. Alternatively, the temperature may be observed as the calorimeter is heated continuously from a low temperature to near the melting point. Here, a marked diminution in warming rate indicates that the sample is absorbing energy as it transposes to a high-temperature polymorph.

Once the properties of the solid have been determined, the heat capacity of the liquid is measured, usually starting at the melting point and working to progressively higher temperatures, with as many overlapping series as necessary.

3. *Crystallization procedures*

Most substances crystallize without much difficulty, but many do so only after considerable undercooling, and for some, the completely crystalline state may be very difficult to obtain. In fact, some relatively simple organic compounds have yet to be crystallized in the pure state. Crystallization usually may be initiated by cooling the sample a few degrees below the melting point; the onset of crystallization is then evidenced by rapid, sometimes almost instantaneous, warming to the melting point. If the sample is rapidly cooled too far below the melting point, a persistent glass or undercooled liquid may be formed. When crystallization will not begin even 20 or 30° below the melting point, it may be necessary to cool the sample to as low as 80°K to form nuclei, which initiate crystallization when the sample is subsequently warmed to just below the melting point.

After crystallization has started, it is continued by removing heat from the calorimeter by whatever means is most convenient with a particular system. The rate of crystallization varies widely from compound to compound. If the rate is slow, too rapid removal of heat from the sample may cause as much as several per cent of the sample to persist as undercooled liquid or glass. Such incomplete crystallization often can be detected in heat capacity measurements by slow warming during an equilibration period; but the warming may be difficult to detect even when a significant amount of material remains uncrystallized. Even when a substance crystallizes fairly well, it is worthwhile to remelt 10 to 50 per cent of the original solid and recrystallize it slowly, removing heat by radiation to the adiabatic shield, which is maintained a few degrees below the melting point. The rate of recrystallization may be adjusted by varying the temperature difference between shield and calorimeter. After the sample has cooled to about 10° below the melting point, it usually may be cooled further as rapidly as desired. In general, the slower a substance crystallizes initially, the slower should be the rate of recrystallization. The process just described has the added advantage of assuring a well annealed, homogeneous solid phase.

The final test for complete crystallization is in the reproducibility of heat capacity and enthalpy of fusion results from repeated crystallizations. For organic compounds, the time required to obtain satisfactory crystallization varies form an hour or so to as much as several weeks. One of the advantages of the adiabatic method of operation is that it provides the convenient and reliable temperature and heat exchange control that is necessary in difficult crystallizations.

4. *Observational procedures*

A. Heat Capacity Measurements

To begin a series of heat capacity measurements, the calorimeter is first cooled to the desired starting temperature. With the apparatus shown in *Figure 2*, for example, the cones of the lower refrigerant tank, adiabatic shield, and calorimeter are brought into contact to cool the sample rapidly. When the desired temperature is reached, contact is broken and the cooling rate slows abruptly. With the calorimeter design shown in *Figure 3*, helium exchange gas is used to cool the sample rapidly and is pumped out when the desired temperature is reached. The shield and ring are then heated to the temperature of the calorimeter to establish adiabatic conditions and prevent further cooling by radiation and conduction along leads. Under most conditions, satisfactory results can be obtained if the temperature of the shield is kept within about 0·01° of the calorimeter temperature. Closer control can be achieved in well designed systems, but to do so manually requires constant attention. Modern automatic temperature controllers can achieve control within 0·001°. With the same calorimeter system, automatic control results in significantly more precise data than manual control. The temperature of the ring—for tempering the gradient in the lead wires (25 in *Figure 2*) —need not be controlled as closely as that of the shield. Fluctuations in the temperature of the ring should not affect the measurements unless they are so large that the shield cannot be controlled properly. Depending upon the design of the shield and the heat transfer characteristics of the leads, the ring may be operated at the same temperature or a few degrees below the temperature of the shield.

(1) Determination of Temperature and Energy Increments. After steady adiabatic conditions have been established, the temperature of the calorimeter is observed at chosen time intervals. As discussed in a following paragraph, systematic deviations from adiabatic conditions may produce a detectable temperature drift. If necessary, the temperature is observed long enough to determine the drift rate, which is used in calculating the initial temperature of the heat capacity measurement.

Following the determination of the initial temperature, electrical energy is supplied to the calorimeter. Normally, the heating rate should not be greater than 1°C/min and, for convenience in measuring the power, may be less if a small temperature increment is desired. The energy input is determined from measurements of the heater current and potential with a precision potentiometer and measurements of the heating interval with a suitable timing device (See Section II—3—B). The energy input is selected in advance to produce a temperature increment small enough to minimize curvature corrections (discussed in Section IV—2—B). Typical temperature increments are: 10 per cent of the absolute temperature up to 50°K; 5 to 6° from 50°K to about 150°K; and 10° at higher temperatures. Near phase changes or regions of high curvature, much smaller temperature intervals may be necessary, but if the heat capacity curve is nearly linear, larger increments may give satisfactory results.

The temperatures of the adiabatic shield and ring must be controlled to follow that of the calorimeter during a heating period. In well designed control systems, the temperature differences indicated by the thermocouple network can be kept as small during a heating period as during an equilibration period, except for insignificant fluctuations when the calorimeter heater is turned on or off. However, when the calorimeter and shield are heated at a rate of about 1°C/min, it is possible that the temperatures of their surfaces may not be as uniform as they are during an equilibration period. If so, unaccounted for heat exchange would occur during a heating period. This effect should be small in properly designed calorimeter systems, but there is no way of detecting its magnitude quantitatively as there is during an equilibration period. One method of detecting the existence of temperature non-uniformities involves repeating heat capacity measurements with just enough exchange gas in the cryostat to provide an additional heat-exchange mechanism[1].

After a heating period, the temperature of the calorimeter again is followed as a function of time. These observations establish the final temperature for the heat capacity measurement and the initial temperature for the next measurements of the series (cf. Section IV–1–B for details).

(2) Effect of Non-Adiabaticity. It is not possible to construct a calorimetric system that can be treated as truly adiabatic at all times and under all conditions. Regardless of how small an apparent temperature difference is maintained between calorimeter and shield, detectable heat exchange may

result from: (*i*) non-uniform temperature of the shield and calorimeter surfaces, and (*ii*) erroneous thermocouple signals caused by stray e.m.fs from inhomogeneities or strains in the thermocouple leads. Normally, the heat exchange is less than 0·0001 W, which is small enough to be negligible except in studies of substances that have low heat capacities or that reach thermal equilibrium very slowly. If desired, the heat exchange can be made even smaller by maintaining the adiabatic shield at an apparent temperature different from that of the calorimeter. However, if heat exchange is significant, it is usually easier to detect the rate of exchange by observing the temperature drift of the calorimeter and apply an appropriate correction. In any particular calorimetric system, the magnitude of this residual heat exchange varies with operating conditions, but in a reproducible way. (cf. Section IV for details.)

(*3*) *Equilibration Rates.* The time required for a sample to reach thermodynamic equilibrium after a heating period varies greatly. At temperatures below about 30°K, almost all substances reach equilibrium within about one min, some within seconds. At higher temperatures, most inorganic and many organic substances reach equilibrium within a few minutes, except near some kinds of thermal anomalies. However, it is not uncommon for organic solids to require an hour or longer to reach equilibrium when they are within 50 to 100° of the melting point. When such long equilibration periods are encountered, it is particularly important to apply corrections for heat exchange as discussed in the preceding section, even when the normal exchange rate is known to be quite small.

Extremely long equilibration periods are not due to a slow rate of reaching temperature equilibrium, that is, a uniform temperature throughout the calorimeter and sample. When hours or days are required for equilibration, the relaxation process responsible for such slow equilibration must be one associated with some change occurring slowly in the molecular crystals.

Careful observations have shown that some organic solids require at least one day to reach equilibrium after a temperature change. Obviously, it is scarcely practical to wait for equilibrium when such behavior occurs over a wide temperature range. Therefore, one expedient employed is to follow the temperature either for the same length of time after each heating period or, if the equilibration rate changes greatly with temperature, until the same temperature drift is reached. If necessary, corrections for heat exchange may be estimated from data obtained in previous studies. Although such measurements are made on a system not in thermodynamic equilibrium, the results should not be seriously in error because the departure from equilibrium temperature is about the same before and after a measurement. An upper limit on the possible error under such conditions may be obtained in the normal course of experiments. For example, if a sample has been maintained at the initial temperature of a series long enough to reach equilibrium, it is often found that the first one or two heat capacity measurements are lower than the curve determined from data obtained as described above. This result is lower because the initial equilibrium temperature is lower than the initial non-equilibrium temperature that had corresponded to the non-equilibrium final temperature. Normally the heat capacity point so determined is only 0·1–0·3 per cent low, and the error, when both initial and final temperatures deviate from equilibrium by about the same amount, can be assumed to be even smaller.

When extremely slow equilibration periods are encountered, the accuracy and precision with which the heat capacity curve is defined must suffer. For this reason, it is often desirable to make a single measurement of the enthalpy increment over the entire temperature range of slow equilibration as a check on the experimental heat capacity curve.

(4) Heat Capacity of the Calorimeter. The heat capacity of an empty (or helium-filled) calorimeter vessel is determined as described above. However, because the heat capacity curves of materials from which calorimeters are constructed are free of anomalies and very flat above approximately 150°K, measurements can be made at higher temperatures with temperature increments as large as 20°. Typical heat capacity curves for a copper and a platinum-gold calorimeter are illustrated in *Figure 16*. With reasonable care, the mass of a

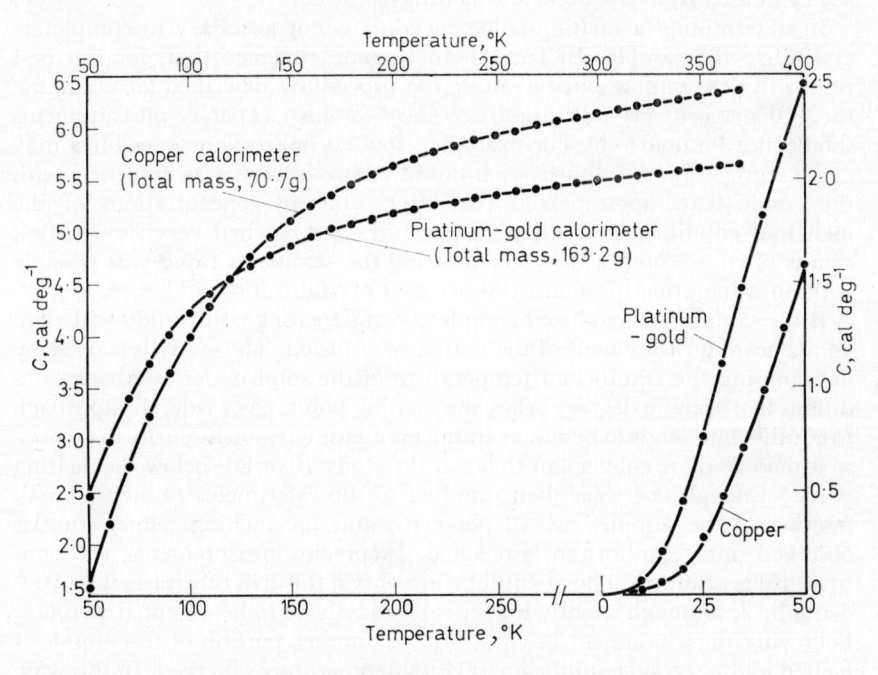

Figure 16. Calibration curves for calorimeter vessels.

calorimeter can be kept the same from one investigation to the next so that repeated calibrations are seldom necessary.

A calorimeter mounted permanently in a cryostat with a fixed filling tube may be calibrated in the same way. Alternatively, two sets of measurements can be made on each substance studied, one with the vessel nearly full and one with it nearly empty[94, 111]. This procedure not only takes into account the tare heat capacity of the calorimeter but also provides data for vaporization corrections (discussed in Section IV–2–D).

B. Enthalpy of Phase Changes

The method of measuring the enthalpy increment of a phase change does not differ in principle from that for measuring heat capacity. The increment in enthalpy from a temperature below the transition point to one above it is measured. If the process is essentially isothermal, e.g. melting, the isothermal increment is calculated from the measured total increment and heat capacity data.

5. *Melting point and purity determinations*

Adiabatic low-temperature calorimetry provides one of the most accurate ways of determining the melting point and purity of a substance. The melting point curve is determined by observing the equilibrium temperature as a function of the fraction of sample in the liquid state. As discussed in Section IV–2–C, the melting point of the actual sample (100 per cent liquid point), triple-point temperature of the hypothetical pure material, and sample purity are evaluated from the observed melting-point curve.

In determining a melting-point curve, it is not necessary to completely crystallize the sample. In fact, there is some evidence that, for the best results in determining purity values, the procedures described for obtaining the 100 per cent crystallization necessary for heat capacity measurements should not be used[83, 84]. For example, slow crystallization procedures may cause segregation of impurity-rich liquid before it freezes as eutectics, with the result that, upon partial remelting, uniform concentrations of the melt and equilibrium melting temperatures are reached very slowly. Best results often are obtained by crystallizing the sample as rapidly as possible without attempting to obtain 100 per cent crystallization.

If the sample is crystallized completely, the melting point study will yield an accurate measurement of the enthalpy of fusion. The study is started by determining the equilbrium temperature of the solid under adiabatic conditions five or more degrees below the melting point. (As a rule, the approach to equilibrium tends to be slower in the premelting region as purity decreases, so it may be more convenient to begin the study 10 or 20° below the melting point.) Enough energy is then supplied to the calorimeter to melt a small fraction of the sample, say 10 per cent, and the melting temperature is observed until equilibrium is reached. In precise measurements, the temperature is assumed to be at equilibrium only if the drift rate is less than 10^{-5} deg min^{-1}, although slightly higher rates may have to be accepted at times. Following the attainment of equilibrium, another portion of the sample is melted and a second equilibrium melting temperature observed. In this way, values of melting temperatures in the two-phase region are determined at a series of fractions melted, for example 10, 25, 50, 70, and 90 per cent. Then the sample is melted completely, and a final equilibrium temperature a few degrees above the melting point is determined. A typical melting curve is illustrated in *Figure 17*.

An alternative method of determining points on the melting curve is to crystallize only a part of the sample and determine the equilibrium temperature. Then the energy required to melt the sample completely is measured; from this the fraction of sample in the liquid state can be calculated. The pro-

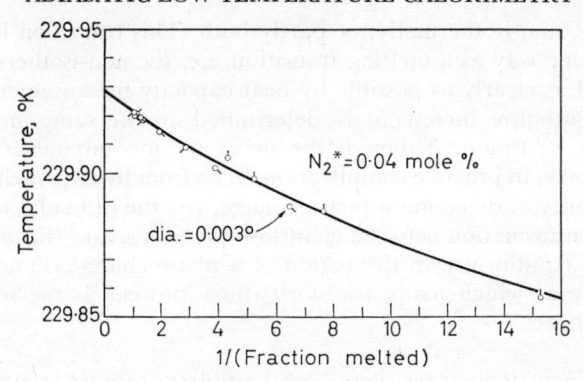

Figure 17. The melting curve of 1-heptanethiol; the different symbols indicate results of four separate determinations.

cedure is repeated for different degrees of crystallization until the melting curve is defined[52, 83].

Although purity determinations (by the calorimetric methods described above) are usually straightforward, careful work has revealed uncertainties involving factors still not clearly understood. Recent studies in the United States by the Bureau of Standards[44] and Bureau of Mines[83, 84] have demonstrated what the limitations of the calorimetric method are and how some of them may be avoided. The cited reports of this work should be consulted for details beyond the scope of this book. One somewhat surprising observation, made apparent by the use of precise automatic shield control, is that more than 24 h may be required for a sample to reach equilibrium in the two-phase melting region.

Most calorimetric systems designed primarily for thermal studies over the entire low-temperature range are more complex and difficult to use than those required for purity determinations alone. For this reason, several investigators have developed specialized adiabatic calorimeters that are convenient to use in melting point and purity determinations[5, 17, 99, 133].

6. *Studies of polymorphic substances*

Many organic and inorganic substances may exist in more than one crystalline state; for example, about one-third of the organic substances studied by low-temperature calorimetry exhibit some kind of polymorphism[82]. The phase behavior of polymorphic substances must be defined accurately when evaluating thermodynamic properties. In fact, nearly all gross errors in reported thermodynamic data determined by low-temperature calorimetry result from failures in detecting or properly defining solid–solid phase changes.

A. Enantiotropic Substances

Substances exhibiting enantiotropy have two or more polymorphic forms, each of which is thermodynamically stable in a definite range of temperature and pressure [37]. The transition between two enantiotropic forms may occur isothermally, non-isothermally, or partly both. This transition is studied in much the same way as a melting transition, i.e. the non-isothermal portion is delineated as clearly as possible by heat capacity measurements, and the isothermal enthalpy increment is determined in the same manner as is the enthalpy of fusion. Although the necessary measurements are simple enough to make, in practice complications arise from three principal sources: (*i*) the difficulty of detecting a phase change, (*ii*) the difficulty of obtaining complete transformation between enantiotropic forms, and (*iii*) the difficulty of obtaining equilibrium in the region of a phase change. Fundamentally, the slow rate at which many transformations proceed is the source of all three difficulties.

(*1*) *Detection of Transitions*. Some solid–solid transitions occur at least as rapidly as the crystallization, or liquid–solid, transitions. Detection and study of such transitions pose no problems unless the associated volume change is large enough and sudden enough to damage a calorimeter vessel. However, when some substances are cooled below a transition temperature, the thermodynamically required transposition to a new stable phase occurs very slowly. It often is possible to cool a high-temperature phase below a transition point rapidly enough to obtain a metastable phase that persists ndefinitely. Reliable heat capacity measurements can be made on such a metastable phase, but as the transition point is approached from below, transformation to the stable low-temperature phase usually begins within 10 to 50° below the transition point. That a transformation is taking place is evidenced by spontaneous warming of the calorimeter as the heat of transition is evolved more or less rapidly. Because the existence of a transition point often is not known in advance, it is particularly important to pay close attention to temperature drift rates in heat capacity determinations. Abnormal warming rates, however small, should be followed long enough to be sure that the rate does not accelerate. In extreme cases, the initial evolution of energy may be at a rate of a few millijoules per minute, and the rate may accelerate very slowly over a period of several days.

(*2*) *Completing Transformations*. Extremely sluggish transformations may warrant special effort to find conditions for maximum rate of transition. At temperatures far below the transition point, the transformation process may cease because of kinetic factors. On the other hand, at temperatures just below a transition point, the process may slow drastically because the free energy difference between the two phases approaches zero. Since the kinetic and thermodynamic factors vary oppositely with temperature, the temperature at which the transition rate is a maximum must be found by trial. Tests for complete transformation to the stable phase below a transition point are made in the same way as tests for complete crystallization; i.e. the sample is held under adiabatic conditions a few degrees below the transition temperature until it is ascertained that no further heat is being evolved.

(*3*) *Equilibration*. Many enantiotropic phase changes involve pretransition effects at temperatures well below the transition point, especially if the transition is at least partly non-isothermal[82, 141]. If a transition is sluggish, long periods of time may be required to reach equilibrium during heat capacity measurements in the pretransition region.

(4) Observational Procedures. Isothermal solid–solid phase changes are studied in the same manner as is the melting process. Non-isothermal transitions are defined by heat capacity measurements. As the heat capacity changes rapidly with temperature near a transition, it is necessary to use smaller and smaller temperature increments in the measurements. If a substance reaches equilibrium quickly enough, increments as small as 0·1° may be used to define a sharp peak in the heat capacity curve. If equilibration is slow, however, it may not be possible to determine the heat capacity accurately in the transition region, regardless of whether the transition is non-isothermal or essentially isothermal. Especially when poor equilibration is found, it is advisable to measure the total enthalpy absorbed in the pretransition and transition regions as a check on the heat capacity and enthalpy of transition measurements.

(5) Hysteresis Effects. The shape of the heat capacity curve of a glassy substance depends both on the rate of cooling through the glass transition and the subsequent rate of making heat capacity measurements. The variations observed in the thermal properties of glasses may be described as the effects of hysteresis. Some early studies of transitions in crystalline substances reported similar hysteresis effects, which undoubtedly were the result of not allowing enough time during the measurements for the substance to reach thermodynamic equilibrium. The procedures just described are designed to allow detection and elimination of such effects of thermal treatment on the measured thermodynamic properties.

B. Monotropic Substances

Substances that have two or more polymorphic forms, only one of which is thermodynamically stable at all temperatures below the melting point, are said to exhibit monotropy[37]. Each of the metastable polymorphs melts at a temperature from a few tenths of a degree to several degrees below the melting point of the stable form. Only rarely are more than two monotropic forms found in substances that are liquid near room temperature.

The metastable (lower melting) form usually crystallizes from the liquid if the sample is cooled only slightly below the melting point. Once crystallized, it is sometimes possible to study the metastable solid down to liquid helium (or hydrogen) temperatures without transposition to the stable form. However, after the metastable form has been cooled to low temperatures, it may transform to the stable form when heated into or near the melting region, probably because nuclei of the stable polymorph are created at low temperatures. This procedure, in fact, is sometimes the only way the stable form can be obtained. On the other hand, a metastable monotropic form may transform to the stable form if cooled more than a few degrees below the melting point. If so, rapid crystallization usually yields the stable form and the metastable form can be obtained only by a slow crystallization procedure involving minimum undercooling. When a stable form is melted and then heated only a few degrees above the melting point, recrystallization may yield the stable polymorph even if the metastable form usually crystallizes first from the liquid. In such cases, the substance must be heated well above the melting point before recrystallization if the metastable form is desired.

From the foregoing discussion, it is apparent that monotropic substances exhibit a wide variety of phase behaviors. Procedures for obtaining the desired crystalline form of a substance must vary accordingly and usually have to be found by trial. Transformations from metastable to stable monotropic forms may occur very slowly, even with a liquid phase present. Therefore, close scrutiny of observed temperature drifts is recommended also to detect the onset of a slow transformation process.

7. *Vapor pressure and enthalpy of vaporization measurements*

Adiabatic calorimeter systems with fixed filling tubes (*Figures 5 and 6*) may be used in direct measurements of both vapor pressure and enthalpy of vaporization[2, 3, 111]. In either type of measurement, the sample is maintained under adiabatic conditions at some temperature below the minimum temperature of the filling tube and associated external lines. The vapor pressures when the sample is at equilibrium at a series of temperatures are read on a mercury manometer attached to the external filling system.

Enthalpy of vaporization measurements are made by introducing a measured quantity of energy to the calorimeter and condensing the vapor produced into an external refrigerated flask of known weight. The calorimeter and sample are kept at essentially constant temperature during the vaporization process by adjusting the power of the calorimeter heater, by adjusting the vapor flow with a throttling valve in the transfer line, or by a combination of power and flow rate adjustments. The principles involved in such measurements are similar in most respects to those described in more detail in Chapter 11. However, it is important to note that no mechanical stirring can be provided in a low-temperature calorimeter vessel, so the rate of vaporization must be kept low enough that temperature gradients and violent ebullition within the vessel are avoided.

IV. Data Reductions

In this section procedures will be described for analyzing the primary observed data to obtain smoothed values of the heat capacity at rounded temperatures in the experimetal range, for extrapolating the smoothed values of heat capacity to $0°K$, and finally for calculating the thermodynamic (thermal) properties.

1. *Treatment of experimental data*

A. Description of Primary Observed Data

The principal measurements to be obtained in adiabatic calorimetry, as described earlier, are heater power, duration of the energy input, and thermometer resistance. Such measurements are made on both the nearly full and either the empty or partially filled calorimeter vessel. Both series of measurements over the same temperature range under conditions as identical as possible are necessary in order to obtain the net heat capacity of the material studied. Considerable experimental data must be analyzed and evaluated in detail before smooth values of heat capacity suitable for calculation of thermodynamic properties are obtained. The data taken during measurements, calibration adjustments that must be made on the observations, and essential calculations should be carefully recorded and examined so that the various associated functions can be sequenced and arranged in the most expedient manner. Printed forms for recording data should be conveniently ordered for subsequent transcription of values, e.g., to punch cards for computer analysis, etc.

B. Adjustment for Heat Leak

The detailed analysis of heat exchange given in Chapter 6 for isothermal calorimetry also applies to adiabatic calorimetry. However, in practice, heat exchange is so small in adiabatic calorimeters that the necessary adjustments can be made almost automatically. Such adjustments may be applied either to the measured electrical energy, q_0, or to the temperature change, ΔT, in calculating the heat capacity.

Because Newton's law of heat transfer is applicable under the operating conditions of the adiabatic calorimeter, the apparent (mean) heat capacity, $q/\Delta T$, may be expressed by the relation:

$$q/\Delta T = \frac{q_0 + \int_{t_1}^{t_m} a_1 C' \, dt + \int_{t_m}^{t_2} a_2 C' \, dt + \int_{t_1}^{t_2} \beta C' \theta \, dt}{T_2 - T_1}. \tag{7}$$

In this expression, q_0 is the measured electrical energy, a is the intrinsic temperature drift (dT/dt) at apparent zero temperature difference between the adiabatic shield and the calorimeter (including the power generated by the current flowing through the thermometer); β is a heat exchange modulus, $d^2T/dt d\theta$, (in terms of the deviations, θ, between the temperatures of the adiabatic shield and the calorimeter vessel); C' is the heat capacity of the calorimeter system (including the sample); t_m is the mid-time of the heating period, and t_1 and t_2 are the times at which T_1, a steady state temperature before heating, and T_2, a corresponding temperature after heating, are observed. The coefficients a and β are temperature dependent and are determined during the course of the heat capacity measurements.

The heat-leak corrections are made to an absolute accuracy compatible with that of the electrical energy measurements. The time dependence of any deviations between the temperatures of shield and calorimeter is determined sufficiently well to calculate $\int \theta \, dt$ with accuracy consistent with the value of β. Adiabatic calorimeters are designed to have small values of a and β, and the total heat exchange correction generally is less than 0·1 per cent. Hence, a, β, and $\int \theta \, dt$ may be determined with low precision. The intrinsic temperature drift, a, is determined at various temperatures by simply observing the temperature drift with the adiabatic shield at control zero (i.e., with $\theta = 0$). The heat-leak modulus, β, is obtained similarly at various temperatures by displacing the temperature control by a fixed number of scale divisions from the control zero and observing the new drift. Then,

$$\beta \approx [(dT/dt)_\theta - (dT/dt)_0]/\theta. \tag{8}$$

In systems with fixed calorimeter filling tubes, α and β do not change much with time, and, once determined, they may be plotted as functions of temperature for routine use with Equation (7). These constants change slightly from experiment to experiment in systems with demountable calorimeters, so they must be redetermined or at least rechecked in each new experiment. However, θ is almost always negligible in well operated systems, and α is automatically determined in the temperature observations. Obviously, in applying values of α and β obtained for a system of heat capacity C' to another of heat capacity C'', these quantities in Equation (7) should be multiplied by the ratio C'/C''.

In an alternative method of adjusting for heat leak, the corrected temperature change, ΔT_{corr}, corresponding to the measured electrical energy, q_0, may be evaluated to obtain the mean heat capacity,

$$q/\Delta T = q_0/\Delta T_{\text{corr}}.$$

$$= q_0 / \left[\left(T_2 - \int_{t_m}^{t_2} \alpha_2 \, dt \right) - \left(T_1 + \int_{t_1}^{t_m} \alpha_1 \, dt \right) - \int_{t_1}^{t_2} \beta \theta \, dt \right]. \tag{9}$$

The α's and β are determined as before. In practice, θ is usually zero during the equilibrium period, so α can be determined easily for each equilibration period. Because α varies slowly with temperature and cannot be measured during a heating period, it is sufficient to evaluate the integrals as $\alpha_2(t_2 - t_m)$ and $\alpha_1 (t_m - t_1)$. These factors are applied to T_2 and T_1 to calculate ΔT_{corr}. This is equivalent to the extrapolation of the temperatures, observed during the equilibration periods before and after heating, to time t_m in order to obtain T_1 and T_2.

Occasionally, when α is unusually large, the temperature of the adiabatic shield may be offset intentionally, and θ is reckoned from the new "zero". If the offset is constant during a heat capacity measurement (or changed at the midpoint of the heating period), the β term in Equations (7) and (9) may still be neglected. In this case appropriate values of α, different from those with the shield in balance, are evaluated from the observed temperature drifts.

2. Reduction of observed data to molal basis

The results obtained as described in the foregoing section are for the calorimeter vessel plus an arbitrary amount of sample. In the following sections are described the calculational procedures needed to obtain thermodynamic data for one mole, or other convenient unit, of the substance studied.

A. Heat Capacity of the Calorimeter Vessel

To compute the net heat capacity of a sample studied in a low-temperature calorimeter, it is necessary to subtract from the gross values the heat capacity of the calorimeter including the thermometer and heater units. This tare heat capacity of the calorimeter must be determined at least as accurately (absolute) as the gross heat capacity of the sample plus calorimeter. If the gross experiments are made under essentially identical conditions for different substances in the same calorimeter system, the tare heat capacity will remain unchanged and need be determined only once. (Corrections for minor changes in calorimeter mass can be made, as described in Section IV–2–E.) However, the tare heat capacity must at least be checked if any significant change is made in operating conditions. The same calorimeter may yield slightly different heat capacities in two essentially identical cryostats because of slightly different systematic errors. Such differences are usually less than 0·1 per cent of the tare heat capacity. At least some systematic errors can be eliminated by careful measurement.

The tare heat capacity may be obtained either from measurements on the empty calorimeter or from two sets of measurements, one with the vessel nearly full and one with it nearly empty. The latter procedure is discussed more fully in Section IV–2–D; the heat capacity of the empty vessel need not be evaluated, although it can be, if desired. In either case, the measurements and data reduction are done in the same way as those in subsequent heat capacity determinations. Helium may be added to the empty vessel to promote thermal equilibration. A correction for the helium content may be necessary, of course, but the amount of helium can be adjusted to make the correction insignificant.

The measurements on the empty vessel are calculated as heat capacity values and corrected for curvature if necessary (Section IV–2–B). The "raw" heat capacity data are smoothed graphically or analytically and a large scale plot or a tabulation of the results prepared. Alternatively, a high speed digital computer may be used for the computations. The plot, table, or computer output is used to find the tare heat capacity at the same mean temperatures for which data are subsequently obtained for the calorimeter plus sample. Another alternative involves preparation of a plot, table, or analytical expression for the enthalpy of the vessel as a function of temperature. This representation is then used to evaluate the energy absorbed by the calorimeter vessel between the initial and final temperatures of subsequent heat capacity measurements.

B. Curvature Corrections

The true heat capacity is given by

$$C = \lim_{\Delta T \to 0} q/\Delta T = \mathrm{d}H/\mathrm{d}T. \tag{10}$$

The result of actual measurements is the mean heat capacity, $C_{\mathrm{mean}} = q/(T_2 - T_1)$, associated with the mean temperature of the interval, $T_{\mathrm{m}} = (T_1 + T_2)/2$. Deviation from linearity in the C versus T curve will therefore require adjustment of the mean heat capacity by a curvature correction to yield the true heat capacity at T_{m}.

In practice, the curvature correction is determined by considering the heat capacity to be represented by a cubic equation in temperature over a limited range

$$C = a_0 + a_1 T + a_2 T^2 + a_3 T^3. \quad (11)$$

On this basis

$$q = \int_{T_1}^{T_2} C\,dT = a_0(T_2 - T_1) + \frac{a_1}{2}(T_2^2 - T_1^2)$$

$$+ \frac{a_2}{3}(T_2^3 - T_1^3) + \frac{a_3}{4}(T_2^4 - T_1^4); \quad (12)$$

and the mean heat capacity is

$$q/(T_2 - T_1) = a_0 + \frac{a_1}{2}(T_2 + T_1) + \frac{a_2}{3}(T_2^2 + T_2 T_1 + T_1^2)$$

$$+ \frac{a_3}{4}(T_2^3 + T_2^2 T_1 + T_2 T_1^2 + T_1^3). \quad (13)$$

The difference between the true heat capacity and the mean heat capacity at T_m is

$$C - C_m = -\frac{(T_2 - T_1)^2}{24}[2a_2 + 3a_3(T_2 + T_1)]. \quad (14)$$

Since at the mean temperature,

$$\frac{\partial^2 C}{\partial T^2} = 2a_2 + 3a_3(T_1 + T_2), \quad (15)$$

the true heat capacity is

$$C = C_m - \frac{(T_2 - T_1)^2}{24}\left(\frac{\partial^2 C}{\partial T^2}\right). \quad (16)$$

Values of $\partial^2 C/\partial T^2$ are obtained by considering the mean heat capacity to be sufficiently close to the true heat capacity and differentiating the smoothed mean heat capacity.

Another approach is useful and valid if the curvature corrections are small, i.e. if the shape of the uncorrected heat capacity curve is nearly the same as that of the corrected curve. With $\Delta T = T_2 - T_1$,

$$q/\Delta T = (1/\Delta T)\int_{T_1}^{T_2} C\,dT$$

$$= (1/\Delta T)[\int_{T_1}^{T_2} C_m\,dT + \int_{T_1}^{T_2}(C - C_m)\,dT]. \quad (17)$$

Because $C - C_m$ is small and constant over the range T_1 to T_2, this expression can be rearranged to give the curvature correction as

$$C - C_m = q/\Delta T - (1/\Delta T)\int_{T_1}^{T_2} C_m\,dT$$

$$= C_m - (1/\Delta T)\int_{T_1}^{T_2} C_m\,dT \quad (18)$$

The integral can be evaluated by Simpson's rule at T_1, T_m, and T_2 to yield

$$C - C_m = 1/3\,[C_m - (C_1 + C_2)/2] \tag{19}$$

in which C_1 and C_2 are the apparent heat capacities at T_1 and T_2. Thus, a simple way to calculate the curvature correction is to lay a straight edge through the apparent heat capacity curve at T_1 and T_2 and take the correction as 1/3 the distance between the straight edge and curve at T_m. The procedure last described is also an easy way to determine if curvature corrections are necessary. In general, such corrections may be made insignificant by using small enough temperature increments in the measurement. By applying curvature corrections routinely, an investigator can use larger temperature increment and reduce time spent on measurements. However, in using large increments, he must take care to avoid missing one of the very small thermal anomalies that sometimes occur in solids.

A literature treatment of curvature correction has been given by Osborne, Stimson, Sligh, and Cragoe[96]. In another section of this chapter (IV–3–B) a method of data analysis is described that requires no curvature correction.

C. Premelting Corrections

Most substances studied today by accurate calorimetric methods are pure enough to render the effect of the impurity on the observed heat capacity data negligible except in the region just below the melting point. Usually, the thermal effect of melting any eutectic formed by a solid-insoluble impurity and major component is not large enough to be detected at the eutectic temperature. As the melting point is approached, however, an increasing proportion of the energy added during a heat capacity measurement is consumed by melting the sample in accordance with the thermodynamics of the solution involved. High values of heat capacity that are manifest in this region are also ascribed to significant changes in the solid density, which are known to occur in certain substances as the melting point is approached[35]. Moreover, abnormal values of heat capacity could be due to non-equilibrium conditions (because of the relatively short time allowed for measurements), the nature of the crystalline state (which depends largely on the rate of crystallization), and the geometry of the calorimeter (see Section III–5). Without additional intensive experimental measurements, the calorimetric data alone do not provide a basis for decision concerning the relative quantitative contributions to the abnormally high heat capacity. Often, because of uncertainty about the nature of the abnormal values, the heat capacity curve is extrapolated "judiciously" from lower temperatures to the melting point and the excess enthalpy above the extrapolated lattice heat capacity curve attributed to premelting. A correction for this effect of premelting must be made in calculating the heat capacity and the enthalpy of fusion of the major component.

(1) Theoretical Formulation. If the liquid solutions formed in the premelting region are ideal and thus obey Raoult's law, an expression for calculating the premelting corrections may readily be derived[53, 56], e.g.

$$F_i \Delta Hm = (x_2^*/x_2)\,\Delta Hm \approx x_2^*\,RT_{TP}^2/(T_{TP} - T_i) \tag{20}$$

in which F_i is the fraction of sample melted at T_i, ΔHm is the enthalpy of fusion (melting), T_{TP} is the triple-point temperature of the pure substance, x_2 is the mole fraction of impurity in liquid solution at T_i, and x_2^* is the mole fraction of impurity in the entire sample. This quantity represents the amount of energy used for partial melting up to T_i. The premelting correction to the energy supplied during a heat capacity measurement between T_1 and T_2 is thus $(F_2 - F_1) \Delta Hm$, which can be calculated from the foregoing equation if x_2^* is known.

Equation (20) is an approximation that applies only if the amount of impurity is quite small and is insoluble in the solid phase. Terms neglected in deriving it are $x_2/2$ and $(1/T_{TP} - \Delta C_S/2\Delta Hm)(T_{TP} - T_i)$ and higher order terms. Neglect of the former term leads to an error in x_2, and therefore in F_i, of 100 $x_2/2$ per cent. For typical organic compounds $\Delta Hm/RT_{TP}^2 (= x_2/(T_{TP} - T_i))$ is 0·04/deg[104], so the error in F_i is $2(T_{TP} - T_i)$. The term $(1/T_{TP} - \Delta C_S/2\Delta Hm)$ is typically 0·004/deg[104], so the corresponding error in F_i is 0·4 $(T_{TP} - T_i)$. These two errors partly compensate for one another. If $x_2 \geqslant 0\cdot01$, higher order terms also become important.

Calculation of the premelting correction for a system that forms a solid solution is considerably more complicated. The distribution coefficient, k, of the impurity between the solid and liquid phase is generally not known. In addition, a shift in equilibrium point (such as that occasioned by increasing the temperature) requires a change in composition of the solid phase. This involves diffusion, which is a slow process requiring an extremely long equilibration time not feasibly measured in laboratory practice. Inasmuch as some of the impurity is frozen out of liquid solution, the impurity composition of the liquid phase does not become large at small values of the fraction melted. Therefore, the simplified equation can be used, for which the enthalpy of fusion is considered constant and $[-\ln(1 - x_2)] = x_2$; moreover, the applicability of the ideal solution concept and Raoult's law is assumed for both phases. If the distribution ratio is greater than one, the composition of the first liquid appearing is purer than that of the overall composition, in contradistinction to the case for which the impurity is liquid-soluble, solid-insoluble.

To calculate the premelting correction for solid-soluble impurity Equation (20) is modified to

$$F_i \Delta Hm = x_2^* RT_{TP}^2/(T_{TP} - T_i) - \Delta Hm\, k/(1 - k), \qquad (21)$$

in which now the distribution coefficient, k, must also be known[81].

(2) *Practical Applications*. The laws of dilute solutions obviously do not hold for the first, very concentrated liquid solution formed at the eutectic temperature. In fact, these laws probably fail until at least a few per cent of the sample has been melted, which for relatively pure materials might be much less than 1° below the triple point. Partly for this reason, it is often found that use of Equation (20) and the value of x_2^* determined in a melting study leads to ridiculous premelting corrections. For example, the "corrected" curve may have a negative slope just below the melting point.

It can be shown that the premelting correction to the heat capacity, ΔCpm, as derived from Equation (20) is:

$$\Delta Cpm = x_2^* \, RT_{\text{TP}}^2/[(T_{\text{TP}} - T_{\text{m}})^2 - (\Delta T/2)^2]. \tag{22}$$

The more exact formulations may give better results, but they do not in all cases and probably are not worth considering in calorimetry on pure substances. It is more practical to use the simple Equation (22), and instead of using the measured value of x_2^*, to select an arbitrary value which will result in a corrected curve that is linear near the melting point. If heat capacity measurements are made close to the melting point, the shape of the corrected curve is highly sensitive to the value chosen as the effective impurity concentration. In this method, x_2^* found in the melting point study must be the upper limit for that to be used in calculating premelting corrections. For some compounds, the corrected heat capacity curve has positive curvature even when the maximum reasonable value of x_2^* is used. For example, the n-paraffin hydrocarbons show this behavior as a result of what has been called homophase premelting[38].

D. Vaporization Corrections

If the substance under investigation exhibits an appreciable vapor pressure in the range of heat capacity measurements, the enthalpy of vaporization (or condensation) associated with the changes in the amount of each phase contributes to the observed gross heat capacity. In a closed calorimeter vessel, the volume of the vapor present at a given temperature is dependent upon the mass of the substance, the densities of both condensed and vapor phases, and the volume of the vessel. To minimize the vaporization correction, the calorimeter vessel is nearly filled so that the vapor volume is very small at the highest temperature of the measurements (on the assumption that the specific volume of the condensed phase of the substance is a monotonic, increasing function of temperature). Overfilling the vessel may result in rupture at higher temperatures. By carrying out two series of measurements, one on the nearly full calorimeter and the other on the partially filled calorimeter (but with sufficient sample to maintain saturation conditions through the measurements) the volume of the calorimeter becomes irrelevant in the analysis of the data because the quantities involved are properties of the substance only. As in any other heat capacity measurements, the accuracy of the final values is dependent upon the difference between the heat capacities of the nearly full and partially filled (or empty) vessel. Therefore, the increment of sample between the nearly full and partially filled vessel should be made as large as practicable. By maintaining the same amount of vapor in the filling tube for corresponding temperatures during measurements on the nearly full and partially filled calorimeter, the reduction of data obtained with the filling-tube calorimeter is comparable to that obtained with the demountable calorimeter. If measurements on the partially filled vessel are not carried out with the filling-tube calorimeter, the reduction of data is considerably more involved and requires correction for sample that has vaporized into the filling tube. In this section, procedures for vaporization correction will be discussed for four combinations of heat capacity experiments with volatile substances:

(*i*) nearly full and empty filling-tube calorimeter,
(*ii*) nearly full and empty demountable calorimeter,
(*iii*) nearly full and partially filled filling-tube calorimeter, and
(*iv*) nearly full and partially filled demountable calorimeter.

Procedures for vaporization correction have been developed by Hoge[61] and by Osborne[94] for the above four types of experiments. Hoge's treatment is couched in terms of the entropy of a two-phase system, and the vaporization correction is applied after the reduction of the experimental data to apparent net (i.e. sample) heat capacity. Osborne treated the energy and thus dealt directly with the experimental observations. Osborne's method is described in detail in Chapter 11; therefore only Hoge's method will be described here.

For the more general case in which a portion of sample is removed from the calorimeter vessel proper (e.g., as in a filling-tube calorimeter)† Hoge[61] obtained the following relations for the heat capacity of a two-phase system:

$$C_{\text{net}} = C_S{}^c (M - M_t) + T(dS'/dT) + \Delta h_v (dM_t/dT). \qquad (23)$$

in which

$$S' = (dp/dT)[V - (M - M_t) v^c]. \qquad (24)$$

The symbols in the equations are defined as follows: M = total mass of sample, M_t = mass of sample in tube, p = vapor pressure, v^c = specific volume of condensed phase, V = volume of calorimeter vessel, Δh_v = specfic enthalpy of vaporization, $C_S{}^c$ = heat capacity per unit mass of the condensed phase, and C_{net} = observed gross heat capacity minus empty heat capacity. Since the above equation has been derived on the basis of an infinitesimal process, curvature corrections must be applied to the heat capacity values prior to their use in the equation. The mass of vapor (assumed to be an ideal gas) in the filling tube at a given calorimeter vessel temperature is

$$M_t = p(W/R) \sum_i (V_i/T_i). \qquad (25)$$

Here W = molecular weight of the substance, R = gas constant, and V_i represents the volume of small slugs of vapor in the tube at temperature T_i. In practice the average temperatures (sensed by thermocouples) of sections of the filling tube are used in the calculation. The term giving the infinitesimal temperature dependence for removal of sample by vaporization is obtained by differentiating Equation (25)

$$dM_t/dT = (W/R)\left[(dp/dT) \sum(V_i/T_i) + p \frac{d}{dT} \sum_i (V_i/T_i)\right]. \qquad (26)$$

† In the operation of a filling-tube calorimeter, in which the design allows the sample vapor to expand into the filling tube external to the calorimeter vessel, the temperature of the filling tube must be higher than that of the calorimeter vessel at all times to avoid condensation of sample therein. Vapor filling-tube calorimeters have been described [cf. Section II–2–B–(2)] in which a relatively short section of thin tube of small diameter joins the calorimeter vessel to a valve which is a part of the adiabatic shield[3, 95]. In the reduction of data obtained with such a calorimeter, one-half of the tube is considered to be a part of the calorimeter vessel.

The last term is evaluated by plotting $\sum_i (V_i/T_i)$ as a function of temperature, then reading values of it from the plot at equally spaced integral temperatures, and finally differencing these values. Values of $\Delta \sum_i (V_i/T_i)/\Delta T$ are generally adequate if the filling tube volume is small and the vapor pressure less than an atmosphere. If more accurate values are needed, numerical differentiation methods are used with the tabulated values of $\sum_i (V_i/T_i)$[86] or with their differences. If the temperature control of the calorimeter shield and filling tube is precisely reproducible, the relationship between the calorimeter temperature and the summation can be considered a property of the calorimeter, independent of the substance being measured. Therefore, the same set of values of $\sum_i (V_i/T_i)$ and $(d/dT) \sum_i (V_i/T_i)$ can be used for different measurements with the same calorimeter. Obviously, the vapor pressure, p, and its temperature derivative, dp/dT, are functions of the substance investigated. Values of dS'/dT are obtained by the same procedure, combining numerical and graphical methods.

The terms of the general Equation (23) may be transposed to show more clearly how the heat capacity of the condensed phase $C_\mathrm{s}{}^\mathrm{c}$ is calculated, once the various adjuvant data are used as described above to obtain the different adjusting terms:

$$C_\mathrm{s}{}^\mathrm{c} = \frac{C_\mathrm{net} - T\, dS'/dT - \Delta h_v\, dM_t/dT}{M - M_t}. \tag{27}$$

In procedure (*iii*), by maintaining the tube temperatures, T_i, during partially filled vessel measurements at settings identical to those during experiments (at the corresponding calorimeter temperatures) on the nearly-full vessel, the adjusting terms containing $\Delta h_v(dM_t/dT)$ and M_t become identical in both series. The general Equation (23) or (26) then simplifies to

$$C_\mathrm{s}{}^\mathrm{c} = \frac{C_\mathrm{net} - T\, d(S_1' - S_2')/dT}{M_1 - M_2}, \tag{28}$$

in which $S_1' - S_2' = -(dp/dT)(M_1 - M_2)v^\mathrm{c}$, and M_1 and M_2 are the masses of sample used in nearly-full and partially-filled experiments, respectively. The values of $d(S_1' - S_2')/dT$ are obtained by a procedure similar to that described for dM_t/dT and dS'/dT. The evaluation of $d(S_1' - S_2')\, dT$ is considerably simpler than that of dS'/dT, for the factor $(M - M_t)$ is temperature dependent whereas $(M_1 - M_2)$ is a constant and the volume of the calorimeter is not involved (cf. Equation 24). In this calculation, C_net is the difference in the values of apparent heat capacity (including that of the calorimeter vessel) between the nearly-full and partially-filled vessel experiments.

In the reduction of experimental data obtained with the demountable calorimeter in accordance with procedure (*ii*), the terms containing the mass of sample in the filling tube are not involved. The transposed general Equation (27) therefore simplifies to

$$C_\mathrm{s}{}^\mathrm{c} = \frac{C_\mathrm{net} - T(dS'/dT)}{M} \tag{29}$$

in which $S' = (\mathrm{d}p/\mathrm{d}T)(V - Mv^c)$ and $M =$ mass of sample. The values of S' and $\mathrm{d}S'/\mathrm{d}T$ are obtained with the procedure outlined earlier. Here C_{net} is the difference in heat capacity obtained for the nearly-full calorimeter experiments and that of the empty calorimeter vessel.

For reducing data from nearly-full and partially-filled vessel experiments with a demountable calorimeter (procedure *iv*), Equation (28) is also applicable and identical procedures are followed. Any strain energy imparted to the calorimeter vessel by the sample is cancelled in procedures (*iii*) and (*iv*), but not in procedures (*i*) and (*ii*). The contribution from this effect is, however, negligible in experiments being considered in this chapter.

E. Other Corrections

(1) Heat Developed in Leads. Unless already compensated for by appropriate attachment of the potential leads, the ohmic heat developed in leads must be taken into account (cf. Section II–1–G–3 and Chapter 4).

(2) Helium Gas Correction. Helium exchange gas is used to improve thermal equilibration in the filled or empty calorimeter. With helium gas, the temperature gradients of the calorimeter for the two measurements also will be more nearly alike. Only a few torr of the gas at 300°K are needed. Ideally, it should be measured volumetrically from a calibrated container into the calorimeter. If V_b is the calibrated glass bulb volume, then the amount of gas (in number of moles) is $n_1 = (P_1 - P_2)V_b/RT$, in which P_1 is the initial pressure and P_2 is the pressure after admitting gas to the calorimeter. To check whether helium gas is adsorbed significantly by the sample, the amount of helium gas in the vessel is calculated from $n_2 = P_2(V_c - m/\rho)/RT$, in which V_c is the volume of the vessel, m is the mass of sample, and ρ is its density. Values of n_1 and n_2 should be equal if the adsorption is not significant.

The correction for helium gas is based upon the assumption that its heat capacity is the heat capacity of a monatomic gas at constant volume, i.e., $(3/2)R$. Although insignificant at 300°K, it is often appreciable at the lowest temperatures.

(3) Mass Corrections. Often the masses of the calorimeter and its parts (such as copper, solder, gold gasket, thermal conductivity grease, etc.) are not the same for the filled and empty measurements. To correct for mass differences, values of heat capacities for solders, thermal conductivity greases, and related substances (based on experimental data or estimates) used in calorimeter construction may be tabulated at small intervals to facilitate linear interpolation for making adjustments. Some materials involved, such as grease used for thermal contact between the thermometer–heater assembly and calorimeter vessel, the cements used for attaching the lead wires, or glass used in sealing a platinum calorimeter vessel, are proprietary products which are not well characterized. Either heat capacity measurements on these substances should be made in individual laboratories or, alternatively, care should be taken to minimize amounts used or to carefully reproduce their amounts. To avoid confusion in the sign of the

adjustment for the difference in mass, the mass difference should be considered relative to the mass used for the tare measurements. Thus

$$C_{net} = C_{apparent} - C_{tare} - c_i(\Delta m_i), \tag{30}$$

in which c_i is the heat capacity per gram (specific heat) of a material of the calorimeter–vessel system which differs in mass between the two sets of measurements.

The heat capacities of solders are frequently estimated from their composition by application of the Kopp–Neuman rule. For 50:50 lead–tin eutectic solder, measurements from 1·8 to 8°K have been reported by Shiffman, et al.[112] from 1·3 to 20°K by De Nobel and du Chatenier[28], and from 20 to 300°K by Ziegler and Mullins[148]. Cerroseal solder (50 per cent indium, 50 per cent tin, by weight) is a very useful low-melting solder (m.p. about 108°C) for calorimetric closures. The heat capacity of "Apiezon-T" high-vacuum grease has been determined from 1 to 10°K by Osborne and Flotow[92] and from 5 to 300°K by Westrum[138]. This material, as well as "Lubriseal" stopcock grease, is employed to facilitate the establishment of thermal contact between the thermometer–heater–calorimeter assembly. Apiezon–T may be used to 350°K without vaporization or fusion, but has a region of high and irreproducible heat capacity between 220 and 260°K.

Other materials used in cryogenic calorimetry of which the heat capacities have been reported are:

Araldite (Type I baked at 180°C); 1·5–20°K, by Parkinson and Quarrington[97].

Formite (Bakelite) baked varnish; 4–90°K, by Hill and Smith[59].

Glyptal (air dried at 25°C); 4–100°K, by Pearlman and Keesom[98].

Wood's Metal (12·5 per cent Sn, 12·5 per cent Cd, 25 per cent Pb, 50 per cent Bi); 1·5–20°K, by Parkinson and Quarrington[97].

3. Smoothing the experimental data

For the evaluation of derived thermodynamic properties, heat capacities over the temperature range to near 0°K are employed. Satisfactory methods for representing the experimental values of $H_2 - H_1$ as a function of temperature are: graphically smoothed heat capacity values at closely and evenly spaced integral temperatures, one (or several) equation(s) expressing the temperature dependence of the heat capacity, or equations in $H_T - H_{T_0}$ (relative enthalpy). (see Section IV–3–B.)

In low-temperature calorimetry, the analysis and correction of experimental data are usually carried out in terms of the heat capacity; therefore, procedures for smoothing this type of experimental data will be described first. The smoothing of the data in terms of the enthalpy increments between the observed temperatures will be discussed in Section IV–3–B. Both procedures can be done best by a high-speed computer.

A. Smoothing Heat Capacity Data

(1) Graphical Methods

The simplest procedure for smoothing the experimental values is plotting the apparent heat capacity on a large scale (e.g., on mm-ruled paper, one meter wide) and drawing a smooth curve that best represents the data with appropriately few constraints (e.g., spline weights). Curvature corrections may be applied graphically (to the gross and to the tare heat capacities) and the plotted values adjusted accordingly. Values of heat capacity may be read at equally spaced integral temperatures from the smooth curve. The smoothness of the tabular values obtained may be checked by examining the regularity of the third or higher differences. Smoothness in the third difference is sufficient in most cases. The differences can be smoothed graphically and adjustments to the original values made accordingly. When the values of heat capacity are at equally and closely spaced temperature intervals, a desk calculator can be used in numerical smoothing. The seven-point least-square approximation and cubic smoothing coefficients[145a] are suitable for this purpose.

The size of the graph can be reduced by plotting the deviation of the observed values from empirical algebraic equations. The type of empirical equation selected depends on the shape of the heat capacity curve. For solids with no transitions, three overlapping cubic equations should cover the complete range—one from 10° or lower to slightly below the inflection point, another in the region of the inflection point, and the third at higher temperatures. The constants of the equation are determined by using selected values of heat capacity in the range. Additional equations may be needed for substances that have phase transitions in the range of measurements. The smooth values are obtained by combining the readings from the smooth deviation curves and the values calculated at the corresponding temperatures from the empirical equations. By overlapping the equations, values that join most smoothly can be selected and smoothed further by numerical methods mentioned above[145a].

(2) Analytical Methods

In an analytical method, the experimental values of heat capacity are represented by an equation or a set of equations. Either a relatively short range or a broad range of temperature may be utilized. A number of different procedures can be devised for either method. Whatever the method, the experimental data must be represented satisfactorily within the experimental accuracy. The short range method will be considered first.

(a) Short Temperature Range Method. Use of the short temperature range method requires that the experimental values be distributed suitably over the complete range of measurements and that values deviating exceptionally from the other values be rejected by using the Chauvenet or other suitable criterion[90]. Fitting quadratic or cubic equations to heat capacity values obtained at random temperatures permits interpolation of values at closely and evenly spaced integral temperatures. Except at the beginning and end of the experimental temperature range, the procedure involves interpolation for values at equally-spaced integral temperature arguments between two experimental values located near the middle of the temperature interval spanned by the 4 to 5 experimental values. For the terminal portions of the experimental range, the same interpolation formula used for the middle of an interval may be used or special formulas that are available

for the purpose may be employed. When the complete table at equally spaced integral temperature arguments is obtained, the smoothness is checked by examining the differences. Smoothing, if needed, can now be done by standard processes previously mentioned for equally spaced arguments.

If the data are highly precise and closely distributed so that the curvature correction is negligible, a simple linear interpolation between the observed temperatures may be used to obtain values at regular temperature intervals. The values are then smoothed by numerical methods previously mentioned.

Methods of interpolation with unequally spaced arguments are described in treatises on numerical calculus[12, 86, 87, 145a]; Newton's, Lagrange's, or Aitken's interpolation methods are suitable.

Interpolation procedures or alternative procedures can be carried out most conveniently by using a high-speed digital computer. Quadratic or cubic equations may be successively fitted by least squares to the experimental values and the coefficients used for interpolation. Further constraints may be imposed; e.g., first and second derivatives may be conformed to those of the previously interpolated values so that all interpolated values would be smooth. This method is related to that of interlaced parabolas[145b]. The values obtained may be checked for smoothness by differencing and, if necessary, smoothed by numerical methods.

(*b*) *Representation of the Data by Equations.* Selection of a single or several equations to best represent the heat capacity data is accomplished most conveniently by means of a high-speed digital computer. Here the random errors of observations are treated statistically. In contrast to the graphical method or the previously described analytical method over short ranges of temperature, the values obtained are smoothed *a priori*. Equations composed of linear combinations of linearly independent functions often have coefficients fitted by least squares. This method employs minimization of the sum of the squares of the deviations (observed values less those calculated) as criteria. Over large temperature ranges in which the heat capacity may vary over several orders of magnitude or for uneven distribution of heat capacity values, the equation obtained by this method may not best represent the heat capacity data; however, this method is still very effective in smoothing the data. The solution of the normal equations to obtain the coefficients of the equation can be simplified considerably if the functions which make up the equation are orthogonal. The terms of the equation may be various heat capacity-like functions such as Debye functions, Einstein functions, certain rational fractions, and polynomials, for example, Equation (31), which has all four kinds of terms.

$$C = \Sigma\, a_i D\,(\Theta_{D_i}/T) + \Sigma\, b_j E\,(\Theta_{E_j}/T) + \Sigma\, c_k\,(\Sigma\, d_l T^l / \Sigma\, e_m T^m) + \Sigma f_n T^n. \qquad (31)$$

The number of terms and the characteristic temperatures of the Debye and Einstein functions are preassigned on the nature of the heat capacity curve. Similarly, the number of rational fractions and their coefficients d_l and e_m are determined *a priori*. The coefficients a_i, b_j, c_k, and f_n are determined by least squares.

B. Smoothing Enthalpy Increments

Experimentally observed enthalpy increments $H_{T_2} - H_{T_1} = \Delta H_{2,1}$, may be smoothed and analyzed by the following procedures. Let the heat capacity C be represented by a polynomial

$$C = \sum_{-m}^{n} a_i T^i \tag{32}$$

The observed enthalpy increment is, therefore (except for $m = 1$)

$$\Delta H_{2,1} = \int_{T_1}^{T_2} C\, dT = \sum_{-m}^{n} \frac{a_i}{i+1} (T_2^{i+1} - T_1^{i+1}). \tag{33}$$

The coefficients of the polynomial are obtained, for example, by the method of least squares from the observed values of $\Delta H_{2,1}$ over the range for which the polynomial represents the heat capacity. As mentioned previously in Section IV-3-A (which deals with the smoothing of heat capacity data) the terms in the various powers of T can be replaced by a series of Debye and Einstein functions, having discrete preassigned characteristic temperatures, and the coefficients determined in the usual manner by least squares.

The relative deviations of the observed values of $\Delta H_{2,1}$ from the polynomial, given by

$$[\Delta H_{2,1} (\text{obs}) - \Delta H_{2,1} (\text{calc})]/\Delta H_{2,1} (\text{calc}) \tag{34}$$

are plotted at the mid-temperatures of the heating intervals to permit examination of the deviations for possible trends. The presence of trends implies the desirability of selecting another equation that will fit the data better. The deviations, moreover, indicate the uncertainties in the determination of $\Delta H_{2,1}$ and the corresponding temperatures.

The evaluation of Equation (32), in the temperature range for which it was fitted, yields the heat capacity directly without adjustment for curvature. Several overlapping equations usually will be required to represent the experimental data over the complete temperature range. In the overlapping region, the values that join most smoothly are selected. The tabular values are differenced to check their smoothness, and, if necessary, further smoothed by a numerical procedure[145a].

C. Application of High Speed Digital Computers to Calorimetric Data

The progress from raw experimental data to final thermodynamic quantities involves much tedious repetitive computations. The process is therefore ideally suited for performing on a high speed digital computer, with the added advantage that the calculation is performed more rapidly and with fewer errors. The computer is capable of handling a large volume of data (experimental observations) to obtain relatively few selected representative values (smoothed heat capacities), or the process reversed to obtain a large number of values from relatively few, e.g., a table of thermodynamic properties from smoothed heat capacities. Inasmuch as the computer is intended for repetitive calculations, the experimental observations to be calculated and analyzed must be arranged and introduced always into the computer in identical prescribed procedures. Automatic digital recording procedures of experimental data on a form directly acceptable to the computer equipment would help achieve the latter requirement of uniformity.

The computer program to be used to obtain the energy and temperature properties of the system under investigation is dependent largely on the instrumentation of the laboratory and the procedures that are followed in the measurements. Before assembling any computer program, a detailed, step-by-step, ordered arrangement of the measurement and calculation procedures should be charted or diagrammed. In the process, the laboratory measurement procedures often become better systematized and organized to meet the requirement for rigid order of high speed digital computer techniques.

The following are typical examples of observations that can be handled by computers. The calculation of energy from voltage observations will be discussed first, followed by the calculation of temperatures. There are many procedures for smoothing the observed heat capacities. A brief account of general approaches is given.

(1) Energy Calculation for a Heat Capacity Determination. The energy of a short duration heat capacity determination is calculated by multiplying the average value of the power by the time duration of the energy input. The variation in the voltage applied to the heater tends to be smaller than the variation in current through it, since the voltage of storage batteries tends to decrease in use, whereas the heater resistance rises with temperature. It is usual to treat the variation of voltage with time as being linear. Gibson and Giauque[48] have developed an argument showing that if the current–time variation is not higher than quadratic, the time-average value is given by averaging the current readings at the 21 per cent (I_{21}) and 79 per cent (I_{79}) elapsed times of the energy input. The input energy is given by $(I_{21} + I_{79}) \bar{E} \tau / 2$. (See Section II-3-B-2 for calculating I and E from the voltage observations.)

The computer input for calculating the input energy for this relatively simple case is as follows: voltages corresponding to I_{21} and I_{79} and to $\bar{E}$ (the calorimeter heater voltage at mid-time of heating); values of circuit parameters for calculating $\bar{I}$ and $\bar{E}$; and τ.

(2) Energy Calculation for an Enthalpy Determination. In an experiment in which determination of the enthalpy increment over a large temperature range is desired, the heater resistance is no longer linear in time, and the calculation of the energy input requires a numerical integration of the power.

It is usually not possible to take voltage and current measurements simultaneously; therefore, readings of I and E are made alternately at points equally spaced in time. These values are then interpolated and extrapolated to give a table of simultaneous values of E and I from which the power ordinate may be calculated. It is usual to start and end with an

I reading, and if it is assumed that the variation of I and E is not higher than quadratic with time, Newton's formula employing second differences may be used to generate the other readings. Such a procedure requires at least three readings of each quantity. Values of E and I are calculated from voltage observations by the computer according to the procedures outlined in Section II–3–B–2.

Values of $f(t_x)$, $f(t_{x+2})$, and $f(t_{x+4})$ are measured. From these values, $\Delta f(t_x, t_{x+2})$ and $\Delta f(t_{x+2}, t_{x+4})$ are calculated and from them $\Delta^2 f(t_x, t_{x+2}, t_{x+4})$. If these difference quantities are represented for convenience by Δf_I, Δf_II, and $\Delta^2 f$, the interpolated and extrapolated values of $f(t)$ may be calculated:

Interpolation:
forward differences $\quad f(t_{x+1}) = f(t_x) + \Delta f_\mathrm{I}/2 - \Delta^2 f/8 \qquad (35)$
backward differences $\quad f(t_{x+3}) = f(t_{x+4}) - \Delta f_\mathrm{II}/2 - \Delta^2 f/8 \qquad (36)$

One step extrapolation ($'$):
backward $\quad f(t_{x-1}) = f(t_x) - \Delta f_\mathrm{I}/2 + 3\Delta^2 f/8 \qquad (37)$
forward $\quad f(t_{x+5}) = f(t_{x+4}) + \Delta f_\mathrm{II}/2 + 3\Delta^2 f/8 \qquad (38)$

Two step extrapolation ($''$):
backward $\quad f(t_{x-2}) = f(t_x) - \Delta f_\mathrm{I} + \Delta^2 f \qquad (39)$
forward $\quad f(t_{x+6}) = f(t_{x+4}) + \Delta f_\mathrm{II} + \Delta^2 f \qquad (40)$

Three step extrapolation ($'''$):
backward $\quad f(t_{x-3}) = f(t_x) - 3\Delta f_\mathrm{I}/2 + 15\Delta^2 f/8 \qquad (41)$
forward $\quad f(t_{x+7}) = f(t_{x+4}) + 3\Delta f_\mathrm{II}/2 + 15\Delta^2 f/8 \qquad (42)$

Interpolation formulas are used to generate the values of E and I between original readings. E must then be extrapolated one step to obtain readings at the first and last I readings. To obtain readings at the start and theoretical end of the energy input, I must be extrapolated one step, whereas E must be extrapolated two steps. The scheme for the values (read, interpolated and extrapolated) is diagramed below:

```
                            Time ———→
Step          ——-1    0     1    2    3    4    5    6    7    8    +1
Reading        —      on    I    E    I    E    I    E    I    off
Interpolate —                    I    E    I    E    I
Extrapolate—  E''' E''                                        E''   E'''
              I''  I'       E'                           E'   I'    I''
```

The current and voltage are assumed to vary with time in a manner not higher than quadratic. Multiplication of these two quantities at a step i gives the value of the power ordinate, P_i, at that step. The variation of $P(t)$ may now be as high as quartic in its variation with time. This power is now integrated numerically by using Simpson's rule, which is exact for systems whose time variation is quadratic or lower, but which may be in error in this case. To obtain an estimate of the magnitude of the error, it is necessary to calculate the fourth difference of the power variation $\Delta^4 P$. For this reason, the voltage and current readings are extrapolated beyond the two ends of the energy input (steps -1, $+1$), to give imaginary power ordinates at these points. For $2n + 1$ (odd) ordinates the integration is performed by using Simpson's rule:

$$\int_a^b P(t)\,dt = \frac{1}{3}\frac{(b-a)}{2n}\left[P(t_0) + P(t_{2n}) + \sum_{i=1}^{n} 4P(t_{2i-1}) + \sum_{i=1}^{n-1} 2P(t_{2i})\right] + E(t). \quad (43)$$

$E(t)$ is the error term, and is given by

$$E(t) = \frac{-2nh^5}{180}\frac{d^4P(t)}{dt^4} \cong \frac{-nh^5}{90}\frac{\Delta^4 P(t)}{h^4}. \quad (44)$$

In Equation (44), h is the interval. The $2n + 3$ points (including those erected beyond the energy input) have $n + 1$ ranges (3 points per range)—and the fourth difference for all $n + 1$ ranges may be expressed as a combination of the ordinates

$$E(t) = -(h/90)\{P(t_{-1}) + P(t_{2n+1}) - 4[P(t_0) - P(t_{2n})] + 7[P(t_1)$$

$$+ P(t_{2n-1})] - 8[\sum_{i=1}^{n-1} P(t_{2i}) - \sum_{i=2}^{n-1} P(t_{2i-1})]\} \quad (45)$$

In this way the integrals for the energy may be evaluated to the theoretical end of the input. A trapezoidal rule error adjustment involving P_{2n} and P_{2n-1} or P_{2n+1} may be made if the input is short of or beyond the "end".

(3) *Temperature.* The raw data for temperature determination by the potentiometric method involves voltage measurements across the thermometer (E) and of that across a standard resistor in series with the thermometer (I). Usually the change in I with time is very slight, and it suffices to take one reading before and one after the calorimetric experiment. The value of E (corresponding to the thermometer resistance) alters in time because of heat leaks, and it is customary therefore to establish the drift by a series of readings so as to be able to extrapolate E forward to the middle of the energy input—or, in the case of enthalpy-type experiments, to the start of the input. (see Section IV–1–B.) Similarly, observations after the input enable one to back-extrapolate E either to the midpoint or to the end of the input. These midpoint E's together with the I readings then yield values for the resistance before, and after the experiment.

The Mueller bridge method yields resistances directly. The resistance of the thermometer is determined from the observations of the resistances in the N and R connections (see Section II–3–A–2) as a function of time.

The resistances are then transformed into temperatures. Above $90\cdot18°K$, the IPTS, defined in terms of the Callendar–Van Dusen equation

$$W = \frac{R_t}{R_0} = 1 + a\left(1 + \frac{\delta}{100}\right)t - \frac{a\delta t^2}{100^2} - \frac{a\beta}{100^4}(t - 100)\,t^3, \quad (46)$$

may be solved for t by an iterative technique with

$$t \simeq (W_{exp} - 1)\left[\alpha\left(1 + \frac{\delta}{100}\right)\right]^{-1} \tag{47}$$

used to give an initial estimate of the temperature. On rearrangement, the Callendar–Van Dusen equation becomes

$W_{exp} - f(t) \simeq 0$, and upon expanding

$$W_{exp} - f(t_{i+1}) = W_{exp} - f(t_i) + \frac{d}{dt}\left(W_{exp} - f(t_i)\right)(t_{i+1} - t_i)$$

$$t_{i+1} = t_i - F(t_i)/F'(t_i)$$

$$F(t_i) = W_{exp} - f(t_i) \quad \text{and} \quad F'(t_i) = \frac{d}{dt}\left(W_{exp} - f(t_i)\right) \tag{48}$$

This expansion by the Newton–Raphson method is continued until the difference between t_{i+1} and t_i is $<10^{-5}\,°C$. The temperature is then converted to °K by the addition of 273·15. Below 90·18°K and down to 10°K, the calibration is given as a table of resistances at 1° intervals. The divided difference method of Newton may be used to interpolate this table. Such a method is equivalent to fitting a polynomial whose solutions are W_0 through W_n through the observed W values near the solution. The degree of the polynomial used is of importance for the error term is not convergent. Usually, a fourth order formula is used with a negligible contribution from the fifth order. When computing the temperature, the W-table is first searched by a binary method for the entry just less than the experimental resistance ratio; this is W_0 for the equation, and the two entries before and after are used to form the table of divided differences, from which the temperature may be calculated.

For temperatures below 10°K, an empirical equation is developed of the form

$$W = \frac{A + BT^2 + CT^5}{R_0}. \tag{49}$$

Values of the resistance at the He^4 boiling point, at 10°K, and of dR/dT at 10°K enable one to produce the constants. Values calculated from Equation (49) are then added to the W-table, and the temperature calculated as above.

Another method for obtaining values of W below 10°K is to take the published resistances of a thermometer as a function of temperature, and from them to generate a W-table for one's own thermometer.

The computer program must obviously interrogate first the resistance value to determine which of the methods for temperature computation to use.

The heat capacity of the calorimeter plus sample system is then calculated from the corresponding $q/(T_2 - T_1)$ with curvature corrections applied according to Section IV-2-B whenever necessary. The observed heat capacity of the sample is obtained by subtracting the heat capacity of the empty vessel obtained by interpolation at $(T_2 + T_1)/2$ in a table of smoothed heat capacity for the empty vessel. Corrections for differences in the mass of the vessel, for helium exchange gas, and for the impurities are also applied. (see Section IV for details and other corrections for special cases.)

(*4*) *Digital Computer Smoothing of the Experimental Data.* Some of the smoothing procedures that are adaptable to computer calculation are given in Section IV–3. The observed heat capacities are subject to experimental error; thus the data are smoothed by fitting a representative function of temperature to the observed points. The usual scheme is to fit the heat capacity as a polynomial in temperature (in principle, certain obvious advantages would accrue from the use of theoretically more amenable functions such as those involving Einstein and Debye terms), by applying "least squares" criteria to evaluate the coefficients of the powers of T. The true least squares procedure is, however, not satisfactory for machine computation if many terms are involved, as the off-diagonal elements in the matrix of the normal equations are not necessarily zero, and much machine error may be introduced in their calculation.

A way around this difficulty is to generate the normal equations by using orthogonal polynomials for which the off-diagonal elements in the normal equation matrix must necessarily be zero. An advantage of the use of orthogonal polynomials is the ease of generating sets of coefficients in terms of ascending degrees of the approximating polynomial. It is then possible for the computer to use some sort of error estimate as a criterion for terminating the curve fitting process. When this scheme of computation is used on good data, an excellent fit (0·05 per cent) is usually obtained by using a polynomial with about one third as many powers as there are points for it to fit.

The mathematics of the procedures outlined above for smoothing the observed heat capacities may be found in books on numerical analysis[39, 55, 64, 106, 109] and elsewhere[68].

Methods for calculating thermodynamic properties are given in Section IV–5.

4. Extrapolation of heat capacity below about 10°K

Although typical applications of thermodynamic properties may involve relatively high temperatures, evaluation of these properties requires inclusion of the often significant heat capacities between 0°K and the lowest experimental determination. If the extrapolation involves only a few degrees, use may be made of the Debye limiting law

$$C_V \cong \frac{12\pi^4 R}{5}\left(\frac{T}{\Theta_D}\right)^3 = \text{constant} \times T^3. \qquad (50)$$

This result has been shown to be approximately true for many substances and is often adequate for chemical thermodynamic purposes for simple solids, e.g., monatomic elements. The heat capacities of complex solids often tend toward zero in a fashion rather similar to that for monatomic elements. In a few simple cases, heat capacities of ionic crystals, such as that of sodium chloride, may be roughly approximated by the Debye theory with each ion taken equivalent to an atom of a monatomic crystal. However, an attempt to apply similar treatment to AgI results in wide divergence with the data. Substances which deviate greatly from isotropic structure also differ significantly in the temperature dependence of their heat capacity. Graphite is a good example of a lamellar solid with striking departure in heat capacity[76, 125-127] from that predicted by the Debye theory; fibrous structures show even more marked deviation[76, 124, 128]. Although the Born and von Kármán theory[10] is potentially more exact than the Debye theory, it requires more detailed information about interatomic forces than is usually available. Current advances in the application of this theory to particular crystals have been reviewed recently[75]. However, the method most generally used (especially over an appreciable range) involves use of a Debye heat capacity function fitted to the values of heat capacity at the lower temperatures (10 to 50°K). This may be refined by terms in the Einstein function. The coefficients of a series of Debye and Einstein functions with discrete pre-selected characteristic temperatures may be obtained by the least-squares method as previously described (see Section IV-3-A-(2)-(b)). When a single Debye function is used, the "degree of freedom" that yields the most constant characteristic temperature with the data at the lower temperatures is selected. This degree of freedom and the characteristic temperature are then used in the extrapolation.

For substances with magnetic or electronic complications, additional contributions may require consideration in extrapolation of the heat capacity. For solids possessing conduction electrons (e.g., metals) the contribution of the electrons may be treated as that due to a degenerate Fermi–Dirac gas. On this basis, the electronic heat capacity at low temperatures rises with the first power of the absolute temperature[115], i.e.:

$$C_{el} = \gamma T \tag{51}$$

The electronic heat capacity may greatly exceed that due to the lattice vibrations at temperatures below 3°K. Consequently, extrapolation for a substance with such electronic contributions might well be made in terms of the equation

$$C_P = aT^3 + \gamma T \tag{52}$$

in which a and γ are the constants representing, respectively, the Debye term for the lattice vibration and the term due to the conduction electrons. It is convenient to plot C_P/T versus T^2 since, on this basis, Equation (52) becomes a straight line with slope a and intercept γ at $T^2 = 0$.

Such a plot is also of considerable utility in making extrapolations to 0°K, even for substances which may not be expected to follow Equation (52) (especially when one uses available data extending to lower temperatures for similar substances from the rapidly growing collection of data). It is to be noted that the intercept of the heat capacity curve is at the origin except for substances with $\gamma \neq 0$ (e.g., those possessing metallic conduction). Heat capacity curves plotted against the suggested coordinates are shown in *Figures 18* and *19* for metals, ionic solids, other inorganic substances, and molecular crystals.

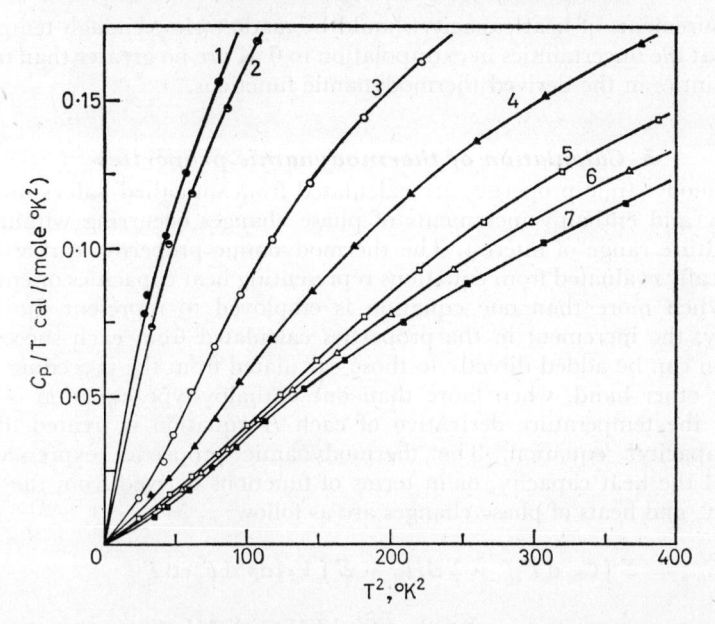

Figure 18. Heat capacity plots in the form C/T vs. T^2 for organic and inorganic molecular crystals [1, tetraphosphorus triselenide; 2, tetraphosphorus decasulfide; 3, tetraphosphorus trisulfide; 4, 3-oxabicyclo[3,2,2]nonane[138]; 5, 2,2,2-bicyclo-octane; 6, quinuclidine; 7, 2,2,2-bicyclooct-2-ene].

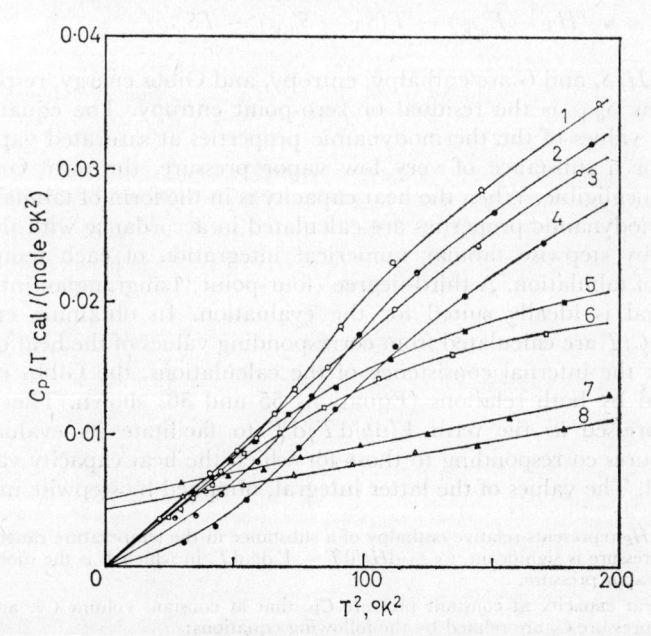

Figure 19. Heat capacity plots in the form C/T vs. T^2 for metals and inorganic solids [1, tellurium metal[138]; 2, γ-uranium trioxide; 3, α-uranium trioxide; 4, β-uranium trioxide; 5, thallium iodide[123]; 6, thallium bromide; 7, hypostoichiometric uranium dicarbide ($UC_{1.90}$)[142]; 8, uranium monocarbide].

Measurements of heat capacity should be made to low enough temperatures that the uncertainties in extrapolation to 0°K are no greater than other uncertainties in the derived thermodynamic functions.

5. *Calculation of thermodynamic properties*

Thermodynamic properties are calculated from smoothed values of heat capacity and enthalpy increments of phase changes occurring within the temperature range of interest. The thermodynamic properties can also be analytically evaluated from equations representing heat capacities or enthalpies. When more than one equation is employed to represent the heat capacity, the increment in the properties calculated from each successive equation can be added directly to those calculated from the preceding one. On the other hand, when more than one enthalpy-type equation is employed, the temperature derivative of each H equation is treated like a heat capacity† equation. The thermodynamic properties expressed in terms of the heat capacity, or in terms of functions derived from the heat capacity, and heats of phase changes are as follow:

$$H_T - E^\circ_{0^\circ K} = \Sigma \int C_{S_i} dT |_0^T + \sum_j \Delta H_{t_j} + \sum_i \int V_i (dp/dT)_i dT |_0^T \qquad (53)$$

$$S_T - S^\circ_{0^\circ K} = \sum_i \int (C_{S_i}/T) dT |_0^T + \sum_j \Delta H_{t_j}/T_t, \qquad (54)$$

$$G_T - E^\circ_{0^\circ K} = \int_0^T (S_T - S^\circ_{0^\circ K}) dT + \sum_i \int V_i (dp/dT)_i dT |_0^T - TS^\circ_{0^\circ K} \qquad (55)$$

$$= (H_T - E^\circ_{0^\circ K}) - T(S_T - S^\circ_{0^\circ K}) - TS^\circ_{0^\circ K}, \qquad (56)$$

in which H, S, and G are enthalpy, entropy, and Gibbs energy, respectively. The terms $S^\circ_{0^\circ K}$ is the residual or zero-point entropy. The equations are given for values of the thermodynamic properties at saturated vapor pressure.‡ For a substance of very low vapor pressure, the term containing dp/dT is negligible. When the heat capacity is in the form of tabular values, the thermodynamic properties are calculated in accordance with the above relation by stepwise tabular numerical integration of each temperature interval of tabulation. A third-degree (four-point) Langrangian integration polynomial is ideally suited for the evaluation. In obtaining entropies, values of C/T are calculated from corresponding values of the heat capacity. To check the internal consistency of the calculations, the Gibbs energy is calculated by both relations (Equations 55 and 56) shown. The Vdp has been expressed as the term $V(dp/dT)dT$ to facilitate its evaluation at temperatures corresponding to those for which the heat capacity values are tabulated. The values of the latter integral, obtained by stepwise numerical

---

† When H_S represents relative enthalpy of a substance in the temperature range in which its vapor pressure is significant, $C_S = dH_S/dT - V dp/dT$, in which V is the molal volume and p the vapor pressure.

‡ The heat capacity at constant pressure C_P, that at constant volume C_V, and that at saturation pressure C_S are related by the following equations:

$$C_P = C_V - T(\partial V/\partial T)^2_P/(\partial V/\partial P)_T$$

and

$$C_S = C_P - T(\partial V/\partial T)_P (dP/dT)_S.$$

integration, are applied in the evaluation of enthalpy and Gibbs energy. The increments in the thermodynamic properties over transition intervals are added as indicated in the equations.

When the interval between the tabulated values of heat capacity cannot be represented by the integrating equation, values of heat capacity should be tabulated at closer temperature intervals or a more complex integrating equation used. The four-point Langrangian integration polynomial is generally adequate for values tabulated at one-degree intervals below about 100°K and at five-degree intervals above this temperature.

V. General Discussion

Although the discussion in this chapter has been concerned largely with techniques and operational procedures employed by the authors and the colleagues in their laboratories, other modes of adiabatic calorimetry have been used and advocated. Moreover, a brief discussion of the advantages of adiabatic calorimetry in comparison with other methods is in order.

1. *Other modes of operation*

In the preceding discussion, the employment of intermittent or discontinuous heating of the calorimeter has been assumed throughout; i.e., after establishment of the quasi-adiabatic foredrift, a measured quantity of electrical energy is added to the calorimeter over a short period (typically 1 to 10 minutes) and the afterdrift followed until an equilibrium value is obtained and the temperature established. The time required for equilibrium may vary from a minute (for heat capacity of metals or macroscopic ionic molecular crystals) to more than twenty-four hours (for transitions or fusion if hysteresis or diffusion of impurities is involved). Because of the limitations of the calorimeter shield controls to maintain the temperature deviations to within about a millidegree, a restriction on the rate of heating is required. In practice this seldom exceeds a degree per minute. The temperature increments also usually do not exceed 10 per cent of the absolute temperature or 10°—whichever is smaller—except in "enthalpy" determinations through transitions or over large temperature intervals.

In an alternative adiabatic procedure, the calorimeter is heated continuously at a constant (or otherwise known) power input and the calorimeter temperature recorded continually as a function of time. (see Section II-3.) Although the method has been relatively little used in adiabatic calorimetry, it was that originally employed by Southard and Andrews[116]. The method is open to concern regarding temperature gradients in the sample, temperature increments between thermometer and sample, lack of thermal equilibration in the sample, etc. These errors may be minimized by reducing the heating rate (at the expense of prolonging the experiment); the inherent simplicity of the method lends itself to automatic control and recording—with a reduction in the tedium of making the measurements. However, it is our opinion that intermittent heating and the attendant wait for thermal equilibration is essential for some substances and clearly preferable for many others.

2. Precision and accuracy

At the present time, the most stringent requirements on calorimetry over the cryogenic range under discussion are (as previously noted) made by chemical thermodynamicists, whose interest centers ultimately in the Gibbs energy increments for chemical reactions. This increment involves the difference between the summations of the thermodynamic properties of the products and the reactants; and because it is small in comparison with the directly measured quantities and derived functions, random errors may have a more significant effect than systematic ones. Consequently, the accuracy of adiabatic calorimetry has been pushed beyond the limits imposed by uncertainties in the temperature scale, especially in the region below 90°K, for which no internationally agreed scale exists. However, these small uncertainties in the temperature tend to be minimized in the integration process for the entropy and Gibbs energy.

The achievement of high reproducibility, not only among the measurements made by one laboratory, but also between laboratories, has been engendered particularly by the Calorimetry Conference through its promulgation of a resolution concerning the publication of calorimetric data[16], encouragement for the preparation of standard samples for thermal measurements, and provision of standardization with respect to platinum- and germanium-resistance thermometers for use at lower temperatures. Because of the aforementioned possibilities of comparison between laboratories of samples of the same compound, the location of systematic errors has been facilitated; it has even become feasible to aim at an experimental consistency of several hundredths of a percent. Indeed, such a figure is claimed over much of the temperature range by a number of laboratories. However, because platinum-resistance thermometry becomes less sensitive at low temperatures, the precision of the data may decrease to an uncertainty approaching 1 per cent as 15°K is approached and to an uncertainty perhaps as much as 5 per cent below 10°K. With germanium resistance or other thermometers more sensitive at low temperatures, improved precision is attainable below 15°K.

3. Calorimetry conference standards

The Calorimetry Conference has been instrumental in arranging for the provision of large batches of highly purified and well-characterized samples for thermal calorimetry. In particular, samples of n-heptane[33, 49, 50], benzoic acid[43, 49, 50], and synthetic sapphire[42, 49, 50], have been prepared by the National Bureau of Standards and distributed internationally to calorimetrists. The synthetic sapphire sample is intended also to serve into the intermediate and high temperature region. More recently, a standard sample of very high purity copper[92] has been prepared by the Argonne National Laboratory for use as a heat capacity standard at the very low end of the temperature range considered in this chapter.

Although some comparisons of the data for n-heptane have been made elsewhere[141], there has, as yet, been relatively little correlation and comparison of the data obtained by the scores of investigators who have made measurements on these samples. Such studies would help to establish definitive values of the heat capacity. However, comparison with the published results does provide a good basis for the elimination of systematic errors between laboratories.

4. Comparison of adiabatic calorimetry with other methods of calorimetry

Because of the omnipresence of heat leak throughout the confines of our most carefully-designed apparatus, the accuracy of any type of calorimetry is likely to be limited by thermal losses. One must ultimately choose between an experiment in which these losses are quite significant, but, to a certain degree, calculable or approximately so, and an experiment in which the losses are minute, but essentially incalculable. Obviously, the method of calorimetry advocated in this chapter follows the latter choice. A brief summary of the relative advantages of adiabatic calorimetry may then be made with respect to isothermal, low-temperature calorimetry and to enthalpy-type calorimetry, employed at the upper end of the cryogenic range.

A. Comparison with Isothermal Low-temperature Calorimetry

Below temperatures of about 250°K it is probable that the so-called isothermal method of low-temperature calorimetry is capable of being operated as reliably as the adiabatic method for normal heat capacity determinations. In this method, the environment of the calorimeter is held essentially constant (by thermal inertia) near the expected mean temperature of the measurement. The heat leak between the calorimeter and surroundings is evaluated from observations of the drift in temperature of the calorimeter before and after the energy input. Above 250°K the accuracy of the method decreases rapidly because of the increasing size of the correction for heat exchange. Discussions of the advantages and reliability of isothermal, low-temperature calorimetry have been presented in Chapter 6 of this book and by Cole, et al.[20]. However, in studies of molecular crystals and other substances with transformations involving considerable thermal hysteresis, the adiabatic method has a distinct advantage over the isothermal because it makes possible reliable measurements, even when thermal equilibration of the samples requires periods of more than a day. Moreover, the simplicity of evaluating the resultant heat capacity data, the economy of refrigerant, and the rapidity with which measurements can be made provide operational convenience in favor of the adiabatic working, especially when automatic temperature control is provided for the adiabatic shield.

B. Comparison with Intermediate Temperature ($> 250°K$) Methods

In the region above about 273°K most existing data on the thermal properties of substances have been obtained by means of the method of mixtures or drop calorimetry, the general techniques of which are described in Chapter 8 of this volume. In this method, the enthalpy increment between some reference temperature near 300°K and a sequence of temperatures spaced by 100° intervals is obtained. The heat capacity is then obtained by differentiation of the enthalpy curve. The enthalpy increment associated with the transition is obtained by extrapolation of the experimental enthalpy curves to the transition temperature. Although this method permits relatively rapid working and quite small samples, it suffers from its inability to delineate the fine structure in the heat capacity curve or the temperature dependence of the heat capacity in the vicinity of a transition and, in some instances, may lead to quite erroneous results because of the "freezing in" of a metastable phase. When transformations occur that are not reversible on cooling, enthalpy data cannot be obtained by the "method of mixtures" alone. However, usually in conjunction with other measurements, such as determination of enthalpies of solution, the difficulty may be circumvented. For transitions in which the thermal effect is small or the temperature dependence of the heat capacity over a small temperature range is of interest, determination as the difference between large quantities results in poor delineation of both nature and magnitude of the thermal effect. For measurements such as the thermal effect in the dehydration of clays on firing, which is not reversible, direct methods must be used. Similarly, in the transition range of vitreous substances, the enthalpy depends on the rate of cooling, and the direct method is necessary for the investigation of these effects. The development of adiabatic calorimeters suitable for operation in the intermediate temperature range may significantly increase the precise calorimetric data obtained by this means.

5. *Future developments*

The importance and vitality of adiabatic calorimetry suggest that improvements in calorimetric technique and thermometry will enable extraction of energetic and thermodynamic data from most substances. Although a significant increase in precision beyond that of the best definitive work today does not seem to be required for current problems, improvements in thermometry especially in the region near 10°K would materially improve precision at this end of the temperature scale. Precision is already high enough so that smoothing of the temperature scales and the development of internationally acceptable temperature scales approaching the thermodynamic temperature scale would provide significant improvement. Improvement in the design of calorimetric vessels might permit more rapid and secure loading of materials with less opportunity for exposure to the atmosphere and consequently less contamination.

With improvements in the precision of measurements and in the theoretical understanding of matter, coupled with the ever-expanding capabilities of high-speed digital computers, samples prepared under more rigid controls will be needed.

Development of techniques that would permit high precision on samples of less than 1 gram size would enable considerable expansion of endeavor in the study of expensive, explosive, and rare materials. Improved cryostat design might provide programmed cooling of the materials whose thermal history is important to the measured properties and also provide the opportunity for very rapid quenching of metastable states. Further development of automatic adiabatic shield control would provide increased precision and reliability. This greater automation of the entire measurement process could result in more efficient use of scientific manpower in the determination of thermodynamic data. The automatic, self-balancing bridge for resistance thermometry installed in the thermodynamics laboratory of the National Bureau of Standards, together with automated punch card data output suitable for direct processing by high-speed digital computers[40], is an example of one trend in the direction of automation.

VI. Acknowledgement

One of us (E.F.W.) wishes to express his appreciation to the Division of Research of the U.S. Atomic Energy Commission for its generous support of his research endeavor, which to a large extent enabled his research in adiabatic calorimetry.

VII. References

[1] Aston, J. G., private communication.
[2] Aston, J. G., and M. L. Eidinoff, *J. Am. Chem. Soc.*, **61**, 1533 (1939).
[3] Aston, J. G., H. L. Finke, G. J. Janz, and K. E. Russell, *J. Am. Chem. Soc.*, **73**, 1939 (1951).
[4] Aston, J. G., H. L. Finke, and S. C. Schumann, *J. Am. Chem. Soc.*, **65**, 341 (1943).
[5] Aston, J. G., H. L. Finke, J. W. Tooke, and M. R. Cines, *Anal. Chem.*, **19**, 218 (1947).
[6] Aston, J. G., E. Willihnganz, and G. H. Messerly, *J. Am. Chem. Soc.*, **57**, 1642 (1935).
[7] Aven, M. H., R. S. Craig, and W. E. Wallace, *Rev. Sci. Instr.*, **27**, 623 (1956).
[8] Beattie, J. A., D. D. Jacobus, and J. M. Gaines, Jr., *Proc. Am. Acad. Arts Sci.*, **66**, 167 (1930).
[9] Beehler, R. E., R. C. Mockler, and C. S. Snider, *Nature*, **187**, 681 (1960).
[10] Born, M., and T. von Kármán, *Physik. Z.*, **13**, 297 (1912); **14**, 15 (1913).
[11] Brickwedde, F. G., H. van Dijk, M. Durieux, J. R. Clement, and J. K. Logan, *J. Res. Natl. Bur. Std.*, **64A**, 1 (1960).
[12] Buckingham, R. A., *Numerical Methods*, rev. ed., Pitman, London, 1962.
[13] Burk, D. L., and S. A. Friedberg, "The Atomic Heat of Diamond at Low Temperatures," in J. R. Dillinger, ed., *Low Temperature Physics & Chemistry*, University of Wisconsin Press, Madison, 1958.
[14] Burns, J. H., D. W. Osborne, and E. F. Westrum, Jr., *J. Chem. Phys.*, **33**, 387 (1960).
[15] Busey, R. H., *U.S. Atomic Energy Commission Report ORNL*-1828, 1955; *Chem. Abstr.*, **50**, 14277c (1956).
[16] Calorimetry Conference, *Physics Today*, **14**, No. 2, 47 (1961).
[17] Clarke, J. T., H. L. Johnston, and W. DeSorbo, *Anal. Chem.*, **25**, 1156 (1953).
[18] Clever, H. L., E. F. Westrum, Jr., and A. W. Cordes, *J. Phys. Chem.*, **69**, 1214 (1965).
[19] Clusius, K., and J. V. Vaughen, *Z. Ges. Kälte-Ind.*, **36**, 215 (1929).
[20] Cole, A. G., J. O. Hutchens, R. A. Robie, and J. W. Stout, *J. Am. Chem. Soc.*, **82**, 4807 (1960).
[21] Corak, W. S., M. P. Garfunkel, C. B. Satterthwaite, and A. Wexler, *Phys. Rev.*, **98**, 1699 (1955).
[22] Cruickshank, A. J. B., *Phil. Trans. Roy. Soc. London*, **A253**, 407 (1960).
[23] Dauphinee, T. M., *Can. J. Phys.*, **31**, 577 (1953).
[24] Dauphinee, T. M., "Potentiometric Methods of Resistance Measurement," in C. M. Herzfeld, ed., *Temperature, Its Measurement and Control in Science and Industry*, Vol. 3, Part 1, Reinhold, New York, 1962.
[25] Dauphinee, T. M., D. K. C. MacDonald, and H. Preston-Thomas, *Proc. Roy. Soc. (London)*, **A221**, 267 (1954).
[26] Dauphinee, T. M., and H. Preston-Thomas, *Rev. Sci. Instr.*, **25**, 884 (1954).
[27] Dauphinee, T. M., and H. Preston-Thomas, *J. Sci. Instr.*, **35**, 21 (1958).
[28] De Nobel, J., and F. J. du Chatenier, *Physica*, **29**, 1231 (1963).
[29] DeSorbo, W., *J. Chem. Phys.*, **21**, 876 (1953).
[30] Din, F., and A. H. Cockett, eds., *Low-Temperature Techniques*, George Newnes Ltd., London, 1960.
[31] Dole, M., W. P. Hettinger, Jr., N. Larson, J. A. Wethington, and A. E. Worthington, *Rev. Sci. Instr.*, **22**, 812 (1951).
[32] Donjon, A., *Proces-Verbaux Seances, Comite Inter-Natl. Poids Mesures*, **25**, 89 (1956).
[33] Douglas, T. B., G. T. Furukawa, R. E. McCoskey, and A. F. Ball, *J. Res. Natl. Bur. Std.*, **53**, 139 (1954).
[34] Dunfee, B. L., *J. Res. Natl. Bur. Std.*, **67C**, 1 (1963).

35. Egan, C. J., and J. D. Kemp, *J. Am. Chem. Soc.*, **59**, 1264 (1937).
36. Eucken, A., *Physik. Z.*, **10**, 586 (1909).
37. Findlay, A., *The Phase Rule and Its Applications*, p. 39, 9th ed., rev. by A. N. Campbell and N. O. Smith, Dover, New York, 1951.
38. Finke, H. L., M. E. Gross, G. Waddington, and H. M. Huffman, *J. Am. Chem. Soc.*, **76**, 333 (1954).
39. Forsythe, G. E., *J. Soc. Ind. Appl. Math.*, **5**, 74 (1957).
40. Furukawa, G. T., private communication.
41. Furukawa, G. T., presented at the 11th Calorimetry Conference, Johns Hopkins University, Baltimore, Maryland, September 14–15, 1956.
42. Furukawa, G. T., T. B. Douglas, R. E. McCoskey, and D. C. Ginnings, *J. Res. Natl. Bur. Std.*, **57**, 67 (1956).
43. Furukawa, G. T., R. E. McCoskey, and G. J. King, *J. Res. Natl. Bur. Std.*, **47**, 256 (1951).
44. Furukawa, G. T., and J. H. Piccirelli, "Calorimetric Determination of the Purity of Benzene (IUPAC-59)," International Union of Pure and Applied Chemistry, Commission on Physicochemical Data and Standards, Cooperative Determination of Purity by Thermal Methods, Report of the Organizing Committee, July 14, 1961.
45. Furukawa, G. T., M. L. Reilly, and W. G. Saba, *Rev. Sci. Instr.*, **35**, 113 (1964).
46. Gaede, W., *Physik. Z.*, **4**, 105 (1902).
47. Giauque, W. F., and R. Wiebe, *J. Am. Chem. Soc.*, **50**, 101 (1928).
48. Gibson, G. E., and W. F. Giauque, *J. Am. Chem. Soc.*, **45**, 93 (1923).
49. Ginnings, D. C., and G. T. Furukawa, *J. Am. Chem. Soc.*, **75**, 522 (1953).
50. Ginnings, D. C., and G. T. Furukawa, *J. Am. Chem. Soc.*, **75**, 6359 (1953).
51. Ginnings, D. C., and E. D. West, *Rev. Sci. Instr.*, **35**, 965 (1964).
52. Glasgow, A. R., Jr., G. S. Ross, A. T. Horton, D. Enagonio, H. D. Dixon, C. P. Saylor, G. T. Furukawa, M. L. Reilly, and J. M. Henning, *Anal. Chim. Acta*, **17**, 54 (1957).
53. Glasstone, S., *Thermodynamics for Chemists*, Van Nostrand, New York, 1947, p. 343.
54. Harris, F. K., *Electrical Measurements*, Wiley, New York, 1952, p. 176.
55. Hildebrand, F. B., *Introduction to Numerical Analysis*, McGraw-Hill, New York, 1956.
56. Hildebrand, J. H., and R. L. Scott, *The Solubility of Nonelectrolytes*, 3rd ed., Reinhold, New York, 1950, p. 27.
57. Hill, R. W., *J. Sci. Instr.*, **30**, 331 (1953).
58. Hill, R. W., "Low-temperature Calorimetry," in K. Mendelssohn, ed., *Progress in Cryogenics*, Vol. 1, Heywood & Company Ltd., London, 1959.
59. Hill, R. W., and P. L. Smith, *Phil. Mag.*, **44**, 636 (1953).
60. Hoare, F. E., L. C. Jackson, and N. Kurti, eds., *Experimental Cryophysics*, Butterworths, London, 1961.
61. Hoge, H. J., *J. Res. Natl. Bur. Std.*, **36**, 111 (1946).
62. Hoge, H. J., *Rev. Sci. Instr.*, **20**, 59 (1949).
63. Hoge, H. J., and F. G. Brickwedde, *J. Res. Natl. Bur. Std.*, **22**, 351 (1939).
64. Householder, A. S., *Principles of Numerical Analysis*, McGraw-Hill, New York, 1953.
65. Huffman, H. M., *Chem. Rev.*, **40**, 1 (1947).
66. Johnson, C. L., *Analog Computer Techniques*, 2nd ed., McGraw-Hill, New York, 1963.
67. Johnston, H. L., J. T. Clarke, E. B. Rifkin, and E. C. Kerr, *J. Am. Chem. Soc.*, **72**, 3933 (1950).
68. Justice, B. H., "Calculation of Heat Capacities and Derived Thermodynamic Functions from Thermal Data with a Digital Computer," Appendix to Ph.D. Dissertation, University of Michigin; *U.S. Atomic Energy Commission Report TID-12722*, 1961.
69. Keesom, W. H., and J. A. Kok, *Physica*, **1**, 770 (1934).
70. Keesom, W. H., and J. N. Van den Ende, *Commun. Kamerlingh Onnes Lab. Univ. Leiden*, 219b (1932).
71. Kostryukova, M. O., and P. G. Strelkov, *Compt. Rend. (Dokl.) Acad. Sci. URSS*, **B90**, 525 (1953).
72. Kunzler, J. E., T. H. Geballe, and G. W. Hull, *Rev. Sci. Instr.*, **28**, 96 (1957).
73. *Landolt-Börnstein Physikalisch-Chemische Tabellen*, Vol. II, 5th ed., Springer, Berlin, 1923, p. 1049.
74. Lange, F., *Z. Physik. Chem. (Leipzig)*, **110**, 343 (1924).
75. de Launay, J., *Solid State Phys.*, **2**, 219 (1956).
76. Lifshits, I. M., *Zh. Eksperim. i Teor. Fiz.*, **22**, 471 (1952); **22**, 475 (1952).
77. MacDonald, D. K. C., "Electrical Conductivity of Metals and Alloys at Low Temperatures," in S. Flügge, ed., *Low Temperature Physics I* (Encyclopedia of Physics, Vol. XIV) Springer, Berlin, 1956, p. 137.
78. Markowitz, W., R. G. Hall, L. Essen, and J. V. L. Parry, *Phys. Rev. Letters*, **1**, 105 (1958).
79. Martin, D. L., *Phil. Mag.*, **46**, 751 (1955).
80. Martin, D. L., *Can. J. Phys.*, **38**, 17 (1960).
81. Mastrangelo. S. V. R., and R. W. Dornte, *J. Am. Chem. Soc.*, **77**, 6200 (1955).

[82] McCullough, J. P., *Pure Appl. Chem.* **2**, 221 (1961).
[83] McCullough, J. P., and G. Waddington, *Anal. Chim. Acta.*, **17**, 80 (1957).
[84] Messerly, J. F., S. S. Todd, G. B. Guthrie, and J. P. McCullough, U. S. Bur. Mines, *RI 6273*, Washington, D.C., 1963.
[85] Meyers, C. H., *J. Res. Natl. Bur. Std.*, **9**, 807 (1932).
[86] Milne, W. E., *Numerical Calculus; Approximations, Interpolation, Finite Differences, Numerical Integration and Curve Fitting*, Princeton University Press, Princeton, New Jersey, 1949, p. 93.
[87] Milne-Thomson, L. M., *The Calculus of Finite Differences*, MacMillan, London, 1951.
[88] Morrison, C. F., Jr., *Generalized Instrumentation for Research and Teaching*, Washington State University Press, Pullman, Washington, 1964.
[89] Mueller, E. F., "Precision Resistance Thermometry," in *Temperature, Its Measurement and Control in Science and Industry*, Vol. I, American Institute of Physics, Reinhold, New York, 1941.
[90] Natrella, M. G., *Experimental Statistics* (U.S. Natl. Bur. Std. Handbook 91), U.S. Department of Commerce, National Bureau of Standards, Washington, D.C., 1963.
[91] Nernst, W., *Sitzber. kgl. preuss. Akad. Wiss.*, **12**, **13**, 261 (1910); *Chem. Abstr.*, **4**, 2397 (1910).
[92] Osborne, D. W., and H. E. Flotow, personal communication.
[93] Osborne, D. W., E. F. Westrum, Jr., and H. R. Lohr, *J. Am. Chem. Soc.*, **77**, 2737 (1955).
[94] Osborne, N. S., *J. Res. Natl. Bur. Std.*, **4**, 609 (1930).
[95] Osborne, N. S., and D. C. Ginnings, *J. Res. Natl. Bur. Std.*, **39**, 453 (1947).
[96] Osborne, N. S., H. F. Stimson, T. S. Sligh, Jr., and C. S. Cragoe, *Sci. Papers (U.S.) Natl. Bur. Std.*, **20**, (Sci. Paper No. 501), 65 (1925); *Chem. Abstr.*, **19**, 2159 (1925).
[97] Parkinson, D. H., and J. E. Quarrington, *Brit. J. Appl. Phys.*, **5**, 219 (1954).
[98] Pearlman, N., and P. H. Keesom, *Phys. Rev.*, **88**, 398 (1952).
[99] Pilcher, G., *Anal. Chim. Acta*, **17**, 144 (1957).
[100] Ramanathan, K. G., and T. M. Srinivasan, *Phil. Mag.*, **46**, 338 (1955).
[101] Rayne, J. A., *Australian J. Phys.*, **9**, 189 (1956).
[102] Rose-Innes, A. C., *Low Temperature Techniques*, English Universities Press Ltd., London, 1964.
[103] Rosengren, K., *Rev. Sci. Instr.*, **32**, 1264 (1961).
[104] Rossini, F. D., K. S. Pitzer, R. L. Arnett, R. M. Braun, and G. C. Pimentel, *Selected Values of Physical and Thermodynamic Properties of Hydrocarbons and Related Compounds* Carnegie Press, Pittsburgh, 1953.
[105] Rubin, T., H. L. Johnston, and H. Altman, *J. Am. Chem. Soc.*, **73**, 3401 (1951).
[106] Rudin, B. D., *Lockheed Math. Computer Res. Rept.*, Missile Systems Div., Palo Alto, California, **1**, No. 3, 8 (1957).
[107] Ruehrwein, R. A., and W. F. Giauque, *J. Am. Chem. Soc.*, **61**, 2940 (1939).
[108] Ruehrwein, R. A., and H. M. Huffman, *J. Am. Chem. Soc.*, **65**, 1620 (1943).
[109] Scarborough, J. B., *Numerical Mathematical Analysis*, 2nd ed., Johns Hopkins Press, Baltimore, 1950.
[110] Scott, R. B., *Cryogenic Engineering*, Van Nostrand, Princeton, New Jersey, 1959.
[111] Scott, R. B., C. H. Meyers, R. D. Rands, Jr., F. G. Brickwedde, and N. Bekkedahl, *J. Res. Natl. Bur. Std.*, **35**, 39 (1945).
[112] Shiffman, C. A., J. F. Cochran, M. Garber, and G. W. Pearsall, *Rev. Mod. Phys.*, **36**, 127 (1964).
[113] Silsbee, F. B., and F. J. Gross, *J. Res. Natl. Bur. Std.*, **27**, 269 (1941).
[114] Sligh, T. S., Jr., *Sci. Papers (U.S.), Natl. Bur. Std.*, **17**, 49 (Sci. Paper No. 407) (1921); *Chem. Abstr.*, **15**, 3004 (1921).
[115] Sommerfeld, A., *Z. Physik*, **47**, 1 (1928).
[116] Southard, J. C., and D. H. Andrews, *J. Franklin Inst.*, **209**, 349 (1930).
[117] Southard, J. C., and F. G. Brickwedde, *J. Am. Chem. Soc.*, **55**, 4378 (1933).
[118] Southard, J. C., and R. T. Milner, *J. Am. Chem. Soc.*, **55**, 4384 (1933).
[119] Staveley, L. A. K., and A. K. Gupta, *Trans. Faraday Soc.*, **45**, 50 (1949).
[120] Sterrett, K. F., D. H. Blackburn, A. B. Bestul, S. S. Chang, and J. Horman, *J. Res. Natl. Bur. Std.*, **69C**, 19 (1965).
[121] Stimson, H. F., *J. Res. Natl. Bur. Std.*, **42**, 209 (1949).
[122] Stull, D. R., *Anal. Chim. Acta.* **17**, 133 (1957).
[123] Takahashi, Y., and E. F. Westrum, Jr., *J. Chem. Eng. Data*, **10**, 244 (1965).
[124] Tarasov, V. V., *Dokl. Akad. Nauk SSSR*, **46**, 22 (1945); *Compt. Rend. (Dokl.) Acad. Sci. URSS*, **46**, 20 (in English); *Chem. Abstr.*, **39**, 5163[4] (1945).
[125] Tarasov, V. V., *Dokl. Akad. Nauk SSSR*, **54**, 803 (1946); *Compt. Rend. (Dokl.) Acad. Sci. URSS*, **54**, 795 (in English); *Chem. Abstr.*, **41**, 6126g (1947).
[126] Tarasov, V. V., *Zh. Fiz. Khim.*, **24**, 111 (1950); *Chem. Abstr.*, **44**, 4742f (1950).
[127] Tarasov, V. V., *Zh. Fiz. Khim.*, **27**, 1430 (1953); *Chem. Abstr.*, **49**, 5952b (1955).
[128] Tarasov, V. V., *Zh. Fiz. Khim.*, **29**, 198 (1955); *Chem. Abstr.*, **50**, 13588h (1956).
[129] Thomas, J. L., *Precision Resistors and Their Measurement* (U.S. Natl. Bur. Std. Circular 470), U.S. Government Printing Office, Washington, D.C., 1948.

[130] Timmerhaus, K. D., "Low Temperature Thermometry," in R. W. Vance and W. M. Duke, eds., *Applied Cryogenic Engineering*, Wiley, New York, 1962.
[131] Todd, L. J., R. H. Dettre, and D. H. Andrews, *Rev. Sci. Instr.*, **30**, 463 (1959).
[132] Trowbridge, J. C., and E. F. Westrum, Jr., *J. Phys. Chem.* **68**, 42 (1964).
[133] Tunnicliff, D. D., and H. Stone, *Anal. Chem.*, **27**, 73 (1955).
[134] Vance, R. W., ed., *Cryogenic Technology*, Wiley, New York, 1963.
[135] Webb, F. J., and J. Wilks, *Proc. Roy. Soc. (London)*, **A230**, 549 (1955).
[136] Wenner, F., *J. Res. Natl. Bur. Std.*, **25**, 229 (1940).
[137] West, E. D., and D. C. Ginnings, *J. Res. Natl. Bur. Std.*, **60**, 309 (1958).
[138] Westrum, E. F., Jr., unpublished data.
[139] Westrum, E. F., Jr., "Application of Cryogenic Calorimetry to Solid-State Chemistry," in K. D. Timmerhaus, ed., *Advances in Cryogenic Engineering*, Vol. 7, Plenum Press, New York, 1962.
[140] Westrum, E. F., Jr., J. B. Hatcher, and D. W. Osborne, *J. Chem. Phys.*, **21**, 419 (1953).
[141] Westrum, E. F., Jr., and J. P. McCullough, "Thermodynamics of Crystals," in D. Fox, M. M. Labes, and A. Weissberger, eds., *Physics and Chemistry of the Organic Solid State*, Vol. I, Interscience, New York, 1963.
[142] Westrum, E. F., Jr., E. Suits, and H. K. Lonsdale, "Uranium Monocarbide and Hypostoichiometric Dicarbide. Heat Capacities and Thermodynamic Properties from 5 to 350°K.," in S. Gratch, ed., *Advances in Thermophysical Properties at Extreme Temperatures and Pressures*, American Society of Mechanical Engineers, New York, 1965.
[143] White, G. K., *Experimental Techniques in Low-temperature Physics,*, Oxford University Press, London, 1959 p. 190.
[144] White, G. K., and S. B. Woods, *Rev. Sci. Instr.*, **28**, 638 (1957).
[145] Whittaker, E. T., and G. Robinson, *The Calculus of Observations. A Treatise on Numerical Mathematics*, 4th ed., Blackie and Son Ltd., London, 1944. [a]p. 291; [b]p. 300.
[146] Yost, D. M., C. S. Garner, D. W. Osborne, T. R. Rubin, and H. Russell, Jr., *J. Am. Chem. Soc.*, **63**, 3488 (1941).
[147] Zabetakis, M. G., R. S. Craig, and K. F. Sterrett, *Rev. Sci. Instr.*, **28**, 497 (1957).
[148] Ziegler, W. T., and J. C. Mullins, *Cryogenics*, **4**, 39 (1964).
[149] "Atomic Frequency Standards," *Natl. Bur. Std. (U.S.)*, *Tech. News Bull.*, **45**, 8 (1961).
[150] *Standard Frequency and Time Services of the National Bureau of Standards* (Misc. Pub. 236—1965 ed.), U.S. Dept. Comm., Natl. Bur. Std., Washington, D. C., 1965.
[151] "World Sets Atomic Definition of Time," *Natl. Bur. Std. (U.S.)*, *Tech. News Bull.*, **48**, 209 (1964).
[152] In the application of constant current power supplies, the time constant should be carefully examined in relation to the switching time (i.e., switch SW_1 of *Figure 11*) and switch action, make-before-break or break-before-make. The energy stored in the filter capacitor of the power supply during the transient period may be significantly different from that of its normal operation and may be a source of large unaccounted energy introduced into the calorimeter.

CHAPTER 6

Low-temperature Calorimetry with Isothermal Shield and Evaluated Heat Leak

J. W. STOUT

Department of Chemistry and James Franck Institute,
University of Chicago, Chicago, Illinois, U.S.A.

Contents

I.	Introduction	216
II.	Description of Typical Calorimetric Apparatus	217
	1. General Description of a Typical Cryostat	217
	2. A Cryostat for Calorimetry of Condensed Gases	221
	3. Calorimeters, Thermometers, and Heaters	224
	A. Calorimeter with Thermometer–Heater on Outer Surface	224
	B. Calorimeter with Internal Thermometer–Heater	226
	(1) Calorimeter	226
	(2) The Thermometer–Heater	227
	4. Equipment for Measurement of Resistance, Energy and Time	229
	5. Experimental Procedures	230
	A. Calorimeter for Solids	230
	B. Calorimeter for Condensed Gases	231
III.	Calculation of Heat Capacity Data	232
	1. Corrections for Heat Leak	232
	A. Newton's Law of Cooling	232
	B. Thermal Relaxation Times	234
	C. Calculation of Heat Capacity Neglecting Thermal Gradients in the Calorimetric System	238
	D. Correction for Thermal Gradients in the Calorimetric System	242
	(1) Calorimeter with External Thermometer–Heater The $(A/E) B_N$ Correction	242
	(2) Calorimeter with Internal Thermometer–Heater. The γB_N Correction	245
	(3) Temperature Gradients between Calorimeter and Sample	247
	E. Representative Values of A/E and B_N	248
	F. Non-constant Heating Correction	249
	2. Electrical Energy Measurements and Calculations	251
	A. Current and Potential during the Heating Period	251
	B. Transient Heating Corrections	252
	C. Corrections for Lead Wires	253
	(1) Four-Lead Thermometer–Heater	253
	(2) Two-Lead Thermometer–Heater	254
	D. Correction for Heating by Thermometer Current	254
	3. Correction for Heat Capacity of Empty Calorimeter, Solder, and Helium	255
	4. Temperature Scales	255
	5. Calculation of $C_P°$ and other Thermodynamic Properties	256
	A. Calculation of $\Delta H/\Delta T$	256
	B. Curvature Corrections	257
	C. Thermodynamic Properties	258
	6. Precision and Accuracy	259
IV.	Acknowledgements	259
V.	References	260

I. Introduction

The accurate determination of the heat capacity of a substance involves the measurement of the energy required to raise the temperature of the substance from an initial temperature, T_1, to a final temperature, T_2. Since the heat capacity varies with temperature, the temperature rise in a single measurement, $\Delta T = T_2 - T_1$, should be small. A thermometer in good contact with the sample and of sufficient precision to measure ΔT with the required accuracy is required. A calorimeter, preferably constructed of a metal of good thermal conductivity to facilitate temperature equilibrium, is needed to contain the sample whose heat capacity is to be determined. Electrical energy, which can be readily determined to high precision through measurements of the potential drop, current, and time interval of direct current passing through a heater in thermal contact with the calorimeter, is a convenient and accurate means of introducing energy. Account must be taken of the exchange of heat between the calorimeter and its environment. To minimize this heat leak, the space between the calorimeter and its surroundings is evacuated to a pressure less than 10^{-5} mm mercury to reduce gaseous thermal conduction; all solid connections between the calorimeter and its surroundings, required for mechanical support and as electrical leads, are made small to minimize heat conduction through them; and the surfaces of the calorimeter and of its surroundings are made of materials of high reflectivity to thermal radiation in order to minimize radiative heat transfer.

Calorimeters incorporating these principles were developed around 1909 in Germany by Eucken[13] and by Nernst[40], and used in the United States by Eastman and Rodebush[12] and by Parks, Gibson and Latimer[45]. In these early calorimeters, the outer wall of the vacuum space surrounding the calorimeter was in thermal contact with a liquid bath of boiling refrigerant, and, when the temperature of the calorimeter was far from that of the bath, the heat leak became large and introduced serious errors in the measurement of heat capacity. These errors may be reduced through the introduction of a metallic thermal shield, whose temperature can be controlled, and which hangs in the vacuum space between the calorimeter and the surrounding bath. Descriptions of calorimeters involving this feature are given by Nernst and Schwers[41] and by Simon[48]. An excellent review of low-temperature calorimetry up to 1928 has been given by Eucken[14].

At the University of California, Berkeley, Giauque and his collaborators[3, 6, 18-24, 28, 34, 39, 43] further developed apparatus for the measurement of heat capacities and of heats of transition, fusion and vaporization at temperatures between 10 and 300°K. In this apparatus, the calorimeter is surrounded by a massive thermal shield made of copper and lead which hangs in the insulating vacuum space. The shield is provided with an electrical heater and thermocouples to measure its temperature. In a heat capacity measurement, the temperature of the calorimeter is never far from that of the shield, and the heat leak correction can thus be made small and accurately calculable. With this apparatus, the errors in the measurement of heat capacity were reduced to about two tenths of a per cent over the temperature range from 35°K to room temperature. Measurements with this type of calorimetric apparatus have sometimes been loosely described by the term "isothermal calorimetry". The word isothermal refers not to the calorimeter whose temperature necessarily changes during a heat capacity measurement, but to the surrounding massive shield whose temperature remains nearly constant during the course of a single measurement. Descriptions of calorimetric apparatus similar to that developed by Giauque and his collaborators have been published by Parks[44], Millar[38], Latimer and Greensfelder[35], Haas and Stegeman[26], Aston and Messerly[1], Hicks[27], Pitzer and Coulter[46], Clusius and Popp[8], Johnston, Clarke, Lifkin and Kerr[31], Johnston and Kerr[32], De Sorbo[11], Taylor, Johnson and Kilpatrick[54], Busey[5] and Cole, Hutchens, Robie, and Stout[9].

Calorimetric apparatus of the "adiabatic" type, in which the thermal shield is of low heat capacity and is maintained close to the temperature of the calorimeter throughout a measurement, is described in Chapter 5. In this chapter, typical apparatus and methods, following principally those developed by Giauque and collaborators, for low-temperature calorimetry employing a massive thermal shield and with careful evaluation of heat leak, will be described. The description will lean heavily on the author's personal experience at Berkeley and at Chicago, and no attempt at a comprehensive survey of the literature will be made.

II. Description of Typical Calorimetric Apparatus

1. *General description of a typical cryostat*

In *Figure 1* is shown a schematic drawing of the cryostat used for low temperature heat capacity measurements on solids[9] in the author's laboratory at the University of Chicago. Many of the features were copied from earlier designs of Giauque and coworkers at the University of California, Berkeley. The outer insulating case is made of sheet metal and wood and is filled with expanded vermiculite insulation. Its outer diameter is 40 cm and the well in the case is 20·5 cm in diameter and 119 cm deep. An alternative and superior design for the outer case, which has been employed in cryostats used for other purposes in the author's laboratory, makes use of a heavy seamless brass tube, 4–6 mm thick, for the outer wall. The inner wall is of thin monel, cupronickel or stainless steel tubing reinforced with brass rings soft soldered to its outside at 15 cm intervals. The inner and outer tubes of the case are soldered to flanges at the top which form part of an O-ring seal. All other joints are soldered. The space between the tubes is filled with Santocel insulation and evacuated at room temperature to a pressure of about 10 micron of mercury. A case of this design provides considerably better thermal insulation than that shown in *Figure 1*.

The Pyrex glass Dewar vessel, of capacity 8·5 liters, is contained in a vacuum tight outer can made of monel metal with a stainless steel ball-vee mechanical joint[19] at the top. In dismantling the apparatus, the outer can assembly is lifted with a hoist, the case removed, and the outer can assembly lowered in place for removal of its lower part which contains the Dewar, and for further dismantling of the apparatus. Liquid nitrogen or liquid hydrogen is added to the glass Dewar through a filling tube. A vacuum

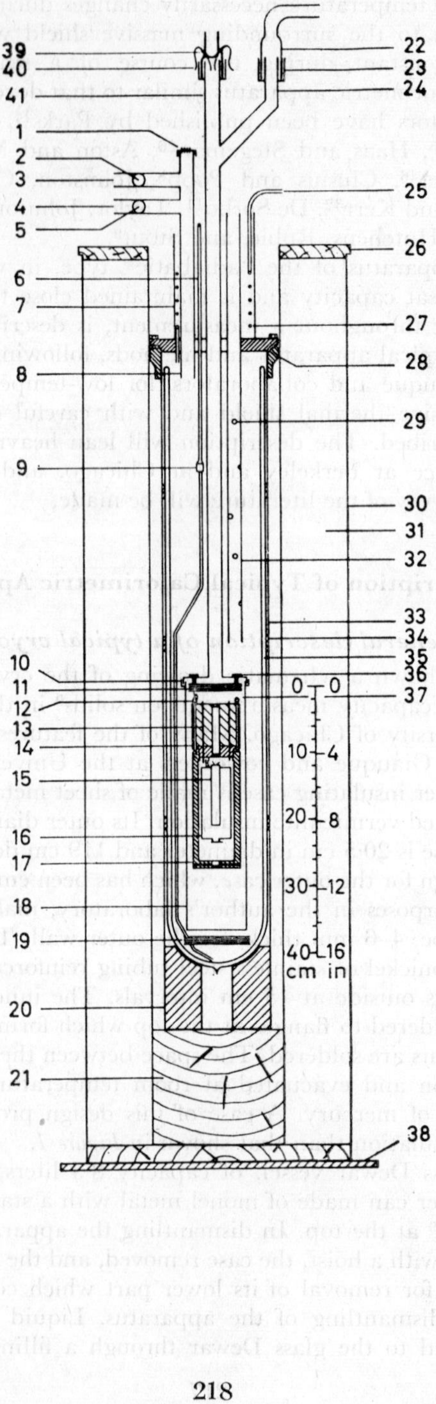
39
40
41
1
2
3
4
5
6
7
8
9
10
11
12
13
14
15
16
17
18
19
20
21
22
23
24
25
26
27
28
29
30
31
32
33
34
35
36
37
38
0—0
10—4
20—8
30—12
40—16
cm in

Figure 1. Cryostat for heat capacity measurements. [A. G. Cole, J. O. Hutchens, R. A. Robie, and J. W. Stout, *J. Am. Chem. Soc.*, **82**, 4807 (1960).]

1, Kovar-ceramic seals. Two seals are used for the Dewar heater leads and two for the thermocouples at the bottom and top of the inner vacuum can. The wires are sealed with wax in Kovar tubes. To prevent cracking, the wax seals should be farther from the cold parts of the cryostat than shown
2, Connection to safety valve
3, Connection to hydrogen gas-holder
4, Connection to vacuum pump
5, Rubber hose outlet to room
6, Blow-out tube
7, Monel cup
8, Gas outlet tube
9, Union joint in blow-out tube
10, Solder cup at top of inner vacuum can
11, Heat station in thermal contact with refrigerant bath
12, Shield supporting screws with supporting string
13, Copper wire leads through top of shield
14, Top of shield
15, Resistance thermometer–heater
16, Calorimeter
17, Bottom of shield with vent holes
18, Inner vacuum can
19, Dewar heater
20, Balsa wood Dewar support
21, Wood block
22, Glass tube to high vacuum pumping line
23, Wax seal
24, Monel pumping tube
25, Blow-in tube
26, Wood top of insulating case
27, Coil of blow-in tube immersed in monel cup
28, Stainless steel ball-vee joint
29, Holes in filling tube
30, Galvanized iron sheet metal forming outside of insulating case
31, Expanded vermiculite insulation in insulating case
32, Filling tube
33, Monel sheet metal forming wall of outer can
34, Brass reinforcement rings soldered to outer can
35, Pyrex Dewar vessel
36, Copper studs for thermal contact between heat station and bath
37, Pushing rods
38, Metal base of insulating case
39, Wax seal
40, Electrical leads and shield thermocouples
41, Monel tube

jacketed transfer tube[51] is inserted for the transfer of liquid hydrogen. The outer can space may be connected to a hydrogen gas holder, to a 200 cubic foot per minute vacuum pump or to the room. A blow-out tube leading to the bottom of the glass Dewar is used to remove unwanted refrigerant. In this operation, the required gas pressure is led in through a blow-in tube which has a coil of several turns in a cup attached to the head of the outer can. When liquid hydrogen is in the Dewar, this cup is filled with liquid nitrogen to minimize heat leak to the hydrogen Dewar. A copper wire coil or a commercial Nichrome resistor, of room temperature resistance about 100 Ω, at the bottom of the glass Dewar, is used to introduce electrical energy. This feature is particularly useful for melting solid nitrogen, which plugs the blow-out tube.

The inner vacuum can hangs from a monel tube 1·9 cm outer diam. and 0·13 cm wall thickness. The thermally insulating vacuum is pumped through this tube, and the various thermocouples and leads to the resistance thermometer and heater run through it. The bottom of the vacuum can is made from monel tubing, 8·9 cm outer diam. and 0·13 cm wall thickness. The vacuum tight seal to the top of the can is made with Rose metal solder (m.p. 95°C), which is contained in a small annular cup on the bottom section of the can. This joint is easy to assemble and disassemble, and little difficulty has been encountered with leaks in it. The massive thermal shield is made in two sections, the top section containing 3·2 kg of lead and 1·0 kg of copper and the bottom section containing 2·4 kg of copper. The joint between the two sections is a cone of 8° included angle, and the weight of the shield assembly is carried on screws fastened into the bottom section and extending through clearance holes in the top section. Thus the weight of the top of the shield is always pushing on the conical joint between the two sections, and upon thermal cycling this joint becomes tighter. There is good thermal contact between the two sections of the joint unless the copper conical joint becomes dirty or scratched. Three brass rods, threaded into the top of the shield and pushing against a flat surface on the bottom, are used to loosen the conical joint when the bottom of the shield is being removed, and to support the upper part of the shield when the bottom is removed. Heaters, made of No. 30 B. and S. gauge constantan wire, are wound in grooves cut on the outside of the top and bottom sections of the shield. The top section has a resistance of 200 Ω and the bottom section 400 Ω. The heater leads are connected to a circuit which enables the currents in the top and bottom heaters to be adjusted separately. Radial holes drilled through the base of the bottom section of the shield provide ports for the passage of gas but prevent direct access of radiation from outside to the calorimeter. The outside of the shield is covered with a thin sheet of German silver, to protect the wires. Aluminum foil, to minimize radiation exchange between the shield and the vacuum can, is wrapped around this sheet. The inside of the shield is gold plated, or it may be lined with aluminum foil. When using aluminum foil, it is essential that none of the radiating surface be covered with an organic cement since such material is an almost black body for room temperature radiation. In fabricating the top of the shield, the copper piece is first machined, and then monel tubes to provide holes for passage of the electrical leads are set in the copper and the lead cast around these tubes. It is important that all wires leading to the calorimeter be brought into good thermal contact with the top of the shield, but they must be electrically insulated from it. No. 16 B. and S. gauge double Formex insulated wire can be imbedded in a low melting indium solder contained in the small monel tubes provided to carry the electrical wires through the top of the shield. Thermal contact between the shield and the No. 16 wires may also be made with stopcock grease although there is danger of melting the grease through accidental overheating of the shield. The electrical leads for the thermometer–heater, of No. 30 B. and S. gauge insulated copper wire, are brought through an Apiezon W wax seal from the room into the vacuum pumping tube. They pass down this tube and then are wrapped around a heat station to bring them to the temperature of the top of the vacuum can. Thermal contact between the heat station and the top of the can is made with silver-soldered copper studs, and the wires, wrapped around the heat station, are brought into thermal contact with it by painting them with clear glyptal varnish. At the top of the shield the No. 30 wires are soldered to No. 16 wires. The three screws supporting the weight of the shield are hung by loops of braided nylon string from hooks attached to the heat station. Copper–constantan thermocouples are used to measure the temperatures of the top and bottom of the shield and the top and bottom of the vacuum can.

2. *A cryostat for calorimetry of condensed gases*

In *Figure 2* is shown the cryostat described by Giauque and Egan[19] for measurements of heat capacity, heats of transition, fusion and vaporization, and vapor pressure of condensed gases, and of substances which are liquid at room temperature. The outer case, outer container and Dewar vessel, and massive thermal shield are similar to those described above. The high vacuum container surrounding the thermal shield is closed at the top by a ball-vee joint made of stainless steel. The upper part of the joint is machined to a spherical surface which makes line contact with the conical surface of the lower part. The bottom of the vacuum container is removable. The vacuum seal is made by tightening nuts on studs which are screwed into holes in the lower part of the joint and extend through clearance holes in the upper part. If care is taken that the mating surfaces are clean and free of scratches, a satisfactory high vacuum joint is obtained.

For measurements on condensed gases, it is necessary to have a tube leading from outside the cryostat to the calorimeter for the introduction and removal of the gas. A soda glass tube, 2 mm inside diam., is connected by a cobalt glass seal to a short platinum tube leading to the inside of the calorimeter. The glass tube is wound with a heater of No. 30 B. and S. gauge insulated constantan wire from a point 1 cm above the connection to the platinum tube to a point, designated 13 in *Figure 2*, near the top of the cryostat. By means of this heater, the tube may be maintained at a higher temperature than the calorimeter, and condensation of material in the tube during measurements at temperatures at which the substance under investigation has appreciable vapor pressure, is avoided. Copper wires leading outside the cryostat are attached to the tube heater at 10 cm intervals.

30 cm

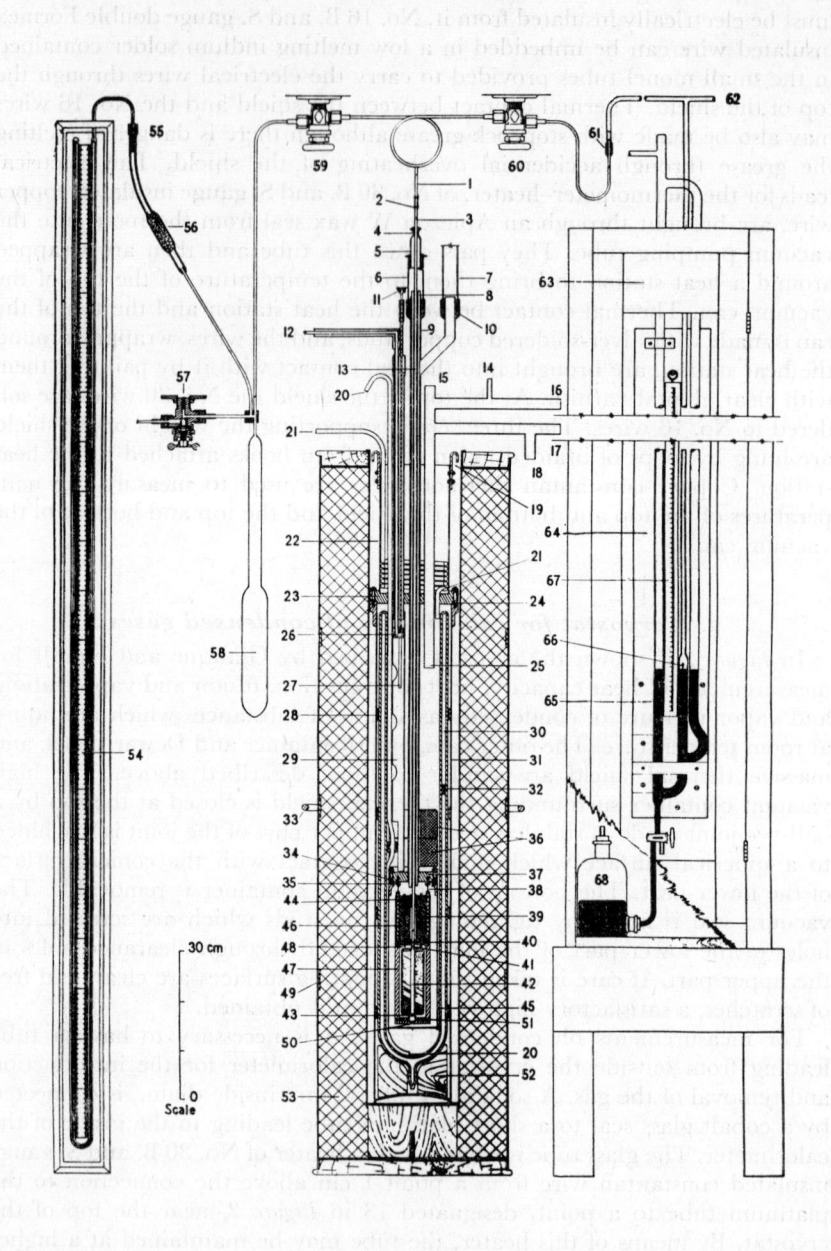

0
Scale

222

Figure 2. Apparatus for low-temperature calorimetry and vapor pressure determinations for condensed gases. [From W. F. Giauque and C. J. Egan, *J. Chem. Phys.*, **5**, 45 (1937)]

1,	Copper tubing
2,	Cobalt glass seal
3,	Soda glass tube
4,	De Khotinsky joint
5 & 6,	Glass tubes
7,	Glass line to vacuum system
8,	Glass tubes containing heater leads and thermocouples
9,	German silver tube
10,	De Khotinsky joint
11,	Valve handle
12,	Vacuum jacketed transfer tube
13,	Top of calorimeter tube heater
14,	Tube for taking out thermocouple from bottom of container
15,	Extra opening
16,	Connection for rubber safety valve
17,	To vacuum pumps
18,	To low pressure hydrogen system
19,	Hooks for raising apparatus
20,	Tube for removing liquid in Dewar
21,	Tube for introducing and precooling hydrogen gas
22,	Monel cup for liquid air
23,	Stainless steel disk
24,	Ball-vee type joint
25,	Outlet tube
26,	Valve in transfer tube
27,	End of transfer tube 12
28,	Steel supporting rings in monel case
29,	Fluted tube
30,	Silvered Pyrex Dewar vessel
31,	Monel case
32,	Asbestos plug in Dewar vessel
33,	Handle on outer case
34,	Ball-vee type joint
35,	Hooks in upper part of container
36,	Lead blocks
37,	Upper part of vacuum container
38,	Suspension cords for thermal shield
39,	Suspension pins screwed into lower part of thermal shield
40,	Cobalt glass seal
41,	Platinum tube
42,	Calorimeter
43,	Thermocouple well; platinum tube filled with Rose metal
44,	Woods metal
45,	Lower part of vacuum container
46,	Upper half of thermal shield
47,	Lower half of thermal shield
48,	Calorimeter heater terminals
49,	Reinforcing rings in outer case
50,	Removable plug in thermal shield
51,	Holes through thermal shield
52,	Balsa wood block
53,	Outer kapok filled case
54,	High pressure manometer
55,	De Khotinsky joint
56,	Tube containing copper foil
57,	Valve to preparation line
58,	Steel bomb
59 & 60,	McKay valves
61,	De Khotinsky joint
62,	Tube to vacuum system
63,	Manometer case
64,	Manometer for vapor pressure measurements
65,	Tube used for levelling purposes
66,	Manometer tube containing fixed point
67,	Standard meter bar
68,	Mercury reservoir

These wires may be used to introduce energy into different parts of the tube and may also be used, in conjunction with the constantan heater wire, as thermocouples to measure the temperature at various positions along the length of the tube. The glass tube is brought into thermal contact with the top part of the thermal shield by means of Rose metal. During measurements of heat of vaporization and of heat capacity at temperatures at which the substance investigated has appreciable vapor pressure, the shield must be maintained at a temperature slightly above that of the calorimeter in order to avoid condensation of material in the tube in the region of thermal contact with the shield. Outside of the cryostat, the tube is connected by a metal–glass seal to a copper tube, which, in turn, is connected through valves to a filling bomb and preparation line, to a high pressure mercury manometer, and to a precision mercury manometer used for the measurement of vapor pressure.

3. *Calorimeters, thermometers, and heaters*

A. Calorimeter with Thermometer–Heater on Outer Surface

A good representative of this type of calorimeter is the gold calorimeter for use in measurements of condensed gases described by Giauque and Wiebe[23], Giauque and Johnston[20], Blue and Giauque[3], Giauque and Egan[19] and Kemp and Giauque[34]. The calorimeter is constructed entirely of gold, with welded joints, except for the 2 mm internal diam. filling tube welded to the top which is made of platinum. The calorimeter is of cylindrical shape, 4·04 cm in outer diam. and 13·3 cm high. The wall thickness is 1 mm and the weight about 440 g. Twelve radial vanes, 2 mm thick and 12·5 cm tall are welded to the inside of the calorimeter wall. These vanes are to improve the thermal contact between the calorimeter and the sample. Calibrated copper–constantan thermocouples, soldered with Rose metal into wells attached to the top and bottom of the calorimeter, are used as the primary temperature standards against which the resistance thermometer is calibrated. A 50 cm length of each thermocouple is wrapped around the calorimeter to minimize temperature gradients at the junction. A resistance thermometer–heater, made of gold wire containing 0·175 per cent silver to increase its low-temperature electrical resistance, covers the outer cylindrical surface of the calorimeter except for 0·5 cm at each end. A typical resistance of the thermometer–heater at room temperature is 400 Ω. The surface of the calorimeter on which the thermometer–heater is to be wound is roughened and covered with thin lens paper. The gold wire, drawn to No. 40 B. and S. gauge and having a double silk insulation, is wound helically around the calorimeter. Copper lead wires of No. 30 B. and S. gauge are attached at either end of the thermometer–heater and near the middle. One inch of No. 40 copper wire is used to connect each of the three copper lead wires to their connections, which are in thermal contact with but electrically insulated from the upper portion of the thermal shield. The use of this smaller wire is necessary to make the thermal conductance between calorimeter and shield small. The thermometer–heater is painted with Bakelite varnish and baked at 120°C. In order to reduce radiation, the outside of the calorimeter is covered with 0·002 mm gold foil cemented on its under side. Care must be taken to avoid having any organic material, which is a nearly black body for thermal radiation, on the outside surface of the calorimeter which can exchange heat by radiation with the surroundings.

A later version of this type of calorimeter, described by Murch and Giauque[39] is illustrated in *Figure 3*. This calorimeter is designed for samples

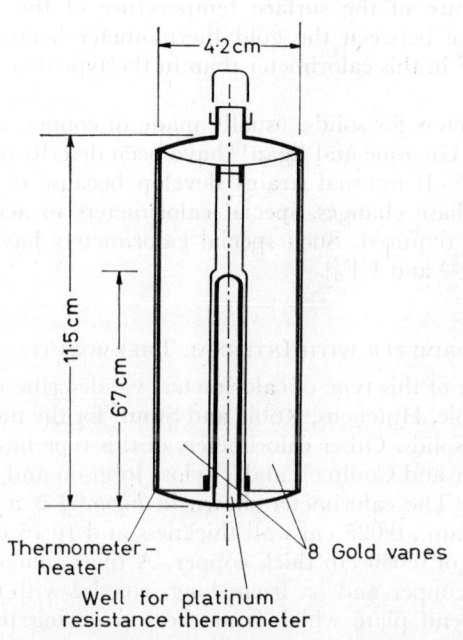

Figure 3. Gold calorimeter for liquids or solids with external thermometer–heater. [From L. E. Murch and W. F. Giauque, *J. Phys. Chem.*, **66**, 2052 (1962).]

which are solid or liquid at room temperature and is, therefore, provided with a filling tube to which a cap is soldered after the sample and helium gas have been introduced into the calorimeter. The calorimeter is made of gold, 1·0 mm thick. There are eight radial vanes 0·25 mm thick. The primary standard for temperature measurement is a Leeds and Northrup strain free platinum resistance thermometer calibrated by the National Bureau of Standards, which is fastened with Rose metal solder to a well within the calorimeter. A thermometer–heater made of 0·0031 inch diam. wire of gold containing 0·1 per cent silver is wound on the external cylindrical surface of the calorimeter. Two layers of China silk, each 0·003 inch thick, are placed on the outer gold surface of the calorimeter and cemented in place with three coats of Formvar varnish, baked at 125°C. The gold wire and an insulating silk thread, 0·004 inch diam., are wound together over the silk cloth to make a closely wound helix of 467 turns. The ends of the gold wire are soldered to leads of 0·010 inch diam. double silk insulated copper wire. The gold–copper junctions are in thermal contact with the calorimeter to avoid thermoelectric effects. The completed thermometer is covered with two layers of silk cloth cemented with Formvar varnish. The bottom copper lead passes between the two outer layers of silk. The outside of the calorimeter is covered with heavy gold leaf to increase the reflectivity for room temperature radiation.

The gold thermometer–heater is compared with the platinum thermometer at the beginning and end of each heat capacity measurement. During the rating periods and the time of energy input the gold thermometer gives a continuous measure of the surface temperature of the calorimeter. The thermal resistance between the gold thermometer–heater and the calorimeter is smaller[17] in this calorimeter than in the type described by Giauque and Egan[19].

Many calorimeters for solids, usually made of copper, and following the general design of Giauque and Egan[19] have been described[24, 44, 38, 35, 26, 18, 21, 32, 6, 11, 5, 16]. If unusual strains develop because of volume changes accompanying phase changes, special calorimeters to accommodate these changes may be required. Such special calorimeters have been described for work on H_2O[22] and UF_6[4].

B. Calorimeter with Internal Thermometer–Heater

As an example of this type of calorimeter, we describe the copper calorimeter used by Cole, Hutchens, Robie and Stout[9] for the measurement of the heat capacity of solids. Other calorimeters of this type have been described by Hicks[27], Pitzer and Coulter[46] and Taylor, Johnson and Kilpatrick[54].

(*1*) *Calorimeter*. The calorimeter shown in *Figure 4* is a copper cylinder, 4·45 cm outer diam., 0·025 cm wall thickness and 10·15 cm long. The end pieces are made of 0·050 cm thick copper. A thermometer–heater well, of 0·025 cm thick copper and *ca*. 9 cm long, coaxial with the calorimeter, is attached to the end plate which is at the top during the measurements. In the calorimeter used for heat capacity measurements of benzoic acid, amino acids, and proteins, 6 radial fins, 0·040 cm thick are attached with 50–50 PbSn solder to the outside of the thermometer–heater well. In other calorimeters[7], used with materials of greater thermal conductivity, the fins have been omitted without seriously lengthening the thermal relaxation time within the calorimeter. All joints except the cap closure in this calorimeter are silver soldered. In the calorimeter with fins the only additional soft soldered vacuum joint is the one connecting the upper end plate to the outside copper cylinder. Three eyelets, through which run loops of nylon thread used to hang the calorimeter from hooks beneath the top of the shield, are soft soldered to the upper end plate of the calorimeter. The outside of the calorimeter is covered with 0·0005 cm thick dull gold plate. A filling tube, made of 0·95 cm outer diam. cupronickel tubing, is silver soldered into the lower end plate. When being filled with a sample, the calorimeter is turned over so the filling tube is on top. After filling with a sample the filling tube is closed with a dished cap made of 0·025 cm thick copper, which is soft soldered (Rose metal) to the end of the filling tube. The cap has a pinhole, 0·05 cm diam., which is left open. After the calorimeter and contents are weighed, the air is removed through the pinhole and replaced with helium at one atmosphere pressure. For this operation the calorimeter is placed in a bell jar connected to a high vacuum pumping line. After the calorimeter is filled with helium, the bottom of the bell jar is lowered and the pinhole closed quickly with indium solder. The calorimeter is again weighed and then replaced in the bell jar and tested for leaks. The empty weight of the calorimeter with fins, including the thermometer–heater, is 124·4 g. The sample volume is 145 cm^3.

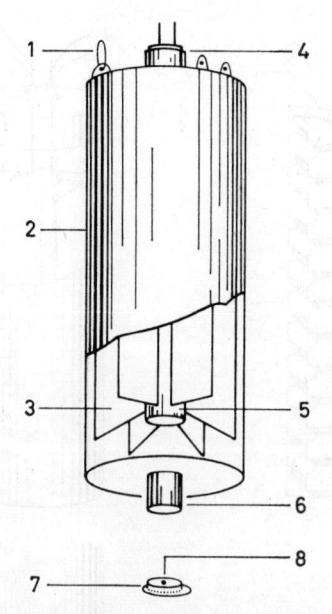

Figure 4. Copper calorimeter for solids with internal thermometer–heater. [A. G. Cole, J. O. Hutchens, R. A. Robie, and J. W. Stout, *J. Am. Chem. Soc.*, **82**, 4807 (1960).]
1, Nylon thread in eyelet
2, Calorimeter
3, Fins soft soldered to thermometer well
4, Resistance thermometer–heater
5, Thermometer well
6, Filling tube
7, Cap
8, Pinhole in cap

(2) *The Thermometer–Heater.* The resistance thermometer–heater is made of pure platinum wire, 0·07 mm diam. It is of strain free construction, similar to that described by Meyers[37] and by Catalano and Stout[7]. The thermometer–heater assembly is shown in *Figure 5*. The platinum wire, in a spiral coil of inner diam. 0·28 mm, is wound on a mica cross notched to receive the coil. The assembly is then placed in a Vycor tube and annealed in an oven, in air, for 19 hours at 500 °C. It is then placed in the thermometer case made of 0·25 mm thick copper, 9·3 cm long and 1·27 cm outer diam. The bottom of the tube contains a pinhole 0·10 cm diam. and the top is open. A two-tube Kovar metal–ceramic seal (Stupakoff No. 95:5007) is soldered (50–50 PbSn) into a copper cap which fits closely inside the thermometer case. Two heavy platinum leads, 0·66 mm diam. and 2·5 cm long

are soldered (50–50 PbSn) inside the Kovar tubes. The ends of the resistance thermometer–heater are welded to the heavy Pt leads, and the cap is then soldered to the case with Rose metal. The thermometer case is sealed to a high vacuum line and, while the case is evacuated, the thermometer is further annealed by passing a current of 0·13 amp for 1 hour through the wire to raise the wire temperature to about 450°C. The water–ice triple

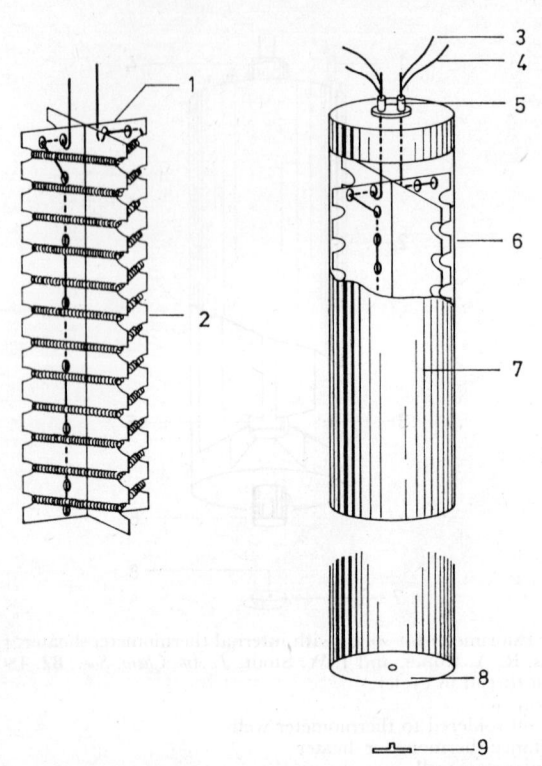

Figure 5. Strain-free platinum thermometer–heater. [A. G. Cole, J. O. Hutchens, R. A. Robie, and J. W. Stout, *J. Am. Chem. Soc.*, **82**, 4807 (1960).]

1, Mica cross
2, Bare thermometer–heater
3, Current lead
4, Potential lead
5, Kovar-ceramic seals
6, Resistance thermometer–heater
7, Thin-walled copper case
8, Bottom of thermometer case
9, Plug for hole in bottom of thermometer case

point resistance of the thermometer is measured before and after several successive annealings. The thermometer case is then filled with helium at 1 atm pressure and the pinhole closed with a small brass plug soldered with Rose metal. The thermometer case is soldered into the well of the calorimeter with an indium solder (0·49 Bi, 0·18 Pb, 0·15 Sn, 0·18 In, by weight) melting at 50°C. The weight of a typical thermometer and case is 11·39 g.

4. *Equipment for measurement of resistance, energy and time*

The resistance of the thermometer is measured by comparing with a high precision potentiometer, such as those designed by White[57] or Wenner[57], the potential drops across the thermometer and a calibrated standard resistance in series with it. The potentiometer should be autocalibrated. With a high sensitivity galvanometer and an illuminated scale at about 10 m distance, a precision of a part per million in the measurement of resistance may be obtained. The same potentiometer may be used to measure the heating current and, by use of a calibrated voltage divider, the

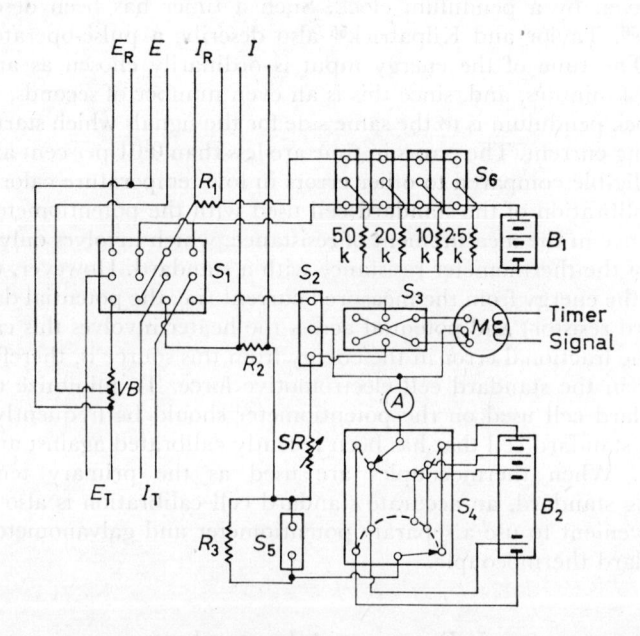

Figure 6. Wiring diagram of circuits for resistance and energy measurements. Wires E_T and I_T are the potential and current leads, respectively, to the thermometer–heater. A selector switch (not shown) enables any one of the four potentials, E_R, E, I_R or I to be read on a potentiometer. E_R is the potential drop across the thermometer–heater during the resistance measurements in the rating periods and I_R, the potential drop across a standard resistor, R_1, either 10 or 100 Ω, is a measure of the corresponding current. E, the potential across a tap of the voltage divider VB, measures the potential drop across the thermometer–heater during energy input, and I, the potential drop across a standard resistance R_2, 1 or 0·1 Ω, is the corresponding current. The four-pole double-throw switches, S_1, is in the up position during the rating periods and in the down position during the heating period. The thermometer current used during the rating periods is provided by a 6 volt battery, B_1. The four single-pole double-throw switches, S_6, permit the selection of any combination of thermometer currents of 0·12, 0·30, 0·60 and 2·4 mA. The heating current is provided by a battery of 15 cells, B_2. Any integral number of cells may be selected by the switches S_4. A series resistance, R_3, may be introduced into the heating circuit by opening the switch S_5. A few seconds before the beginning of the heating period, the reversing switch S_3 is set so the heating current is passing through the substitute resistance, SR, and the single-pole double-throw switch S_2 is opened. S_1 is thrown to the down position. The signal from the timer activates the msec relay, M, and the current is switched to the thermometer–heater. S_2 is then closed in the up position and S_3 opened. A few seconds before the end of the heating period, S_3 is closed so as to pass current through the heater and S_2 is opened. The time signal then causes the millisecond relay to transfer the current from the heater to the substitute resistance. The milliammeter A is used to adjust SR so as to obtain a stabilizing current matching that during energy input.

potential across the heater. The heating current and the current for measuring the thermometer resistance during the rating period are provided by Edison cells or low discharge lead cells. To avoid drifts in the electromotive force of these cells, they should be in a constant temperature environment. A typical arrangement of wiring and switches in the external electrical circuits is shown in *Figure 6*. Various thermometer currents may be selected by switches without altering the total drain on the cells supplying these currents. A substitute resistance matching the resistance of the calorimeter heater is used to stabilize the cells supplying the heating current. The heating current is turned off and on at preset times by means of a mechanical timer driven by a pendulum clock. Such a timer has been described by Johnston[30]. Taylor and Kilpatrick[55] also describe a pulse-operated timing system. The time of the energy input is ordinarily chosen as an integral number of minutes, and, since this is an even number of seconds, the swing of the clock pendulum is to the same side for the signals which start and stop the heating current. The errors in time are less than 0·01 per cent and, therefore, negligible compared to other errors in low temperature calorimetry.

The calibration of the standard cell used with the potentiometer is of no consequence in the measurement of resistance, which involves only the comparison of the thermometer resistance with a standard. However, the calculation of the energy from the measured current (i.e. the potential drop across a standard resistor) and potential across the heater involves this calibration twice. The fractional error in the energy from this source is, therefore, twice the error in the standard cell electromotive force. To minimize this error, the standard cell used on the potentiometer should be frequently checked against a standard cell that has been recently calibrated against an absolute standard. When thermocouples are used as the primary temperature measuring standard, an accurate standard cell calibration is also required. It is convenient to use a separate potentiometer and galvanometer to read the standard thermocouple.

5. *Experimental procedures*

A. Calorimeter for Solids

In this section procedures used with the cryostat and calorimeter described in Sections II–1 and II–3–B are given. The calorimeter is filled with sample and with helium gas at 1 atm pressure and then sealed off. Careful account is kept of the weights during the filling of the calorimeter in order to obtain an accurate weight of the sample and to know the amounts of solder and helium in the filled calorimeter compared to the empty calorimeter. The heat capacity of the empty calorimeter is determined in a separate series of experiments. The calorimeter is hung from the top of the shield and electrical connections from the heavy Pt leads of the thermometer–heater to the four No. 16 gauge copper wires embedded in the top of the shield, which are used for current and potential leads, are made with four Pt leads about 2·6 cm long. The two current leads are 0·10 mm diam. and the two potential leads are 0·07 mm diam. The length of the current leads is measured in order to permit correction for the energy developed in them during the heating periods. The cryostat is then assembled and the insulating vacuum space pumped for several days at room temperature or slightly above. The pressure read on an ionization gauge with a liquid N_2 trap is less than 10^{-6} mm Hg. An extended pumping at the maximum safe temperature is desirable in order to remove water and other condensable impurities, which are not detected by the trapped pressure gauge but which will increase the thermal conductance between calorimeter and shield.

In order to cool the calorimeter, liquid nitrogen is added to the Dewar and a pressure of about 1 mm Hg of helium gas admitted to the insulating vacuum space. By pumping on the nitrogen in the Dewar, the temperature of the calorimeter can be lowered to about 50°K. The inner can is then evacuated and a series of heat capacity measurements may be started. If measurements are to be made at liquid hydrogen temperatures, the calorimeter is cooled as above to slightly above the melting point of nitrogen, and the helium is then pumped from the insulating vacuum space. The bath pressure is then raised to one atm, and the liquid nitrogen is removed through the blow-out tube. The Dewar space is flushed by alternate evacuation and addition of hydrogen gas, liquid nitrogen is put in the cup, and liquid hydrogen is transferred to the Dewar. A small pressure of helium gas is again put in the insulating vacuum space and, by pumping on the hydrogen bath, the temperature of the calorimeter can be lowered to between 10 and 11°K. The insulating vacuum space is then pumped out and a series of measurements begun. An alternative procedure which is more economical in the use of liquid hydrogen is to cool the calorimeter to 50°K by pumping on solid nitrogen before pumping the insulating vacuum. After standing overnight, the solid nitrogen in the Dewar will have melted and the liquid may be removed through the blow-out tube. The calorimeter and shield, protected by the insulating vacuum, will not rise in temperature by more than a degree in this period.

In an individual measurement, the temperature of the shield is adjusted a little above the midpoint between the expected initial and final temperatures of the calorimeter. A series of measurements of the resistance of the thermometer is made in the fore rating period, the electrical energy introduced during a heating period, and the resistance again measured in an after rating period. The times used for the heating periods vary from 3–15 min (generally 6–10 min). The total time for an individual measurement varies from 30–60 min. In addition to measurements of the electrical resistance in the rating periods and the current and potential during the heating period, thermocouple readings to determine the temperatures of the top and bottom of the shield and top and bottom of the vacuum can are taken. During a series of measurements with no refrigerant present the temperature of the can is adjusted so as to minimize the temperature drift of the shield.

B. Calorimeter for Condensed Gases

In the type of cryostat and calorimeter described in Sections II–2 and II–3–A, the sample is introduced by distillation through the glass tube. For measurements at low temperatures at which the sample has negligible vapor pressure, a small pressure of helium gas is added to the calorimeter for thermal conductivity. At higher temperatures the helium is removed, since the vapor pressure of the sample provides sufficient gas for thermal conductivity. In all measurements with an appreciable vapor pressure of the sample and no helium gas present, it is necessary to maintain the shield and all parts of the glass tube at a temperature higher than that of the sample in the calorimeter in order to prevent distillation of some of the sample to colder regions.

During the measurement of heat of vaporization, electrical energy is introduced only into the lower half of the heater in order to avoid possible superheating of the gas leaving the calorimeter. The temperature of the surface of the evaporating liquid or solid is known from measurements of the pressure during the vaporization. The pressure drop because of gas flow in the tube is negligible. The temperature of the outer surface of the calorimeter during a heat of vaporization measurement is needed for the heat leak corrections. The temperature of the lower half of the heater is determined by its resistance calculated from the electrical current and potential during the time of energy input; that of the upper half of the heater is inferred by extrapolation of a series of measurements of its resistance beginning 10 sec or less after the electrical energy is turned off; and the thermocouples on the top and bottom of the calorimeter may be used to measure the temperatures at these positions.

In a heat of vaporization experiment, the amount of material vaporized is required and, in order to obtain this amount accurately, the amount of gas in the calorimeter and filling tube before and after energy input must be known. This amount may be calculated from the measured pressures and temperatures and from the equation of state of the vapor. During evaporation, the pressure within the calorimeter is maintained constant, and the amount of material removed from the filling tube should be accurately measured. Giauque and Johnston[20] describe an apparatus for this purpose consisting of a large thermostated glass bulb in which the pressure is maintained constant by adjusting the amount of mercury in it. The amount of gas collected in the bulb is calculated from measurements of the pressure and temperature and the calibrated volume of the bulb. When a substance may be completely absorbed by a chemical reaction in a weighed absorption bulb, the constant pressure during vaporization may be obtained by the atmospheric pressure on the outlet of the absorption bulb. Giauque and Wiebe[23] describe absorption equipment used in measurements on hydrochloric acid. For substances of sufficiently high boiling point, the material evaporated may be condensed in a bulb cooled by liquid nitrogen and the bulb weighed at room temperature before and after collection of the sample. The pressure in the collecting bulb at liquid nitrogen temperatures is essentially zero, and a capillary tube[53] chosen to have the proper resistance to gas flow is used to maintain the desired calorimeter pressure during vaporization.

III. Calculation of Heat Capacity Data

1. *Corrections for heat leak*

A. NEWTON'S LAW OF COOLING

The transfer of heat by conduction through gases at low pressures, by conduction through solids, and by radiation can be described by Newton's law of cooling. This law states that the heat transferred per unit time, $\dot{q}$ from a body at temperature T_2 to one at temperature T_1, is given by

$$\dot{q} = K(T_2 - T_1) \qquad (1)$$

in which K is the thermal conductance between the two bodies. The contribution to K from conduction through solid bodies and through gases at low pressures may be calculated from thermal conductivities and geometrical factors. The variation with temperature of the thermal conductivity of gases and solids is sufficiently small that negligible error is introduced by assuming the thermal conductance constant over the range of temperature involved in a single heat capacity measurement. The thermal conductance will, however, change by a significant amount in a series of measurements covering a large temperature range.

The heat transferred by radiation between two bodies at temperatures T_2 and T_1 is proportional to

$$T_2^4 - T_1^4 = (T_2 - T_1)(T_2^3 + T_2^2 T_1 + T_2 T_1^2 + T_1^3) \qquad (2)$$

The contribution to the thermal conductance, K, arising from radiation is, therefore, proportional to

$$T_2^3 + T_2^2 T_1 + T_2 T_1^2 + T_1^3 = 4 T_{Av}^3 \left[1 + \left(\frac{T_2 - T_1}{2 \; T_{Av}} \right)^2 \right] \qquad (3)$$

In calorimetric apparatus described in this chapter, in which the calorimeter is surrounded by a massive radiation shield, the difference in temperature between the calorimeter and the shield, $T_2 - T_1$, does not exceed 5°K. The thermal conductance by radiation between the shield and the calorimeter, therefore, varies essentially as the cube of their mean absolute temperature. In a single measurement, the mean absolute temperature changes at most by 5°K, which, at 300°K, corresponds to a variation of 5 per cent in K. The substitution of a constant value of K equal to an average over the temperatures encountered in a single measurement is, therefore, permissible. At low temperatures, although the fractional variation in the radiation contribution to K is larger, the effect of this variation on the calculated heat capacity is small since the radiation contribution to the heat exchange varies as the cube of the absolute temperature.

At low temperatures, the principal contribution to the thermal conductance between a calorimeter and its surrounding shield arises from the conductivity of metallic wires in thermocouples and as leads to thermometers and heaters on the calorimeter. In the temperature range from 10 to 40°K, the thermal conductivity of pure metals decreases considerably with increasing temperature[47] but again the variation over the temperatures encountered in a single measurement is small enough that it is sufficiently accurate to replace the varying thermal conductance by a mean value taken as constant over a single measurement.

Normally the space between a calorimeter and its surrounding shield is evacuated to a pressure of less than 10^{-5} mm Hg. At this pressure, the contribution of the thermal conductivity of the gas to the total thermal conductance is negligible. Occasionally higher pressures are encountered, because of leaks in the vacuum system or inadequate pumping times at low temperatures, and because of desorption of condensed material at higher temperatures. Although the variation of gas conductivity with temperature at constant pressure is small enough that its contribution to the conductance may, over the temperature variation of a single measurement, be replaced by an average value, this conductivity is, at low pressures, proportional to the pressure and, if inadequate pumping time has been allowed or desorption is occurring, the pressure may change by a large factor during the course of a single measurement. The conductance is then a function of time. For accurate work it is desirable to avoid such variations of thermal conductance. In order to evaluate a correction, described in section III–1–D–(2), a constant pressure of about 2×10^{-2} mm Hg of helium gas is sometimes introduced into the vacuum space. In this case, the gas pressure is essentially constant and the gaseous conductivity contribution to the conductance between calorimeter and shield may be adequately approximated by a constant over the temperature variation encountered in a single measurement. It should be pointed out that, at gas pressures in the neighborhood of 1 atm, the major mechanism of heat transfer by a gas may be convection. Convective heat transfer may not obey Newton's law of cooling and depends in a complicated way on the geometry and temperature distributions. The contribution of convection to the overall heat transfer is, however, negligible at the gas pressures present in the insulating vacuum space of the type of apparatus described in this chapter.

B. Thermal Relaxation Times

To illustrate the nature of the solutions of the linear differential equations governing heat flow which obeys Newton's law of cooling, we will examine the simplified model illustrated in *Figure 7*. The calorimeter temperature is designated by T_c and its heat capacity by C_c. The heater has temperature T_h and heat capacity C_h. Electrical power, $P(t)$, in which t is the time, is introduced into the heater. The calorimeter is connected to the heater by

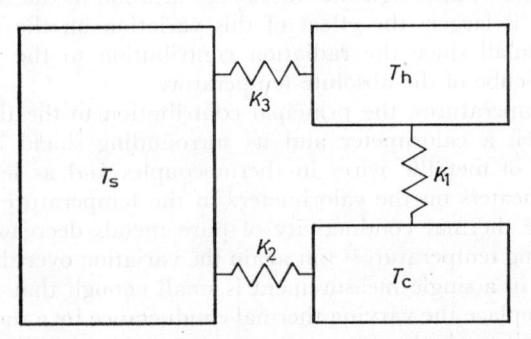

Figure 7. Schematic diagram of thermal connections between calorimeter at temperature T_c, heater at temperature T_h, and external heat reservoir (thermal shield) at temperature T_s.

the thermal conductance K_1 and to an external heat reservoir at temperature T_s by the thermal conductance K_2. The heater is in contact with the heat reservoir through the thermal conductance K_3. All thermal conductances are taken as constants independent of time and temperature. For simplicity we take T_s to be independent of time and define $\theta_h = T_h - T_s$ and $\theta_c = T_c - T_s$. The equations of heat conduction are then

$$\frac{d\theta_h}{dt} + \frac{K_1 + K_3}{C_h}\theta_h - \frac{K_1}{C_h}\theta_c = \frac{P(t)}{C_h} \qquad (4)$$

$$\frac{d\theta_c}{dt} - \frac{K_1}{C_c}\theta_h + \frac{K_1 + K_2}{C_c}\theta_c = 0 \qquad (5)$$

The solutions of this pair of linear, simultaneous, first order differential equations may be obtained by standard methods[49]. If the power input, P, is a constant independent of time and if the heat capacities and thermal conductivities are constants, the solutions are

$$\theta_h = c_{1h} \exp(m_1 t) + c_{2h} \exp(m_2 t) + P \frac{K_1 + K_2}{K_1 K_2 + K_2 K_3 + K_3 K_1} \qquad (6)$$

$$\theta_c = c_{1c} \exp(m_1 t) + c_{2c} \exp(m_2 t) + P \frac{K_1}{K_1 K_2 + K_2 K_3 + K_3 K_1} \qquad (7)$$

in which m_1 and m_2 refer to the positive and negative signs, respectively, in the expression

$$m = -\tfrac{1}{2}\left(\frac{K_1 + K_3}{C_h} + \frac{K_1 + K_2}{C_c}\right)$$

$$\pm \tfrac{1}{2}\left[\left(\frac{K_1 + K_3}{C_h} + \frac{K_1 + K_2}{C_c}\right)^2 - 4\frac{K_1 K_2 + K_2 K_3 + K_3 K_1}{C_h C_c}\right]^{\frac{1}{2}} \qquad (8)$$

Since the thermal conductances and heat capacities are essentially positive, both m_1 and m_2 are negative, and the time dependence of both θ_h and θ_c is described in terms of a constant plus two exponentially decreasing functions of time with time constants $-m_1^{-1}$ and $-m_2^{-1}$ respectively. The constants c_{1h}, c_{2h}, c_{1c} and c_{2c}, only two of which are independent, are determined from the initial conditions.

In order to obtain an accurate measurement of the heat capacity of the calorimeter, it is necessary that nearly all the electrical energy introduced into the heater be used to heat the calorimeter, and that the energy transferred to the heat reservoir, the heat leak, be a small fraction of the total. This requires that $K_1 \gg K_3$ and $K_1 \gg K_2$. Furthermore it is ordinarily true that

$C_\text{h} \ll C_\text{c}$. Then, dropping terms of higher order than the first in the small quantities K_2/K_1, K_3/K_1 and C_h/C_c, one obtains

$$m_1 = -\frac{(K_2 + K_3)}{C_\text{c} + C_\text{h}} \tag{9}$$

$$m_2 = -\frac{K_1}{C_\text{h}}\left(1 + \frac{C_\text{h}}{C_\text{c}} + \frac{K_3}{K_1}\right) \tag{10}$$

There are, therefore, two time constants of vastly different magnitude. The first, $-m_1^{-1}$, is the ratio of the total heat capacity of the calorimeter plus heater to the total conductance between them and the surroundings and ranges in magnitude from 10 min in the least favorable cases to over 10 hours in the most favorable cases. The second time constant, $-m_2^{-1}$, is essentially the ratio of the heat capacity of the heater to the thermal conductance between heater and calorimeter and is of the order of a few seconds. Because of the large difference between the two time constants, their effects are readily separated. The term involving $\exp(m_2 t)$ is a quickly dying out transient superimposed on the slowly varying term containing $\exp(m_1 t)$.

If, for simplicity, we take as the initial conditions $\theta_\text{h} = \theta_\text{c} = 0$ at $t = 0$, the solutions of the differential equations become

$$\theta_\text{c} = \frac{PK_1}{C_\text{h} C_\text{c} (m_1 - m_2)} \left(\frac{1 - \exp(m_2 t)}{m_2} - \frac{1 - \exp(m_1 t)}{m_1}\right) \tag{11}$$

$$\theta_\text{h} = \frac{K_1 + K_2}{K_1} \theta_\text{c} + \frac{P\,[\exp(m_1 t) - \exp(m_2 t)]}{C_\text{h}\,(m_1 - m_2)} \tag{12}$$

For times when $|m_1 t| \ll 1$, and to the first order in the small quantities K_2/K_1, K_3/K_1 and C_h/C_c, these expressions become

$$\theta_\text{c} = \frac{P}{C_\text{c}}\left(1 - \frac{C_\text{h}}{C_\text{c}} - \frac{K_3}{K_1}\right) t - \frac{P\,C_\text{h}}{K_1 C_\text{c}}\left(1 - \frac{2 C_\text{h}}{C_\text{c}} - \frac{2 K_3}{K_1}\right)[1 - \exp(m_2 t)] \tag{13}$$

$$\theta_\text{h} = \frac{P}{C_\text{c}}\left(1 - \frac{C_\text{h}}{C_\text{c}} - \frac{2 K_3}{K_1}\right) t + \frac{P}{K_1}\left(1 - \frac{2 C_\text{h}}{C_\text{c}} - \frac{K_3}{K_1}\right)[1 - \exp(m_2 t)] \tag{14}$$

The temperature of the heater initially rises at the rapid rate P/C_h but exponentially relaxes with the time constant $-m_2^{-1}$ to the much slower rate, essentially P/C_c. The temperature of the calorimeter initially has zero slope against time but, after a brief transient, again with time constant $-m_2^{-1}$, rises linearly with time, at a rate only slightly greater than that of the heater. After the transients have died out, the temperature head of the heater relative to the calorimeter is

$$\theta_\text{h} - \theta_\text{c} = \frac{P}{K_1}\left(1 - \frac{C_\text{h}}{C_\text{c}} - \frac{K_3}{K_1} - \frac{K_3 t}{C_\text{c}}\right) \tag{15}$$

which is essentially equal to the constant P/K_1.

Let the power, P, be reduced to zero at some time t_1. Ordinarily $|m_1 t_1| \ll 1 \ll |m_2 t_1|$. We will designate the temperature of the calorimeter at t_1 as θ_1 and that of the heater by $\theta_1 + \Delta\theta_1$. At times later than t_1 the solutions of the differential equations (4) and (5) with $P = 0$ are

$$\theta_c = \frac{-K_1 \Delta\theta_1 + K_2 \theta_1}{C_c (m_2 - m_1)} \{\exp [m_1(t - t_1)] - \exp [m_2(t - t_1)]\}$$
$$+ \theta_1 \left(\frac{m_2 \exp [m_1(t - t_1)] - m_1 \exp [m_2(t - t_1)]}{m_2 - m_1}\right) \quad (16)$$

$$\theta_h = \frac{(K_1 + K_3) \Delta\theta_1 + K_3 \theta_1}{C_h (m_2 - m_1)} \{\exp [m_1(t - t_1)] - \exp [m_2(t - t_1)]\}$$
$$+ (\theta_1 + \Delta\theta_1) \left(\frac{m_2 \exp [m_1(t - t_1)] - m_1 \exp [m_2(t - t_1)]}{m_2 - m_1}\right) \quad (17)$$

These equations describe a rapid attainment of temperature equilibrium between the heater and calorimeter, with time constant $-m_2^{-1}$, and thereafter a slow drift downward of both calorimeter and heater, essentially as $\theta_1 \exp[m_1(t - t_1)]$. At times large compared to $-m_2^{-1}$, the small temperature difference between thermometer and heater is given, to the first order in small quantities, by

$$\theta_h - \theta_c = \theta_1 \left[-\frac{K_3}{K_1}\left(1 - \frac{C_h}{C_c}\right) + \frac{C_h K_2}{C_c K_1} \right] \exp [m_1(t - t_1)] \quad (18)$$

If the heater is on the outside of the calorimeter, so that K_2 is small compared to K_3, then the heater is intermediate in temperature between the calorimeter and the surroundings, and a correction must be applied to the temperature of the heater to obtain that of the calorimeter. If, on the other hand, the calorimeter surrounds the heater, so that K_3 is zero, then the steady state temperature of the calorimeter is between the shield temperature and that of the heater. The temperature difference is, however, very small in this case because of the factor C_h/C_c.

In a real calorimetric apparatus, there are additional complications not considered in the simplified model analyzed above. The surroundings of the calorimeter may slowly change temperature with time, and more important, there are finite relaxation times associated with the attainment of thermal equilibrium within the calorimeter shell and between this shell and the contents of the calorimeter. The essential criterion for accurate calorimetric work is that the thermal relaxation times within the calorimetric system, the times for the attainment of thermal equilibrium between the calorimeter, heater, thermometer, and sample, be short compared to the thermal relaxation time between the calorimetric system and its surroundings. When this criterion is satisfied, it is possible to treat the calorimetric system in the first approximation as a unit and then make small corrections for the steady state temperature differences existing when energy is being introduced or during the rating periods.

C. Calculation of Heat Capacity Neglecting Thermal Gradients in the Calorimetric System

In this section we derive the expressions for the corrections for heat leak in a heat capacity measurement with the approximations that thermal equilibrium is instantaneous within the calorimetric system comprising the calorimeter, sample, heater and the thermometer, and that the power input is constant. In later parts of Section III–1, the corrections necessary because of the difference between a real calorimeter and the approximation will be considered.

The electrical energy introduced into the heater in a measurement of heat capacity may readily be measured to an accuracy of a few hundredths of one per cent. Let the times of the beginning and end of the period during which electrical energy is being introduced into the calorimeter be designated as $t = -\tau$ and $t = \tau$, respectively. The zero of time, therefore, corresponds to the midpoint of the heating period. The total electrical energy added to the calorimeter is

$$E = \int_{-\tau}^{\tau} P(t)\,dt \tag{19}$$

in which $P(t)$ is the electrical power. In evaluating E, the change of P with time is taken into consideration (see Section III–2–A), but to a good approximation it is sufficient in evaluating the heat leak correction to take P as a constant. The correction made necessary by this approximation is described in Section III–1–F.

The heat interchange between the calorimeter and its surrounding thermal shield is estimated from the variation with time of the thermometer readings in the fore and after rating periods. Ordinarily the temperature drift of the calorimeter is a constant during each of these rating periods, and the correction for heat leak is made by linearly extrapolating the observed fore and after period temperatures to the midpoint of the heating period. However, when the thermal conductance between the calorimeter and surrounding shield is unusually large or when the heat capacity of the calorimeter is unusually small, the calorimeter drifts become exponential, and a more elaborate analysis, given below, is required. *Figure 8* illustrates the observed temperature–time curves found in such an extreme case, the measurement of the heat capacity of an empty calorimeter in the temperature range between 12 and 13°K[9].

Let T_s indicate the temperature of the thermal shield, T that of the calorimetric system, and C the heat capacity of the calorimetric system. Then

$$C\,dT/dt = P + B_N(T_s - T). \tag{20}$$

B_N is the Newton's law coefficient of heat interchange, or the thermal conductance between the thermal shield and the calorimetric system. Equation (20) may be written

$$dT/dt = \beta + \alpha(T_s - T) \tag{21}$$

in which $\alpha = B_N/C$ and $\beta = P/C$. α^{-1} is the thermal relaxation time between

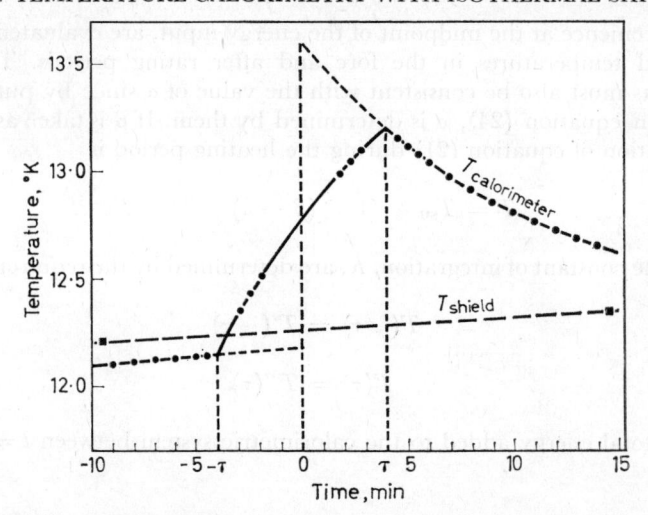

Figure 8. Temperature–time curve in a heat capacity measurement of an empty calorimeter. [From A. G. Cole, J. O. Hutchens, R. A. Robie, and J. W. Stout, *J. Am. Chem. Soc.*, **82**, 4807 (1960).] The time of the heating period, 2τ, is eight minutes.

the calorimetric system and the thermal shield. To a sufficient approximation the temperature of the shield may be represented by a linear function of time,

$$T_s = T_{s0} + kt. \tag{22}$$

The constant k may be determined from the readings of a thermocouple on the shield which serve to determine the temperature drift even though the absolute temperature is not accurately known. The temperature T_{s0}, which would involve an absolute calibration of the shield thermocouple, is not required. Let T' be the temperature of the calorimetric system in the fore rating period (including the extrapolation of this temperature to the midpoint of the heating period), T'' be the corresponding calorimeter temperature in the after rating period, and T be the actual temperature of the calorimetric system during the heating period. During the rating periods β is zero. The solution of equation (21) is then

$$T' = T'_0 + kt + (1/a)\,[(\mathrm{d}T'/\mathrm{d}t)_0 - k]\,(1 - e^{-at}) \tag{23}$$

and a similar expression in which T'' is everywhere substituted for T'. The value of a, taken as constant, is an average over the temperature rise of the heating period and is given by the expression, following from equations (21) and (22),

$$a = [(\mathrm{d}T'/\mathrm{d}t)_1 - (\mathrm{d}T''/\mathrm{d}t)_2]/[T''_2 - T'_1 - k(t_2 - t_1)] \tag{24}$$

in which t_1 is a time in the fore period near the beginning of the energy input and t_2 a time in the after period soon after the end of the energy input. The constants T'_0, T''_0, $(\mathrm{d}T'/\mathrm{d}t)_0$ and $(\mathrm{d}T''/\mathrm{d}t)_0$ in equation (23), chosen for convenience at the midpoint of the energy input, are evaluated from the observed temperatures in the fore and after rating periods. These four constants must also be consistent with the value of a since by putting $t_1 = t_2 = 0$ in equation (24), a is determined by them. If β is taken as constant, the solution of equation (21) during the heating period is

$$T = T_{s0} + (1/a)(\beta - k) + kt + Ke^{-at} \qquad (25)$$

β and the constant of integration, K, are determined by the requirements that

$$T(-\tau) = T'(-\tau)$$

$$T(\tau) = T''(\tau) \qquad (26)$$

The total energy added to the calorimetric system between $t = -\tau$ and $t = \tau$ is

$$E + \int_{-\tau}^{\tau} B_N(T_s - T)dt = E + aC \int_{-\tau}^{\tau} (T_s - T)dt \qquad (27)$$

$B_N = aC$ is experimentally found to be sufficiently constant with temperature so a mean value may be taken outside the integral sign. It is convenient to calculate the energy needed to heat the calorimeter instantaneously from T_0' to T_0''. This energy is

$$E + aC[\int_{-\tau}^{0}(T' - T)dt + \int_{0}^{\tau}(T'' - T)\,dt] \qquad (28)$$

The integrals may be evaluated by combining equations (21), (23), (25), and (26) to give

$$a[\int_{-\tau}^{0}(T' - T)dt + \int_{0}^{\tau}(T'' - T)dt] =$$

$$(T_0'' - T_0')[1 - (a\tau/\sinh a\tau)] \qquad (29)$$

Substituting in equation (28), one obtains for the mean heat capacity of the calorimetric system

$$C = \frac{E}{T_0'' - T_0'} \frac{\sinh a\tau}{a\tau} \qquad (30)$$

The approximation $(\sinh a\tau)/(a\tau) = 1 + (a\tau)^2/6$ is valid to 0·01 per cent for $a\tau < \frac{1}{3}$ and may be used for all except the lowest temperature measurements.

When the thermal relaxation time, a^{-1}, is large compared to the times of the rating periods, the exponentials in equation (23) may be expanded leading to

$$T' = T_0' + (dT'/dt)_0\, t + [k - (dT'/dt)_0]\,(at^2/2) \qquad (31)$$

and a similar expression with T' replaced by T''. In most heat capacity measurements, the temperature drift of the shield, k, and the reciprocal of the thermal relaxation time, a, are sufficiently small so that only the linear term in these expressions need be considered, and the heat capacity may be calculated from equation (30), by employing simply the electrical energy and a temperature rise calculated by linear extrapolation of the temperatures measured in the fore and after periods to the midpoint of the heating period. At very low temperatures at which the heat capacities are low and at high temperatures at which the B_N term becomes large because of radiation, the values of a become large enough so that the observed temperature drift of the calorimeter is not constant and an exponential extrapolation is desirable. There is a limited range in which the quadratic approximation to an exponential, equation (31), is sufficient, but if the observed calorimeter drifts are not constant because of a large a, it is better to make an exponential extrapolation rather than a quadratic one. If, however, the observed calorimeter drifts are not constant because of a changing shield temperature, then a quadratic extrapolation adequately approximates equation (23).

In the derivation of equation (30), it was assumed that a and β were constants independent of temperature during a single heat capacity measurement. A small correction for the variation of β is described below in section III–1–F. The maximum variation of a occurs at low temperatures at which the heat capacity of the calorimeter varies approximately as T^3 and the thermal conductance decreases with increasing temperature[47]. The variation of a in a calorimeter for which the thermal conductance to the shield arises from conductivity along platinum wires and the heat capacity varies as T^3 is about ten per cent per degree. The value of a is calculated from the thermometer drifts in the fore and after periods, combined with measurement of the drift in shield temperature, and therefore represents some average value between those at the initial and final temperatures of the calorimeter. One can represent the observed values of a over a range of temperature as a Taylor's series expanded about a mean temperature and carry out the integration of equation (21). The details of this tedious calculation will not be presented here, but the result is that, for a temperature rise not exceeding ten per cent of the absolute temperature, the error in replacing the varying a by a mean value is small compared to other errors arising from insensitivity of resistance thermometers. As the temperature increases the errors, because of the variation of a, decrease rapidly.

Typical values of a for an empty copper calorimeter with platinum lead wires[9] are 0·11 min^{-1} at 12°K, 0·022 min^{-1} at 20°K, 0·0043 min^{-1} at 30°K, 0·0006 min^{-1} at 100°K, 0·0029 min^{-1} at 200°K, and 0·013 min^{-1} at 300°K. The values of a for a filled calorimeter are smaller by the ratio of the heat capacity of the empty calorimeter to that of the full.

It should be emphasized that the extrapolation of the fore and after temperatures by equation (23) for use in calculating the heat capacity with equation (30) is only a mathematical device for calculating the heat interchange between the calorimeter and shield during the period between the last temperature measurement, T_1', at a time t_1 in the fore rating period and a temperature measurement, T_2'', at a time t_2 in the after rating period when the calorimetric system is in internal thermal equilibrium after the introduction of energy. As can be seen from *Figure 8*, the actual calorimeter temperature may differ widely from the extrapolation. The heat gained by the calorimeter from the shield between the times t_1 and t_2, plus the electrical energy introduced, is the energy needed to heat the calorimetric system from T_1' to T_2''. In assigning a mean temperature to a heat capacity measurement the average of the temperatures T_1' and T_2'', corresponding to equilibrium temperature measurements in the fore and after periods, should be used rather than an average of the extrapolated temperatures T_0' and T_0''.

In order to minimize the corrections for heat leak it is desirable that the heating time be short. This is particularly important at temperatures in the liquid hydrogen range when the thermal relaxation time within the calorimetric system is very short and the calorimeter drifts are large and non-linear. Because of the insensitivity of metal resistance thermometers in this range, the possible accuracy of a heat capacity measurement is less than at higher temperatures, and it is wise to take fewer measurements of the heating current and potential in order to decrease the time of the heating period. A heating period of five or six minutes is convenient in this temperature range. At higher temperatures when the heat leak correction is smaller, the time of energy input is increased to ten minutes or more with a slight gain in accuracy in the measurement of the electrical energy.

Over most of the temperature range, the resistance of the thermometer is nearly linear in temperature, and it is convenient to express the heat leak corrections and the temperatures of the thermometer in terms of thermometer resistance rather than converting all resistance measurements to degrees. At temperatures in the range of liquid hydrogen, however, the temperature coefficient of resistance is changing rapidly with temperature, and all resistances and drifts should be converted to degrees before calculating the corrections for heat leak.

D. Correction for Thermal Gradients in the Calorimetric System

(1) *Calorimeter with External Thermometer–Heater. The $(A/E)B_N$ Correction.* This correction was first described by Giauque and Wiebe[23]. In this type of calorimeter, the temperature rise in a heat capacity measurement is calculated from measurements of the resistance of the thermometer wound on the external surface of the calorimeter. During the heating period, the resistance of the thermometer–heater is also measured, and thus the surface temperature of the calorimeter is known. The course of the surface temperature (taken as the temperature of the thermometer–heater) and the mean temperature of the calorimetric system is shown schematically in *Figure 9*. The mean temperature of the calorimetric system at any time is the temperature that would be obtained if the system were allowed to come to internal equilbrium at constant energy. For clarity the difference between the heater temperature and the calorimeter temperature in the rating periods is exaggerated in *Figure 9*. The variation with time of the heater temperature and the calorimeter temperature is given by equations similar to those developed in section III–1–B. If the heater completely covered the surface of the calorimeter, K_2 defined in that section would be zero. In a calorimeter of the type

described by Giauque and Egan[19], about 6/7 of the calorimeter surface is covered by the heater and K_2 is small compared to K_3.

During the heating period, the difference between the surface temperature and that of the calorimeter is, after a short relaxation time of a few seconds,

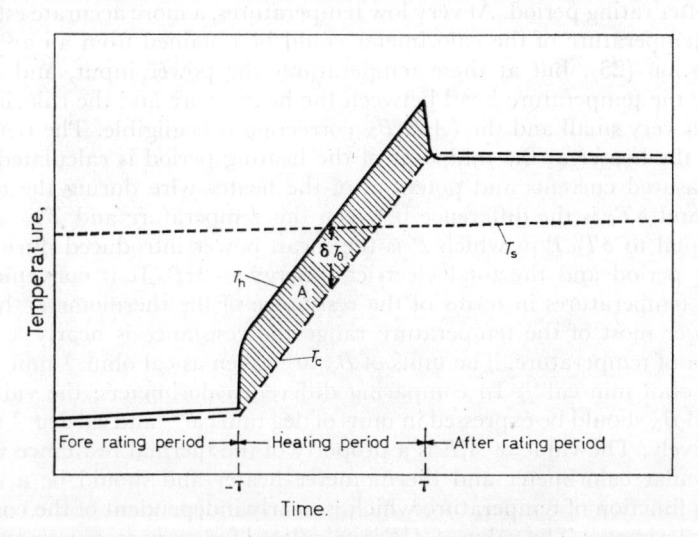

Figure 9. Schematic drawing of the shield temperature, T_s, calorimeter temperature, T_c, and thermometer–heater temperature, T_h, during a heat capacity measurement with a calorimeter with external thermometer–heater. For clarity the temperature difference between thermometer-heat and calorimeter in the rating periods has been exaggerated.

given by equation (15), or to sufficient accuracy by $\delta T = P/K_1$, in which P is the power input to the heater and K_1 is the thermal conductance between heater and calorimeter. In the first order calculation of the correction for heat interchange between the calorimeter and shield, the surface temperature of the calorimeter has been taken as the mean temperature of the calormetric system given by equation (25). The actual surface temperature is higher than assumed and, therefore, the heat leak to the calorimeter has been over-estimated by an amount $B_N \int_{-\tau}^{\tau} \delta T \, dt = B_N A$. A is the shaded area in *Figure 9*. To a sufficient approximation, A is equal to the product of a measured δT_0 at the midpoint of the heater period and the time of heating, 2τ. A correction $-B_N A$ must be added to the energy introduced into the calorimeter, which is approximately equal to the electrical energy, E, introduced into the heater. A fractional correction

$$1 - (A/E)B_N \tag{32}$$

must, therefore, be made to the first order calculation of the heat capacity to account for higher surface temperature during energy input.

The quantity $(A/E) = (2\tau \delta T_0)/E$ is, to the first order, equal to the reciprocal of the thermal conductance, K_1, between the calorimeter and the heater wire. It should, therefore, be independent of the power input or the heating time and should vary slowly with temperature. The calorimeter temperature, T_{Av}, is taken as the mean of the temperatures $T''(-\tau)$ and $T'''(\tau)$ obtained by extrapolation to the beginning and end of the heating period of the temperature drifts observed in the fore rating period and after equilibrium is attained in the after rating period. At very low temperatures, a more accurate estimate of the temperature of the calorimeter could be obtained from an analogue of equation (25), but at these temperatures the power input, and consequently the temperature head between the heater wire and the calorimeter, becomes very small and the $(A/E)B_{\mathrm{N}}$ correction is negligible. The temperature of the heater at the midpoint of the heating period is calculated from the measured currents and potential of the heater wire during the energy input, and δT_0 is the difference between this temperature and T_{Av}. A/E is then equal to $\delta T_0/P$ in which P is the mean power introduced during the heating period and the total electrical energy is $2\tau P$. It is convenient to express temperatures in terms of the resistance of the thermometer–heater, since over most of the temperature range the resistance is nearly a linear function of temperature. The units of B_{N} are taken as cal ohm^{-1} min^{-1} and of A/E ohm min cal^{-1}. In comparing different calorimeters, the values of A/E and B_{N} should be expressed in units of deg min cal^{-1} and cal deg^{-1} min^{-1} respectively. The value of A/E is a property of the thermal resistance within a particular calorimeter and thermometer–heater and should be a slowly varying function of temperature which is nearly independent of the contents of the calorimeter. The value of A/E is calculated for every measurement, and a point lying off the smooth curve for the calorimeter is a sensitive indication of an error in the measurement or calculation of the current or potential during the heating period.

In the rating periods, the temperature of the thermometer–heater will, as discussed in section III–1–B, lie between the temperature of the shield and the calorimeter although close to the latter. Let T_{h} be the temperature of the thermometer–heater, T_{c} the temperature of the calorimeter, and T_{s} that of the thermal shield. Then, from equation (18), neglecting small quantities and putting $K_2 = 0$, one obtains for the temperatures at the beginning $(t = -\tau)$ of the heating period

$$T_{\mathrm{h}}(-\tau) - T_{\mathrm{c}}(-\tau) = -(K_3/K_1)[T_{\mathrm{c}}(-\tau) - T_{\mathrm{s}}(-\tau)]$$

$$= (A/E)B_{\mathrm{N}}[T_{\mathrm{s}}(-\tau) - T_{\mathrm{c}}(-\tau)] \tag{33}$$

At the end of the heating period $(t = \tau)$, the temperatures extrapolated back from a time after equilibrium has been obtained are

$$T_{\mathrm{h}}(\tau) - T_{\mathrm{c}}(\tau) = (A/E)B_{\mathrm{N}}[T_{\mathrm{s}}(\tau) - T_{\mathrm{c}}(\tau)] \tag{34}$$

Combining these two equations, and defining $\Delta T_{\mathrm{h}} = T_{\mathrm{h}}(\tau) - T_{\mathrm{h}}(-\tau)$, $\Delta T_{\mathrm{c}} = T_{\mathrm{c}}(\tau) - T_{\mathrm{c}}(-\tau)$, $\Delta T_{\mathrm{s}} = T_{\mathrm{s}}(\tau) - T_{\mathrm{s}}(-\tau)$, one obtains

$$\Delta T_{\mathrm{h}} = \Delta T_{\mathrm{c}}\left[1 - \frac{A}{E}B_{\mathrm{N}}\frac{\Delta T_{\mathrm{c}} - \Delta T_{\mathrm{s}}}{\Delta T_{\mathrm{c}}}\right] \tag{35}$$

In the first order calculation of the heat capacity, equation (30), one uses the measured temperatures of the resistance thermometer–heater to describe the calorimeter temperature. Actually the temperature rise of the thermometer–heater is slightly less than that of the calorimeter and a correction

$$1 - \frac{A}{E} B_N \frac{\Delta T_c - \Delta T_s}{\Delta T_c} \tag{36}$$

must be applied to the first order calculation of the heat capacity because of this difference. In most measurements, the drift in the shield temperature during the time of energy input is sufficiently small such that $(\Delta T_c - \Delta T_s)/\Delta T_c$ may be put equal to one. Then the total correction to the first order calculation of heat capacity for non-equilibrium effects in the calorimeter is the factor

$$1 - 2(A/E) B_N \tag{37}$$

The maximum value observed for $(A/E)B_N$ is of the order of 0·01 near 300°K, and terms of the order of $[(A/E)B_N]^2$ are, therefore, negligible. Near room temperature, it is sometimes desirable to purposely introduce a constant upward drift in shield temperature in order to decrease $\Delta T_c - \Delta T_s$ in equation (36) and thus decrease the $(A/E)B_N$ correction.

In a calorimeter of the type described by Giauque and Egan[19], about one seventh of the calorimeter surface is not covered by the thermometer–heater. A correction is sometimes made to allow for this but since the correction is difficult to make accurately because of lack of detailed knowledge of the temperature distributions throughout the calorimetric system, and since the correction is small, it is dubious whether the accuracy is increased by applying it.

(2) *Calorimeter with Internal Thermometer–Heater. The γB_N Correction.* In this type of calorimeter, the thermal conductance between the thermometer–heater and the thermal shield, K_3, is small compared to that between the calorimeter and shield, K_2 (see *Figure 7*). The temperature of the outer surface of the calorimer is, therefore, closer to the mean temperature of the calorimetric system during the heat input period and the rating periods than when the thermometer is wound on the surface. It is nonetheless desirable to make a quantitative correction for the temperature difference between the thermometer and the mean temperature of the calorimetric system. By observing the effect of the shield temperature upon the resistance of the thermometer and using a simplified model for the thermal connections between different parts of the calorimetric system Catalano and Stout[7] estimated this correction. A superior method, described by Cole, Hutchens, Robie and Stout[9] involves the measurement of the heat capacity with different values of B_N and extrapolating the measured points to $B_N = 0$ to obtain the true heat capacity. Regardless of the details of the temperature distribution, the temperature differences between various parts of the calorimetric system in the steady state will be proportional to $B_N(T_c - T_s)$ and will vanish as B_N approaches zero. One may, therefore, relate the

observed temperature rise of the thermometer–heater ΔT_h, to that of the calorimetric system at equilibrium by the equation

$$\Delta T_\mathrm{h} = \Delta T_\mathrm{c}\,(1 - 2\gamma B_\mathrm{N}) \tag{38}$$

in which γ is a constant to be experimentally determined. If C is the heat capacity calculated by taking the temperature of the thermometer–heater to represent that of the calorimetric system and C_0 is the true heat capacity of the calorimetric system, then

$$C_0 = C_\mathrm{h}\,(1 - 2\gamma B_\mathrm{N}) \tag{39}$$

By measuring the apparent heat capacity, C_h, at varying values of B_N but at the same mean temperature one may determine both the true heat capacity, C_0, and the correction factor, γ. In practice some helium gas, sufficient to increase B_N by a factor of two or three, is admitted to the insulating vacuum space separating calorimeter and shield, and a series of measurements of heat capacity is made. The values of γ calculated are plotted against temperature, and points read off the curve are used to correct the heat capacity measurements taken with a good insulating vacuum. The maximum $2\gamma B_\mathrm{N}$ correction is found with an empty calorimeter. At 300°K it amounts to about one per cent. With a filled calorimeter the correction is smaller, not exceeding 0·3 per cent at 300°K. The correction diminishes rapidly at lower temperatures, since B_N varies approximately as T^3 and γ varies slowly with temperature. In an empty calorimeter, the mean temperature of the calorimeter is intermediate between that of the shield and thermometer during the rating periods, and during the heating period the temperature of the outer surface lags slightly behind the mean calorimeter temperature. The value of γ for an empty calorimeter is, therefore, negative. When the calorimeter is filled with sample, however, the temperature of the thermometer during the rating periods may be either slightly closer to or further from the shield temperature than is the mean temperature of the calorimetric system, depending on the heat capacities and thermal conductivities within the calorimetric system. Likewise, during energy input, the presence of the sample lowers the mean temperature of the calorimetric system relative to the surface temperature. The addition of sample to the calorimeter makes the value of γ more positive, which in practice often reduces its magnitude almost to zero.

In a calorimeter with internal thermometer–heater, the value of A/E is calculated as described in section III–1–D–(1) for each measurement. This quantity is a measure of the thermal resistance between the thermometer–heater and the calorimeter, and for the type of strain free platinum thermometer–heater described in section II–3–B–(2) varies inversely as the thermal conductivity of the gas in the thermometer case. At 0°C the thermal conductivity of helium gas is 6·3 times that of air, and the contamination of the helium gas in the thermometer by a small amount of air is readily observed as an increase in the A/E value. During the initial filling of the thermometer case with helium, the value of A/E is measured to make sure that air is not accidently introduced when the pinhole in the case is sealed.

After the thermometer is attached to the calorimeter, a record of A/E values is kept during all measurements of heat capacity. With even a very small leak in the case, the A/E will gradually rise over a long period of time as helium is pumped out of the case during times when the calorimeter is in a vacuum and replaced by air when the calorimeter is exposed to the atmosphere. In such a situation, the thermometer case should be removed from the calorimeter, evacuated and all joints checked with a sensitive leak detector, refilled with pure helium and resealed.

The routine calculation of a value of A/E for every measurement is also needed in the transient heating correction described below in section III–2–B and further serves as a sensitive detector of errors in the heating current or potential difference.

(3) Temperature Gradients Between Calorimeter and Sample. At low temperatures, the time of establishment of thermal equilibrium within the calorimetric system is very short, and a measurement of the resistance of the thermometer one minute after the end of the heating period represents an equilibrium temperature. At higher temperatures, non-equilibrium thermometer drifts are normally observed for 3 to 6 min after energy input. With a sample of large heat capacity and poor thermal conductivity, the time for equilibrium may rise to 10 min at 150°K and 15 min near room temperatures. As long as the time for internal equilibrium in the calorimeter is short compared to the thermal relaxation time, a^{-1}, between the calorimetric system and the shield, the effect of the steady state temperature gradients is taken into account by the $(A/E)B_N$ or γB_N corrections described above. Near room temperature, the thermometer drifts are not constant in the rating periods even after the calorimetric system is in the steady state because the large drifts change the temperature difference between calorimeter and shield. It is important to separate this effect from the change in the drifts as the steady state is approached. This separation may be done by observing whether the drifts are consistent with the equations

$$\mathrm{d}T/\mathrm{d}t = a(T_s - T) \tag{40}$$

and

$$\mathrm{d}^2 T/\mathrm{d}t^2 = a(\mathrm{d}T_s/\mathrm{d}t - \mathrm{d}T/\mathrm{d}t) \tag{41}$$

with the same value of a throughout the fore and after rating periods. During the initial relaxation after the introduction of energy, the observed drift, $\mathrm{d}T/\mathrm{d}t$, and its change with time are larger in magnitude than those obtained from these equations with the value of a appropriate to the fore period and the after period in the steady state. If B_N, and consequently a, is changed during the heating period by desorption of gas, then the drifts are also large, and it is difficult to separate this effect from the attainment of equilibrium in the calorimetric system. For this reason it is very important to avoid the presence of water or other condensable gases which may condense on the calorimeter and desorb during the heating period. Data taken under such circumstances are best discarded.

There is no single relaxation time to describe the attainment of a steady state temperature distribution within a filled calorimeter.

E. Representative Values of A/E and B_N

In *Figures 10* and *11* are shown typical values of A/E and B_N as functions of temperature for a calorimeter with internal thermometer–heater of the type described in section II–3–B. The data are taken from measurements on a copper calorimeter filled with $CrCl_2$[52]. These data are chosen to illustrate various anomalies which may occur in the temperature variation of A/E and B_N and which would not be present under ideal conditions. At high temperatures, the principal mechanism of heat transfer between the

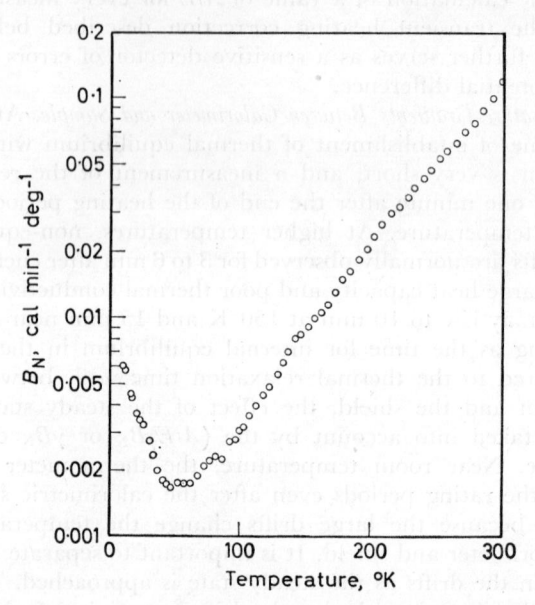

Figure 10. Thermal conductance between calorimeter and shield, B_N, (logarithmic scale) *versus* temperature. Data from measurements on $CrCl_2$, J. W. Stout and R. C. Chisholm, *J. Chem. Phys.*, **36**, 979 (1962).

calorimeter and shield is radiation, which contributes a term varying as T^3 to B_N. The plot of $B_N\ T^{-3}$, in *Figure 11*, shows that this quantity slowly increases between 200 and 300°K, an effect probably due to gas conduction by condensable gases, water and perhaps organic vapors, which have been incompletely removed from the vacuum space. The removal of such material is facilitated by lengthy pumping at a temperature of about 35°C, and the B_N values at the higher temperatures are lowered by this procedure. Small leaks in the vacuum system can also, over an extended period of time with the cryostat cold, cause condensed water and CO_2 to collect in the vacuum space. The slightly high values of B_N near 140°K in *Figure 11* are probably due to a small amount of CO_2 which has collected in the vacuum system. Below about 100°K the $B_N\ T^{-3}$ curve rises, since the contribution to B_N of the thermal conductance of the platinum wires connecting the shield and calorimeter, which is approximately constant above 50°K, becomes important compared to that of radiation. *Figure 10* shows that the total B_N passes through a minimum near 50°K and rises at lower temperatures because of the increasing thermal conductivity[47] of the wires.

The major contribution to A/E is the thermal resistance of the helium gas in the thermometer shell. The thermal conductivity of helium gas[33] decreases with temperature. The conductivity at 300°K is 7·3 times that at 20°K and 2·2 times that at 100°K. The observed A/E's increase as expected at lower temperatures but by smaller factors. At temperatures below 25°K,

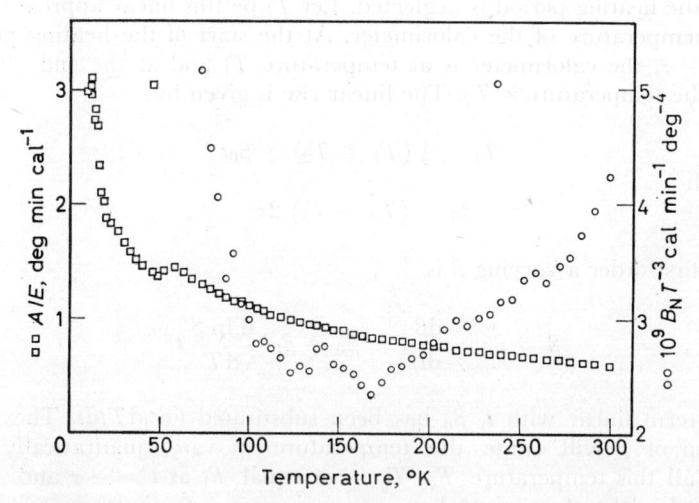

Figure 11. Thermal conductance between calorimeter and shield divided by cube of the absolute temperature, $B_N T^{-3}$, (circles, right scale), and thermal resistance between heater and calorimeter, A/E (squares, left scale), versus temperature. Data from measurements on CrCl$_2$, J. W. Stout and R. C. Chisholm, *J. Chem. Phys.*, **36**, 979 (1962).

the measured values of A/E are relatively inaccurate since at the small rates of heat input needed for heat capacity measurements in this temperature range the temperature difference between heater and calorimeter becomes small. There is a small bump in the curve of A/E near 60°K. This is because of the presence of a small amount of air in the helium in the thermometer case. At lower temperatures, the air is condensed and the thermal resistance is that of pure helium. As the temperature rises, the air evaporates, decreasing the thermal conductivity of the gas mixture and raising the value of A/E. When all the air has evaporated, the A/E drops with increasing temperature as the thermal conductivity of the gas mixture increases. The thermal effect of the vaporization of the quantity of air needed to produce the observed bump in the A/E curve is so small as to be undetectable in the heat capacities. When care is taken to have pure helium in the thermometer case, the bump in the A/E curve near 60°K disappears.

In a calorimeter with external heater, of the type described in section II-3-A, the A/E at room temperature is smaller by a factor of about twenty than that shown in *Figure 11*. It also increases at lower temperatures, at 12°K being 5 to 10 times as large as at room temperature.

F. Non-constant Heating Correction

In deriving equation (30), β, the ratio of the electrical power to the heat capacity of the calorimetric system, was taken as constant. Actually the power input will vary because of the change in electrical resistance of the thermometer–heater during energy input and the heat capacity of the calorimeter and contents will also change. This small correction may be estimated to sufficient accuracy by calculating the correction to the linear rise in temperature with time that occurs if β is constant and the heat leak during the heating period is neglected. Let T_l be this linear approximation to the temperature of the calorimeter. At the start of the heating period, at $t = -\tau$, the calorimeter is at temperature T_1 and at the end, at time $t = \tau$, the temperature is T_2. The linear rise is given by:

$$T_l = \tfrac{1}{2}(T_1 + T_2) + \beta_0 t \tag{42}$$

in which

$$\beta_0 = (T_2 - T_1)/2\tau \tag{43}$$

To the first order a varying β is

$$\beta = \beta_0 + \frac{d\beta}{dt} t = \beta_0 + \beta_0^2 \frac{d \ln \beta}{dT} t \tag{44}$$

In the term linear with t, β_0 has been substituted for dT/dt. The linear variation of β will cause the temperature to vary quadratically with time. Call this temperature T_q. T_q must equal T_1 at $t = -\tau$ and T_2 at $t = \tau$ and is, therefore, given by

$$T_q = \tfrac{1}{2}(T_1 + T_2) + \beta_0 t + \beta_0^2 (d \ln \beta/dT)(t^2 - \tau^2)/2 \tag{45}$$

The fractional correction to the energy, and to the heat capacity, is

$$\frac{B_N}{E} \int_{-\tau}^{\tau} (T_l - T_q)\, dt = \frac{a\tau \Delta T}{6} \frac{d \ln \beta}{dT} \tag{46}$$

Here $\Delta T = T_2 - T_1$, $a = B_N/C = B_N \Delta T/E$ is the reciprocal of the thermal relaxation time between calorimeter and shield (see section III–1–C), and the approximation $\Delta T = 2\beta_0 \tau$ has been used. Let the resistance of the heater be R and the heating current be kept constant. The power developed in the heater is $I^2 R$. Since $\beta = P/C$,

$$(d \ln \beta/dT) = (d \ln R/dT) - (d \ln C/dT) \tag{47}$$

and for constant heating current the heat capacities must be multiplied by a correction factor

$$1 + \frac{a\tau \Delta T}{6}\left[\frac{d \ln R}{dT} - \frac{d \ln C}{dT}\right] \tag{48}$$

This correction reaches a maximum value of about 0·3 per cent between

10 and 15°K, and is down to 0·02 per cent at 30°K and above. Above about 100°K the heater voltage is kept constant during energy input. In this case the power decreases as the resistance increases and the factor by which the heat capacities are multiplied is

$$1 + \frac{\alpha\tau\Delta T}{6}\left[-\frac{d\ln R}{dT} - \frac{d\ln C}{dT}\right] \tag{49}$$

In this range the correction never exceeds 0·02 per cent. This correction is proportional to α and, therefore, to B_N and so is already included in the γB_N correction described in section III–1–D–(2) which is applied at the higher temperatures.

2. Electrical energy measurements and calculations

A. Current and Potential during the Heating Period

The electrical energy, E, introduced into the heater during the heating period is given by

$$E = \int_{-\tau}^{\tau} \varepsilon I \, dt \tag{50}$$

in which I is the current through the heater, ε is the electrical potential across the heater and the integral over time extends from the beginning of the heating period at $t = -\tau$ to the end at $t = \tau$. It is desired to measure E to a few hundredths of one per cent, and it is necessary to make sufficient readings of ε and I during the heating period to determine E to this accuracy. Because the resistance of the thermometer–heater, $R = \varepsilon/I$, changes during the heating period it is not possible to maintain both ε and I constant. If the current to the heater is supplied through a low resistance circuit from a battery of cells, then ε will drift only slightly during the heating period and may be adequately represented by a linear function of time whose average value is that read at the midpoint of the heating period. If ε is constant, I will vary inversely as the resistance, R, of the heater. The variation of I during the heating period is too large to be represented by a linear function of time but, except for a short relaxation period, discussed in section III–2–B below, it is represented to sufficient accuracy by a quadratic function. The mean value of a quadratic function of time is[24]

$$\frac{1}{2\tau}\int_{-\tau}^{\tau} I \, dt = \tfrac{1}{2}\left[I\left(-\frac{\tau}{\sqrt{3}}\right) + I\left(\frac{\tau}{\sqrt{3}}\right)\right] \tag{51}$$

If ε is a linear function of time and I is quadratic then

$$\int_{-\tau}^{\tau} \varepsilon I \, dt = \frac{(2\tau)\,\varepsilon_0}{2}\left[I\left(\frac{-\tau}{\sqrt{3}}\right) + I\left(\frac{\tau}{\sqrt{3}}\right) + \frac{(2\tau)^2}{12}\left(\frac{dI}{dt}\right)_0\left(\frac{d\varepsilon}{dt}\right)_0\right] \tag{52}$$

In this equation the subscript zero refers to values at $t = 0$, the midpoint of the run. The last term in equation (52) ordinarily contributes less than

0·01 per cent and may be neglected. The measurements are scheduled so that I is read at $t = -\tau/\sqrt{3}$ and $t = \tau/\sqrt{3}$ (corresponding to fractions 0·211 and 0·789 of the total heating time) and ε is read at the midpoint of the heating period. In a six or more minute heating period there is time for six readings of I which are scheduled in two groups of three centered about $t = -\tau/\sqrt{3}$ and $t = \tau/\sqrt{3}$ respectively and three readings of ε centered around $t = 0$. There is thus obtained a drift rate and galvanometer sensitivity measurement from each set of three readings. In order to verify that with a particular calorimeter ε is represented by a linear function of time and I by a quadratic function with the required accuracy, occasional measurements, not part of a series of heat capacity measurements, are taken in which only ε or I is read at short intervals throughout the heating period. When an unusually large amount of energy is to be introduced into the calorimeter over an extended period of time, such as occurs during the measurement of a heat of fusion or transition, the measurements are scheduled as a series of shorter intervals, each covering a sufficiently short time so the assumptions of a quadratic variation of I and linear variation of ε are valid. During such a measurement the heating current is on continuously.

At the lowest temperatures, the resistance of the thermometer–heater becomes small, and a single cell may provide too large a power input into the heater. It is then convenient to supply the heating current from 10 or 15 cells in series with an adjustable external resistance. During energy input the current through the heater is nearly constant and may be adequately represented by a linear function of time whereas the potential, ε, varies as the resistance of the heater changes and requires a quadratic function of time to represent it. The roles of ε and I are then interchanged from those in the discussion above. In a calorimeter with internal heater such as is described in section II–3–B, "constant current" measurements are made at temperatures below about 80°K and "constant voltage" measurements at higher temperatures.

Since the potential across the heater may be as large as 30 volts, it is necessary to use a voltage divider to measure ε with a potentiometer whose upper limit is 0·1 volt. A calibrated "volt box" with taps which reduce the potential by factors ranging from 20 to 500 is used. For accurate work, the resistance of the copper wires leading from the voltage divider to the thermometer–heater must be considered in the voltage divider calibration. The current, measured by the potential drop across a calibrated standard resistance, is the sum of the currents through the heater and through the voltage divider. The current through the voltage divider, ε/R_{VB}, in which R_{VB} is the resistance of the voltage divider, must be substracted from the measured current to obtain the current through the heater. The current through the voltage divider is less than one per cent of the total current.

B. Transient Heating Corrections

During the heating period the temperature of the heater exceeds the mean temperature of the calorimeter by

$$\delta T = T_h - T = (A/E)\,P \tag{53}$$

or, if the temperature is expressed in terms of the resistance of the thermometer, by

$$\delta R = R_h - R = (A/E)' P \tag{54}$$

The units of $(A/E)'$ are ohm min cal^{-1}. At the beginning of the heating period, the heater is at essentially the temperature of the calorimeter and it will acquire the steady-state temperature head, δT, exponentially with a relaxation time

$$t_w = (A/E) C_h \tag{55}$$

in which C_h is the heat capacity of the heater (see Section III–1–B). This relaxation time is a few seconds and is not taken into account by the measurements of ε and I described above. In a measurement at constant voltage, additional energy above that previously accounted for is introduced during the transient period and the heat capacities must be multiplied by the factor

$$1 + (\delta R/R_h)(t_w/2\tau). \tag{56}$$

Here R_h is the value of the thermometer–heater resistance at the midpoint of the heating period. In a measurement at constant current the power during the transient period is less than in the steady state and the heat capacities must be multiplied by the factor

$$1 - (\delta R/R_{Av})(t_w/2\tau) \tag{57}$$

in which R_{Av} is the resistance corresponding to T_{Av} defined in Section III–1–D–(1). When the case enclosing the thermometer–heater is filled with pure helium the maximum value of t_w observed is about 5 sec, and the correction does not exceed 0·1 per cent. If, however, the helium in the case is contaminated by air, A/E and t_w may increase by a factor of three or four with a corresponding increase in the correction.

In order to determine t_w, special measurements with an ammeter of short time constant are made of the current in the first 15 sec of a heating period. From these measurements, made at two or three temperatures, one calculates by equation (55) the value of the effective heat capacity of the heater. This heat capacity is found to be somewhat larger than that of the platinum wire and presumably includes some of the heat capacity of the mica support. A smooth curve is drawn through the measured values of C_h, using as a guide the known heat capacity of platinum and an estimated heat capacity of mica. Values of C_h from this curve, together with the observed values of A/E are then used to calculate t_w for the correction to each heat capacity measurement.

C. Corrections for Lead Wires

(1) Four-Lead Thermometer–Heater. During the heating period, energy is

developed in the two platinum wires, 0·01 cm in diameter and about 2·5 cm long, through which the heating current is led from the shield to the calorimeter. The potential drop across these wires is not included in the measured ε, and a correction must be made for the energy developed in them. Half of this energy goes into the calorimeter and half to the shield. The heat capacities must be multiplied by a correction factor

$$1 + r/R \tag{58}$$

in which r is the resistance of *one* wire lead and R is the resistance of the heater. If the wire lead is of the same material as the thermometer–heater, the correction is independent of temperature. It amounts to about 0·2 per cent. This correction is unnecessary if one potential lead is connected at the shield terminal of one current lead and the other potential lead is connected at the calorimeter terminal of the other current lead.

(2) *Two-Lead Thermometer–Heater.* In this case, the measured potential during the heating period includes that across the wire leads, and correction must be made for the half of the heat developed in them which flows to the shield. The heat capacity must be multiplied by the correction factor

$$1 - r/R \tag{59}$$

A second correction arises because the resistance of the wire leads is included in the resistance of the thermometer–heater measured during the rating periods. The mean temperature of a lead will be half-way between the temperatures of the calorimeter and shield, and the observed resistance change will, therefore, be less than that which would be found if the lead wires were at the calorimeter temperature. The heat capacities must be multiplied by the correction factor

$$1 - \frac{dr}{dR} \frac{(\Delta T_c - \Delta T_s)}{\Delta T_c} \tag{60}$$

in which dr/dR is the ratio of the temperature coefficient of resistance of *one* wire lead to that of the thermometer and ΔT_c and ΔT_s are the temperature rise of the calorimeter and shield, respectively, during the time of energy input.

D. Correction for Heating by Thermometer Current

During the rating periods, a small amount of heat is produced in the thermometer–heater by the current needed to measure the resistance. During the heating period, this current does not pass through the heater. An amount of electrical energy equivalent to that which would have been produced by the thermometer current must, therefore, be subtracted from the energy measured during the heating period. The heat capacities should be multiplied by a factor

$$1 - i^2/I^2 \tag{61}$$

in which i is the thermometer current and I the current during the heating period. Ordinarily this correction is less than 0·01 per cent and may be ignored, but, at the lowest temperatures at which a large thermometer current is used to increase the sensitivity and at which the heating current may be small, the correction may amount to 0·1 per cent.

3. *Correction for heat capacity of empty calorimeter, solder, and helium*

The heat capacity of the empty calorimeter is measured over the entire temperature range. Careful account is kept of the weight of the empty calorimeter and its wire leads as it is used for measurements of the heat capacity of various substances. It is convenient to construct a graph or table of the heat capacity of the empty calorimeter at a reference weight and, when the calorimeter is used in the measurement of the heat capacity of a substance, to make small corrections for the heat capacity of the difference in weight of solder between that present in the filled calorimeter and the reference empty. The calorimeter, sample, and cap is weighed filled with air or dry nitrogen and then weighed again in air after filling with helium gas at a measured temperature and pressure and the pinhole in the cap has been sealed. The amount of helium in the calorimeter is calculated from the pressure, temperature, and the difference in volume between the calorimeter and sample. The difference in weight of the calorimeter when filled with air or nitrogen and with helium is calculated, and any small difference between the calculated weight change and that observed is used to estimate changes in the weight of solder during the sealing operation. The heat capacity of the helium gas, 2·98 cal deg^{-1} mole^{-1} times the number of moles of helium in the calorimeter, and the heat capacity of the empty corrected for solder changes, are subtracted from the measured heat capacity to give that of the sample.

4. *Temperature scales*

The resistance of the thermometer–heater must be calibrated in terms of a standard thermometer. In the calorimeter described by Giauque and Egan[19] the resistance thermometer was calibrated during the fore and after rating periods of each heat capacity measurement against a standard copper–constantan thermocouple, which had in turn been calibrated by comparison with a gas thermometer. Murch and Giauque[39] substituted a strain-free platinum resistance thermometer calibrated by the National Bureau of Standards for the standard thermocouple. In the calorimeter described in Section II–3–B, the thermometer–heater is calibrated in a separate experiment against a strain-free platinum resistance thermometer calibrated by the National Bureau of Standards. The calibration of the thermometer–heater is routinely checked at the boiling point of hydrogen and occasionally against other fixed points such as the boiling points of oxygen or nitrogen or the triple point of water. The apparatus described by Stimson[50] is very convenient for the triple point calibration.

A smooth table of resistance versus temperature at one degree intervals is constructed from the calibration measurements, and this table is used to calculate the temperatures from the resistances measured during heat capacity measurements. The thermometer calibrations of the National Bureau of Standards are based on the International Temperature Scale of 1948 between the boiling points of sulfur and of oxygen and on the scale of Hoge and Brickwedde[29] below the oxygen boiling point. There is a discontinuity in the derivative of resistance *versus* temperature at the junction of the International Temperature Scale and that of Hoge and Brickwedde. In order to avoid an artificial kink in curves of measured heat capacities, this discontinuity should be smoothed out in constructing the resistance–temperature table for the thermometer–heater.

5. *Calculation of $C_P°$ and other thermodynamic properties*
A. Calculation of $\Delta H/\Delta T$

In a sealed calorimeter, the change in internal energy of the entire calorimetric system, ΔU_{total}, as the calorimeter temperature changes from T to $T + \Delta T$, is by the first law of thermodynamics equal to q, the heat absorbed by the system from its surroundings, plus E, the electrical energy (electrical work) introduced into the system. The mean heat capacity as calculated by equation (30), together with the various corrections listed in Sections III–1–D–E–F and III–2, is, therefore, $\Delta U_{\text{total}}/\Delta T$. The change in energy of the empty calorimeter under identical conditions, and of the helium gas are subtracted as discussed in Section III–3. The change in energy of the helium gas is its molal heat capacity at constant volume, $(3/2)R$, times the number of moles of helium and times ΔT. The change in energy with pressure of the gas is negligible. We are left with $\Delta U/\Delta T$ of the sample corresponding to changing its temperature from T to $T + \Delta T$, its pressure from P to $P + \Delta P$ and its volume from V to $V + \Delta V$.

$$\frac{\Delta U}{\Delta T} = \frac{H(T+\Delta T, P) - H(T,P) - P\Delta V + U(T+\Delta T, P+\Delta P) - U(T+\Delta T, P)}{\Delta T} \tag{62}$$

Here $H = U + PV$ is the enthalpy.

$$U(T + \Delta T, P + \Delta P) - U(T + \Delta T, P) = \int_P^{P+\Delta P} [-T(\partial V/\partial T)_P - P(\partial V/\partial P)_T]\, dP = (-T\alpha + P\beta)\, V\Delta P \tag{63}$$

in which $\alpha = (\partial \ln V/\partial T)_P$ is the cubical coefficient of expansion at constant pressure and $\beta = -(\partial \ln V/\partial P)_T$ is the isothermal compressibility. Then

$$\Delta U/\Delta T = (\Delta H/\Delta T)_P - PV\alpha + (-T\alpha + P\beta)\, V(\Delta P/\Delta T) \tag{64}$$

To a very good approximation, the change in pressure may be calculated by taking the helium as an ideal gas at constant volume, or $\Delta P/\Delta T = P/T$. Then

$$\Delta U/\Delta T = (\Delta H/\Delta T)_P - 2PV\alpha + P^2V\beta/T \qquad (65)$$

The magnitude of the two correction terms in equation (64) is ordinarily so small as to be negligible compared to the experimental errors in measuring $(\Delta H/\Delta T)_P$. If the calorimeter contains helium gas at 1 atm at room temperature then PV is 1 cal for a molal volume of 41 cm^3. For solids, α ranges from 3×10^{-4} to 10^{-5} deg^{-1} so the term $2PV\alpha$ usually lies between 10^{-3} and 10^{-5} cal mole^{-1} deg^{-1}, 0·01 per cent or less of the heat capacity at room temperature. As the temperature is lowered, the fractional correction to $\Delta U/\Delta T$ to obtain $(\Delta H/\Delta T)_P$ decreases approximately as the absolute temperature, since α varies roughly as the heat capacity and P as T. The compressibility, β, is about 10^{-4} atm^{-1} for liquids and 10^{-6} atm^{-1} for solids and the term $P^2V\beta/T$ is, therefore, less than 10^{-6} cal mole^{-1} deg^{-1}. It also decreases roughly proportional to T and is always negligible.

With a calorimeter used for condensed gases such as that described in Section II–3–A, the helium gas introduced into the calorimeter for heat conduction at temperatures when the vapor pressure of the sample is small is at a low pressure, and the corrections described above are negligible. At higher temperatures, measurements are often made with no helium gas present, and the pressure in the calorimeter is determined by the vapor pressure of the sample. In this case, work is done by the expanding solid or liquid sample on the gas in the filling tube, and this work must be included in the energy balance of the calorimeter. The result is that the mean heat capacity calculated as described above is the heat capacity under the saturated vapor pressure, C_S, rather than C_P. When the vapor pressure of the sample is appreciable, a correction to the energy must be made for the heat used to vaporize material into the gas space in the calorimeter and filling tube.

During a series of measurements with a sealed calorimeter, each value of $(\Delta H/\Delta T)_P$, which, in the limit as ΔT becomes small, is the constant pressure heat capacity, C_P, corresponds to a different pressure. A correction to the heat capacity, $C_P°$, in the standard state ($P = 1$ atm for solids and liquids) may be made by means of the thermodynamic formula

$$(\partial C_P/\partial P)_T = -T(\partial^2 V/\partial T^2)_P \qquad (65)$$

For measurements in which the maximum difference from the standard pressure is 1 atm and for ordinary substances, the correction to convert C_P to $C_P°$ is less than 0·01 per cent and is negligible compared to the experimental errors.

B. Curvature Corrections

The measured heat capacities are converted to $(\Delta H/\Delta T)_P$, the ratio at constant pressure of the change of enthalpy to a finite change in temperature. If it is desired to calculate the differential heat capacity,

$C_P = (\partial H/\partial T)_P$, a correction for the non-linearity of C_P versus T must be applied. The heat capacity may be expanded in a Taylor's series around the mean temperature of a measurement, T_{Av}, to terms in T^3

$$C_P(T) = C_P(T_{Av}) + (dC_P/dT)(T - T_{Av}) +$$
$$(1/2)(d^2C_P/dT^2)(T - T_{Av})^2 + (1/6)(d^3C_P/dT^3)(T - T_{Av})^3 \quad (66)$$

The derivatives are evaluated at T_{Av}.

$$\frac{\Delta H}{\Delta T} = \frac{1}{\Delta T}\int_{-\Delta T/2}^{\Delta T/2} C_P d(T - T_{Av}) = C_P(T_{Av}) + (1/24)(d^2C_P/dT^2)(\Delta T)^2 \quad (67)$$

The curvature may be determined from a plot of $\Delta H/\Delta T$ versus T with sufficient accuracy for this correction. The curvature, d^2C_P/dT^2 is $8/(\Delta T)^2$ times the difference between the average of the values of $\Delta H/\Delta T$ at $T_{Av} + \Delta T/2$ and $T_{Av} - \Delta T/2$ and the value of $\Delta H/\Delta T$ at T_{Av}. The value of C_P is then obtained by subtracting from $\Delta H/\Delta T$ at T_{Av} one-third of the difference between the average of $\Delta H/\Delta T$'s at temperatures corresponding to the beginning and end of a measurement and $\Delta H/\Delta T$ at T_{Av}.

C. Thermodynamic Properties

In many investigations, the entropy or some other derived thermodynamic quantity is of greater interest than the heat capacities themselves. Procedures for calculating thermodynamic properties as recommended[56] by the Calorimetry Conference are described in Chapter 5, Section IV–5.

The calculation of the derived thermodynamic quantities requires an extrapolation of the heat capacity from about 10°K to the absolute zero, and, in order that this extrapolation be valid, it is necessary that it involve only a small contribution to the thermodynamic quantities arising from the lattice vibrations and perhaps the low-temperature tail of magnetic ordering effects. Many paramagnetic substances possess at 10°K appreciable amounts of entropy arising from disorder of the magnetic moments. In such cases, there is a temperature coefficient of the magnetic susceptibility at 10°K, and special consideration of the magnetic contributions to the thermodynamic functions must be made. In nearly all substances, the heat capacity arising from the ordering of nuclear spins occurs at temperatures below 1°K. For this reason, it is customary to ignore the contribution of the nuclear spin to the entropy and other thermodynamic functions and to tabulate the thermodynamic functions relative to a reference state at the absolute zero which includes the entropy of a random nuclear spin orientation. In a like fashion, the entropy of mixing of isotopes is not included in the tabulated thermodynamic functions. In chemical reactions at higher temperatures, the contributions of the nuclear spins and of isotope mixing to the thermodynamic functions are equal on both sides of a chemical equation.

In diamagnetic solids with no electronic contribution to the heat capacity, the extrapolation of the heat capacity from 10°K to the absolute zero involves the estimation of the lattice heat capacity in this range. At sufficiently low temperatures the lattice heat capacity will follow the Debye T^3 law, but in many substances there are significant deviations[2] from the Debye[10] heat capacity equation in the neighborhood of 10°K. A convenient procedure is to plot the observed values of C_P/T^2 versus T in the temperature range 10–20°K. A curve is drawn through these points and through the origin. The portion of the curve near the origin is made a straight line of finite slope so the heat capacity is made to vary as T^3 in the low-temperature limit.

6. *Precision and accuracy*

Between 40 and 250°K, the precision of measurement of heat capacity by calorimetric equipment of the type described in this chapter, as measured by the standard deviation of experimental points from a smooth curve, corresponds to an error of less than 0·1 per cent. The scatter is greater than this above 250°K because of increased heat leak and greater below 40°K because of the decreasing sensitivity of the thermometer and the increasing importance of the heat leak correction. The precision does not include an allowance for systematic errors which are present in a particular apparatus, such as errors in the standard temperature scale, errors in electrical calibration, errors in weighing, etc. These errors are difficult to allow for but, in one attempt[9] to take them into account, it was estimated that the total error in heat capacity does not exceed 5 per cent at 10°K, 2 per cent at 15°K, 1 per cent at 20°K and 0·2 per cent between 40 and 250°K. Above 250°K the error increases, reaching about 0·4 per cent at 300°K.

Systematic errors that vary from one laboratory to another may be detected by the comparison of the heat capacities of the same standard substance measured in various laboratories (see Chapter 1). The heat capacity of a National Bureau of Standards sample of benzoic acid[25] has been measured by Busey[5] and by Cole, Hutchens, Robie and Stout[9] in calorimeters with isothermal shield of the type described in this chapter, and by Furukawa, McCoskey and King[15] and by Osborne, Westrum and Lohr[42] with an "adiabatic" calorimeter of the type described in Chapter 5. Comparison of the results of the various investigators shows that there is agreement to within the limits of error estimated above except at the lowest temperatures near 10°K. The National Bureau of Standards temperature scale[29] was used by all investigators, and systematic errors in it would not be disclosed by the comparison of data from the different laboratories.

IV. Acknowledgements

I thank Professor W. F. Giauque for helpful correspondence, for supplying me with a copy of his unpublished "Notes on the Calculation of Heat Capacity" which he has prepared for the use of his students, and for permission to reproduce *Figures 2* and *3*. Discussions with Professor J. O. Hutchens and with graduate students and fellows who have collaborated in low-temperature heat capacity measurements have served to clarify many questions. The partial support of this work by the Office of Naval Research and by the National Science Foundation is gratefully acknowledged.

V. References

[1] Aston, J. G., and G. H. Messerly, *J. Am. Chem. Soc.*, **58**, 2354 (1936).
[2] Bijl, D., Chapter XII in C. J. Gorter, ed., *Progress in Low Temperature Physics*, Vol. II, North Holland Publishing Co., Amsterdam, 1957.
[3] Blue, R. W., and W. F. Giauque, *J. Am. Chem. Soc.*, **57**, 991 (1935).
[4] Brickwedde, F. G., H. J. Hoge, and R. B. Scott, *J. Chem. Phys.*, **16**, 429 (1948).
[5] Busey, R. H., *J. Am. Chem. Soc.*, **78**, 3263 (1956); ORNL-1828, Office of Technical Services, Dept. of Commerce, Washington 25, D.C.
[6] Busey, R. H., and W. F. Giauque, *J. Am. Chem. Soc.*, **74**, 4443 (1952).
[7] Catalano, E., and J. W. Stout, *J. Chem. Phys.*, **23**, 1284 (1955).
[8] Clusius, K., and L. Popp, *Z. Physik. Chem.*, **B46**, 63 (1940).
[9] Cole, A. G., J. O. Hutchens, R. A. Robie, and J. W. Stout, *J. Am. Chem. Soc.*, **82**, 4807 (1960).
[10] Debye, P., *Ann. Physik*, (4), **4**, 553 (1901).
[11] De Sorbo, W., *J. Am. Chem. Soc.*, **75**, 1825 (1953).
[12] Eastman, E. D., and W. H. Rodebush, *J. Am. Chem. Soc.*, **40**, 489 (1918).
[13] Eucken, A., *Physik Z.*, **10**, 586 (1909).
[14] Eucken, A., *Energie- und Wärmeinhalt*, Vol. VIII, Part I, of Wien-Harms *Handbuch der Experimentalphysik*, Akadem. Verlagsgeschellschaft, Leipzig, 1929.
[15] Furukawa, G. T., R. E. McCoskey, and G. J. King, *J. Res. Natl. Bur. Std.*, **47**, 256 (1951).
[16] Gerkin, R. E., and K. S. Pitzer, *J. Am. Chem. Soc.*, **84**, 2662 (1962).
[17] Giauque, W. F., private communication.
[18] Giauque, W. F., and R. C. Archibald, *J. Am. Chem. Soc.*, **59**, 561 (1937).
[19] Giauque, W. F., and C. J. Egan, *J. Chem. Phys.*, **5**, 45 (1937).
[20] Giauque, W. F., and H. L. Johnston, *J. Am. Chem. Soc.*, **51**, 2300 (1929).
[21] Giauque, W. F., and P. F. Meads, *J. Am. Chem. Soc.*, **63**, 1897 (1941).
[22] Giauque, W. F., and J. W. Stout, *J. Am. Chem. Soc.*, **58**, 1144 (1936).
[23] Giauque, W. F., and R. Wiebe, *J. Am. Chem. Soc.*, **50**, 101 (1928).
[24] Gibson, G. E., and W. F. Giauque, *J. Am. Chem. Soc.*, **45**, 93 (1923).
[25] Ginnings, D. C., and G. T. Furukawa, *J. Am. Chem. Soc.*, **75**, 522 (1953).
[26] Haas, E. G., and G. Stegeman, *J. Am. Chem. Soc.*, **58**, 879 (1936).
[27] Hicks, J. F. G., Jr., *J. Am. Chem. Soc.*, **60**, 1000 (1938).
[28] Hildenbrand, D. L., and W. F. Giauque, *J. Am. Chem. Soc.*, **75**, 2811 (1953).
[29] Hoge, H. J., and F. G. Brickwedde, *J. Res. Natl. Bur. Std.*, **22**, 351 (1939).
[30] Johnston, H. L., *J. Opt. Soc. and Rev. Sci. Instr.*, **17**, 381 (1928).
[31] Johnston, H. L., J. T. Clarke, E. B. Lifkin, and E. C. Kerr, *J. Am. Chem. Soc.*, **72**, 3933 (1950).
[32] Johnston, H. L., and E. C. Kerr, *J. Am. Chem. Soc.*, **72**, 4733 (1950).
[33] Keesom, W. H., *Helium*, Elsevier, Amsterdam, 1942, p. 103.
[34] Kemp, J. D., and W. F. Giauque, *J. Am. Chem. Soc.*, **59**, 79 (1937).
[35] Latimer, W. M., and B. S. Greensfelder, *J. Am. Chem. Soc.*, **50**, 2202 (1928).
[36] Lewis, G. N., and M. Randall, *Thermodynamics*, revised by K. S. Pitzer and L. Brewer, McGraw-Hill, New York, 1961, p. 65.
[37] Meyers, C. H., *J. Res. Natl. Bur. Std.*, **9**, 807 (1932).
[38] Millar, R. W., *J. Am. Chem. Soc.*, **50**, 1875 (1928).
[39] Murch, L. E., and W. F. Giauque, *J. Phys. Chem.*, **66**, 2052 (1962).
[40] Nernst, W., (a) *Sitzb. Kgl. preuss. Acad. Wiss.*, **12**, **13**, 261 (1910); *Chem. Abstracts*, **4**, 2397 (1910), (b) *Ann. Physik*, **36**, 395 (1911).
[41] Nernst, W., and F. Schwers, *Sitzb. Kgl. preuss. Akad. Wissenschaften*, **1914**, 355.
[42] Osborne, D. W., E. F. Westrum, Jr., and H. R. Lohr, *J. Am. Chem. Soc.*, **77**, 2737 (1955).
[43] Ott, J. B., and W. F. Giauque, *J. Am. Chem. Soc.*, **82**, 1308 (1960).
[44] Parks, G. S., *J. Am. Chem. Soc.*, **47**, 338 (1925).
[45] Parks, G. S., G. E. Gibson, and W. M. Latimer, *J. Am. Chem. Soc.*, **42**, 1533 (1920); **47**, 338 (1925).
[46] Pitzer, K. S., and L. V. Coulter, *J. Am. Chem. Soc.*, **60**, 1310 (1938).
[47] Powell, R. L., and W. A. Blanpied, *Thermal Conductivity of Metals and Alloys at Low Temperature*, Natl. Bur. Standards Circular 556, 1954.
[48] Simon, F., *Ann. Physik*, **68**, 241 (1922).

[49] Sokolnikoff, I. S., and R. M. Redheffer, *Mathematics of Physics and Modern Engineering*, McGraw-Hill, New York, 1958, Chapter I.
[50] Stimson, H. F., *J. Res. Natl. Bur. Std.*, **42**, 209 (1949).
[51] Stout, J. W., *Rev. Sci. Instr.*, **25**, 929 (1954).
[52] Stout, J. W., and R. C. Chisholm, *J. Chem. Phys.*, **36**, 979 (1962).
[53] Stout, J. W., and L. H. Fisher, *J. Chem. Phys.*, **9**, 163 (1941).
[54] Taylor, R. D., B. H. Johnson, and J. E. Kilpatrick, *J. Chem. Phys.*, **23**, 1225 (1955).
[55] Taylor, R. D., and J. E. Kilpatrick, *Rev. Sci. Instr.*, **26**, 1132 (1955).
[56] Westrum, E. F., Jr., *Science*, **123**, 522 (1956).
[57] White, W. P., in *Temperature, Its Measurement and Control in Science and Industry*, Vol. I, Am. Inst. of Physics, Reinhold, New York, 1941, pp. 265–278.

CHAPTER 7

CALORIMETRY BELOW 20°K

R. W. HILL
Clarendon Laboratory, Oxford, England

DOUGLAS L. MARTIN
Division of Pure Physics, National Research Council, Ottawa, Canada

DARRELL W. OSBORNE
Argonne National Laboratory, Argonne, Illinois, U.S.A.

Contents

I.	Introduction	263
II.	Cryostats for Calorimetry below 20°K	264
	1. Cryostats for Use between 4° and 20°K (with possible extension in each direction)	265
	2. Cryostats for Use in the Liquid Helium Range	266
	3. Cryostats for Measurements below 1°K	267
	A. ^{3}He Cryostats	267
	B. Cyclic Demagnetization (Magnetic Refrigerator)	268
	C. ^{3}He Dilution Refrigerators	269
	D. Adiabatic Demagnetization	269
III.	Thermal Isolation of Calorimeter	272
IV.	Methods of Cooling Calorimeters	273
	1. With Helium Exchange Gas	273
	2. With a Condensing Pot	273
	3. By Waiting	274
	4. With Mechanical Heat Switches	274
	5. With Superconducting Heat Switches	278
	6. Comparison of Heat Switches	279
V.	Thermal Contact with Solid Samples	279
VI.	Thermometry	282
	1. Germanium Thermometers	282
	2. Measuring Circuits for Germanium or Carbon Resistance Thermometers	283
	3. Magnetic Thermometry	284
VII.	Isothermal versus Adiabatic Calorimetry	287
VIII.	Special Techniques below 1°K	288
IX.	References	290

I. Introduction

There are three basic difficulties inherent in calorimetry below 20°K which do not arise at higher temperatures. The first is that no accurately calibrated stable secondary thermometer has been available for temperatures less than about 14°K (although with the advent of germanium thermometers the position is now changing), and therefore no practical temperature scale had been established. The experimenter has had to calibrate his own secondary thermometers against vapor pressures of liquid H_2, liquid ^{4}He, or liquid ³He, or against a gas thermometer, or against a magnetic thermometer, or against some combination of these, as discussed in Chapter 2. The second difficulty is that all specific heats become vanishingly small as the absolute zero is approached, and great attention must therefore be given to stray heat influxes. In particular, heat influx due to vibration and to stray radio-frequency sources is a very serious problem which is not encountered at higher temperatures. The third difficulty is that of obtaining thermal contact between the calorimeter and the sample, and between the individual particles of the sample. At higher temperatures adequate thermal contact is often achieved by means of helium exchange gas in the calorimeter, but at temperatures below 10°K, particularly with powdered samples, adsorption of the helium may result in complete loss of thermal contact, or the change in adsorption with temperature may cause a spurious heat effect that is appreciable compared to the heat capacity being measured.

For these reasons neither the accuracy nor the precision of calorimetric results below 20°K has been as good as can be obtained at higher temperatures. However, in the present state of the art it seems that an overall accuracy of 0·2 per cent should be possible, although few results published to date are better than ±1 per cent.

The use of pure copper as a calorimetric standard substance below 20°K was suggested at the 1964 Calorimetry Conference, and suitable samples may now be obtained from Argonne National Laboratory (Attention: D. W. Osborne or H. E. Flotow), Argonne, Illinois[55]. A better picture of accuracies will emerge when these samples are in general use. (It should be noted that the specific heat of copper is very sensitive to the presence of small amounts of transition metal impurities[22]. Samples made from ordinary commercial copper are therefore quite unsuitable for checking the performance of a calorimeter.)

There are, however, several simplifications which are characteristic of work at very low temperatures. One is that heat exchange by radiation ceases to be appreciable when the surfaces concerned are below 20°K. (It is necessary, however, to have radiation traps in the pumping tubes in order to exclude radiation from higher temperatures.) Also, the high vacua necessary to limit heat transfer by gas conduction are relatively easy to obtain. The availability of superconducting materials is a further advantage; for example, the use of superconducting leads (below 7°K) eliminates the correction for heat developed in heater leads.

In discussing techniques for this temperature range it is important to avoid being dogmatic. Heat capacities vary enormously in magnitude, and a method appropriate to one substance may fail for another simply because the heat capacity is several orders of magnitude greater or smaller. This is in contrast to the position at room temperature, where specific heats of condensed substances are the same within about one order of magnitude.

II. Cryostats for Calorimetry below 20°K

The design of cryostats and calorimeters for use below 20°K has been strongly influenced by the need for calibration of the thermometer. Because most secondary thermometers used heretofore below 14°K (e.g. carbon thermometers and phosphor-bronze thermometers) change calibration on warming it has usually been necessary to provide for calibration *in situ* each time the apparatus is cooled from room temperature. This facility can be eliminated if a stable germanium thermometer, calibrated in a separate apparatus, is available, but most workers prefer to include some provision for checking the calibration occasionally *in situ*.

Since most calorimetric experiments are rather lengthy, it is desirable that the required temperatures should be maintained for long periods of time without having to refill with refrigerants. The refrigerants may be contained within glass or metal Dewar vessels or in metal cans within the vacuum space of the cryostat. A combination of these techniques may also be used. It should be noted that some provision must be made for the occasional repumping of Pyrex glass Dewars that are used to contain helium, because of the slow diffusion of helium gas through borosilicate glass in the warm parts of the apparatus. Further information about the construction of cryostats is given by Hoare, Jackson, and Kurti[33], White[75], and Rose-Innes[67].

The temperature range under consideration can conveniently be subdivided further into the following ranges: 4°–20°K, 1°–4°K, and below 1°K.

1. *Cryostats for use between 4° and 20°K (with possible extension in each direction)*

Cryostats for this temperature range may either contain both liquid hydrogen and liquid helium stages or helium only. Of these arrangements, the former has the particular virtues that the helium consumption is less and that the thermometers may readily be calibrated against the vapor pressures of both liquids, whereas the latter has the virtue of simplicity. The outer stage could alternatively contain liquid nitrogen. Typical cryostats are sketched in *Figure 1*. In each of these cryostats, the specimen is surrounded by a shield whose temperature can be raised above that of liquid helium by

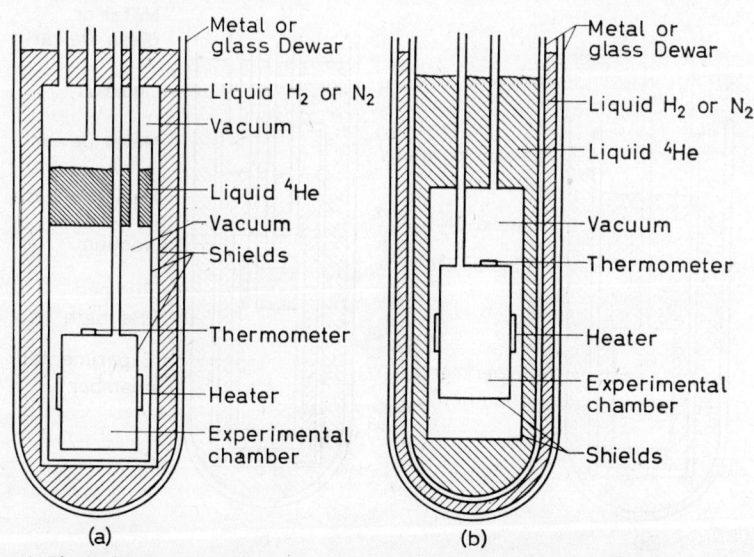

Figure 1. Typical cryostats (schematic) for calorimetry between 4° and 20°K. Often the separate pumping line for the experimental chamber is omitted.

electrical heating. Only a very small heating rate is necessary to maintain a steady temperature if the shield is well insulated from the helium bath. Because of the high thermal diffusivity (in this temperature range) of copper and other metals of which the shield is usually constructed, and also because the effect of radiation is negligible, the detailed disposition of the shield heaters and thermometer is not important. For example, it is usually not necessary to have separate heaters and separate thermometers for the top, middle, and bottom of the shield, contrary to the practice at higher temperatures (see Chapter 5).

Two methods of working are commonly used: one is the adiabatic method, and the other is often loosely called the isothermal method. In the former the shield temperature is adjusted, either manually or automatically, so as to follow faithfully the specimen temperature while measurements are being made. In this method the thermometer shown in *Figure 1* is usually a difference thermocouple between the shield and the specimen. In the latter method it is the shield, not the specimen, which is isothermal, for the shield temperature is maintained constant during a heating period. Further discussion of these two methods is given in section VII of this chapter, as well as in Chapters 5 and 6.

2. *Cryostats for use in the liquid helium range*

Temperatures in this range are produced by pumping over liquid helium, and cryostats are divided into two classes, depending on whether the main helium bath or an auxiliary bath is used. The two types are shown schematically in *Figure 2*. In these diagrams the liquid nitrogen is shown in a separate Dewar surrounding the helium Dewar; often, particularly in metal cryostats, the liquid helium is surrounded by a shield attached to a

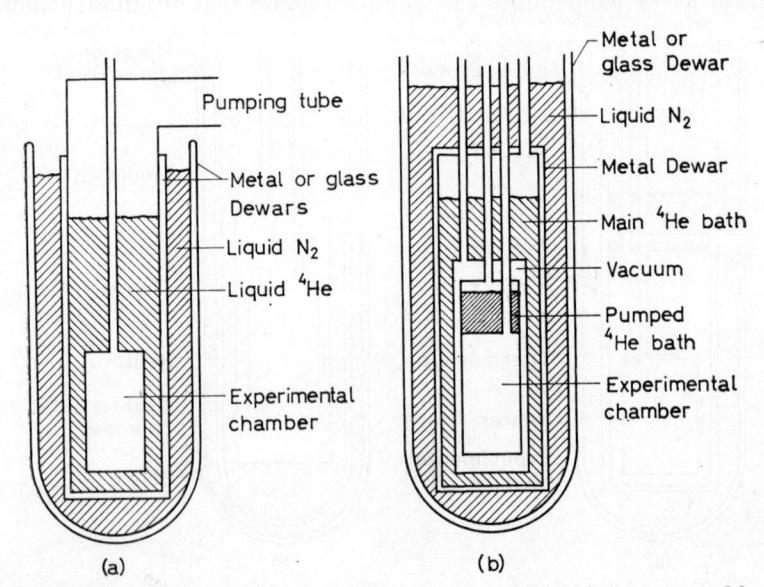

Figure 2. Typical cryostats (schematic) for calorimetry in the liquid helium range. Often the separate pumping line shown in (b) for the experimental chamber is omitted.

liquid nitrogen container in the same vacuum space as the liquid helium container. The cryostat in which the main bath is pumped is clearly much simpler but has the serious disadvantage that the relatively large evaporation rate of the pumped bath makes it difficult to achieve temperatures close to 1°K. Special pumps and large diameter pumping tubes and valves are necessary. These difficulties are effectively avoided in the more complicated cryostats, where, with proper design of the pumping system, temperatures down to 0·9°K can be reached by the use of diffusion and rotary pumps of modest size. If a small constriction is used at the lower end of the main pumping tube, superfluid film flow can be limited without undue limitation on the pumping speed[11]. In this way, for small cryostats, the evaporation rate of the pumped helium can be reduced to around 1 cm³ of liquid per hour, compared with about 50 cm³ per hour from the simple cryostat. The corresponding difference in the volume of gas to be pumped becomes very important when the pressure is small. A choice between these two types of cryostat may, therefore, depend critically on the lowest temperature which it is desired to obtain.

In all helium cryostats the consumption of helium may be materially reduced by using the cold gas evaporating from the liquid to cool the lead wires and support tubes. Thus in *Figure 2* it would be more economical to bring the electrical leads out through the ⁴He bath rather than through a highly evacuated tube. Another method is to have an "economizer" or heat exchanger to which the lead wires are thermally anchored[73].

These types of cryostat have become very popular because the thermometers can be calibrated directly against the vapor pressure of liquid helium and the necessity for gas thermometry is thereby avoided. However, there has been a regrettable tendency to attempt to derive too much information from measurements made only in this restricted temperature range.

3. *Cryostats for measurements below* 1°K

Four types of cryostat are in use at the present time.

A. ³He Cryostats

These employ ³He as the final refrigerant. The cost of this material is relatively high (currently about $0·15 per cm³ of gas at STP), and so the quantities used are quite small, ranging from a few cm³ of liquid down to a few tenths of 1 cm³. Because this liquid, in contrast to liquid ⁴He, does not exhibit superfluidity, evaporation rates are readily reduced to around 0·1 cm³ of liquid per hour. In the past, when the price of ³He was considerably greater, some cryostats used a continuous circulation system to obtain long running time with a small quantity of ³He or to cool samples of high thermal capacity. This technique now seems to be an unnecessary complication.

Typical non-circulating ³He cryostats are sketched in *Figure 3*. They differ from the ⁴He-range cryostats of *Figure 2* only by the addition of an extra stage. The ³He is liquefied by condensation at the temperature of the pumped ⁴He bath. Because the ³He pumping tube is usually made fairly large so that the lowest temperatures may be obtained, an appreciable fraction of the gas "store" may be left in this tube unless the condensing temperature is kept low. Usually a temperature of 1·2°K, at which the vapor pressure of ³He is 20 mm, is adequate.

The pumping system for the ³He must be carefully sealed to avoid loss of gas. Sealed rotary pumps are available, e.g. from the Welch Scientific Company (Skokie, Illinois, U.S.A.)[66]. These pumps can be used either alone

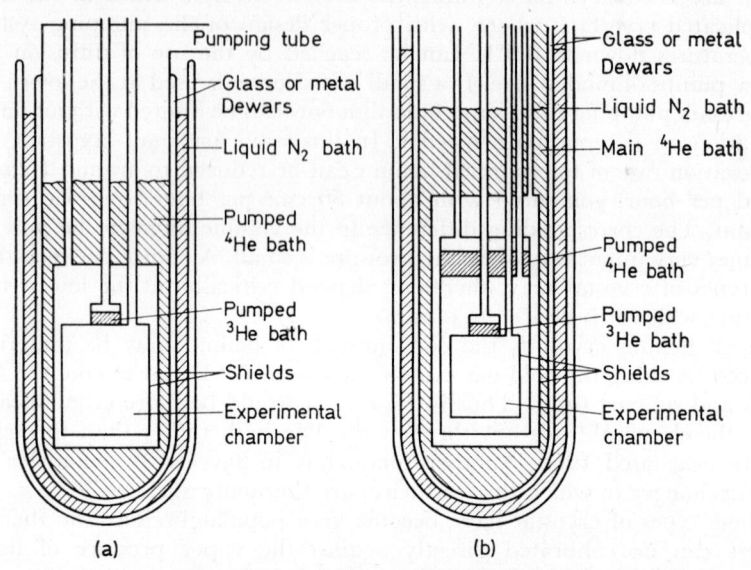

Figure 3. Typical ³He cryostats (schematic).

or in conjunction with a diffusion pump, depending on the lowest temperature desired. Alternatively, a diffusion pump with high critical backing pressure, e.g., the Speedivac 2M4A made by Edwards High Vacuum, Ltd. (Crawley, England) may be used without a rotary pump, in which case the gas is either pumped into a large storage vessel or returned to the cryostat and condensed in a vessel attached to the pumped ⁴He bath. By a suitable arrangement of valves the same pump may be used for moving the ³He from the storage vessel to the ³He cryostat and vice versa.

The lowest temperature which can be reached with a ³He cryostat is in the vicinity of 0·3°K. Thus it may be said that the use of this isotope extends the temperature range accessible with liquid refrigerants by about a factor of three. This extension has proved to be of great value in the study of electronic specific heats of metals, particularly superconductors, and has rendered accessible the anomalous "hyperfine" specific heats associated with the ordering of nuclear magnetic moments in ferromagnetic metals and other ordered magnetic systems. Current interest in these fields has led to the construction of quite a large number of ³He cryostats for calorimetry in the last few years[43, 47, 59, 71].

B. Cyclic Demagnetization (Magnetic Refrigerator)

This interesting technique was developed[29] at a time when ³He was still too scarce and expensive to be used as a refrigerant. However, the ³He supply position improved greatly a few years later, and so the magnetic refrigerator technique has never been widely adopted. The lowest temperature achieved so far by this method[30] is approximately 0·2°K.

C. ³He Dilution Refrigerators

These new devices will doubtless be widely used. Cooling is produced by the transfer of ³He from a liquid phase rich in ³He into the dilute solution of ³He in ⁴He coexisting at low temperature. Several cryostats of this type have been described[80, 82, 84], the lowest temperature continuously maintained being about 0·020°K[84].

D. Adiabatic Demagnetization

In this method of cooling, the working substance is a paramagnetic salt. Magnetization of such a salt reduces the magnetic disorder and hence the entropy; it follows that isothermal magnetization requires the rejection of heat to the bath. Conversely, isothermal demagnetization is accompanied by absorption of heat by the salt. If the salt is thermally isolated so that the demagnetization is adiabatic, cooling results. This effect exists at all temperatures, but its magnitude depends on the magnetic susceptibility, which increases with decrease of temperature. At the same time, heat capacities become smaller (aside from "humps" in the heat capacity curves for some substances), so that a given heat effect produces a bigger temperature effect.

The experimental realization of this type of cooling is sketched in *Figure 4*, which shows a paramagnetic salt pill suspended in a 1°K cryostat and surrounded by mutual inductance coils for temperature measurement. Efficient cooling from 1°K requires the use of magnetic fields of 10 kilogauss or more, and the physical size of the cryostat may have to be limited if it is to fit into an available magnet. To meet this requirement while still retaining a reasonably large liquid helium capacity, Dewar vessels of fairly large

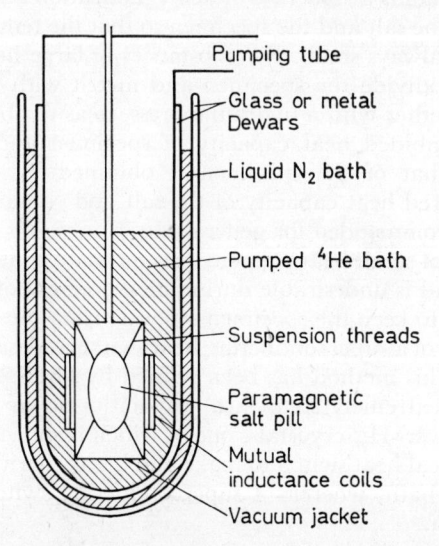

Figure 4. An adiabatic demagnetization apparatus (schematic).

diameter may be fitted with "tails" of smaller diameter. Only the tail need go into the magnetic field. Alternatively, a superconducting magnet located in the liquid helium bath could be used. Another variation is to include a ^{3}He stage in the cryostat; this has the advantage of reducing the magnetic field required.

The simplest method of using this type of cryostat is as follows:

(i) Establish thermal contact between salt and bath by admitting a little helium exchange gas to the vacuum jacket.
(ii) Magnetize the salt; it will overheat, and some minutes must be allowed for it to regain the temperature of the bath.
(iii) Pump away the exchange gas from the vacuum jacket.
(iv) Demagnetize.

In order to measure the temperatures achieved (between 0·1 and 0·001°K, depending on the choice of salt and the conditions of the experiment) use is again made of the magnetic properties of the salt. The magnetic susceptibility, which varies with temperature roughly as T^{-1}, can conveniently be measured by means of mutual inductance coils which form part of either an a.c. or a ballistic bridge.

Magnetic thermometry is briefly discussed in Chapter 2 and in Section VI of this chapter. For a fuller account of all aspects of magnetic cooling, including magnetic thermometry, the review by de Klerk[16] should be consulted. The review by Ambler and Hudson[2] is also quite informative.

The first heat capacity measurements to be made below 1°K were those on paramagnetic salts, with the salt acting as coolant, specimen, and thermometer. In order to make wider use of the method it is essential to be able to cool substances other than paramagnetic salts. This cooling may be done by maintaining thermal contact between specimen and salt during the demagnetization. The cooling will be most efficient if it is reversible; this circumstance demands a slow rate of demagnetization and intimate thermal contact between the salt and the specimen so that the temperature difference between them is always small. For substances of large heat capacity it may be possible to subdivide the specimen and mix it with the salt, the whole being pressed together with a hydraulic press, so as to obtain a large area of contact. The combined heat capacity of specimen and salt will then be measured, and that of the specimen is obtained by subtraction of the previously measured heat capacity of the salt and container. However, this method is not recommended for general use.

For specimens of small heat capacity this intimate contact is not necessary during cooling and is undesirable during measurement of the heat capacity. It is much better to keep the specimen separate from the salt and to connect the two by means of a superconducting heat switch as described in Section IV of this chapter. This method has been refined by Phillips[53, 60] to permit the measurement of extremely small heat capacities down to temperatures of 0·1°K or even lower. His cryostats, one of which is shown in *Figure 5*, incorporate a mechanical heat switch to make and break thermal contact between the salt and the bath, whereas a superconducting switch is used between salt and specimen.

The present lower limit of temperature for calorimetry is about 0·05°K,

CALORIMETRY BELOW 20°K

depending on the substance to be measured, even though temperatures ten times lower are fairly easily achieved by adiabatic demagnetization. One principal reason for this limit is the increasing difficulty of establishing

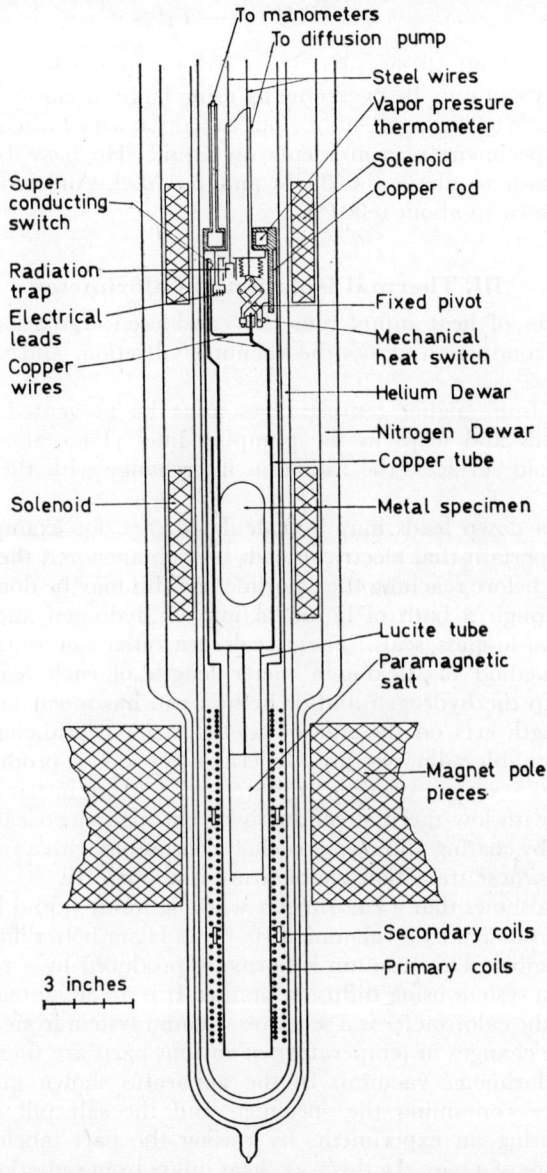

Figure 5. An adiabatic demagnetization apparatus used by Phillips[60] for heat capacity measurements on metals between 0·1° and 4°K. In a later modification[53] the solenoids for applying a magnetic field to the superconducting heat switch and to the specimen are smaller, superconducting coils located in the helium bath rather than in the nitrogen bath. Also, the magnetic thermometer is a spherical crystal of cerium magnesium nitrate located below the cooling salt and connected to the specimen by a copper wire interrupted by the superconducting heat switch. (By courtesy of the Editor, *Physical Review.*)

thermal contact between salt and specimen as the temperature is reduced. This can be seen from the formula which has been given[5, 52] for the rate of heat transfer between a paramagnetic ionic crystal and copper:

$$\dot{q} = DA(T_1^4 - T_2^4) \tag{1}$$

in which D is about 10^5 ergs sec^{-1} cm^{-2} deg^{-4} and A is the contact area. This difficulty can only be overcome by using larger areas of contact, which may be difficult to arrange. An obvious exception arises when liquid ^{3}He or ^{4}He is the specimen; measurements on liquid ^{3}He have been made by Peshkov[58] down to about 0·0033°K and by Abel, Anderson, Black, and Wheatley[1] down to about 0·004°K.

III. Thermal Isolation of Calorimeter

Five sources of heat influx must be considered: radiation, conduction down leads, conduction across the vacuum, vibration, and stray electrical heating.

Radiation from higher temperatures must be prevented by means of suitable shields and traps in the pumping lines. The calorimeter should only "see" cold surfaces, the radiation interchange with these being negligible.

Conduction down leads may be calculated (see, for example, White[75]). It is most important that electrical leads be well anchored thermally at low temperatures before reaching the calorimeter. This may be done by bringing the leads through a bath of liquid helium or hydrogen and leading out through metal-to-glass seals. These seals are often not reliable, and the alternative method of sticking a meter length of each lead wire, non-inductively, to the hydrogen and/or helium can has much to commend it. (A meter length errs on the side of safety. A few centimeters is often insufficient.) Suitable adhesives are General Electric Co. products No. 7031 and No. G 9825, surface coating Araldite, etc. For use in the liquid helium range, leads with low thermal conductivity but high electrical conductivity can be made by coating alloy wires such as constantan with a superconductor such as lead, whose transition temperature is about 7°K.

Calculation shows that a calorimeter working in the liquid helium region should be surrounded by a vacuum of 10^{-9} mm Hg or better if heat exchange is to be negligible. This vacuum is normally produced by a relatively high speed vacuum system using diffusion pumps. It is an advantage if the space surrounding the calorimeter is a separate vacuum system from the rest of the cryostat since changes in temperature of various parts are then less likely to affect the calorimeter vacuum. In the apparatus shown in *Figure 5* the vacuum space containing the specimen and the salt pill is completely closed off during an experiment, by raising the part labeled "radiation trap" by means of a wire. In this way, heat influx from radiation or from hot molecules entering through the pumping tube is prevented.

For work below 4°K some workers[63] have attempted to produce a suitable vacuum by freezing the air out of the space around the calorimeter which was sealed at room temperature. However, the helium content of the atmosphere[35] is 1 in 200 000 and the resultant pressure of helium at 4°K

($\sim 5 \times 10^{-5}$ mm) would be much too high for safety. The apparatus should, therefore, be evacuated to a rough vacuum before freezing out the residual gases.

Vibration and stray electrical heating become significant as the temperature is decreased because of the rapidly decreasing heat capacity of most samples. Vibration heating arises from inelastic stretching of fibers supporting the calorimeter and can be as high as 1000 ergs/min because of pump vibrations, etc. The problem becomes acute below 1°K, and special precautions may be necessary[2, 47, 74]. With care the vibration heating can be reduced to about 1 erg/min.

Heating effects may also occur because of stray radiofrequency fields. This effect may be particularly serious when semiconductor resistance thermometers are used[3, 24]. The cure is adequate electrical screening; in some locations, unfortunately, it may be necessary to screen the entire room. Heating sometimes occurs because of the operation of relays, etc. in the calorimeter control circuits. Condensers placed at strategic locations normally effect a cure.

Care must be taken, of course, that the current in the thermometer circuit is not causing undue heating. Some experimenters have found that the switching in of one type of double potentiometer may cause momentary interconnection of the measuring circuits, with a resultant undesired heating of the calorimeter.

IV. Methods of Cooling Calorimeters

1. *With helium exchange gas*

In this method the vacuum between calorimeter and surroundings is broken by a pressure of helium gas of the order of 10^{-1} mm Hg. This is a fast method of cooling, but if this method is used to cool as far as 4°K, it is very difficult to pump off all the exchange gas in a reasonable time because of adsorption on walls. If all the gas is not removed, then it may be given off intermittently during measurements. Thus the heat of desorption of the gas may be measured, but the chief effect is due to spoiling the vacuum and causing an unknown heat exchange between calorimeter and surroundings. Therefore, unless very great care is taken, the use of exchange gas is fraught with dangers. It has been used quite successfully on cooling as low as about 1°K by using high speed pumping systems and monitoring the vacuum with a mass spectrometer leak detector[12, 23]. As a general rule, however, it is recommended that the upper temperature limit of the measurements should be less than the temperature at which the exchange gas was previously evacuated.

2. *With a condensing pot*

In this method helium or hydrogen is condensed into a pot on the calorimeter. If two tubes are connected to the pot, cold helium gas may be circulated to cool the calorimeter from relatively high temperatures[46]. The use of a condensing pot is very convenient because the pot serves as a vapor pressure bulb for the calibration of the secondary thermometer. After use for this purpose and before calorimetric measurements are commenced, the liquid must be pumped away, and a good vacuum established within the pot. This is not always easy; good thermal insulation of the calorimeter requires the connecting tube to be small, and the pumping speed is inevitable small. Thus there is a possibility of gas being desorbed during measurements, although this may not be too serious because the main vacuum cannot be affected. However, the heat capacity of even a small amount of ^{3}He gas or ^{4}He gas may be appreciable compared to the heat capacity being measured. With ^{4}He below 2·2°K there is the possibility of high heat leak because of superfluid film flow if all the helium gas is not pumped out.

3. *By waiting*

In this method the calorimeter is cooled by conduction of heat along the wiring. This method is very slow but does avoid possible errors mentioned under (1) and (2) above.

4. *With mechanical heat switches*

These switches provide the currently preferred method of cooling calorimeters to liquid helium temperatures and will, therefore, be discussed in some detail. They have also been used below 1°K, but here they meet strong competition from superconducting switches (see below). One type of mechanical heat switch involving mating conical surfaces has been described in Chapter 5. This type, which was first used by Keesom and Kok[36], has been used extensively to cool from room temperature to 14°K[68] or to 4°K[73]. It is usually not satisfactory for operation below 4°K because of the frictional heat generated on releasing the contact, although it has been used to 1·3°K for samples of relatively high heat capacity[10, 38]. There are two basic types in common use which are satisfactory in this respect. In one the calorimeter is pressed against its surrounding jacket or against a cooled plate to make thermal contact and is lifted clear to obtain insulation, whereas in the second the calorimeter is permanently suspended and a system of movable jaws is used to grip a wire or foil attached to it. In the first type the contact may be obtained by external application of force, as in *Figure 6*, or simply by the weight of the specimen, as in *Figure 7*.

Most of the switches of the second type are based on the design of Webb and Wilks[72]. This switch, which is sketched in *Figure 8*, was originally designed for measurements of small heat capacities from 1·5° to 4·0°K. The emphasis, therefore, lay on small frictional heating on release, and for this reason the switch is well suited to operation at lower temperatures.

For the original switch, the mean thermal resistance at liquid helium temperatures is given as 6000 deg watt^{-1}, and the heat generation in the specimen on release as 10^3 ergs. The switch has been further developed by Manchester[45], who used indium coated jaws instead of copper ones and Teflon bearings on the moving parts. Very complete details are given about the operating procedure, and the thermal resistance of the switch was measured over a wide temperature range. When the pressure at the contact was small enough to give "negligible" heating on release (100 to 200 ergs) the thermal resistances at 4° and 1·6°K were found to be 10^4 and 10^5 deg watt^{-1}, respectively. Smaller resistances could be obtained at the expense of a greater heating on release.

A variation due to Phillips[60] (see *Figure 5*) uses a more complicated system of levers. With plain copper jaws, thermal resistances of 500 and 2500 deg watt^{-1} at 4·2° and 1·3°K, respectively, were observed, together with a heating on release ranging from about 50 to several hundred ergs. Phillips uses a rigid support for the specimen; this reduces the heating on release by preventing the conducting link from chattering between the jaws during

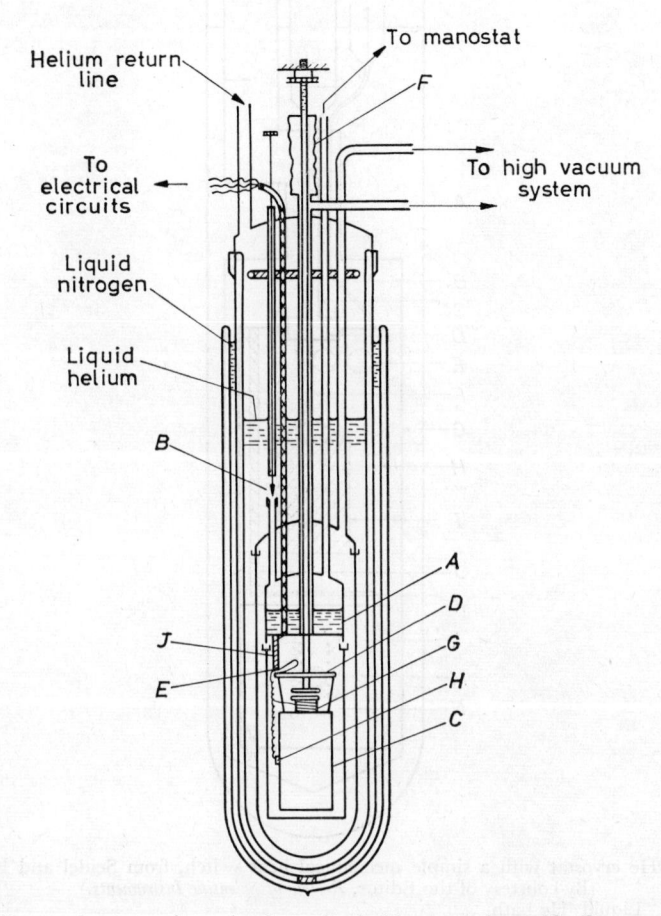

Figure 6. Cryostat incorporating a simple mechanical heat switch, used by Rayne and Kemp[65] for metals below 4·2°K. (By courtesy of the Editor, *Philosophical Magazine*.)

- A, Pumped ^{4}He bath
- B, Valve for filling A from the main ^{4}He bath
- C, Specimen of metal
- D, Disc from which the specimen is suspended by Nylon threads
- E, Copper strap between disc D and a copper rod J, attached to the pumped ^{4}He bath
- F, Screw and bellows mechanism for moving disc and specimen and for pressing the specimen against the flat bottom of the container attached to A
- G, Heater on a copper former screwed into the specimen
- H, Carbon resistance thermometer
- J, Copper rod to which the electrical lead wires and strap E are anchored thermally

opening. Many authors agree on the importance of skillful adjustment of the contact pressure if rapid cool-down is to be combined with small heating on release.

The other type of switch, in which the specimen is cooled by being pressed against the surrounding shield has been widely used[63, 65], mainly for

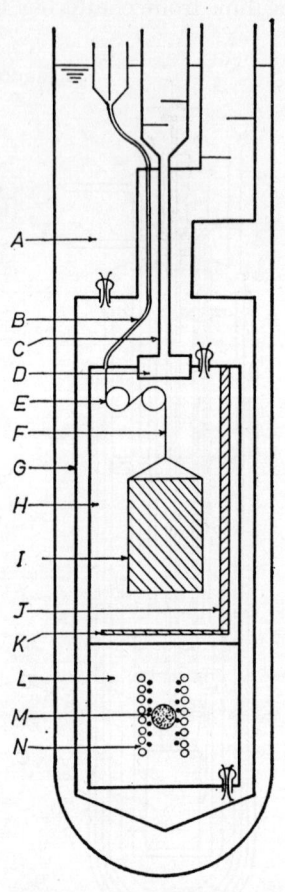

Figure 7. ³He cryostat with a simple mechanical heat switch, from Seidel and Keesom[71]. (By courtesy of the Editor, *Review of Scientific Instruments*.)

- A, Liquid ⁴He bath
- B, Thin-walled capillary
- C, Thin-walled, German silver tube
- D, Pumped liquid ³He bath
- E, Pulley system for moving the specimen
- F, 0·12 mm diameter steel wire
- G, Outer vacuum can
- H, Evacuated chamber, made of copper
- I, Specimen (suspended from a Nylon thread and the steel wire F)
- J, Copper rod
- K, Copper plate
- L, Lower chamber, made of copper, coated on the inside with 50:50 lead–tin solder to reduce eddy currents and containing ⁴He (1 atm at room temperature) for thermal conductivity
- M, Paramagnetic salt thermometer
- N, Inductance coils

work on metals in which the calorimeter takes the form of a solid block. Quite large contact pressures can be used with this type of switch, and rapid cooling is possible, though the heat generation on release may be considerably bigger than with the Webb and Wilks type. In this connection it is important that the movement of the surfaces be purely translational; if rotation is not prevented the heat generation may be unmanageably large.

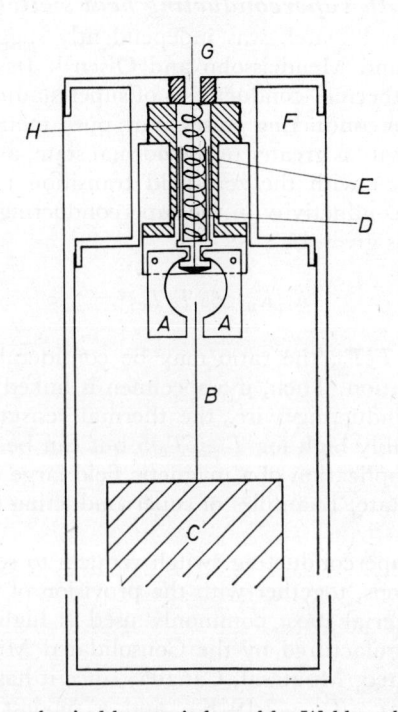

Figure 8. Diagram of a mechanical heat switch used by Webb and Wilks[72] for the measurement of low heat capacities below 4·2°K. (By courtesy of the Editor, *Proceedings of the Royal Society*.)

- A, Copper jaws
- B, Copper wire projecting from the specimen C
- D, Copper cylinder, which moves smoothly inside a copper turning E screwed to the cryostat F
- G, Piano wire (0·12 mm diam.) to a bellows at the top of the apparatus, by means of which the jaws A are operated
- H, Spring to open the jaws when the tension on the wire G is released. The specimen is suspended by fine Nylon threads (not shown)

When the long cool-down times are not a matter of concern, the application of pressure is unnecessary. The specimen can be lowered so as to sit on the bottom of the shield and lifted clear for insulation[43, 71]. Though slow, this method is very simple, and it has proved possible to use it to surprisingly low temperatures. Thus, Lounasmaa[42] has cooled rare earth metal specimens of large (hyperfine) heat capacity to around 0·3°K in a ³He cryostat of large capacity. The specimens were normally left to cool overnight. With some substances for which the hyperfine heat capacity was particularly large, e.g. terbium[44], a little stopcock grease was used between the bottom of the sample and the cooling platform in order to reach the lowest temperatures in a reasonable time. The grease contact was forcibly broken by lifting the sample before beginning the experiments. In order to repeat the measurements the grease contact was re-established by lowering the sample and warming the apparatus to room temperature.

5. *With superconducting heat switches*

This type of thermal switch was independently suggested by Gorter[25], Heer and Daunt[31], and Mendelssohn and Olsen[51]. Its operation depends on the fact that the thermal conductivity of superconductors is different in the normal and superconducting states. For pure metal superconductors, the thermal conductivity is greater in the normal state, and for temperatures not too low compared with the zero-field transition temperature T_c the ratio of the thermal conductivity in the superconducting state K_s to that in the normal state K_n is given by

$$K_s/K_n \simeq (T/T_c)^2. \tag{2}$$

For small values of T/T_c, the ratio may be considerably smaller than is predicted by this relation. Then, if a specimen is linked to the appropriate bath with a superconducting wire, the thermal resistance between them can be made reasonably high for $T < T_c/5$ but can be reduced to a fairly small value by the application of a magnetic field large enough to keep the wire in the normal state. Examples of superconducting switches are shown in *Figures 5* and *10*.

The design of a superconducting switch reduces to selecting the type of wire and its dimensions, together with the provision of the necessary magnetic field. The material most commonly used is high purity lead (e.g., Tadanac brand, manufactured by the Consolidated Mining and Smelting Co. of Canada, Limited, Montreal, Canada) since it has a reasonably high transition temperature ($T_c = 7\cdot19°K$), and a modest threshold field of about 800 oersteds when $T \ll T_c$. The dimensions of the lead wire or tape used must be chosen for a particular problem to give the most acceptable compromise between cool-down rate in the normal condition and heat leak in the superconducting condition. It is advisable to "solder" the lead wire or tape with pure lead.

The switching field may be obtained in a variety of ways: from a conventional solenoid either external to the cryostat or immersed in the main helium bath, or from a superconducting solenoid. When designing the solenoid it may be important to be sure that the stray field at the specimen is small, particularly when metal specimens are under investigation; otherwise considerable eddy current heating may occur during switching. The heat generation in the superconductor during switching is negligible. A double solenoid may be used to advantage. This typically consists of two concentric and coaxial coils of equal length. If one has n turns and cross-sectional area A, and the other has n/a turns and area aA, and if the coils carry the same current but in opposite senses, then the field at points remote from the solenoid falls off as the inverse fifth power of distance instead of the third power as for a simple solenoid. The field at the center of the coil is naturally reduced also but by a smaller factor. As an alternative the sample may be shielded with a superconducting screen.

Some problems have arisen[40, 48] from a combination of thermal contact difficulties and the heat leak through a superconducting heat switch in the "open" position. Quite significant errors in specific heat results may occur when such problems are present. For a discussion of methods which may be used (perhaps with some modification) to correct for thermal gradients in the calorimetric system see Chapter 6, Section III–1–D.

6. *Comparison of heat switches*

The principal properties of the two main types of heat switches may be summarized as follows. In the case of the superconducting switch we have

(*i*) Very small heat generation on release.
(*ii*) Simplicity of construction and operation.
(*iii*) Relatively poor ratio of resistance in the "open" to "closed" condition.
(*iv*) Unsatisfactory operation above about 1·5 °K.

For the mechanical switch the position is rather different:

(*i*) Relatively large heat generation on release.
(*ii*) Relatively difficult to construct and operate satisfactorily.
(*iii*) Good ratio of resistance in the "open" to "closed" condition.
(*iv*) Can be used to indefinitely higher temperatures.

It is then clear that a choice between the two types will be determined largely by the temperature range of the experiment and by the heat capacity of the specimen at the lowest temperatures.

V. Thermal Contact with Solid Samples

At higher temperatures thermal contact between calorimeter and sample is usually made by means of an exchange gas. At low temperatures, however, the exchange gas may become adsorbed on the sample and then be given off during a heating period. Since the heat of desorption may be large, considerable errors may arise in this way. (The errors are not compensated by measuring an empty calorimeter since the surface areas are different.) It is, therefore, advisable to avoid using exchange gas within the calorimeter for measurements extending into the liquid helium region. When the sample is in the form of a powder the problem becomes acute. However, in measurements on grey tin (which can only be obtained in the form of powder) Webb and Wilks[72] relied on the contact between loosely packed grains of the metal to establish thermal equilibrium. The same technique was used by De Sorbo[18] on diamonds. It may be possible to improve thermal contact sufficiently by pressing or sintering the powder into a compact whole, or by casting it in a suitable oil or resin. For example, Craig, Massena, and Mallya[13] have used glycerol for this purpose.

If exchange gas is not to be used, the calorimeter need no longer be a vacuum-tight container and may be nothing more than a holder of heater and thermometer. But it is still necessary to provide for proper thermal contact with the specimen, and the heat capacity of the "calorimeter" must be known, preferably by direct measurement. Specimens which can be machined present the least problem, for heater and thermometer units can be screwed into place[37, 43]. Alternatively, the heater–thermometer assembly can fit snugly into a hole in the specimen[12, 26], or can be carried on a clamp which grips the specimen firmly[22]. The latter method is illustrated in *Figure 9*. If excessive local strain is to be avoided, both specimen and

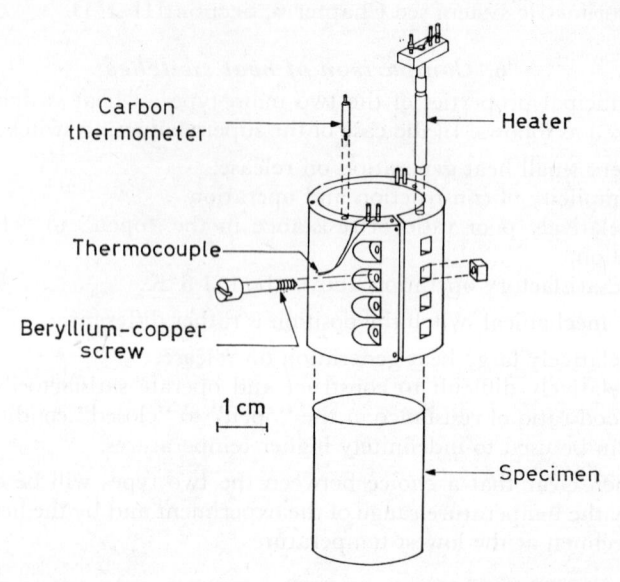

Figure 9. Sketch of a "clamp" type calorimeter[22]. The carbon thermometer has now been replaced by a germanium thermometer. (By courtesy of the Editor, *Proceedings of the Royal Society.*)

clamp must be accurately cylindrical and should be of the same material or have similar thermal expansion. A little high vacuum grease can be used to improve the thermal contact but is unnecessary with the type of clamp shown in *Figure 9* [22].

When machining is not possible, pieces of the sample may be stuck to the calorimeter, which may be simply a light tray[46] carrying the thermometer and heater, by means of an adhesive or grease. Some greases which have been used for this purpose are Dow-Corning silicone high vacuum grease, Apiezon N, Apiezon T, and Lubriseal. Martin[46] has found that the silicone grease forms a stronger bond than the Apiezon greases at low temperatures, but with any grease it is important to have similar thermal expansion for both the tray and the sample or to use small pieces of the sample, in order to avoid fracturing the grease bond on cooling. It is necessary to meter the adhesive or grease accurately so as to correct for its heat capacity, which may be surprisingly large[32, 56, 57]. The metering can be done by weighing, or grease can be metered volumetrically from a hypodermic syringe with sufficient accuracy for most measurements. Since the thermal diffusivity of the grease is not very large, the grease layer should be as thin as possible to avoid long time effects connected with stray lumps of grease.

CALORIMETRY BELOW 20°K

It is possible to turn differential contraction to advantage if the specimen can be cast into a calorimeter in such a way as to make it adhere on cooling. An example of this is to cast the specimen (an alkali metal, for example) around a metal wire of lower thermal expansion and to attach the wire to the calorimeter[64]. Another possibility is still to rely on grease but to avoid

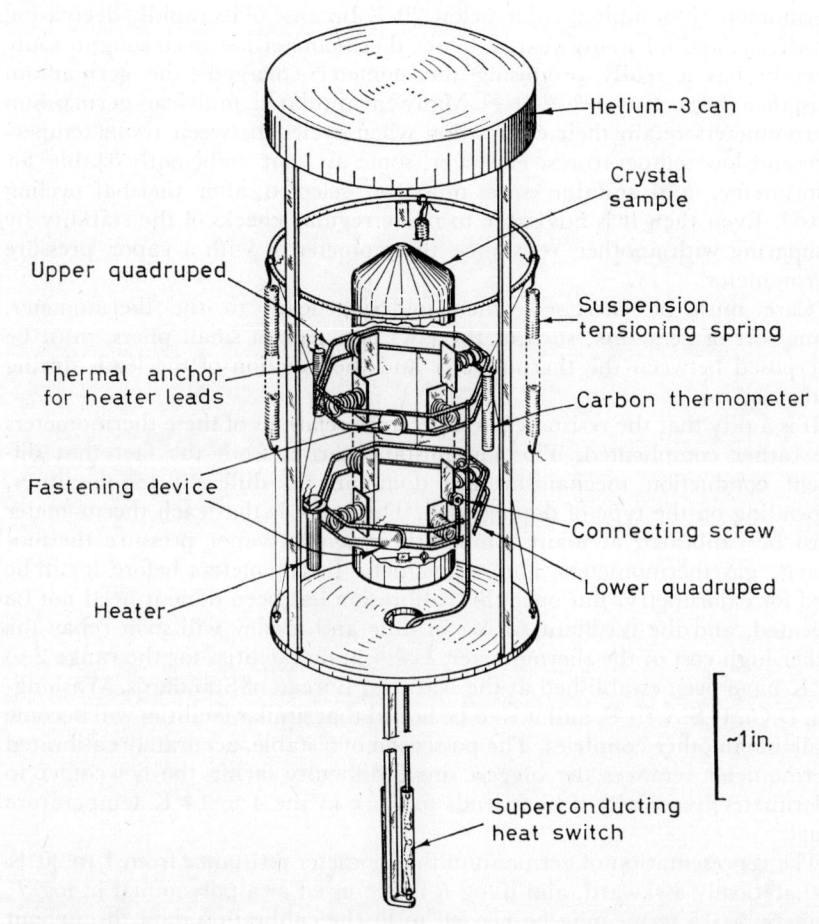

Figure 10. Calorimeter with spring-loaded contacts for samples with large thermal contractions, as used by Martin[48] on alkali halide crystals from 0·5° to 1·5°K. A measured amount of silicone high vacuum grease (totalling 0·02 cm³) is applied to the spring-loaded feet before measuring the empty calorimeter as well as when a crystal is inserted. The superconducting heat switch shown is protected from breakage by a loosely fitting glass tube. (By courtesy of the Editor, *Proceedings of the Physical Society*.)

stresses by making the calorimeter partly flexible, e.g., by using several spring loaded contacts, each of fairly small surface area[48], as shown in *Figure 10*. In devising systems of this sort it is not always necessary to attempt to minimize the heat capacity of the calorimeter but rather to be sure that it can be measured accurately.

281

VI. Thermometry
1. *Germanium thermometers*

Accurate calorimetry over wide ranges of temperature was first made possible by the introduction of the platinum resistance thermometer. The special virtues of this device were its stability and the fact that its behavior could be established by calibration at a few fixed points. This excellent thermometer is of limited value below 20°K because of its rapidly decreasing sensitivity, and for many years a better thermometer has been sought. Only recently has a really promising thermometer emerged—the germanium resistance thermometer[20, 39, 41, 55]. Many encapsulated, four-lead germanium thermometers retain their calibration when cycled between room temperature and low temperatures. However, some are not sufficiently stable for calorimetry, and suitable ones must be selected after thermal cycling tests[79]. Even then it is advisable to make regular checks of the stability by comparing with another resistance thermometer or with a vapor pressure thermometer.

Care must be exercised when soldering leads to the thermometer. Some sort of heat sink, such as the jaws of a pair of small pliers, must be interposed between the thermometer and the junction of the leads during soldering.

It is a pity that the resistance–temperature relations of these thermometers are rather complicated. This circumstance arises from the fact that different conduction mechanisms are dominant at different temperatures, depending on the type of doping used. The result is that each thermometer must be calibrated at many temperatures against vapor pressure thermometers, gas thermometers, and/or magnetic thermometers before it can be used for calorimetry. But once this calibration has been done it need not be repeated, and the resultant saving in time and money will soon repay the rather high cost of the thermometer. Calibration facilities for the range 2 to 20°K have been established at the National Bureau of Standards, Washington, D.C., U.S.A.[77, 83], and it is to be hoped that similar facilities will become available in other countries. The possession of a stable, accurately calibrated thermometer removes the biggest single difficulty facing the newcomer to calorimetry, especially if he intends to work in the 4 to 14°K temperature range.

The representation of germanium thermometer resistance from 1 to 20°K is analytically awkward, and if $\log R$ is expressed as a polynomial in $\log T$, as many as 15 terms may be needed to fit the calibration data throughout to 0·001°K or better[55]. But if the data are to be handled by a computer, as it normally would be even for simpler formulae, the presence of many terms is not a matter of any great concern.

There is a limit to the temperature range in which a given thermometer should be used; one which has a good sensitivity at 20°K will probably have an excessive resistance below about 1°K. Another thermometer will keep a resistance of manageable size down to 0·1°K, but it will be too insensitive for use much above 4°K. One consideration is the heating effect of the thermometer current at the lowest temperature, where the resistance is a maximum and the heat capacity usually a minimum. The pertinent resistance is, of course, that between the current leads; this resistance should normally not be more than twice the resistance between the potential leads. For any particular application, a suitable resistor must be picked from manufacturers' catalogues or arrangements must be made for construction of a special thermometer with suitable characteristics.

Another point which should be kept in mind when selecting a germanium resistance thermometer is that many thermometers contain helium gas to improve the response time. If such a thermometer is used at too low a temperature, the helium will condense to a liquid and there will be an appreciable absorption of heat on warming through the temperature range in which the liquid vaporizes[78]. Thus, a thermometer containing 1 atm of ^{4}He gas at room temperature will have a "hump" in its heat capacity near 1·5°K. However, the helium gas can be eliminated entirely, because the thermometer will be rapidly equilibrated by conduction along the lead wires, especially if copper wires with thin electrical insulation are soldered to the thermometer leads and cemented to the calorimeter. (In any case all of the electrical leads to the calorimeter, which usually are of some alloy with low thermal conductivity such as manganin or constantan, should be cemented to the calorimeter for a length of 10 cm or so before they are connected to the heater or the thermometer.)

If measurements are to be made in a magnetic field, e.g. when measuring the heat capacity of a superconductor in its normal state, it will be necessary to take account of magnetoresistance effects. All semiconducting thermometers show relatively large magnetoresistance, usually with an increase of resistance proportional to the square of the field and increasing with reduction of temperature. Thus additional calibration data will be needed, mainly at the lower temperatures.

When calibrating a germanium thermometer or other secondary thermometer against a vapor pressure, magnetic, or other type of thermometer, great care must be taken that no temperature gradient exists between the two thermometers or else the gradient must be measured[47]. Some experimenters have attempted to calibrate a thermometer mounted on the calorimeter by comparison with a primary or calibrated secondary thermometer carried on a shield surrounding the calorimeter. Equality of temperatures cannot be assured by maintaining a difference thermocouple between shield and calorimeter at zero e.m.f., because of the possibility of thermal e.m.f.s from inhomogeneities in the lead wires, nor by maintaining the shield at a temperature such that the calorimeter drift is zero. In the latter method unsuspected vibration heating of the calorimeter may give rise to appreciable temperature differences. Even the use of exchange gas to couple the calorimeter and shield is fraught with danger because the surface of the shield may not be isothermal and because there may be vibration heating of the calorimeter.

2. *Measuring circuits for germanium or carbon resistance thermometers*

Germanium and carbon thermometers are frequently measured potentiometrically, the potential drop between the potential leads to the thermometer being compared with the potential drop across a standard resistor. One advantage of this method is that the lead resistances are not included in the measured resistance, and another is that the range of resistances that can be measured is wider than with a Mueller or Smith bridge. If possible, the current should be the same as that used in the calibration (usually 0·1 to 10 μA); if a different current is used, the effect of the current on the calibration should be determined. A six or seven decade, thermofree potentiometer with the last decade capable of steps of 0·01 μV (obtainable from Rubicon, Leeds and Northrup, Guildline, Sensitive Research Instrument Corp., or Tinsley, for example) is recommended. A sensitive null detector such as a photoelectric galvanometer or a chopper-type amplifier is required. Provision should be made for reversing the current through the thermometer, in order to eliminate the effect of thermal e.m.f.s.

Another circuit that has been successfully used in calorimetry with carbon thermometers[47] and that can be used as well for germanium thermometers is the thermal-free isolating potential comparator circuit of Dauphinee and Mooser[15]. The output from the circuit can be fed into a potentiometer recorder. With this method only one measurement is needed to determine the resistance, independent of the lead resistances and of any thermal e.m.f.s. Suitable commercial instruments for this circuit are the Type 9744 four-section mechanical chopper and the Type 9801 four-terminal variable resistor (10^5 Ω maximum in steps of 0·001 Ω, 0·01 per cent accuracy) manufactured by Guildline (Smith's Falls, Ontario, Canada).

At the lowest temperatures, at which the resistance of a semiconductor becomes large, it may be advisable to use an a.c. bridge to keep the power dissipated in the thermometer sufficiently low. (The circuit mentioned in the preceding paragraph has also been used successfully for calorimetry at the rather low power level of 10^{-10} W.) The low power level is desirable for two reasons: not only to keep the heating of the calorimeter low but also to keep the temperature of the thermometer from rising appreciably above that of the calorimeter. Thermal e.m.f.s, of course, do not affect measurements with an a.c. bridge. If the resistance and the change of resistance of the thermometer is large, as is usually the case when these thermometers are used, the resistances of the leads can be cancelled with sufficient accuracy by connecting three of the leads to the bridge as shown in Chapter 2, *Figure 2(a)*, provided the ratio of the fixed arms of the bridge is unity. A 33-c/s bridge similar to the one designed by Blake, Chase, and Maxwell[9] has been used in conjunction with a phase-sensitive detector to measure carbon thermometers down to 0·02°K with a power dissipation in the thermometer of less than 10^{-12} watt[8]. A less sensitive a.c. resistance bridge for cryogenic thermometry that uses measuring powers as low as 10^{-9} watt is manufactured by Cryonetics, Inc. (Clinton, N.J., U.S.A.).

3. *Magnetic thermometry*

Although this subject is discussed briefly in Chapter 2 and more thoroughly by de Klerk[16], a few remarks regarding methods of measurement may be helpful. The paramagnetic salt used as the magnetic thermometer should preferably be a single crystal and should preferably have a spherical shape. Thermal contact with the magnetic thermometer can be achieved in a variety of ways: by the use of grease, and in particular by gluing flat crystals of cerium magnesium nitrate to flat copper or brass plates by means of

CALORIMETRY BELOW 20°K

Apiezon N[5, 52], by varnishing a copper wire to the crystal[53], by varnishing flat crystals to copper plates[74], by growing the crystal around wires[14, 19, 70], by the use of liquid ^{4}He in a closed container[54, 71], or by the use[1, 43] of liquid ^{3}He.

The mutual inductance of a set of coils containing the paramagnetic salt (the magnetic thermometer) is usually measured with a Hartshorn bridge[27] (see *Figure 11*). In order to minimize eddy current heating of metal

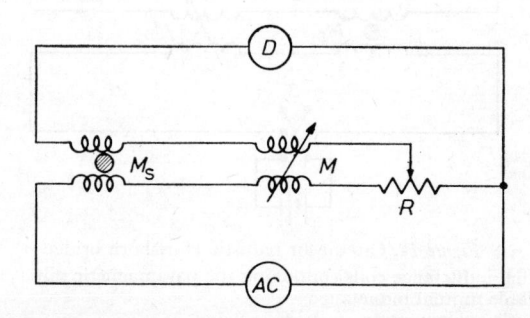

Figure 11. Circuit for alternating current Hartshorn bridge.
AC, Constant frequency current generator
M_s, Mutual inductance coils containing the paramagnetic salt
M, Variable mutual inductance
R, Resistive voltage divider
D, Detector

parts of the apparatus the bridge is operated ballistically or at low frequencies of 13 to several hundred c/s. The construction of suitable variable mutual inductances has been described by de Klerk and Hudson[17], by Erickson, Roberts, and Dabbs[21], and by McKim and Wolf[50]. H. Tinsley and Co., Ltd. (London, England) manufacture an astatic variable mutual inductance (Type 4229) with 3 decades and a continuously variable dial that has been used in conjunction with a transistorized tuned amplifier and null detector (Type 1232–A, General Radio Co., Concord, Mass., U.S.A.). It is necessary to calibrate the inductor for accurate work, however. A very good detector, which also has a built-in output for exciting the primary circuit, is a Princeton Lock-In Amplifier and Pre-Amplifier (Princeton Applied Research Corp., Princeton, N.J., U.S.A.) plus an oscilloscope. The resistive voltage divider labeled R in *Figure 11* is used to balance out the out-of-phase component. It can be assembled from a.c. resistance boxes, or an a.c. 4-decade voltage divider (for example, Model 1454–A, General Radio Co.) can be used. An important characteristic of R is that the inductances and capacities should be small and independent of the resistance settings.

For ballistic operation the generator a.c. in *Figure 11* is replaced by a reversing switch connected to a battery, the detector is a ballistic galvanometer, and the connection between the primary and secondary is eliminated, as shown in *Figure 12*. It is convenient for this mode of operation to have auxiliary resistances which keep the resistance in the secondary constant as M is varied[17].

An electronic a.c. mutual-inductance bridge designed by Pillinger, Jastram, and Daunt[61] has been widely used in low temperature experiments. Commercial models of this bridge are available from Cryonetics, Inc.

Wiebes, Hulscher, and Kramers[76] have recently described an accurate a.c.

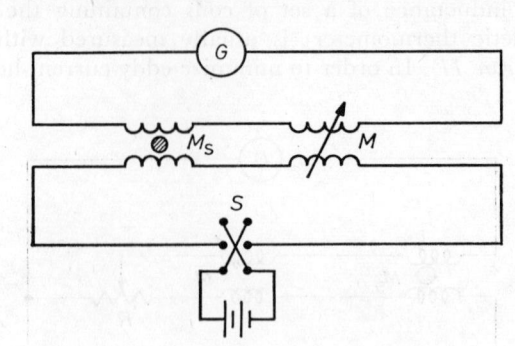

Figure 12. Circuit for ballistic Hartshorn bridge

M_s, Mutual inductance coils containing the paramagnetic salt
M, Variable mutual inductance
G, Ballastic galvanometer
S, Reversing switch

mutual-inductance compensator which can be easily assembled from components that are commercially available. They also have given[76] a critique of other mutual inductance bridges.

An excellent modification of the a.c. Hartshorn bridge has recently been described by Maxwell[49]. This circuit, shown in *Figure 13*, uses a precise 5-, 6-, or 7-decade ratio transformer that is available commercially (for example, from Gertsch Products, Inc., Los Angeles, Calif., U.S.A., or from Electro Scientific Industries, Portland, Oregon, U.S.A.). The recommended detector D is again a Princeton Lock-In Amplifier and Pre-Amplifier plus an oscilloscope, and the voltage divider R should have the characteristics noted above.

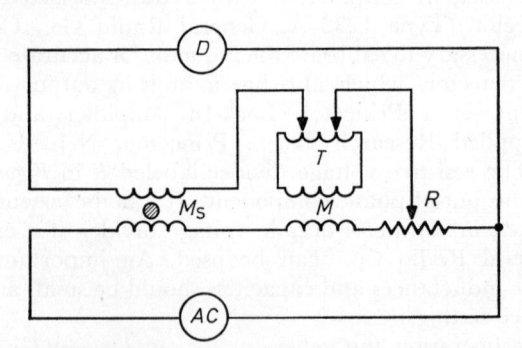

Figure 13. Modified a.c. Hartshorn bridge using a ratio transformer[49]

AC, Constant frequency current generator
M_s, Mutual inductance containing the paramagnetic salt
M, Fixed mutual inductance
R, Resistive voltage divider
T, 5-, 6-, or 7-decade ratio transformer
D, Detector

A final remark concerning magnetic thermometry is that the accuracy of the temperature measurements is usually limited by the accuracy of the T^*-T correction, in which T^* is the magnetic temperature calculated from the mutual inductance measurements, and T is the thermodynamic temperature. Therefore it is preferable to use a paramagnetic salt for which this correction is small. At present the best salt in this respect is cerium magnesium nitrate, $Ce_2Mg_3(NO_3)_{12} \cdot 24\ H_2O$, for which the difference $T^* - T$ is zero within experimental error down to 0·006°K[16, 81]. This substance is currently used by the majority of experimenters doing accurate magnetic thermometry below 1°K.

VII. Isothermal versus Adiabatic Calorimetry

When designing a cryostat for calorimetry, it is necessary to decide whether adiabatic or isothermal operation is to be used. The factors to be taken into account depend on the temperature range, and it is therefore appropriate to discuss the 1 to 20°K range at this point.

Many of the substances which are studied at low temperatures are blocks of pure metal or alloy which have very good thermal diffusivity and not too small a heat capacity. It has already been noted that good thermal insulation is relatively easily achieved in this range, and consequently the isothermal method can often be used under ideal circumstances where the drifts are small. On the other hand, if the substances to be studied have a poor thermal diffusivity, as would be the case with a powder, or if the thermal contact between sample and calorimeter is poor, the long relaxation times would at once point to the desirability of adiabatic operation. Similarly, if the specimen may undergo some slow process such as annealing, with emission or absorption of heat, the adiabatic method is preferable. The fact that such effects might not be expected and might be difficult or impossible to identify when using an isothermal calorimeter is a good reason for using an adiabatic calorimeter when possible.

In both methods the calorimeter is usually heated discontinuously. In the isothermal method the periods between heatings are used to evaluate the heat leak after equilibrium has been reached (see Chapter 6), and in the adiabatic method one simply waits for equilibrium. If the relaxation times are all sufficiently short it is possible to use either continuous or discontinuous heating with adiabatic operation[78].

Relatively few workers have thought it worthwhile to set up adiabatic calorimeters for this range, probably because it has often been the case that the accuracy of experiments has been limited by uncertainties in the temperature scale rather than by heat leak corrections. A special difficulty is the lack of a really suitable material for a difference thermocouple. Alloys of gold with cobalt or iron[6, 7, 62] can be used against "normal" silver (silver containing 0·37 at. % gold) or chromel, but their sensitivity becomes small at the lowest temperatures and is strongly temperature dependent through most of the range. Another difficulty is that inhomogeneities in the thermocouple wires in the regions where there are temperature gradients may give rise to thermal e.m.f.s amounting to several μV[62]. Consequently, the difference thermocouple does not read zero when the temperature difference is zero.

One solution is to operate in a quasi-adiabatic fashion[73]. In this method the adiabatic shield is controlled at an arbitrary galvanometer deflection or e.m.f. of the difference thermocouple. The control point is chosen to give nearly zero drift of the calorimeter at equilibrium before beginning a sequence of heat capacity measurements. The small heat leaks before and after the heating period (after equilibrium has been reached) are evaluated by measuring the drifts of the calorimeter temperature, and a correction is made for the heat leak by extrapolating the temperatures to the middle of the heating period—exactly as in isothermal calorimetry. The thermal e.m.f.s from the inhomogeneities must either be steady or vary linearly with time and linearly with the temperature of the calorimeter over the time and temperature range of one heat capacity measurement.

VIII. Special Techniques below 1°K

The most accurate results in this range have been obtained by the conventional techniques. However, there may be situations which call for one of the special techniques (or some modification thereof) discussed below.

The temperature-wave method of Howling, Mendoza, and Zimmerman[34] was intended for the measurement of the specific heats of metals below 1°K, but the application seems to be limited and the accuracy poor.

An interesting technique was used by Samoilov[69] to measure the specific heats of cadmium from 0·3° to 0·9°K in the normal and superconducting states. The sample of cadmium was linked to a paramagnetic cooling salt by means of a thin copper wire and copper vanes around which the salt was pressed. The dimensions of the wire were chosen so that about an hour was required to cool the cadmium to the temperature of the salt after demagnetization (approximately 0·1°K). On the other hand the heater and thermometer (a phosphor bronze resistance thermometer) were in very much better thermal contact with the cadmium sample. It was therefore possible to use short heating periods of ten seconds and to determine the temperature rise of the cadmium before much of the heat flowed to the salt. The corrected temperature rise was determined in the usual way by extrapolating the temperatures measured before and after heating to the middle of the heating period. To reduce the drift of the sample at temperatures above 0·5°K the sample was connected to a heater through a copper wire identical to the one used for the connection to the salt. The flow of heat from this auxiliary heater offset the flow of heat to the salt (which had a very high heat capacity) and thus raised the equilibrium temperature of the sample.

These techniques are not mentioned because of their usefulness between 0·3° and 0·9°K, where a ^{3}He cryostat and mechanical and/or superconducting switches can now be used to better advantage, but because the ideas involved may be useful at the lowest temperatures that can be reached by adiabatic demagnetization[1]. Thus Anderson, Salinger, Steyert, and Wheatley[4] have measured the heat capacity of liquid ^{3}He between 0·008° and 0·04°K with the calorimeter containing liquid ^{3}He and cerium magnesium nitrate coupled as tightly as possible to a chromium potassium alum cooling salt. Nevertheless the time for attaining equilibrium in the calorimeter was so short that it was possible to heat the calorimeter and reach equilibrium before too much heat was lost to the cooling salt.

CALORIMETRY BELOW 20°K

A more useful method has been developed by Haseda and Miedema[28] that is applicable to single crystals of various salts between 1° and 0·03°K. The sample (see *Figure 14*) is linked to a heat sink at about 0·01°K through

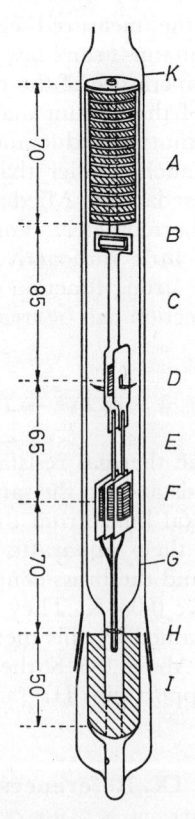

Figure 14. Haseda and Miedema's apparatus for measuring specific heats and susceptibility down to 0·03°K[28]. The dimensions shown are in mm. (By courtesy of the Editor, *Physica*.)

- A, CrK-alum cooling salt
- B, γ-Ray thermometer
- C, Heat link (a strip of brass, 60 × 6 × 0·3 mm)
- D, Sample, glued to brass plates with Apiezon N grease
- E, Copper connection
- F, CeMg-nitrate thermometer
- G, Glass thermal shield
- H, Support pins made from photographic film
- I, CrK-alum thermal guard
- K, Outer glass tube, immersed in a liquid helium bath.

a known thermal resistance. The temperature of the sample falls continuously during an experiment, and measurements of the temperature are made about every ten seconds. Then the heat capacity of the sample and thermometer system can be calculated from the rate of change of the temperature and the known heat flow through the thermal link.

289

L

The thermal link is a piece of brass, and the heat flow through it is given by

$$\dot{q} = a(T_h^2 - T_c^2) \tag{3}$$

In this equation the constant a (84·7 erg sec^{-1} deg^{-2} in one particular apparatus) is determined from the measured electrical resistance at helium temperatures and the Wiedemann–Franz law, T_h is the temperature of the "hot" end, and T_c is the temperature of the cold end. T_h is obtained from the magnetic susceptibility of the cerium magnesium nitrate thermometer (measured with the aid of mutual inductance coils wound on the outer glass tube K). Since T_c is much smaller than T_h it does not need to be known very accurately; Haseda and Miedema obtain it with adequate accuracy from a γ-ray thermometer. For this, a single crystal of $MgSiF_6 \cdot 6\,H_2O$ containing a little radioactive ^{54}Mn is used. The angular distribution of the γ-rays is a strong function of temperature, and therefore the γ-ray intensity in one direction can be used as a thermometer. The heat capacity is then given by

$$C = \dot{q}/(dT_h/dt) = -a(T_h^2 - T_c^2)/(dT_h/dt) \tag{4}$$

It is important to have the thermal resistance between the sample and the metal and the thermal resistance in the sample itself small relative to the thermal resistance of the metal link. From the available data Haseda and Miedema estimate that in their apparatus the difference between the temperatures of the sample and the brass contact plate is only 2 per cent of the sample temperature above 0·05°K. They estimate that the accuracy of the heat capacity results obtained by this method is better than 5 per cent between 0·05°K and 0·5°K. Above 0·5°K the accuracy is reduced by some extra cooling through the support pins H.

IX. References

[1] Abel, W. R., A. C. Anderson, W. C. Black, and J. C. Wheatley, *Phys. Rev. Letters*, **14**, 129 (1965); *Physics*, **1**, 337 (1965); *Phys. Rev.*, **147**, 111, (1966).
[2] Ambler, E., and R. P. Hudson, *Rept. Progr. Phys.*, **18**, 251 (1955).
[3] Ambler, E., and H. Plumb, *Rev. Sci. Instr.*, **31**, 656 (1960).
[4] Anderson, A. C., G. L. Salinger, W. A. Steyert, and J. C. Wheatley, *Phys. Rev. Letters*, **6**, 331 (1961).
[5] Anderson, A. C., G. L. Salinger, and J. C. Wheatley, *Rev. Sci. Instr.*, **32**, 1110 (1961).
[6] Berman, R., J. C. F. Brock, and D. J. Huntley, *Cryogenics*, **4**, 233 (1964).
[7] Berman, R., and D. J. Huntley, *Cryogenics*, **3**, 70 (1963).
[8] Black, W. C., Jr., W. R. Roach, and J. C. Wheatley, *Rev. Sci. Instr.*, **35**, 587 (1964).
[9] Blake, C., C. E. Chase, and E. Maxwell, *Rev. Sci. Instr.*, **29**, 715 (1958).
[10] Burns, J. H., D. W. Osborne, and E. F. Westrum, Jr., *J. Chem. Phys.*, **33**, 387 (1960).
[11] Cooke, A. H., and R. A. Hull, *Nature*, **143**, 799 (1939).
[12] Corak, W. S., M. P. Garfunkel, C. B. Satterthwaite, and A. Wexler, *Phys. Rev.*, **98**, 1699 (1955).
[13] Craig, R. S., C. W. Massena, and R. W. Mallya, *J. Appl. Phys.*, **36**, 108 (1965).
[14] Dabbs, J. W. T., L. D. Roberts, and S. Bernstein, *Phys. Rev.*, **98**, 1512 (1955).
[15] Dauphinee, T. M., and E. Mooser, *Rev. Sci. Instr.*, **26**, 660 (1955).
[16] de Klerk, D., in S. Flügge, ed., *Handbuch der Physik*, Vol. 15, Springer-Verlag, Berlin, 1956, pp. 38–209.
[17] de Klerk, D., and R. P. Hudson, *J. Res. Natl. Bur. Std.*, **53**, 173 (1954).
[18] De Sorbo, W., *J. Chem. Phys.*, **21**, 876 (1953).
[19] Dugdale, J. S., D. K. C. MacDonald, and A. A. Croxon, *Can. J. Phys.*, **35**, 502 (1957).

[20] Edlow, M. H., and H. H. Plumb, in C. M. Herzfeld, editor-in-chief, *Temperature, Its Measurement and Control in Science and Industry*, Vol. III, Part 1 (F. G. Brickwedde, ed.), *Basic Concepts, Standards, and Methods*, Reinhold, New York, 1962, pp. 407–411.
[21] Erickson, R. A., L. D. Roberts, and J. W. T. Dabbs, *Rev. Sci. Instr.*, **25**, 1178 (1954).
[22] Franck, J. P., F. D. Manchester, and D. L. Martin, *Proc. Roy. Soc.* (*London*), **A263**, 494 (1961).
[23] Garfunkel, M. P., and A. Wexler, *Rev. Sci. Instr.*, **25**, 170 (1954).
[24] Gniewek, J. J., and R. J. Corruccini, *Rev. Sci. Instr.*, **31**, 899 (1960).
[25] Gorter, C. J., in *Les Phénomènes Cryomagnétiques*, Collège de France, Paris, 1948, p. 76.
[26] Griffel, M., and R. E. Skochdopole, *J. Am. Chem. Soc.*, **75**, 5250 (1953).
[27] Hartshorn, L., *J. Sci. Instr.*, **2**, 145 (1925).
[28] Haseda, T., and A. R. Miedema, *Physica*, **27**, 1102 (1961); Commun. Kamerlingh Onnes Lab. Univ. Leiden, No. 329c.
[29] Heer, C. V., C. B. Barnes, and J. G. Daunt, *Phys. Rev.*, **91**, 412 (1953).
[30] Heer, C. V., C. B. Barnes, and J. G. Daunt, *Rev. Sci. Instr.*, **25**, 1088 (1954).
[31] Heer, C. V., and J. G. Daunt, *Phys. Rev.*, **76**, 854 (1949).
[32] Hill, R. W., and P. L. Smith, *Phil. Mag.*, **44**, 636 (1953).
[33] Hoare, F. E., L. C. Jackson, and N. Kurti, eds., *Experimental Cryophysics*, Butterworths, London, 1961.
[34] Howling, D. H., E. Mendoza, and J. E. Zimmerman, *Proc. Roy. Soc.* (*London*), **A229**, 86 (1955).
[35] Keesom, W. H., *Helium*, Elsevier, Amsterdam, 1942, p. 8.
[36] Keesom, W. H., and J. A. Kok, *Physica*, **1**, 770 (1933–34); Commun. Kamerlingh Onnes Lab. Univ. Leiden, No. 232d.
[37] Keesom, W. H., and J. N. van den Ende, *Proc. Kon. Akad. Amsterdam*, **35**, 143 (1932); Commun. Kamerlingh Onnes Lab. Univ. Leiden, No. 219b.
[38] Kostryukova, M. O., and P. G. Strelkov, *Doklady Akad. Nauk S.S.S.R.*, **90**, 525 (1953).
[39] Kunzler, J. E., T. H. Geballe, and G. W. Hull, Jr., *Rev. Sci. Instr.*, **28**, 96 (1957).
[40] Lien, W. H., and N. E. Phillips, *Phys. Rev.*, **133**, A1370 (1964).
[41] Lindenfeld, P., in C. M. Herzfeld, editor-in-chief, *Temperature, Its Measurement and Control in Science and Industry*, Vol. III, Part 1 (F. G. Brickwedde, ed.), *Basic Concepts, Standards, and Methods*, Reinhold, New York, 1962, pp. 399–405.
[42] Lounasmaa, O. V., *Phys. Rev.*, **126**, 1352 (1962).
[43] Lounasmaa, O. V., and R. A. Guenther, *Phys. Rev.*, **126**, 1357 (1962).
[44] Lounasmaa, O. V., and P. R. Roach, *Phys. Rev.*, **128**, 622 (1962).
[45] Manchester, F. D., *Can. J. Phys.*, **37**, 989 (1959).
[46] Martin, D. L., *Phil. Mag.*, **46**, 751 (1955).
[47] Martin, D. L., *Proc. Roy. Soc.* (*London*), **A263**, 378 (1961).
[48] Martin, D. L., *Proc. Phys. Soc.* (*London*), **83**, 99 (1964).
[49] Maxwell, E., *Rev. Sci. Instr.*, **36**, 553 (1965).
[50] McKim, F. R., and W. P. Wolf, *J. Sci. Instr.*, **34**, 64 (1957).
[51] Mendelssohn, K., and J. L. Olsen, *Proc. Phys. Soc.* (*London*), **A63**, 2 (1950).
[52] Miedema, A. R., H. Postma, N. J. v. d. Vlugt, and M. J. Steenland, *Physica*, **25**, 509 (1959); Commun. Kamerlingh Onnes Lab. Univ. Leiden, No. 315 b.
[53] O'Neal, H. R., and N. E. Phillips, *Phys. Rev.*, **137**, A748 (1965).
[54] Osborne, D. W., B. M. Abraham, and B. Weinstock, *Phys. Rev.*, **94**, 202 (1954).
[55] Osborne, D. W., H. E. Flotow, and F. Schreiner, *Rev. Sci. Instr.*, **38**, 159 (1967).
[56] Parkinson, D. H., and J. E. Quarrington, *Brit. J. Appl. Phys.*, **5**, 219 (1954).
[57] Pearlman, N., and P. H. Keesom, *Phys. Rev.*, **88**, 398 (1952).
[58] Peshkov, V. P., *J. Exptl. Theoret. Phys.* (*U.S.S.R.*), **46**, 1510 (1964); *Soviet Phys. — JETP*, **19**, 1023 (1964).
[59] Peshkov, V. P., K. N. Zinov'eva, and A. I. Filimonov, *J. Exptl. Theoret. Phys.* (*U.S.S.R.*), **36**, 1034 (1959); *Soviet Phys. — JETP*, **9**, 734 (1959).
[60] Phillips, N. E., *Phys. Rev.*, **114**, 676 (1959).
[61] Pillinger, W. L., P. S. Jastram, and J. G. Daunt, *Rev. Sci. Instr.*, **29**, 159 (1958).
[62] Powell, R. L., M. D. Bunch, and R. J. Corruccini, *Cryogenics*, **1**, 139 (1961).
[63] Ramanathan, K. G., and T. M. Srinivasan, *Phil. Mag.*, **46**, 338 (1955).
[64] Rayne, J., *Phys. Rev.*, **95**, 1428 (1954).
[65] Rayne, J. A., and W. R. G. Kemp, *Phil. Mag.*, **1**, 918 (1956).
[66] Reich, H. A., and R. L. Garwin, *Rev. Sci. Instr.*, **30**, 7 (1959).
[67] Rose-Innes, A. C., *Low Temperature Techniques; The Use of Liquid Helium in the Laboratory*, English Universities Press, London, 1964.
[68] Ruehrwein, R. A., and H. M. Huffman, *J. Am. Chem. Soc.*, **65**, 1620 (1943).
[69] Samoilov, B. N., *Doklady Akad. Nauk S.S.S.R.*, **86**, 281 (1952).
[70] Schroeder, C. M., *Rev. Sci. Instr.*, **28**, 205 (1957).
[71] Seidel, G., and P. H. Keesom, *Rev. Sci. Instr.*, **29**, 606 (1958).
[72] Webb, F. J., and J. Wilks, *Proc. Roy. Soc.* (*London*), **A230**, 549 (1955).

[73] Westrum, E. F., Jr., J. B. Hatcher, and D. W. Osborne, *J. Chem. Phys.*, **21**, 419 (1953).
[74] Wheatley, J. C., D. F. Griffing, and T. L. Estle, *Rev. Sci. Instr.*, **27**, 1070 (1956).
[75] White, G. K., *Experimental Techniques in Low-Temperature Physics*, Clarendon Press, Oxford, 1959.
[76] Wiebes, J., W. S. Hulscher, and H. C. Kramers, *Appl. Sci. Research*, **B11**, 213 (1964); Commun. Kamerlingh Onnes Lab. Univ. Leiden, No. 341c.
[77] Cataland, G., and H. H. Plumb, *J. Res. Natl. Bur. Std.*, **70A**, 243 (1966).
[78] Cochran, J. F., C. A. Shiffman, and J. E. Neighbor, *Rev. Sci. Instr.*, **37**, 499 (1966).
[79] Edlow, M. H., and H. H. Plumb, *J. Res. Natl. Bur. Std.*, **70C**, 245 (1966).
[80] Hall, H. E., P. J. Ford, and K. Thompson, *Cryogenics*, **6**, 80 (1966).
[81] Hudson. R. P., and R. S. Kaeser, *Physics*, **3**, 95 (1967).
[82] Neganov, B., N. Borisov, and M. Liburg, *J. Exptl. Theoret. Phy. (U.S.S.R.)*, **50**, 1445 (1966); *Soviet Phys.—JETP*, **23**, 959 (1966).
[83] Plumb, H. and G. Cataland, *Metrologia*, **2**, 127 (1966).
[84] Vilches, O. E., and J. C. Wheatley, *Physics Letters*, **24A**, 440 (1967).

CHAPTER 8

High-temperature Drop Calorimetry

Thomas B. Douglas

National Bureau of Standards, Washington D.C., U.S.A.

Edward G. King

*Albany Metallurgy Reasearch Centre, Bureau
of Mines, Albany, Oregon, U.S.A.*

Contents

I. Introduction	294
1. General Requirements	294
2. Advantages and Disadvantages of the Drop Method	295
A. Cases for which the Drop Method is Unsuitable	295
II. The Furnace	296
1. Furnace Design and Operation	296
A. General Type of Furnace Most Commonly Used	297
(1) Materials of Construction	298
(2) The Heaters	298
(3) Heat Transfer and Temperature Gradients	299
B. Furnaces with a Good Heat-Conducting Core	300
C. Furnaces Operating Above 1500°C	302
D. Automatic Temperature Control of the Furnace	303
2. Temperature Measurement	304
A. The Thermocouple	304
(1) Calibration	304
(2) Common Sources of Thermocouple Error	305
B. Use of Other Temperature-Measuring Instruments	305
3. The Sample in the Furnace	306
A. The Container	306
(1) Choice of Material	306
(2) Container Design; Enclosing the Sample	307
B. Suspending and Dropping the Sample	308
(1) Suspension and Dropping Mechanisms	308
(2) Time in the Furnace	309
III. The Calorimeter	310
1. The Isothermal Calorimeter	310
A. Design of the Ice Calorimeter	310
B. Assembly	312
C. Calibration Factor	313
D. General Procedure of Operation	313
E. A Typical Experiment	314
F. Tests for Accuracy	316
G. Hunting Causes of Trouble	317
2. The Isothermal-Jacket, Block-Type Calorimeter	318
A. Design and Construction	318
B. Additional Measuring Equipment	319
C. A Typical Experiment	320
D. Calibration Factor	322
IV. Treatment of the Data	322
1. Correcting to Standard Conditions	322
A. Correction to a Basis of a Pure Sample	322
(1) Correcting for Insoluble Impurities	322

  (2) Correcting for Soluble Impurities 323
 B. Correcting for Vaporization Inside the Container; Correcting
  to a Different Pressure 323
2. Smoothing and Representing Enthalpy Values 325
 A. Graphing 325
  (1) Choice and Treatment of Measurements 326
 B. Representation by Equations 326
  (1) Theoretical and Semi-Theoretical Equations. 326
  (2) Empirical Equations 326
 C. Treatment of Fusions and Transitions 328
3. Derivation of Other Thermodynamic Properties from Relative
 Enthalpy 328
4. Precision and Accuracy 329
V. References 330

I. Introduction

The determination of accurate high-temperature enthalpies provides a valuable tool in the study of materials and processes at elevated temperatures. As interest in high-temperature research increases, impetus is given to experimental methods that will extend the temperature range of measured enthalpies.

The enthalpies of the simpler, ideal gases can be determined much more accurately from spectroscopic data than by measurement. Calculation methods may also be accurately applied for less simple gases with molecules in a single electronic state and without internal rotations. In these cases molecular-constant data must be available, *i.e.*, molecular-structure and vibrational-frequency data.

At present, enthalpies of solids and liquids must be determined experimentally; and most of the accurate data for solids and liquids, above 100°C, have been provided by drop calorimetry. With this method a sample is heated to a known temperature and is then dropped into a calorimeter (usually operating near room temperature), which provides measurement of the heat evolved by the sample in cooling to the calorimeter temperature.

Currently the major applications of enthalpy and enthalpy-related data are to problems in microstructure, heat transfer, and chemical thermodynamics. In particular, the last application is of great importance. If room-temperature values of enthalpy, Gibbs energy, or entropy have been obtained by other means, the values may be extended to the limit of the high-temperature measurements. In many cases equilibrium vapor pressures can be calculated more accurately than they can be directly measured. If vapor pressure measurements do exist, the measured enthalpies and derived data can be used in a complementary way.

1. *General requirements*

The general requirements for drop-calorimetry operation of high accuracy may be considered in terms of (*i*) the sample, (*ii*) the furnace, and (*iii*) the calorimeter.

(*i*) The substance measured must not change irreversibly to a significant extent during the measurements. Such a change can occur through decomposition, reaction with the container, or deterioration caused by container leakage. In passing from the higher (furnace) temperature to the lower (calorimeter) temperature, the amount of heat evolved by the sample is meaningful only if the forms of the sample at the two temperatures are defined.

(*ii*) The accurate determination of the average temperature of the sample referred to some well defined temperature scale (usually the International Practical Temperature Scale) is the most important requirement of the furnace. The heat lost by the sample as it drops through the region between the furnace and the calorimeter must cancel out through empty-container measurements. This requirement dictates a high and reproducible rate of fall (most easily accomplished by a close approximation to free fall) and constancy of surface emissivity, because these factors strongly affect radiation losses which predominate at high temperatures.

(*iii*) The calorimeter must be capable of measuring the heat delivered by the sample with high accuracy, a requirement which entails the accurate determination of the heat losses from the calorimeter itself.

In designing calorimetric apparatus for high accuracy and choosing materials of construction, intuition is a poor substitute for relatively simple calculations, even ones which may be in error by as much as a factor of two or three. This is especially true of questions involving heat transfer. For considerably lower requirements in accuracy, a much simpler furnace and type of calorimeter can replace those subsequently described in this chapter; in this case, simple calculations or tests may indicate an otherwise unsuspected degree of simplification in construction and operation which may be safely adopted.

2. *Advantages and disadvantages of the drop method*

Advantages and disadvantages of the drop method compared with other calorimetric methods are discussed in Chapter 1, Section V. Perhaps the greatest advantage of the drop method over the adiabatic method is that the precise calorimetric measurements are made at or near room temperature where suitable conditions are most easily maintained. Above 500°C or so the drop method at present excels applications of the adiabatic method in accuracy as well as simplicity. With high-temperature adiabatic and isothermal-jacket calorimeters there is considerable difficulty in temperature control and in accounting for heat losses by radiation.

With the drop method, enthalpies of transition and fusion are obtained by the difference between two measured quantities of heat. If these quantities are large, the errors in the enthalpies of transition or fusion may be correspondingly large. This may be regarded as a weakness of the drop method; however, this effect is partially cancelled by systematic procedure.

A. Cases for which the Drop Method is Unsuitable

The success of the drop method depends upon complete return to the same, defined, thermodynamic reference state after each measurement. The method inherently involves the rapid cooling of the sample. Sometimes, during cooling, a sample fails to complete a solid–solid transition or fails to completely crystallize upon returning from a fused state. If the calorimeter is normally precise, such cases will usually make themselves known through lack of precision or through continued heat evolution to the calorimeter over a longer period of time than usual (sometimes over a period of many hours). If the time is too long or the results inconsistent, the drop method is unsuitable and must be abandoned. If the behavior is consistent and measurable, the difficulty may be resolved through use of auxiliary calorimetry such as solution calorimetry by which enthalpies are referred to a desired reference state.

Particularly uncertain are drops from the liquid state if the solid-composition line is not vertical. Changes of composition of solid phases will occur upon cooling, with resultant indefinite final states. Such cases can usually be predicted if phase diagrams are available.

II. The Furnace

Invariably the furnace involves a vertical cylindrical core surrounding the sample. Sometimes the core is heated by means of its own electrical resistance, but more commonly through the electrical resistance of one or more suitable wire windings on it. Sometimes at the higher temperatures inductive heating is used instead.

Reference may be made to published descriptions of some of the furnaces designed specifically for drop calorimetry. Important contributions to precision in operation were made by White[36, 37]. Southard[33] described a wire-wound furnace suitable for use up to 1500°C. His design has been used and described more fully by Kelley, Naylor, and Shomate[19], and its essential features have been adopted by numerous other investigators, such as Ginnings and Corriccini[10]. The U.S. National Bureau of Standards[6] improved the temperature profile of a furnace for use to 900°C by means of thick silver tubes surrounding the core. Nickel was used for the same purpose by Oriani and Murphy[28], Dworkin and Bredig[3], Gilbert[7], and others. Graphite was used for this purpose for the temperature range 900–1650°C by Lucks and Deem[24]. Douglas and Payne[2] have presented design calculations for a furnace operating up to 1500°C. Fomichev, Kandyba, and Kantor[5] used a solid tube of platinum as a resistance heater, and nickel tubes to equalize the temperature. These last workers also describe a vacuum furnace for the range 900–3000°K with a graphite heater, together with a system of coaxial graphite and metal shields. Levinson[23] used a graphite resistance furnace for measurements in the range 1000–2400°C. Olette's[27] furnace follows the design of Southard's[33], but gives measurements to as high as 1900°C. The furnace of Kirillin, Sheindlin, and Chekhovskoi[20, 21] is heated by tungsten windings and has been used up to 2400°C. Hoch and Johnston[12] have described a furnace in which the sample is heated by radio frequency induction in the temperature range 700–2700°C.

1. *Furnace design and operation*

In designing and constructing a furnace, ease of operation and ready access for repairs are certainly important (along with superior performance) if the purpose of the undertaking is a satisfactory rate of production of a body of data. For example, the ability to easily swing or hoist the furnace away from the calorimeter is an advantage. It is convenient to be able to exchange heater windings and thermocouples without disturbing other parts. Easy and

HIGH-TEMPERATURE DROP CALORIMETRY

rapid return of the specimen from the calorimeter to the furnace and readying conditions for a succeeding measurement deserve consideration.

A. General Type of Furnace Most Commonly Used

Most accurate drop calorimetry to date has been performed with wire-wound, resistance-heated furnaces at temperatures of 1500°C or lower. The cross-section of such a furnace is diagrammed in *Figure 1* and may be considered typical. However, the following discussion will present several alternative features.

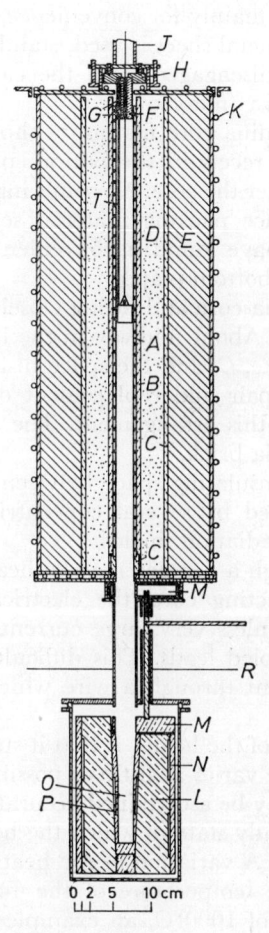

Figure 1. Diagram of furnace and copper-block calorimeter

A, Alumina winding tube
B, Alumina tube
C, Mullite
D, Alumina powder
E, Magnesia brick
F, Refractory plug
G, Porcelain tube
H, Stainless steel holder
J, Brass dropping tower
K, Copper cooling coil
L, Copper block
M, Gates
N, Resistance thermometer
O, Receiving "well"
P, Heater
R, Oil level
T, Thermocouple

297

(1) Materials of Construction. For operation up to 1500°C, the furnace core (or cores, if there are two or more coaxial ones) is usually a tube of alumina. The core, A, shown in *Figure 1* (inside diameter 2·5 cm, wall thickness 0·5 cm, length 58 cm) is a porous tube made of alumina. It is sometimes desirable to pass an inert gas through the core, partly to prevent oxidation of the heater winding, sample container, or suspension wire. This protection is not essential when platinum (or platinum–rhodium alloy) is used throughout for these purposes.

The top and bottom plates of the furnace shown are of ordinary steel plate. The outside boundaries of the furnace are water-cooled through use of copper tubing. This is mainly for convenience, so metal sheet all around should be satisfactory. If metal sheet is used, stainless steel is recommended. The furnace pictured is disengaged from the calorimeter by means of a hand-operated, geared hoist (not shown).

The bottom of the alumina core fits into a short section of mullite tube which in turn fits into a recess in the bottom plate of the furnace. This arrangement serves to center the core. No centering device is used for the top of the furnace; the furnace packing material serves to hold it in place. However, some workers have found it convenient to silver-solder stainless steel rings to the top and bottom plates.

Surrounding the alumina core and fitting closely to it is a second tube of alumina, B, 0·6 cm thick. About midway in the insulation space is located a tube of refractory mullite, C, and the space inside it is packed with pure alumina powder. The repair and replacement of the furnace winding is easily accomplished with this arrangement. The space outside the mullite tube is filled with magnesia brick.

As an alternative, the insulating space between the core and the outside container may be occupied by several concentric, highly polished metal cylinders which serve as radiation shields.

(2) The Heaters. Although a furnace may be heated by passing an electric current through a conducting core, the electrical resistance of the core must be relatively high unless very large currents are used, requiring unusually large or water-cooled leads. This difficulty is usually avoided by passing the heating current through a wire which may be wound on the furnace core.

The main requirement of the heater is that it surround the sample with a region whose temperature varies as little as possible so that the mean temperature of the sample may be ascertained accurately. Constancy of furnace temperature implies a steady state in which the heat losses are balanced by the output of the heaters. A variation of the heating current by 1 per cent will vary the steady-state temperature of the furnace by approximately 20° in the neighborhood of 1000°C, an example which illustrates the fact that approximate constancy of furnace temperature depends more upon the thermal inertia (large heat capacity) of the furnace than is often realized.

The furnace power is nearly always supplied by one or more heaters wound on the core, which may be purchased with grooves on the outside for uniform heater spacing. Frequently more power is supplied to the ends of the core than the middle to compensate for the heat losses from the top and bottom of the furnace. This may be accomplished by shunting part of the current from the center portion of a single winding or by having three separate windings with independent controls. In practice, the center heater usually extends over the middle 65–80 per cent of the core. Care must be taken that the temperature profile of the furnace does not have or develop unwanted humps or hollows. Although a single winding has a narrower constant-temperature zone, it is free from this defect.

When massive cores are employed *within* the main heaters, additional secondary heaters are advisable to prevent temperature lags and to aid in controlling the core temperature. These heaters may be wound around a separate core immediately surrounding the sample, or if the massive core is a sufficiently good heat conductor, be located within the conductor itself. The use of the secondary heaters can be more important if automatic temperature control is employed.

Power is conveniently supplied by means of variable autotransformers. If more than one heater section is used, it is advisable to feed each section by its own transformer.

Nichrome (Ni–Cr 80:20) wire is used most often for heater windings at lower temperatures. It has a fairly low temperature coefficient of resistivity, but cannot be used in air much above 1000°C because of excessive oxidation. Platinum–rhodium windings are suitable for continuous use up to about 1500°C, but in air they deteriorate rather rapidly above this temperature. Platinum containing 10 per cent rhodium has about three times the resistivity at 1500°C as at room temperature, but this is no real disadvantage. Platinum containing 20 per cent rhodium is sufficiently flexible for winding on the core if annealed; its advantage is mainly a slightly higher melting point. For still higher temperatures a more refractory winding is needed. Lucks and Deem[24] used molybdenum windings up to 1650°C, surrounding the wire with a protective atmosphere. Main heater wire should not be too fine, because of decreased mechanical strength and shorter life. A diameter of from 0·05–0·1 cm has been commonly used. Care should be taken to wind the heaters firmly and uniformly.

The single-winding heater used in the furnace shown in *Figure 1* is made up of 58 ft. of *B* and *S* No. 18 (diameter, 0·1 cm) platinum containing 20 per cent rhodium on the spirally grooved alumina core. AC power is supplied by means of a voltage stabilizer and two 7·5 kVA variable autotransformers in series. This scheme permits sensitive manual control. The power consumption at 1500°C is about 1700 W.

(*3*) *Heat Transfer and Temperature Gradients*. Despite the practical importance of enthalpy measurements at high temperatures and the simplicity of the drop method, the results of different investigators on the same materials have often varied by several per cent. This type of calorimetry is thus seen to be less developed than techniques at low temperatures. The principal reason for disagreement undoubtedly lies in measuring the sample temperature in the furnace. The singly wound furnace shown in *Figure 1* provides an axial temperature gradient of less than 0·1 per cent of the temperature (Celsius) per inch over the range 100–1500°C. Refinement may be made to improve and lengthen this 'hot zone'. The use of multiple heaters (Ginnings and Corruccini[10]) provided a few tenths of a degree per inch at 1100°C in a furnace 46 cm long. The use of innermost, heat-distributing tubes of silver provided the U.S. National Bureau of Standards[6] with virtually no temperature gradient within the masses of silver. This furnace is fully described in a subsequent section. Fomichev et al.[5] used nickel tubes and obtained a precision of measurement of 0·1° up to 1500°K.

It is desirable to reduce the temperature difference between the inner furnace wall and the sample as much as possible. This is particularly important if the temperature of the sample is measured by means of a thermocouple whose junction is attached to the wall. Also, the presence of a gas surrounding the sample (a few mm Hg suffices) will minimize the wall-to-sample gradient, as will fusing the sample in its container if this is possible. Heat transfer between the container and wall is increased by reducing the gap between them. Consideration should be given also to the heat-transfer coefficient of the gas to be used in the furnace. When the conditions are favorable and the nearly isothermal part of the core is not too short, the sample will radiate an insufficient amount of power from the bottom of the furnace to invalidate its temperature measurement.

B. Furnaces with a Good Heat-Conducting Core

A good heat-conducting core is provided by a material of high thermal conductivity and by using a large cross-section of the material. The result is an unusually large total heat capacity. The major consequences in the performance of the furnace compared with that described in Section II-1-A are, first, that axial temperature gradients and errors in measuring the sample temperature can be made smaller or alternatively that they can be made equally small with considerably less effort, and second, that the inertia of the furnace to temperature change becomes so large that the temperature can easily be kept constant to 0·01° or better by simple control by hand.

A few examples of such furnaces and their major differences from the more commonly used type will be discussed. The U.S. National Bureau of Standards[6] constructed a furnace having a core of silver, which is resistant to corrosion, has the highest thermal conductivity of any metal, and can be used up to its melting point, 962°C. A cross-section of the furnace is shown in *Figure 2*. The main furnace heater was made in three separate sections corresponding in elevation to the three silver cylinders, J, K, and L, which were located inside the alumina. The silver cylinders are supported and separated by porcelain spacers, Y, having a far lower thermal conductivity than silver, so that the end silver cylinders, J and L, need to be maintained at a temperature no less than a few tenths of a degree of that of the central silver cylinder, K, for the gradients in cylinder K to be negligible. Coaxially with the silver and porcelain cylinders are Inconel tubes which enclose the sample container, D, with its suspension wire and shields, S. Helium flows upwards past the container and escapes from a small orifice at the top. (Care must be taken that cold air cannot fall into the lighter helium in the furnace.)

Figure 2 shows some of the vertical holes, N, drilled through the silver and porcelain and placed 90 degrees apart azimuthally. These holes contain the platinum resistance thermometer, the platinum–rhodium thermocouple, and the differential thermocouples between the end silver cylinders, J and L, and the central cylinder, K. In one of these holes are placed three small

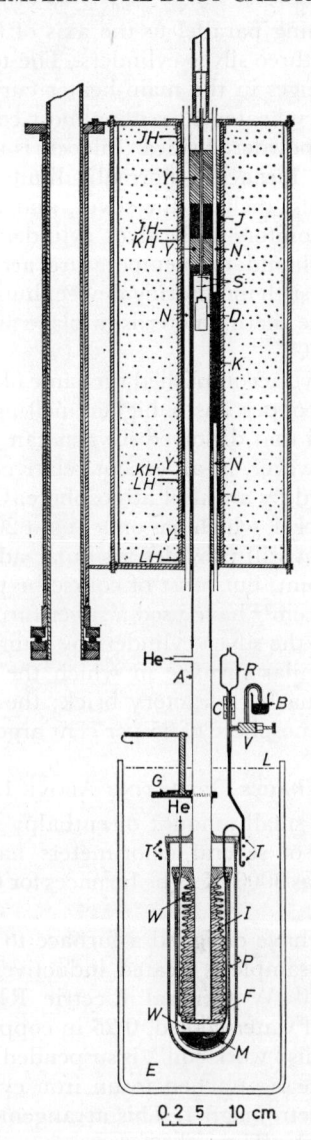

Figure 2. Diagram of silver-core furnace and ice calorimeter

A, Calorimeter "well"	M, Mercury
B, Beaker of mercury	N, Inconel tubes
C, Glass capillary	P, Pyrex vessels
D, Sample container	R, Mercury reservoir
E, Ice bath	S, Platinum shields
F, Copper vanes	T, Mercury "tempering" coil
G, Gate	V, Needle valve
J, Ice mantle	W, Water
	Y, Porcelain spacers

JH, KH, LH, Furnace heater leads
J, K, L, Silver cylinders

secondary heaters, running parallel to the axis of the furnace and located at the elevations of the three silver cylinders. The temperature of the silver responds to sudden changes in the main-heater currents in 3–6 min, but to changes in the secondary-heater currents almost completely within 1 min. For furnace insulation, powdered silica (Silocel) is used.

The operation of the furnace is normally limited to 900°C, to provide a margin of safety against melting the silver, and a thermocouple-reading meter (Sym-ply-trol) monitors each silver cylinder to shut off the heating current automatically should the temperature accidentally become a few degrees higher. A fourth such monitoring device shuts off the heating currents if the platinum resistance thermometer is in place and the furnace temperature exceeds about 625°C.

Some investigators have accomplished the same objectives by using copper instead of silver. Pure copper has a higher melting point (about 1083°C, but it has been reported that dissolved oxygen can lower the melting point by as much as 20°); however, at all except relatively low temperatures the copper must be protected by an inert atmosphere. Other investigators have replaced the silver by nickel, which has only about 20 per cent of the thermal conductivity of silver but still provides a similar advantage; the nickel has a still higher melting point, but must of course be protected from excessive oxidation. Lucks and Deem[24] have used a silver furnace of the above design up to 900°C, separating the silver cylinders by alumina spacers. From 900–1650°C they used a similar furnace in which the silver was replaced by graphite and the alumina by refractory brick; the graphite was protected from oxidation by an atmosphere of 95 per cent argon–5 per cent hydrogen.

C. Furnaces Operating Above 1500°C

Although a relatively small amount of enthalpy data above 1500°C has been reported, designs of several calorimeters have been published for measurement to as high as 3000°K. The furnaces for two of these calorimeters will be described briefly.

Hoch and Johnston[12] have designed a furnace to be used in drop calorimetry to 3000°K. The sample is heated inductively, the radio frequency being supplied by a 20 kW General Electric RF generator through a "work coil" consisting of water-cooled, 0·25 in copper tubing. The sample, heated in the center of this "work coil", is suspended by means of a tantalum wire. The tantalum wire is attached to an iron cylinder which is held in place by an external electromagnet. This arrangement provides a dropping mechanism for the sample. Thus, the furnace is essentially a vacuum-tight shell containing the coil and sample. The sample temperature is measured by sighting holes in the sample container with a disappearing-filament optical pyrometer. The furnace and attached calorimeter are sealed vacuum tight with O-rings and the system is evacuated with an oil diffusion pump. A small amount of helium is introduced after dropping the sample to facilitate heat distribution. The furnace walls are wound on the outside with copper tubing. The tubing is used to help degas the furnace when heated by steam and then to cool it during a run. The furnace is about 34 cm long and 25 cm in diameter.

Fomichev et al.[5] describe a furnace for use at 1000–3000°K. Heating is accomplished by means of a graphite tube with an internal diameter of 4·5 cm, a length of 60 cm, and a wall thickness of 0·3 cm. Electrical contact is made with water-cooled flanges of copper or brass. These flanges are attached to springs to allow for expansion of the graphite tube. The heater is surrounded by nine coaxial shields; the first is made of graphite, the second and third of tantalum, the fourth, fifth and sixth of molybdenum, and the remaining three of stainless steel. All outer surfaces of the furnace are water-cooled. In addition, two water-cooled screens are provided above the top and below the bottom of the central furnace openings. These screens can be moved to one side by means of bellows. Temperature is measured by optical pyrometry, utilizing a reflecting prism which is located along the central axis and between the sample and the calorimeter. The prism is moved out of the way with a bellows. All parts of the apparatus are vacuum-tight, and different parts may be sealed off so that samples may be placed in the furnace without disturbing the furnace temperature. The sample is suspended in the furnace by a 0·1 mm tungsten wire which is melted to release the sample. An inert gas (at a pressure of 10–20 mm Hg) is used during the measurements, as it improves the general performance. Power is supplied through stepdown transformers. At 3000°K the power consumption is 40 kW.

D. Automatic Temperature Control of the Furnace

The two furnaces illustrated by *Figures 1* and *2* use manual control of heaters. The resulting temperature fluctuations are well within the limits of other uncertainties involved in the measurement of the sample temperature. Manual control satisfactorily provides for the "over-shooting" necessary for overcoming the thermal lag of the furnace parts, thus providing a quicker approach to thermal equilibrium. Automatic controls have been found to be of limited value for these furnaces.

The detailed treatment of automatic-control instrumentation can be found elsewhere. Specific applications to calorimetry are discussed in Chapters 5 and 9. A few general comments may be made here relating to automatic stabilization of furnaces in drop calorimetry.

The important point is to have the response time between the sensing element and the heater it controls as small as practicable. This sensing element may reflect the temperature of a point or the temperature of a region. The former may be sensed by a single-junction thermocouple, the latter by a multiple-junction thermocouple or an element whose electrical resistance varies sufficiently with temperature. When the furnace winding is of platinum–rhodium, the temperature coefficient is such that its resistance may be used as the temperature-sensing element.

Furnaces with low thermal inertia, such as those using reflection shields, are more sensitive to small current changes and are best served by automatic control devices. Furnaces with massive metal inner tubes are very sluggish to temperature change and will benefit least from automatic controls.

The response of the best controls on the market is dependent on three factors. Briefly, they are: how far the temperature differs from the desired value, how long it has been so, and how fast it is changing. If the refinements of control are adequate for the time-lag of response, the major remaining problem is likely to be securing a wide enough range of control to handle all likely variations in heater power and furnace heat losses.

2. *Temperature measurement*
A. The Thermocouple

The most common temperature-measuring device up to 1500°C has been the thermocouple. Some base-metal thermocouples (such as Chromel–Alumel and iron–constantan) have the appeal of large sensitivities. However, with modern potentiometers reading to one microvolt or less, this sensitivity is not needed, and for the most accurate work, the base–metal thermocouples are not preferred over the more stable Pt *versus* Pt–Rh thermocouple.

Thermocouples of 94 Pt–6 Rh *versus* 70 Pt–30 Rh[38] and 80 Pt–20 Rh *versus* 60 Pt–40 Rh[15] have the advantage of maintaining their calibrations for comparable periods of time at temperatures some 300° higher than the conventional Pt *versus* 90 Pt–10 Rh couple. Additional investigation of these and other thermocouples particularly adapted to high temperatures may recommend them for measurement in calorimetry. At the present time the reliable Pt *versus* 90 Pt–10 Rh thermocouple (or the Pt *versus* 87 Pt–13 Rh thermocouple) is the best for accurate work. This couple shows very little change up to 1200°C, and can be used up to about 1500°C with frequent calibrations. Roeser and Wensel[34] have stated that with favorable calibration the uncertainty in interpolated values of temperature should be as small as 0·3° near 1100°C and 2° near 1500°C.

(1) *Calibration*. Details of thermocouple calibration *outside* the furnace are dicussed in Chapter 2. The ideal Pt *versus* 90 Pt–10 Rh thermocouple is completely homogeneous and free from strains so that its reading depends solely on the temperatures of its junctions. As the thermocouple is used, the e.m.f. changes are mainly due to diffusion through the junction and to oxidation. These effects can be made negligible if the working couple is frequently calibrated *in situ* in the furnace. Calibration below the point of thermocouple deterioration (about 1500°K) may be accomplished by comparison with a calibrated thermocouple. A special method of calibration makes the comparison with the hot junction of a "standard" thermocouple firmly attached (preferably in a well) to the sample container which is hanging in place in the furnace.

Otherwise calibration *in situ* at one temperature is generally made by melting a short length of gold wire (at least 1 mm) welded between the elements of the thermocouple. The e.m.f. at the melting point corresponds to the halt in the temperature rise. The assumed temperatures of all fixed points used for thermocouple calibration should be stated because such information has sometimes permitted later revision of old data to a basis closer to the true thermodynamic temperature scale. Roeser and Wensel[34] report that in most cases curves of differences between actual thermocouples and standard reference tables are linear over the entire range 0–1700°C.

The use of Pt *versus* 90 Pt–10 Rh thermocouples above about 1500°K necessitates frequent recalibrations. It has been found that, in typical circumstances, the rates of change per 3 h of heating in an oxidizing atmosphere are about 0·2° at 1600°K, 0·4° at 1700°K, and 1·0° at 1800°K.

(2) *Common Sources of Thermocouple Error.* The temperature measurement of the sample suspended in the furnace requires that the thermocouple junction and the sample be at the same temperature. Further, the temperature gradient of the thermocouple along the leads to the junction should be small. In the furnace shown in *Figure 1*, the thermocouple leads are brought into the furnace through a 5 cm-long refractory plug, F, by means of porcelain tubes, G. The plug is made of zirconium silicate tubing for high thermal shock resistance. The porcelain tubes are fixed in position with alumina and cement. The thermocouple leads run along near the furnace wall to the sample container. The junction touches the top wall of the suspended container at temperatures below 800°K, and it is placed within 1 mm of the top above this temperature to prevent welding or other interaction between the two surfaces. Proximity of the leads to the solid parts of the furnace is highly desirable to prevent conduction into or away from the junction; a similar result can be achieved by welding the junction to a band of a highly conducting metal such as platinum which lies firmly against the inside wall of the furnace. The problem here is the difficulty in making frequent calibrations of the couple. If, as is normally to be expected, heat is being conducted *away* from the junction, an error from this effect will result in an enthalpy that is too high. This error is equivalent to associating the net enthalpy of the sample with a furnace temperature that is too low, and in this sense the same effect in the empty-container run affords no compensation. The furnace of *Figure 2* has its thermocouple tubes inserted directly into the thick inner wall of silver.

A possible source of error in thermocouple readings may arise if there is electrical leakage from the heaters. The heaters are usually operated with a.c. to minimize this, but such errors might arise from some rectifying effect in the insulator materials. Momentary shutting off of the furnace power is the standard technique for testing this source of error. Other sources of error are contamination through container leakage and contamination from impurities in the insulating inlet tubes.

When differential thermocouples are used, leads to the outside should be of platinum rather than platinum–rhodium to avoid changes of composition arising from passage through regions of large temperature gradients.

B. Use of Other Temperature-Measuring Instruments

Even at the lowest furnace temperatures, the error in temperature measurement with a thermocouple may easily be as much as 0·1 °C. The platinum resistance thermometer defines the International Practical Temperature Scale up to 630·5°C and, because an accuracy between 0·01° and 0·1°C is not difficult to obtain with it, may be substituted in this temperature range. If the thermometer uses mica for insulation, as has been customary in the past, it cannot be used for higher temperatures. The thermometer must be encased in some material which does not soften at the highest temperatures of use and at the same time allows a snug fit inside a solid part of the furnace so that conduction along the leads is not a source of appreciable error. Such a mica-insulated thermometer, encased in 90 per cent Pt–10 per cent Rh, has been used in the silver-core furnace at the U.S. National Bureau of Standards[6]. This thermometer had an ice-point resistance of about the conventional 25Ω, which is better than thermometers of much lower resistance that have been constructed. Periodic reading of the ice-point value, particularly after the thermometer has been to its highest temperatures, is highly desirable to detect the need for recalibration. If the resistance element becomes stretched, readings at all temperatures should be displaced by the same percentage as at the ice point. However, if other changes occur, *e.g.* through contamination of the platinum, the readings at all temperatures may tend to be changed by more nearly the same *amounts* as at the ice point.

If a platinum resistance thermometer and thermocouple are both used in the furnace, it becomes possible not only to calibrate the thermocouple by the thermometer (up to 630°C) or to compare their independent calibrations, but also to compare them regularly as a means of detecting any large, otherwise unsuspected, change in the calibration of either measuring instrument. Compared with a thermocouple, a platinum resistance thermometer has the disadvantages of slower response to temperature changes (offset by the temperature sluggishness of a massive core), and difficulty in construction, particularly in mounting and insulating the long fine wire.

The optical pyrometer is usually used to measure furnace temperatures above the range of Pt–Rh thermocouples. It affords less precision than the Pt *versus* 90 Pt–10 Rh thermocouple, but has the advantage that it defines the International Practical Temperature Scale above 1063°C.

3. *The sample in the furnace*
A. The Container

(1) *Choice of Material.* The choice of a suitable container material is usually not difficult at lower temperatures, but it well may be the greatest single factor hindering the extension of data to higher temperatures. Important considerations for container materials are mechanical strength, melting point, emissivity, reactivity, and reproducibility of their thermodynamic states.

The most common factor limiting the choice of a suitable container material is reactivity with the sample, through either chemical reaction or miscibility. If the container is exposed to an oxidizing atmosphere, this is also a factor. A useful generalization is that a metallic sample is more likely to be inert to a non-metallic container and vice versa. Since reactivity often varies greatly with temperature, a calculation or estimation of the free energy changes of all possible reactions is often useful. Platinum–rhodium (10 or 20 per cent) most generally satisfies the requirements for a metallic container to about 1500°C. Gold and silver have similar advantages below their melting points (1063° and 962°C). The high purity and reproducible state attainable with these metals allow container replacement with no appreciable change in enthalpies. If a non-metallic container is needed silica glass has frequently been used. Devitrification to cristobalite above 1300°C limits its use to about 1450°C.

Measurements above 1500°C are sparse, so that container information is also lacking. Tantalum has been used to 2600°C and tungsten to 2500°C. Graphite, surrounded by a shell of metal to improve the radiation characteristics or to afford an extra seal against possible escape of the sample, could sometimes be used. It must be recognized that in many cases a high container enthalpy will have to be tolerated.

In the absence of any suitable container material, the problem has been solved by using the sample itself in mechanically cohesive form. In this case the heat lost in the dropping process must be estimated or else determined by varying the ratio of mass to the surface exposed. Beryllium oxide was measured to 2200°K[16] by using a sintered sample of BeO surrounded by a 0·2 mm-thick sheath of molybdenum metal which protected the sample from excessive heat loss and provided a hollow for temperature measurement.

Many laboratories have adopted transition-metal or base-metal containers for use to moderately high temperatures. The ready availability and workability of these metals and alloys have made them a logical choice. However, it is wise to avoid the use of alloys that undergo transitions during cooling. If the degree of transformation and the consequent heat evolved are the same in measurements when empty and when full, the transition cannot cause any error. However, with a full container the slower rate of cooling *can* result in the evolution of more of the heat of transition. The net heat attributed to the sample is consequently too high. Ginnings[8] has critically examined three sets of drop calorimetry data involving containers of Nichrome V (80 Ni–20 Cr) and stainless steel, types 430 and 446, all of which undergo transitions between 500° and 700°C. He detected abrupt increases in the enthalpy–temperature functions derived for the samples from the measurements and showed that, by attributing the effect to the above cause, the data above the transition temperature became much smoother. The enthalpy correction in these cases ranged from 0·05–1·8 cal/g of sample and lowered the calculated heat capacities by 0·3 to several per cent.

(2) *Container Design; Enclosing the Sample*. The container is usually cylindrical and has a capacity of the order of 10 cm³. The capacity is determined primarily by the requirements of the receiving medium after the drop. The wall thickness of metal containers is usually 0·2–0·3 mm, but a thicker wall is needed for non-metallic containers.

Five designs of sample containers are shown in *Figure 3*. *Figure 3(a)* shows a single cylinder with snug-fitting end caps. When made of a base metal, the edges can be sealed by induction welding if gas-tightness is needed, with the sample remaining cold and surrounded by any desired gas. If the container is of noble metal (Pt–Rh, Au, or Ag) the edges can be welded with a gas–oxygen torch. *Figure 3(b)* shows a variation used by the U.S. Bureau of Mines in which the top is connected to a narrow neck. After the sample has been introduced, the container may be evacuated and filled with helium gas. The neck may be pinched shut near the top and sealed gas-tight by gas–oxygen welding. A container sealed in this manner will usually remain so to 1500°C. If volatility during welding is a problem, the neck can be sealed shut by gold soldering; this, of course, limits the measurements to temperatures below the melting point of gold. In extreme cases of volatility, placing the container in a shallow dish of water keeps the temperature down during the soldering process.

Figures 3(c) and *3(d)* show two containers closed by hard metal cones against gold gaskets. These have been used extensively at the U.S. National Bureau of Standards, container in *Figure 3(c)* being of Nichrome V, and container in *Figure 3(d)* being of Monel for measuring hydrocarbons up to 20 atm vapor pressure. The threads of the screw cap are pre-oxidized to

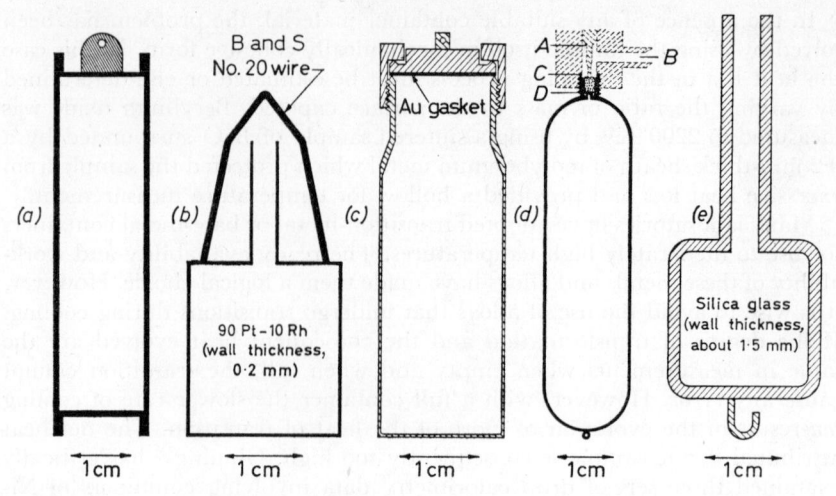

Figure 3. Designs of sample containers
    a, Of base or noble metal and with welded end-caps
    b, Of 90% Pt–10% Rh, sealed by welding or soldering
    c, Of Nichrome V (80% Ni–20% Cr), sealed by a gold gasket
    d, Of Monel, sealed by a gold gasket *D* (for volatile liquids) (filling device: *A*, screw driver; *B*, filling tube; *C*, tin gasket)
    e, Of silica glass, sealed by welding

prevent sticking by self-welding at high temperatures. Such mechanical closure is convenient, but gas-tight seals are difficult to maintain.

Figure 3(e) shows a type of silica glass container. Upon filling, the tube is attached to a vacuum pump and "pulled off" with a hydrogen–oxygen torch close to the body of the container, retaining the vacuum inside.

Accurate knowledge of the masses of sample and container materials is especially important because the series of runs could have a high inter-consistency that would not reveal large constant or systematic errors in mass. A simple test for a gas leak in the container is to pump around it, then surround it with air one time and helium (or hydrogen) another; with a few cubic centimeters of gas space inside, the difference in weight when there is a leak is easily detectable.

B. Suspending and Dropping the Sample

(1) *Suspension and Dropping Mechanisms*. The sample is suspended in the furnace by means of a suitable wire, the kind and diameter of which depend upon the temperature and the load. Platinum–rhodium (10 or 20 per cent) has commonly been used up to 1500°C. Diameters range between 0·1 mm and 0·4 mm. Tungsten and tantalum have been used at higher temperatures. Southard[33] and many later investigators left the suspension wire attached to the container as it fell into the calorimeter, enabling easy return of the sample to the furnace. The enthalpy of the wire is assumed to cancel out in the measurements of the container when full and when empty. The other common procedure is to melt or cut the wire. One investigator used a magnetic release[12]. Another had an arrangement whereby the wire was made to slip off the container[4].

The presence of two or three thin circular disks of bright platinum attached to the suspension wire just above the sample will reduce radiation losses out of the top of the furnace. In the furnace shown in *Figure 2* such shields are more important in preventing serious heat losses by convection from the hot sample out the calorimeter[10], because there is only one gate and it must be outside the calorimeter proper to dissipate radiation from the furnace. However, in the furnace shown in *Figure 1* the lower gate is inside the calorimeter and therefore prevents such losses of heat from the sample.

If the sample falls unattached into the calorimeter, the impact with the calorimeter may be cushioned by falling on to a deformable body. Southard's procedure of leaving the entire suspension wire attached includes the use of a heavy metal plunger which, attached to the suspension wire, falls through a vertical tube above the furnace. Through the use of holes in the tube there is nearly free fall until the plunger reaches a position corresponding to the point where the sample has entered the calorimeter, when it is slowed by an air cushion. The air escapes through a small hole in the tube near the bottom of the plunger travel. The entire fall (including deceleration) takes less than 0·5 sec. To minimize stretching of the suspension wire, Douglas and Payne[2] arranged for the container to decelerate uniformly at the end of the fall by stretching a pre-set spring that had a stop at the end of the fall.

The addition of weights to the plunger when the empty container is used (to keep the total falling mass constant) will help to prevent a systematic error between the full and empty experiments. It is more important to minimize friction within the plunger tube in order to approach as nearly as possible the acceleration of free fall. Ginnings and Corruccini[10] have estimated that in free fall at 900°C, a typical container with sample would lose not more than one calorie more than the empty container.

(2) *Time in the Furnace*. The length of time that the sample is in the furnace is important and varies with the conductivity and size of the sample. A simple test for temperature equilibrium is to compare several experiments with different times in the furnace and then to select a time which is clearly adequate. Under conditions such that the sample is heated mainly by conduction, it may be shown that approximately

$$k = t/\log_{10}[q_0(q_0 - q)] \qquad (1)$$

in which q is the heat found with time t in the furnace, q_0 is the heat when the time is adequate for reaching virtual equilibrium, and k is a constant representing the time for the temperature difference between the furnace and sample to be reduced to 10 per cent of the original value. A time in the furnace can then be adopted of at least $4k$, since according to equation (1) this time brings heat transfer to the sample to within 0·01 per cent of completion. Actually, the contribution of radiation to the heating of the sample increases rapidly with temperature, with the result that the use of equation (1) gives an upper limit to the time required in the furnace at higher temperatures.

Additional factors, such as proximity to temperatures of melting, transition, or other complex structural changes, may make equation (1) unsuitable.

III. The Calorimeter

1. *The isothermal calorimeter*

An isothermal calorimeter absorbs or evolves heat without a change in temperature, and the amount of heat is measured by an accurately measurable change in some quantity such as volume. Although calorimeters involving vaporization are more sensitive, those involving fusion (particularly of ice) have been more commonly used in drop calorimetry. If the volume change is determined from the mass of displaced mercury, the ratio of heat absorbed to the mass of mercury (K, the calibration factor of the calorimeter) is given by

$$K = \Delta H_m / (v' - v) d_M \qquad (2)$$

in which ΔH_m is the enthalpy of fusion of the melting substance per unit mass, v' and v are its specific volumes in the less and more dense states respectively, and d_M is the density of mercury (all at the calorimeter temperature).

Compared with other calorimeters, the isothermal type has certain advantages and disadvantages well exemplified by the ice calorimeter. Its calibration factor is a universal constant, no temperature or electrical measurements are required in its operation, and when surrounded by an ice bath the heat leak can be made very small and quite constant. However, such a calorimeter is restricted to use at a fixed temperature and must be completely leak-free and so rigid that its volume capacity is extremely constant. The isothermal type of calorimeter is not inherently more precise nor more accurate than the best of the other types, and its construction for accurate work requires a fair amount of time and care; but once properly built, it can yield measurements rapidly and accurately by relatively simple operation.

In this chapter detailed discussion of the isothermal calorimeter will be limited to the ice calorimeter. Isothermal calorimeters employing organic liquids in place of water have also been used. The diphenylether calorimeter, which has been applied to the determination of the enthalpy of metals[14] as well as in reaction calorimetry, differs from the ice calorimeter principally in operating at a higher temperature (26·87 °C), expanding during absorption of heat, and being over three times as sensitive.

A. Design of the Ice Calorimeter

A cross-sectional diagram of an ice calorimeter which has been in use in high-temperature drop calorimetry at the U.S. National Bureau of Standards for many years[6] is shown in the bottom half of *Figure 2*. The calorimeter "well", A, which receives the sample from the furnace, is constructed of some low-thermal-conductivity metal such as Nichrome, with a wall thickness of approximately 0·25 mm. (The slight taper in the tube shown in *Figure 2* is really unnecessary.) The hot sample comes to rest near the bottom of the well, and conduction of unmeasured heat out of the well is negligible. Dry gas (He) flows slowly up the well at all times to keep out moisture, and the top of the well is stoppered loosely when not attached and sealed to the furnace.

The well is really in two sections joined by a gate, G, which is opened briefly to admit the sample but otherwise shields the calorimeter from radiation from the hot furnace. The gate is a tinned copper disk with a hole and a wide-angle slot which moves the sample and its suspension wire to the side of the well when the gate is closed after dropping in the sample.

Around the bottom end of the calorimeter well are two coaxial cylindrical vessels, P, the inner one enclosing the "calorimeter proper" and the outer one constituting the calorimeter jacket. These vessels are commonly of Pyrex glass to permit visual observation as ice is being frozen; however, to reduce somewhat the heat transfer by radiation between them, some experimenters have silvered the surfaces except for a narrow vertical strip.

The two glass vessels are attached at their tops to metal "caps". At the U.S. National Bureau of Standards the outside top of each glass vessel was first ground to true roundness, and the seal to the metal was then made with a thin intervening layer of Apiezon W wax. If the wax layer is too thick, the calorimeter may sag or fall apart when allowed to warm up, but the thickness of the hard wax must be several times the difference in shrinkage of the glass and the metal when cooled. In one case[2], the wax gap was successfully reduced to 0·15 mm by using a low-expansion iron alloy called "Therlo" (29 per cent Ni, 17 per cent Co, 0·3 per cent Mn), which, however, must be protected against corrosion. Some experimenters have replaced the wax joints by less fragile seals. Leake and Turkdogan[22] interposed a winding of string impregnated with Araldite synthetic resin, with compression by a screw collar outside the metal flange. Glass-to-metal seals[28] and O-rings have also been used.

Figure 2 shows a series of horizontal copper disks or vanes, F, which provide good heat paths inside the calorimeter vessel. These disks are conveniently separated by short metal sleeves. In soldering this assembly to the well and subsequently tinning or silver-plating the surfaces to render them inert to water, it is believed desirable to avoid all narrow crevices which might be filled to an irreproducible extent by the water in the calorimeter.

Several experimenters have provided the top metal cap of the calorimeter vessel with a filling tube to introduce the water or replace it. Such a tube (necessarily closed at the top) entails the danger that during the freezing of an ice "mantle", I, the water in it may freeze solid at the level of the cap and burst the tube owing to the expansion accompanying freezing. The use of only one opening to the outside (the "mercury inlet tube"), which normally is completely filled with mercury, obviates this danger. This inlet tube connects the mercury pool, M, in the calorimeter with the mercury accounting system (C, B, V and R) outside. The center portion, T, of this tube, wound around the outside of the outer metal cap, has preferably a capacity for as much ice-cooled mercury (perhaps 10 cm^3) as will ever enter the calorimeter in a single run. The ends of the tube, however, should be of smaller bore, partly to minimize the effect of varying room temperature on the volume of the mercury outside the ice bath. Since solder which would dissolve in the mercury must be avoided, one can use a single tube and draw down its ends.

The mercury-accounting system has two glass capillary branches, one leading through a valve, V, to a weighed beaker of mercury, B, and the other (a capillary, C, of 0·6 mm bore which is preferably calibrated after assembly) being provided with a scale for observing small volume changes when the valve is closed. The capillary is sealed to a mercury reservoir, R, above. For the valve, a hard-steel needle seating *precisely* on soft steel gives satisfactory closure, but a rotationless stem and a softer seat (platinum or Kel-F) may be preferable[2]. Oriani and Murphy[28] substituted a precision-bore glass stopcock. The capacity of the valve must be small and reproducible because one cubic millimeter of mercury is equivalent to almost one calorie of heat.

The calorimeter heat leak can be made so small that it does not require frequent or accurately timed readings. The scale for the capillary outside the calorimeter is located some 20 cm above the mercury level inside the calorimeter so that the pressure head on the water in the calorimeter lowers its freezing point to approximately that of the air-saturated water in the ice bath outside, the heat leak therefore being conveniently small. Some investigators have evacuated the jacket space of the ice calorimeter to reduce heat leak. However, keeping the space filled with a *dry* gas (such as CO_2) at atmospheric pressure has the advantages of decreasing the net downward force on the inside (calorimeter) vessel and of making the volume capacity of the latter nearly independent of changes in atmospheric pressure.

In designing rigid mechanical supports for the calorimeter (not shown in *Figure 2*), some compromise must be made between low flexibility and low-conduction heat paths from the room to the calorimeter. It is also wise to provide sufficient adjustability to facilitate making the calorimeter well and the furnace core coaxial and strictly vertical.

Although modern techniques of thermoregulation permit control of the temperature of a bath to $\pm 0 \cdot 001\,°C$, a simple ice bath, E, accomplishes the same result with no necessary temperature fluctuations and no instrumentation. If wax seals are present, it is convenient to replace the ice bath by circulated cold water during idle periods, but such a bath is likely to be inadequate to preserve any ice mantle present in the calorimeter.

B. Assembly

The two glass vessels are attached and tested for the slightest leak; a helium leak detector is convenient. Pure gas-free water, W, which must be kept in a closed system to prevent absorption of air, is then allowed to fill, by gravity, through the mercury inlet tube if no special filling tube has been incorporated, the dry and thoroughly evacuated inner glass vessel. Unless freezing is begun immediately, some mercury is then drawn into the reservoir to seal the calorimeter water from air.

A suitable amount of mercury is next substituted for an equal volume of water in the calorimeter by alternately freezing ice to expel water and melting the ice to draw in mercury in its place. A closed-bottom tube inserted into the calorimeter well can serve to hold crushed 'Dry Ice' (solid CO_2) to freeze an ice mantle or to contain a stream of warm water to melt it.

When the mercury level inside the calorimeter is above the bottom of the mercury inlet tube, one more ice mantle must be frozen, and this must be large enough to expel a combined volume of water and mercury equal to the inlet-tube capacity plus the largest amount (30 cm³ or more for the calorimeter shown in *Figure 2*) that will be expelled in freezing any future ice mantle. Owing to a lens effect, the ice mantle will soon appear to occupy the entire cross section inside, even when it is far from doing so, and an estimate of the total volume (and hence average diameter) of the ice present is far more reliably provided by the total volume of mercury that has been expelled.

High purity of the water inside the calorimeter is essential for calorimetric accuracy for two reasons. First, any dissolved gas forced out of solution by freezing would tend to redissolve slowly and in so doing vitiate the accurate volume determinations. Second, if the water is impure the calorimeter will absorb, without measuring, heat of the order of a calorie for every 0·001 °C that the calorimeter temperature rises as the ice melts. It is believed possible, however, to remove as much as 99·9 per cent of the dissolved air by a single fractional distillation of the water, and tests of electrical conductivity can be used to detect electrolytic solutes. The mercury used should be of very high purity. As long as the mercury is kept out of contact with metals which it dissolves, it can be separated from surface impurities by the usual process of fractional filtering and can be reused repeatedly in the calorimeter.

C. Calibration Factor

The calibration factor used for the ice calorimeter should not be a value calculated by substitution into equation (2), but one determined by direct electrical measurement. Nevertheless, the same calibration factor should be applicable to all precise ice calorimeters. The following electrically-determined values are "ideal" in the sense that they have been corrected to the basis of no pressure change through a change in the mercury level inside the calorimeter when heat is absorbed.

Ginnings and Corruccini[9], in an extensive investigation of optimum conditions, found 270·42 ± 0·06 abs. J/g Hg; and later Ginnings, Douglas, and Ball[11], using an improved ice calorimeter and more favorable conditions, found from about 100 determinations a mean value which, with a slight correction[6], is 270·48 ± 0·03 abs. J/g Hg. (The tolerances stated are estimates of absolute accuracy.) Leake and Turkdogan[22] later obtained a value of 270·54 abs. J/g Hg, with a standard deviation of ± 0·17. These values agree within their uncertainties, but the value of Ginnings, Douglas, and Ball is believed to be the most reliable.

D. General Procedure of Operation

For heat measurements, the first step is to freeze an ice mantle. There is usually more or less supercooling, so that the first ice appears suddenly, and may branch out so as to *appear* to fill the calorimeter. For precise calorimetry it is recommended that most of this initial ice be melted. A thin layer of ice in the top region of the calorimeter is needed to trap small amounts of heat that might otherwise escape. If the calorimeter is to be operated the same day, a very small layer of ice on all the metal surfaces should be deliberately melted first. The inside of the well is then wiped dry and the entire ice bath is replenished (to the level, *L*, in *Figure 2*). Since ice–water interfaces in the bath must be kept close to the calorimeter, freshly ground clear ice is used, and the interstices are filled with pure water. When the heat leak becomes "normal" and steady (usually within 2 h), enthalpy measurements may begin.

It is believed that ice mantles may be used in accurate heat measurements up to several days after they are frozen, provided the melting of ice at the top of the ice bath during periods of standing (especially overnight) has not allowed much of the upper, protective part of the ice mantle to melt. To maintain a constant heat leak when the calorimeter is in actual operation, it sometimes proves necessary to pack occasionally a little fresh ice in close contact with the emerging calorimeter well. If the ice bath is well insulated (Styrofoam of a few cm thickness serves excellently), one packing of ice in the bath should be sufficient for 12 h or so if the top few cm of ice are replenished often enough. Once an ice mantle is partly melted, it is considered mechanically safer to melt all the ice attached to the inside calorimeter surfaces before freezing a new ice mantle.

When the beaker of mercury is removed from the submerged tip, the mercury in the latter should not recede owing to the presence of air in the valve. If the packing of the stem in the mercury valve is not airtight, it may be necessary to replace the air in the valve with mercury daily. Keeping the inside surfaces of the glass capillary and mercury reservoir clean may prove troublesome, but a few mm of water above the mercury meniscus helps to maintain its proper shape. Oriani and Murphy[28] used a precision-ground glass taper-joint on the capillary to permit easy removal for cleaning. The top part of the mercury in the capillary, where foreign particles tend to accumulate, may be periodically removed by suction. Before each reading of the mercury meniscus, it is advisable to apply enough pressure to the reservoir to depress the mensicus slightly (about 1 mm) so that it will rise to a reproducible level.

To allow mercury from the weighed beaker to enter the calorimeter in a heat measurement, the valve may simply be opened and checked against possible clogging. However, at first, some samples deliver heat to the calorimeter so rapidly that the meniscus would fall below the capillary and allow air to enter the calorimeter unless the pressure in the reservoir were first lowered sufficiently. An alternative procedure is to draw more than enough mercury for the experiment from the weighed beaker into the reservoir and then close the valve. (One of the important precautions in using the ice calorimeter is to avoid introducing heat to the well, even from an object at room temperature, unless the inlet tube is connected to a supply of mercury in the accounting system.) The gate of the calorimeter is then opened long enough to admit the sample.

E. A Typical Experiment

The data and calculations involved in a typical enthalpy measurement with an ice calorimeter are given in *Table 1* and *Figure 4*. Examples of calorimeter misbehavior and numerous small corrections are introduced to show how they are handled.

In this example, a sample of aluminum oxide in a 90%Pt–10%Rh container was measured. (The empty container, also with air sealed in at 600°C and 1 atm, had already been measured in separate runs.) In *Figure 4*

HIGH-TEMPERATURE DROP CALORIMETRY

Table 1. Data and calculations for a typical drop calorimetry experiment with a platinum resistance thermometer and ice calorimeter

Sample ("Al$_2$O$_3$")		Furnace Temperature	
weight	16·8299 g	*Time*	*Thermometer ohms*
buoyancy correction	+0·0026	10:55	77·9094
0·14% SiO$_2$ by analysis	−0·0236	10:57	77·9083
		11:01	77·9078
mass, pure Al$_2$O$_3$	16·8089 g		

Mean: 77·9085 (found)
−77·8733 (=600°C)

0·0352

$dR/dT = 0{\cdot}0802\ \Omega/\deg$ at 600°C
$600 + 0{\cdot}0352/0{\cdot}0802 = 600{\cdot}44$°C
(sample temperature)

Heat measurement

+126·9382 g	(weight of beaker + Hg before drop)
−74·0812	(weight of beaker + Hg after drop)
−13·6333	(corrected mass of Hg, empty container, 600·44°C)
−0·0027	[buoyancy correction for net Hg (sample + container)]
+0·0069	[offset correction, g Hg (see *Figure 4*)]
−0·0006	(correction for 0·0021 g Pt − 10% Rh extra with sample, g Hg)†
+0·0035	(correction for air, 600°C and 1 atm, displaced by sample, g Hg)†
−0·0524	(correction for SiO$_2$ in sample, g Hg)†
−0·0324	(correction to exactly 600°C, g Hg)†
39·1460 g	(corrected mass of Hg for sample at 600°C)

Enthalpy calculation

$(H_{600°C} - H_{0°C})$ (final, pure Al$_2$O$_3$)
= (39·1460) (270·48 + 0·01‡)/16·8089.
= 629·94 abs J/g
($\div 4{\cdot}1840 = 150{\cdot}56$ defined cal/g)

†Computed from reference 17.
‡Calculated correction for rise of mercury level in this calorimeter.

the mercury capillary readings taken when the valve was closed are plotted against time, with a new line through the points whenever the mercury valve was opened and the capillary meniscus reset. (The initial abnormal steepness was traced to moisture in the well that had to be removed.)

The temperature of the furnace was read frequently, just before the drop, with a platinum resistance thermometer. The sample had too much thermal inertia to follow the small short-time drift indicated by the three readings in *Table 1*, so the average was taken.

Eight minutes after the sample was dropped, the mercury valve was closed and the meniscus read, but the negative slopes up to 11:30 suggested that the sample was still evolving heat. The decreased slope after 12:00 was traced to a failure to keep the ice bath well packed on top, and illustrates the error that would have resulted had this later part of the graph been extrapolated back to the time of the drop. The "reset" correction was computed from an empirical calibration factor of 6·3 mg Hg/scale mm, and was subtracted algebraically from the *loss* of mercury in the beaker. To show readily their importance relative to the precision of duplicate

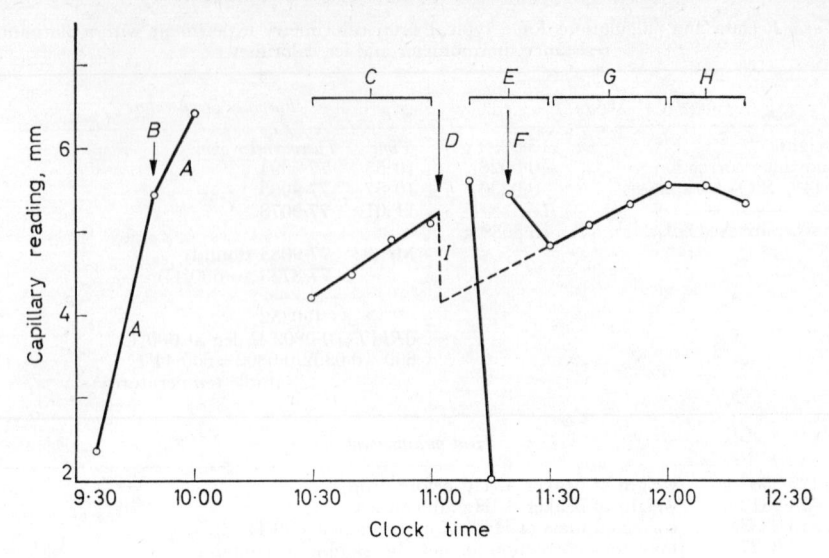

Figure 4. Ice-calorimeter heat leak and reset correction in a typical experiment
          A, Calorimeter "well" wet
          B, Gas flow through well reduced
          C, Normal heat-leak rate before drop
          D, Time of sample drop
          E, Sample not at thermal equilibrium with calorimeter yet
          F, Valve opened and capillary meniscus reset
          G, Near-normal heat-leak after drop
          H, Ice-bath packing deficient
          I, Reset correction = $-1 \cdot 1$ mm.

experiments, which was of the order of 5 mg of mercury, it was convenient to express the corrections in terms of mass of mercury.

In applying the correction for the impurity (SiO_2), it was assumed on the basis of the high-temperature preparation of the sample that the silica was present in the form of an aluminum silicate, and the correction was accordingly computed by subtracting from the enthalpy[17] of Al_2SiO_5 that of an equivalent amount of Al_2O_3.

F. Tests for Accuracy

Overall checks of the accuracy of a particular ice calorimeter used in high-temperature drop calorimetry are afforded by comparing enthalpy measurements on a suitable material with the best published values. Highly pure α-Al_2O_3 (corundum) is currently considered the best single material for comparisons up to above 1500°C; the published values have been referenced and reviewed recently by Kelley[17], and more recent values by Hoch and Johnston[13] (up to 2000°C) are probably the most accurate available at the highest temperatures. However, the use of such differences to establish an empirical calibration of a new apparatus is not recommended in work of high accuracy.

Several comparisons of enthalpy changes obtained by precise drop and adiabatic calorimetry have been made at the U.S. National Bureau of Standards. Values for water over the range 0–250°C agreed within an average of 0·05 per cent by the two methods[6], whereas the results for aluminum oxide over the range 25–107°C were about 0·2 per cent higher by the drop method than by low- and high-temperature adiabatic calorimeters[6]. (The two adiabatic calorimeters agreed with each other to 0·01 per cent on the average[35].) At these relatively low temperatures the adiabatic calorimeters were believed to be several times as accurate as the drop method apparatus. At somewhat higher temperatures the precision of adiabatic calorimetry decreases considerably because of increased radiation effects, and for this reason the two methods were believed to give comparable accuracy at these higher temperatures; however, for the range 242–416°C the difference found for aluminum oxide averaged only 0·05 per cent[35]. At still higher temperatures, the drop method is the more accurate.

G. Hunting Causes of Trouble

The possible causes behind a perplexing symptom of trouble are often numerous. If the cause is likely to be a single one, it is economical to eliminate from initial consideration those causes which can produce effects only of the opposite sign or of a different order of magnitude from that observed. If the only symptom is unsatisfactory precision or accuracy of the enthalpy values, it may not be immediately clear whether the sample, the furnace, or the calorimeter is responsible. Errors arising from the sample and the furnace are discussed elsewhere in this chapter.

When the trouble is in the ice calorimeter itself, one source of error and poor precision is the melting of a hole in the ice mantle. This allows unmeasured heat to escape, but fortunately, such a hole can usually be observed visually, or at least it is usually manifest by an abnormally slow equilibration after the drop. (Occasionally two successive runs have relatively large errors of comparable magnitude but opposite sign, presumably because of a slight irreproducible collapse of the calorimeter like that of the bottom of an oil can.)

A large positive or negative heat-leak rate of the ice calorimeter, apart from its indication of a condition which is likely to be highly variable, is too great to be followed accurately. If there is apparently a large heat leak *into* the calorimeter (falling meniscus), there may be a hole in the ice mantle, the ice bath may need more careful packing, further time may be required for the calorimeter to equilibrate with respect to heat and volume changes, or there may be air bubbles in the calorimeter which are slowly dissolving in the water. (In the last case the compressibility of the calorimeter may be found to have increased considerably.) On the other hand, if there is apparently a large heat leak *out* of the calorimeter (rising meniscus), the calorimeter well may be wet, or so large an ice mantle may have been frozen that some water has replaced mercury in the inlet tube (causing an abnormally high calorimeter temperature).

There are other possible causes of an apparently abnormal heat leak. Fluid leaks of the calorimeter deserve special mention. If water is leaking out of the top of the inner glass vessel, this will be visible in time but will immediately cause a falling (or more rapidly falling) mercury meniscus. On the other hand, if there is leakage at a level higher than the meniscus, fluid will be entering the calorimeter and the meniscus will tend to rise.

Fluid leaks may be verified by standard procedures. A time-consuming procedure is to follow the meniscus readings of a thoroughly temperature-steady ice-free calorimeter. A more rapid method involves brief evacuation above the mercury meniscus (when the mercury column will break), but prolonged evacuation may lead to volume hysteresis which simulates a fluid leak.

2. *The isothermal-jacket, block-type calorimeter*

The isothermal-jacket, block-type calorimeter absorbs or evolves heat with a change of temperature, the change usually not exceeding 4–5°. Although block calorimeters have been designed to operate adiabatically[23], the usual type employs isothermal surroundings. The following discussion is limited to a model of this type.

A. Design and Construction

The block calorimeter shown in *Figure 1* is the one designed by Southard[33], and is being used at present by the U.S. Bureau of Mines. It is essentially a cylindrical block of copper, L, 12·6 cm in diameter and 20·3 cm high. The mass is large enough that the largest heat absorption does not cause a temperature rise of greater than about 5°. Aluminum blocks have been used successfully in place of copper. The block shown is gold-plated and rests on plastic knife-edges that are glued to the bottom of the surrounding jacket. The interior surface of the jacket is also gold-plated and both jacket and block are kept polished to minimize the heat interchange between them. A brass tube (2·5 cm diameter) through which the sample drops, connects the jacket to the furnace. The top of the jacket is removable and is flange-connected to the lower part with machine screws and a Tygon gasket. The jacket and calorimeter are immersed to level R in a stirred oil bath maintained at 29·70 ± 0·005°C. It is more convenient to maintain the temperature of a bath by heating than by cooling or a combination of the two; thus, if room temperature conditions are such, the bath may be more conveniently maintained at a reference temperature higher than the standard temperature of 25°C.

The receiving "well", O, is a removable, tapered copper plug bearing a 100Ω manganin wire heater, P, which is used in the electrical calibration of the heat capacity of the calorimeter. The well is covered by a circular copper gate, M, about 2 cm thick. The gate (solid, except for a 2·5 cm hole) is closed except for the brief interval necessary for dropping the capsule. The bottom surface is machined to fit into a similarly machined recess in the block itself, thus providing good thermal contact and guarding against heat losses by convection. The gate rotates on a shaft eccentric with the center line of the receiving well and is operated manually. The shaft is also connected to a similar, but hollow, gate located just under the furnace. This gate is cooled by a slow stream of water, preventing the intrusion of heat into the calorimeter from the furnace. Both gates have a thin slot of the proper radius to allow the suspension wire to pass through when the gate is closing.

A slow stream of CO_2 (about 50 cm³/min) is introduced at the bottom of the jacket and flows continually around the calorimeter and through the furnace. The procedure prevents condensation of atmospheric moisture on the calorimeter and furnace refractory. It also maintains an oxidizing atmosphere around the furnace thermocouple, which is located within the furnace core. The heat exchange rates between the block and jacket are reduced by the CO_2 to about 60 per cent of those in air. This rate is about 0·002°/min for a temperature difference of one degree between the block and the jacket and is reproducible to about 1 per cent. There is a slight departure from this figure for large temperature rises. This variation is taken into account in the calculations.

The calorimeter resistance thermometer, wound on recess N, is of the transposed bridge type described by Maier[25]. The thermometer is wound so as to cover about one-half of the outer surface of the block and is coated with Bakelite varnish. The winding is then covered by a tapered copper sleeve that makes a driving fit on to the similarly tapered block. It consists of four windings, two of copper and two of manganin wire alternately arranged in the bridge. The two copper coils are wound together in one operation, as are the two manganin coils. The current passing through the thermometer is then exactly divided at all temperatures and the bridge is in balance at only one temperature. The four windings are approximately 210Ω each and have a balance-point at 20·8°C.

The use of this kind of thermometer reduces the number of decades of a potentiometer—in this case three—necessary for temperature measurements, and provides greater temperature sensitivity than a single winding. In this particular case the temperature coefficient is 410 μV/deg with a current of 1 mA. The use of a Leeds and Northrup White 100 000 μV potentiometer with a high-sensitivity galvanometer gives a precision of 0·03 μV or about 0·0001°C. Heat is generated by the thermometer at 0·003 cal/min, which is a negligible quantity. (If a singly wound resistance thermometer of either platinum or copper is used, a bridge such as the Leeds and Northrup G-2 type Mueller bridge should be used for comparable precision.) The voltage drop across the thermometer is 3860 μV at 30·00°C. The heat capacity of the block is 3·691 defined calories per microvolt change in the thermometer. There is a small heat capacity change of the block with temperature.

B. Additional Measuring Equipment

The same potentiometer and galvanometer are used in the electrical calibration of the calorimeter and in measuring the resistance of the manganin heater. Two standard resistances are used, both calibrated by the U.S. National Bureau of Standards. A 0·1Ω resistance is used in the energy-measuring circuit, and a 100Ω resistance is used in determining, and in periodically checking, the resistance of the manganin heater. The standard cell used with the potentiometer is checked periodically against several standard cells used only for this purpose.

The potentiometer working cells and the cell for supplying the thermometer current are 200 ampere-hour lead cells of the charge-retaining type. The energy source for heating the block is composed of a battery of 23 Edison cells with taps to use any number. Experience has shown that the lead cells and Edison cells are equally suited to any of these three uses.

Energy input is made by using a stop watch which is checked at regular intervals against a standard chronometer. If electrical timing is used, care should be taken to see that the line frequency is maintained at 60 cycles; otherwise a proper correction must be applied. Time for attainment of thermal equilibrium by the calorimeter is around 10 min for a calibration and 10–60 min for a calorimetric measurement, depending mainly upon the thermal conductivity of the sample.

C. A Typical Experiment

A typical experiment will serve to illustrate the method of operation of the calorimeter and the calculations involved. During the time that the container and contents are coming to thermal equilibrium in the furnace, the block is set at some desired temperature. This is conveniently done with dry ice and the heater used in the calibration measurements. During the last 20 min or so preceding the drop, the temperature of the calorimeter is read at two-minute intervals and the furnace temperature is checked for constancy. The gates are then momentarily opened and the plunger falls, dropping the sample into the calorimeter. This operation takes about 2 sec and correction is made for the radiation into the calorimeter from the furnace. (This heat gain is negligible below 750°C and does not exceed 0·1 per cent at higher temperatures.) During the time that the calorimeter is heated, its temperature is observed, first at one-minute intervals and then at longer intervals, until equilibrium is established. Finally, readings are taken again at two-minute intervals. *Table 2* gives the record of a typical experiment.

The sample temperature in the furnace is given by the thermocouple reading at 23 min and 23·5 min after the start of the experiment—just prior to dropping the capsule. The constant resistance thermometer current of 1000 μA is checked several times during the warming and cooling rate periods. Adjustment is made with a resistance box connected in series with the thermometer to provide constancy to $\pm$ 0·01 μA. During the equilibration period the change of current is more rapid, necessitating a significant adjustment before the final equilibrium cooling period.

Correction for heat interchange of the block with the thermostat is made with the equation $R = dT/dt = -0\cdot002027\,(T - 3734)$. In this equation, dT/dt is the measured rate in microvolts per minute, $-0\cdot002027$ is Newton's constant, T is the calorimeter temperature in microvolts, and 3734 is the temperature, also in microvolts, at which the heat interchange rate is zero. The magnitude of Newton's constant depends mainly upon the heat capacity of the block, the thermal conductivity of the gas and metal leads between the block and jacket, the exposed area of the block or jacket, and the distance between them.

The remaining calculations of *Table 2* are for the most part self-explanatory. The conversion from microvolts to joules is made by using the heat capacity of the calorimeter (15·445 J/μV) as determined by separate calibration measurement. The conversion from electrical units to defined calories is made by using the relation 1 cal = 4·1840 J. The heat value of the experiment is corrected from the equilibrium temperature (3823 μV) to 30°C (3860 μV) by using 1°C = 410 μV. The heat capacity value of 1·23 cal/deg. for the container and contents is taken from Kelley and King[18]. The enthalpy of the container was obtained from a separate series of measurements. The

HIGH-TEMPERATURE DROP CALORIMETRY

Table 2. Data and calculations for a typical drop-calorimetry experiment with a thermocouple and copper block calorimeter (Sample, La_2O_3)

Thermostat temperature	$29.70 \pm 0.005°C$
Wt. of La_2O_3 (corrected to vacuum)	10.4440 g
Wt. of Pt–10% Rh container (corrected to vacuum)	10.9640 g
Time in furnace (at temperature) before start of experiment	1 h
Resistance thermometer maintained at	1000 ± 0.01 μamp

Record of Experiment

Time (min)	T or RT e.m.f.† (μV)	Time (min)	T or RT e.m.f.† (μV)	Time (min)	Resistance thermometer e.m.f. (μV)
0	9976 (T)	22	3443.78‡(RT)	38	3822§
2	3431.75‡(RT)	23	9980(T)	42	(3823.21)¶
4	3432.93‡(RT)	23.5	9980(T)	44	3822.87¶
6	3434.04‡(RT)	24	(3444.98)*(RT)	46	3822.52¶
8	3435.39‡(RT)	24.5	3613§(RT)	48	3822.15¶
10	3436.57‡(RT)	25	3685§(RT)	50	3821.81¶
12	3437.78‡(RT)	26	3749§(RT)	52	3821.48¶
14	3439.07‡(RT)	27	3781§(RT)	54	3821.15¶
16	3440.19‡(RT)	28	3799§(RT)	56	3820.75¶
18	3441.36‡(RT)	30	3814§(RT)	58	3820.40¶
20	3442.55‡(RT)	34	3821§(RT)	60	3820.02¶
				62	3819.68¶

Calculations

Temperature of container and contents in furnace (9980μV)	1308.5°K
Final resistance thermometer e.m.f.	3823.21 μV
Initial resistance thermometer e.m.f.	3444.98 μV
Thermometer rise	378.23 μV
Correction for time gates were open	-0.20μV
Correction for heat interchange with thermostat, by using e.m.f. vs. time plot for equilibration period, as indicated below	2.27 μV

Time interval (min)	Resistance thermometer e.m.f. (μV)	Heat interchange rate (μV/min)	Correction (μV)
24–24.5	3542	0.390	0.195
24.5–25	3652	0.167	0.084
25–26	3720	0.029	0.029
26–27	3766	−0.064	−0.064
27–28	3790	−0.113	−0.113
28–30	3807	−0.147	−0.294
30–34	3818	−0.170	−0.680
34–42	3822	−0.178	−1.424
			Total −2.27

Corrected temperature rise		380.30μV
Conversion to calories	$\dfrac{(380.30 \times 15.445)}{4.1840}$	1403.86 cal
Correction to 30.00°C	$\dfrac{[(3860 - 3823)1.23]}{410}$	−0.11 cal
Enthalpy of empty container		−419.61 cal
Net enthalpy of La_2O_3		984.14 cal
Enthalpy per mole	$\dfrac{(984.14 \times 325.82)}{10.4440}$	30 700 cal
Correction to 25°C (5×25.9)		130 cal
Final value, La_2O_3 ($H_{1308.5} - H_{298.15}$)		30 830 cal/mole

*Drop Capsule
†T = Thermocouple, RT = Resistance thermometer
‡Warming rate, $R_1 = 0.601$ μV/min
§Period of equilibration
¶Cooling rate, $R_2 = -0.176$ μV/min

final correction of the La_2O_3 enthalpy from 30 °C to 25 °C utilizes the room temperature heat capacity data of Kelley and King[18]. If this datum is not available from some compilation such as the one cited, the measured enthalpy data must be extrapolated. Very little error results from this procedure.

D. Calibration Factor

The calibration factor of the block calorimeter described here has remained substantially unchanged for several years. A single calibration is made for each substance, and its constancy serves as an overall check on the precision measuring equipment.

A block calibration experiment is conducted in the same manner as the run, except that measured electrical energy is supplied in place of the heat from the container. This heat is supplied through the calibrated manganin heater coil, P, surrounding the receiving well. The current passing through the heater is measured on the White potentiometer by measuring the voltage drop across a standard resistor ($0 \cdot 1 \Omega$) in series with the heater. These readings are made alternately with those of the resistance thermometer. A substitute external resistance of the same magnitude as the heater is used to "exercise" the energy battery for about an hour before introducing energy into the block. This procedure provides a total change in current during a calibration of about $0 \cdot 03$ per cent, so that the current values may be averaged to determine the energy input. In case a precision bridge is used to measure temperature, a less precise potentiometer such as the Leeds and Northrup type K-3 is entirely suitable for the measurement of the energy of calibration.

As can be seen from the sample experiment, there is no need to convert from microvolts to degrees. Calibrations are calculated in terms of joules per microvolt and applied to the experiment in that form.

IV. Treatment of the Data

1. *Correcting to standard conditions*

A. Correction to a Basis of a Pure Sample

Frequently a sample contains from several tenths to several per cent of impurities. The thermal behavior of these impurities can be the source of significant error if ignored. A spectrographic analysis will identify and estimate the orders of magnitude of the elements of these impurities if they are different from those of the main sample. An X-ray diffraction analysis will show the crystal form and indicate impurities, including uncombined compound constituents, usually to the extent of about 1 per cent or more. Microscopic examination also is an aid in determining the presence of an impurity phase, particularly amorphous content if this possibility exists. In most cases a quantitative chemical analysis is highly desirable to determine the chemical composition with certainty.

(1) *Correcting for Insoluble Impurities.* For all impurities which are completely immiscible with the sample the thermal corrections are strictly additive. Sometimes such an impurity will melt or undergo transition in the measured range. Proper enthalpy correction removes the resulting apparent discontinuity of enthalpy of the main sample. At high temperatures substances

usually become more miscible, so that a misjudgement based on assumed immiscibility may result in significant error. This consideration stresses the need for taking pains to secure samples of high purity.

Some compilation such as that of Kelley[17] is useful in computing corrections for insoluble impurities.

(2) *Correcting for Soluble Impurities.* Enthalpy corrections for impurities that are wholly or partly in solution in the main substance are almost never strictly additive. There is a variety of combinations which can occur regarding the states of the impurity and the main sample, as well as the solubility of the impurity at the two temperature extremes of the measurement. Corrections for the heat effects involved are often relatively small and frequently neglected, because they are difficult even to estimate. The heat corrections are most important when the impurity is in liquid solution only at the upper temperature. If no heat-of-solution data are available, a good approximation is to apply a correction based on the heat of fusion of the impurity at its melting point. Other cases are estimated by treating the impurity as though it were insoluble. If the correction needs to be refined, special information is required, *i.e.*, information that provides the degree of solubility at the two temperatures of the measurement and the accompanying heat effects.

When impurities are dissolved in a sample that melts in the range of measurements, they cause a lowering of the melting point often called "pre-melting". This subject is discussed at some length in other chapters of this volume. If the amount of dissolved impurities is low, the solution may be considered ideal. In this case the enthalpy correction is approximately

$$\Delta H_{\text{corr}} = -N_2 R T \, T_m/(T_m - T), \tag{3}$$

in which N_2 is the total mole fraction of impurities, R is the gas constant, T is the absolute temperature in question, and T_m is the melting point of the pure sample. If the solution is not ideal so that equation (3) is not applicable, or if the analysis for the proportion of impurities cannot be relied on, the value of N_2 in equation (3) is better determined empirically as the value that will give corrections to make the enthalpy–temperature relation the most plausible one.

This matter of judgement often introduces considerable uncertainty in the corrections and the resulting corrected heat of fusion, even for samples as pure as 99·5 per cent. A sensitive way to examine the results for pre-melting effects is to plot mean heat capacities for successive intervals of furnace temperature and then to look for excessive upturn in the curve below the melting point. The heat capacity–temperature curves of some pure solids show an upward turn as the melting point is approached. This upward curvature can be attributed to lattice vacancies or anharmonic vibrations, but this effect is more gradual than that due to pre-melting. If the empirical application of equation (3) leads to a curve which shows a heat capacity decreasing with temperature, the value chosen for N_2 is too great.

B. Correcting for Vaporization Inside the Container; Correcting to a Different Pressure

If even a small fraction of the sample or one of its components vaporizes inside the sample container in the temperature range of the measurements, the heat involved in the vaporization process may be an appreciable fraction of the total heat measured, and correction for it is usually desired. If the necessary subsidiary information is available, the total correction can be computed from an exact thermodynamic relation given by Osborne[29].

$$[mH]_1^2 = [q]_1^2 + [PV]_1^2 - [(V - mv)L/(v' - v)]_1^2. \tag{4}$$

In equation (4) the superscript 2 and the subscript 1 indicate the value of the bracketed quantity at the higher temperature less that at the lower temperature; m is the total mass of sample, including the vapor; H is the enthalpy, per unit mass, of the condensed phase (solid or liquid) at pressure P, which is its vapor pressure; $[q]_1^2$ is the heat evolved by the sample (after correction for the container and its parts); V is the inside volume of the container; L is the heat of vaporization (or sublimation, if solid) per unit mass at the temperature in question; and v' and v are the volume per unit mass of the vapor and condensed phase, respectively. It is usually more convenient to make a substitution in equation (4) from the exact Clapeyron equation, to obtain

$$[mH]_1^2 = [q]_1^2 + [PV]_1^2 - [(V - mv)T\,dP/dT]_1^2. \tag{5}$$

If the vapor pressure and its temperature coefficient are known with sufficient accuracy (which must be high if P and consequently dP/dT are large), it is sufficient to apply equation (5) to measurements on a single sample, with the precaution to minimize the gas space $(V - mv)$ (after due allowance for thermal expansion of the sample) in order to minimize the last correction term. However, an alternative procedure is available if the vapor pressure data are not sufficiently reliable. The most uncertain quantity in equation (5) can be eliminated by making heat measurements on both a "high" and a "low" filling of the container and applying the equation to each. Sufficient sample must be used in the "low filling" experiments to ensure that not all the condensed phase is evaporated.

As an example, Douglas et al.[1] measured the enthalpy (relative to 0°C) of n-heptane at 250°C, at which temperature the vapor pressure is about 21 atm, using "high" and "low" fillings in which the liquid occupied 83 per cent and 27 per cent of the container, respectively. Although in the case of the "low" filling experiments the net correction calculated from equation (5) amounted to about 12 per cent of the total heat measured (the correction was far smaller for the "high" filling), the two separately corrected liquid enthalpies differed by only 0·1 per cent. Each value was the mean from several duplicate experiments.

If the pressure on the sample varies with temperature either because of a rapidly changing vapor pressure or because inert gas is sealed in at some temperature and pressure, it may be desirable to correct the enthalpy increment for the condensed sample, calculated from equation (4) or (5), to the basis of a given fixed pressure, such as a standard state of 1 atm. The correction may be computed for both terminal temperatures from the exact thermodynamic relation

$$(\partial H/\partial P)_T = V - T(\partial V/\partial T)_P \tag{6}$$

in which V and H apply to the same quantity of sample. For large pressure differences, equation (6) requires integration if the thermal expansion data are accurate enough to justify it. For a change of pressure of only a few atmospheres, however, the correction is very small (less than 0·1 per cent). The calculation of C_V from C_P, which are thermodynamically related by

$$C_P - C_V = a^2 V T / \beta \qquad (7)$$

in which a is the volume coefficient of thermal expansion and β is the compressibility, is a special case of correction to different pressures.

2. Smoothing and representing enthalpy values

Measurements of relative enthalpies over a range of temperature may be presented by graph, equation, or tabulation.

A. Graphing

Although graphing is not generally considered to be a satisfactory way to represent the final data, it is a good aid in smoothing the data and finding a suitable form of empirical equation or its coefficients. For these purposes it is not suitable to graph the enthalpy data directly; instead, functions of enthalpy are plotted. These function plots must be sensitive enough to show deviations from smooth continuity which are within the precision attained by the measuring technique.

Generally, some form of heat capacity function is most convenient for plotting the measured data. This procedure not only provides a good basis for smoothing the data, but also shows continuity with any existing low-temperature heat capacity data directly and provides a precise means of comparison with other work.

The heat capacity curve derived from the smoothed enthalpy data is often plotted together with unsmoothed heat capacity values corresponding to mean heat capacities given directly by $\Delta H / \Delta T$ for pairs of successive furnace temperatures. If the heat capacity does not vary linearly with temperature, a curvature correction is needed to convert mean values to true heat capacities; the curvature correction is given by the infinite series according to Osborne et al.[30]

$$C - (q/\Delta T) = -(\partial^2 C/\partial T^2)(\Delta T)^2/24 - (\partial^4 C/\partial T^4)(\Delta T)^4/1920 - \ldots \qquad (8)$$

in which C is the true heat capacity, q is the heat interval for the sample over the temperature interval ΔT, and C and its derivatives apply at the mean temperature of this interval. Even for a temperature interval as large as 100 °C, the first term of the right-hand side of equation (8) is normally only a very small fraction of the total heat capacity and the remaining terms are completely negligible. If this is not true because the curvature is unusually great, measurements should be made at one or more intervening temperatures. If the enthalpy–temperature relation is represented analytically, it is convenient to compute the derivatives in equation (8) from that relation. If the heat q is that for the sample at constant pressure, q may be replaced by ΔH and C by C_P.

Another method is to plot the quotients of the individual enthalpies divided by the corresponding temperature intervals of the measurements. The value of the ordinate is the mean heat capacity between the furnace temperature and the standard reference temperature. Thus if 298°K is the reference temperature, the ordinate of this curve at 298°K is the true heat capacity at this temperature. Any existing low-temperature data can easily be calculated in this form to show continuity with these data. This comparison is important, because almost always the low-temperature heat capacity data are more accurate than the high-temperature data. If $(H_T - H_{298})/(T - 298)$ is the mean heat capacity as defined here, then the true heat capacity is given by the relation

$$C_P = (H_T - H_{298})/(T - 298) \\ + (T - 298)d\,[(H_T - H_{298})/(T - 298)]/dT. \qquad (9)$$

(1) *Choice and Treatment of Measurements.* In the process of smoothing data, the choice of the number of measurements over the same temperature interval requires some decision as to how such duplicates will be handled. It is the usual practice to repeat measurements for which agreement is unsatisfactory. This raises the question as to which individual measurements to discard, and of those remaining, which will receive the most weight. Statistical methods have been developed to show when a single measurement can be discarded objectively. Most of these methods depend upon the assumption that the probability of encountering a single measurement with such a large deviation from the mean is small. This mean can be considered as the mean value of several duplicate experiments or as a smooth curve such as the heat capacity curves discussed. Of the measurements remaining for each temperature range, it is a good procedure to treat them equally in getting the mean value, and then to treat these mean values equally in deriving smooth values or an empirical equation to represent them. Additional measurements for any one temperature range should be chosen primarily so as to give statistical precision to the mean value of enthalpy or its derivatives. However, the existence of systematic errors usually makes a large number of repetitive experiments of limited value.

B. Representation by Equations

(1) *Theoretical and Semi-Theoretical Equations.* For the crystalline state, the Debye treatment for intermolecular vibrations is often coupled with the Einstein treatment for intramolecular vibrations. This procedure has given a representation of the C_V–temperature curve so closely approximated by experimental results for many substances as to be useful, especially at low temperatures for which the difference between C_P and C_V is generally small. However, the limiting value $(3R)$ of the Debye and Einstein functions is generally exceeded over the high-temperature range. In this case calculations of heat capacities are so complicated by such factors as anharmonic vibrations, electronic heat capacity, and the difference between C_P and C_V that such calculations are seldom made.

(2) *Empirical Equations.* Empirical equations for enthalpy may necessitate more labor in obtaining values at odd temperatures than from tables, and

they are usually less precise than the smooth values given in tables; but they have the advantage of affording an easy means of deriving related thermodynamic properties such as heat capacity and entropy increments. In addition, if the form of the enthalpy equation is the standard one suggested by Maier and Kelley[26], the constants in the calculation of chemical reactions may be expeditiously combined. This equation is

$$H_T - H_{298\cdot 15} = aT + bT^2 + cT^{-1} + d \tag{10}$$

A procedure for evaluating the constants of this equation was given by Shomate[31]. If the heat capacity at 298·15°K was accurately known from low-temperature calorimetry, Shomate used the following equations

$$C_{P,298\cdot 15} = a + 596\cdot 30b - c/(298\cdot 15)^2 \tag{11}$$

$$0 = 298\cdot 15a + (298\cdot 15)^2 b + c/298\cdot 15 + d \tag{12}$$

$$[(H_T - H_{298\cdot 15}) - C_{P,298\cdot 15}(T - 298\cdot 15)]\,T/(T - 298\cdot 15)^2$$
$$= bT + c/(298\cdot 15)^2 \tag{13}$$

The function on the left side of equation (13) is evaluated for each measured high-temperature enthalpy value (or for evenly spaced, smooth-curve values), and the results are plotted against T. The slope and intercept of what is judged to be the best straight line give values for b and c respectively. If c is nearly zero, the line is sometimes drawn to make it exactly so. The other constants are then supplied by equations (11) and (12). If the substance has a melting point or transition in the measured temperature range, the same procedure can be applied above and below the temperature of discontinuity. Otherwise, four simultaneous equations are used to determine the constants. Frequently the equation form can be simplified above a temperature of discontinuity. This is particularly true if the discontinuity is above about 600°K.

The general adoption of equation (10) for all substances has the advantage of fitting data reasonably well while minimizing the number of terms. In addition, the error in using the equation to extrapolate the data to higher temperatures, while generally considerable, is often less than with many other forms of equation. However, there are the disadvantages that in many cases this equation form does not fit the precise data at all temperatures within the precision of the measurements or does not give weight to the temperature derivative of C_P at 298·15°K indicated by precise low-temperature calorimetry.

A better fit of an equation to the data may be obtained for many specific cases by choosing a different temperature function with the same or a greater number of constants. The number of constants should not be increased to the extent that implausible points of inflexion appear. The labor of fitting coefficients of an empirical equation by the least squares method is greatly reduced by the present-day availability of high-speed electronic computers and the ability to develop general-use codes, each of which includes any desired combination of temperature terms. A least squares method has the advantage of mathematical objectivity, but loss is encountered through the difficulty of selecting the temperature range of "best fit".

Unless an equation is found which fits the enthalpy measurements within their precision at all temperatures, tabular values should be used for the most precise calculations.

C. Treatment of Fusions and Transitions

After correcting for any pre-melting, the enthalpy of a pure substance as a function of temperature is extrapolated to the melting point for both the solid and the liquid. If the enthalpies of both the solid and the liquid are expressed relative to the same state at the same temperature, the difference is the heat of fusion. The accuracy of the heat of fusion so obtained will depend upon the uncertainties of these extrapolations, the correction for "pre-melting", and the melting point. Since the heat capacities of the solid and liquid are not too far different at the melting point, an error of several degrees in the melting point has little effect on the resulting heat of fusion.

Unless the melting point is already known, it may be determined directly in the furnace by the cooling curve method, or by using the enthalpy data as a basis. In the latter case the melting point is taken as the lowest temperature at which the enthalpy measurement of the pure material falls on the liquid curve.

Solid–solid transitions are classed as first and second order according to whether, at equilibrium, they are abrupt and isothermal or gradual over a range of temperature. A first order transition is sometimes so sluggish that it is difficult to distinguish between the two types by drop calorimetry. These cases are spread out as an enhanced heat capacity. Sluggish heats of transition may sometimes be assigned arbitrarily as isothermal transitions for convenience in deriving the empirical equations. In any case, if the range of interest is above the range of transition, the enthalpies will be unaffected by the interpretation of the transition range, and the entropy calculations are likely to suffer little error. In efforts to measure enthalpies of substances in the neighborhood of either transition points or melting points, it may be necessary to greatly increase the length of time in the furnace. Also it is sometimes necessary to "overshoot" the temperature of transformation and then lower the temperature in order to increase the rate of transformation.

If a latent heat of transition is quite small, its existence may be hard to detect. An enthalpy-function plot such as the $\Delta H/\Delta T\, vs\, T$ plot mentioned earlier is of considerable value in detecting and defining them. Shomate[32] describes another type of plot used for this purpose; it is based on equation (13).

3. Derivation of other thermodynamic properties from relative enthalpy

The following common thermodynamic functions can be readily derived by the indicated mathematical operations on the enthalpy equation; the values are those at temperature T unless otherwise indicated

$$C_P^\circ = [\partial(H^\circ - H^\circ_{298.15})/\partial T]_P \tag{14}$$

$$S° = \int_{T'}^{T} (C_P°/T) dT + S_{T'}° \qquad (15)$$

$$-(G° - H°_{298\cdot 15})/T = S° - (H° - H°_{298\cdot 15})/T. \qquad (16)$$

Sometimes $(H°_{298\cdot 15} - H°_0)$ is known from low-temperature calorimetry; in such cases the enthalpy and free-energy functions are often expressed relative to 0° instead of 298·15°K. For high-temperature thermodynamic applications, however, it is much more important to be able to substitute into equation (15) an accurate value of the integration constant $S_{T'}°$ (at some temperature T' within the range of the high-temperature enthalpy measurements). In this case, equation (16) will yield free-energy functions, the use of which predicts equilibrium relations at high temperatures without the need for their direct measurement. A value for $S°_{298\cdot 15}$ is often available from low-temperature calorimetry on the substance in question, and assumption of the Third Law of Thermodynamics.

If for some reason the enthalpy is represented only approximately by an empirical equation, the deviations of the unsmoothed values from the equation may be plotted and an empirical curve drawn through them. A smooth value of enthalpy at any temperature in the range can then be obtained by adding algebraically the ordinate of the deviation plot to the value given by the equation. A value of the heat capacity can also be obtained by adding the slope of the deviation plot to the temperature derivative of the (approximate) enthalpy equation. A variation for determining heat capacity is given by equation (9). Usually the contribution of the second term is only a few per cent of the first term, and the error in its graphical determination thus contributes only a small percentage error to the C_P so evaluated.

Kelley[17] describes a correction in the calculation of entropy based on the standard equation form. The corrections are applied incrementwise every 100°C as $q/T_{av.}$, in which q is the enthalpy deviation of the equation from the smooth values for the temperature interval, and $T_{av.}$ is the mean temperature of the interval.

4. *Precision and accuracy*

Values derived from drop measurements are often given accompanied by tolerance figures, but it is sometimes not clear whether precision or estimated accuracy is meant. In drop calorimetry the systematic errors are the important ones, so that the precision, as calculated by standard statistical procedures, is not as large as the absolute uncertainty of the results. Some estimate of the systematic error should be made to combine with the precision in arriving at a value of estimated uncertainty. Examples of experimenters who have presented such detailed analysis of their errors are Ginnings and Corruccini[10], Furukawa et al.[6], and Hoch and Johnston[12]. Particular sources of error are measurement of the temperature of the sample; calibration of the calorimeter and other instruments such as standard cells, potentiometers, thermometer bridges, thermocouples, and pyrometers; the calorimetric measurement, including allowance for heat leak corrections; sample impurities and mass changes; empty-container measurements; and the loss of unmeasured heat by the sample during the drop and in the calorimeter.

Some investigators will combine possible errors statistically and then multiply the result by some arbitrary factor such as two, whereas others will combine possible errors with the signs taken to give the largest overall error. Comparison with work of different laboratories is an aid in error estimation.

The uncertainty of the heat capacity is difficult to estimate. The process of differentiation of enthalpy may multiply the percentage uncertainty characteristic of the enthalpy by a factor of two to much more if the heat capacity itself is changing rapidly with temperature. The uncertainty of the derived entropy may be calculated from the equation

$$S_T - S_{298.15} = (H_T - H_{298.15})/T + \int_{298.15}^{T} [(H_T - H_{298.15})/T^2]\,dT. \qquad (17)$$

By substituting in the uncertainties of the enthalpy measurements and integrating the second part graphically, the entropy uncertainty may be determined. It is a close approximation to assume the same percentage uncertainty for the entropy increment as for the measured enthalpy increment.

Precise calorimetric data gain maximum value not only when the uncertainties can be estimated and stated, but also when the results are reported in sufficient detail to permit future corrections to less arbitrary procedures of data treatment or to more accurate physical constants.

V. References

[1] Douglas, T. B., G. T. Furukawa, R. E. McCoskey, and A. F. Ball, *J. Res. Natl. Bur. Std.*, **53**, 139 (1954).
[2] Douglas, T. B., and W. H. Payne, in S. F. Booth, ed., Natl. Bur. Standards Handbook 77, *Precision Measurement and Calibration*, Vol. II, *Selected Papers on Heat and Mechanics*, U.S. Govt. Printing Office, Washington, 1961, pp. 241–276.
[3] Dworkin, A. S., and M. A. Bredig, *J. Phys. Chem.* **64**, 269 (1960).
[4] Ferrier, A., *J. Sci. Instr.* **39**, 233 (1962).
[5] Fomichev, E. N., V. V. Kandyba, and P. B. Kantor, *Izmeritel. Tekhn.* **1962**, No. 5, 15.
[6] Furukawa, G. T., T. B. Douglas, R. E. McCoskey, and D. C. Ginnings, *J. Res. Natl. Bur. Std.*, **57**, 67 (1956).
[7] Gilbert, R. A., *J. Chem. Eng. Data*, **7**, 388 (1962).
[8] Ginnings, D. C., *J. Phys. Chem.*, **67**, 1917 (1963).
[9] Ginnings, D. C., and R. J. Corruccini, *J. Res. Natl. Bur. Std.*, **38**, 583 (1947).
[10] Ginnings, D. C., and R. J. Corruccini, *J. Res. Natl. Bur. Std.*, **38**, 593 (1947).
[11] Ginnings, D. C., T. B. Douglas, and A. F. Ball, *J. Res. Natl. Bur. Std.*, **45**, 23 (1950).
[12] Hoch, M., and H. L. Johnston, *J. Phys. Chem.*, **65**, 855 (1961).
[13] Hoch, M., and H. L. Johnston, *J. Phys. Chem.*, **65**, 1184 (1961).
[14] Hultgren, R., P. Newcombe, R. L. Orr, and L. Warner, *N.P.L. Symposium* No. 9 (1958), *Met. Chem.*, H.M.S.O. (1959), p. 1H.
[15] Jewell, R. C., E. G. Knowles, and T. Land, *Metal Ind. (London)* **87**, 217 (1955).
[16] Kantor, P. B., R. M. Krasovitskaya, and A. N. Kisel, *Fiz. Metal. i Metalloved.* **10**, No. 6, 835 (1960).
[17] Kelley, K. K. "Contributions to the Data on Theoretical Metallurgy. XIII. High-Temperature Heat-Content, Heat Capacity, and Entropy Data for the Elements and Inorganic Compounds", *U.S Bur. Mines Bull.* 584 (1960).
[18] Kelley, K. K., and E. G. King, "Contributions to the Data on Theoretical Metallurgy. XIV. Entropies of the Elements and Inorganic Compounds", *U.S. Bur. Mines Bull.* 592 (1961).

[19] Kelley, K. K., B. F. Naylor, and C. H. Shomate, *U.S. Bur. Mines Technical Paper*, No. 686 (1946).
[20] Kirillin, V. A., A. E. Sheindlin, and V. Y. Chekhovskoi, *Dokl Akad. Nauk SSSR* **135**, No. 1, 125 (1960).
[21] Kirillin, V. A., A. E. Sheindlin, and V. Y. Chekhovskoi, *Dokl. Akad. Nauk SSSR*, **139**, No. 3, 645 (1961).
[22] Leake, L. E., and E. T. Turkdogan, *J. Sci. Instr.*, **31**, 447 (1954).
[23] Levinson, L. S., *Rev. Sci. Instr.*, **33**, 639 (1962).
[24] Lucks, C. F., and H. W. Deem. "*Thermal Properties of Thirteen Metals*", ASTM Special Technical Publication No. 227, American Society for Testing Materials, Philadelphia, Pa. (1958).
[25] Maier, C. G., *J. Phys. Chem.*, **34**, 2860 (1930).
[26] Maier, C. G., and K. K. Kelley, *J. Am. Chem. Soc.*, **54**, 3243 (1932).
[27] Olette, M., *Compt. rend.*, **244**, 1033 (1957).
[28] Oriani, R. A., and W. K. Murphy, *J. Am. Chem. Soc.*, **76**, 343 (1954).
[29] Osborne, N. S., *Natl. Bur. J. Res.* **4**, 609 (1930).
[30] Osborne, N. S., H. F. Stimson, T. S. Sligh, and C. S. Cragoe, *Bur. Standards Sci. Pap.*, **20**, 65 (1925).
[31] Shomate, C. H., *J. Am. Chem. Soc.* **66**, 928 (1944).
[32] Shomate, C. H. *J. Phys. Chem.* **58**, 368 (1954).
[33] Southard, J. C., *J. Am. Chem. Soc.*, **63**, 3142 (1941).
[34] Wensel, H. T. and W. F. Roeser. In *Temperature, Its Measurement and Control in Science and Industry*, Vol. I, American Institute of Physics, Reinhold Publishing Corp., New York, 1941, pp. 284–314.
[35] West, E. D., and D. C. Ginnings, *J. Res. Natl. Bur. Std.*, **60**, 309 (1958).
[36] White, W. P., *The Modern Calorimeter*. Chemical Catalog Co., New York, 1928.
[37] White, W. P., *J. Am. Chem. Soc.*, **55**, 1047 (1933).
[38] Wisely, H. R., *Thermocouples for Measurement of High Temperature*, Ceramic Age, Newark, N.J. (1955).

CHAPTER 9

Adiabatic Calorimetry From 300 to 800°K

E. D. WEST

National Bureau of Standards, Washington, D.C., U.S.A.

EDGAR F. WESTRUM, JR.

University of Michigan, Ann Arbor, Michigan, U.S.A.

Contents

I.	Introduction	333
	1. Methods of Calorimetry above 300°K	333
	2. Comparison with Low-Temperature Adiabatic Calorimetry	335
II.	Heat Exchange in Adiabatic Calorimetry	335
III.	A High-Precision Adiabatic Calorimeter Utilizing Intermittent Heating over the Range 300 to 800°K	338
	1. The Calorimeter (Sample Container)	339
	2. The Calorimeter Heater	341
	3. The Adiabatic Shield	342
	4. Adiabatic Shield Temperature Control	345
	5. Environmental Control	347
	6. Measurement of Electrical Power	348
	7. Experimental Procedures	349
	8. Determination of Enthalpies of Phase Transitions	350
	9. Determination of Transition Temperature	352
IV.	An Adiabatic Calorimeter for Thermal Measurements from 250 to 600°K	353
	1. Intermediate-Temperature Thermostat	353
	2. The Silver Calorimeter	356
	3. Operational Method	358
	4. Calibration of Calorimeter	359
V.	General Discussion	360
VI.	Survey of Adiabatic Calorimeters in the Intermediate and Higher Temperature Ranges	361
VII.	Acknowledgment	365
VIII.	References	365

I. Introduction

This chapter is concerned primarily with a general discussion of adiabatic calorimetry above room temperature. Two modern adiabatic calorimeters are described here in some detail, and a terse summary of the characteristics of other adiabatic calorimeters described in the literature is presented in tabular form.

1. *Methods of calorimetry above* 300°*K*

Two methods are widely used for making heat capacity measurements above room temperature. One, the *method of mixtures*, or *drop calorimetry*, discussed in Chapter 8, obtains the enthalpy of a sample as a function of temperature t. The enthalpy data are usually represented by a power series in t and differentiated to obtain the heat capacity and thence the free energy and entropy. This method, which can be relatively simple and accurate and which is applicable to very high temperatures, depends on return of the calorimetric sample to the same initial state after each drop from the furnace. This method, however, may be inapt for a sample which undergoes one or more transitions in cooling, for it may not be completely transformed in the calorimeter; that is, a variable amount of energy may be frozen into the sample.

In the study of transitions and samples which do not reach a reproducible state in a reasonable time, *adiabatic calorimetry*, the second general method, may be used to advantage. Ideally, the adiabatic shield is always maintained at the temperature of the calorimeter so that there is no temperature differential and hence no heat exchange between calorimeter and surroundings. In such a calorimeter, temperature changes take place comparatively slowly, not more than one transition need be involved in an experiment, and equilibration takes place at the highest temperature of the experiment, at which the rate of conversion should be most favorable.

Two theoretically equivalent but practically different methods of operation are used in adiabatic calorimetry. In the *continuous-heating method*, the calorimeter and shield are heated at constant power input and the times recorded at which the thermometer indicates selected temperatures. The energy input is determined from concurrent measurements of time and power. Division of this energy by the temperature rise gives the average heat capacity of the calorimeter plus the sample; the average heat capacity of the sample is obtained by subtracting the experimentally-determined heat capacity of the empty calorimeter. The chief advantages of this method are that more data can be taken in a given time, and the problem of controlling the adiabatic shield at the temperature of the calorimeter is somewhat simplified.

In the *intermittent-heating method*, the temperature of the calorimeter is measured at equilibrium with no power input. The calorimeter is then heated at a constant power for a known time interval. After power is turned off, the temperature is again measured when the transient effects of heating have become negligible. Because of the sudden change in the power requirements of the adiabatic shield at the beginning and end of the heating period, the shield control is somewhat more difficult. Intermittent heating is particularly useful, however, when the sample undergoes transitions or fusion, or when a slow process, such as annealing, is to be observed. It has the considerable advantage that the melting point and purity of the sample can often be estimated from fractional melting, as described in Chapter 5. Such data provide information about the sample as it was actually used and may be used to check on possible contamination by reaction with the container. A considerable advantage of the intermittent- over the continuous-heating method is that the "zero" heat leak is determined for each experiment at the time of the experiment. This technique guards against day-to-day changes in the "zero" heat leak which might affect the precision and accuracy of the data.

In contrast to the adiabatic method, in which the exchange of heat to the surroundings is minimized, is the *heat-leak calorimeter*, in which thermal exchange is used directly in evaluating absorption or evolution of the energy of the calorimetric sample. Both *radiation* and *conduction* types of calorimeters have been described[1, 26, 55, 71, 75, 77, 99]. Here the rate of change of calorimeter temperature is inversely proportional to the heat capacity of the calorimeter and its contents, provided that a constant temperature increment between calorimeter and surroundings is maintained, that thermal gradients within the calorimeter and shield may be neglected, that the thermal leakage modulus is constant, and that processes occurring in the calorimeter have a constant relaxation time. The extreme simplicity of these calorimeters is their main virtue; thermal effects are obtained from the integration of the calculated heat leak.

The heat-leak principle has been exploited further in the ingenious microcalorimeters of Tian[78, 79], developed more extensively by Calvet[16] for the calorimetry of very slow processes, and recently modified for use at high temperatures by Bros[15], Kleppa[38], and others.

2. *Comparison with low-temperature adiabatic calorimetry*

Theoretically, an adiabatic calorimeter for use above room temperature should not differ from that used below room temperature, but there is a considerable practical difference. Low-temperature calorimeters work in vacuum, which eliminates heat transfer by gas conduction; however, this advantage is nullified at higher temperatures by the increased heat transfer by radiation. The larger coefficient for radiant heat exchange between the calorimeter and the adiabatic shield requires correspondingly more precise control of the temperature difference for comparable uncertainty in the heat capacity data.

The choice of construction materials is more severely limited as the maximum operating temperature is increased. The melting point of some materials, such as soft solder, may be exceeded. Similarly, varnish or a substitute (for obtaining good thermal contact of heaters and thermocouples with the metallic surfaces on which they are mounted) cannot be used to high temperatures. The lack of such a material usually results in an increased lag in the response of these elements to temperature changes and in the increased difficulty of adiabatic shield control. Oxidation must be avoided, partly because it changes the emissivity of a surface and thereby the heat transfer coefficient. A few of the low-temperature problems, however, can be avoided: large variations in the heat capacity of the sample, larger thermal conductivity of copper leads at low temperatures, and refrigerant baths of liquid nitrogen, hydrogen, or helium.

The techniques common to both low and high ranges are the energy and temperature measurements, automatic shield control, and the treatment of data discussed in Chapter 5.

II. Heat Exchange in Adiabatic Calorimetry

The energy measurements for adiabatic calorimetry — voltage, current, and time — and the increase in the calorimeter temperature can usually be made with less uncertainty than the uncertainty in heat exchange between the calorimeter and adiabatic shield. A qualitative understanding of heat flow in an adiabatic calorimeter is therefore important in understanding the design and operation of the apparatus.

If the outer surface of the calorimeter and the inner surface of the shield are good thermal conductors, the temperature can be treated as approximately constant over each surface. If, in addition, the condition is imposed (by manual or automatic control) that some point on the inner surface of the shield has the same temperature as some point on the outer surface of the calorimeter, the two surfaces will be at the same temperature and no heat will flow between the calorimeter and the shield. If these conditions were exactly satisfied, the calorimeter would be truly adiabatic.

In practice, the temperature of such a calorimeter changes slowly with time though there is no electrical power input to the calorimeter, even when the temperature difference between the shield and the calorimeter is controlled to zero. One explanation of this heat leak is that there are gradients on the calorimeter and shield so that the average temperatures are not matched even when the observed temperature difference is zero.

It is not difficult to imagine exaggerated gradients in an adiabatic calorimeter such that, during the heating period, heat will be transfered from some part of the calorimeter to some part of the shield. Since that part of the shield receives heat from the calorimeter, its heat capacity is logically part of that of the calorimeter, The calorimeter boundary is even more nebulous in the case of electrical leads and supports between the calorimeter and the adiabatic shield. Dickinson[20] has discussed this ambiguity in connection with an isoperibol calorimeter. According to Bridgman[13], if a calorimeter boundary is defined, there must be a corresponding set of operations for measuring the boundary. No such operations are set forth for calorimetry, but operations for determining the energy equivalent are rather carefully prescribed. Actual practice, then, emphasizes the fact that the energy equivalent is the important quantity; the "boundary" of a calorimeter is a meaningless term in Bridgman's sense.

If the gradients on the calorimeter and shield are the same in the experiment and in the corresponding determination of the energy equivalent, any heat exchange due to the gradients is accounted for. The determination of the energy equivalent is therefore a powerful technique for accounting for small amounts of heat which cannot be detected in any other way. In heat capacity work, measurements with the empty calorimeter account for heat exchange due to temperature gradients. The design criteria for calorimeters which have equivalent gradients in the full and empty experiments can be deduced from the overall heat flow problem discussed below.

In an ideal calorimeter with no power input, there is no heat leak from the calorimeter to the adiabatic shield when the temperature difference is zero. In real calorimeters under these conditions, there is usually a small but finite heat exchange which we will call the "zero" heat leak to distinguish it from heat exchange due to other causes and to point up its paradoxical nature. Two possible causes of "zero" heat leak are (i) there may be a spurious e.m.f. in the shield control circuit which causes the shield temperature to be slightly different from that of the calorimeter, or (ii) the power supplied to the shield to maintain its temperature above that of the environment flows from the shield heater through the shield to the surface, where it is lost to the environment. The gradient thus established causes the temperature of the inner surface of the shield to be generally different from that of the control thermocouple. Thus, even with no spurious e.m.f., the real shield temperature is not, on the average, the same as the calorimeter temperature, and a net heat flow is the result. As the temperature difference between the shield and the environment increases, the power in the shield increases so that the "zero" heat leak should increase. Before the correction for this heat leak is considered, some comments are in order regarding the partial differential equations describing heat flow.

Heat flow equations have been formulated by West[87] describing calorimeters composed of a number of different materials in which heat is transferred only by conduction and radiation. With the approximations that the heat capacity per unit volume and thermal conductivity are independent of temperature, the equations form a linear system to which the principle of superposition of solutions (discussed in Chapter 4) may be applied. The temperature can then be treated as a sum of "temperatures" which are due to the heat flow from the shield, constant power in the calorimeter heater, or such other effects as must be taken into account.

Most calorimetric methods are based on the assumption, expressed or implied, that the principle of superposition can be used in describing the heat flow. For instance, the rate of change of the calorimeter temperature in fore and after periods is used to make a heat leak correction for the reaction or heating period. Such a correction assumes that there is no interaction of the heat flow due to either reaction or heating with the heat flow due to the heat leak.

In the mathematical analysis[87], the principle of superposition is used to express the temperature as a sum of contributions from various sources. One term is used to describe the temperature distribution present throughout the experiment due to sources other than electrical heating, which causes the "zero" heat leak; another describes the temperature distribution due to heating at a constant rate. In the case of intermittent heating, two additional terms are used to describe the effects of the initial and final transients when the calorimeter heater is turned on and off.

When equations with initial and boundary conditions are set down for each temperature term, it is found that the energy contributions from the transient terms in the intermittent heating are equal and opposite. If the experiment is sufficiently long that transients are damped out, their integral effect on the observed heat capacity is negligible.

The spurious e.m.f.s in the control circuit can have two origins. If they originate outside the constant-temperature environment, they contribute either constant or random effects (or both) to the heat transfer between the calorimeter and jacket. The random effects will appear as scatter in the heat capacity data, but the constant effects will appear as constant contributions on the "zero" heat leak and are allowed for in the correction for this heat flow. Spurious e.m.f.s caused by inhomogeneities within the environment can be treated as temperature differences summed from the various terms; the contribution of each term can be considered with the equations for that term alone and are accounted for by the heat-leak corrections and by the full- and empty-calorimeter measurements insofar as they are the same for both. Since the transients cancel, heat exchange between calorimeter and shield due to the continuous-heating term is the same for the intermittent- and continuous-heating methods. It is found that this contribution to the temperature is proportional to the heating rate, whereas the time required to heat the calorimeter over a given temperature interval is inversely proportional to the heating rate. Since heat exchange is proportional to the product of the (constant) heating term and the time, the heating rate cancels out; consequently, the heat exchange due to continuous heating is not directly determinable by changing the heating rate.

This unknown heat exchange is accounted for by experiments with the empty calorimeter only to the extent that the experimental conditions for the full calorimeter are reproduced. Absence of the sample necessarily changes the response time, but the transients cancel for each experiment. The requirement remains that the sample should not alter appreciably the temperature distribution on the calorimeter surface. Because the power is changed in proportion to the heat capacity, this requirement also extends to the calorimeter heater.

The correction for "zero" heat leak is different for the two methods. In the intermittent-heating method, the average of the observed fore- and after-period "zero" heat leaks is applied to the intervening heating period. In continuous heating, this procedure is not possible; however, the heat transfer corresponding to the "zero" heat leak may be determined by varying the heating rate or by separate determinations. The continuous-heating method is subject to the disadvantage that the causes of the "zero" heat leak are assumed constant over a period of a day or more, in contrast to about an hour in the intermittent-heating method.

The preceding analysis does not apply unless the experiment takes proper account of the transient terms. In the intermittent-heating method, the final temperature is measured only after the observer is certain that the transients have become negligible. Likewise, in the continuous-heating method, heat-capacity data should not be taken unless the observer is certain that the transients have become negligible. In the former method the "zero" heat leak observations can be extended for some of the experiments to determine that the transients are negligible. In the latter method, comparison of experiments from series having different starting temperatures should reveal deviations due to the transients. The transients will change with temperature and when the state of the sample is altered, as in fusion, but otherwise they should be reproducible. Once the time for the disappearance of the transients is established for a particular temperature and experimental condition, it should not be necessary to check it for replicate measurements of the heat capacity.

III. A High-Precision Adiabatic Calorimeter Utilizing Intermittent Heating over the Range 300 to 800°K

This calorimeter was designed specifically to measure the heat capacity of sulfur up to its normal boiling point. Sulfur is a good example of a material for which the adiabatic method is preferable to the drop method. The solid consists of S_8 rings, undergoes a transition at 368·6°K, and melts at 388·33°K to a liquid consisting partly of long chains. The concentration and length of the chains is a function of the temperature. When the liquid is quenched, as occurs in the "drop" method, the chains and other forms peculiar to the liquid do not have time to revert to S_8 rings, and, thus, varying amounts of energy are frozen in, so that the requirement that the sample come to a reproducible state in the calorimeter is violated.

The intermittent-heating method is preferable to the continuous-heating method for this material because liquid sulfur comes to equilibrium slowly, and data are required for the heats and temperatures of transition and fusion, and for an estimate of purity from the melting curve.

The construction and operation of the original calorimetric apparatus have been described in some detail[89, 90]. In the interim, several improvements have been made to obtain more reliable or simpler operation or better accuracy. Making the thermopile of noble-metal alloys[86] has apparently reduced thermocouple failures due to simple mechanical breakage. Better control of the environment beyond the adiabatic shield has improved the reproducibility of the "zero" heat leak correction, although the precision of measurement still appears to be about 1 part in 10 000, as in the original apparatus.

1. *The calorimeter (sample container)*

A schematic vertical section of the apparatus (which is generally cylindrical about the vertical axis) is shown in *Figure 1*.

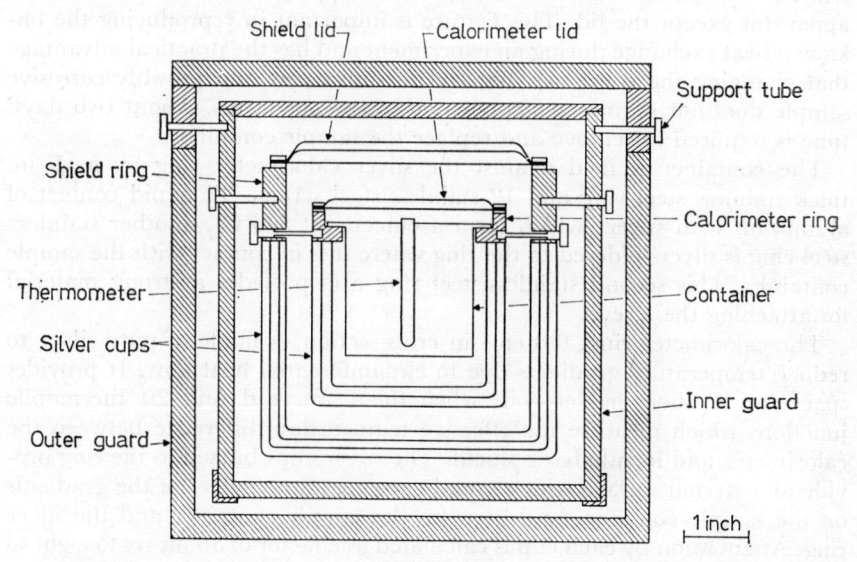

Figure 1. Schematic vertical section of the adiabatic calorimeter and associated thermostat[89, 90].

The calorimeter proper (as defined in Chapter 1) consists of two main parts: a removable container and a permanent outer "surface" consisting of a silver ring and thin silver radiation shields. This construction is adopted as the best way to reproduce the unknown heat exchange discussed in Section II above and to avoid the related systematic errors.

The sample container should be constructed of materials compatible with the sample at the highest experimental temperature anticipated and having the largest thermal diffusivity. A system of metallic vanes reduces the time required for the sample to come to thermal equilibrium, thus reducing the uncertainty in the heat-leak correction. In addition, this construction serves to keep the temperature gradients in the container small for both the full- and the empty-calorimeter experiments. The effects of these gradients on the surface of the calorimeter will be correspondingly small.

The volume occupied by the container is about 100 ml. The aluminum container used in the measurements on sulfur has a space-consuming construction which reduces the useful volume to 70 ml; the "dead" volume includes the thermometer and heaters. These elements are separated sufficiently from one another so that the thermometer is not at the hottest part of the calorimeter, a condition which permits measurement of the calorimeter temperature to about 0·05°K during heating. The initial and final temperatures can thus be selected to within a few thousandths of a degree, so the observer is enabled to reproduce experiments and make accurate comparisons of the data.

Three coaxial silver cups of 0·25 mm wall are silver alloy-brazed to a silver ring. A lid of similar construction completes the enclosure for the container. It is apparent from the figure that the container can be removed without disturbing the thermocouples or supports or other parts of the apparatus except the lid. This feature is important in reproducing the unknown heat exchange during an experiment and has the practical advantage that changing the container material to suit a new and possibly corrosive sample does not require reconstruction of the apparatus. About two days' time is required to remove and replace the sample container.

The container is held against the silver calorimeter ring by a $\frac{1}{16}$ in. thick stainless steel ring and 12 stainless steel screws. To avoid contact of aluminum with silver (which form a eutectic at 566°C), another stainless steel ring is silver-soldered to the ring where it is in contact with the sample container. This second stainless steel ring also provides a strong material for attaching the screws.

The calorimeter ring, 0·8 cm^2 in cross section, is made of pure silver to reduce temperature gradients due to circumferential heat flow. It provides space for 12 short heater coils when these are used and 20 thermopile junctions which measure the effective temperature difference between the calorimeter and its adiabatic shield. The silver cups brazed to the ring provide an external surface which greatly attenuates the effect of the gradients on the sample container and between the sample container and the silver ring. Attenuation by each cup is calculated as a factor of about six to eight so that the attenuation by three cups should be a factor of several hundred. Significant temperature differences on the surface of the sample container are expected only during the heating interval. They are calculated to be not greater than 0·02°K for the empty calorimeter and 0·04°K for the full calorimeter, so that their average effect on the outer calorimeter surface should be considerably less than one millidegree. In the original apparatus, the temperature difference between the sample container and the ring was made small by distributing the calorimeter heater windings between them approximately in proportion to their heat capacity. By various combinations of the heating coils, allowance could be made for the full or empty sample container. During heating, the temperature difference between the sample container and the ring was calculated to be about 0·05°K, which is less than 10 per cent of what it would have been if the heater were all on the ring. If the uncertainty in the reproducibility of this gradient is assumed to be as much as the calculated gradient itself, the variation in the average temperature of the outer surface of the calorimeter due to variations on the sample container is estimated to be only about 0·0001°K. Some experiments have been carried out without the heaters in the ring; the gain in simplicity of construction is apparent. Eliminating the ring heaters also eliminates one source of electrical trouble, as these heaters have a tendency to develop grounds to the calorimeter, but this elimination results in slower response of the ring to the heater and to greater excursions of the shield control. The excursions could probably be reduced by suitably programming the uncontrolled power in the shield heater (which is increased and decreased stepwise at the beginning and end of the heating period). To overcome the probable decrease in performance due to the increased temperature difference between the container and the ring, an additional silver cup might be provided.

More critical than the thermal contact between the container and the ring is the contact between the lid and the ring, because variations in this unattenuated contact directly affect temperature over almost one-sixth the outer surface of the calorimeter. This contact is made between the silver of the lid, backed by a stainless steel ring, and a smooth stainless steel surface. Although the heat capacity of the lid assembly is small, all the heat required to raise its temperature during the heating interval must come from the ring through this mechanical contact. In disassembling the calorimeter, contact is usually found to be so good that the lid and ring must be forced apart, an indication that some heat is transferred by direct metal-to-metal conduction. However, to arrive at an estimate of how large the temperature difference between the ring and the lid may be, it is assumed that heat transfer is only by gaseous conduction through an effective spacing of 0·0002 in. With these assumptions, the temperature difference between lid and ring is calculated to be 0·001°K. A single-junction thermocouple installed between the ring and the center of the lid was not sensitive enough to determine variations in this thermal contact, but it did serve to set an upper limit of a few thousandths of a degree. To improve on this feature of the apparatus, several thermocouple junctions might be installed permanently between the lid and the silver ring and an arrangement made for merely tipping the lid out of the way to remove the sample container. If the junctions are disturbed in loading the calorimeter, comparisons for full and empty measurements are in error because of the change in thermal contact.

2. *The calorimeter heater*

The calorimeter heater has a resistance near 70 Ω and a small heat capacity. Originally the coils for this heater were made by winding helices of oxidized

38 B & S gauge Nichrome V wire to fit inside porcelain tubes with an inner diameter of 1 mm. The lead from the lower end of each coil was brought up through a small porcelain tube inside the helix. To reduce the heat developed in the leads between coils, the helices were arc-welded in helium to larger nickel leads. The total resistance of all calorimeter heaters in series was about 640 Ω, but the sections were connected in series–parallel combinations to give a net 77 Ω for the empty calorimeter and 60 Ω for the full calorimeter. The heat capacity of the heater was about one $J°K^{-1}$, which is less than 0·3 per cent of the heat capacity of the full calorimeter. Simplified heaters were constructed by threading Nichrome V through four-hole porcelain tubes of about 1·5 mm outer diameter. They have a smaller surface than the heater coils, so a higher temperature increment is required to deliver the same power, and a slightly lower heat capacity, so that the time constant is somewhat greater.

The voltage taps for the heater must be so placed that they account for the electric power delivered to the calorimeter. In this apparatus, the current leads are 0·2 mm gold wires, and the lengths of wire between the shield and the calorimeter are about 4 cm. The combined resistance of the two current leads between the calorimeter and the shield is 0·06 Ω, or 0·1 per cent of the resistance of the calorimeter heater. On the assumption that one-half of the heat generated in each lead will flow to the calorimeter and that the two leads are of equal length and have equal thermal contacts on the shield and calorimeter, the voltage taps have been placed with one on the calorimeter junction of one current lead and one on the shield junction of the other lead. This problem is considered in more detail in Chapter 4.

3. *The adiabatic shield*

The accuracy of adiabatic calorimetry depends to a large extent on the effectiveness of the adiabatic shield in keeping the heat exchange with the calorimeter small and the same in both the full and empty calorimeter experiments. Both calorimeter and shield have temperature gradients on their surfaces which prevent perfect matching of their temperatures. On the assumption that the heat-transfer coefficient is constant over the surface, control thermocouples can be distributed to average the surface temperatures. Alternatively, the shield may be controlled in sections—the shield top matched to the calorimeter top, etc. In this arrangement, some allowance is made for differences in the heat-transfer coefficients of different parts of the calorimeter. To a considerable extent it can avoid averaging radiant heat transfer with thermal conduction of leads and supports. The design of the present calorimeter shifts the emphasis from matching temperatures as closely as possible to reproducing in measurements with empty calorimeter the heat exchange in the measurements with the full calorimeter. To this end, the apparatus is designed so that the thermal contact between the shield and calorimeter is disturbed as little as possible when the container is removed for filling.

The adiabatic shield is constructed in a fashion similar to that of the calorimeter, with a massive silver ring with silver cups and lid. It is held as closely as possible to the calorimeter temperature to minimize heat flow from the calorimeter. Around the adiabatic shield is a closed cylinder of silver about

¼ in. thick which is called the *inner guard*. It is controlled a few tenths of a degree below the shield temperature to provide a good reproducible environment for the shield. Surrounding the inner guard is a cylindrical aluminum cup which is called the *outer guard*. Its function is to supply the major part of the heat flow to the room. It is insulated with six thin aluminum shields which reduce radiation and convection. A small tube runs from outside the calorimeter to the bottom of the inner guard to bring carbon dioxide to that region and flush outward any air which may diffuse into the apparatus. Because of our difficulties with this system, we would recommend a closed system using an argon atmosphere.

The uncertainty in heat exchange between calorimeter and shield is the sum of uncertainties in the correction for the "zero" heat leak, which is made for each experiment, and the correction for the (unknown) heat exchange during the heating period, which depends on the reproducibility between the full and the empty experiments. Both corrections depend on the product of the various heat-transfer coefficients and an associated temperature difference. It is unlikely that a change in the heat transfer coefficient causes an error in the "zero" heat leak correction, which depends on measurements made relatively frequently. A change in heat-transfer coefficients between measurements with the full and empty calorimeter will, however, affect the unknown heat exchange during the heating period.

The heat transfer coefficient was kept small in this calorimeter so that fractional changes would have a smaller absolute effect. For the original apparatus, the heat-transfer coefficient at 670°K was calculated to be 0·18 w °K^{-1} temperature difference between shield and calorimeter. The calculation ascribed 0·1 w°K^{-1} to radiation, 0·08 w°K^{-1} to carbon dioxide conduction, and 0·006 w°K^{-1} to conduction through leads and supports. By deliberately offsetting the shield control and observing the change in "zero" heat leak, the heat-transfer coefficient was found to be 0·23 w°K^{-1}. Changes might conceivably occur by a change in the thermal contact of the leads and supports, but these were not disturbed except for the lids. It is necessary to remove the shield lid when the container is removed to change the sample. Here, as with the calorimeter, the change in temperatures over the lid cannot be checked for a particular perturbation, except by estimating the effect by comparative measurements in which the only change is the removal and replacement of the lid. (Installation of several thermocouples to provide direct indication of the temperature drop across the mechanical contact between lid and ring might be considered.) The original supports for the calorimeter were three Nichrome strips 6 mm wide, 0·025 mm thick, and 2 cm long. Although these strips conducted very little heat, they allowed considerable motion of the calorimeter relative to the adiabatic shield. This motion caused occasional grounding of thermocouples or heater and thermometer leads and made the position of the calorimeter relative to the shield somewhat ambiguous. Since a change of relative position might change the unknown heat leak, the strips have been replaced by three radial tubes of Inconel with outer diameter of 3 mm and 0·1 mm wall, which conduct more heat, but hold the calorimeter rigidly in place.

If the carbon dioxide were diluted, the contribution from gas conduction might change. Therefore, the carbon dioxide flow is monitored several times a day, and no important measurements are made after resealing the calorimeter until the apparatus has been flushed overnight with carbon dioxide. (The possibility of operator error could be eliminated by making the apparatus gas-tight.) Although the heat-transfer coefficient could be reduced by the amount of the gas conduction term if the system were evacuated, the gain would be more than offset by the adverse effects on thermal contact, i.e., at the lids, heaters, thermocouple junctions, and thermometer.

The radiant heat-transfer coefficient differs from the others in that the emissivity obviously changes by formation of a coating of silver sulfide on the calorimeter and shield surfaces when the apparatus is open for any appreciable time. After heating to about 525°K, the sulfide coating disappeared and left a clean silver surface. The silver surfaces were polished in the original assembly, but during the first heating the silver on the surface recrystallized to give a diffusely-reflecting surface. After a period of disuse, this surface was restored by heating to remove the sulfide. Heat-capacity measurements were not detectably affected by the sulfide coating, perhaps because it disappeared below the temperature at which radiation becomes predominant. One pronounced effect was attributed to the slow oxidation of the sulfur coating by traces of oxygen present in the carbon dioxide. The "zero" heat leak in the range 420 to 520°K was positive and larger than usual the first time the calorimeter was heated to this range after being open for a time. This effect would have been attributed to the heat of reaction offsetting the usually small, negative "zero" heat leak. The effect was constant and small enough so that when "zero" heat leak corrections were made, the corrected data were indistinguishable from regular data.

It is not feasible to avoid the formation of the sulfide layer by plating the silver with gold because gold at the higher temperatures of operation diffuses into silver. The "zero" heat leak correction for the improved apparatus is very simple to apply because the apparatus satisfies the approximation of theory[87]. It is sufficiently constant from fore to after periods that the time between initial and final temperature readings is multiplied by the rate of change of the calorimeter temperature. The correction is applied to increase the temperature difference to what it would have been if the "zero" heat leak were truly zero. The "zero" heat leak is influenced by the temperature difference between the shield and the guard and by fluctuations in this temperature difference. Better guard control in the improved calorimeter makes the "zero" heat leak more reproducible. At 743, 753, and 763°K, the "zero" heat leak for one series was 2·80, 2·74, and 2·84 $\times 10^{-4}$°K min^{-1} with a heat capacity[9] of about 300 J°K^{-1}. For a forty-minute experiment with a 10°K temperature rise, the correction is a few parts in 10000 on the heat capacity, with the uncertainty smaller by an order of magnitude.

The temperature difference between the adiabatic shield ring and the calorimeter ring is measured by two, ten-junction thermopiles made of gold–40 per cent palladium vs. platinum–10 per cent rhodium wire 0·125 mm dia. A series connection of the two gives 1200 μV°K^{-1} in the upper temperature range. With the original apparatus, two thermopiles were compared frequently by opposing their e.m.f.s to detect circumferential gradients on the silver rings. These gradients were found to be related to the way the glass fiber insulation was packed around the guards. The new system for insulating the apparatus from the room has eliminated the need for this test. The junctions are made at "thermal tie-downs," which consist of small gold tabs sandwiched between mica washers for electrical insulation, and squeezed against the silver rings by stainless steel nuts. These thermal tie-downs are similar to, but larger than, those previously described[58]. At the usual heating rate of $0.5°K$ min^{-1}, tests indicate that junctions of the thermopile lag about $0.02°K$ below the rings. The lag in junctions on the calorimeter is compensated for by the lag in those on the shield, so that the error in the observed temperature will be considerably less than $0.02°K$. However, it is the reproducibility and not the absolute value which is important. It is necessary to have this temperature difference the same to better than $0.001°K$ for measurements with the full and with the empty calorimeter. It does not seem likely that the thermal tie-downs could be rebuilt to this accuracy, so, to insure maximum reproducibility, the thermopile junctions are not disturbed when the sample container is removed for filling. If a thermocouple junction must be replaced for some reason, it is necessary to show by repeating some heat-capacity experiment that its thermal contact is essentially the same as before or to be able to allow for the difference in thermal contact. To prevent cooling of thermopile junctions by conduction along the leads to the outside, an extra thermal tie-down is installed in each of the four leads to bring them to the shield temperature prior to the first junction.

The leads from the thermopile, as well as those from the thermometer and heater, are 0.2 mm gold wires, which practically eliminate spurious e.m.f.s[64]. Just outside the calorimeter, the junction is made to copper wires which go to the control and observation stations. The gold–copper junction contributes very little e.m.f., and the other junctions in the circuit are mechanical copper-to-copper joints, or made with solder having a low thermal e.m.f. against copper.

4. *Adiabatic shield temperature control*

Control of the adiabatic shield is of critical importance to the accuracy of the heat-capacity measurements. Fortunately, it is not necessary to control the absolute temperature difference between the shield and the calorimeter at zero. It is only necessary that the deviation from zero be reproducible for the length of the experiment. If, for example, a spurious e.m.f. in an external circuit produces a constant offset in the shield control, the resulting heat exchange between the calorimeter and shield will be observable in the fore and after periods. Such constant contributions to the "zero" heat leak are accounted for in each individual experiment, and constancy is not necessary over the larger time interval between measurements with the empty and full calorimeter.

A useful corollary to the accidental constant offset of the control is the deliberate offset. By this means the rate of change of the calorimeter temperature can be made very small, resulting in more precise temperature measurement. A zero rate of change does not necessarily mean that there is no heat exchange between the calorimeter and the shield but only that heat flow into one part of the calorimeter is balanced by heat flow out at another part.

If the observer is waiting for the slow equilibration of a sample, this device is useful to eliminate the change in temperature which may shift the equilibrium.

A similar argument is applicable to small constant deviations in the electronic controlling equipment. The control system might suffer an offset related to the increased power levels during the heating period. Such an offset might be caused, for example, by increased 60-cycle pickup from a heater causing a zero shift in an amplifier, or by an increased d.c. current leak from a heater circuit to a thermopile circuit. Correction for such an effect would depend on reproducing it for both the full and empty calorimeter.

The problem remains of correcting for observed deviations of the controlled temperature. In the original apparatus[89], the deviations were integrated over the time of the experiment and a correction obtained by multiplying the integral with the observed heat-transfer coefficient determined as described in the preceding section. As the control was improved, the only observable deviations came at the beginning and end of the heating period. These deviations were ordinarily about 0·005°K in amplitude for about 20 sec before returning to the control point and were approximately equal and of opposite signs; hence, the corrections were insignificant and not applied. More recently, some experiments were carried out without heaters in the calorimeter ring. The response time of the thermopile was increased, and suitable changes in the control system restored the usual control. If desired, a careful adjustment of the control for a particular temperature range can reduce excursions to less than 0·002°K maximum amplitude.

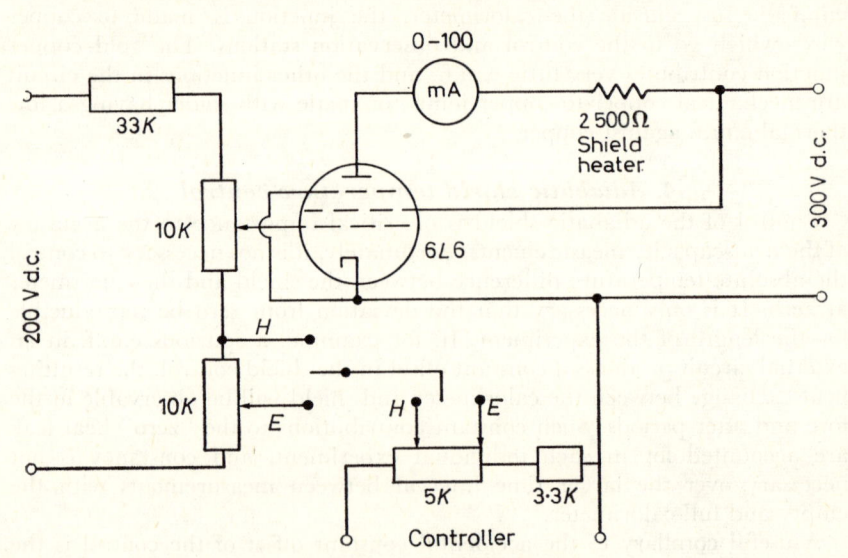

Figure 2. Adiabatic shield control circuit[89].

The shield control circuit is shown in *Figure 2*. The shield heater is 2500 Ω of Nichrome V wire placed in the plate circuit of a 6L6 electron tube. The heater current varies from about 5 mA during an equilibration period to about 40 mA when heating at $0.5°K$ min^{-1}. In the grid circuit of the tube are a bias supply and the output from the electronic controller. Two relays operate at the beginning and end of the heating period to aid the controller. Contacts H are made for the heating period, E for equilibration. At the beginning of the heating period the upper relay steps up the power level to the approximate requirement for the chosen heating rate. Simultaneously, the lower relay increases the influence of the controller on the current in the shield heater (decreases the proportional band). The two relays are thrown by a simple electron tube relay which is actuated synchronously with the calorimeter heater switch. Reactance can be used in the grid circuit of this tube to delay the time at which the power is increased.

The remainder of the control circuit is made up of commercially-available components[89]. The three-mode controller supplies 0–5 mA at 0–10 V. The reset or integrating mode is particularly useful in eliminating zero offset due to changes in the power supply requirement. A strip chart recorder is used which will respond within one second to a signal corresponding to a full scale deflection. The recorder contains a slidewire bridge which provides the controller with the sign and magnitude of deviations from the control point. This bridge circuit is electrically isolated from the microvolt-level amplifier, a useful feature in dealing with high temperature apparatus in which resistances to ground and between circuits decrease with temperature. The low-level amplifier has an effective input filter to remove 60-cycle a.c. pickup.

Numerous checks on the zero stability of the control system have been made by occasional observations of the thermopile with a microvolt potentiometer during equilibration periods. The null-point is stable to ± 0.2 μV, which is somewhat better than the rating on the components. Stability is required only over the time of an experiment since offsets are accounted for in "zero" heat leak measurements.

A useful change is now practical for this control system. The electron tube control element may be replaced by a transistor, to permit the use of lower voltages and higher currents in the shield heater, so that its resistance might be a small fraction of the resistance of the present heater. The smaller resistance should simplify the construction considerably.

5. *Environmental control*

The environment for the adiabatic shield is provided by an inner guard of 6 mm silver and an outer guard of 10 mm commercially pure aluminum. This silver inner guard responds quickly to changes in shield temperature, and the aluminum outer guard supplies the bulk of the heat flowing to the room. The temperature of the cylindrical portion of the silver inner guard is normally controlled about $0.3°K$ below the temperature of the adiabatic shield, so that the controller can cool the shield by reducing the shield heater current. The top and bottom of the silver inner guard are controlled relative to the cylindrical portion. The aluminum outer guard is made in two parts, each controlled about $5°K$ below the silver guard.

The control thermopiles for these parts of the apparatus are two junctions of gold–40 per cent palladium *vs.* platinum–10 per cent rhodium. The shield-to-inner-guard control uses a commercially available, low-level, microvolt amplifier directly connected to a 3-mode controller. The other controls, made in the laboratory, have microvolt sensitivity but are subject to zero shifts from a.c. pickup and temperature changes. Adjusting the zeros of these controllers is a considerable part of the operation. To facilitate control, each power level is stepped up at the beginning of the heating period by the electron tube relay. Power to these heaters is controlled by means of type 6080 (6 AS 7) electron tubes. For reasons cited in connection with the shield heater discussion, control of the power with transistors appears to be preferable.

The effect of the environment on the calorimeter can be determined by changing the temperature difference between the silver inner guard and the adiabatic shield. As the difference is increased, the shield power increases, and a larger gradient with a corresponding increase in the magnitude of the "zero" heat leak is set up. In the original apparatus the "zero" heat leak was 3 J h^{-1} for 0·25°K offset and 9 J h^{-1} for 0·75°K offset. The temperature difference is controlled within $\pm$ 0·012°K, which corresponds to an uncertainty in the "zero" heat leak of less than $\pm$ 0·15 J h^{-1}.

The apparatus is insulated from the room by seven thin aluminum shields 0·5 mm thick and spaced 1 cm apart. Originally, the apparatus had only one guard cylinder which was insulated with glass fiber. Packing the glass fiber in various ways was found to alter the temperature distribution in the adiabatic shield. At that time, the outer guard was installed to reduce the effect. The aluminum cylinders have several advantages over the glass fiber insulation, for they respond more quickly to temperature changes because of their relatively lower heat capacity. In addition, they are much cleaner to handle when work is being done on the apparatus and make it possible to use the same distribution of "insulation" for measurements with the calorimeter full and empty. Insulated in this way, the aluminum outer guard requires 130 W to maintain the temperature at 720°K and 220 W at 820°K.

6. *Measurement of electrical power*

The measurement of electric power follows the basic techniques for d.c. power measurements described in Chapter 3. In addition to variable series resistors, the circuit contains dummy (ballast) resistors, which are adjusted to the resistance of the calorimeter heater. The current is switched between the dummy resistors and the calorimeter heater by relays with mercury-wetted contacts, which respond in about 1 msec. The current relays and the timer clutch relay are operated sychronously with a one-cycle per second signal. The dummy and series resistors, together with a solid-state power supply which is variable in small steps, facilitate the maintenance of constant power in the calorimeter through the transient period when the calorimeter heater is turned on[29]. The current is evaluated from the e.m.f. drop across a 0·5 Ω standard resistor in series with the calorimeter heater. The voltage is evaluated from the e.m.f. measured across 1 per cent of a 20 000 Ω voltage divider connected to the heater potential leads as described above. Corrections are applied for the current in the voltage divider and its leads. All e.m.f.s are referred to calibrated, saturated standard cells. The duration of the heating period is measured with a 60-cycle timer powered by a crystal controlled oscillator. The d.c. timer clutch is switched by one of the mercury relays; the usual heating interval is about 1200 sec. When the apparatus is operated properly, the agreement between the 1 c/s relay and the timer is about 0·01 sec. For times of the order of 10 sec or less, the timer and 1 c/s signal do not agree. The timer reads high by 0·04 sec for a 4 sec interval and 0·06 for 2 sec.

The temperature is measured by a platinum resistance thermometer sealed in a special glass capsule to extend the useful temperature range[46]. It behaves well at the sulfur point, and even at 873°K the effect of the electrical conduction of the glass causes an uncertainty of only 0·001°K. The thermometer is calibrated initially at the ice, steam, and sulfur points, but its resistance is checked in an ice triple-point cell each time the calorimeter is opened to change the sample. Resistance measurements are made with a Mueller bridge.

7. Experimental procedures

Experimental procedure for heat-capacity measurements is reduced to routine as much as possible. When one day's measurements are concluded, all heaters are shut off except that of the aluminum outer guard. Control is switched to a single-junction thermocouple on the outer guard referred to an ice bath. The control temperature is selected by means of a 10-turn potentiometer in the thermocouple circuit. When the e.m.f. of the thermocouple falls below the reference potential of the potentiometer, the heater is turned on automatically to maintain a temperature several degrees below the desired starting temperature for the following day. To begin measurements, the controls are set up and the calorimeter heated to the desired starting temperature, with the heater current adjusted to obtain about 0·5°K min^{-1} heating rate. The rate is not critical, but the use of a very different rate requires readjustment of all control circuits. Except for measurement in regions of transition and fusion, the heating rate is maintained at 0·5°K min^{-1} within about 3 per cent, although observations at 0·25°K min^{-1} give the same heat capacity to 0·01 per cent. Of course, experiments at different rates have different uncertainties, because the uncertainty in heat leak correction increases with the length of the experiment.

The experiment is begun by determining the "zero" heat leak, which is observed as the change in the resistance of the platinum thermometer, until the rate of change becomes constant. The calorimeter heater current is then switched on at an integral second. Several measurements of current and voltage are made during the heating period. Thermocouples, including those used for control, are checked with the potentiometer or by galvanometer deflection; in this way, operation of the various controllers is checked. Near the end of the heating period the thermometer current is turned on and the resistance followed with time. From knowledge of the thermometer lag behind the average calorimeter temperature, the final temperature of the experiment can be selected to within 0·05°K. Such an arrangement allows close duplication of the temperature ranges in succeeding experiments. An experienced operator can carry out the calculations of average heat capacity for the completed experiment simultaneously with measurements on the following experiment. This arrangement allows quick and accurate comparisions of data.

8. *Determination of enthalpies of phase transitions*

To measure an enthalpy of transition or fusion, about one half the usual power is used. The experiment is started well below the transition temperature, so that the first part of the experiment is just like a heat capacity measurement. When the transition begins, the controls must be adjusted manually to allow for the decrease in the rate of rise of the calorimeter temperature. The reverse occurs at the end of the transition. At the usual heating rate, the problem of manual adjustment is aggravated. In either case, corrections are required for control deviations.

Heating is carried well beyond the transition temperature to make sure that all of the sample is transformed. As the transformation takes place, the interface between phases recedes from the relatively better heat-conducting, metallic surfaces, which consequently must increase in temperature to maintain the heat flux through the transformed material to the interface. The magnitude of this temperature increase is greater for greater heating rates, so that the temperature at the thermometer may be well beyond the transition temperature. However, an adequate amount of untransformed sample remaining may bring the average temperature back to the transition temperature, leaving an ambiguous amount of sample transformed.

The data for such an experiment are corrected to the transition temperature by use of experimental data for the heat capacity of the appropriate phase. These data can be obtained on either side very close to the transition temperature and extrapolated a short distance. For samples melting in the temperature range studied, the melting point and purity of the sample can be estimated from plots of the calorimeter temperature against the reciprocal of the fraction of the sample melted. These curves are obtained by adding a known amount of electrical energy to the calorimeter and observing the temperature after the rate of change of the temperature becomes constant and approximately equal to the magnitude expected for that temperature. The fraction melted is calculated from the excess energy added, divided by the total enthalpy of fusion for the sample in the calorimeter. For these curves it is not necessary to know accurately the fraction melted, and adjustments for the small control deviations may be omitted. For example, in the melting curve for sulfur (*Figure 3*) the square point at reciprocal fraction melted $1/F = 5.3$ lies off the curve. The question of whether adjustment of the "zero" heat leak might improve the fit may be answered by noting that the heat flow from the calorimeter is less than that at $673°K$ and is hence less than 3 J h^{-1} or 24 J for the eight-hour equilibration. This amount of energy corresponds to *freezing* 0.5 per cent of the sulfur present, and besides being too small, the adjustment has the wrong sense.

Shmidt and Sokolov[69] have suggested that such deviations and perhaps a major part of the temperature change in the melting curves are caused by temperature gradients on calorimeters. They do not present a detailed argument, but the line of reasoning must be somewhat as follows: An adiabatic calorimeter is not perfect and gains (or loses) a small amount of heat to its surroundings during equilibration periods. If the calorimeter gains heat during equilibration with both solid and liquid present, the heat must flow from the surface of the calorimeter to the interface between the two phases. The thermometer is located in good thermal contact with the metallic

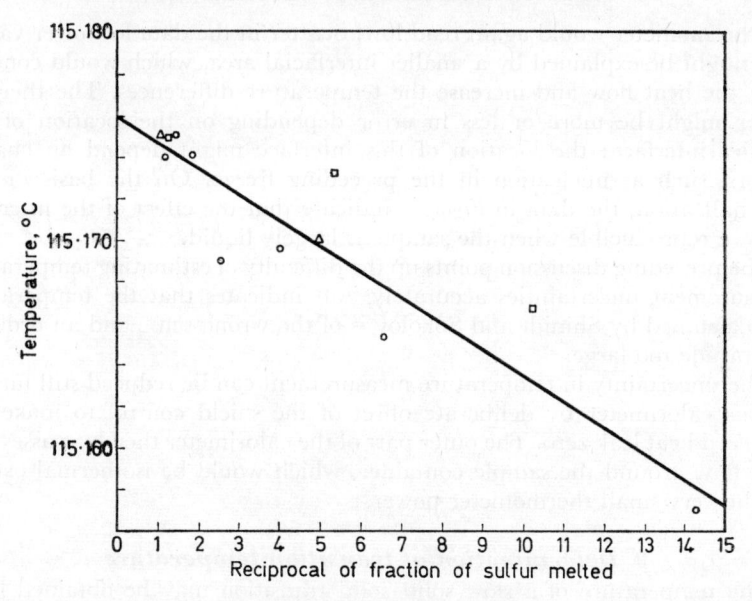

Figure 3. Melting curve for sulfur[85].

parts of the calorimeter and not at the interface; therefore, its indicated temperature (independent of its own excess temperature) will lie above that of the interface. As more solid is melted, the interface recedes from the metallic parts of the calorimeter and the thermometer reads still higher relative to the interface temperature. The net result is an increase in the thermometer temperature as the sample is progressively melted, independent of any impurity. The geometry of the calorimeter does indeed have an influence on the usual plot of temperature *vs.* the inverse of fraction melted; moreover, the direction of the deviation from the theoretically straight line depends on the sign of the zero heat leak.

To estimate the magnitude of this effect for the calorimeter in *Figure 1*, some assumption must be made as to the location of the liquid–solid interface. If we assume that the solid melts away from the aluminum walls and drops to the bottom of the container, heat flow from the interface to the top of the container would make the thermometer read too low. As more sulfur melted and the interface approached the bottom of the container, the thermometer would read still lower, which is contrary to observation but might be possible if it were superimposed on an impurity effect larger than that indicated in *Figure 3*. Because of the sign of the heat leak, heat flows out of the container, so that its coldest region is at the contact with the silver ring. If we assume that the solid sticks to the aluminum surface and that the interface is here, the temperature gradient at the thermometer is only that due to the thermometer power and is an order of magnitude smaller than the heat leak. If we assume that melting takes place radially in each of the cylindrical spaces in the aluminum container and that, in the worst case, the solid remains symmetrically distributed in the space, the heat flow from the liquid sulfur would set up a temperature difference of 3×10^{-4}°K for $1/F = 4/3$ and the thermometer would again read low. Scatter in the data for lower values of F might be explained by a smaller interfacial area, which would concentrate the heat flow and increase the temperature differences. The thermometer might be more or less in error depending on the location of the smaller interface; the location of this interface might depend on chance factors, such as nucleation in the preceding freeze. On the basis of this rationalization, the data in *Figure 3* indicate that the effect of the interface is more reproducible when the sample is largely liquid.

The preceding discussion points up the difficulty of estimating temperature measurement uncertainties accurately, but indicates that the temperature drop assumed by Shmidt and Sokolov is of the wrong sense and an order of magnitude too large.

The uncertainty in temperature measurement can be reduced still further in this calorimeter by deliberate offset of the shield control to make the observed heat leak zero. The outer part of the calorimeter then by-passes any heat flow around the sample container, which would be isothermal except for the very small thermometer power.

9. *Determination of transition temperature*

The temperature of a slow solid–solid transition may be obtained by a somewhat different method. Since there is no impurity effect here, the transition temperature is assumed to be independent of the fraction transformed, although Shmidt and Sokolov[69] observe in their calorimeter a variation

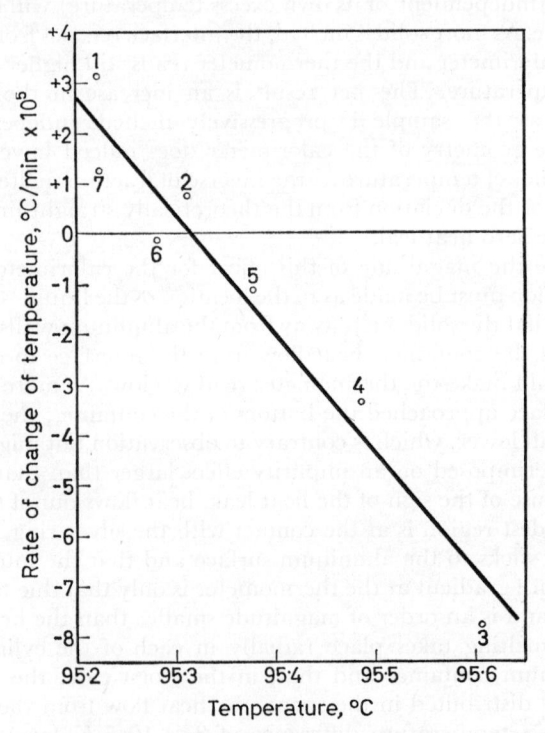

Figure 4. Determination of the transition temperature for rhombic sulfur[85].

which they ascribe to a change in the temperature distribution in the calorimeter. The temperature uncertainties of a few tenths of a millidegree calculated above apply also to the present case, but other experimental difficulties cause greater uncertainty.

The determination of the temperature for the transition from rhombic to monoclinic sulfur is shown in *Figure 4*. The calorimeter with rhombic sulfur was heated until the transformation started at about 373°K. The power was then shut off and the calorimeter was cooled by heat transfer to the adiabatic shield until the temperature was thought to be slightly below the transition temperature. The adiabatic shield was then brought up to the temperature of the calorimeter and the rate of change of the calorimeter temperature was observed for 0·5 h. The calorimeter was then heated for a short time and a new rate observed at a slightly higher temperature. This procedure was repeated, with equilibrium approached from both higher and lower temperatures to obtain both positive and negative rates. The results are plotted in *Figure 4* numbered in chronological order. The method is based on the assumption that, within a few tenths of a degree of the transition temperature, the rate of transformation of one phase into the other is proportional to the difference from the equilibrium temperature. Determinations of the rate of transformation as a function of temperature can be plotted to find the temperature at which the rate is zero, which is taken as the transition temperature. The uncertainty in the transition temperature is believed to be about 0·1°K, with allowance for uncertainty in the heat leak.

IV. An Adiabatic Calorimeter for Thermal Measurements from 250 to 600°K

Obtaining data on the transition and fusion behavior and thermodynamic properties of molecular crystals into the liquid state requires adiabatic calorimetry to temperatures of about 600°K, with reasonable convenience in loading and unloading the calorimeter. The calorimeter of Westrum and Trowbridge[81,82,91] described in this section was devised to meet these problems and provides an instrument of considerable utility, operable from conventional low-temperature measuring and adiabatic shield control circuits, and yet of very simple construction. Compared to the adiabatic calorimeter described in Section III, it is less precise, but far more convenient in operation. The calorimeter described has already been used in the measurement of thermal properties of nearly two score molecular crystals and ionic substances.

1. *Intermediate-temperature thermostat*

The adiabatic thermostat depicted in *Figure 5* in schematic cross section is essentially a modification of the aneroid, low-temperature cryostat described in Chapter 5 of this volume. That part of the apparatus shown in *Figure 5* is contained within an additional enveloping ("floating." i.e., not heated), copper radiation shield in a high vacuum cylindrical brass casing similar to that shown in Chapter 5 (except for the elimination of the refrigerant tanks). The "floating" radiation shield significantly reduces the electrical power required to maintain the highest temperatures, for, in contrast to

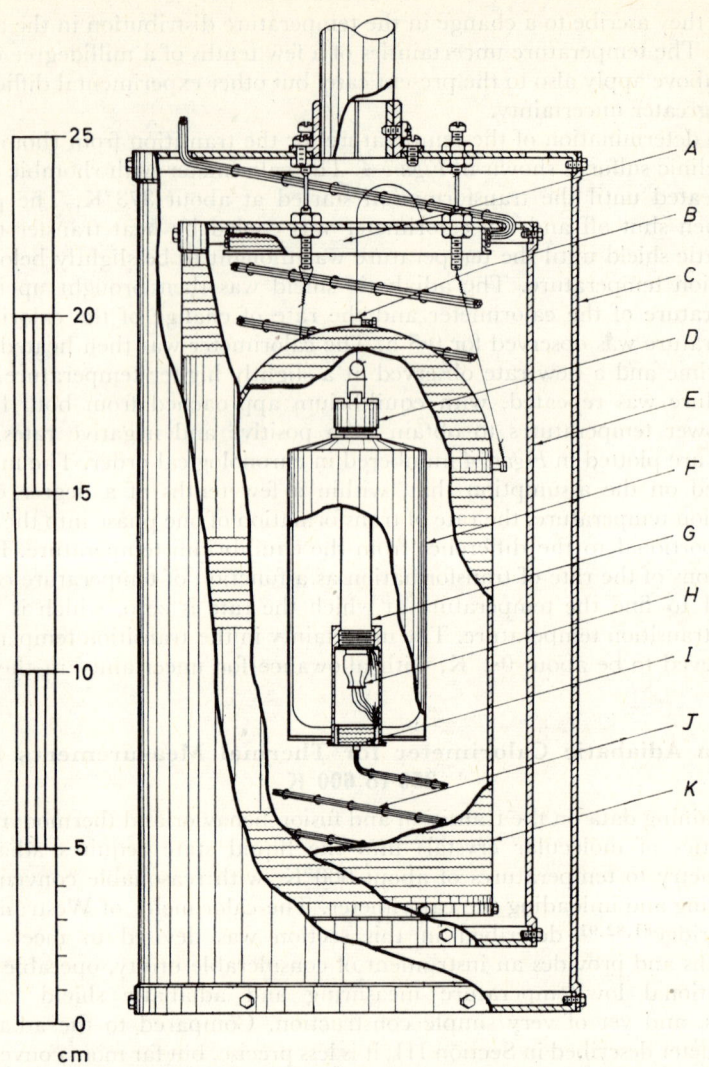

Figure 5. Cross-sectional diagram of intermediate temperature adiabatic calorimeter[91]
A, Guard shield ring
B, Calorimeter suspension collar
C, Primary radiation shield
D, Calorimeter closure assembly
E, Guard shield
F, Calorimeter assembly
G, Thermometer–heater assembly
H, Thermometer
I, Thermal equilibrium spool
J, Lead bundle
K, Adiabatic shield

most calorimeters made for operation in this range, no external furnace is provided. For some purposes it has been found convenient to add a liquid nitrogen tank to permit cooling the calorimeter into the cryogenic range in order to obtain a larger overlap with cryogenic, adiabatic calorimetry. The stainless steel tube shown at the top of *Figure 5* has 1·91 cm inner diameter,

354

0·25 mm wall and serves to support the entire thermostat–calorimeter unit within the evacuated brass casing. The top of the primary radiation shield, C, is secured to the lower end of the large stainless steel tube. Situated between this radiation shield and the top plate is a tank of 1·5 liters capacity for liquid nitrogen coolant. A filling tube of 1·25 cm inner diameter, thin-wall stainless steel passes through the top plate to the tank, and a variable heat leak in the form of a stainless steel tube provides thermal contact between the liquid nitrogen tank and the thermostat itself. In this manner both thermostat and calorimeter are cooled by admitting less than 0·1 torr helium gas into the vacuum chamber to provide a heat transfer mechanism.

The guard shield, E, supported from the primary radiation shield by three machine screws (reduced to a small diameter over most of their length), functions as the main buffer between the adiabatic shield, K, and ambient (room) temperature. Temperature control of this 1·3 kg shield is maintained by two independent heaters, each of about 500 Ω resistance and each extending over all three parts (top, cylinder, and bottom). The bulk of the heating is done through manually preset variable auto-transformers and step-resistors. The temperature maintained by an electronic control channel is about 0·6°K below that of the adiabatic shield; the sensing element used is a copper–Constantan differential thermocouple buried in the enamelled-lead bundle on the top plate. On the inside top of the guard shield is a ring, A, which provides a thermal tie-down junction for the lead bundle, J.

The adiabatic shield, K, spun from 1·27 mm copper, has a cylindrical middle section capped with a hemispherical top. The open lower end of the shield is closed with a closely fitting copper plate held in place by six machine screws, but readily removable for loading. The entire surface of the middle-top section of the shield is covered with a 300 Ω, fiber glass-covered, Advance wire heater. The bottom section is covered with 500 Ω of wire. The temperature differentials of both sections of the adiabatic shield are separately maintained within ± 0·001°K by two automatic electronic shield controls provided with proportional, rate, and reset control actions; copper–Constantan differential thermocouples are used to provide the sensing e.m.f. A schematic diagram of one channel of the electronic shield control circuit is presented in Figure 6 and described in more detail by Carlson[17], as modified from a prototype developed by Furukawa[23]. The middle-top section responds to a difference in temperature between the calorimeter, F, and the shield, whereas the bottom plate and the guard shield are controlled with respect to the middle-top section. Each channel consists of a stabilized d.c. microvolt amplifier to amplify the output from the differential thermocouple (or thermopile). The amplifier output is displayed on a Speedomax Type-H microvolt recorder where the degree of imbalance from a preset control point is sensed and compensated with a Series 60 Current Adjusting Type (CAT) control system. The d.c. CAT output is converted to a.c. for amplification, and after impedance matching, is fed directly to the heater. Both heaters on the adiabatic shield have their own control channels and are regulated independently. A third channel is used to maintain the temperature of the guard shield about 0·6°K below that of the adiabatic shield. The completely automatic controls for maintaining adiabatic conditions enable a single operator to make all measure-

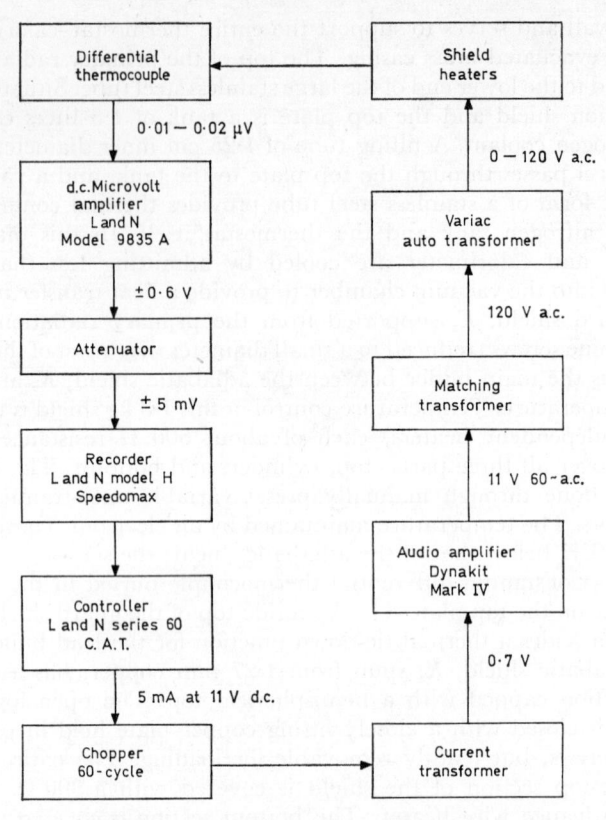

```
          ┌─────────────────┐                    ┌─────────────────┐
          │  Differential   │◄───────────────────│     Shield      │
          │  thermocouple   │                    │     heaters     │
          └─────────────────┘                    └─────────────────┘
                 │                                        ▲
             0·01 – 0·02 μV                          0 – 120 V a.c.
                 ▼                                        │
          ┌─────────────────┐                    ┌─────────────────┐
          │ d.c. Microvolt  │                    │                 │
          │   amplifier     │                    │     Variac      │
          │    L and N      │                    │ auto transformer│
          │   Model 9835 A  │                    │                 │
          └─────────────────┘                    └─────────────────┘
                 │                                        ▲
              ± 0·6 V                                120 V a.c.
                 ▼                                        │
          ┌─────────────────┐                    ┌─────────────────┐
          │   Attenuator    │                    │    Matching     │
          │                 │                    │   transformer   │
          └─────────────────┘                    └─────────────────┘
                 │                                        ▲
              ± 5 mV                                11 V 60~ a.c.
                 ▼                                        │
          ┌─────────────────┐                    ┌─────────────────┐
          │    Recorder     │                    │ Audio amplifier │
          │ L and N model H │                    │    Dynakit      │
          │   Speedomax     │                    │    Mark IV      │
          └─────────────────┘                    └─────────────────┘
                 │                                        ▲
                                                       0·7 V
                 ▼                                        │
          ┌─────────────────┐                    ┌─────────────────┐
          │   Controller    │                    │                 │
          │ L and N series 60│                   │                 │
          │    C. A. T.     │                    │                 │
          └─────────────────┘                    └─────────────────┘
                 │                                        ▲
            5 mA at 11 V d.c.                             │
                 ▼                                        │
          ┌─────────────────┐                    ┌─────────────────┐
          │    Chopper      │───────────────────►│    Current      │
          │   60-cycle      │                    │  transformer    │
          └─────────────────┘                    └─────────────────┘
```

Figure 6. Typical electronic automatic shield control channel[17].

ments and to provide continuous control over a period which may extend over several days for studying equilibrium transition phenomena.

2. *The silver calorimeter*

The silver sample container or calorimeter used in the intermediate temperature adiabatic cryostat may also be seen in *Figure 5*. It is similar in size and shape to those used in low-temperature work but with two marked deviations: (*i*) regarding the method of sealing the sample compartment, and (*ii*) regarding the installation of the thermometer–heater assembly. Except for the sealing mechanism, the calorimeter was machined from two portions of pure silver rod. The bottom, entrant well, and six vertical, radial vanes (which provide thermal conductivity and are tapered radially to economize mass) were fabricated from one rod. In a recent modification, further mass was removed and movement of liquid samples facilitated by V-shaped notches in the periphery of the vanes. The body of the calorimeter is a cylinder 3·9 cm in diameter and 0·4 mm thick. All seams for the completed assembly were force-fitted and sealed vacuum-tight by silver alloy brazing with APW #355 alloy. The calorimeter has an internal volume of 82 ml and originally weighed 120 g; recently it has been reduced in mass to 105 g.

ADIABATIC CALORIMETRY FROM 300 TO 800°K

The axial, entrant well, *G*, is so constructed as to accept the thermometer–heater assembly shown in *Figure 7*. A Leeds and Northrup capsule-type, platinum resistance thermometer, *H*, is gripped in a slender, beryllium–

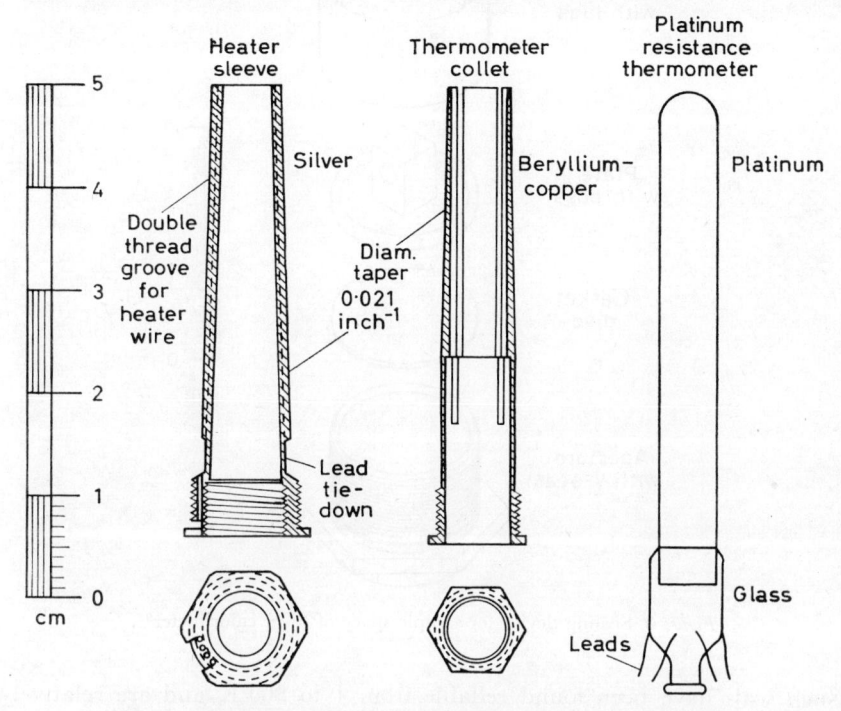

Figure 7. Thermometer-heater assembly of adiabatic calorimeter[91].

copper collet arrangement. This collet in turn is screwed into the tapered heater sleeve, to place the thermometer in contact with the heater sleeve by means of the beryllium–copper. The heater contains 250 Ω of 38-gauge B & S Karma wire wound bifilarly. A recessed groove of the heater sleeve is used to "tie down" the heater leads before they enter small holes under the threads and pass out to the calorimeter spool. Good thermal contact is achieved between thermometer, heater, sample and calorimeter since thermal equilibration is achieved typically within 15 min after an energy input. The spool, *I*, around which all lead wires pass ensures that the leads achieve the surface temperature of the calorimeter.

Figure 8 is an enlarged view of the closure used in place of the usual soldered closure. The vacuum-tight closure is achieved by forcing the annealed gold gasket against a circular knife edge. Gasket and plate are prevented from turning during closure to assure a clean cut into the gold. The aperture is 1 cm in diameter and constructed of stainless steel so as to avoid locking of the threads with the Monel cap. A glass chamber encasing a remotely operating wrench permits evacuation of the calorimeter to 10^{-6} torr, subsequent addition of helium, and seating of the closure. The lugs on the cap fit into recesses of the externally-operated wrench arrangement.

Cap
with lugs

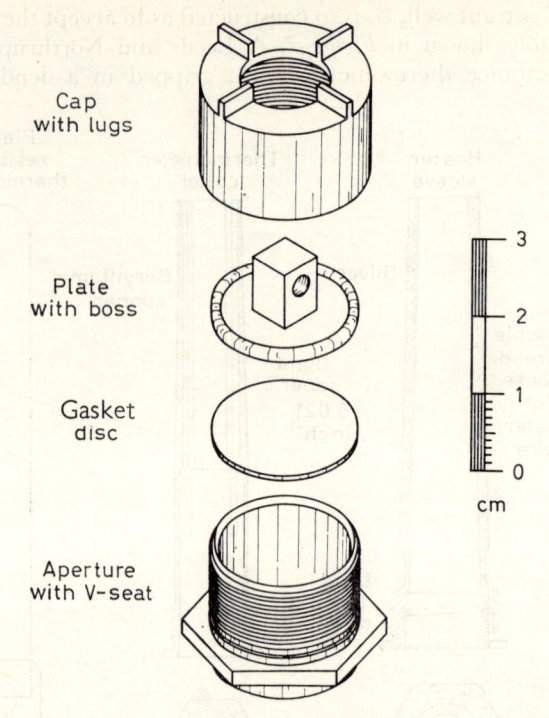

Plate
with boss

Gasket
disc

Aperture
with V-seat

Figure 8. Sealing device for sample space of silver calorimeter[91].

Such seals have been found reliable from 1 to 800°K and are relatively trouble free. A recently developed version of the same aperture is provided with a demountable operating device that enables coupling the calorimeter directly to a vacuum line, removing the cap and gasket, evacuating, distilling in the sample, adding helium gas for thermal conduction, and sealing the valve. Further details concerning the demountable thermometer-heater assembly, the nature of the seal for the sample space, and the mechanism for closing this seal under vacuum may be found elsewhere[81]. The convenience, reliability, and rapidity of handling permit loading of the calorimeter and reassembly of the cryostat in about 4 hours.

In use, the calorimeter, F, is suspended by a #40 Advance wire through a set-screw-activated fixture, B, on the adiabatic shield.

3. *Operational method*

The capsule-type, platinum resistance thermometer was calibrated by the National Bureau of Standards and its resistance checked in a special cell at the triple-point of water.

All measurements are made by the intermittent-heating method with the "zero" heat flow determined during suitable fore and after periods, as described in Section III. Occasionally, enthalpy-type determinations are made over larger ranges of temperature as a check on the integration of the heat capacity data. In all other respects, the operation of the calorimeter and the evaluation of the data parallel procedures previously described in conjunction with low-temperature adiabatic calorimetry in Chapter 5.

Although the two calorimeters already described have been designed primarily for thermal rather than for thermochemical measurements, it is obvious that with suitable modifications they might also be adapted to the latter use.

4. *Calibration of calorimeter*

Although independent calibrations of the thermometer and electrical measuring circuits had been performed, it was considered desirable and indeed necessary to check all calibrations by studying the heat capacity of a well characterized substance. For this purpose the heat capacity of a Calorimetry Conference standard sample of synthetic sapphire[81], Al_2O_3, was studied from 260 to 540°K. The sample (187·9954 g *in vacuo*) was loaded as previously described and a few torr of helium gas added for thermal conduction. The calorimeter represents about 20 per cent of the total measured heat capacity at the lower temperature and decreases to 15 per cent at higher temperatures. The data from this research agree well with those of West and Ginnings[88], whose data are shown also on the deviation plot of *Figure 9*. The zero deviation line represents the best smooth curve

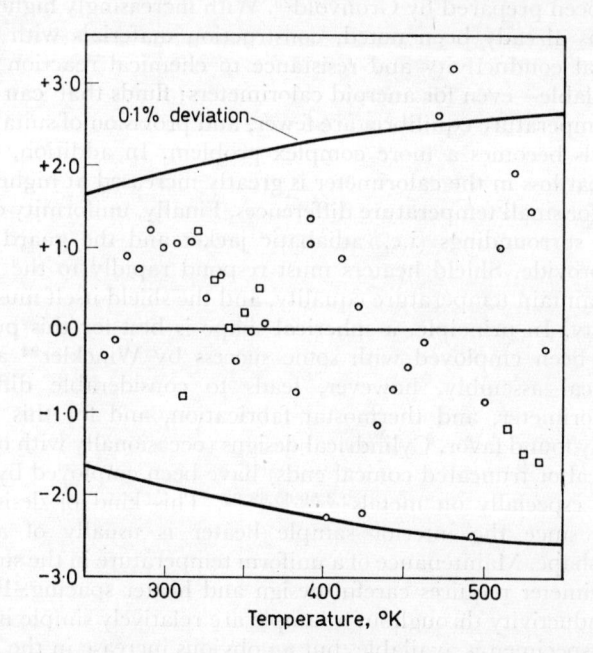

Figure 9. Deviation of the heat capacity of the Calorimetry Conference standard sample of synthetic sapphire (aluminum oxide) taken by Westrum and Trowbridge[81, 91], (○), and of corresponding data by West and Ginnings[88], (□), from the smooth curve of Westrum and Trowbridge.

through the data of this research, with most experimental points falling well within 0·1 per cent deviation limit. Both sets of heat capacity data above

500°K are more than 0·1 per cent higher than earlier published values of the National Bureau of Standards[81].

To ascertain the thermal interaction with the shield, a thermal head was caused by controlling the top-middle and bottom portions of the adiabatic shield at a small temperature displacement from zero. Under such conditions the heat-leak coefficient, λ, defined as the ratio of the time rate of change of temperature to the thermal head, may be computed. For this calorimeter, λ takes on a linear dependence with temperature which can be best expressed by the relation, $\lambda(\min^{-1}) = 4\cdot 3 \times 10^{-3} - 8\cdot 17 \times 10^{-5} T$ (T in °K). From these results it was possible to assure a nearly zero drift over the entire temperature range by judiciously offsetting the adiabatic shield control point. A 0·001°K uncertainty in the adiabaticity amounted to 0·03 per cent of the Al_2O_3 heat capacity for a typical 50-min experiment at 400°K. This 0·03 per cent is comparable to the observed scatter (cf. deviation plot, *Figure 9*) of the data in this temperature region and serves to emphasize the importance of strict adiabaticity.

V. General Discussion

The general difficulties with high-temperature calorimetry have been discussed by Kingery[37] and in depth by White[92]; an extensive, more recent survey has been prepared by Grønvold[25]. With increasingly higher temperatures, as has already been noted, construction materials with sufficiently high thermal conductivity and resistance to chemical reaction are not as readily available—even for aneroid calorimeters; fluids that can be stirred to obtain temperature equilibria are fewer, and provision of suitable insulating materials becomes a more complex problem. In addition, the rate of radiative heat loss in the calorimeter is greatly increased at higher temperatures, even for small temperature differences. Finally, uniformity of temperature in the surroundings (i.e., adiabatic jacket and the guard shields) is harder to provide. Shield heaters must respond rapidly to the controlling device to maintain temperature equality, and the shield itself must be of low heat capacity. In principle, a spherical shape is best for this purpose and indeed has been employed with some success by Winckler[94] and others. The spherical assembly, however, leads to considerable difficulties in sample, calorimeter, and thermostat fabrication, and for this reason has not generally found favor. Cylindrical designs (occasionally with one or more hemispherical or truncated conical ends) have been employed by a number of workers, especially on metals[2,3,59,65,68,93]. This kind of design is more convenient, since the interior sample heater is usually of an over-all cylindrical shape. Maintenance of a uniform temperature in the surroundings of the calorimeter requires careful design and heater spacing. Problems of thermal conductivity throughout a sample are relatively simple if a massive, conducting specimen is available, but an obvious increase in the complexity results when an ionic or molecular crystalline material in a finely-divided state is involved.

For electrically-conducting materials which can be fabricated in the form of rods, a relatively long cylindrical furnace has occasionally been employed[59]. In the limit, this design can be extended to the form of thin wires[6,22,96] or by the use of a capillary to enclose a thin "rod" of liquid metal.

Occasionally, essentially adiabatic conditions are sought by using impulses of a few seconds duration[49], giving a very rapid temperature rise [e.g., 100°K sec^{-1} or even an order of magnitude higher[36]]. Such methods have provided reasonably good agreement at temperatures below about 800°C.

Essentially adiabatic methods for directly determining enthalpies of reaction at elevated temperatures have also been employed by Kingery[37] and others[18,25,41,42].

VI. Survey of Adiabatic Calorimeters in the Intermediate and Higher Temperature Ranges

Information concerning high-temperature adiabatic calorimeters developed for general or specific purposes is contained in *Table 1*. Because of the difficulties involved, this information has not been evaluated or reviewed critically, but simply presented for coverage and to facilitate noting their most interesting and relevant features, their ranges of operation, and the claimed precision of data, together with comments concerning special applicability of these instruments.

Table 1. Survey of intermediate temperature adiabatic calorimeters

Author, temperature range, and precision	Construction, nature of sample, and operating technique	References, primary and derived
Armstrong 670 to 870°K ~0.5% at 670° ~1% at 870°	Rigid aneroid, in-furnace-type; samples may be machined conical cylinder of metal, loose particles, or powder sealed in helium. Continuous and intermittent heating.	2
Awbery and Griffiths 320 to 1170°K ~1%	Concentric stainless-steel radiation shields and low-thermal inertia heaters in water-cooled vessel at 0.1 torr pressure; cylindrical iron sample with entrant well for heater. Continuous heating.	3
Backhurst 870 to 1870°K 3 to 5%	Large, water-cooled tank containing successive rectangular enclosures of ceramic materials, molybdenum, and rhodium platinum radiation shields on rigid support. Massive, solid or liquid metallic sample in molybdenum or aluminum crucible. Continuous heating. Largely automatic adiabatic control.	4
Barger, Bolze, Hughes, and Medved 300 to 530°K ~5%	Simplified, semiautomatic for rapid heat capacity determination on lubricants.	5
Belaga, Coddington, and Marcus 310 to 530°K ~2%	Modified continuous-heating calorimeter for determination of heat capacities of 500 ml samples of organic liquid.	7
Berge and Blanc 80 to 800°K Several percent	Aneroid, pendant calorimeter for solid or powdered samples.	8
Bláha 270 to 520°K ~2%	Calorimetric thermostat in Dewar vessel for determination of heat capacity of liquids.	10

Table 1—continued

Author, temperature range, and precision	Construction, nature of sample, and operating technique	References, primary and derived
Braun and Kohlhaas 300 to 1700°K ~2%	Calorimeter for massive metal samples.	12 (cf. Braun and Kohlhass[11])
Dench 800 to 1670°K $\Delta Hr \pm 150$ cal (g atom)$^{-1}$	Multi-layered molybdenum sheet furnace and shield system of low thermal inertia; massive or divided metallic samples in unsealed tantalum capsule. Determination of heat contents, heat capacities, heats of transformation and fusion, and small endo- or exothermic heats of reaction between condensed phases.	19
Dole, Hettinger, Larson, Wethington, and Worthington 250 to 550°K ~0.6% (accuracy ~1.5%)	Aneroid calorimeter with automatic adiabatic control and temperature recording for heat capacity measurements on high polymeric materials. Samples placed on "trays" inside rectangular chambers.	21 (cf. Worthington, Marx, and Dole[97])
Hellwege, Knappe, and Wetzel 300 to 450°K ~1%	Calorimeter vacuum-jacketed by submarine in stirred, fluid, temperature-regulated bath.	27
Johnston, Sterrett, Craig, and Wallace 300 to 550°K ~0.2%	In-furnace-type of calorimeter for measurement of heat capacities and transitions in metals and alloys.	32 (cf. Sterrett and Wallace[73])
Jost, Oel, and Schneidermann 290 to 820°K ~0.3%	Rigid calorimeter of Armstrong-type for heat capacity of ionic crystals.	33
Kangro 300 to 1100°K	Calorimeter to study enthalpies of inorganic reactions and of mixing.	34
Kay and Loasby 300 to 970°K ~5%	Silver-plated tantalum crucible mounted in silica cradle surrounded by spherical copper furnace and spherical silver radiation shields for heat capacity determinations of radioactive plutonium metal. Self-heating provided by sample.	35
Krauss 960 to 1600°K ~5%	Quasi-adiabatic calorimeter for determining the heat capacity of cylindrical specimens of metal.	39
Krauss and Warncke 350 to 870°K ~0.5 to 1%	Quasi-adiabatic calorimeter for the determination of the heat capacity of a massive nickel specimen by a continuous heating method.	40
Kubaschewski and Walter 830 to 930°K In ΔHr ~2.5%	Massive shield, in-furnace-type calorimeter for the determination of enthalpy increments in alloy formation.	42
Lyusternik 320 to 1270°K ~1%	In-furnace-type calorimeter for determination of heat capacity and enthalpy effects in ferrous alloys. Automatic regulation by thyratron photo relays. Continuous heating.	43

Table 1—continued

Author, temperature range, and precision	Construction, nature of sample, and operating technique	References, primary and derived
Martin 300 to 570°K ~1%	Spherical stainless-steel calorimeter with heater and stirring fan surrounded by spherical adiabatic shield and vaccum jacket for determination of constant volume heat capacity of gases.	44 (cf. de Nevers and Martin[50]; and Hwang[31])
Martin and Rixon 310 to 770°K ~3%	Variation of Armstrong design for heat capacity of flat cylinders of metals.	45
Moser 320 to 970°K ~0.5%	In-furnace-type calorimeter for heat capacities of metals, oxides, etc., within a silver vessel.	47
Nagasaki and Takagi 270 to 1170°K ~1%	Armstrong-type calorimeter for heat capacity of metals and alloys by continuous heating at about 2°K min^{-1}	48 (Hirabayashi, Nagasaki, and Kono[28])
Nölting 320 to 920°K ~0.3%	Rigid, Armstrong-type, in-furnace calorimeter on samples of low thermal conductivity (e.g., ionic crystals) in entrant-well container. Intermittent heating due to slow rate of attainment of equilibrium.	51
North 370 to 1070°K ~2%	Rigid, Sykes-type, water-cooled-jacket, with automatic shield control and continuous heating. Uranium sample in steel calorimeter.	52
O'Neal and Gregory 320 to 730°K ~1%	Suspended, aneroid calorimeter with thin, automatically-controlled, radiation shields. Sample of transition metal halide in 60 ml space between concentric uranium glass spheres.	57 (O'Neal[56]; Gregory and O'Neal[24]; Oetting and Gregory[53, 54]; and Wydeven and Gregory[98])
Osborne, Stimson, and Ginnings 370 to 650°K ~0.2%	Calorimeter vessel capable of withstanding high pressure for determination of thermal properties of liquid and gaseous water. Continuous heating technique.	58
Phipps 270 to 373°K ~0.5%	Calorimeter designed for determination of heat capacity of liquids (e.g., milk) with avoidance of local overheating by heater design and stirring by oscillation of sample about vertical axis. Intermittent heating.	60
Popov and Gal'chenko 420 to 1070°K ~1%	In-furnace-type calorimeter for determination of the thermal properties of refractory oxides in a silica-glass tube coated with silver foil. The continuous heating method was employed.	61 (Popov, Gal'chenko, and Senin[62])
Raine, Richards, and Ryder 300 to 430°K ~5%	Williams- and Daniels-type calorimeter for measurement of heat capacity of polythene cylinder in copper can with external fins for stirring oil in silver calorimeter vessel. Fluid in outer bath used to provide adiabatic operation.	63

Table 1—continued

Author, temperature range, and precision	Construction, nature of sample and operating technique	References, primary and derived
Scheil and Normann 300 to 840°K ~6%	Cylindrical design calorimeter with massive shields for heat capacities in solid and liquid phases, enthalpies of transition and mixing in metallic systems contained in graphite crucible.	66
Schmidt and Leidenfrost 290 to 670°K ~0.2%	Spherical calorimeter and spheroidal shield system in a water bath for determination of the heat capacity of divided metals (e.g., nickel). The calorimeter is of gold-plated copper with a central spherical heater. Slow (2°K h^{-1}) continuous heating is employed.	67
Shmidt and Sokolov 300 to 1020°K ~0.4%	Rigid calorimeter for determination of heat capacities of fragmented samples of low thermal conductivity utilizing automatically regulated low-heat capacity shields with separately mounted shield heaters. Metallic sample vessel has axial heater and eccentrically mounted calorimeter. Intermittent heating.	69
Sinel'nikov 300 to 970°K ~1.5%	Moser-type calorimeter in silica-glass envelope for study of the $\alpha \rightleftharpoons \beta$ transition in quartz.	70
Stansbury, McElroy, Picklesimer, Elder and Pawel 320 to 1270°K ~0.5%	Calorimeter-in-tube-furnace type for heat capacity, enthalpies of transition, and mixing in bulk.	72 (cf. Brooks and Stansbury[14])
Stow and Elliott 300 to 570°K ~0.5% (accuracy ~1%)	Rugged, corrosion resistant submarine-type with internal aneroid adiabatic shield and surrounding oil bath for heat capacity of solids and liquids, especially rosins. Intermittent heating.	74 (cf. Horowitz and Stow[30])
Sykes and Jones 290 to 720°K ~0.5% (accuracy ~1%)	Calorimeter in silica tube within horizontal-tube-furnace for study of thermal properties of cylindrical bulk-metal samples with internal heater within massive copper block by continuous heating technique.	76
Tret'yakov, Troshkina, and Khomyakov 370 to 1120°K ~1%	Cylindrical, suspended calorimeter-in-furnace-type for thermal properties of high magnetic coercivity alloys. Massive steatite block for continuous heating method.	80
Trowbridge and Westrum 250 to 600°K (see text)	Calorimeter for thermal properties of poorly conducting samples. (See text.)	82 (cf. Trowbridge[81])
Verhaeghe, Robbrecht, and Bruynooghe 290 to 620°K ~1%	Vacuum, Nernst-type calorimeter for cylindrical specimens of ferrite with external heater imbedded in helical grooves and surrounding cylindrical copper adiabatic shield.	83
Warfield, Petree, and Donovan 300 to 430°K ~2%	Nernst-type calorimeter for rapid measurement of thermal properties of polymeric materials.	84

Table 1—continued

Author, temperature range, and precision	Construction, nature of sample, and operating technique	References, primary and derived
West and Ginnings 300 to 800°K ~0·05 to 0·14%	High-precision calorimeter for thermal properties of poorly conducting samples (see text).	90 (cf. West[85])
Williams and Daniels 300 to 380°K ~1%	Submarine-type calorimeter for determination of heat capacity of liquids, immersed in ferric chloride solution in glycerol heated both by immersion heater and by direct passage of current through the solution.	93
Worthington, Marx, and Dole 250 to 550°K ~0·1% below 430° ~0·5% at 550°	Modified aneroid, cylindrical calorimeter for heat capacities of high polymeric materials.	97

VII. Acknowledgment

One of us (E.F.W.) wishes to express his appreciation to the Division of Research of the U.S. Atomic Energy Commission for its generous support of his research endeavor, which to a large extent enabled his research in adiabatic calorimetry.

VIII. References

[1] Andrews, D. H., *J. Am. Chem. Soc.*, **48**, 1287 (1926).
[2] Armstrong, L. D., *Can. J. Res.* **28A**, 44, 51 (1950).
[3] Awbery, J. H. and E. Griffiths. *Proc. Royal Soc. (London)*, **A174**, 1 (1940).
[4] Backhurst, I., *J. Iron Steel Inst. (London)*, **189**, 124 (1958).
[5] Barger, J. W., C. C. Bolze, R. L. Hughes, and T. M. Medved, *Lubrication Eng.*, **16**, 105 (1960).
[6] Behrens, W. U. and C. Drucker, *Z. Phys. Chem.*, **113**, 79 (1924).
[7] Belaga, M. W., D. M. Coddington, and H. Marcus, *Rev. Sci. Instr.*, **27**, 948 (1956).
[8] Berge, P. and G. Blanc, *J. Phys. Radium*, **21**, No. 7 (Suppl.), 129 A (1960).
[9] Birky, M. M., unpublished data.
[10] Bláha, O., *Cesk. Casopis Fys.*, **12**, 216 (1962).
[11] Braun, M. and R. Kohlhaas, *Z. Naturforsch.*, **19a**, 663 (1964).
[12] Braun, M. and K. Kohlhaas, *Arch. Eisenhüttenw.*, in press.
[13] Bridgman, P. W., *Logic of Modern Physics*, Macmillan, New York, 1946.
[14] Brooks, C. R. and E. E. Stansbury, *Acta Met.*, **11**, 1303 (1963).
[15] Bros, J. P., in *Les Dèveloppements Récents de la Microcalorimétrie et de la Thermogenèse*, Editions du Centre National de la Recherche Scientifique, Paris, 1967, p.275.
[16] Calvet, E. and H. Prat, *Microcalorimétrie — Applications Physico-Chimiques et Biologiques*, Masson, Paris, 1956.
[17] Carlson, H. G. "Thermodynamic Properties of Methyl Alcohol, 2-Methyl- and 2,5-Dimethylthiophene and 2-Methylfuran," Ph.D. Dissertation, University of Michigan; U.S. Atomic Energy Commission Report TID-15153, 1962.
[18] Chipman, J. and N. J. Grant, *Trans. Am. Soc. Metals*, **31**, 365 (1943).
[19] Dench, W. A., *Trans. Faraday Soc.*, **59**, 1279 (1963).
[20] Dickinson, H. C., *Natl. Bur. Std. (U.S.), Bull.*, **11**, 189 (1915).
[21] Dole, M., W. P. Hettinger, Jr., N. Larson, J. A. Wethington, and A. E. Worthington, *Rev. Sci. Instr.*, **22**, 812 (1951).
[22] Förster, F. and G. Tschentke, *Z. Metallk.*, **32**, 191 (1940).
[23] Furukawa, G. T., personal communication.
[24] Gregory, N. W. and H. E. O'Neal, *J. Am. Chem. Soc.*, **81**, 2649 (1959).
[25] Grønvold, F., in *Thermodynamic*, Vol. I, "International Atomic Enery Agency," Vienna, 1967, p. 35.

[26] Haworth, E. and D. H. Andrews, *J. Am. Chem. Soc.*, **50**, 2998 (1928).
[27] Hellwege, K. H., W. Knappe, and W. Wetzel, *Kolloid-Z.*, **180**, 126 (1962).
[28] Hirabayashi, M., S. Nagasaki, and H. Kono, *J. Appl. Phys.*, **28**, 1070 (1957).
[29] Hoge, H. J., *Rev. Sci. Instr.* **20**, 59 (1949).
[30] Horowitz, K. H. and F. S. Stow, Jr., *Anal. Chem.*, **21**, 1571 (1949).
[31] Hwang, Yu-Tang, "The Constant Volume Heat Capacities of Gaseous Tetrafluoromethane, Chlorodifluoromethane, Dichlorotetrafluoroethane, and Chloropentafluoroethane," Ph.D. Dissertation, University of Michigan, 1961; *Dissertation Abstr.*, **22**, 2322 (1962).
[32] Johnston, W. V., K. F. Sterrett, R. S. Craig, and W. E. Wallace, *J. Am. Chem. Soc.*, **79**, 3633 (1957).
[33] Jost, W., H. J. Oel, and G. Schniedermann, *Z. Physik. Chem. (Frankfurt)*, **17**, 175 (1958).
[34] Kangro, W., *Z. Elektrochem.*, **34**, 253 (1928).
[35] Kay, A. E. and R. G. Loasby, *Phil. Mag.*, **9**, 37 (1964).
[36] Khotkevich, V. I. and N. N. Bagrov. *Dokl. Akad. Nauk SSSR*, **81**, 1055 (1951).
[37] Kingery, W. D., *Property Measurements at High Temperatures*, Wiley, New York, 1959, Chapter 10.
[38] Kleppa, O. J., *J. Phys. Chem.*, **64**, 1937 (1960).
[39] Krauss, F., *Z. Metallk.*, **49**, 386 (1958).
[40] Krauss, F. and H. Warncke, *Z. Mettalk.*, **46**, 61 (1955).
[41] Kubaschewski, O. and H. Villa, *Z. Elektrochem.*, **53**, 32 (1949).
[42] Kubaschewski, O. and A. Walter, *Z. Elektrochem.*, **45**, 630 (1939).
[43] Lyusternik, V. E., *Pribory i Tekhn. Eksperim.*, **1959**, 127 (1959); *Instr. Exptl. Tech. (USSR)*, **2**, 647 (1959, Publ. 1960).
[44] Martin, J. J., "Constant Volume Heat Capacity of Gases," in J. F. Masi and D. H. Tsai, eds., *Progress in International Research on Thermodynamic and Transport Properties*, American Society of Mechanical Engineers, Academic Press, New York, 1962.
[45] Martin, W. H., and J. Rixon, *J. Sci. Instr.*, **36**, 179 (1959).
[46] Meyers, C. H., personal communication.
[47] Moser, H., *Physik. Z.*, **37**, 737 (1936).
[48] Nagasaki, S. and Y. Takagi *J. Appl. Phys. (Japan)*, **17**, 104 (1948).
[49] Nathan, A. M., *J. Appl. Phys.*, **22**, 234 (1951).
[50] de Nevers, N. H. and J. J. Martin, *A. I. Ch. E. Journ.*, **6**, 43 (1960).
[51] Nölting, J., *Ber. Bunsenges. Physik. Chem.*, **67**, 172 (1963).
[52] North, J. M., *At. Energy Res. Estab. (G. Brit.) Rept.*, M/R 1016 (1956); *Chem. Abstr.*, **50**, 10506f (1956).
[53] Oetting, F. L. and N. W. Gregory, *J. Phys. Chem.*, **65**, 138 (1961).
[54] Oetting, F. L. and N. W. Gregory, *J. Phys. Chem.*, **65**, 173 (1961).
[55] Ohlmeyer, P., *Z. Naturforsch.*, **1**, 30 (1946).
[56] O'Neal, H. E., "A New Adiabatic High Temperature Calorimeter and the Heat Capacity of Iron(II) Bromide," Ph.D. Dissertation, University of Washington, 1957; *Dissertation Abstr.*, **18**, 1276 (1958).
[57] O'Neal, H. E. and N. W. Gregory, *Rev. Sci. Instr.*, **30**, 434 (1959).
[58] Osborne, N. S., H. F. Stimson, and D. C. Ginnings, *J. Res. Natl. Bur. Std.*, **18**, 389 (1937).
[59] Pallister, P. R., *J. Iron Steel Inst.*, **161**, 87 (1949).
[60] Phipps, L. W., *J. Sci. Instr.*, **32**, 109 (1955).
[61] Popov, M. M. and G. L. Gal'chenko, *Zh. Obshch. Khim.*, N.S., **21**, 2220 (1951).
[62] Popov, M. M., G. L. Gal'chenko, and M. D. Senin, *Zh. Neorg. Khim.*, **3**, 1734 (1958); *Russ. J. Inorg. Chem.*, **3**, 18 (1958).
[63] Raine, H. C., R. B. Richards, and H. Ryder, *Trans. Faraday Soc.*, **41**, 56 (1945).
[64] Riggs, F. B., Jr., *Rev. Sci. Instr.*, **32**, 366 (1961).
[65] Roberts, H. S., *Am. J. Sci.*, **35A**, 273 (1938).
[66] Scheil, E. and W. Normann, *Z. Metallk.*, **51**, 159 (1960).
[67] Schmidt, E. O. and W. Leidenfrost, "Adiabatic Calorimeter for Measurements of Specific Heats of Powders and Granular Materials at 0 to 500°C.," in J. F. Masi and D. H. Tsai, eds., *Progress in International Research on Thermodynamic and Transport Properties*, American Society of Mechanical Engineers, Academic Press, New York, 1962.
[68] Seekamp, H., *Z. Anorg. Allgem. Chem.*, **195**, 345 (1931).
[69] Shmidt, N. E. and V. A. Sokolov, *Russ. J. Inorg. Chem.*, **5**, 797 (1960).
[70] Sinel'nikov, N. N., *Dokl. Akad. Nauk SSSR*, **92**, 369 (1953).
[71] Smith, R. H. and D. H. Andrews, *J. Am. Chem. Soc.*, **53**, 3644 (1931).
[72] Stansbury, E. E., D. L. McElroy, M. L. Picklesimer, G. E. Elder, and R. E. Pawel, *Rev. Sci. Instr.*, **30**, 121 (1959).
[73] Sterrett, K. F. and W. E. Wallace, *J. Am. Chem. Soc.*, **80**, 3176 (1958).
[74] Stow, F. S., Jr. and J. H. Elliott, *Anal. Chem.*, **20**, 250 (1948).
[75] Stull, D. R., *J. Am. Chem. Soc.*, **59**, 2726 (1937).
[76] Sykes, C. and F. W. Jones, *J. Inst., Metals*, **59**, 257 (1936).
[77] Thomas, S. B. and G. S. Parks, *J. Phys. Chem.*, **35**, 2091 (1931).

[78] Tian, A., *Bull. Soc. Chim. France*, **33**, 427 (1923).
[79] Tian, A., *J. Chim. Phys.*, **30**, 665 (1933).
[80] Tret'yakov, Yu. D., V. A. Troshkina, and K. G. Khomyakov, *Zh. Neorg. Khim.*, **4**, 5 (1959); *Russ. J. Inorg. Chem.*, **4**, 2 (1959).
[81] Trowbridge, J. C., "Thermodynamic Functions and Phase Transitions including Fusion for Perchloric Acid, Pentaerythrityl Fluoride and Triethylenediamine," Ph.D. Dissertation, University of Michigan, 1963; U.S. Atomic Energy Commission Report TID–17352, 1963.
[82] Trowbridge, J. C., and E. F. Westrum, Jr., *J. Phys. Chem.*, **67**, 2381 (1963).
[83] Verhaeghe, J. L., G. G. Robbrecht, and W. M. Bruynooghe, *Appl. Sci. Res. Sect. B*, **8**, 128 (1960).
[84] Warfield, R. W., M. C. Petree, and P. Donovan, *SPE (Soc. Plastics Eng.) J.*, **15,** 1055 (1959).
[85] West, E. D., *J. Am. Chem. Soc.*, **81**, 29 (1959).
[86] West, E. D., *Rev. Sci. Instr.*, **31**, 896 (1960)
[87] West, E. D., *J. Res. Natl. Bur. Std.*, **67A**, 331 (1963).
[88] West, E. D. and D. C. Ginnings, *J. Phys. Chem.*, **61**, 1573 (1957).
[89] West, E. D. and D. C. Ginnings, *Rev. Sci. Instr.*, **28**, 1070 (1957).
[90] West, E. D. and D. C. Ginnings, *J. Res. Natl. Bur. Std.*, **60**, 309 (1958).
[91] Westrum, E. F., Jr. and J. C. Trowbridge, unpublished data.
[92] White, W. P., *J. Phys. Chem.*, **34**, 1121 (1930).
[93] Williams, J. W. and F. Daniels, *J. Am. Chem. Soc.*, **46**, 903 (1924).
[94] Winckler, J. R., *J. Am. Ceram. Soc.*, **26**, 339 (1943).
[95] Wittig, F. E., *Z. Elektrochem.*, **54**, 288 (1950).
[96] Worthing, A. C., *Phys. Rev.*, **12**, 199 (1918).
[97] Worthington, A. E., P. C. Marx, and M. Dole, *Rev. Sci. Instr.*, **26**, 698 (1955).
[98] Wydeven, T. J. and N. W. Gregory, *J. Phys. Chem.*, **68**, 3249 (1964).
[99] Ziegler, W. T. and C. E. Messer, *J. Am. Chem. Soc.*, **63**, 2694 (1941).

CHAPTER 10

Vapor-flow Calorimetry

JOHN P. MCCULLOUGH[†]

*Bartlesville Petroleum Research Center, Bureau of Mines,
Bartlesville, Oklahoma, U.S.A.*

GUY WADDINGTON

*Office of Critical Tables, National Academy of Sciences–National
Research Council, Washington, D.C., U.S.A.*

Contents

I.	Introduction	370
	1. Brief Survey	370
	2. Constant Flow Methods	370
II.	Non-adiabatic Calorimeters	372
	1. Apparatus	373
	A. The Vaporizer Vessel	374
	(1) Construction of Vessel	374
	(2) Vaporizer Heater	375
	B. The Vapor Calorimeter Vessel	375
	(1) Construction of Vessel	375
	(2) Heating Element	376
	(3) Thermometers	376
	C. The Flow System	377
	(1) Flow Path	377
	(2) Manostat	377
	(3) Sample Collection System	377
	(4) Liquid Introduction System	377
	D. The Thermostats	378
	2. Operating Procedures	379
	A. Establishment of Flow Conditions	379
	B. Power and Temperature Measurements	379
	C. Heat of Vaporization Experiments	380
	(1) Procedures	380
	(2) Sequence of Observations	381
	D. Heat Capacity Experiments	381
	(1) Procedures	381
	(2) Sequence of Observations	382
	(3) Observational Techniques	383
	3. Calculation of Experimental Results	383
	A. Heat of Vaporization	383
	B. Heat Capacity	385
	(1) Practical Method	386
	(2) General Method	387
	4. Derived Results	388
	A. Second Virial Coefficient and Equation of State	389
	B. Ideal Gas Heat Capacity	389
	C. Other Useful Information	389

[†]Present address: Central Research Division Laboratory, Mobil Research and Development Corporation, Princeton, New Jersey, U.S.A.

JOHN P. McCULLOUGH and GUY WADDINGTON

III. Discussion 390
   1. Accuracy of Results 390
     A. Accuracy of Heat of Vaporization Measurements .. 390
     B. Accuracy of Heat Capacity Measurements 391
     C. Accuracy of Second Virial Coefficient Values 391
   2. Reference Substances 392
     A. Heat of Vaporization Measurements 392
     B. Heat Capacity Measurements 392
IV. References 393

I. Introduction

1. *Brief survey*

One of the most important properties in the thermodynamic characterization of a pure substance is its heat capacity in both the real and ideal gas states. The reasons for its importance are twofold: (*i*) accurate knowledge of the heat capacity of the real gas as a function of temperature at constant pressure, C_P, and of its derivative with respect to pressure at constant temperature, $(\partial C_P/\partial P)_T$, are important parts of the thermodynamic description of the real gas; and (*ii*) knowledge of the value of $C_P°$, the heat capacity at constant pressure for the ideal gas state, provides a powerful guide in establishing vibrational assignments from spectroscopic data, in determining the heights of barriers to internal rotation, and in deciding questions of rotational isomerism. It is not surprising, then, that experimental determination of gas heat capacity has received much attention in recent years and has been brought to the state of a precision measurement.

In the past half century many experimental methods, both direct and indirect, have been described in the literature for determination of C_P, C_V or the ratio of these two, commonly designated γ. The early ones were reviewed thoroughly in Partington and Shilling's "Specific Heat of Gases" in 1924[28]. Methods used since that date, both new and modifications of early methods, have been described briefly and referenced by Masi[16]. More recently Rowlinson[31] has outlined the various methods that have been used in the past half century.

The title of this Chapter is "Vapor-flow Calorimetry". The reason for selecting this title is that of all the methods described in the literature for determining the heat capacity of vapors only variants of the constant flow method have been used consistently to accumulate substantial bodies of data of high accuracy. True, other methods have been effective for special situations. For example Kistiakowski and Rice[14] made good use of the isentropic expansion method to determine the heat capacities of C_2H_6 and C_2D_6 gases near room temperature. Brinkworth[3, 4], and Eucken and von Lude[9] have also used the same method. Michels and Strijland[23] have used the constant volume method to obtain C_V for CO_2 under pressure. But neither these methods, nor many others reviewed by Masi[16], with the exception of constant flow methods, can be applied to a sufficient variety of compounds, over a useful range of temperature and pressure, and give accurate results.

2. *Constant flow methods*

The constant flow method, originated by Callendar and Barnes[5] for liquids, was adapted to gases by Swann[37]. In this method a steady flow of gas, in a suitable tubular calorimeter, is caused to flow past a thermometer, a heater, then over a second thermometer. With a constant flow of gas, and constant power input to the electrical heater, a steady state is established in the calorimeter. By measurement of the rate of the flow of gas, of power input to the heater, and initial and final temperature of the gas, the apparent heat capacity of the gas may be calculated. Heat losses occur along the path of the flowing gas for which corrections must be made to obtain the true heat capacity. The design of the particular calorimeter determines the way in which the correction for heat loss is made.

Among the several general designs of constant flow calorimeters to be found described in the literature, a differentiation may be made according to the dependence of heat loss on flow rate of the gas. In calorimeters described by Scheele and Heuse[32], and by Bennewitz and coworkers[1, 2], the fractional loss is inversely proportional to the square of the flow rate. This design has not been made into a modern precision method. In a design initiated by Pitzer[30] and improved by U.S. Bureau of Mines workers (hereafter designated as the non-adiabatic method[21, 40]), fractional heat loss varies inversely as the flow rate. A third variant is the "nearly-adiabatic" flow calorimeter in which, by heated shielding, the heat loss is reduced to a very low level, although small corrections for residual heat loss must still be made. Examples of the "nearly-adiabatic" design are to be found in the work of Osborne, Stimson and Sligh[25], and Scott and Mellors[36] both of the U.S. National Bureau of Standards (NBS).

Both the non-adiabatic and nearly adiabatic methods have been the basis for the determination of substantial amounts of accurate data. Each method has its advantages and disadvantages.

The NBS design is adapted to the study of gases of low molecular weight. Thus C_P for ammonia[25], oxygen[38], carbon dioxide[15], butadiene[36] and others have been studied by this method. Because of the rather small diameter of the tubing in the flow path, and its length, considerable pressure drop occurs as gas flows through the calorimeter, and this fact limits its application to studies at modest pressures and, as stated, to gases of low molecular weight. The generation of constant flow is by throttling from a cylinder or other temperature-controlled vessel, and the determination is by condensation and weighing of the effluent vapor on the downstream side of the calorimeter. This method of producing, controlling, and measuring flow is best suited to low boiling fluids. The upper attainable temperature limit of the nearly adiabatic method is that of the melting point of the solder used in fabrication of the calorimeter. This method in expert hands has yielded excellent results, but its limited range of applicability and the difficulties of fabrication and operation of the equipment have caused it to fall into disuse. For this reason it will not be described in detail here. Any reader wishing to consider the method seriously is referred to references 15, 25, and 36.

The third variant of the constant flow method, the "non-adiabatic" type due to Pitzer[30] and Bureau of Mines workers[21, 40], is the only one which has been used widely in the last decade, hence will be the only one to be described in detail in these pages. It meets the following criteria of a good flow calorimeter: (*i*) a constant accurately measured rate of flow; (*ii*) a heater to introduce an accurately measured amount of heat into the flowing gas; (iii) thermometers that measure the temperature increase of the gas caused by the known power input; and (iv) a design that prevents excessive heat losses from the flowing gas to the environment and that permits accurate corrections for the loss occurring.

The essential parts of the non-adiabatic calorimeter are (i) a cycling vaporizer for electrical evaporation of liquid to produce the constant flow of gas; (ii) the calorimeter containing a platinum resistance thermometer, heater and two more resistance thermometers, all in the path of the flowing gas; and (iii) a condensing system that can either return the condensed gas to the vaporizer for recycling *or* divert the flowing gas to a cold trap for a measured time interval, thus providing a measure of the flow rate and incidentally a value of the heat of vaporization of the liquid for the conditions of the experiments.

As with the "nearly adiabatic" type of calorimeter, the method of cycling tends to impose some limitations on the classes of substances that can be studied. Very low and high boiling fluids cannot be handled with existing designs, but the very large number of liquids with normal boiling points between about 15°C and a little above 100°C can be investigated. Heat capacity at constant pressure can be determined at pressures from about 0·1 atm to about 2 atm. The resulting values of C_P, plotted against pressure, permit an accurate extrapolation, nearly linear, to zero pressure thus providing a value of $C_P°$, a very important quantity in statistical thermodynamics. The slope of the linear plot provides a value of $(\partial C_P/\partial P)_T$, also a very useful quantity.

Regarding the possible temperature range of heat capacity measurements attainable with the non-adiabatic method, this range is limited at the low end by the boiling point of the liquid of interest, at the minimum operable temperature of the cycling vaporizer (e.g., a temperature corresponding to a vapor pressure of about 0·1 atm) and at the high end by materials of construction of the calorimeter, the functioning of the thermostat in which the calorimeter is situated, and the increasing heat losses by the silvered surfaces of the calorimeter encountered at high temperature. It should also be mentioned that the materials of construction in the path of the flowing gas may be subject to corrosive action by chemically active gases or, in rare cases, may cause catalytic decomposition of the gas. After these general introductory remarks, attention will now be given to a detailed account of a typical non-adiabatic calorimeter system, its operation, and the results obtained.

II. Non-adiabatic Calorimeters

In calorimeter systems of the type developed by Pitzer[30] and others[40], no attempt is made to establish adiabatic conditions. The calorimeter is designed so that heat loss occurs only by radiation, and the experiments are made in a way that allows accurate corrections for radiative heat loss. In principle, this method can be used to measure the heat capacity of any gaseous substance, but all modern systems using the method have been designed to operate with substances that are normally liquid at room temperature. Such systems include a calorimetric vaporizer, or boiler, which produces a steady flow of vapor and which allows measurements of both heat of vaporization and vapor heat capacity in the same system. The system described in the following is that developed in the United States by the Bureau of Mines[21, 22, 40]. Similar systems have been used by other investigators in the United States[20, 29, 41], and one incorporating a number of significant modifications was developed recently by the National Physical Laboratory in England[11].

1. *Apparatus*

A complete vapor-flow calorimeter system is illustrated schematically in *Figure 1*. The principal parts of the system are: (i) the calorimetric vaporizer; (ii) the vapor-flow calorimeter vessel; (iii) the flow system, including the sample collection, liquid introduction, and pressure regulation devices;

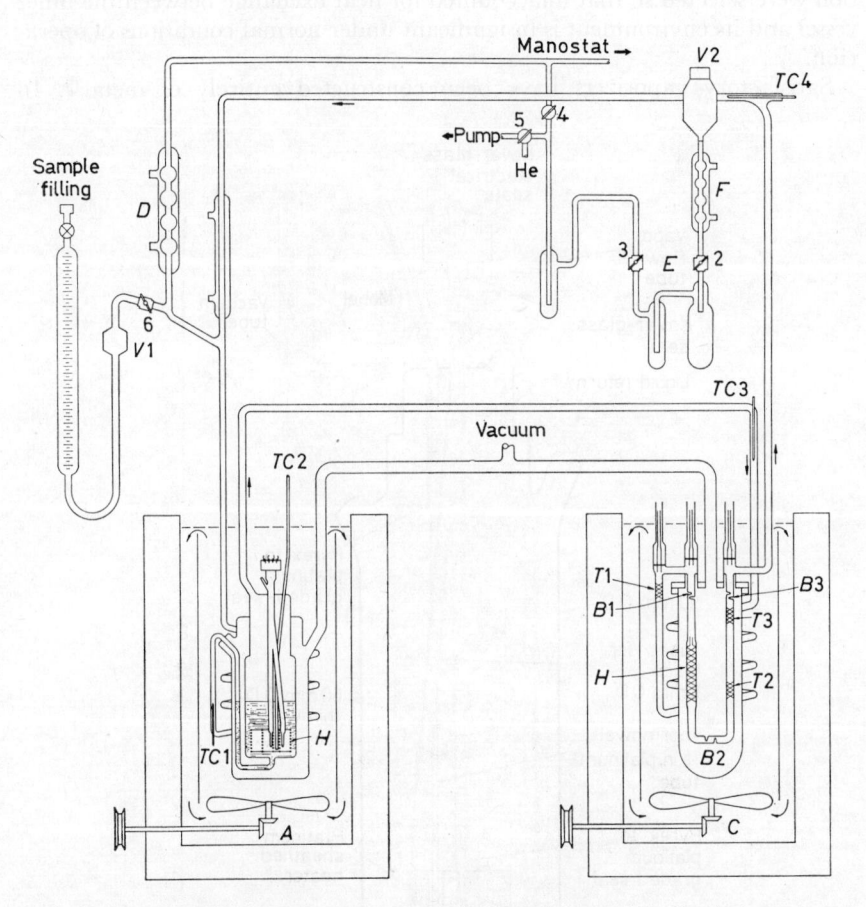

Figure 1. Vapor flow calorimeter system. A, vaporizer vessel thermostat; C, calorimeter vessel thermostat; D, condensers; E, sample receiver; 2 and 3, sample receiver valves and connections to system; 4, 5, and 6, valves; $V1$ and $V2$, solenoid-operated valves; $T1$, $T2$, and $T3$, platinum resistance thermometers (not shown: thermometers $T4$ and $T5$ in calorimeter and vaporizer vessel thermostats, respectively); H, heaters; $TC1$, $TC2$, $TC3$, and $TC4$, thermocouples; $B1$, $B2$, and $B3$, indentations for baffles.

373

and (*iv*) the thermostats. These parts are described in detail in the following paragraphs.

A. The Vaporizer Vessel

(*1*) *Construction of Vessel.* A calorimetric vaporizer designed for use with corrosive organic liquids is shown in detail in *Figure 2*. The inner glass vessel contains the liquid sample, a heating element, and a thermocouple well. It is constructed from thin-walled glass (1 mm thick) to reduce its heat capacity and thermal lags. Glass tubulation is provided for entry of liquid and exit of vapor. The inner vessel is surrounded by an outer glass vessel to provide an insulating vacuum jacket. The surfaces within the vacuum jacket are silvered to reduce radiative heat exchange. Other details of the vessel's construction are given in *Figure 2*. The dimensions and materials of construction were selected so that unaccounted for heat exchange between the inner vessel and its environment is insignificant under normal conditions of operation.

Satisfactory vaporizers have been constructed entirely of metal[39]. In

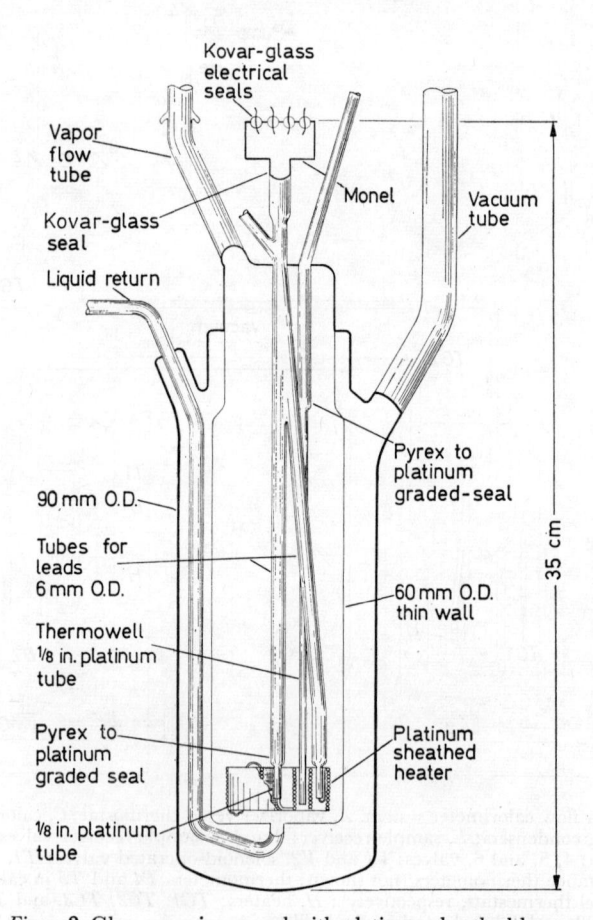

Figure 2. Glass vaporizer vessel with platinum-sheathed heater.

principle, a metal vaporizer has the advantages of smaller heat capacity and faster thermal response than a glass vaporizer. However, these advantages are not important when the operating procedures described below are followed. Glass vaporizers are recommended for general purpose systems because they are more inert chemically and less susceptible to leak troubles.

(2) *Vaporizer Heater.* The heating element of the vaporizer, one of the most critical parts in the design of a flow calorimeter, is used to generate a steady flow of vapor. For highest accuracy, the flow rate should not fluctuate more than about 0·01 per cent. The design of this element must be such that superheating of the liquid is minimized and that entrainment of liquid droplets in the vapor stream is avoided. If the electrical element can be exposed directly to the substance being studied, heaters of the kind described in references 11 and 40 give satisfactory results. The platinum-sheathed element shown in *Figure 2* has been found to give excellent results and can be used with substances that are corrosive or electrically conducting. This element consists of glass-insulated manganin wire (10 Ω resistance, No. 24 AWG) sheathed by a platinum tube, 0·040 in. dia., 0·005 in. wall. In constructing the heater assembly, the element is wrapped with glass thread and then wound in three concentric helices, which are spaced and supported by a Mycalex† framework (not shown). The ends of the platinum sheath are gold-welded to $\frac{1}{8}$ in. dia. platinum tubes (before the resistance wire is installed) which, in turn, are joined through graded glass seals to borosilicate glass tubes. The heater current leads (not shown) are brought through the platinum and glass tubes and out of the vaporizer through a ring seal. Potential leads are attached to the current leads about 5 mm above the platinum–glass seal. The current and potential leads are No. 16 and No. 24 AWG silver wire. With this arrangement, corrosive compounds come into contact with inert materials only, and the electrical circuit is insulated from the boiling substance. In operation, very small bubbles of vapor form over the entire surface of the heating element, and because each coil of the element is spaced from the others, the bubbles rise to the liquid surface without coalescing. Experience has shown that bubbles much larger than 1 mm dia. cause more turbulent ebullition and entrainment of liquid droplets in the vapor stream leaving the vaporizer. The nichrome heater protected by glass coated with carborundum powder used by the National Physical Laboratory[11] is also designed to eliminate entrainment.

B. THE VAPOR CALORIMETER VESSEL

(1) *Construction of Vessel.* A typical vapor calorimeter is shown in *Figure 3*. As shown in the figure, vapor enters the vessel and passes through a thin-walled U-tube of borosilicate glass containing a heating element and two platinum resistance thermometers. The outer surface of the tube is silvered. Semi-circular indentations in the U-tube mix the vapor and stop radiation along the axis of the tube. The U-tube is surrounded by a metal vessel to provide an insulating vacuum. A glass outer jacket is equally satisfactory calorimetrically, but the metal jacket facilitates construction and repair of

†Mycalex, compression molded glass-bonded mica, is manufactured by Mycalex Corp. of America.

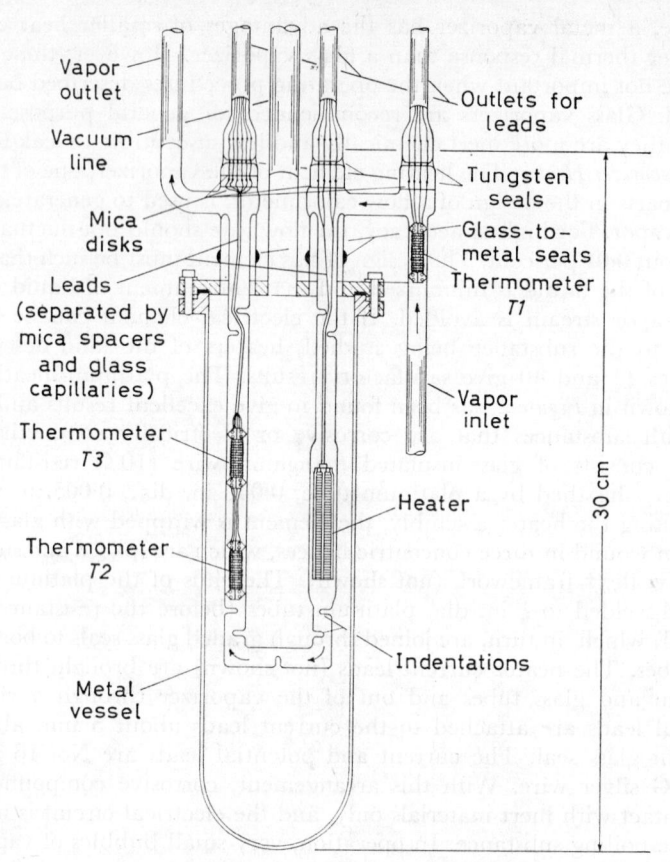

Figure 3. Metal-jacketed flow calorimeter vessel.

the calorimeter. The inner surface of the jacket is also silvered, either electrolytically when a metal jacket is used or chemically with a glass jacket.

(2) *Heating Element*. The heating element shown in *Figure 3* consists of six 3 mm dia. helices of No. 30 AWG nichrome wire connected in a parallel-series arrangement to give an effective resistance of 25 Ω. The six helices are grouped about a central glass tube and insulated from each other by mica strips. Two glass rings hold the assembly together and space it from the U-tube wall. Current and potential leads of No. 24 gold are attached at the upstream side of the heater. They are brought out of the calorimeter against the vapor stream to eliminate conduction heat losses along the leads. The leads are brought out of the flow system into the protective leads tube by tungsten seals.

(3) *Thermometers*. The two platinum resistance thermometers are placed in the other leg of the U-tube to isolate them from radiation from the heater. The baffles in the end of the U-tube also stop radiation as well as mix the heated vapor. These thermometers, *T2* and *T3*, and also *T1* outside the calorimeter, are of the coiled-filament strain-free type with ice-point resistances of about 25 Ω. Such thermometers have been found capable of with-

standing for many years the high vapor velocities normally met in operation. Because of the high vapor velocities, the thermometers are fixed in the tube by shrinking the glass to the mica cross at four points of contact at each end. Thermometers $T2$ and $T3$ are connected in series, and the two current and four potential leads of platinum wire are brought out of the calorimeter in the direction of the flowing gas. They are spot-welded or silver-soldered to tungsten wires sealed through the glass walls.

Because the effluent gas is hotter than the thermostat fluid surrounding the vapor exit tube and thermometer leads tube, relatively large temperature gradients can occur in the vicinity of the tungsten seals, leading to undesirable residual e.m.f.s in the thermometer circuit. To minimize such thermal e.m.f.s a close fitting mica disk, with holes for passage of platinum leads, is placed just above the vapor exit tube so that the vapor surrounding the platinum–tungsten joint is stagnant. Also, the tungsten seals are made as far above the vapor exit as possible, and the vapor outlet tube is led away from the seals, so as to reduce temperature gradients as much as possible in the vicinity of the platinum–tungsten junctions.

C. The Flow System

(1) Flow Path. The flow calorimeter is a cyclic system. Vapor from the vaporizer passes to the calorimeter thermostat through 8 mm inner dia. glass tubing, which is wrapped with a resistance-wire heating element and glass wool insulation. It then passes through a helical tempering coil made from 2 m of 8 mm inner dia. glass tubing, past thermometer $T1$, and into the calorimeter vessel. From the calorimeter vessel, the vapor passes to the three-way, solenoid valve[22], which also is heated. After leaving the valve, the vapor is condensed and returns to the vaporizer through another tempering coil, which is made of 2 m of 6 mm inner dia. glass tubing. The temperatures of the two vapor flow lines and the three-way valve are individually controlled by the use of thermocouples located as shown in *Figure 1*.

(2) Manostat. The system shown in *Figure 1* was designed to operate at pressures between 0·1 and 2 atm. For satisfactory operation, the pressure in the flow system must be controlled to $\pm$ 0·1 mm or better. The manostat located in a helium-filled part of the system is connected to the main flow system through a refrigerated trap (not shown). A bellows-type (Wallace and Tiernan Company, U.S.A.) manostat is used in the system described here. A mercury manostat with a nitrogen leak has been used successfully by the National Physical Laboratory[11].

(3) Sample Collection System. During heat of vaporization experiments, the material generated in the vaporizer is collected in the system shown beneath the three-way valve. The vapor is diverted to the collection system by the valve, condensed, and the liquid is collected in the double receiver shown. The receiver, equipped with stainless steel valves[22] is detachable. The left side of the receiver is connected to the manostat and flow system through a refrigerated trap, which prevents a volatile substance from entering the receiver by diffusion through the back way.

(4) Liquid Introduction System. The liquid introducer shown in *Figure 1* is used both to admit a sample to the system initially and to meter liquid into the vaporizer during a heat of vaporization experiment at the same rate

at which material is withdrawn into the collection system. The metering apparatus consists of a buret to which is attached a length of capillary tubing and a water-cooled, stainless steel, solenoid valve. The rate of flow of liquid is controlled by adjusting the pressure of helium gas above the liquid in the buret by means of a bellows-type manostat. Hales, et al.[11] control the rate of flow of liquid by adjusting the number of drops per unit time from a special orifice to accomplish the same end.

D. The Thermostats

The vaporizer and calorimeter vessels are contained in separate thermostats, as shown in *Figure 1*. The temperature of the vaporizer thermostat should be controlled within limits of ± 0·01°, and that of the calorimeter thermostat within ± 0·005°. For best results and greater ease of operation, temperature control within ± 0·002° is desirable. Thermostats of the type shown in *Figure 4* can be controlled within these close limits over the range

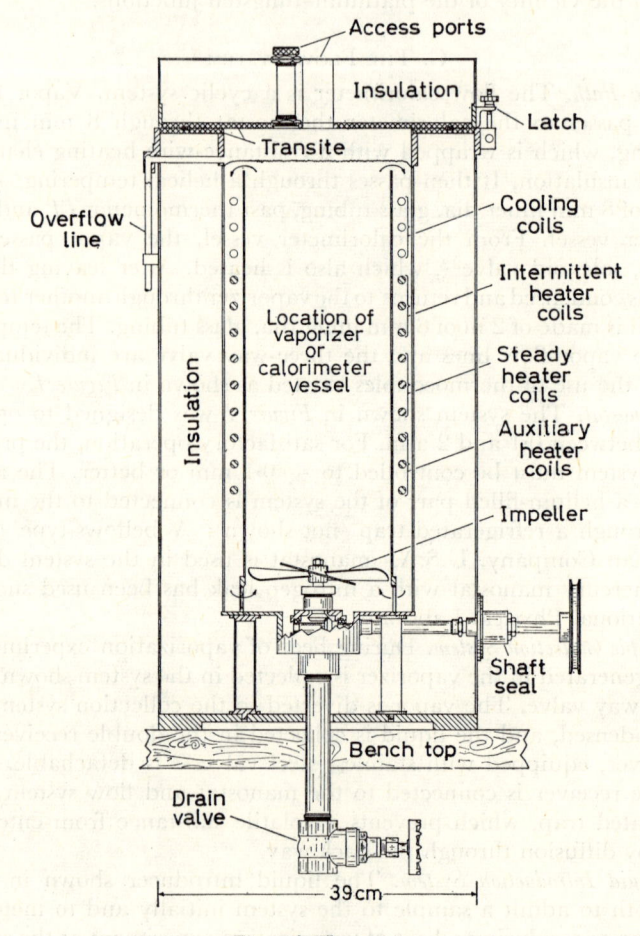

Figure 4. Thermostat.

0 to 250°C. The impeller located at the bottom of the thermostat forces a uniform stream of fluid (such as a silicone oil) up through the central cylinder, which contains the experimental vessel. The fluid flows over the top of the inner cylinder and returns to the bottom of the thermostat through the annulus, which contains heating elements and a cooling coil. With this arrangement, temperature gradients within the inner cylinder are about $\pm$ 0·001°. The thermostat temperature can be controlled within $\pm$ 0·002° by using an electronic regulator that provides both proportional and reset control features.

2. *Operating procedures*

A. Establishment of Flow Conditions

In a complete investigation of a substance, the flow calorimeter system is operated at several temperatures and pressures. The vaporizer vessel usually will be operated at 3 or more temperatures corresponding to the substance's saturation pressures in the range 0·1 to 2 atm. Normally, the calorimeter vessel will be operated at a series of 5 or more temperatures for each condition of the vaporizer. In starting up the system, the two thermostats are heated to the desired operating temperatures, and the flow lines are heated above the condensation temperature of the sample. Then, the pressure within the flow system is set approximately, and electrical energy from a bank of storage batteries is supplied to the sample in the vaporizer.

When boiling starts, the pressure on the system is adjusted until the boiling temperature within the vaporizer is about 0·1° below the thermostat temperature, as determined by the differential thermocouple, $TC1$, shown in *Figure 1*. The calorimeter thermostat is operated at least 5° above the boiling temperature. The temperature of the flowing vapor is adjusted to within $\pm$ 2° of the calorimeter bath temperature before it enters that bath. The temperature of the vapor leaving the three-way valve is maintained about 20° above the condensation point. These adjustments of vapor temperature are monitored with the differential thermocouples, $TC2$, $TC3$, and $TC4$. The flow of water (or air) in the condensers above the vaporizers is adjusted so that the condensate returning to the vaporizer thermostat is nearly at saturation condition. A heating element on the liquid return line is used to keep the temperature of the condensate just below the boiling point.

When the rate of vaporization is altered, all of the controls mentioned in the foregoing paragraph must be changed to maintain the same conditions throughout the flow system. As the rate of flow of vapor is increased, the pressure drop through the system increases, and the manostat pressure at the top of the condensers must be decreased to maintain the same boiling temperature in the vaporizer. It is important to be sure that all controls are changed in a way that prevents premature condensation of vapor in the flow path.

B. Power and Temperature Measurements

The essential electrical measurements in vapor-flow calorimetry involve determinations of the resistance of five platinum resistance thermometers ($T1$ through $T5$, *Figure 1*), the power input of two heating elements, and the e.m.f. of one differential thermocouple (*TC1*). The precision of the resistance thermometer measurements should be equivalent to about ± 0·0002°, and the accuracy of the power measurements should be about ± 0·01 per cent. The temperature difference of about 0·1° measured by *TC1*, should be determined with an accuracy of about ± 0·001°.

Any suitable electrical measuring instruments, such as precision bridges and potentiometers (Chapters 2 and 3), may be used. However, because of the number and variety of observations that must be made in sequence, there is a definite advantage to using a single measuring instrument, especially if the observations are to be made by a single investigator. A precision potentiometer with two independent measuring circuits, such as the White double potentiometer, is particularly convenient for this application.

It is almost impossible to eliminate thermal e.m.f.s in the thermometer circuits of the flow calorimeter vessel. These e.m.f.s are almost constant for a particular flow condition, but change significantly with changes of condition. When the thermometer resistances are measured potentiometrically, the thermal e.m.f. of each thermometer can be measured with the thermometer current source turned off, and a correction for the residual e.m.f. can be applied accurately.

C. Heat of Vaporization Experiments

(*1*) *Procedures*. The first measurements of an investigation are of the heat of vaporization, the results being used to calculate the vapor flow rate in subsequent heat capacity experiments. The heat of vaporization is determined at a series of temperatures at which the vapor pressure corresponds to the series of pressures chosen for heat capacity measurements.

In preparation for heat of vaporization experiments, the system is brought to steady operating conditions, with vapor cycling through the normal flow path, condensing, and returning to the vaporizer. The flow calorimeter plays no part in these experiments, and its thermostat is controlled about 20° above the vaporizer temperature. An empty, previously weighed receiver, *E, Figure 1*, is attached to the collection system at *2* and *3*, and a liquid nitrogen bath is placed around the trap at the left of the receiver. To test the diversion valve, *V2*, for leaks, the collection system, still isolated from the manostat by valve *4*, is evacuated for about 5 min by a mechanical pump. Any substance leaking through the valve is easily detected as frost in the trap. Helium is then admitted to the collection system, refrigerant baths are placed around the two bulbs of the receiver, the pressure is adjusted to that in the flow system, and the collection system is opened to the manostat. Liquid nitrogen is used as a refrigerant for the left hand bulb, and ice water, dry ice slurry, or liquid nitrogen bathes the right hand bulb. From calibration data, obtained while originally introducing the sample to the system, the pressure on the liquid introducer is set to provide flow of liquid into the vaporizer at the same mass rate as vapor is to be removed in the collection system. The system is then ready for a determination, which consists of measuring the electrical energy required to vaporize a given mass of substance (collected in the receiver), the amount of substance in the vaporizer being held constant by admitting liquid from the sample-filling buret through valve *V1*.

After a sample is collected in the receiver, a liquid nitrogen bath is placed around the right hand bulb, which contains the bulk of the material. When the sample is thoroughly cooled, the system is isolated from the manostat and evacuated for about 2 min, or long enough to ensure that all of the material is condensed inside the receiver. The valves of the receiver are then closed, and it is removed for final weighing.

To return the sample to the liquid introducer, the receiver is attached to the fitting shown at the top of the buret. A small hole in the tube just below the ring seal on the right bulb of the receiver allows the bulb to drain completely.

(2) *Sequence of Observations.* With steady flow conditions in the flow system and the collection system in readiness, the following initial observations are made in the order named: (*i*) Temperature of the vaporizer thermostat, $T5$; (*ii*) electrical power supplied to the vaporizer heater; and (*iii*) the difference in temperature between the vapor within the vaporizer and the thermostat, $TC1$. While observing the differential thermocouple at 10 to 30 sec intervals, the operator throws a single control switch that simultaneously starts an electric stopclock (0·01 sec accuracy) and opens the diversion valve, $V2$ and the liquid introducer valve, $V1$ (both of which are solenoid operated). When the vapor is diverted to the collection system, the boiling temperature may change by as much as $0·1°$ because the pressure drops through the alternate paths differ slightly. Therefore, the e.m.f. of $TC1$ is recorded at 30 sec intervals during a collection period. Normally, the collection period is of 5 to 10 min duration. When enough sample has been collected, the control switch is turned off, stopping the clock and returning the flow system to its original state. The initial observations are then repeated in reverse order to complete the electrical measurements.

The precision of measurements made with a system of this kind should be at least $\pm$ 0·05 per cent, usually better. Once such a system has been thoroughly tested, three determinations at each temperature are enough to define the heat of vaporization. Under good conditions, 3 to 6 determinations can be made in an 8-hour day. The measurements are usually made at a vapor flow rate of about 0·05 mole min^{-1} because the collection of a sample at higher flow rates, up to 0·25 mole min^{-1}, causes increasing disturbances in the flow system. However, in testing new apparatus, measurements should be made over as wide a range of flow rates as possible. Such experiments will allow detection of liquid entrainment from the vaporizer and some kinds of systematic error.

D. Heat Capacity Experiments

(*1*) *Procedures.* Vapor heat capacity measurements normally are made at two or more pressures at each of a series of temperatures. Once the heat of vaporization has been determined at boiling temperatures corresponding to the pressures chosen for heat capacity measurements, the rate of flow of vapor can be calculated directly from the measured power input to the vaporizer heater. The system is prepared for heat capacity measurements as described in the foregoing, except that the collection system and liquid introducer are not needed. At a particular pressure at one temperature, the heat capacity is measured at four vapor flow rates covering the widest possible range of velocities, normally from a high rate of about 0·25 to a low of about 0·05 mole min^{-1}. Such a series of experiments is made continuously in a 6- to 8-hour period.

(2) *Sequence of Observations.* The principal determinations made in a heat capacity measurement are of the power input to the vaporizer and calorimeter heating elements and the temperature increment of the vapor as it passes through the calorimeter. The power measurements yield the rate of flow of vapor through the calorimeter and the net enthalpy increment as the vapor passes the calorimeter heater. These factors and the temperature increment allow calculation of an *apparent* heat capacity value.

Accurate determination of the temperature increment requires observations of the four thermometers, $T1$ through $T4$, under two conditions: First, a "no energy" experiment with vapor flowing through the calorimeter but with no energy supplied by that vessel's heater; and second, an "energy" experiment under identical flow conditions but with energy supplied to heat the flowing vapor. There are two reasons for following this procedure. First, it is desired to measure the temperature increments, ΔT_2 and ΔT_3 (about 8°), with a precision of about $\pm$ 0·0002°, but such precision cannot be obtained by taking the difference indicated by two resistance thermometers, particularly when the thermometers are mounted in a rapidly moving gas stream. Such precision can be obtained, however, by subtracting the temperature indicated by $T2$ (or $T3$) during a no energy experiment from that of the same thermometer indicated during an energy experiment. A correction for drift in the thermostat temperature between the two experiments, usually less than 0·01°, can be made accurately from the corresponding observations on $T1$ and $T4$ (only one of which is strictly necessary). Second, the procedure also provides data needed to compute a correction for the Joule–Thompson cooling that occurs when the gas flows through the calorimeter, as described in the following section.

The observations for a no energy experiment provide a temperature base for the corresponding energy experiment at the same vapor flow rate, so only temperatures need be measured. Therefore, the four resistance thermometers are observed in the following sequence: $T4$, $T1$, $T2$, $T3$, $T1$, $T4$, residual thermometer e.m.f.s (if necessary).

Observations for the energy experiments require power as well as temperature measurements. The sequence of observations is as follows: $TC1$, $T5$, $T4$, $T1$, P_C, P_V, $T2$, $T3$, $T2$, P_V, P_C, $T1$, $T4$, $T5$, $TC1$, residual thermometer e.m.f.s (if necessary). P_C and P_V denote the power input to the calorimeter and vaporizer heaters. As some point during the experiment, the pressure indicated by the manometer is recorded to the nearest 0·05 mm.

A pair of no energy and energy experiments, made at the same vapor flow rate, provides the data needed to calculate the *apparent* heat capacity at a particular condition of pressure, temperature, and flow rate. Normally, the *apparent* heat capacity is determined at four flow rates, beginning with the highest rate. When the flow rate is changed and when the calorimeter energy is turned on or off, re-establishment of steady flow conditions may require 30 min or longer. The following sequence of experiments minimizes the time required for equilibration: No energy (i), energy (i); energy (ii), no energy (ii); no energy (iii), energy (iii); and energy (iv), no energy (iv).

The apparent heat capacity values as a function of flow rate provide the information needed to compute the *true* heat capacity at the particular temperature and pressure.

(*3*) *Observational Techniques.* The vapor-flow calorimeter is a complex and dynamic system in which the observables are never quite as constant as one would like them to be. Some of the changes that occur in the system during an experiment are uniform and unidirectional. For example, the power inputs to the vaporizer and calorimeter heaters, supplied by storage batteries, change linearly with time and seldom by as much as 0·05 per cent in the course of an experiment. Also, the thermostat temperatures and system pressure may drift slightly during an experiment. The effects of such uniform drifts are made negligible by taking observations in the symmetrical way recommended in the foregoing, and averaging all results to correspond to the midtime of an experiment.

Superimposed on the linear changes are fluctuations, which may be more or less erratic, that arise from three principal sources. The first are short-term fluctuations of a few thousandths degree in the thermostat temperatures. The effect of these fluctuations can be reduced by observing $T1$ through $T5$ for about 2 min each and averaging the results of repeated observations on each thermometer. A second and usually minor source of fluctuations is the manostat. Pressure fluctuations are evidenced most directly as short term fluctuations in the boiling point, usually less than 0·01°, which also may be taken care of by averaging the reading of $TC1$ over a 2 min period. The third and most troublesome source of fluctuations is in the boiling characteristics of the sample itself. Many organic liquids boil very smoothly in the vaporizer, but others are a continual source of trouble, usually because of a tendency for the liquid to gain and lose superheat erratically. This is why the design of the vaporizer heater is so important in flow calorimetry. If erratic results are obtained (and the electrical circuitry is in good order), the source of the difficulty almost always is in the poor boiling characteristics of the sample. (When poor boiling is encountered, about the only remedy is to supply a spurt of excess electrical energy to the vaporizer heater in the hope that the "boiling chip" action of its surface will be regenerated.)

All fluctuations and uniform drifts are reflected (as an algebraic sum) in the temperature of the heated vapor in the calorimeter. For this reason, observations of $T2$ and $T3$ are made with special care at the midtime of a no energy or energy experiment. When the system is operating smoothly, the readings of these thermometers during an energy experiment are recorded at 10 sec intervals, first $T2$ for 90 sec, then $T3$ for 180 sec, and finally $T2$ for another 90 sec. When the system is not operating smoothly, observation of $T2$ and $T3$ while energy is being supplied to the calorimeter is one of the best ways of detecting troubles so they may be corrected.

3. *Calculation of experimental results*

A. Heat of Vaporization

If both the temperature and liquid level within the vaporizer remain constant throughout a heat of vaporization measurement, the heat of vaporization can be calculated directly from the relationship

$$\Delta Hv = q'_V M/m_S \quad (1)$$

in which ΔHv is the heat of vaporization, q'_V is the heat input to the vaporizer during a collection period, m_S is the mass of sample collected, and M is the molecular weight. In practice, corrections ranging from 0·01 to about 0·2 per cent must be applied in calculating ΔHv. These corrections are described in the following paragraphs.

(*i*) A small change in the temperature within the vaporizer almost always occurs during a heat of vaporization measurement. Thus, q'_V must be corrected for the heat gained or lost by the inner parts of the vaporizer plus contents during the timed sample collection period. This correction can be calculated as

$$[m_L(\text{ave.}) \, c_L + \Sigma(m\,c)_V] \, [t_F - t_I] \quad (2)$$

in which m_L and c_L are the average mass and specific heat of the liquid *within* the vaporizer during a collection period; $\Sigma(mc)_V$ is the sum of the mass—specific heat products for all solid parts of the inner vaporizer vessel, heater assembly, and thermocouple well (with sufficient accuracy, a constant for a given vessel); and t_I and t_F are the vaporizer temperatures at the beginning and end of a collection period. This correction is normally less than 0·1 per cent of ΔHv.

(*ii*) A second correction is required to account for the total enthalpy change in the vaporizer due to a change in liquid level during the collection period. The need for this correction can be seen schematically from *Figure 5*. As liquid is removed from the vessel under a constant boiling pressure, the temperature profile within the vaporizer changes as shown. It is assumed that the temperature increases below the boiling surface because of increased hydrostatic head, but that the temperature below the top of the heater is

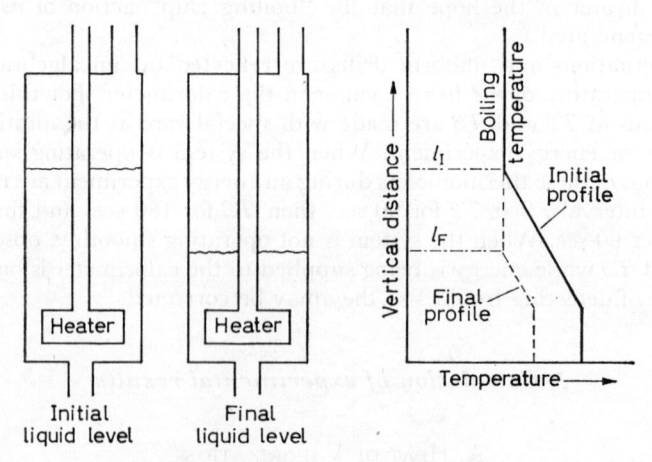

Figure 5. Vaporizer vessel before and after change of liquid level at constant boiling pressure and resulting change in the temperature profile.

constant because of more vigorous mixing in that region. The resulting changes in temperature and mass of sample cause a corresponding change in the net enthalpy of the vaporizer and contents. The correction for change in liquid level is calculated from equations of the following form:

Vaporizer components: $\quad -[a_V + b_V(l_I + l_F)] [m_\delta] [dt/dp] \quad (3)$

Sample: $\quad -[a_L + b_L(l_I + l_F)][m_\delta\, c_L\rho] [dt/dp] \quad (4)$

The factors a_V and a_L are related to the amount of solid components and liquid sample below the top of the heater, and b_V and b_L are similar factors per unit height above the top of the heater; l_I and l_F are the initial and final heights of the liquid surface above the heater; m_δ is the change in mass of sample in the vaporizer; c_L and ρ are the specific heat and density of the sample; and dt/dp is the derivative of t with respect to p for the sample, obtained from vapor pressure data. With the liquid introduction system shown in *Figure 1*, m_δ may be made very small so that the total correction from equations 3 and 4 is negligible. However, the small correction may be of either sign depending on the sign of m_δ. If the liquid introduction system is not used, $-m_\delta = m_S$, and the total correction may be as large as 0·5 per cent of ΔHv, depending on the fullness of the vaporizer and the magnitudes of m_δ and dt/dp.

(*iii*) When the liquid introduction device is used during a collection period, liquid at the temperature of the thermostat flows into the vaporizer. This liquid normally is about 0·1° above the chosen boiling temperature, and a correction must be applied to account for the energy it carries into the vaporizer.

After these three corrections have been applied to q'_V, the heat of vaporization may be calculated by the equation

$$\Delta Hv = q_V [1 + (m_\delta/m_S)(V_L/V_G)]M/m_S \quad (5)$$

in which q_V is the corrected energy, V_L and V_G are molal volumes of liquid and gas at vaporizer conditions. If the liquid introduction system is not used, $-m_\delta = m_S$, and the factor in brackets becomes $(1 - V_L/V_G)$, which is the same as that used in Chapter 11. This factor accounts for the fact that when the liquid level in the vaporizer falls or rises during a collection, the amount of vapor condensed in the receiver is either less than or greater than the amount actually generated.

B. Heat Capacity

The flow-calorimeter vessel is designed so that heat loss from the inner U-tube occurs only by radiation from the hotter inner member to the cooler jacket wall. If the temperature profile along the inner member is not affected significantly by changes in vapor flow rate, the absolute magnitude of the radiative heat loss will be the same at all flow rates. Therefore, at a particular pressure and temperature, both the heat loss and true heat capacity can be calculated from two or more determinations of the apparent heat capacity. This condition is true for nearly all substances that can be studied in the apparatus described, and it provides the basis for the practical method of calculation discussed next. For gases with very low *specific* heat, the radiative heat loss may depend slightly on flow rate, and a more exact method of calculation for such gases is given later.

(1) Practical Method. An energy balance on the flow-calorimeter vessel during a heat capacity measurement yields

$$q_C = C_P F \Delta T + k_1 \Delta T \qquad (6)$$

in which q_C is the energy supplied by the calorimeter heater per unit time; C_P is the true heat capacity of the flowing gas; F is the gas flow rate in moles per unit time; ΔT is the temperature increment of the gas within the calorimeter; and $k_1 \Delta T$ is the radiative heat loss per unit time. (k_1 is a constant involving emissivities of the calorimeter surfaces and a geometrical factor to account for the linear temperature increase along the heater.) Rearranging equation 6 yields

$$q_C/F\Delta T = C_P + k_1/F = C_P(\text{app.}) \qquad (7)$$

in which $C_P(\text{app.})$ is the apparent heat capacity. In practice, C_P (app.) is measured at four flow rates in the range 0·05 to 0·20 moles min⁻¹. The true heat capacity is found by extrapolating a plot of $C_P(\text{app.})$ vs. $1/F$ to $1/F = 0$. A typical plot of results is shown in *Figure 6*. Two independent values of $C_P(\text{app.})$ are obtained at each flow rate from observations of

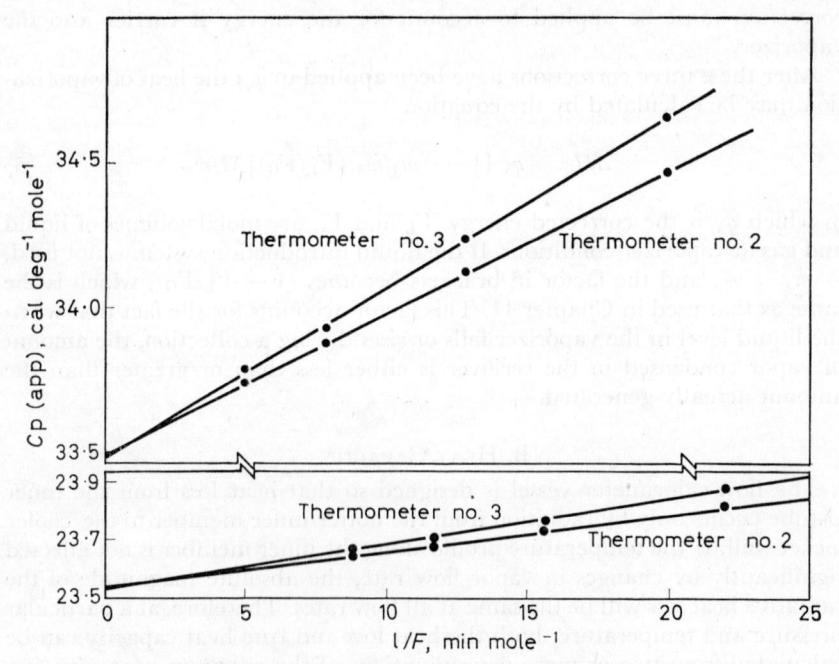

Figure 6. Apparent heat capacity *vs* reciprocal of flow rate. Upper: benzene at 500·17°K and 1 atm. Lower, benzene at 348·16°K and 0·125 atm.

thermometers 2 and 3, so two lines are determined in each series. The assumptions made in deriving equation 7 are verified by the facts that the points fall well within 0·05 per cent of a straight line and that the lines for $T2$ and $T3$ intersect (within 0·05 per cent) at $1/F = 0$. The slopes of these lines are almost independent of pressure but proportional to the cube of the absolute temperature (T^3).

As is true of most calorimetric methods, small corrections to the observed data must be applied in the calculations. These corrections, usually totaling less than 0·2 per cent, are:

(*i*) The flow rate, F, is calculated from the previously determined $\Delta H v$ and the power input to the vaporizer heater. The energy carried into the vaporizer by liquid continuously returning from the condensers must be added to the measured electrical energy input, as in heat-of-vaporization measurements described in Section II–3–A.

(*ii*) As determined by direct measurement, about 0·03 per cent of the energy supplied to the calorimeter heater is generated in the leads and not measured in normal power observations. This small leads correction factor is determined in separate experiments as a function of temperature.

(*iii*) During heat capacity measurements at a series of flow rates, the pressure within the vaporizer is maintained constant. Because the pressure drop through the system is a function of flow rate, the pressure in the calorimeter changes with flow rate. In final calculations of C_P, each value of C_P(app.) is corrected by adding $(P_V - P_C) (\partial C_P/\partial P)_T$, in which P_V and P_C are the pressures in the vaporizer and calorimeter. P_C is not measured directly, but $P_V - P_C$ is a nearly constant fraction of the total pressure drop. This fraction is determined in original tests on the system, and the total pressure drop is observed in each determination of C_P(app.).

(*iv*) The method of measuring ΔT eliminates most of the effect of Joule–Thompson (J–T) cooling caused by the pressure drop through the calorimeter vessel. However, a small correction to C_P(app.) must be made because the J–T cooling is slightly smaller for an energy measurement than for the corresponding no energy measurement. The data needed to apply this correction are derived from the no energy observations of $T1$, $T2$, and $T3$ obtained at each temperature, pressure, and flow rate of an investigation.

(2) *General Method.* For vapors of low specific heat, it is necessary to account for a small dependence of total heat loss on flow rate. The origin of a heat loss term inversely proportional to F can be seen from *Figure 7*, which is a schematic representation of the temperature profile in the calorimeter. The temperature of the entering vapor is measured by a platinum-resistance thermometer at position A, electrical energy is supplied by a heating element between points B and $C(F)$, and the temperature is measured at points D and $E(G)$ before the vapor leaves the calorimeter. Equation 7 applies strictly only when the temperature increment and profile at all flow rates is that given by the solid line $ABCDE$. For this condition, the total heat loss up to the thermometer at D is proportional to the area of $BCDH$, which usually is assumed to be independent of flow rate if ΔT is held constant. Actually, the heat losses that occur in the calorimeter will cause the temperature of the vapor to fall after it passes the heater, as shown on a much exaggerated scale by the dashed profile $BFDG$. Therefore, the heat loss up to point D is

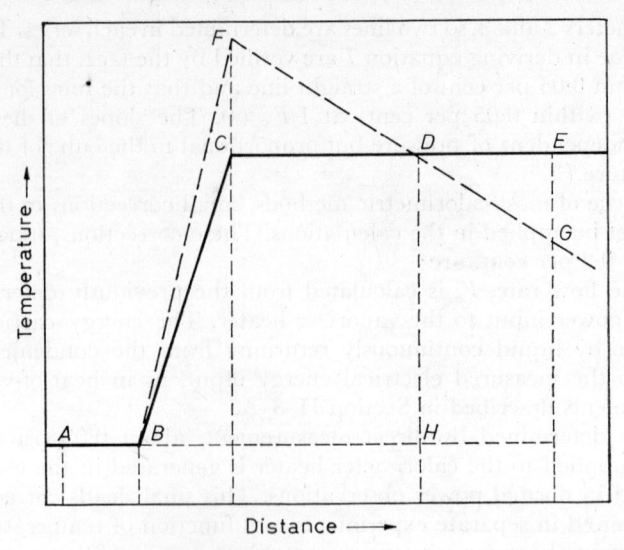

Figure 7. Schematic representation of the temperature profile in a vapor flow calorimeter.

proportional to the area $BFDH$, the area of $BFDC$ being inversely proportional to the flow rate. An energy balance on the calorimeter up to point D then gives

$$q_C = C_P(\text{app.})\, F\Delta T = C_P F\Delta T + k_1 \Delta T + (k_2/F)\Delta T \tag{8}$$

In this equation, $k_1\Delta T$ is related to the area $BCDH$ and $(k_2/F)\Delta T$ is related to the area $BFDC$. In the experiments, the slope of the line FDG is determined from the temperatures measured at points D and G, and the profile BFD is established from the results. The equation for $C_P(\text{app.})$ including the flow-dependent term for heat loss is

$$C_P(\text{app.}) = C_P + k_1/F + k_2/F^2 \tag{9}$$

In applying equation 9, $C_P(\text{app.}) - C_P$ is evaluated approximately by equation 7, and k_2/F^2 is calculated as [area $BFDC$/area $BFDH$] $[C_P(\text{app.}) - C_P]$, the ratio of areas being calculated from the known temperature profile. A plot of $C_P(\text{app.}) - k_2/F^2$ vs. $1/F$ is then used to evaluate C_P at $1/F = 0$. Actually, the k_2/F^2 term is negligible for most organic vapors, but verges on significance for such vapors as nitromethane and carbon disulfide because of their unusually low *specific* heats. For example, neglect of the term would have introduced an error of 0·025 cal deg^{-1} mole^{-1} (0·2 per cent) in the value of $C_P°$ for carbon disulfide at 502·25°K, a smaller error at 453·35°K, and entirely negligible errors at the lower temperatures[39].

4. Derived results

The direct results of a vapor-flow calorimetric investigation are: $\Delta H v$ as a function of temperature; and C_P as a function of both pressure and temperature. Other useful thermodynamic data that may be derived from these results and accurate vapor pressure data are discussed in the following sections.

A. Second Virial Coefficient and Equation of State

In the low pressure region in which the flow calorimeter is normally operated, most substances obey the equation of state

$$PV = RT(1 + B/V + \ldots) \tag{10}$$

in which B is the second virial coefficient and higher virial coefficients make a negligible contribution. Values of B may be calculated from ΔHv and vapor pressure data by use of a rearranged form of the Clapeyron equation:

$$V_G = (\Delta Hv/T)(dT/dP) + Vl \tag{11}$$

$$B = V_G(PV_G/RT - 1) \tag{12}$$

B. Ideal Gas Heat Capacity

Both the effect of pressure on heat capacity and the ideal gas heat capacity, $C_P°$, may be expressed in terms of the foregoing equation of state and measured values of C_P. From standard texts on thermodynamics,

$$(\partial C_P/\partial P)_T = - T(\partial^2 V/\partial T^2)_P \tag{13}$$

In terms of this relationship and equation 10, the vapor heat capacity is given by

$$C_P = C_P° - RT^2(d^2B/dT^2)/V + R[B - T(dB/dT)]^2/[V^2 + 2BV] \tag{14}$$

or, with slight approximations, in the more convenient form

$$C_P = C_P° - PT(d^2B/dT^2) + 2[P^2/R][B(d^2B/dT^2)][1 - 2BP/RT] \tag{15}$$

Equation 15 is fitted by an iterative process to the experimental values of C_P to obtain values of $C_P°$ and d^2B/dT^2 at each temperature[22].

The mathematics of the foregoing correlations is greatly simplified if the equation of state $PV = RT + BP$ is used instead of equation 10. However, in its closed form, this pressure explicit equation implies a linear relationship between C_P and P. Accurate C_P data show that the relationship actually is non-linear, as required by equations 14 and 15. It should be noted that numerical values of B calculated from V_G (equation 11) will differ slightly depending on whether the volume explicit or pressure explicit equation of state is used.

C. Other Useful Information

Values of $C_P°$ are the primary objective of flow calorimetric investigations, but the other information discussed in the foregoing has many useful applications. For example, the values of B and d^2B/dT^2 may be used to evaluate the constants in a semi-empirical equation for B[12, 35].

$$B = a + b \exp c/T \tag{16}$$

Thus, from vapor pressure data, measured values of ΔHv, equations 10 and 16, and well known thermodynamic relationships, values of $\Delta Hv°$, $\Delta Sv°$, and $\Delta_G v°$ may be calculated for use with experimental values of ΔHf and S for the liquid state. Experimental values of d^2B/dT^2 also provide a unique test of assumed intermolecular potential functions, as shown by Douslin, et al.[7, 8]

III. Discussio

1. *Accuracy of results*

A. ACCURACY OF HEAT OF VAPORIZATION MEASUREMENTS

There is no direct way of establishing the absolute accuracy of heat of vaporization data determined as described in Section II. However, precision of the measurements is ± 0·05 per cent or better; ± 0·1 per cent is a conservative estimate of accuracy because all physical measurements are accurate to about 0·01 per cent, and the possibility of important systematic errors is small. To obtain such accuracy requires close control of and constant attention to all elements of the dynamic flow system. Condensation in flow lines, superheating of the boiling liquid, small pressure fluctuations, poor condensing characteristics, and other, sometimes obscure, malfunctions can cause errors if undetected. Fortunately, such effects lead to poor precision in replicate experiments, so the investigator knows something is wrong and can look for the difficulty.

Table 1 gives comparisons of results obtained in a flow calorimeter with those measured in other types of apparatus. With one exception (benzene

Table 1. Comparison of experimental values of heat of vaporization

Compound	T (°K)	Flow Calor.† ΔHv, cal mole⁻¹	Ref.	Other	Ref.
Water	338·2	10108	19	10099	26
	373·2	9714	19	9717	26
Benzene	298·2	8089‡	35§	8090	27
	323·2	7755	35§	7756	10
	353·2	7343	35§	7353	10
Nitromethane	298·1	9138‡	22	9147	13
2-Thiapropane	291·1	6696	18	6688	24

† Values determined in the flow calorimeter described in Section II.
‡ Extrapolated values.
§ Value given revised slightly from that in ref. 35 on the basis of more recent unpublished results.

at 353·2°K), all of these results agree within about 0·1 per cent. The comparisons in *Table 1* are typical of all those for which experimental data of comparable accuracy are available.

Comparisons of experimental values with those calculated from other thermodynamic data support the foregoing statements about accuracy. For example, the standard heat of vaporization of 3,3-dimethyl-2-thiabutane[34] is, $\Delta Hv°_{298.15} = 8\,560$ cal mole^{-1}, as obtained by an extrapolation of data measured in the range 330–372°K. By use of the Clapeyron equation, vapor pressure data, and gas-imperfection data, $\Delta Hv°_{298.15}$ for this substance is calculated to be 8 550 cal mole^{-1}. From a value of $S°_{298.15}$ calculated by the methods of statistical mechanics, an experimental value of $S_{298.15}$ for the liquid, and vapor pressure data, $\Delta Hv°_{298.15}$ is found to be 8 550 cal mole^{-1}. Such consistent results have been obtained for many other compounds for which the necessary thermodynamic data are available.

B. Accuracy of Heat Capacity Measurements

The method of measuring vapor heat capacity described in detail in this chapter is capable of yielding results precise to $\pm\,0.05$ per cent or better. From consideration of possible systematic errors, the absolute accuracy uncertainty is estimated to be $\pm\,0.2$ per cent or less. Independent verification of this accuracy claim can be made by comparing experimental values of $C_P°$ with those calculated by the methods of statistical mechanics. However, to calculate values of $C_P°$ with accuracy approaching that claimed for the experimental values requires very complete and detailed knowledge of the molecular energy levels. Such knowledge is available for only a few polyatomic molecules, none of which is more than triatomic. In *Table 2*, results

Table 2. Comparison of experimental and calculated values of gas heat capacity

Compound	T, °K	$C_P°$, cal deg^{-1} mole^{-1} Expt.	Calc.	δ†(%)	Ref.
Carbon dioxide	243.2	8.243	8.243	0.00	15
	273.2	8.593	8.595	−0.02	
	323.2	9.149	9.141	+0.09	
	363.2	9.546	9.539	+0.08	
Water	361.8	8.107	8.113	−0.07	19
	381.2	8.146	8.148	−0.02	
	410.2	8.206	8.206	0.00	
	449.2	8.291	8.292	−0.01	
	487.2	8.380	8.383	−0.04	
Carbon disulfide	325.6	11.175	11.172	+0.03	39
	367.6	11.568	11.565	+0.03	
	407.1	11.880	11.883	−0.03	
	453.4	12.218	12.209	+0.07	
	502.2	12.484	12.509	−0.21	

† Experimental − observed, %

obtained with an adiabatic calorimeter[15] for CO_2 and with a non-adiabatic calorimeter for H_2O and CS_2 are compared with calculated values. In each case, the agreement is within the claimed accuracy uncertainty of the experimental values and probably within the accuracy uncertainty of the calculated values.

C. Accuracy of Second Virial Coefficient Values

The values of B and d^2B/dT^2 derived from data obtained in flow calorimetric experiments are more reliable than most values obtained by other, sometimes more direct methods. Values of B calculated from the most accurate heat of vaporization and vapor pressure data (equations 11 and 12) are probably accurate to 0·1 to 0·2 per cent of the molal volume of the saturated vapor, V_G. Table 3 gives a comparison of values obtained in this

Table 3. Comparison of experimental values of second virial coefficient

Compound	T, °K	V_G, cm^3 mole^{-1}	$-B$, cm^3 mole^{-1} Flow Calor.†	PVT‡	δ, %V_G §
Pyridine[6, 17]	346·6	112 450	1 292	1 301	+0·01
	366·1	58 900	1 157	1 107	−0·08
	388·4	30 850	990	923	−0·22
Toluene[6, 33]	341·3	110 240	1 717	1 753	+0·03
	361·0	57 740	1 472	1 468	−0·01
	383·8	30 170	1 267	1 203	−0·21
	410·1	15 660	1 095	955	−0·9

† Determined from heat of vaporization and vapor pressure data.
‡ Determined from compressibility measurements.
§ B (flow calor.) $-B$(PVT), expressed as %V_G.

way with those determined from vapor compressibility measurements. Below a saturation pressure of 1·0 atm, the two sets of data for both pyridine and toluene agree within about 0·2 per cent of V_G, but the data for toluene at 2·0 atm pressure disagree by 0·9 per cent of V_G. The deviations for each compound show a distinct trend with temperature.

2. Reference substances

A. Heat of Vaporization Measurements

Water and benzene are two useful reference substances for heat of vaporization measurements. Several highly accurate studies have been made of each (see for examples: water[27], and earlier references cited therein; and benzene[35], and earlier references cited therein), and both are easily obtained in pure form. Water may cause experimental trouble in some systems because of its poor boiling characteristics[19] or because of corrosion and electrical conduction. For these reasons, benzene is probably the most convenient reference substance to use in a system such as that described in Section II, or any similar system.

B. Heat Capacity Measurements

Water, carbon disulfide, and benzene are important reference substances for heat capacity measurements with the non-adiabatic method. All three are easily obtained in pure form. Accurate calculated values of $C_P°$ for water and carbon disulfide are available for checking experimental data (Section III-1-B). However, as discussed in Section III-2-A, water may present experimental problems. Also, carbon disulfide is slightly corrosive[39]. For these reasons, benzene is probably the most convenient reference substance to use in apparatus similar to that described in Section II. Benzene is easy to handle experimentally; most investigators working in the field have reported at least some experimental data for it; and it can also be used as a reference substance for heat of vaporization measurements.

Carbon dioxide is a useful reference gas for the nearly adiabatic method. It is easy to purify in large amounts; the heat capacity is relatively high; and highly accurate experimental and calculated values of $C_P°$ have been reported[15]. However, its use in the non-adiabatic method would require a different method for measuring the rate of vapor flow.

IV. References

[1] Bennewitz, K., and O. Schulze, *Z. physik. Chem.*, **A186**, 299 (1940).
[2] Bennewitz, K., and W. Rossner, *Z. physik. Chem.*, **B50**, 143 (1941).
[3] Brinkworth, J. H., *Proc. Roy. Soc. (London)*, **107A**, 510 (1925).
[4] Brinkworth, J. H., *Proc. Roy. Soc. (London)*, **111A**, 124 (1926).
[5] Callendar, H. L. and H. T. Barnes, Report of Br. Assoc., 552–553, 1897 and *Phil. Trans.*, **199A**, 55–263 (1902).
[6] Cox, J. D., and R. J. L. Andon, *Trans. Faraday Soc.*, **54**, 1 (1958).
[7] Douslin, D. R., and Guy Waddington, *J. Chem. Phys.*, **23**, 2453 (1955).
[8] Douslin, D. R., R. T. Moore, J. P. Dawson, and Guy Waddington, *J. Am. Chem. Soc.*, **80**, 2031 (1958).
[9] Eucken, A., and K. von Lude, *Z. physik. Chem.*, **B5**, 413 (1929).
[10] Fiock, E. F., D. C. Ginnings, and W. B. Holton, *J. Res. Natl. Bur. Stand.*, **6**, 881 (1931).
[11] Hales, J. L., J. D. Cox, and E. B. Lees, *Trans. Faraday Soc.*, **59**, 1544 (1963).
[12] Hirschfelder, J. O., F. T. McClure, and I. F. Week, *J. Chem. Phys.*, **10**, 201 (1942).
[13] Jones, W. M., and W. F. Giauque, *J. Am. Chem. Soc.*, **69**, 983 (1947).
[14] Kistiakowsky, G. B., and W. W. Rice, *J. Chem. Phys.*, **7**, 281 (1939).
[15] Masi, J. F., and B. Petkof, *J. Res. Natl. Bur. Stand.*, **48**, 179 (1952).
[16] Masi, J. F., *Transactions of the ASME*, 1067–1073, October 1954.
[17] McCullough, J. P., D. R. Douslin, J. F. Messerly, I. A. Hossenlopp, T. C. Kincheloe, and Guy Waddington, *J. Am. Chem. Soc.*, **79**, 4289 (1957).
[18] McCullough, J. P., W. N. Hubbard, F. R. Frow, I. A. Hossenlopp, and Guy Waddington, *J. Am. Chem. Soc.*, **79**, 561 (1957).
[19] McCullough, J. P., R. E. Pennington, and Guy Waddington, *J. Am. Chem. Soc.*, **74**, 4439 (1952).
[20] McCullough, J. P., W. B. Person, and R. Spitzer *J. Am. Chem. Soc.*, **73**, 4069 (1951).
[21] McCullough, J. P., D. W. Scott, H. L. Finke, W. N. Hubbard, M. E. Gross, C. Katz, R. E. Pennington, J. F. Messerly, and Guy Waddington, *J. Am. Chem. Soc.*, **75**, 1818 (1953).
[22] McCullough, J. P., D. W. Scott, R. E. Pennington, I. A. Hossenlopp, and Guy Waddington, *J. Am. Chem. Soc.* **76**, 4791 (1954).
[23] Michels, A., and J. C. Strijland, *Physica*, **16**, 813 (1950).
[24] Osborne, D. W., R. N. Doescher, and D. M. Yost, *J. Am. Chem. Soc.*, **64**, 169 (1942).
[25] Osborne, N. S., H. F. Stimson, and T. S. Sligh, *Natl. Bur. Standards Sci. Technol. Papers*, **20**, 119 (1925).
[26] Osborne, N. S., H. F. Stimson, and D. C. Ginnings, *J. Res. Natl. Bur. Stand.* **23**, 261 (1939).
[27] Osborne, N. S., and D. C. Ginnings, *J. Res. Natl. Bur. Stand.* **39**, 453 (1947).
[28] Partington, J. R., and W. G. Shilling, *The Specific Heat of Gases*, Ernst Benn, London, England, 1924.
[29] Pennington, R. E., and K. A. Kobe, *J. Am. Chem. Soc.* **79**, 300 (1957).
[30] Pitzer, K. S., *J. Am. Chem. Soc.*, **63**, 2413 (1941).
[31] Rowlinson, J. S., *The Perfect Gas*, Vol. V of Topic 10, The Fluid State (Series: The International Encyclopedia of Physical Chemistry and Chemical Physics). Pergamon Press, London, 1963.
[32] Scheele, K., and W. Heuse, *Ann. Physik*, **37**, 79 (1912).
[33] Scott, D. W., G. B. Guthrie, J. F. Messerly, S. S. Todd, W. T. Berg, I. A. Hossenlopp, and J. P. McCullough, *J. Phys. Chem.*, **66**, 911 (1962).
[34] Scott, D. W., W. D. Good, S. S. Todd, J. F. Messerly, W. T. Berg, I. A. Hossenlopp, J. L. Lacina, Ann Osborn, and J. P. McCullough, *J. Chem. Phys.*, **36**, 406 (1962).
[35] Scott, D. W., Guy Waddington, J. C. Smith, and H. M. Huffman, *J. Chem. Phys.*, **15**, 565 (1947).
[36] Scott, R. B., and J. W. Mellors, *J. Res. Natl. Bur. Stand.*, **34**, 243 (1945).
[37] Swann, W. F. G., *Proc. Roy. Soc.*, **82A**, 147 (1909).
[38] Wacker, P. F., R. K. Cheney, and R. B. Scott, *J. Res. Natl. Bur. Stand.*, **38**, 651 (1947).

[39] Waddington, Guy, J. C. Smith, K. D. Williamson, and D. W. Scott, *J. Phys. Chem.*, **66**, 1074 (1962).
[40] Waddington, Guy, S. S. Todd, and H. M. Huffman, *J. Am. Chem. Soc.*, **69**, 22 (1947).
[41] Williamson, K. D., and R. H. Harrison, *J. Chem. Phys.*, **26**, 1409 (1957).

CHAPTER 11

Calorimetry of Saturated Fluids Including Determination of Enthalpies of Vaporization

D. C. GINNINGS

National Bureau of Standards, Washington, D.C., U.S.A.

H. F. STIMSON

National Bureau of Standards, Washington, D.C., U.S.A. (retired)

Contents

I.	Introduction	395
II.	Theory of the Method	396
III.	General Calorimetric Design and Procedures	399
IV.	Calorimeters for Measuring Properties of Water	403
	1. Calorimeter for 0–270°C	403
	2. Calorimeter for 100–374°C	404
	3. Calorimeter for 0–100°C	407
	A. Accounting for Mass of Sample	411
	B. Accounting for Energy	411
	(1) Electrical Energy	411
	(2) Pump Energy	412
	(3) Heat Leaks	412
	C. Accounting for Change in State	412
	4. Comparison of Results with the Three Calorimeters	413
V.	Calorimeter for Hydrocarbons	413
VI.	Other Calorimetric Methods for Measuring Enthalpy of Vaporization	418
VII.	Summary	419
VIII.	References	419

I. Introduction

Calorimetry of saturated fluids is the measurement of the heat required for changes of state in them (liquids and vapors). The term "saturated", as used in this chapter, restricts the states of the fluid to those for which the liquid and the vapor phases are both present in equilibrium within the calorimeter. The relative amounts of the two phases, however, may differ by large factors; i.e., the calorimeter may be nearly full of liquid in a high-filling experiment or may have only a little liquid in a low-filling experiment.

In heat capacity determinations both high and low fillings can be used. Sometimes such experiments are called gross and tare experiments. The method of saturation calorimetry described in this chapter differs from calorimetric methods in some other chapters of this book in which tare experiments are made with an empty calorimeter. In 1917 Osborne[5] described his experiments on the specific and latent heats of ammonia in which he made tare experiments with an empty calorimeter. These required him to make some troublesome corrections about whose accuracy he felt somewhat uncertain. In 1924 he developed[6] the method described here, to be used for determining the thermal properties of water and steam. This method of saturation calorimetry not only simplifies the computations but also, in principle, eliminates certain errors.

In saturation calorimetry the temperatures depend only on the pressures along the vapor pressure line; thus, at any instant, the temperature at all liquid–vapor interfaces in the calorimeter is essentially uniform and depends only on the pressure of the vapor there. Enthalpy of vaporization experiments, therefore, can be performed at a constant temperature, by adding heat and withdrawing fluid from the calorimeter while controlling the flow rate to keep the pressure constant.

In heat capacity experiments, the power input can be chosen so that the temperature may be raised at essentially the same rate in both the high and the low fillings, thus making the pressure changes occur in the same way in both. This procedure has the advantage that, by taking differences, one can eliminate the correction for fluids in the filling tubes and the correction for work done by strain of the calorimeter as the pressure is raised.

By Osborne's method the enthalpy values of both the saturated liquid and the vapor can be obtained directly from three types of experiments made in one apparatus without precise knowledge of the volume in the calorimeter or other thermal properties of the fluid. This calorimetric method was first applied to water and steam[7,9,10] and later to hydrocarbons[12]. A review of the theory of this method will be given, together with experimental applications to determine some thermal properties of water in the temperature range from 0–374 °C. In addition, there will be some discussion of other methods of measuring enthalpies of vaporization.

II. Theory of the Method

The method makes use of a single calorimetric apparatus. A sample (both liquid and vapor) is so isolated (in the calorimeter) that its quantity, state and energy can be accounted for. With this apparatus, a system of measurements can be made at progressively higher temperatures to determine values of enthalpy of both the liquid and the vapor, referred to values at some reference temperature such as 0 °C.

The calorimeter is provided with two outlet tubes, one at the bottom for introducing or withdrawing liquid and the other at the top for withdrawing vapor. Valves on these tubes either shut the fluid in the calorimeter or control the rate at which fluids are withdrawn. Provisions are made for observing the mass of fluid put into the calorimeter, the heat added, and the temperature of the calorimeter with its contents. The calorimeter is surrounded by a shield whose temperature is controlled so as to minimize net exchange of heat between them. All experiments are made with both the liquid and vapor phases present in the calorimeter.

The essential equations involved in the method follow; for more detail on the derivations of these equations the reader is referred to Osborne's original papers[6]. First consider two heat capacity experiments with the calorimeter, (*i*) one with a large volume of liquid and some vapor, and (*ii*) the other with a small volume of liquid and the rest vapor. In these two experiments the calorimeter contains the sample masses m_A and m_B,

CALORIMETRY OF SATURATED FLUIDS

respectively. The corresponding quantities of energy, q_A and q_B, which are necessary to heat the calorimeter and its contents from an initial temperature, T_1, to a higher final temperature, T_2, are found. From these experiments it is concluded that the difference of the energies, $q_A - q_B$, is used to heat the difference of the masses of the sample, $m_A - m_B$, from T_1 to T_2. The quotient of these differences would represent the change of the enthalpy, H, of the saturated liquid, were it not for the larger volume of vapor in experiment (*ii*). This requires a "vapor correction", $\Delta H_v[V/(V' - V)]$ in which ΔH_v is the enthalpy of vaporization and V and V' are the specific volumes of the saturated liquid and vapor respectively. The equations are

$$\Delta q/\Delta m = (q_A - q_B)/(m_A - m_B) = \left[H - \Delta H_v[V/(V' - V)] \right]_{T_1}^{T_2}$$
$$= [a]_{T_1}^{T_2} \qquad (1)$$

The quantity in brackets, $[H - \Delta H_v[V/(V' - V)]]$, is a specific thermodynamic property of the sample itself and is *not* dependent on the calorimeter. This property, which Osborne called a, has the dimensions of enthalpy.

Equation (1) shows that this property, a, is less than H by a vapor correction term $\Delta H_v[V/(V' - V)]$ which Osborne called β and which is also a specific thermodynamic property of the sample. Hence

$$a = H - \Delta H_v[V/(V' - V)] = H - \beta. \qquad (2)$$

Now consider a second type of experiment which can be used for measuring the enthalpy of vaporization ΔH_v. Let heat be supplied to evaporate liquid while withdrawing vapor. A throttle valve is operated to control the rate of withdrawal so as to keep the pressure, and consequently the temperature of evaporation, constant throughout the experiment. The vapor thus removed is essentially saturated at the temperature of the experiment. The theory shows that the heat added, Δq, divided by the mass withdrawn, Δm, is equal to the enthalpy of vaporization ΔH_v plus the same vapor correction β described above in the a experiments. The quantity $\Delta q/\Delta m$ used for this vaporization experiment, and called γ by Osborne, is also a specific thermodynamic property of the sample so that

$$\gamma = \Delta q/\Delta m = \Delta H_v + \Delta H_v[V(V' - V)] = \Delta H_v + \beta. \qquad (3)$$

Now consider a third type of experiment in which liquid is withdrawn through the lower tube as heat is added. Here it is obvious that all energy goes to producing vapor to fill the space whence liquid is removed. For unit mass of liquid withdrawn, there is $[V/(V' - V)]$, of unit mass of liquid vaporized in the calorimeter, so that

$$\Delta q/\Delta m = \Delta H_v[V(V' - V)] = \beta. \qquad (4)$$

Here then is an experiment which *directly* evaluates the vapor correction β used to obtain H and ΔH_v from the a and γ experiments.

The relations between Osborne's measured a, β, and γ, and conventional enthalpy (at saturation) are summarized by the equations 5 to 7.

$$H = \alpha + \beta \tag{5}$$
$$\Delta H_v = \gamma - \beta \tag{6}$$
$$H' = H + \Delta H_v = \alpha + \gamma \tag{7}$$

in which H' is the enthalpy of the saturated vapor.

So far these derivations have involved only the first law of thermodynamics. If the Clapeyron relation is used, which involves the second law of thermodynamics, the following equations result

$$\beta = \Delta H_v[V/(V' - V)] = VT(\mathrm{d}P/\mathrm{d}T) \tag{8}$$
$$\gamma = \Delta H_v + \Delta H_v[V/(V' - V)] = \Delta H_v[V'/(V' - V)] = V'T(\mathrm{d}P/\mathrm{d}T) \tag{9}$$

in which T is the absolute thermodynamic temperature and P is the vapor pressure. The ratio of these equations reduces to

$$\frac{\beta}{\gamma} = \frac{V}{V'} \tag{10}$$

at any temperature. From this it is seen that calorimetric measurements can lead directly to the ratio of the specific volumes of the saturated fluids. From equations (8) and (9) it is also seen that the calorimetric measurements can lead to values of the specific volumes if the vapor pressure derivative, $\mathrm{d}P/\mathrm{d}T$, is known.

In order to visualize the relative magnitudes of these specific thermal properties (α, β, γ, and H), *Figure 1* shows the formulated results of the experiments which were performed to obtain values of the thermodynamic properties of saturated water and water vapor. The α is shown starting from zero at 0°C and increasing almost linearly up to the critical point ($\sim$374°C). The γ is greatest at 0°C and decreases with increasing rapidity until its slope becomes $-\infty$ at the critical point. The β, on the other hand, is very small at 0°C, but its slope increases with temperature. Above 300°C the slope increases faster than exponentially and β meets γ at the critical point where the slope is ∞. The enthalpy of the saturated liquid, H, is the sum of α and β. The enthalpy of the saturated vapor, H', is the sum of α and γ and it meets H at the critical point. At any temperature the difference $H' - H$, which is the same as the corresponding $\gamma - \beta$, is the enthalpy of vaporization ΔH_v. The mean diameter is defined as $(\gamma + \beta)/2$.

Figure 1 shows that for water, β is relatively small except at the higher temperatures and pressures. At 100°C, the normal boiling point of water, β is only about $\gamma/1600$ and about $\alpha/300$. For liquids below their normal boiling points it is usually more accurate to calculate the values of β, by using the equation $\beta = VT(\mathrm{d}P/\mathrm{d}T)$, than to measure them calorimetrically.

The saturation enthalpy is $H = \alpha + \beta$, so that

$$(\partial H/\partial T)_\mathrm{s} = \mathrm{d}\alpha/\mathrm{d}T + \mathrm{d}\beta/\mathrm{d}T. \tag{11}$$

For most liquids near their normal boiling points $\mathrm{d}\beta/\mathrm{d}T$ is as much as 1 per cent of $\mathrm{d}\alpha/\mathrm{d}T$ so some care must be taken there when using the Clapeyron relation for calculating β and $\mathrm{d}\beta/\mathrm{d}T$ accurately.

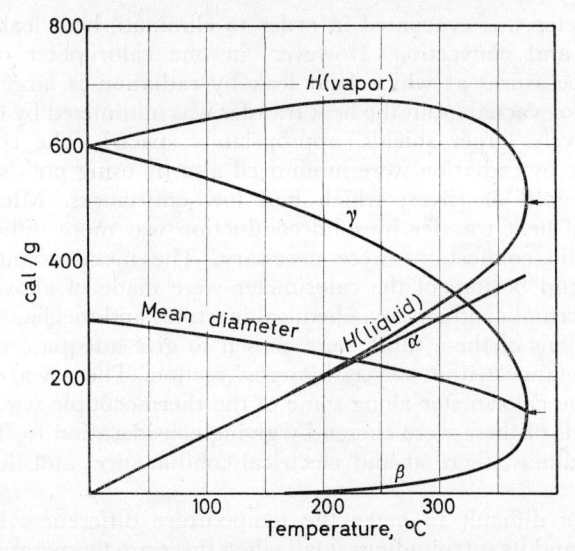

Figure 1. Thermodynamic properties of water.

Values of the specific heats, $C_S = (\partial q/\partial T)_S$ and $C_P = (\partial q/\partial T)_P$, are often needed. The value of C_S can be obtained from the relation

$$C_S = (\partial H/\partial T)_S - V(\partial P/\partial T)_S = \mathrm{d}\alpha/\mathrm{d}T + \mathrm{d}\beta/\mathrm{d}T - \beta/T. \qquad (12)$$

C_P may be obtained from C_S using the relation

$$C_P = C_S + T(\partial V/\partial T)_P(\partial P/\partial T)_S. \qquad (13)$$

As a general rule $(\partial H/\partial T)_S$, C_S, and C_P differ from each other by less than 0·1 per cent at temperatures below the normal boiling points.

III. General Calorimetric Design and Procedures

The design of apparatus for the calorimetry of saturated fluids depends upon such factors as the amount of the sample available, the temperature and pressure range, the accuracy desired, the thermal properties of the sample, and the state of calorimeter design at the time. In the calorimetry of Osborne and his collaborators, several general principles were adopted at the start. Some of these will now be discussed.

Early in the program it was accepted that heat leak uncertainties could be the largest item of error in accurate calorimetry and that it was necessary to minimize and evaluate the heat leaks. The solution to this problem involved three different procedures, (*i*) minimization of the heat transfer coefficient between calorimeter and surroundings, (*ii*) minimization of the temperature differences between calorimeter and surroundings, and (*iii*) measurement of all unavoidable temperature differences in order to correct for the resulting heat leak. The application of these three procedures follows.

In three of the calorimeters described in this chapter the space surrounding the calorimeter was evacuated in order to eliminate heat leak by gaseous conduction and convection. However, in one calorimeter operating at higher temperatures at which heat leak by radiation is large, it was not convenient to evacuate, but the heat transfer was minimized by using a series of progressively larger shields, appropriately spaced. The coefficients of heat transfer by radiation were minimized also by using polished reflecting surfaces of gold or silver, which had low emittances. Minimizing the coefficient of heat transfer by solid conduction was more difficult because some metallic connections were necessary. The metal connecting tubes at the top and bottom of the calorimeter were made of alloys chosen for their low thermal conductivity, chemical inertness and mechanical strength. The dimensions of these tubes were chosen to give adequate strength and fluid conductance without excessive cross section. There was also a heat leak from the calorimeter along some of the thermocouple wires. The sizes and materials of these were chosen by giving consideration to the combination of sturdiness, thermal and electrical conductance, and thermoelectric power.

It is more difficult to make the temperature differences between the calorimeter and its surroundings small when there are temperature gradients on the surface of the calorimeter and the shield. The first step, therefore, was to minimize temperature gradients. This was accomplished in several different ways, depending upon the particular calorimeter. In two calorimeters, mechanical stirring was used to distribute heat and thus reduce the temperature differences over the surface of the calorimeter. In the calorimeter designed for higher pressures and temperatures, for which mechanical stirring would have been difficult, bubbles of vapor mixed the fluid as in a coffee percolator and heat was distributed further by conduction in a system of silver vanes. In a fourth small calorimeter, many copper vanes distributed the heat by conduction.

It is not only important to consider the temperature gradients in the calorimeter but also the gradients in its surroundings. Although it is true that calorimetric procedures usually reduce the effect of errors which are due to temperature gradients in a shield, this reduction of errors is more effective if the gradients are small and essentially independent of the environment. One calorimeter used a stirred liquid bath to distribute heat over a shield (shell) surrounding the calorimeter. In another calorimeter at higher temperatures, for which it was impractical to use a stirred liquid bath, a heavy silver shield was used to distribute heat from a heater wound on it. In still another designed for the maximum accuracy at moderate temperatures, a novel heat distribution system using a saturated vapor bath was very effective over a large surface.

In addition to minimizing heat transfer coefficients and temperature gradients, the remaining unavoidable temperature differences were recorded and summed algebraically so corrections could be made for the net amount of heat transfer. The evaluation of the overall heat leak in an experiment was performed by dividing the heat leak into four parts, (i) the heat leak to the shield from most of the surface area of the calorimeter, usually measured by means of a combination of thermocouples, (ii) the heat leak along the upper tube connecting to the calorimeter, (iii) the heat leak along the lower tube and (*iv*) a heat leak called the "residual" heat leak, which is that heat leak occurring when there is no indicated temperature difference between the calorimeter and its surroundings. In order to evaluate the various heat leaks adequately, it was necessary to give systematic attention to experiments designed for this purpose. For example, the residual heat leak was usually evaluated before and after each experiment. Experiments to evaluate the various heat transfer coefficients at different temperatures were made daily by deliberately exaggerating the temperature difference and measuring the change of temperature produced in the calorimeter.

The most difficult problem in measuring and evaluating heat leak is the measurement of the effective temperature differences along the various paths for heat flow between the calorimeter and its surroundings, especially when the calorimeter is not at thermal equilibrium. One obvious solution of this problem would be to use a great many thermocouples located at representative places. In one calorimeter[9], 38 thermocouples were used to obtain information on temperature distribution. Of these, 31 had measuring junctions located on the calorimeter and its immediate surroundings. Experience had shown that the design of the attachment of a thermocouple junction to a calorimeter was vital to the validity of the temperature measurement. When the thermojunction could be soldered directly to metal, it was assumed to be in good thermal contact. In this case, if the wires were small and made of material having low thermal conductivity, the temperature of the junction would represent the temperature of the metal to which it was attached. In most cases, however, the thermocouple junctions needed to be electrically insulated from the metal so that they could be connected in series for averaging temperatures or be used for measuring temperature differences. Since most good electrical insulators are also good thermal insulators, a special type of thermocouple attachment was developed. This type of attachment, called a "thermal tie-down" and shown in *Figure 2*, was used extensively in all calorimeters developed by Osborne and his colleagues. It has been used subsequently in a calorimeter described in Chapter 9. Basically, this thermojunction tie-down consists of a gold terminal, T, insulated by thin mica, M, all pressed firmly against the metal part by a nut on a threaded stud. Two or more gold terminals were used together to connect thermocouples with each other. If the thermocouples were to be used in a region where there would be large temperature gradients and heat flow along the thermocouple wires which would affect the temperature of the thermojunctions, thermal tie-downs were used on the lead wires before they reached the thermojunction. In some calorimeters, second tie-downs were used for this purpose. This type of thermal tie-down was found to be effective even in an evacuated space. The general theory of heat tempering along wires is discussed in Chapter 4.

An alternative method of minimizing heat leak uncertainties, which avoided the use of a multitude of thermocouples, was developed for one calorimeter[12]. The principle of this method seems so important to calorimetry that some discussion will now be given. In accurate calorimetry, it is desirable to utilize the comparison method as much as possible for minimizing errors. In heat capacity calorimetry, the comparison method is used by

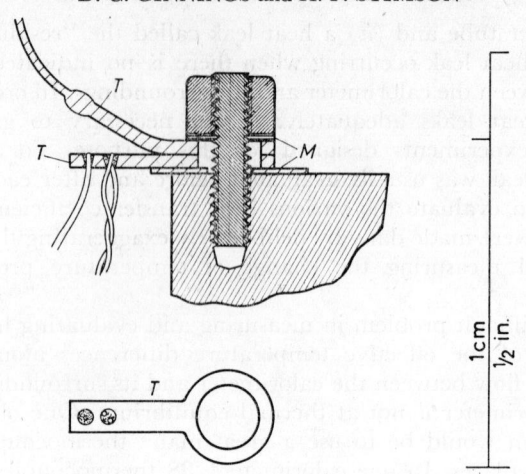

Figure 2. Details of thermal tie-downs.

making two series of experiments, one gross series with the calorimeter nearly full and one tare series with the calorimeter nearly or completely empty. If the two series of experiments are made in the same manner, the gradients on the shield (immediate surroundings) should be very nearly identical in the two series of experiments. If the gradients on the calorimeter are also the same in the two series, the heat leak will be the same. Even though its magnitude is not known, the heat leak should be balanced out by differences when the tare measurements are subtracted from the gross measurements. The difficulty with this assumption is that different amounts of the sample in the calorimeter practically always give rise to different temperature gradients in the calorimeter, so that the above assumption is not usually valid. It is possible, however, to design a calorimeter which validates this assumption reasonably. The principle of the design is to add an outer calorimeter surface whose temperature–time function is essentially independent of the contents of the calorimeter, so that temperature gradients on it are the same in the gross and tare series. This can be accomplished by providing an outer calorimeter surface which is thermally separated from the sample container except in one chosen zone whose temperature is measured. Examples of this type of construction have been given in Chapter 9. and further examples are given later in this chapter.

Another characteristic of all four of the calorimeters used by Osborne and his collaborators in the study of water and hydrocarbons was the use of a "reference block" to contain the platinum resistance thermometers which determine the temperature on the international scale. Many experimenters using a resistance thermometer place it inside the calorimeter and use its temperature at equilibrium to represent the temperature of the calorimeter. In the calorimeters used by Osborne and his collaborators, the resistance thermometers were placed in a reference block separate from the calorimeter. The block temperature was kept at about the same temperature as the calorimeter, and all temperatures on the calorimeter and its surroundings were referred to the reference block temperature by means of thermocouples. The use of a reference block has several advantages. (*i*) The heat capacity of the empty calorimeter can be less. (*ii*) It facilitates the use of two or more resistance thermometers instead of one, so that a failure of one resistance thermometer may be detected before it invalidates the experimental results. (*iii*) By the use of a silver or copper block of appropriate dimensions, the reference block can conveniently provide an isothermal region for both the resistance thermometers and the reference junctions of thermocouples. By using thermocouples differentially so that the thermocouple wires do not pass through large temperature gradients, most of the effects of thermocouple inhomogeneities are avoided. While the use of a reference block has the above advantages, it adds two complications which may have prevented its general use. These are (*i*) that it may require a more complex design and also (*ii*) that the temperature of the reference block may have to be controlled separately, depending on the design.

IV. Calorimeters for Measuring Properties of Water

1. *Calorimeter for 0–270°C*

The first calorimeter described here was used to measure values of the thermodynamic properties α, β, and γ of water up to 270°C. *Figure 3* shows a diagram of the calorimeter and its accessory parts. In the bottom of the calorimeter there was a double centrifugal pump which served two functions, (*i*) to circulate liquid over almost the entire inner wall of the calorimeter and (*ii*) to flow water in a thin stream over the heater, located near the top of the calorimeter. The pump capacity was sufficient to circulate the entire contents of the calorimeter in about 5 sec, so that temperature gradients on the calorimeter surface were minimized. Furthermore, in the vaporization experiment, the use of a thin flowing layer of water over the heater, H, permitted quiet evaporation, thus producing essentially saturated vapor. The calorimeter was held in place by the upper and lower tubes, ST, inside a shell, E, whose temperature was controlled by a circulating oil bath. In the upper part of this bath was located the reference block, R, containing three resistance thermometers, T, and various reference junctions of thermocouples, J. The upper and lower tubes, connecting to the calorimeter, led to throttle valves used for controlling the flow of vapor or liquid, respectively. In the vaporization (γ) experiments, vapor flow was diverted from one receiving container to another by valves actuated by time signals. In the β experiments (withdrawal of liquid), however, only one receiving container was used, so that the flow of liquid was started and stopped during an experiment.

In this calorimeter, α experiments were made over the entire range 0–270°C, but β and γ experiments were made only over the range 100–270°C. The values of α from these experiments were converted to H, enthalpy at saturation pressure referred to 0°C, by using measured values of β above 100°C. Below 100°C, the values of β are so small that values calculated from the relation $\beta = TV\,dP/dT$ were better. The values of ΔH_v were obtained directly from the observed values of β and γ between 100 and 270°C. The accuracy of the derived values of H, C, and ΔH_v was usually

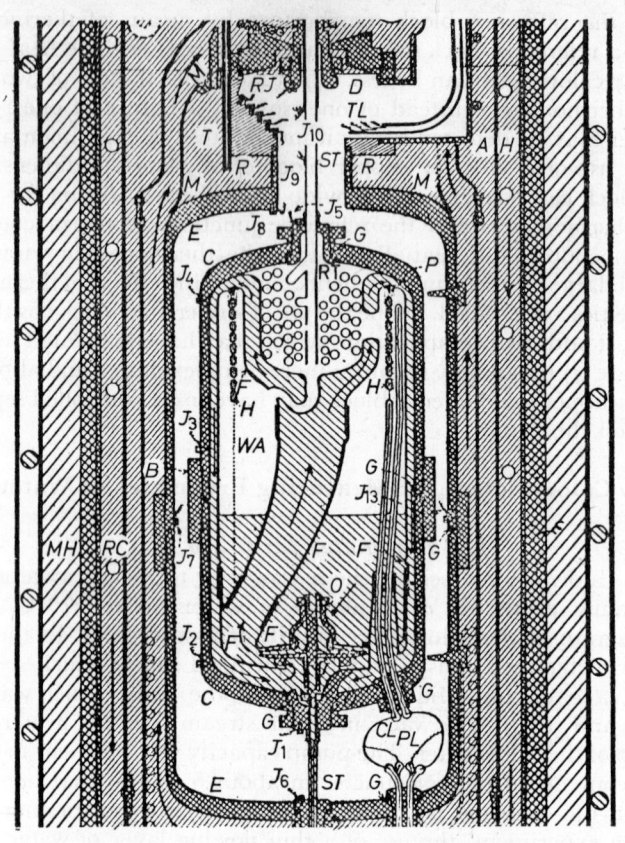

Figure 3. Calorimeter for 0–270°C

C,	Calorimeter shell	CL,	Current lead
E,	Envelope shell (shield)	PL,	Potential lead
B,	Threaded band	R,	Reference block
G,	Gold gaskets	M,	Mantle
ST,	Support tubes	J_1, J_2, etc., Measuring junctions of thermocouples	
I,	Pump impeller		
F,	Pump casing	RJ,	Reference junctions of thermocouples
O,	Ball bearings		
P,	Water port	TL,	Thermocouple leads
H,	Calorimeter heater	T,	Platinum resistance thermometer
WA,	Gauze apron	AH,	Auxiliary heater
		MH,	Main heater
		RC,	Refrigerating coil

better than 0·1 per cent, as confirmed by experiments with two later calorimeters.

2. *Calorimeter for* 100–374°C

This calorimeter was designed for measurements on water up to the critical temperature (about 374·1°C) and pressure (about 218 atm). Because of the higher pressure and temperature, the design required more mechanical strength and corrosion resistance. Although in the earlier calorimeter (0–270°C) a copper–nickel alloy had sufficient strength for the calorimeter shell, it was necessary in the 100–374 °C calorimeter to use a special non-corrosive alloy steel to avoid mechanical creep at the higher temperatures. *Figure 4* shows a diagram of this calorimeter, together with some of its

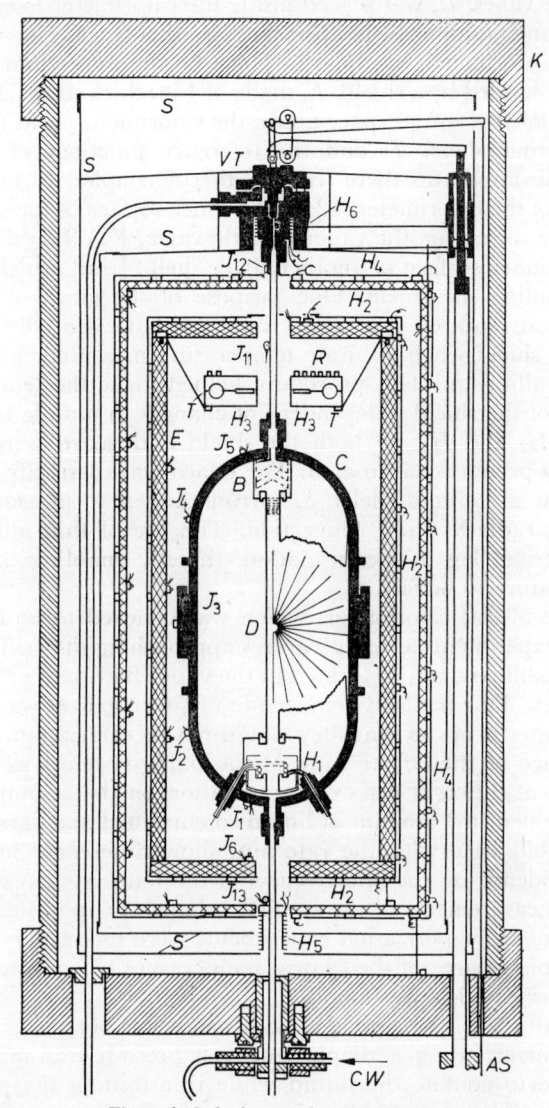

Figure 4. Calorimeter for 100–374 °C

C,	Calorimeter shell	H_2,	Shield heater
H_1,	Calorimeter heater	H_4,	Guard heater
D,	Heat diffusion system	H_5,	Lower tube heater
R,	Reference block	CW,	Cooling water
H_3	Reference block heater	S,	Aluminum radiation shields
T,	Platinum resistance thermometers	VT,	Vapor throttle valve
J_1, J_2, etc. Measuring junctions of thermocouples		H_6,	Heater for valve
		AS,	Adjusting screw for valve
E,	Shield	K,	Outer casing

accessory parts. The calorimeter heater, H_1, was located in the bottom of the calorimeter, C, and was concentrated so as to produce bubbles of vapor for convective mixing by use of the principle of the coffee percolator. A system of radial silver vanes, D, was placed inside the calorimeter to speed thermal equilibrium and minimize temperature gradients. As in the previous example, the calorimeter was held in place by the upper and lower tubes, in a space surrounded by a shield, E, made of $\frac{1}{4}$ in. thick silver. The reference block, R, also located in this space above the calorimeter, held two platinum resistance thermometers, T, and the reference junctions of the thermocouples. In this apparatus there were 38 thermocouples, 13 located on the outer surface of the calorimeter, 12 on the inner surface of the shield ($\frac{1}{4}$ in.), 2 on the upper tube, 1 on the vapor throttle valve, VT, 3 on the lower tube and 7 on the inner surface of another silver shell ($\frac{1}{8}$ in.) which served as a guard surrounding the shield. One purpose of the guard was to supply most of the heat required at a steady state, so that essentially no heat was needed in the shield when accurate temperature measurements were made at thermal equilibrium. This procedure helped make the temperatures on various parts of the shield independent of changes in outside temperatures. The heaters, H_2 and H_4, on both the shield and guard were distributed approximately proportional to area. The guard was thermally insulated by successive thin aluminum shells, S, surrounding it to impede the loss of heat both by radiation and convection. The use of thin aluminum also minimizes thermal lag in the insulation, thereby enabling the control of guard temperature to be fast.

One feature of this calorimeter, which was believed to be important in vaporization experiments at temperatures approaching the critical temperature, was the baffle system, B, located in the vapor flow path at the exit from the calorimeter. This baffle system, made of silver gauze, was designed to catch small water drops so that they would not be carried out of the calorimeter. Evidence of the effectiveness of the baffle system was obtained by measurements at different rates of vaporization on the assumption that, if drops were formed, the amount of liquid so entrained and carried out with the vapor would vary with the rate and show differences in the results. Additional evidence on the effectiveness of the baffle system was obtained from special measurements of the specific volume of vapor, described later. In this calorimeter, vaporization experiments were successfully made up to within about one degree of the critical temperature before any evidence of entrained water droplets was found.

The apparatus was also provided with means for observing the pressure in the calorimeter, using a diaphragm-type pressure-transmitting cell in the liquid line to confine the liquid while transmitting the pressure to a pressure gauge. The use of this calorimeter for accurate measurement of the vapor pressure of water is described in a separate publication[8]. In the calorimetric experiments for β and γ, this provision for pressure measurement proved very useful. At thermal equilibrium before starting an experiment, the pressure in an outside container was made equal to that inside the calorimeter, as indicated by the null position of the diaphragm in the pressure transmitting cell. During the γ experiment, the fluid withdrawal rate was controlled so that the pressure in the calorimeter was the same as before the experiment. In this way, the surface of the liquid in the calorimeter was maintained essentially at the starting temperature, even though most of the liquid became somewhat warmer during evaporation. The throttle valves used to control the fluid withdrawal rate were specially designed to provide continuous adjustment without back-lash.

Since this calorimeter was used for accurate pressure measurements, it was possible to provide an independent check on β and γ at high temperatures where the possibility of mixing of liquid and vapor phases might exist, even though there had been no evidence for it in the experiments changing the rate of flow. As stated earlier in section II of this chapter using the Clapeyron relation we have

$$\gamma = TV'\mathrm{d}P/\mathrm{d}T$$
$$\text{and } \beta = TV\mathrm{d}P/\mathrm{d}T.$$

If V' and V are known, values of γ and β can be calculated from the vapor pressure data. Although it was not convenient to use the calorimeter as a volume-measuring device for extensive measurements on the specific volumes of saturated liquid and vapor, a method was devised to make such measurements at 370°C in order to check the calorimetrically measured γ and β. For these measurements it was possible to establish a reasonably sharp boundary for the volume of the calorimeter, by means of the thermocouples and heater on the lower tube from the calorimeter. From a calibration of this volume it was possible to measure the specific volumes of both the saturated liquid and vapor at 370°C, and to calculate values of γ and β by using the vapor pressure data. These calculated values of γ and β agreed with the calorimetrically measured values of γ and β to within about 0·3 per cent, which could be attributed to the error of these specific volume measurements.

3. *Calorimeter for* 0–100°C

Measurements of the heat capacity of water in the range 0–100°C have special significance because water has been used for a heat capacity standard in this range. The calorimeter described here was designed to obtain as high an accuracy as possible, using the techniques available in 1938, on both heat capacity and enthalpy of vaporization. Following Osborne's method, this meant measurements of α, β, and γ. However, at temperatures below 100°C, β is so small that values calculated by using the Clapeyron relation were more accurate than values measured calorimetrically. At 100°C, β is less than 0·1 per cent of γ, and at lower temperatures it is a still smaller fraction. Therefore, no measurements of β were made with this calorimeter.

In order to strive for accuracies of 0·01 per cent for heat capacities, it was decided to use a calorimeter with a capacity larger than the earlier ones, namely almost 1200 cm³. It was thought that a large sample would be more favorable for higher accuracies for several reasons. (*i*) The mass accounting should be more reliable. (*ii*) The volume of a calorimeter of a given shape increases as the cube of the linear dimension, whereas the surface increases only as the square and hence the area subject to heat leak is proportionately less. (*iii*) A sphere has the least surface for its volume and the spherical shape deforms less under pressure. At temperatures seldom exceeding 100°C the pressure would be moderate so the calorimeter could be thin-walled.

It was recognized that the larger surfaces of the calorimeter would cause larger temperature differences which would be less favorable for heat leak accounting, so it was decided to stir the liquid contents by using a synchronous motor for its constant speed. By these means it was possible to keep the effective heat capacity in the low-filling experiments down to less than 5 per cent of the effective sample size, which is the difference between masses of samples in high and low fillings. (With many calorimeters the empty calorimeter heat capacity is more than 20 per cent of the sample heat capacity.)

A study was made of the method of the stirring to provide the maximum amount of mixing with a minimum of pump power. The system chosen is shown in *Figure 5*. The circulating system consisted of two screw propellers on one shaft, and a system of guides to direct the water flow. The larger (upper) propeller, P_1 did most of the circulating, and the smaller (lower) propeller, P_2, was used to avoid stagnation in the extreme bottom of the calorimeter. The liquid went downward next to the wall of the calorimeter shell and was directed by radial guides, PG, at the bottom so that the liquid flowed without swirl past the heater, H_1, into the propeller. The water from the propellers was guided axially upward by a circulator pump casing, PC, and curved guide vanes served to remove the swirl imparted to the water by the propellers. A synchronous motor kept the pump speed very constant at 70 rev/min which gave adequate circulation while dissipating only about 0·004 W in the calorimeter. Such low pump power was possible in this design because the water level was not raised as it had been in the 0–270°C calorimeter. It was found, however, that even with this low power, there was a random variation in the dissipated pump power which may have been a major contribution to the variation in the experimental results.

The calorimeter shell, C, was made of copper, all surfaces of it and parts inside were gold-plated, and the outer calorimeter surface was polished to minimize heat transfer by radiation. A baffle, B, was installed near the vapor outlet tube to prevent passage of liquid drops during the vapor withdrawal experiments.

For accurate evaluation of electric power in the calorimeter, particular attention was given to the design of the calorimeter heater. The earlier calorimeter heaters had resistances of about 10 Ω with the usual four electrical leads, i.e., two current leads and two potential leads attached to the current leads at points midway between the calorimeter and its surrounding shield. This selection of points for attaching the potential terminals seemed reasonable for measuring the power to the calorimeter. The choice of size and resistance of the current leads between the calorimeter and shield was a compromise between the amount of heat developed in the leads and the heat transferred along these leads. With calorimeter heaters having 10 Ω resistance, this compromise was difficult to make when accuracies of 0·01 per cent were sought. If, however, the heater resistance was increased from 10 Ω to 100 Ω, the relative heat developed in the leads compared to that in the calorimeter heater was reduced by a factor of 10. A discussion of this problem is given in Chapter 4. The resistance of this

CALORIMETRY OF SATURATED FLUIDS

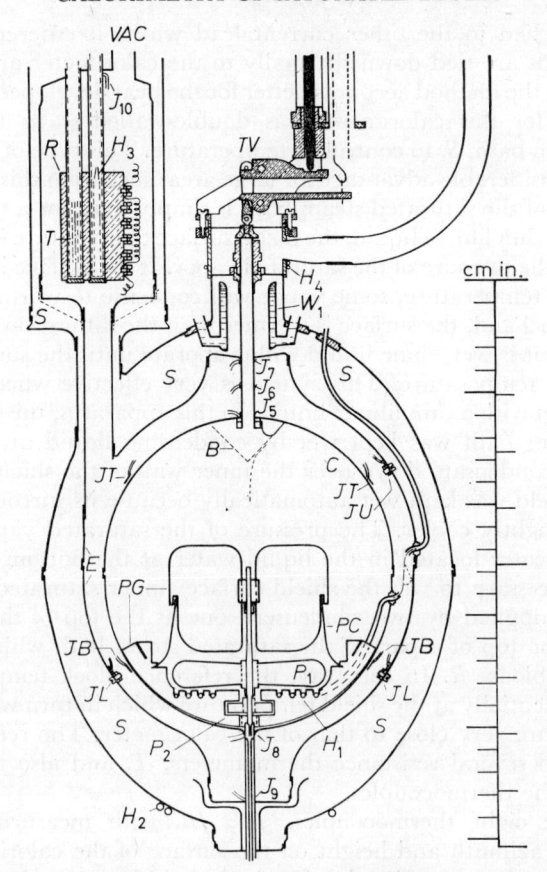

Figure 5. Calorimeter for 0–100°C

C,	Calorimeter shell	H_2,	Steam bath heater
H_1,	Calorimeter heater	R,	Reference block
$P_1, P_2,$	Pump propellers	H_3,	Heater for block
PC,	Pump casing	T,	Platinum resistance
PG,	Flow guides		thermometers
B,	Gauze baffle	W_1,	Condenser
JT, J_5, etc.,	Measuring junctions	TV,	Throttle valve
	of thermocouples	H_4,	Heater for valve
S,	Saturated steam bath	VAC,	Vacuum line

calorimeter heater was made about 107 Ω so that the problem of accounting for heat developed in the leads was considerably simplified. Actually the entire heat developed in both current leads between calorimeter shell and shield amounted to only about 0·03 per cent of the heat developed in the calorimeter heater.

In this calorimeter a different method of attaching potential leads was used. The mid-points of the current leads between the calorimeter and shield became hot, so that it seemed preferable to avoid attaching potential leads to the mid-points if possible. One potential lead was attached to one current lead where it entered the calorimeter shell and the other potential lead was attached to the other current lead where it entered the shield. If the two leads are tied down thermally to the calorimeter and the shield symmetrically, the method accounts better for the heat developed in the leads.

The shield for this calorimeter was double-walled so as to contain a saturated steam bath, S, to control its temperature. This type of temperature control has considerable advantage for large areas such as in this calorimeter. The principle of the saturated steam bath is simply that, for a metal surface covered with a thin film of liquid, the metal surface temperature is determined essentially by the pressure of the saturated vapor. If the surface is cooler than the saturation temperature, some vapor will condense to warm the surface. If, on the other hand, the surface is warmer than the saturation temperature of the vapor but is wet, some liquid will evaporate until the surface cools to the saturation temperature. This system is very effective when the surface can be kept wet with a thin film of liquid. In this apparatus, the surface of the saturated steam bath was kept wet by condensing liquid at the top and directing the condensate down over the inner wall of the shield. The outer wall of the shield was kept wet automatically because its surroundings were always kept slightly cooler. The pressure of the saturated vapor was controlled by a heater located in the liquid water at the bottom of the bath. The liquid necessary to wet the shield surface (inner saturated steam bath surface) was supplied by two condensers, one at the top of the shield and the other in the top of a part of the saturated steam bath which contained the reference block, R. In this way, the reference block temperature was maintained essentially at the shield temperature which in turn was controlled to a temperature very close to that of the calorimeter. The reference block accommodated several resistance thermometers, T, and also the reference junctions for the thermocouples.

There were eight thermocouples, JT, JB, with measuring junctions distributed in azimuth and height on the surface of the calorimeter, in an attempt to give them equal weighting for heat leak accounting. Individual leads to these thermocouples were brought out either for check of the temperature uniformity, when used individually, or for connection in series for evaluation of the average temperature. All eight were used in series to measure the temperature difference between calorimeter and reference block, or the group of eight could be opposed to eight corresponding thermocouples distributed over the inside area of the shield for maintaining adiabatic conditions. In addition to these, there were three thermocouples, J_5, J_6, J_7, on the upper tube and two thermocouples, J_8, J_9, on the lower tube. The thermocouple, J_6, located at the mid-point of the upper tube, furnished data for calculation of the temperature of the vapor leaving the calorimeter during the vaporization experiments. The thermocouples used in all the calorimeters described here were made of Chromel P versus Constantan (or their equivalent) because these materials give a high thermoelectromotive force (~ 60 μV/°C near room temperature) and have reasonable thermal and electrical resistivities.

Since this calorimeter was intended to give results consistent with the best techniques available in 1938, some discussion will now be given of the precautions taken to insure highest accuracy. This discussion will be in three parts as follows.

CALORIMETRY OF SATURATED FLUIDS

A. Accounting for Mass of Sample

Precautions were taken to insure pure samples of water and proper accounting for the masses. In the heat capacity measurements, the difference in masses of water in the calorimeter in the high and low fillings was approximately 1 kg. In all experiments, the amount initially put into the calorimeter was compared with the amount finally taken out. Usually these amounts agreed to 0·02 g, which corresponded to about 1 part in 50 000 in the heat capacity experiments. When the mass accounting was not this accurate, the experimental results were discarded. In preparing the sample, precautions were taken to reduce the impurities (including dissolved gases) to a negligible amount. For transferring samples, the connecting tubes were evacuated before the transfer. In the vaporization experiments, the mass taken out in a typical experiment was about 20 g. Systematic precautions were taken to insure reproducible weighing conditions for the sample container. It is believed, therefore, that the accounting for mass was not the principal limitation on the overall accuracy of the results.

B. Accounting for Energy

The energy supplied to the calorimeter can be considered in three principal parts (*i*) the electrical energy supplied, (*ii*) the energy supplied by the circulating pump, and (*iii*) the heat leak from the calorimeter. These three parts will now be considered, with emphasis on their effects on the overall accuracies of the experiments.

(*1*) *Electrical Energy*. With modern techniques the measurement of electrical energy does not seem to be a limitation on the accuracy of calorimetric experiments. For standards of e.m.f., a group of saturated standard cells was maintained in a specially insulated aluminum box whose temperature was controlled constant within 0·01 °C[4]. These standard cells and also the resistance standards were calibrated periodically. The temperatures of both the standard cells and the standard resistors were usually checked at both the beginning and the end of the day. Two specially constructed volt boxes were used in combination so that if any part of either volt box changed resistance between calibrations, the daily check on them would detect it. A check was made at least twice a day for any change in the potentiometer factor. This check consisted essentially of comparing a given potentiometer reading (which is effectively a ratio) with the ratio of two standard resistors.

The accounting for heat developed in the current leads to the calorimeter heater has already been discussed. Another small effect which was considered was the "starting effect" of the change in calorimeter heater resistance immediately after the current was started. With the high resistance of the calorimeter heater which required a high voltage storage battery, it was not convenient to use an external resistor having the same resistance in order to minimize the effect on power of change in resistance of the heater[3]. Consequently, because of a significant change in resistance of the heater during the first 30 sec before the first power reading, the calorimeter heater power could not be extrapolated perfectly from regular periodic observations starting at 30 sec. A study was made of this starting correction, observing the change in power during the first 30 sec. The result was that a correction was made which amounted to only 0·01 per cent of the whole energy input, even for the extreme case of a short experiment.

In addition to measurements of power, measurements of time intervals were necessary in order to calculate energies. These time interval measurements were perhaps the easiest of all measurements because they were determined by using accurate standard time signals (seconds). The time signals were used to trip a switch in the heater circuit. The time for operation of this switch was determined to be negligible.

(2) *Pump Energy.* As mentioned previously, the pump power in the heat capacity experiments was approximately 0·004 W which, with the pump running continuously, amounted to about 0·015 per cent of the energy put into the sample during a typical 10° heat capacity experiment. Although this is small, separate measurements of this power were made near the time of the experiment in order to estimate the correction. Even though the pump speed was very constant, the mechanical power dissipated was found to vary at random by amounts larger than could be accounted for by the uncertainty in temperature measurement. This variation, however, was usually not more than would correspond to about 0·005 per cent for the value of a.

(3) *Heat Leaks.* The accurate control and measurement of heat leak is vital in accurate calorimetry. Heat leaks in this calorimeter are classified, like those in earlier calorimeters, into shield heat leak, upper tube heat leak, lower tube heat leak, and residual heat leak. Measurements of these heat leaks were made daily throughout the experiments, and calibration experiments were made for determinations of the various heat transfer coefficients. In spite of the large calorimeter surface it was possible to keep the shield heat leak down to a very low value by use of the evacuated space and polished gold-plated surfaces. The use of numerous thermocouples on this surface, in addition to the effective use of the circulating pump to reduce temperature gradients, made it possible to reduce the actual shield heat leak to a small value. The residual heat leaks, such as could have resulted from imperfect weighting of the controlling thermocouples, were almost negligible and were automatically included with the pump energies in the experiments. The upper and lower tube heat leaks were small and were accurately evaluated by the thermocouples located on them. In the heat capacity experiments, the net corrections applied for all measured heat leaks averaged less than 0·001 per cent of the energy put into the sample.

C. Accounting for Change in State

The temperature of the calorimeter was determined by measurements with platinum resistance thermometers, calibrated on the International Temperature Scale, combined with thermocouple measurements. The temperature of the calorimeter was observed at essentially thermal equilibrium at the beginning and end of each experiment. For this temperature measurement, the temperature of the reference block containing the resistance thermometers was brought very close to the temperature of the calorimeter, as indicated by the eight thermocouples (in series) having measuring junctions on the calorimeter and reference junctions on the reference block. Thus, the temperature measured by the resistance thermometers was essentially the average temperature of the calorimeter, except for a small difference which was calculated from the thermocouple readings. For this calculation, the sensitivity of the thermocouples was determined in place directly against the resistance thermometers.

In all the experiments, the change in proportion of liquid and vapor in the calorimeter is accounted for in the theory of the method as described earlier. In the vaporization experiments, the calorimeter was usually about half-full. Under this condition, the temperature of the vapor leaving the calorimeter was determined by means of the group of four thermocouples on the upper part of the calorimeter, representing its average temperature, in conjunction with the three thermocouples giving temperatures along the upper tube. During the actual vaporization and withdrawal of vapor through the upper tube, this tube was tempered by the outgoing vapor so there was no correction for upper tube heat leak during this period. In the vaporization experiments, the shield heat leak accounting was slightly less accurate than in the heat capacity experiments because of the larger temperature gradients on the calorimeter. These gradients were larger on account of the sharp temperature gradient in the liquid next to the surface where evaporation was taking place. These gradients were large even when the circulating pump speed was increased to 106 rev/min. A small correction, sometimes amounting to as much as 0·002 per cent and depending on the location of the liquid level in the calorimeter, was applied to account for the change in liquid level.

4. *Comparison of results with the three calorimeters*

Of the three calorimeters described, the 0–100°C calorimeter was designed to have the highest accuracy, probably about 0·01 per cent. It is interesting to compare the results from this calorimeter with the results from the others where measurements were made at the same temperatures. In all three calorimeters, the measurements were made of the enthalpy of vaporization at 100°C. If the vaporization results with the 0–100°C calorimeter are used as a basis for comparison at 100°C, the 0–270°C calorimeter gave a value 0·06 per cent high whereas the 100–374°C one gave a value 0·02 per cent high. Measurements of enthalpy of vaporization at 50°, 70°, and 90°C with the 0–270°C calorimeter[2] gave values averaging less than 0·02 per cent higher than the one used for 0–100°C. Heat capacities from measurements made with the 0–270°C calorimeter could also be compared with those from the one used for 0–100°C over its entire range. These values were found to be about 0·07 per cent higher.

V. Calorimeter for Hydrocarbons

As part of a program for the investigation of the properties of hydrocarbons, accurate measurements of enthalpies of vaporization at 25°C of a large number of pure samples were needed. The calorimeter described here was designed particularly for measuring these although it was used for a few times at other temperatures above and below 25°, both for vaporization and heat capacity experiments.

In designing this calorimeter, there were three principal aims. (*i*) The calorimeter shell should be made small because some of the pure hydrocarbons were available in relatively small amounts. (*ii*) The measurements (especially vaporization) should be accurate to at least 0·1 per cent. (*iii*) The apparatus should permit relatively rapid measurements of vaporization so that a large number of materials could be investigated. Making the calorimeter small has certain other advantages besides favoring the use of small samples. In a small calorimeter, it is possible to distribute heat by conduction with a system of many copper vanes and dispense with mechanical stirring. These vanes minimized the time required for thermal equilibrium and contributed both to accuracy and ease of operation. Mechanical stirring was purposely avoided after having found in the 0–100°C calorimeter that there were unpredictable variations in the dissipated pump power. Another advantage of a small calorimeter is that it is possible to use effectively unheated metal shields for the control and evaluation of heat leak, and thus practically eliminate one common source of error in adiabatic heat capacity calorimetry. This important feature will be discussed while describing details of the apparatus.

Because measurements on most of the materials were at temperatures considerably below the normal boiling point, values of the quantity β were negligible in some cases, and in other cases small enough so that they could be calculated adequately from known thermal properties of the material. Therefore, this calorimeter was not provided with a lower tube for β experiments. All samples were vaporized into and out of the calorimeter. A schematic drawing of the apparatus is shown in *Figure 6*, and a scale drawing of essential features is shown in *Figure 7*. The calorimeter, holding approximately 100 ml of sample, had the heat distributing system shown in *Figure 8*.

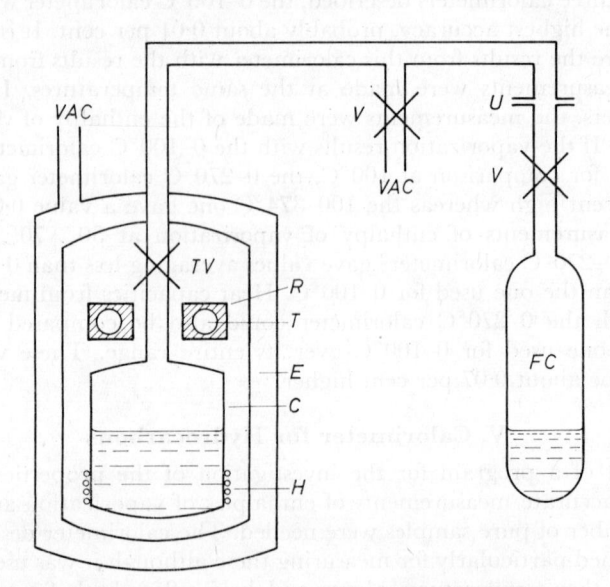

Figure 6. Schematic diagram of calorimeter for hydrocarbons

C,	Calorimeter shell	R,	Reference block
E,	Adiabatic shield	T,	Resistance thermometer
FC,	Fluid container	TV,	Throttle valve
H,	Calorimeter heater	U,	Union
V,	Valve	VAC,	Vacuum line

CALORIMETRY OF SATURATED FLUIDS

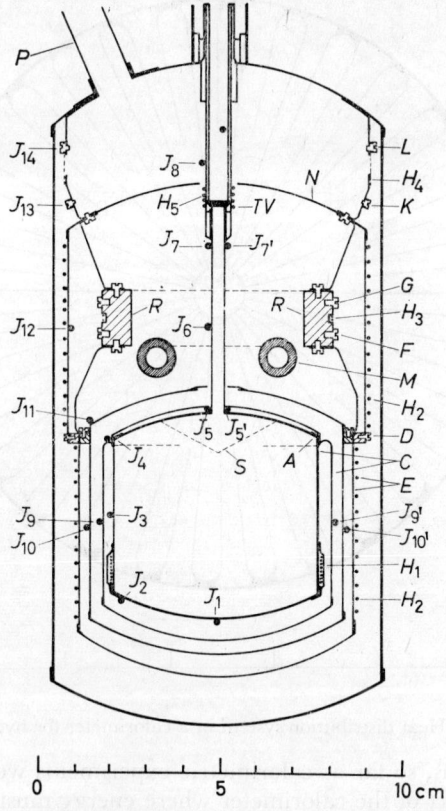

Figure 7. Scale drawing of calorimeter for hydrocarbons

- A, Attachment of calorimeter shell to its shield
- C, Calorimeter shell and shield
- D, Attachment of adiabatic shield to its shield
- E, Adiabatic shield with its shield
- F, G, K, L, Zones of thermal attachment of thermocouples
- H_1, Calorimeter heater
- H_2, Adiabatic shield heater
- H_3, Reference block heater
- H_4, Heater
- H_5, Throttle valve heater
- J_1, J_2, etc., Measuring junctions of thermocouples
- M, Receptacles for platinum resistance thermometers
- N, Deck of adiabatic shield
- P, Lead wire and vacuum duct
- R, Reference block
- S, Gauze baffle
- TV, Throttle valve seat

The tinned copper vanes were arranged so that no point in the liquid sample was more than 3 mm from a conducting metal part. The heater, H_1, (100 Ω) was located on the bottom part of the calorimeter shell, as shown in *Figure 7*. The heater was covered with a copper sheath soldered to the calorimeter shell. Thermocouple measuring junctions, J, were located as shown, with reference junctions on the reference block. The calorimeter shell was provided with a thin copper 'heaterless' shell surrounding it and attached to it

415

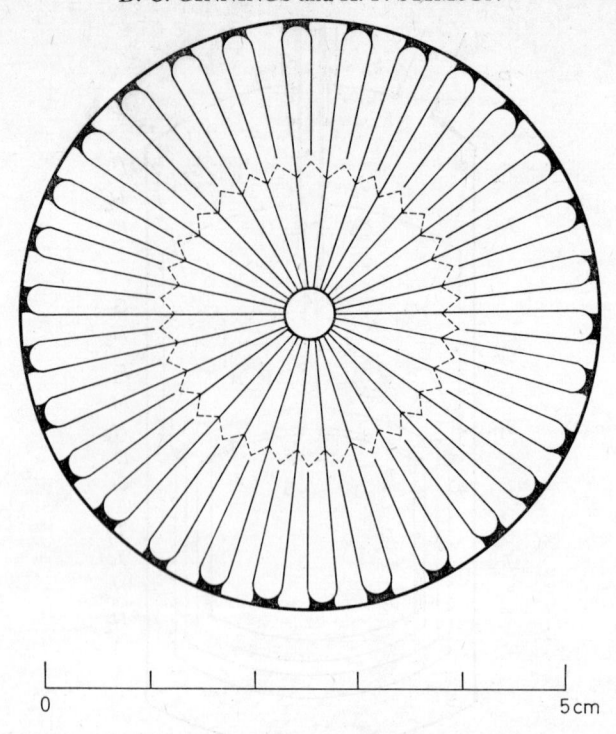

Figure 8. Heat distribution system in a calorimeter for hydrocarbons.

at one zone. Thus, so far as calorimetric experiments were concerned, this thin shell was part of the calorimeter where energy must be accounted for. The calorimeter was surrounded with an adiabatic shield, E, provided with heaters. This adiabatic shield, similar to the calorimeter shell, was also provided with a thin heaterless copper shell which was attached inside the shield at a zone at the same height as that on the calorimeter. Thus the outer calorimeter surface was one heaterless thin copper shell and the inner shield surface another. The calorimeter shell, the adiabatic shield, the heaterless copper shells and the reference block, R, were all gold plated and polished to minimize radiation emittance. The spaces between and around these parts were evacuated to eliminate gaseous conduction and convection. There were therefore two paths for heat leak from the calorimeter, one by radiation from the polished gold surface of the calorimeter heaterless thin shell, and the other by solid conduction along the upper tube and electrical leads from the calorimeter which were thermally attached to the tube. The heat leak along the upper tube was controlled and measured by means of the thermocouples (J_7-J_5) and $(J_{7'}-J_{5'})$ attached to it. The heat leak from the calorimeter was controlled and measured using the thermocouples $(J_{10}-J_9)$ and $(J_{10'}-J_{9'})$.

In the vaporization experiments, even though there was a significant temperature gradient on the calorimeter shell, the heaterless copper shell attached to the calorimeter shell was nearly isothermal owing primarily to the very low heat transfer coefficient to the calorimeter at all points except at the chosen attachment zone. The thermocouples J_9 and $J_{9'}$ measured effectively the temperature of the calorimeter surface relative to the temperature of the reference block. Similarly, in the vaporization experiments, J_{10} and $J_{10'}$ measured the temperatures of the surrounding surface.

During the heat capacity experiments, however, there were necessarily temperature gradients on both of these heaterless shells. Although the two thermocouples were located so as to represent approximately the average temperature of these two surfaces when heating, it is not necessary that they do so. When using the method of two series of heat capacity experiments, a gross series and a tare series, the only requirement is that both surfaces have the same temperature distributions at corresponding times in both series of experiments. This means that to have a valid heat leak accounting, any change in temperature gradients on the surface of the calorimeter shell itself must not affect the temperature distribution on its heaterless shell. With the design shown, the temperature distribution on this heaterless shell was essentially independent of any temperature gradients on the calorimeter shell due to either the presence of the sample or the type of experiment conducted.

Similarly, the temperatures on the heaterless shell which was attached to the adiabatic shield were independent of the temperature gradients on the adiabatic shield. Thus with this design in the heat capacity experiments, the thermocouples $(J_{10}-J_9)$ were not required to measure the true heat leak in any experiment, although these thermocouples probably did so.

It is believed that the above principle of design is vital to all calorimetry and that many of the differences in results in accurate calorimetry can be attributed to the effects on heat leak caused by temperature gradients in the calorimeter and by changes in these gradients between the gross and tare series in heat capacity experiments. A much more comprehensive discussion of this and related questions is given in Chapter 9.

Since further details of the apparatus can be obtained from the original publication, further discussion will be limited to significant points in its use and the accuracies obtained. The effectiveness of the overall heat accounting was remarkable. It was proved possible to use this calorimeter for accurate vaporization experiments with samples of less than 3g in the calorimeter. In various vaporization experiments, temperature differences on the calorimeter varied by about 1°, depending upon liquid level and rate of vaporization. In spite of this there was no evidence of change in measured enthalpy of vaporization due either to liquid level or to rate of vaporization. This latter independence is also evidence of the withdrawal of essentially saturated vapor, rather than of either superheated vapor or vapor containing liquid droplets. The independence of the measured enthalpies of vaporization upon temperature differences in the calorimeter shell was evidence of the effectiveness of the method using unheated shields.

Possibly the most convincing evidence of the accuracy of this calorimeter is the comparison of results on the same material used both in this calorimeter and the earlier 0–100°C calorimeter. For this purpose, vaporization measurements at 25°C were made on water before, during, and after the series of experiments on 59 hydrocarbons. The average of the measurements, with the present calorimeter, for enthalpy of vaporization at 25°C differed by about 0·001 per cent from the tabulated value in the previous work[11], and the average deviation of values from individual experiments from this tabulated value was about 0·01 per cent.

It is interesting to note that this calorimeter was used also on materials having relatively low vapor pressures, such as n-octane. With this material, the mass rate of withdrawal was limited to about 0·15 g/min. In spite of these unfavorable conditions, it is believed that the results justify the estimate of accuracy of 0·1 per cent in the vaporization experiments, which was made when the calorimeter was designed.

Although the heat capacity experiments with this calorimeter were incidental to the vaporization experiments, it is interesting to examine the heat capacity results for evidence of their accuracy. Here again, possibly the best evidence lies in a comparison of results on water used to check the performance of the calorimeter. Heat capacity measurements with water were made at the beginning and near the end of this series of experiments. It was found that the 30 experiments with water gave results about 0·025 per cent different from the previous results in the 0–100°C calorimeter. Additional evidence was also available in the measurements on n-heptane, because some experiments were made on a sample of heptane in the 0–100°C calorimeter. In spite of the use of a sample from a different source, heat capacity measurements in the range 10–35°C gave an average deviation of 0·037 per cent from a least-square formulation of the results with the 0–100°C calorimeter. It is of interest that these earlier experimental measurements gave an average deviation of 0·028 per cent from the same formulation. The evidence from the results seems to indicate that the hydrocarbon calorimeter, although intended for vaporization experiments, is capable of an accuracy better than 0·1 per cent for heat capacity experiments in its temperature range.

VI. Other Calorimetric Methods for Measuring Enthalpy of Vaporization

The calorimeters described have been used in vaporization measurements in the range 0–374°C and at pressures up to 218 atm. No attempt will be made to describe other vaporization calorimeters which use similar methods in this temperature range. At temperatures below 0°C heat capacity calorimeters have also been used for vaporization measurements. Some of these are described in Chapters 5 and 6. At high temperatures at which accurate direct measurements of enthalpy of vaporization are very difficult or impossible, indirect measurements can be made by using the Clapeyron relation $\Delta H_v = T(dP/dT)(V' - V)$. For this evaluation the determination of dP/dT is probably the most difficult.

In the direct methods described, the heat required to vaporize the liquid has usually been supplied by electrical heaters because of the convenience and accuracy of the electrical measurements. Some measurements, mostly of historic value, have measured the heat of condensation, rather than that of vaporization.

Another direct method which has proved useful in certain vaporization measurements is the transpiration method, sometimes known as the carrier gas method. In this method, which has application mostly below the normal boiling temperatures of liquids, vapor is removed from the calorimeter with an inert carrier gas. Measured electric power is usually used to evaporate the liquid. An advantage of this method is the relative ease in controlling the rate of vaporization by controlling the rate of flow of the carrier gas; no elaborate throttle valve is required for this control. A recent example of the application of this method at 25 °C is the work of Wadsö[13], who measured enthalpies of vaporization of samples as small as 0·15 g to an accuracy better than 0·5 per cent. His calorimeter was designed for samples having vapor pressures at 25 °C in the range 1–100 mm Hg.

In the transpiration method there are several precautions that should be considered in addition to the usual heat leak measurement. As in the method used by Osborne and colleagues it is necessary, in principle, to make a β vaporization correction, that is, to correct for the vaporization into the space occupied by the liquid evaporated. As pointed out earlier, this vaporization correction is small and sometimes negligible at temperatures below the normal boiling point. Another possible correction, which is usually small, allows for departures from the ideal process of evaporation at saturation pressure, such as might occur at large rates of gas flow.

Another interesting vaporization calorimeter using the transpiration method is that of Coon and Daniels[1]. They made measurements on small samples of carbon tetrachloride incidental to its use as a refrigerating material in microcalorimetry. By using twin calorimeters and compensating unknown heat with the vaporization of carbon tetrachloride they were able to measure exothermic reactions at room temperature which gave heat rates of about 4 J/h.

VII. Summary

Discussion has been mainly of calorimetry of saturated fluids, as carried out at the National Bureau of Standards over the period 1928–1942. From these experimental measurements values of enthalpy of vaporization, heat capacities and derived thermodynamic quantities were obtained. Four calorimeters have been described, three for water to pressures of 218 atm in the temperature range 0–374 °C, and one for hydrocarbons near 25 °C. An effort has been made to discuss the advantages and disadvantages both of method and various experimental features. Although in general this calorimetry was carried out with as high an accuracy as the "state of the art" permitted at the time, it must be remembered that the calorimeters were developed about three decades ago. With advances in technology since that time, and with the benefit of the experience with these calorimeters, it seems certain that improvements could now be made in both design and operation.

VIII. References

[1] Coon, E. D., and F. Daniels, *J. Phys. Chem.* **37**, 1 (1933).
[2] Fiock, E. F., and D. C. Ginnings, *J. Res. Natl. Bur. Std.* **8**, 321 (1932).
[3] Hoge, H. J., *J. Res. Natl. Bur. Std.*, **36**, 111 (1946).
[4] Mueller, E. F., and H. F. Stimson, *J. Res. Natl. Bur. Std.*, **13**, 699 (1934).
[5] Osborne, N. S., *Bu. Stds. Bull.*, **14**, Scientific Papers 301, 313, and 315 (1917).
[6] Osborne, N. S., *J. Opt. Soc. Am.*, **8**, 519 (1924); *J. Res. Natl. Bur. Std.* **4**, 609 (1930); *Trans. Am. Soc. Mech. Engrs.*, **52**, 221 (1930).
[7] Osborne, N. S., H. F. Stimson, and E. F. Fiock, *J. Res. Natl. Bur. Std.* **5**, 411 (1930); *Trans. Am. Soc. Mech. Engrs.*, **52**, 191 (1930).

[8] Osborne, N. S., H. F. Stimson, and D. C. Ginnings, *J. Res. Natl. Bur. Std.*, **10**, 155 (1933); *Mech. Eng.*, **54**, 118 (1932).
[9] Osborne, N. S., H. F. Stimson, and D. C. Ginnings, *J. Res. Natl. Bur. Std.*, **18**, 389 (1937).
[10] Osborne, N. S., H. F. Stimson, and D. C. Ginnings, *J. Res. Natl. Bur. Std.*, **23**, 197 (1939).
[11] Osborne, N. S., H. F. Stimson, and D. C. Ginnings, *J. Res. Natl. Bur. Std.*, **23**, 261 (1939).
[12] Osborne, N. S., and D. C. Ginnings, *J. Res. Natl. Bur. Std.*, **39**, 453 (1947).
[13] Wadsö, I., *Acta Chem. Scand.*, **14**, 566 (1960).

CHAPTER 12

Heat Capacity of Liquids and Solutions Near Room Temperature

A. J. B. CRUICKSHANK

University of Bristol, Bristol, England

TH. ACKERMANN

Institut für physikalische Chemie der Universität Münster, Münster, Germany

P. A. GIGUÈRE

Université Laval, Québec, Canada

Contents

I. Survey of Experimental Methods	423
1. General Considerations	423
2. Methods of Limited Precision	425
A. Indirect Methods	425
(1) The Piezo-Thermometric Method	425
(2) The Compressibility Method	428
B. Dynamic Methods	431
C. Flow Calorimetry	434
II. Adiabatic Heating Methods	435
1. Theoretical Introduction	435
A. Derived Thermodynamic Quantities and Experimental Precision	435
B. Principles of Comparison Methods	437
(1) Comparison of Heating Rates	438
(2) Comparison of Masses of Test and Reference Specimens	438
(3) Comparison of Heating Powers	438
(a) Variable-Resistance Heaters	439
(b) Independently Controlled Heaters	439
(c) Auxiliary Heaters	439
C. Principles of the Direct Methods	440
(1) Continuous Heating	441
(2) Intermittent Heating	442
D. Applicability of the Comparison and Direct Methods	443
E. Theory of the Recording Adiabatic Twin-Vessel Calorimeter	444
2. Calorimeter Vessels	449
A. General Requirements	449
(1) Shape of the Vessel Proper	450
(2) Materials of Construction	451
B. Filling the Calorimeter Vessel	454
(1) Single-Vessel Calorimeters	455
(2) Twin-Vessel Calorimeters	456
C. Sealing the Calorimeter Vessel	457
(1) Glass Vessels	458
(2) Metal Vessels	458
D. Heating the Calorimeter Vessel and the Liquid Specimen	460
(1) Heat Transport in the Liquid	461
(2) Heating Elements	462
(a) Heating Elements for Metal Vessels	463
(b) Heating Elements for Glass Vessels	465
E. Accommodating the Thermal Expansion of the Liquid Specimen	465
(1) The Vapor Space Method	465
(2) The Variable Volume Method	468

F. Measuring the Temperature of the Calorimeter Vessel and Contents .. 470
   (1) Choice of Method 470
   (2) Resistance Thermometers 472
      (a) Siting the Resistance Thermometer 472
      (b) Mounting Internal Resistance Thermometers 472
      (c) Measuring the Resistance 473
G. Details of Representative Calorimeter Vessels 474

3. Adiabatic Shields and Regulators 477
  A. Regulator Theory and Shield Control 477
   (1) Definitions 477
   (2) Control Requirements 478
      (a) Comparison Methods 478
      (b) Direct Methods 479
   (3) Regulator Systems 479
      (a) The Two-Valued Response Regulator 479
      (b) The Linear Response Regulator 481
      (c) The Linear-Plus-Rate Response Regulator 482
      (d) The Regulator with Reset Response 482
   (4) Discussion 482
  B. Characteristics of Wet and Dry Shields 482
   (1) Dry Shields 483
   (2) Wet Shields 485
  C. Choice of Shield and Regulator 487
   (1) The Twin-Vessel Calorimeter Heated Intermittently 487
   (2) The Twin-Vessel Calorimeter Heated Continuously 488
   (3) The Single-Vessel Calorimeter Heated Intermittently 488
   (4) The Single-Vessel Calorimeter Heated Continuously .. 489
  D. Dry Shields for Continuously Heated Adiabatic Calorimeters .. 489
   (1) The Single-Vessel Calorimeter 489
   (2) The Twin-Vessel Calorimeter 491
  E. A Wet Shield Controlled by Liquid Flow 494
   (1) Shield Vessel and Controlling Bath 494
   (2) Water Circulation System 497
   (3) Regulators 498

4. Control Thermocouples for Adiabatic Shields 500
  A. Heat Transfer between Vessel and Shield 500
  B. Radiation and Conduction along Supports and Electrical Leads .. 502
  C. Thermocouple Layout and Materials 504
   (1) Design Criteria 504
   (2) Layout Characteristics 504
      (a) The Rate of Response 504
      (b) The Efficiency 505
   (3) Discussion 505
   (4) Thermocouple Materials 507
  D. Mounting the Thermocouple Junctions 509
   (1) Fixed Junctions 509
   (2) Free-Contact Junctions 511
  E. The Optimum Number of Junctions 512

5. Experimental Procedure 515
  A. Continuously Heated Calorimeters 515
   (1) Measuring Procedures 515
   (2) Calculations 519
  B. Intermittently Heated Calorimeters 519
   (1) Measuring Procedures 519
   (2) Calculations 521

III. Isothermal Drop Calorimetry 521
1. Introduction 521
2. Isothermal Drop Calorimeter Receivers 523
3. Drop Calorimeter Furnaces and Vessels 529
4. Procedure and Calculations 531

IV. References 533

I. Survey of Experimental Methods

1. *General considerations*

In the present context the most important differences between the measurement of the heat capacities of liquids and solutions, and the measurement of other heat capacities are those arising from the precision necessary in the former case, due to the nature of the information to be derived from the results. Thus in order to be able to evaluate either the partial molal heat capacities of solutes or the thermodynamic excess heat capacity of mixing (see Section II–1 in this chapter) it is necessary to make the primary heat capacity measurements with the greatest possible precision. Obviously this requirement of very high precision does not apply to all heat capacity measurements on liquids, and indeed for entropy evaluation up to high temperatures the heat capacity over the liquid range needs to be measured only with the same precision as over the solid and gaseous ranges. If high precision is not required, the only experimental problems which are peculiar to the study of liquids are those associated either with the accommodation in a specimen vessel of a one-phase liquid sample, or with making allowance for changes in the liquid–vapor equilibrium consequent upon the change in temperature.

A general consideration peculiar to the determination of the heat capacity of liquids near room temperature is that in addition to the various procedures involving direct heating of the specimen and measurement of the resulting temperature increment, and the method of mixtures and its more recent refinements (including the use of isothermal calorimeters as receivers), it is possible also to use indirect methods based on the thermodynamic relations between the coefficients of entropy and those of volume. The last mentioned have the advantages, first that the problem of accommodating the volumetric expansion of liquid specimens is avoided altogether, and second that only simple apparatus is required and that no elaborate temperature or energy measuring equipment is needed; but the limited precision of these methods restricts their application to exploratory work only.

The steady-state flow calorimeter also avoids the problem of accommodating thermal expansion. It has the further advantage that independent calibration experiments are not required. Although the method is in principle applicable to high-precision work, it presents major difficulties in connection with the flow control and measurement. On the other hand, quite simple apparatus is capable of 1 per cent precision, so that this method also is well suited to exploratory investigations, especially in the high pressure range.

The long familiar method of mixtures also gives this order of precision, but the use of an isothermal calorimeter—incorporating a two-phase (liquid + solid) reference system as the receiver—enables the precision to be greatly improved without losing the advantage of experimental simplicity, since the energy changes are determined in terms of changes of volume or mass in the reference system of the isothermal calorimeter. The precision of this method is certainly adequate for entropy evaluation and for determining the integral enthalpy change over a temperature interval of the order of tens of degrees.

If the heat capacity measurements are to be used to study the detailed properties of liquids or solutions (evaluation of partial molal heat capacities, etc.) it is probably true that only the direct heating methods are capable of the required precision (see Section II–1 below)†. This category properly includes the various dynamic methods of calorimetry based on steady-state heating or on cooling curve analysis in an isoperibol calorimeter, but as the precision of these methods also is limited, it is convenient to include them in this introductory Section I together with the indirect methods. There are left the *adiabatic* heating methods in which heat leak is eliminated completely or nearly completely. The two main techniques using adiabatic heating, namely the comparison methods (twin-vessel calorimeters) and the direct methods (single-vessel calorimeters) are discussed in Section II below where, apart from the special problems of designing calorimeter vessels for liquids, we confine ourselves to those particular problems of calorimeter design which arise from the high precision necessary when the heat capacity data are to be used to study the detailed properties of liquid mixtures and especially the properties of solutes in dilute solution. Section III deals primarily with the isothermal-receiver version of the method of mixtures (isothermal drop calorimetry).

The procedure adopted in Section II (Adiabatic Heating Methods) is to give a general account of the various versions of the two main methods, with some indications as to the types of investigation to which each appears suited, and to examine in detail the major design problems relating to each of the main components of the calorimeter. In Section III (Isothermal Drop Calorimetry) we concentrate on the use of the near-to-room-temperature isothermal calorimeter as receiver with only passing reference to the use of other types of calorimeter for this purpose, but we discuss questions of detailed design in relation to several instruments of broadly comparable precision.

One important consideration that applies to the subject matter of all three sections is that, in designing calorimeters to work near room temperature, a variety of versatile materials which are not usable at extreme temperatures behave quite satisfactorily between 250°K and 400°K. These include, for example, thermosetting resin coatings and adhesives, and rubber-like gasket materials. We therefore pay particular attention to the ways in which such materials may be used, and to the experimental precautions their use necessitates.

---

† The partial molal heat capacity of a solute at high dilution may be obtained very precisely by adding to the partial molal heat capacity at relatively high concentration (greater than 1 mole liter^{-1}) the temperature derivative of the heat of dilution per mole of solute[111]. This method requires two calorimeters of comparable precision, but the final result is then significantly more precise than if it had been determined directly using the heat capacity calorimeter alone. A second, and much simpler, indirect method[35] is to obtain the partial molal heat capacity of the solute at infinite dilution as the temperature derivative of the heat of solution at infinite dilution. This method is reliable only for strong electrolytes, for which the heat of solution varies linearly with the square root of concentration.

2. Methods of limited precision

A. Indirect Methods

For normal liquids in the temperature range between 250°K and 400°K the magnitudes of the temperature and pressure coefficients of volume are such that they may be measured fairly accurately by relatively simple apparatus. These coefficients are much larger than those of solids, so that the necessary calibration of the specimen vessel need not involve significant losses of accuracy, but (unlike those of gases) they are small enough not to require special arrangements to accommodate the volume changes. Consequently it may, in some cases, be more convenient experimentally to use the thermodynamic relations between the coefficients of entropy and those of volume to *calculate* the heat capacity rather than to measure the latter directly, especially where great accuracy is not necessary.

Two classes of thermodynamic relations are appropriate to this kind of calculation. The first includes

$$\left(\frac{\partial P}{\partial T}\right)_V = \left(\frac{\partial P}{\partial S}\right)_V \cdot \left(\frac{\partial S}{\partial T}\right)_V \tag{1}$$

and

$$\left(\frac{\partial V}{\partial T}\right)_P = \left(\frac{\partial V}{\partial S}\right)_P \cdot \left(\frac{\partial S}{\partial T}\right)_P. \tag{2}$$

Equation (1) enables C_V to be calculated from $(\partial P/\partial T)_V$ and $(\partial T/\partial V)_S$, but neither of these coefficients is amenable to direct measurement. Equation (2), on the other hand, provides the basis of the piezo-thermometric method.

The second class of thermodynamic relations which may be used in this way includes

$$\left(\frac{\partial V}{\partial P}\right)_T = \left(\frac{\partial V}{\partial P}\right)_S + \left(\frac{\partial V}{\partial S}\right)_P \cdot \left(\frac{\partial S}{\partial P}\right)_T. \tag{3}$$

Combining Equations (2) and (3) and making use of the Maxwell relation

$$\left(\frac{\partial S}{\partial P}\right)_T = -\left(\frac{\partial V}{\partial T}\right)_P$$

leads to

$$\left(\frac{\partial S}{\partial T}\right)_P = \left(\frac{\partial V}{\partial T}\right)_P^2 \bigg/ \left[\left(\frac{\partial V}{\partial P}\right)_S - \left(\frac{\partial V}{\partial P}\right)_T\right] \tag{4}$$

which provides the basis of the compressibility method.

(1) *The Piezo-Thermometric Method*

Equation (2) transforms into

$$C_P = T\left(\frac{\partial V}{\partial T}\right)_P \bigg/ \left(\frac{\partial T}{\partial P}\right)_S, \tag{5}$$

a form first derived by Thomson[133]. The molal thermal expansion $(\partial V/\partial T)_P$ is simply the molal volume times the thermal expansivity,

$$a = \frac{1}{V}\left(\frac{\partial V}{\partial T}\right)_P,$$

so that Equation (5) may be written

$$C_P = TaV/(\partial T/\partial P)_S, \qquad (6)$$

whence C_P may be calculated from independent measurements of density, thermal expansivity and the coefficient $(\partial T/\partial P)_S$. The last of these properties may be measured directly in terms of the temperature change consequent upon adiabatic decompression since it is nearly independent of pressure, at least over the range 0–50 atm. Such measurements have been done by several groups of workers, most recently by Burlew[18] and by Duff and Everett[46]. The liquid specimen is contained in a spherical glass vessel (50 to 100 ml) provided with a horizontal filling limb which includes a U-bend. After filling the specimen vessel with the liquid to be studied, a small quantity of mercury is introduced into the U-bend of the filling limb. The mercury serves both to confine the specimen liquid in the vessel and to transmit pressure to it. The filled specimen vessel is then immersed in an oil-filled pressure bomb fitted with a pressure-release valve. A predetermined pressure is applied slowly to the oil inside the pressure bomb (which is mounted inside a thermostat) and the whole allowed to come to thermal equilibrium at thermostat temperature. The pressure in the bomb is then released suddenly and the consequent change of temperature (on adiabatic decompression) is measured by means of a single thermocouple junction set up inside the specimen vessel. A fast-response galvanometer is a suitable indicating instrument.

Alternatively, a small thermistor may be used as the thermometric element. In either case the most important design requirement is that there is a sufficient length of leads inside the liquid specimen to give adequate tempering. In operation, if a fast-response galvanometer is used to indicate the temperature change, the deflected galvanometer image exhibits a slight arrest just short of the final position; this is due to the thermocouple or thermistor material registering first its own $(\Delta T/\Delta P)_S$ and then quite rapidly coming to thermal equilibrium with the liquid. Provided that the temperature-sensitive element is sited near the center of the liquid specimen, the galvanometer deflection remains steady at its second arrest during at least some tens of seconds before the effects of thermal conduction between the specimen and the oil in the pressure bomb become apparent. For details of the technique the reader is referred to the two papers cited. Duff and Everett claim a precision of about 0·1 per cent in $(\partial T/\partial P)_S$.

The thermal expansivity and the density of the specimen liquid may both be measured by a combined dilatometer and pyknometer in which volume changes are measured in terms of the amount of mercury which has to be extracted from the dilatometer in order to keep the total volume of specimen and mercury constant. This technique was developed by Owen, White and

Smith[104] to study water, and it has recently been developed for use with liquid hydrocarbons[36]. The modification consists in replacing the glass tap used for filling by Owen, White and Smith by a plunger seal, using Teflon (polytetrafluoroethylene) as the sealing material. This modification also enables the dilatometer to be filled under vacuum, ensuring air-free specimens. The modified dilatometer is shown schematically in *Figure 1*.

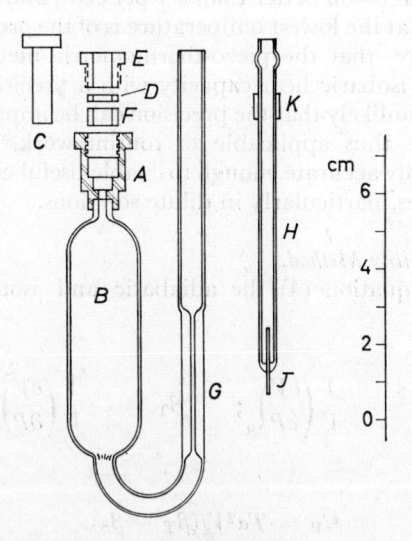

Figure 1. Pyknometer–dilatometer[36] for measuring density and thermal expansion of non-aqueous liquids. The body B and the pipette H are Pyrex, and the cap A and fittings are stainless steel, with a Teflon sealing ring F. The platinum capillary J is 1 mm o.d.

The stainless steel cap A is fixed to the body of the dilatometer B by an epoxide resin cement. The bottom of the sealing plunger C locates accurately in the cap A and the spacer D is forced down by the gland ring E on to the Teflon packing ring F to make the seal. After introducing an accurately weighed amount of mercury, just sufficient to fill the dilatometer to the level of the lower end of the capillary section of the side arm G (1·5 mm bore capillary), the dilatometer is connected to the filling apparatus by a flexible connection to the top of the side arm G and by a hollow plunger otherwise similar to C which is sealed into the cap A in the normal way; the top end of the filling plunger is sealed rigidly into the filling apparatus. The dilatometer is then evacuated and filled with the liquid sample, keeping the pressures inside A and G equal. The dilatometer is detached from the filling apparatus after aerating the former, and the plunger C is then replaced and the dilatometer re-weighed. After thermal equilibration at the lowest temperature the pipette H is inserted into the side arm G and the accessible mercury extracted through the platinum capillary (1 mm outside diameter) J. This mercury collects in the base of H and can be weighed. The thermostat temperature is then raised to the next value and the procedure repeated. The capillary J should project about 3 mm into the capillary section of G when accurately located in G by the collar K which is a parallel-ground glass ring cemented to the outside of H. The level at which mercury extraction ceases is usually constant within 0·01 mm. The thermal expansion of the dilatometer itself has to be found by calibration with water; this serves also to check the technique, since the volumetric expansivity of the glass vessel so obtained should be very close to three times the linear thermal expansivity of the glass used. This method enables the thermal expansivity of the liquid to be determined with a precision better than 0·1 per cent, and the precision of the density as measured at the lowest temperature is of the order of 0·01 per cent.

It seems, therefore, that the piezo-thermometric method might be developed to give the isobaric heat capacity with a precision close to 0·1 per cent, but it appears unlikely that the precision can be improved much beyond this. The method is thus applicable to routine work, especially on pure liquids, but it is hardly accurate enough to enable useful evaluation of partial molal heat capacities, particularly in dilute solutions.

(2) *The Compressibility Method.*
Introducing into Equation (4) the adiabatic and isothermal compressibilities, defined by

$$\beta_S = -\frac{1}{V}\left(\frac{\partial V}{\partial P}\right)_S; \quad \beta_T = -\frac{1}{V}\left(\frac{\partial V}{\partial P}\right)_T,$$

leads to

$$C_P = Ta^2 V/(\beta_T - \beta_S). \tag{7}$$

The method based on Equation (7) has the advantage that no rapid temperature changes have to be measured, but experimental errors are magnified both by the occurrence of the second power of a and by the fact that $\beta_T - \beta_S$ is for some liquids smaller than $\frac{1}{4}\beta_T$.

The technique of determining β_T from observations of compression as a function of applied pressure by either the piston-displacement or the 'sylphon' methods does not seem to have been significantly improved since this field was reviewed by Bridgman[15] in 1946; the attainable precision is about 0·05 per cent. Various recently developed piezometric methods, see Staveley and Parham[124], Diaz-Peña and McGlashan[43], Coleman and Cruickshank[34], are a little less precise, but require pressures of only 10 or 20 atm. The adiabatic compressibility may be determined either from velocity of sound measurements, since

$$\beta_S = 1/\rho u^2, \tag{8}$$

or piezometrically. The interferometric methods for determining the velocity of sound in liquids described by Freyer, Hubbard and Andrews[57] and by Low and Moelwyn-Hughes[96] might give β_S to about 0·1 per cent. The piezometric method entails direct measurement of the expansion on adiabatic decompression, and when precautions are taken to ensure adiabaticity the attainable precision is probably about 0·2 per cent, even though only low pressures (less than 5 atm) can conveniently be used.

HEAT CAPACITY OF LIQUIDS AND SOLUTIONS

Thus it seems unlikely that the best available technique for measuring β_T and β_S can give C_P values more precise than about 1 per cent. On the other hand, the use of a piezometer to measure both adiabatic and isothermal compressibilities in a single experiment, giving C_P values with a precision about 1–2 per cent may be useful for exploratory work, and so we describe briefly the apparatus developed by Coleman and Cruickshank[34].

The piezometer itself is shown schematically in *Figure 2*; it is made entirely of Pyrex glass.

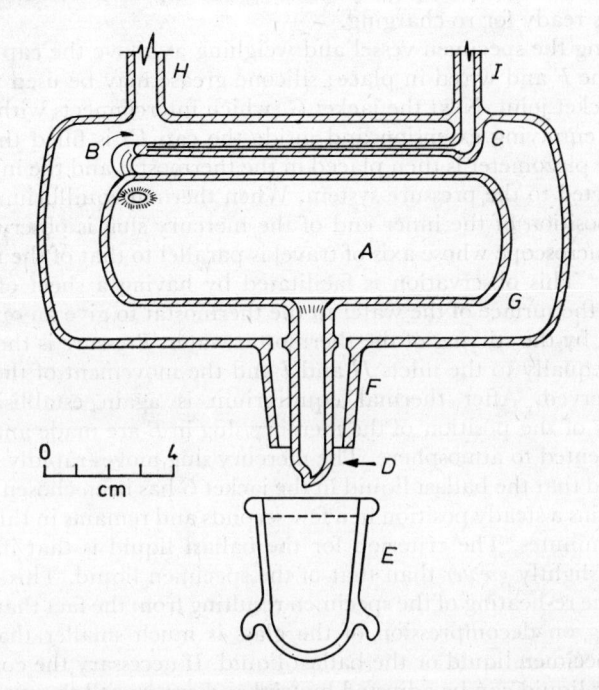

Figure 2. Piezometer[34] for measuring adiabatic and isothermal compressibilities of liquids or solutions up to 5 atm. The specimen is confined by a mercury slug in the graduated capillary B; the pressure jacket G is filled with liquid whose $(dT/dP)_S$ is similar to that of the specimen.

The specimen vessel A (volume 150 ml) has joined to it a measuring capillary B (precision bore, 0·5 mm nominal) which is etched at 1 mm intervals. The actual bore of the selected capillary must be determined before assembly. This is done by observing the apparent lengths of mercury slugs of known mass at some set temperature. The apparent bore corresponding to each slug is then plotted against the reciprocal of slug length; the intercept corresponding to infinite slug length then gives the true bore. In order to fill the specimen vessel it is first evacuated and then the liquid specimen is forced through the measuring capillary B by means of a large vacuum-tight syringe made of precision-bore Pyrex glass with a nylon plunger. When the vessel A and the capillary B are quite full of liquid a small amount of mercury (less than 0·5 g) is introduced into the well C at the entrance to the capillary B. By slightly heating the whole piezometer and then cooling to just above the experimental temperature a slug of mercury about 1 or 2 cm long is drawn into the outer part of the capillary B. The piezometer is then disconnected from the filling apparatus and, after removal of the excess specimen liquid and mercury from C by means of a Polythene catheter tube attached to a suction line, the mass of the specimen is determined by weighing the whole piezometer (the mass of the mercury slug is calculated from its length and the known diameter of the capillary B). On completion of a series of experiments the specimen vessel is emptied by cutting off the tip D of the emptying extension. After thorough cleaning and drying the tip D is resealed and the apparatus is ready for re-charging.

After filling the specimen vessel and weighing as above the cap E is fitted over the cone F and wired in place; silicone grease may be used to seal the cone and socket joint. Next the jacket G (which interconnects with the space around the emptying extension and inside the cap E) is filled through the inlet H. The piezometer is then placed in the thermostat and the inlets H and I are connected to the pressure system. When thermal equilibrium is established the position of the inner end of the mercury slug is observed with a travelling microscope whose axis of travel is parallel to that of the measuring capillary B. This observation is facilitated by having a sheet of Plexiglas 'floating' in the surface of the water in the thermostat to give an optical path undisturbed by the stirring of the thermostat water. Pressure is then applied slowly and equally to the inlets H and I and the movement of the mercury slug is observed. After thermal equilibrium is again established, final observations of the position of the mercury slug in B are made and then the pressure is vented to atmosphere. The mercury slug moves rapidly outwards, and provided that the ballast liquid in the jacket G has been chosen correctly, the slug attains a steady position in a few seconds and remains in this position for several minutes. The criterion for the ballast liquid is that its value of $(\partial T/\partial P)_S$ is slightly *greater* than that of the specimen liquid. This difference just offsets the re-heating of the specimen resulting from the fact that the adiabatic cooling on decompression of the glass is much smaller than that of either the specimen liquid or the ballast liquid. If necessary the composition of the ballast liquid can be adjusted by trial and error until the mercury slug in the measuring capillary B moves steadily to its final position after decompression, without overshooting and then retreating, which is what happens if $(\partial T/\partial P)_S$ for the ballast liquid is too large.

After the piezometer has again attained the temperature of the thermostat (this takes 1 or 2 hours) the whole sequence of compression and decompression is repeated. By using a series of increasing applied pressures a series of results is obtained from which both β_T and β_S can be calculated in terms of the slopes of plots of displacement of the mercury slug against applied pressure for both isothermal compression and adiabatic decompression. A series of 10 to 20 measurements at applied pressures ranging from about 100 torr to 1500 torr is recommended.

The most likely source of error in the isothermal measurement is loss of specimen liquid by creeping past the mercury slug during the application of the pressure. This loss is minimized by the use of a relatively long mercury slug, and by the greatest care in eliminating grease-like contamination from the interior of the measuring capillary. In some cases it may be better to fill

HEAT CAPACITY OF LIQUIDS AND SOLUTIONS

the whole of the space C with mercury, in which case a known mass of mercury must be introduced. This mass may be found by differential weighing of the reservoir from which the mercury is added. A disadvantage of this procedure is that it reduces the mobility of the mercury-to-liquid interface and so may reduce the precision of the adiabatic decompression measurements.

B. Dynamic Methods

The analysis of heating or cooling curves for obtaining the heat capacity has been described in Chapter 6 in connection with the low temperature isoperibol calorimeter. The early work of Ferguson and Miller[52] on organic liquids in the range 40°C to 70°C is a typical application of the method. Further refinements were evolved by Leech[92], who showed that the integral power law of cooling, $dT/dt = kT^n$, holds only approximately, and developed a new method for getting the correct values of dT/dt. The experimental procedure consists in tracing the complete cooling curve for an initial calorimeter temperature about 50 deg above that of the enclosure, and then measuring the electric power necessary to maintain the temperature differential at a preselected constant value. With the particular calorimeter used by Leech, rather large quantities of liquid were needed (100 to 300 ml) and the accuracy, in the region of 0·25 per cent, depended mainly on the determination of dT/dt. Application of the graphical methods to the cases of liquids and solutions near room temperature presents no special problems except possibly when the approach to thermal equilibrium is very slow, or when time lags in the measurements become appreciable, as a result either of low thermal conductivity of the specimen, or of slow response in the measuring device (e.g. a galvanometer). Dynamic methods of analysis must be resorted to in such cases.

In the case of a slow approach to thermal equilibrium, Barieau and Giauque[9] plotted values of $-\ln(dT/dt)$ against time and fitted these data by a linear equation of the form

$$-\ln(dT/dt) = A - Bt. \tag{9}$$

They integrated this equation, giving

$$\ln B(T_t - T_\infty) = A - Bt, \tag{10}$$

and then added the integral to the value of T at the given time t. This plot of the logarithm of the drift, corrected for the heat leak, gave them the true equilibrium temperature for each determination. Two other similar methods have been proposed by Egan[47]. In the first one the cooling curve is taken as a section of a hyperbola,

$$T = T_k - (t - t_k)/(A + Bt) \tag{11}$$

where T_k is the temperature at the selected time t_k. Solving the equation for large values of t gives the equilibrium value of T. In the other method T is plotted against $1/t$ and a straight line is drawn intercepting the T-axis at $t = \infty$.

Correction for time lags is more involved. Kouvel[89] has analyzed the dynamic response of an isoperibol calorimeter in the case both of thermal gradients in the specimen and of a too sluggish galvanometer. It seems unlikely, however, that either situation need arise in measurements on liquids near room temperature.

A typical application of the cooling curve method to the determination of the heat capacity of a liquid is that of Calvet[23], using his conduction microcalorimeter. This instrument, which was originated by Tian[135], and

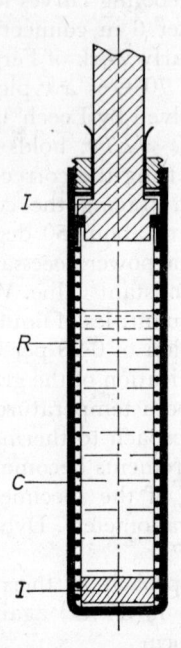

Figure 3. Sample cell[27] for measuring the heat capacity of liquids with the Tian–Calvet conduction microcalorimeter (see ref. 27, p. 60)

C, Cell proper
R, Electrical heating resistance
I, Thermal insulation.

brought to a high degree of perfection by the Marseilles school[22, 26, 27], is intended primarily to follow the evolution of very small heat effects over relatively long periods of time. Consequently it is not particularly well suited to the problem at hand. Indeed, its best features (such as its remarkable sensitivity) are not really taken advantage of in heat capacity measurements. Furthermore, the process of obtaining the data from the integration of differential cooling curves is subject to appreciable errors, hence the limited precision (about 1 per cent).

Briefly the principle is as follows: two microcalorimetric elements, as nearly identical as possible, are mounted in opposition inside the calorimeter proper. Each one contains an identical cell (see *Figure 3*) made of a length of metal tube, wound with an electric heating resistance, and closed at both ends with plugs of some suitable insulating material (for instance a plastic).

An electric current is passed in the two cells connected in series, raising their temperature to the same extent ΔT above that, T, of the surrounding container. It is noteworthy that at equilibrium this increment in temperature is proportional only to the electric power dissipated, P, and is independent of the heat capacity of the cells. The governing factor here is the thermal resistance between cells and container, which is constant.

Three operations are required for the calibration of the instrument and the determination of the heat capacity of a substance. In the first one, both cells are filled with the same mass of water, m, and the heating current is passed for a while. On cooling, the difference in electric current generated between the two sets of thermocouples is recorded. The area under this curve, A, is the correction to be applied to all other measurements with the same apparatus due to the fact that the two elements are not absolutely identical. In the second operation the water in one of the cells is replaced by an equal mass of the liquid to be measured, and the area under the differential cooling curve A' is obtained as above. Lastly a third datum A'' is needed to solve the equation, and this is achieved by replacing the liquid in the sample cell by a different mass of water, m'. The specific heat of the sample, with respect to that of water taken as unity, is given by the relationship

$$c_P = 1 - \left[\frac{A' - A}{A'' - A}\right]\left[\frac{m - m'}{m}\right]. \qquad (12)$$

This method is equally applicable to solids, although for poor heat conductors another correction is required, due to the thermal disequilibrium between the sample and the thermocouple junctions[24]. In a recent improvement of the method[25] the heating (or cooling) of the sample is realized by the Peltier effect, using the conduction thermocouples. The external electric heating (by the Joule effect) is thus dispensed with and, furthermore, the two microcalorimetric elements need not be exactly alike. In principle the microcalorimeter makes it possible to work within an extremely small temperature interval; consequently the correction for evaporation of the liquid sample becomes negligible, and the heat capacity measured is the true value at the temperature of the calorimeter, not the average over a range of temperature. Actually these advantages are somewhat illusory, because of the limited precision of the measurements. The microcalorimeter has been used successfully on several liquids, in particular vinyl chloride[26], but no actual heat capacity data from it seem to have been published so far.

Finally, mention should be made in this section of an application of differential cooling curve analysis to a simple twin-vessel calorimeter. This technique, developed by Spear[122], permits the measurement of the specific heats of liquids to within about 5 per cent. It is based upon the assumption that when two similar calorimeter vessels, one containing a reference liquid and the other the test liquid, are pre-heated and then cooled towards the temperature of the surrounding thermostat the ratio of the differential cooling times of the two calorimeter vessels is a measure of the ratio of their total heat capacities. For measurements of this kind, only a minimum of electrical equipment is required. The necessary equipment includes a controlled-temperature bath, two calibrated thermocouple-recorder units, and two simple (commercially available) magnetic stirrers. The apparatus is particularly suitable for heat capacity measurements on liquid heat-transfer agents, where great precision is not needed. It must be noted, however, that the precision of the results obtained depends on the differences between the physical properties of the sample and of the reference liquid.

C. Flow Calorimetry

The liquid flow calorimeter of Callendar and Barnes[21] exemplifies the advantages of the method. The calorimeter proper is a tube of uniform diameter (2 mm, say) about 50 cm long. Each end widens smoothly into a chamber about 2 cm in dia. and 5 or 10 cm long. Side arms attached to these chambers accommodate the liquid inflow and outflow. Two matched thermometric elements (platinum resistance thermometers in the original) are sited centrally in the end chambers, and the heating resistance is a single platinum wire stretched between the outer shields of the thermometric elements so that it passes accurately along the axis of the flow tube. The complete assembly is surrounded by an evacuated jacket, which is in turn immersed in a liquid-bath thermostat. Under steady conditions of liquid flow and heat input, the thermometric elements indicate a constant temperature difference, ΔT. The differential form of the heat-balance equation is then

$$c\Delta T \frac{dm}{dt} = \frac{dQ}{dt} - \frac{dq}{dt}, \qquad (13)$$

in which c is the specific heat of the flowing liquid and dm/dt is the rate of mass flow; dQ/dt is the power input and dq/dt is the heat loss. The heat loss cannot be made zero because the temperature gradient along the tube is finite. In a well designed apparatus, however, dq/dt is nearly independent of flow rate and is primarily dependent upon ΔT. Thus if the flow rate is changed and the heating power adjusted to maintain again the same ΔT, the heat loss term may be eliminated to give

$$c\Delta T = \frac{(dQ/dt)_1 - (dQ/dt)_2}{(dm/dt)_1 - (dm/dt)_2}. \qquad (14)$$

Alternatively, and better, a series of flow rates are used and the quantity $(dQ/dt)/(dm/dt)$ is plotted against the reciprocal of (dm/dt). Extrapolation to zero, corresponding to infinite flow rate, then gives $c\Delta T$. In practice the main difficulties are (i) ensuring that the radial temperature profiles at the inlet and outlet ends of the flow tube are the same, (ii) keeping the flow rate sufficiently constant over relatively long periods, so that a true steady state is attained, and (iii) measuring the flow rate with a precision comparable to that of the measurements of heating power and temperature increment.

One way of improving the uniformity of the temperature profile is to use a wide bore flow tube, with the heating filament wound helically on an axial glass rod, so that the flow cross-section is an annulus[55]. This reduces the heater-to-wall distance for a given flow range, and at the same time imparts some helical motion to the liquid.

The adaptation of the flow method for high-precision heat capacity measurements does not appear to have been attempted recently, and the apparatus used by Koch[88] to measure the specific heat of water is still one of the more elegant examples. The high pressure flow calorimeter used by Forrest, Brugmann and Cummings[55] to investigate the specific heat of diphenyl (average deviation of results 0·63 per cent) provides a better indication of the precision attainable with fairly simple apparatus.

II. Adiabatic Heating Methods

1. *Theoretical introduction*

A. Derived Thermodynamic Quantities and Experimental Precision

It is characteristic of the liquid state that it exhibits a high degree of short range molecular order while possessing a high degree of molecular mobility. Thus the molal entropies of liquids are, in general, more strongly temperature dependent than those of gases or solids, and the same is broadly true also of the molal heat capacities (if we exclude solids at low temperatures). Moreover, the variation with temperature of the molal isobaric heat capacity shows relatively large differences between different liquids. Thus from the theoretical viewpoint it is important to be able to obtain accurate information about the temperature variation of both the entropy and the isobaric heat capacity itself. This consideration applies especially to the study of solutions, for which the thermodynamic effects of specific interactions, solvation structure, etc. are significant. Most theories of liquid mixtures and solutions lead, in the first instance, to statements about the partial molal free energies (chemical potentials) of solute and solvents. Consequently attention is focussed also on the partial molal enthalpies and the partial molal heat capacities, particularly that of the solute in the case of a dilute solution. The *partial* molal heat capacity of the solute (dilute species) is defined as the derivative with respect to the number of moles of solute, of the total isobaric heat capacity of the sample of solution:

$$\bar{c}_{P,2} = \left(\frac{\partial C_{soln}}{\partial n_2}\right)_{T,P,n_1}. \tag{15}$$

The *apparent* molal heat capacity of the solute, denoted by the symbol[93] ϕ_{C_P}, is directly related to the difference between the total isobaric heat capacity of the solution specimen and the isobaric heat capacity of a quantity of the pure solvent equal to that in the solution specimen. This is, therefore, the quantity most frequently quoted in the literature. If a solution contains n_1 moles of solvent and n_2 moles of solute, the apparent molal heat capacity of the solute is defined by

$$\phi_{C_P} = \frac{C_{soln} - n_1 c°/_{P,1}}{n_2} \tag{16}$$

in which $c°_{P,1}$ is the molal isobaric heat capacity of the pure solvent. Combining Equations (15) and (16) leads to

$$\bar{c}_{P,2} = {}^\phi c_P + n_2 \left(\frac{\partial {}^\phi c_P}{\partial n_2}\right)_{T,P,n_1} \quad (17)$$

If n_1 is chosen to correspond to 1 kg of pure solvent, n_2 becomes identical with the molal concentration, m, of the solute, and Equation (17) can be written

$$\bar{c}_{P,2} = {}^\phi c_P + m \left(\frac{\partial {}^\phi c_P}{\partial m}\right)_{T,P} \quad (18)$$

Thus the *partial* molal heat capacity of the solute is just the sum of the *apparent* molal heat capacity and the molality times the slope of the apparent molal heat capacity as a function of the molality. For electrolyte solutions it may be better to plot ${}^\phi c_P$ against $m^{\frac{1}{2}}$, when Equation (18) becomes[93]

$$\bar{c}_{P,2} = {}^\phi c_P + \tfrac{1}{2} m^{\frac{1}{2}} \left(\frac{\partial {}^\phi c_P}{\partial m^{\frac{1}{2}}}\right)_{T,P} \quad (19)$$

Especially in dilute solutions ($m < 0.05$) the difference ($C_{soln} - n_1 c°_{P,1}$) is small in comparison with C_{soln} itself, and this circumstance determines the precision necessary in the calorimetric measurements. For example, if ${}^\phi c_P$ for an electrolyte in 0·05 molal aqueous solution is of the order of -100 joule deg^{-1}mole^{-1}, then to obtain ${}^\phi c_P$ within 5 per cent it is necessary to measure C_{soln} with a precision better than 0·01 per cent.

The precision requirement is similar also in the study of liquid mixtures where, for example, information about the temperature-dependence of the 'interchange energy'[74] may be obtained from the integral heat capacity of mixing, defined by

$$c_P{}^M = \left(\frac{\partial H^M}{\partial T}\right)_{P,x} = c_{mix} - x_1 c°_{P,1} - x_2 c°_{P,2} \quad (20)$$

whence

$$c_P{}^M = x_1(\bar{c}_{P,1} - c°_{P,1}) + x_2(\bar{c}_{P,2} - c°_{P,2}) \quad (21)$$

where $\bar{c}_{P,i}$ is the partial molal heat capacity of species i in the mixture x_1, x_2 and $c°_{P,i}$ is the molal heat capacity of pure species i. The value of c_{mix} for an equimolar mixture of benzene and diphenylmethane[46], for example, is about 195 joule deg^{-1} mole^{-1}, whereas $\tfrac{1}{2}(c°_{P,1} + c°_{P,2})$ is about 200 joule deg^{-1} mole^{-1}, so that the magnitude of $c_P{}^M(x = 0.5)$ is about 3 per cent of that of c_{mix}; the ratio of $c_P{}^M(x = 0.5)$ to c_{mix} has a similar magnitude in the case of the system hexane plus hexadecane[97]. Thus in order to obtain $c_P{}^M$ accurately to 1 per cent of its value in equimolar mixture (this is about the greatest error acceptable when partial molal heat capacities are to be evaluated by differentiation) it is again necessary to make individual measurements (of c_{mix}, $c°_{P,i}$) with a precision about 0·01 per cent.

Consequently it seems reasonable, in the present context, to take the figure 0·01 per cent as a starting point in discussing high-precision adiabatic calorimeters for the measurement of heat capacities of liquids and solutions near room temperature. The adoption of even such an arbitrary criterion does not, however, suffice to make an unambiguous choice of experimental method, because the detailed requirements of individual projects differ so widely, and further, many of the high-precision adiabatic calorimeters which have been described in recent years have broadly comparable performance. It is nevertheless true that for a particular project, a choice between the comparison method and the direct method can be made on the grounds of experimental convenience. We therefore devote the remainder of this sub-section to outlines of the theory of each of the main variations of these two basic methods, together with some indications as to the types of investigation to which each appears best suited.

The following sub-sections cover in detail the component parts of adiabatic calorimeters, and finally we outline typical procedures for recording and analyzing experimental data. Our approach here is to concentrate on the design problems associated with each component rather than to give detailed descriptions of particular calorimeters. This arrangement enables due emphasis to be given to the most important aspects of individual calorimeters while keeping to the fore the major problems encountered in all calorimeters designed for liquid specimens. Since many of these problems are associated with the calorimeter vessel, we give an extended discussion of the physical theory underlying the design of calorimeter vessels for liquids.

Those major practical difficulties of calorimetry which are simplified by working near room temperature are covered in other chapters and so do not get detailed attention here. For example we ignore the problem of tempering external leads, because it is much simpler than in high-temperature calorimetry, while still following the same principles. Again, it is usually necessary to interpose only one controlled temperature stage between the room and the shield of an adiabatic calorimeter operating between 0°C and 120°C, so we omit completely the designing of the environmental thermostat and concentrate on the attainment of very precise control of the adiabatic shield itself.

B. Principles of Comparison Methods

The term *comparison method* is sometimes used, in a broad sense, to cover the whole field of calorimetric techniques using standard reference substances for calibration. Many experimental determinations of the heat capacities of liquids and solutions have been applications of the comparison method in the sense that they were based upon calibration experiments using water or another reference substance. Of the very few reported measurements of heat capacities of liquids between 0°C and 120°C which can be regarded as direct, or "absolute" measurements, the work of Osborne, Stimson and Ginnings[103] on the heat capacity of water is an outstanding example. In a more limited sense, the meaning of the term comparison method is restricted to the technique pioneered by Joule[86, 105], that is to say the use of twin calorimeter vessels, one containing a reference substance (the value of whose specific heat is either assumed or known independently), and the other containing the substance under investigation. The scope of this subsection is limited according to this restricted meaning of the term.

The early history of the comparison method was reviewed by Richards and Gucker[110]; their paper includes an extensive bibliography. The high-precision twin-vessel calorimeters described in recent literature fall into one or other of three categories according to which parameter it is whose values for the twin calorimeter vessels are compared. Each of these three categories is discussed in turn below.

(1) Comparison of Heating Rates

This method[75, 108, 110] involves the well known and very simple procedure of measuring the difference between the temperature changes in the specimen vessel and the reference vessel during the heating period with the carefully matched heating resistances in the twin calorimeter vessels connected in series so that the heating power is the same in both vessels. The method is not easily adapted to permit continuous heating. For intermittent heating, the operating equation takes the form

$$\frac{C_r - C_s}{C_r} = \frac{\Delta T_s - \Delta T_r}{\Delta T_s}, \qquad (22)$$

where the C_i are the total heat capacities of the specimen and reference calorimeter vessels and the ΔT_i are the corresponding temperature increments. In high-precision work the temperature difference $T_s - T_r$ is usually measured directly (by thermocouple for example) and independently of T_s and T_r. If C_r is also determined, as the quotient of the total energy input to the reference vessel over its temperature increase ΔT_r, the difference $C_r - C_s$ can be evaluated from a single experiment.

(2) Comparison of Masses of Test and Reference Specimens

In this method also[75] equal heating powers are applied to the twin calorimeter vessels. The basis of the method is similar to that of an accurate weighing procedure; that is to say the amount of the reference material in the "tare" or reference vessel is either increased or decreased, by trial and error, until both vessels exhibit the same heating rate when their heating powers are equal. The *specific heats* of the test and reference materials are then in inverse ratio to the calorimetrically balanced masses. This method is in principle highly accurate since all thermal effects are the same for both vessels, but the experimental procedure is time consuming and tedious.

(3) Comparison of Heating Powers

This method[3, 5, 50, 51, 71, 72] involves the adjustment of the heating power applied to one or other of the twin calorimeter vessels to equalize their heating rates. It is particularly well suited to continuous heating procedures and hence to recording twin-vessel calorimeters, since the electrical power input to the heaters can be controlled continuously during a long heating period so as to keep the instantaneous temperature difference between the two vessels negligibly small. It is essential to this method to be able to determine precisely the small corrections to the heating power of one of the twin vessels, and this may be done in any of three ways, namely (i) the use of variable-resistance heaters, (ii) independent adjustment and measurement of the current passing in each of the heaters, and (iii) the use of a low-power auxiliary heater in the calorimeter vessel containing the liquid of higher total heat capacity, with matched main heaters connected in series.

(a) *Variable-Resistance Heaters.* When the heaters of the twin calorimeter vessels are connected in series, the ratio of the heating powers is given exactly by the ratio of the heater resistances. Thus when the heater resistances are adjusted to give equal heating rates, their ratio gives immediately the ratio of the total heat capacities of the twin vessels, each with its contents. The operating equation is then

$$\frac{C_r - C_s}{C_r} = \frac{R_r - R_s}{R_r} \qquad (23)$$

in which the R_i are the adjusted heater resistances and the other symbols have the same significance as in Equation (22). The main difficulty is to determine the ratio R_r/R_s with sufficient accuracy; the adjustment of one or other of the heater resistances also presents difficulties in practice. In the twin-vessel calorimeter developed by Gucker's group[71, 72] the heaters were made up of low-resistance elements connected in series, so that the total heater resistance could be varied in steps of 0.03 per cent of the maximum value by changing the heater connections. With this type of adjustment the difference in temperature between the specimen and reference vessels after a 1 deg heating could always be made smaller than 10^{-3} deg. The adjusted heater resistances were then compared precisely, by means of a bridge network. In this method, just as with method 1 above, $C_r - C_s$ has to be evaluated in terms of C_r, so that both the total heat input and the total temperature increment have to be measured in each experiment. Heat capacity data for aqueous electrolyte solutions in the room-temperature range[72, 114] have been obtained with a reproducibility about 0.1 per cent by this method.

(b) *Independently Controlled Heaters.* This method requires very precise control and measurement of the current passing in each of the calorimeter vessel heaters. It is also essential that the resistances of the heaters in the specimen and reference vessels be exactly matched. As the power realized in an electrical resistance depends on the square on the current intensity, very fine adjustment of the current intensity is necessary in order to compensate for small differences in the total heat capacities of the two vessels. For example, if the total heat capacities of the specimen and reference vessels (including their contents) differ by 0.1 per cent, then the ratio of the current intensities in the two matched heaters will be close to 1.0005. The difficulty of attaining the implied precision of current control is probably the reason why this method does not seem to have been used for accurate work on liquids and solutions.

(c) *Auxiliary Heaters.* The basis of this method is the separation of the additional heating power required for the liquid specimen of the greater heat capacity in a fully balanced twin-vessel calorimeter. The main practical advantage of this method is that it is comparatively simple to determine with adequate precision the current intensity in the auxiliary heater. Thus, considering again the example quoted above, if both the main and the auxiliary heaters have the same resistance, then an extra heat capacity of 0·1 per cent in one vessel is balanced by a current in the auxiliary heater which is smaller than the current in the main heater by a factor of only about 30. The operating equation in this case, with the same nomenclature used as above, is

$$C_s - C_r = I^2{}_a R_a/(\mathrm{d}T/\mathrm{d}t), \tag{24}$$

in which the subscript a refers to the auxiliary heater in the specimen vessel and $\mathrm{d}T/\mathrm{d}t$ is the common heating rate in deg sec^{-1}. It is obvious from Equation (24) that this is the only comparison method which gives the difference between the total heat capacities of the filled specimen and reference calorimeter vessels directly, without additional information about either C_r or C_s. The detailed theory of one application of this method is set out in Section II–1–E below.

C. Principles of the Direct Methods

In order to avoid restricting the use of the term 'absolute measurement', which generally describes any measurement made in terms of the absolute units of electrical energy, we use the term 'direct measurement' to describe measurements with a single-vessel adiabatic calorimeter, as opposed to comparison measurements made with a twin-vessel calorimeter. In all the direct methods so defined, the primary experiment is the determination of the total heat capacity of the calorimeter vessel and its contents *directly*, by measuring the rise in temperature consequent upon the addition of a measured quantity of electrical energy (corrected for heat leak and other secondary effects). In order to obtain the heat capacity of the liquid specimen itself, the heat capacity of the empty calorimeter vessel must be determined independently. In principle the simplest way of doing this is to carry out a calibration experiment with the calorimeter vessel evacuated[102, 131, 137]. This means that two complementary experiments have to be done with significantly different conditions inside the calorimeter vessel; in particular the conditions governing the approach to thermal equilibrium are widely different in the two experiments, and this is probably the major source of difficulty in the direct methods. The main advantage of these methods is that a direct measurement on a standard liquid, whose molal heat capacity is accurately known, gives an unambiguous check on the overall *accuracy* of the apparatus.

In order to minimize the difficulties of measuring the heat capacity of the empty calorimeter vessel, the procedure has been adopted of filling it, either with a light gas (hydrogen or helium) at low pressure, or with a "reference" liquid whose heat capacity is known accurately and whose thermal conduction characteristics are similar to those of the test liquid. The second procedure is in fact the basic comparison method (comparison of heating rates), but carried out with the maximum of experimental inconvenience. The first procedure, on the other hand, can be treated as a direct measurement, because the use of an estimated value for the very small heat capacity of the filling gas involves only negligible errors.

It is inherent in the direct methods that all the electrical energy liberated must be strictly accounted for, together with the secondary energy effects such as the heat of stirring, residual heat leak, etc. In terms of an overall precision of 0·01 per cent, if the residual heat leak exceeds 0·1 per cent of the average heating power, the former must itself be known to better than 10 per cent. Since these limits are beyond the capabilities of any except a vacuum-jacketed adiabatic calorimeter, we assume, in the following discussion, that the calorimeter vessel is surrounded by an evacuated insulating space and by an adiabatic shield whose temperature is adjusted to match that on the outer surface of the calorimeter vessel. In this context the term *adiabatic* means zero net heat transfer between the calorimeter vessel and the shield, rather than merely that the temperature of the shield is kept more or less equal to that of the calorimeter vessel.

Now, whatever thermal gradients exist within the material of the shield, it is obviously the temperature at its inner surface (i.e. that facing the calorimeter vessel) which, together with the average temperature at the outer surface of the calorimeter vessel, determines the rate of heat transfer between the two by radiation. It is consequently a fundamental point of design (see Section II–3 below) that the temperature at these two surfaces shall control all the other heat transfer processes also, and that the thermal power of these processes shall be calculable with satisfactory precision from a knowledge only of the two surface temperatures.

Even when this condition is fulfilled, however, it is still necessary to know the time integral of the difference between these two surface temperatures at least approximately. Thus if the adiabatic shield is manually controlled, it is necessary to record continuously the temperature difference. When automatic regulation is used, the time integral of the temperature difference across the insulating space is usually similar in similar experiments, and if it is neither zero nor simply related to the total temperature increment, it may still be computed from the observed behavior of the regulator without recourse to continuous recording. We assume in this section that the adiabatic shield regulator is automatic.

The direct method may be realized in terms either of continuous heating of the calorimeter vessel, or of intermittent heatings which are separated by relatively long equilibration periods.

(1) Continuous Heating

This method consists in setting up a constant rate of heating or, more usually, a constant heating power, and measuring both the power and the rate of heating. The latter is usually done by determining the time taken for the temperature to increase by a predetermined amount[2, 11, 48, 102, 137] just as in the twin-vessel calorimeters, Section II–1–E. The operating equation is then

$$C = (\dot{Q} - \dot{q})\,(dt/dT); \qquad (25)$$

the heat-leak rate $\dot{q}$ is known as a function of the regulator behavior and of the *heat-leak coefficient*, governing the transfer between the calorimeter vessel and the shield. This coefficient usually changes only slowly with temperature, so that $\dot{q}$ may in some cases be kept effectively constant over heating intervals of several degrees. The heat capacity of the calorimeter vessel plus the liquid specimen is then compared to that of the empty vessel, measured under equivalent conditions. Provided only that uniformity of temperature is maintained over the whole extent of the outer surface of the calorimeter vessel during heating, this is in principle the most accurate way of realizing the direct method of calorimetry, since differences in the temperature distribution inside the calorimeter vessel under different conditions of filling become irrelevant in steady-state heating. The major difficulty in this technique arises from the comparatively large thermal expansivity of most liquids. Thus it is difficult to keep a one-phase liquid specimen under constant pressure throughout a heating which continues long enough to ensure the establishment of steady conditions; the alternative here is to allow a vapor space, but this entails the computation of the heat of vaporization per degree of temperature increase, and this may easily introduce a significant error.

(2) *Intermittent Heating*

The essential point of this version of the direct method is that the temperature increment is measured between two uniform-temperature states of the calorimeter vessel and its contents. This fact means incidentally that the primary measurement is of ΔT rather than of dT/dt. The most important practical consideration is the impossibility of avoiding a relatively long equilibration period between successive heatings. Thus the apparatus persists alternately, during comparable periods of time, in each of two states which differ significantly in the conditions which influence the processes of heat transfer between the calorimeter vessel and its adiabatic shield. Consequently with this method it is of primary importance to minimize the coefficient of heat transfer between calorimeter vessel and shield. There are two more or less equivalent ways of dealing with the residual heat leak. The first is to evaluate it directly [cf. Equation (40) below] as

$$\delta q = \int \dot{q}\, dt = \int (\xi \zeta)\, dt, \qquad (26)$$

in which ζ is the instantaneous temperature difference between the outer surface of the calorimeter vessel and the inner surface of the adiabatic shield, and ξ is the heat-leak coefficient (watt deg^{-1}) characteristic of the calorimeter vessel and the shield. In this case ξ has to be measured independently, whereas ζ, or its mean value, is found from a knowledge of the functioning of the regulator. Alternatively, the *cooling coefficient* ξ^* (sec^{-1}) may be measured directly in each experiment by setting up a known (and relatively large) value of ζ. Integration with respect to time of ξ^* gives the temperature error δT. The two corresponding forms of the basic equation for the method are then

$$C = (\Delta Q + \delta q)/\Delta T = \Delta Q/(\Delta T - \delta T). \qquad (27)$$

This method is especially suitable if very precise knowledge of the heat capacity is required at a single temperature or over a relatively small range of temperature; it is also useful at the other extreme, at which the calorimeter vessel is filled (without vapor space) at the working temperature for measurements at widely different temperatures. Although this method has advantages with solid specimens (for which the establishment of steady conditions is slow), these considerations are significant with liquids only if no vapor space can be allowed.

D. Applicability of the Comparison and Direct Methods

As noted in Chapter 1 (Section V) the use of the twin-vessel calorimeter can have important advantages in the study of small heat effects. It is well suited to the accurate determination of small differences in heat capacity, as between isomeric hydrocarbons, for example, for which the differences may be of greater interest than the heat capacities themselves, and especially to the study of dilute solutions, for which the difference between the molal heat capacity of the pure solvent and that of the solution may be only a few parts per thousand. For the study of liquids differing widely from the chosen reference liquid, only methods 2 and 3a of Section II–1–B above are suitable, but it is debatable whether, in such cases, the comparison method is so much more accurate than the direct method as to justify the effort of making two calorimeter vessels with effectively identical thermal characteristics. In fact, despite its long history, the comparison method has not been much used for the study of widely different liquids. Thus, in general, twin-vessel calorimeters have been used primarily for the precise determination of small differences in molal heat capacity, for which it is easier to make replicate calorimeter vessels than to attain the extreme overall precision necessary to use the direct method for comparative studies. A problem to which the comparison method is particularly well suited is the determination of the apparent and partial molal heat capacities of aqueous electrolytes, and several different versions of the method[3, 5, 70, 71, 75, 82, 108, 110] have been applied to this problem.

One important practical advantage of the twin-vessel calorimeter when applied to the study of dilute aqueous solutions is that since the relative lowering of the solvent vapor pressure above the solution is small, a vapor space can be tolerated in the specimen vessels. This is because only negligible error arises in the calculation of the small *difference* between the latent heat changes in the two vessels. The inclusion of a vapor space simplifies the design of the vessels and greatly facilitates filling them. The inclusion of a vapor space also greatly extends the temperature increase which can be explored without refilling the specimen vessels, and in particular the method of continuous heating may conveniently be employed. This permits the realization of another advantage which should not be overlooked, namely the time saving which can be effected by the use of a modern recording twin-vessel calorimeter. There is still a notable scarcity of calorimetric heat capacity data pertaining to electrolyte solutions, and this is only partly due to the difficulty of attaining the requisite accuracy; the main reason seems to be the long time required to obtain experimental results of attested precision Consequently any method which eliminates time consuming manual operations and numerical calculations represents a considerable advance. The use of direct-recording control devices does achieve this, and even though it may involve a slight loss of accuracy, the gain in the rate at which thermodynamic information can be obtained certainly outweighs the loss, at least at the current stage of development of the art. Thus for the purpose of relative measurements, giving immediately the apparent molal heat capacity of the solute as a function of temperature, the automatic recording twin-vessel calorimeter has obvious and compelling advantages.

In the study of non-aqueous liquids and liquid mixtures on the other hand, the net advantage probably rests with the direct methods, even for the determination of the heat capacity of mixing. Thus although $c_P{}^M$ appears as a small difference between two large quantities, one of these [the heat capacity of the ideal reference system, see Equation (20)] has no experimental reality, and the simplest way of using the comparison method is to take one or other of the pure components as reference liquid. The heat capacity of mixing may then be calculated from the relative heat capacity of the mixture and that of the second pure component. In neither measurement are the latent heat changes in the two vessels simply related, so that it is not practicable to allow vapor spaces. A major advantage of the comparison method—simplicity of construction of the calorimeter vessels—is thus lost. Moreover, the difficulty of handling calorimeter vessels without vapor spaces is such that it is easier in practice to concentrate on adiabatic control and thermal equilibration than to make two matched calorimeter vessels of this type. It must be stressed, however, that when there is no vapor space in the calorimeter vessel the temperature range accessible without adjusting the amount of specimen is limited (20 or 30 deg perhaps) so that while steady-state heating is possible, its advantages over intermittent heating are greatly reduced, especially in relation to the rate at which data can be accumulated.

E. Theory of the Recording Adiabatic Twin-Vessel Calorimeter

Before we proceed to the discussion of the design problems associated with the calorimeters chosen for detailed examination, it is necessary to outline the principle of operation of the recording adiabatic twin-vessel calorimeter. This apparatus[5] is intended primarily for the determination of apparent molal heat capacities of solutes in relatively dilute solutions, and the operating principle is based upon method 3c, comparison of heating powers by auxiliary heater, of Section II–1–B above. It is convenient to start from the definition of the apparent molal heat capacity of the solute as given in Section II–1–A. In relation to a twin-vessel calorimeter, Equation (16) can be written in the form

$$\phi_{C_P} = \frac{C_s - C_r}{n_2} = (C_s - C_r)\frac{1000}{w_1 m_2} \qquad (28)$$

in which:

C_r is the total heat capacity of the reference vessel and its contents,
C_s is the total heat capacity of the specimen vessel and its contents,
m_2 is the molality of the solute in the specimen vessel,
w_1 is the mass of solvent, which must be the same in both vessels.

HEAT CAPACITY OF LIQUIDS AND SOLUTIONS

The calorimeter includes an automatically controlled adiabatic shield, and since the heating rates (and hence, to a good approximation, the temperatures) of the specimen and reference vessels are always equalized, the residual heat leaks due to imperfectly adiabatic conditions are virtually identical for the two calorimeter vessels, and Equation (24),

$$C_s - C_r = I_a^2 R_a/(\mathrm{d}T/\mathrm{d}t), \qquad (24)$$

may be applied directly. That is to say if the current intensity, I_a, in the auxiliary heater is regulated so as to keep the heating rate, $\mathrm{d}T/\mathrm{d}t$, of the specimen vessel equal to that of the reference vessel, then provided that the resistance, R_a, of the auxiliary heater and $\mathrm{d}T/\mathrm{d}t$ are known and remain constant, I_a^2 gives a measure of $C_s - C_r$ and hence, by Equation (28), of ϕ_{CP}. It follows that a continuous record of I_a^2 over an extended heating gives the value of the apparent molal heat capacity of the solute at any point in the temperature range covered. In using a record of I_a^2 in this way the assumption is that in practice an average value of $\mathrm{d}T/\mathrm{d}t$ over a complete heating will be determined in terms of the total temperature increment and the total heating period, as $\Delta T/\Delta t$, in which ΔT is of the order of tens of degrees. This procedure has the great advantage that the effects of small fluctuations of temperature gradient in the two vessels can be neglected. Over this sort of temperature range the resistance of a constantan element may be regarded as being independent of temperature since the total change of resistance between 300°K and 400°K is only a few tenths of one per cent. Thus the main factor limiting the reliability of the value of ϕ_{CP} determined from a record of I_a^2 is the extent to which $\mathrm{d}T/\mathrm{d}t$ changes as the temperature increases. For a given current intensity in the main heaters of the two vessels (connected in series), $\mathrm{d}T/\mathrm{d}t$ depends on the heat capacity of the calorimeter vessel filled with the liquid having the lower total heat capacity. In general the total heat capacity of the solution will be greater than that of the pure solvent in the reference vessel. With aqueous electrolyte solutions, on the other hand, since the apparent molal heat capacity of the solute is negative, the total heat capacity of the solution will be smaller than that of the corresponding quantity of pure water, and the auxiliary heating will be supplied to the reference vessel containing the pure water. Up to about 400°K the temperature coefficient of the specific heat of water and of dilute solutions is very small. The total change in the specific heat of water between 300°K and 400°K is less than 2 per cent. Thus the greatest error in the apparent molal heat capacity of the solute, calculated according to the above procedure, is unlikely to greatly exceed 1 per cent (at the upper and lower ends of the temperature range covered). With any other method, since the apparent molal heat capacity is determined as the small difference between two large heat capacities, it is difficult to achieve results for aqueous electrolytes at high dilutions which are accurate even to a few per cent. For example, by using the direct method, the heat capacity of a 0·2 molal solution of sodium chloride in water must be determined with an accuracy of 0·01 per cent in order to obtain the apparent molal heat capacity of the sodium chloride accurately to 3 per cent. For most purposes, however, it is sufficient to obtain ϕ_{CP} with an accuracy of about 5 per cent, and in this context the influence of the temperature-variation of the specific heat of water on the results obtained from the simpler form of the recording twin-vessel adiabatic calorimeter may be neglected.

In the case of non-aqueous solvents, however, for which the temperature dependence of the specific heat is greater than in the case of water, it is necessary to keep the heating rate constant, or to monitor it continuously throughout the experiment. In practice it is probably easier to adopt the first alternative, using a special regulator to keep the heating rate in the reference vessel constant. Since this requirement is almost peculiar to this type of investigation, we describe here the principle of an effective temperature–program regulating device. It is worth noting that this scheme is capable of adaptation to a continuously heated single-vessel calorimeter, in which it is important to keep the heating rate *roughly* constant in order not to alter significantly the steady-state temperature gradients within the calorimeter vessel. The electronic circuit of the device is illustrated in *Figure 4*. If the temperature of the apparatus is controlled by a resistance

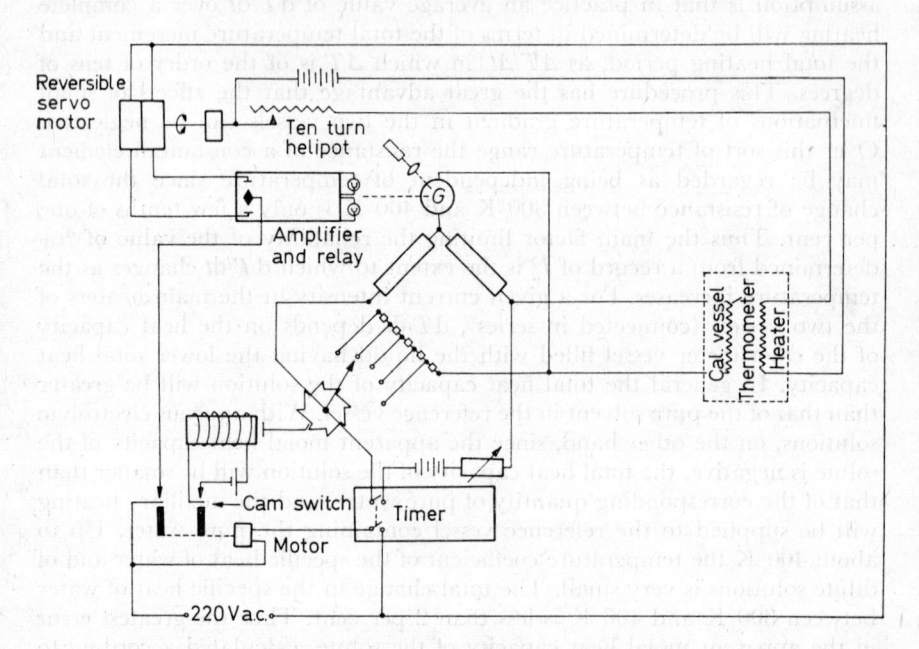

Figure 4. Automatic temperature-program regulator circuit (schematic) for the recording twin-vessel continuously heated calorimeter[5]. This regulator controls the heating rate in the calorimeter vessel whose heat capacity is the smaller.

thermometer (a platinum thermometer of 100 Ω nominal resistance, say) the variable reference rheostat of the bridge circuit can be operated by means of a step-switch relay such as that commonly used in telephone dialling systems. The current passing through the coil of the mirror galvanometer depends on the deviation of the instantaneous resistance of the platinum thermometer from the reference value corresponding to the setting of the dial resistance box. When d.c. voltage from a battery is applied to the bridge circuit, an out-of-balance current, corresponding to a deviation of the instantaneous temperature of the platinum thermometer from the corresponding programmed reference value, will result in a deflection of the galvanometer light beam from its zero position between the two photocells. The photocells are connected to an amplifier and relay circuit which actuates a reversible servo motor; the sense of rotation of the servo motor is determined by the sign of the galvanometer deflection, but its speed is fixed. The servo motor operates a ten-turn precision-wound helical potentiometer which controls the current in the main heater circuit, reducing the power input to the main heaters whenever the instantaneous temperature indicated by the platinum thermometer is higher than the programmed reference value, and vice versa. The whole process operates intermittently, the d.c. supply being connected to the thermometer bridge for a short interval every 30 seconds, say, while the servo motor relay circuit includes a cam switch which cuts off the a.c. supply after a fixed short interval, so that the servo motor rotates through a fixed angle at each cycle of operation, and correspondingly the power input to the main calorimeter heaters changes by one small (positive or negative) step every 30 sec. On completion of each cycle of operation the resistance indicated by the dial resistance box is automatically increased by one step ($0.1\ \Omega$, say). Within a few minutes of the regulator commencing to operate the heating rate of the calorimeter vessels attains a constant value (within 0.5 per cent), which is determined by the pre-selected operating cycle period of the temperature–program regulator and the step value of the dial resistance box in the thermometer bridge circuit. Since this resistance box consists of decadic units (not shown in *Figure 4*) connected in series, continuous operation is possible, so that the pre-selected constant heating rate can be maintained throughout the entire range up to $400°K$.

In the twin-vessel calorimeter, the temperature–program regulator controls the heating rate in the upper calorimeter vessel, which is always filled with the liquid having the smaller total heat capacity. Thus the platinum resistance thermometer in the upper calorimeter vessel is connected into the bridge circuit of the temperature–program regulator; the corresponding platinum thermometer in the lower calorimeter vessel is a dummy, included to ensure that the total heat capacities of the twin vessels are as nearly equal as possible. The auxiliary heating is then always applied to the lower calorimeter vessel, and the second heating coil in the upper calorimeter vessel is also a dummy and is not connected to the power circuit. The details of this layout are shown schematically in *Figure 5*, which also shows the arrangement of the auxiliary heater regulator and the adiabatic shield regulator. The latter is described in Section II–3–D below.

The auxiliary heater regulator keeps the current intensity in the auxiliary heater in the lower calorimeter vessel at the value which is necessary to maintain the twin-vessel system in thermal balance. As may be seen from *Figure 5*, this regulator is essentially similar to the temperature–program regulator, except that the sensing element of the auxiliary heater regulator is a multi-junction thermocouple mounted between the two calorimeter vessels. This is connected to the galvanometer through a circuit arranged to give maximum sensitivity[5]. The light beam from this galvanometer

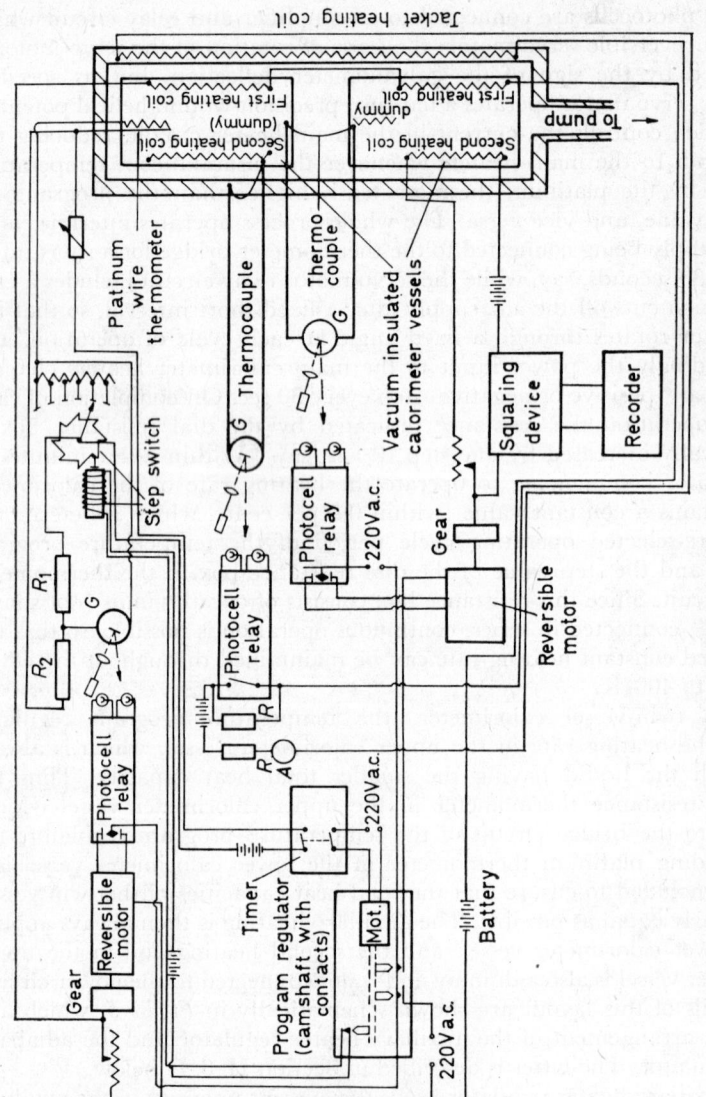

Figure 5. Regulator circuits (schematic) for the recording twin-vessel continuously heated calorimeter[5], showing the adiabatic shield regulator and the regulator for the auxiliary heater in the calorimeter vessel whose heat capacity is the larger.

448

actuates the photocell relay which in turn switches a reversible servo motor resistor unit similar to that in the automatic temperature–program regulator. This unit is also operated intermittently by the timing device which controls the operation of the temperature–program regulator, so that the auxiliary heating power in the second heater element in the lower vessel is also varied in definite steps. In other words, each heating rate check on the upper vessel is followed by a check on the temperature balance between the two vessels, and the current intensity I_a in the auxiliary heater in the lower vessel is automatically adjusted to minimize the temperature difference between the two vessels. To be effective, the thermocouple–galvanometer unit must be capable of indicating temperature differences of the order of 10^{-5} deg.

Finally, the square of the auxiliary heating current, I_a^2, is recorded so that a plot of the apparent molal heat capacity of the solute in the upper vessel is obtained. If an X—Y recorder (not shown in *Figure 5*) is available, the X-scale can be operated by an additional thermocouple measuring the actual temperature of the upper vessel, so that the recorder gives directly a plot of apparent molal heat capacity against temperature.

The final operating equation for this apparatus then takes the form:

$$\phi_{\mathrm{CP}} = \frac{I_a^2}{n_2} R_a \cdot \frac{\Delta R_\mathrm{T}}{\Delta T} \cdot \frac{\Delta t}{\Delta R_\mathrm{T}} = \frac{1000\, I_a^2 R_a}{w_1 m_2} \cdot \frac{\mathrm{d} R_\mathrm{T}}{\mathrm{d} T} \cdot \frac{\Delta t}{\Delta R_\mathrm{T}}, \qquad (29)$$

which may be derived from Equations (24) and (28), where $\mathrm{d}R_\mathrm{T}/\mathrm{d}T$ is the temperature coefficient of the platinum thermometer and $\Delta t/\Delta R_\mathrm{T}$ is the time programmed for a 1 Ω resistance increment in the platinum resistance thermometer. The temperature derivative of R_T does of course vary with temperature (changing by about 5 per cent between 300°K and 400°K), so that the recorded value of ϕ_{CP} has to be corrected to allow for this variation whenever the maximum accuracy is required.

2. Calorimeter vessels

A. General Requirements

The pioneers of calorimetry usually called the central part of their apparatus, containing the specimen under investigation, "the calorimeter" or, occasionally, "the calorimeter proper". Recently it has generally been referred to as "the calorimeter vessel". In discussing the comparison method it is necessary to distinguish between two calorimeter vessels, and these are identified as the specimen vessel and the reference vessel. From the design point of view another important distinction is that between the calorimeter vessel (including heaters, thermometers, stirrers, etc.) on the one hand, and its outer casing or shell, which is normally the main structural unit, on the other. Where necessary we refer to the latter as "the vessel proper".

The basic requirements for a calorimeter vessel for liquids are that it shall contain the liquid specimen absolutely without leakage, and that its internal and external geometry shall be compatible with the operation of an automatically controlled adiabatic shield. The second requirement implies, among other things, that the external area of the vessel proper be minimized with respect to the internal volume available to the specimen, and that the temperature be uniform over the whole external surface. These requirements are common to both the direct and the comparison methods. A third requirement, particularly important in the direct method, is that the total heat capacity of the calorimeter vessel must be small in comparison with that of the liquid specimen. It is easily shown that, irrespective of whether it is the temperature rise or the energy input which is kept the same for both the main (vessel full) measurement and the calibration (vessel empty) measurement, the proportional error on the heat capacity of the specimen is greater than that on the total heat capacity of the vessel and specimen by a factor $(a + 1)(a^2 + 1)^{\frac{1}{2}}/a^2$, in which a is the ratio of the total heat capacity to that of the empty vessel.

It is remarked in Chapter 1 that ease of operation is an important requirement for successful calorimetry. It is pointless to develop an elaborate and sophisticated apparatus if repairs and breakdowns extend unreasonably the time lapse between effective measurements. The most common causes of breakdown include electrical circuit failures, so that an additional requirement is that all parts of the calorimeter vessel be easily removable for testing and repairing.

The design problems arising from the first two requirements may conveniently be grouped into four categories:

(i) filling the calorimeter vessel with an accurately measured mass of specimen and sealing it,
(ii) heating the specimen and disseminating the heat uniformly throughout the specimen and vessel,
(iii) accommodating the thermal expansion of the liquid specimen, and
(iv) measuring the actual temperature of the specimen and also the effective temperature difference between the vessel proper and its surroundings (i.e. the adiabatic shield and, in twin-vessel calorimeters, the reference vessel).

The first two categories are interdependent in that the choice of the method of heat dissemination depends both on the shape of the vessel proper and on the material of which it is made. Both these factors are important also in determining the ratio of the heat capacity of the specimen to that of the empty calorimeter vessel. In fact only a limited range of shapes is compatible with the rapid attainment of thermal uniformity, and this range is further restricted by the need to simplify the fabrication of the vessel as far as possible. The choice of materials also is severely limited. It is convenient, therefore, to discuss the questions of shape and constructional material of the vessel proper before examining in turn the four categories of design problems listed above.

(1) *Shape of the Vessel Proper*

It is argued in Chapter 4, Section II–1, that the ideal shape for a calorimeter vessel is spherical, since this shape minimizes the external area with respect to the internal volume, and gives the greatest strength for a given wall thickness; but in addition to the objections to this shape cited in Chapter 4 we have to consider the following. If, as with experiments on solutions of different concentrations, the calorimeter vessel has to be emptied and cleaned out between each series of experiments, it is necessary to be able easily to remove the calorimeter vessel from the main apparatus. This requirement is incompatible with a regular distribution of thermocouple junctions over the external surface of a spherical vessel, and the incorporation of a resistance thermometer to monitor the whole of a spherical surface is extremely difficult. Consequently it seems reasonable to assume that calorimeter vessels for liquids or solutions will normally be cylindrical, generally with rounded or conical ends. In the latter case it is important to avoid sharp edges, as these accentuate local temperature gradients, thereby causing errors in the radiative heat transfer as computed from the temperatures of the main surfaces. For applications that make it necessary to be able completely to disassemble the vessel, the simplest design is a cylinder (with its axis vertical) having a well sealed lid. Such a vessel may easily be withdrawn upwards from a fixed thermocouple assembly or supporting device. This arrangement is also suitable where only a small movement of the vessel can be achieved without disturbing attached thermocouple junctions.

The optimum ratio of length to diameter for a cylindrical calorimeter vessel will depend on the method of heating, and in particular on the means used to disseminate the heat. If this dissemination is done primarily by conduction through the liquid and by free convection, Sturtevant[128] has shown that a long thin calorimeter vessel will have a shorter equilibration time than a short thick one. In several recently designed calorimeters the heat is disseminated either by forced convection (stirring) or by a combination of free convection and metallic conduction through internal fittings, and in these designs the ratio of length to diameter is usually between $2\frac{1}{2}$ and $3\frac{1}{2}$.

It is probably sound practice first to design the heater and thermometer units and decide whether or not to use mechanical stirring before deciding the shape of the vessel.

(2) *Materials of Construction*

Maintaining temperature uniformity over the external surface is obviously facilitated by making the vessel proper from a material having a high thermal conductivity. Further, since thermal conduction between the liquid specimen and the fabric of the vessel is limited by the boundary-layer effect at the liquid-to-solid interface, it is important to minimize the heat capacity of the vessel proper. The same criterion applies to the complete calorimeter vessel. This means, in general, that the wall thickness of the vessel proper should be as small as strength requirements allow. This is a severe limitation when the specimen liquid has a high vapor pressure at the working temperature. The estimation of the minimum wall thickness for a given internal pressure is discussed in Chapter 4, Section II–1.

In order to compare some of the materials which have been used for calorimeter vessels in recent years, their relevant properties are set out in *Table 1*.

In *Table 1*, $c_P{}^*$ is the *volumetric* specific heat, i.e. the heat capacity per unit volume. This is the simplest property to use for comparing the total heat capacities of machined components, since the strengths of materials are

A. J. B. CRUICKSHANK, TH. ACKERMANN and P. A. GIGUÈRE

Table 1. Thermal properties and strengths of some materials used in calorimetric vessels

Material	c_P (J deg^{-1} g^{-1})	ρ (g ml^{-1})	c_P^* (J deg^{-1} ml^{-1})	Safe stress (dyne cm^{-2} × 10^9)	λ (watt deg^{-1} cm^{-1})
Stainless steel	0·5	7·8	3·9	6·0	0·46
Duralumin	0·85	3·0	2·5	2·3	1·29
Brass	0·38	8·6	3·3	1·2	1·09
Copper	0·38	8·9	3·4	0·7	3·84
Silver	0·24	10·5	2·5	0·5	4·07
Glass	0·9	2·2	2	0·2	0·007–0·010
Araldite†	2·7–4·5	1·1	3–5	0·2	

† Epoxide resin made by Ciba-ARL Ltd., Duxford, Cambridge, England; representative of many similar materials. The smaller c_P figures refer to the range 0 to 100°C and the larger to 140 to 160°C.

necessarily computed in terms of linear dimensions. The *safe stress* quoted in *Table 1* corresponds roughly to the elastic limit of the metallic materials; the figure quoted for glass applies to boro-silicate glasses which have been carefully annealed after manufacture. The figures for thermal conductivity, λ, are averages over the range 0°C to 100°C.

A rough indication of the way in which the total heat capacity of a vessel will vary according to the construction material for a given internal pressure is obtained by dividing the volumetric specific heat by 10^{-9} times the quoted safe stress. The best material from this point of view is then that for which this quotient is the smallest. The figures quoted in *Table 1* indicate that for the same internal pressure, the total heat capacity of a glass vessel will be nearly 16 times that of a stainless steel vessel, 10 times that of a duralumin vessel and twice that of a copper one. Although stainless steel is undoubtedly the best material from the point of view of total heat capacity its relatively low thermal conductivity is a disadvantage. Duralumin is nearly as good as stainless steel from the point of view of total heat capacity, it has a much higher thermal conductivity, and it is much easier to machine. Neither copper nor silver rate highly because of their low safe stress values.

The temperature differences *across* the walls of different calorimeter vessels, all designed to contain the same internal pressure, can be correlated by multiplying the thermal conductivity of the material by 10^{-9} times the safe stress. Taking duralumin as datum at unity, the corresponding temperature differences for the other materials are: stainless steel, 1·1; brass, 2·2; copper, 1·1; silver, 1·5; glass, 1500. Only in the case of glass is the difference really significant. In practice, however, these conclusions are valid only when the internal pressure is relatively high (greater than about 10 atm), since otherwise the wall thickness will be determined less by considerations of strength than by ease of manufacture, etc. In any case the temperature difference *across* the outer wall of the vessel proper is not the most important parameter related to the thermal conductivity of the material, since only in the case of glass is this temperature difference likely to be comparable with that across the liquid-to-wall boundary layer. More important is the rapidity of thermal equilibration in the direction parallel to the wall surface and around the corners of the vessel proper. This implies that stainless steel is quite suitable for a spherical vessel or for a cylindrical vessel with nearly hemispherical ends having a more or less uniform wall thickness. The use of duralumin, on the other hand, will lead to only a small increase in the total heat capacity relative to stainless steel, but the much higher thermal conductivity makes duralumin the more suitable material for designs involving variations in wall thickness.

A second important consideration in selecting the material for the vessel proper is resistance to corrosion. Duralumin is not at all good in this respect, since not only is its surface resistance poor, but it is a very difficult metal to plate or clad. Copper is quite easy to plate (with gold or rhodium, for example) and its high thermal conductivity makes it suitable for cases where the internal pressure does not exceed a few atmospheres, provided that unusually good corrosion resistance is not required. For the study of highly corrosive electrolyte solutions, glass is normally used as the main structural material for the calorimeter vessel[2, 48, 102]; it is then important to get the best possible internal distribution of heat, and to use the smallest practicable wall thickness. In special cases gold-plated metal vessels have been used to study highly corrosive liquids[117]. Franck and Spalthoff[56] used a steel bottle internally plated with silver to measure the heat capacity of pure hydrogen fluoride in the liquid–vapor state. Since soldered joints are very difficult to protect against corrosion by plating, the vessel was made in one piece with a wall thickness sufficient to withstand internal pressures up to 300 atm. An internal heating element was used, consisting of a resistance wire sheathed in a silver capillary tube. The heating element was wound into a coil whose ends were soldered to the heavy conical lid used to close the opening at the top of the steel bottle. Sealing was effected by a soft silver gasket attached to the conical lid. The temperature of the calorimeter vessel was measured by a platinum resistance thermometer wound on to the nickel-clad outer surface of the bottle.

Another factor which must be taken into account, especially in the direct method, is the radiation emittance (see Chapter 4, Section III–2) of the outer surface of the calorimeter vessel. Radiative heat transfer will usually account for a major part of the heat leak from the calorimeter vessel, see Section II–4 below, so it is necessary to make the emittance as low as possible. Highly polished copper has a very low total radiation emissivity (0·02, taking full black-body emissivity as 1·00) comparable to that of silver, but copper tarnishes rapidly so that it is suitable only if it is possible to re-polish between each series of experiments. Duralumin has a fairly low emissivity (0·03 to 0·05) but this increases in air owing to surface oxidation. Gold-plated copper (emissivity 0·03) is a good compromise here; rhodium plating is tougher than gold, but has a higher emissivity (about 0·05).

If the calorimeter vessel has to have a detachable lid, to allow repair of internal plating, for example, or to permit the calibration of a resistance thermometer made integrally with the lid, the choice of the method of sealing the lid to the body depends on how frequently the vessel has to be dismantled. If this is only occasionally, for repair or calibration, the lid may be sealed with soft solder, provided that proper precautions are taken, see Chapter 4, Section II–2. Alternatively, a renewable flange weld may be used, the weld being machined off to open the vessel. When frequent dismantling is necessary, e.g. for cleaning, a gasket seal is to be preferred.

Soft metal sealing rings give good thermal contact, but require heavy flanges to exert the necessary sealing pressure. It may be better to use non-metallic gaskets (Nylon, Polythene, Teflon, etc.) but the seating must then be designed to give some metal-to-metal contact to ensure proper heat distribution. There is, however, a very important limitation which applies especially to the rubber-like materials. Although these are the most convenient to use, the latent heat effects of stress relaxation processes can cause large errors unless great care is taken to make these effects reproducible. An example of this phenomenon is that measuring the specific heat of moderately strained rubber between 20°C and 80°C by drop calorimetry gives values much smaller, depending on the strain, than those found with an isoperibol calorimeter; the latter results agree with those for unstrained rubber[14]. A plausible inference is that the rubber absorbs latent heat as it is cooled to 20°C. If so, it is to be expected that when rubber which has been cooled under strain is reheated, latent heat will be evolved, giving a spuriously low apparent specific heat. In fact, if a Neoprene ring under unidirectional compression is cycled between 20°C and 60°C, the apparent specific heat at 20°C shows successive reductions, the first being as much as 0·5 joule deg^{-1} g^{-1}. Subsequent reductions are smaller, but the total reduction after three or four cycles may exceed 1 joule deg^{-1} g^{-1}. The effect diminishes with increasing temperature, and changes sign at about 50°C[38]. In practice, if it is necessary to duplicate a series of heat capacity measurements on a liquid specimen covering a temperature range of more than a few degrees, then after cooling to the initial temperature the sealing ring must be removed and either normalized or replaced by a nearly identical one (a weight-matched set from the same manufacturing batch may be used), and the vessel refilled.

B. Filling the Calorimeter Vessel

To ensure that the heat capacity measurements refer unambiguously to the selected liquid or solution it is necessary that the operations of filling and sealing the calorimeter vessel neither contaminate the specimen nor, in the case of a solution, alter its composition. In this context it is important that the physical properties of the specimen liquid correspond to a thermodynamically reproducible state. It follows that the method of filling will depend on the nature of the specimen liquid as well as on the calorimeter technique.

Before discussing filling techniques in detail it is pertinent to make some general points. In making what are, in effect, measurements of specific heat it is obvious that the mass of the specimen must be known at least as accurately as the energy input and the resulting temperature increase. The mass of specimen in the calorimeter vessel will usually be determined by weighing either the vessel itself or the reservoir from which it is filled, so that it is important to ensure that all weighing operations are free from systematic error. This means in practice that the total mass of the parts to be weighed must be kept small enough to permit the use of a sensitive balance, discriminating to better than about 2 mg, and that strict attention must be paid to the evaluation of buoyancy corrections, etc.

HEAT CAPACITY OF LIQUIDS AND SOLUTIONS

(1) Single-Vessel Calorimeters

If the calorimeter vessel is designed to be easily removable from the rest of the apparatus without disturbing the electrical wiring, it can best be filled by attaching it to a separate filling apparatus, the mass of the specimen being determined by weighing the vessel before and after filling. Such a calorimeter vessel may conveniently be filled with pure liquids by connecting it directly to a vacuum still; this completely avoids atmospheric or other gaseous contamination. If the calorimeter vessel is not completely filled with liquid (see Section II–2–E below) the same technique may be used for solutions of involatile solutes by first introducing the solute, weighing the calorimeter vessel, and then distilling in the solvent. Great care is necessary, however, to ensure that the resulting solution is homogeneous, and in our experience this is very difficult when the calorimeter vessel is completely filled.

An alternative filling method, especially suitable for liquid mixtures, is first to make up a mixture of the desired composition by weight from outgassed components, and then to transfer a portion of this mixture to the previously evacuated calorimeter vessel. This method involves the use of elaborate liquid handling apparatus, including reservoirs for each of the pure components, and a detachable mixing vessel (with a vacuum-tight valve) which may be weighed separately. A quantity of one component is first vacuum distilled into the mixing vessel and the latter is re-weighed at room temperature. Then, with the first component frozen, the second component is vacuum distilled into the mixing vessel. The sealed mixing vessel is again warmed to room temperature and weighed. The two components are then thoroughly mixed with the mixing vessel valve still closed. A portion of the mixture is finally transferred as liquid to the previously evacuated calorimeter vessel. If the connection to the calorimeter vessel is made by a siphon, the vapor space above the mixture remains constant throughout the filling process, and no net alteration of liquid-phase composition occurs. A siphon connection may be operated by heating the upper part of the mixing vessel slightly to increase the vapor pressure, and then cooling again. It is necessary to know roughly the total vapor volume available in order to estimate the composition of the specimen transferred to the calorimeter vessel, but it should be noted here that the *composition* of the specimen in the calorimeter vessel has, in general, to be known only with a precision similar to that of the derived quantity obtained from the heat capacity measurements (e.g. the excess heat capacity of mixing) whereas the *mass* of the specimen has to be known to better than the overall calorimetric precision, i.e. to better than 0·01 per cent.

If the calorimeter vessel is to be filled completely with liquid it is usual, after filling, to raise the applied pressure to about 1 atm. Foreign gas contamination can be avoided by applying the pressure directly to the liquid. A device for doing this[38] is illustrated in *Figure 6*.

A variation of the above method is to displace the liquid into the calorimeter vessel by running mercury into the mixing vessel. This variation enables the vapor space to be minimized throughout the filling process, so that the specimen composition corresponds closely to that given by the masses of the two components which were distilled into the mixing vessel.

This variation is not recommended for filling temperatures above about 30°C, where contamination of the mixture by mercury might possibly be significant. This is because the solubility of mercury often increases rapidly with increasing temperature, especially in the case of hydrocarbon mixtures

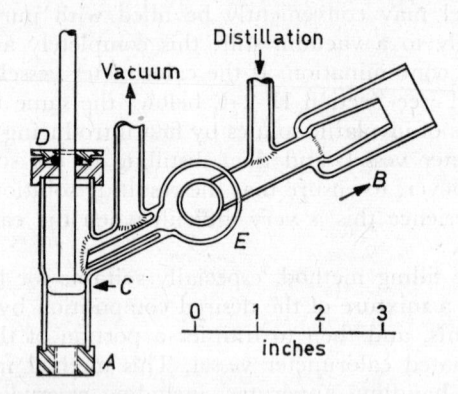

Figure 6. Calorimeter vessel filling apparatus[38] to give an air-free liquid specimen at about 1 atm. The calorimeter vessel is attached at *A* and, after evacuation, liquid is run in from the mixing vessel *B* up to the side arm. The Nylon piston *C* is lowered into the liquid (*D* is a Wilson shaft seal), the diaphragm valve *E* is closed and the vacuum line opened to atmosphere, when the piston transmits this pressure to the liquid.

(see Kuntz and Mains[91]) and because the vapor-phase interactions between hydrocarbons and mercury are very strong, especially at pressures greater than atmospheric (see Jepson and Rowlinson[84]).

With very small calorimeter vessels, if the mixing vessel and the calorimeter vessel are both fitted with serum caps, the transfer may be effected by means of a large hypodermic syringe. The serum cap on the calorimeter vessel must of course be replaced by a vacuum-tight seal after filling is completed.

(2) *Twin-Vessel Calorimeters*

If the calorimeter includes multiple adiabatic and radiation shields, with a large number of thermocouple junctions fixed to the calorimeter vessel, removing the calorimeter vessel for filling would disturb the whole arrangement and the necessary reproducibility of experimental conditions would be very difficult to achieve. This consideration applies especially to twin-vessel calorimeters in which, for example, the relative positions of the specimen and reference vessels must not be changed by the filling process. Consequently special arrangements must be made for filling the twin vessels. Several arrangements involving permanently connected filling pipes have been described[72, 83, 137, 143]. The main disadvantage of such arrangements is that the filling pipes inevitably increase the quantity of heat exchanged between the calorimeter vessel and its surroundings. Another disadvantage is that the masses of the specimen and reference liquids have to be calculated from indirect weighings. One way of avoiding these difficulties is to mount the two calorimeter vessels, together with the two sets of control thermo-

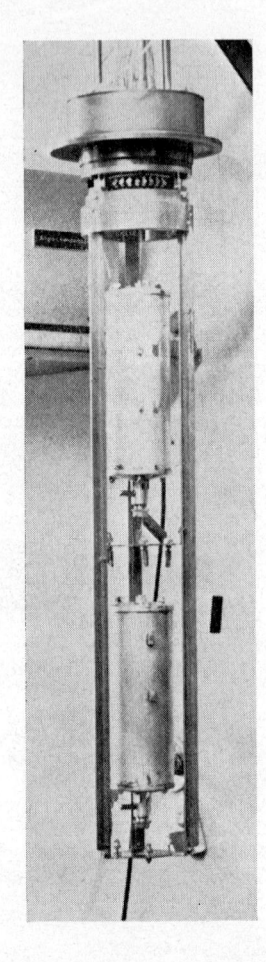

Figure 7. Mounting frame with twin calorimeter vessels[5]. The silver vessels are fixed in the frame, which forms part of the adiabatic shield. The filling nipples can be seen at the lower ends of the calorimeter vessels.

couples, in a metal frame as indicated in *Figure 7*. In this apparatus[5] the frame is designed to be a good sliding fit into the metal cylinder forming the inner part of the adiabatic shield (not shown in *Figure 7*), so that good thermal contact is achieved. The frame is thus effectively part of the adiabatic shield. The top ring of the frame, carrying all the electrical terminals in insulated clamps, can be screwed onto the underside of the lid of the adiabatic shield which is, in turn, fixed to the mounting-plate of the apparatus. The assembling of the apparatus is completed by drawing the cylindrical part of the adiabatic shield up over the frame and fixing it to the shield lid. Thus, by first removing the shield cylinder, the entire inner part of the calorimeter may be removed for filling, and then replaced, without disturbing either of the thermocouple circuits.

The filling connection is made through flexible Teflon hoses, each of which has a threaded metal nozzle which screws into the top of the filling nipple, see Section II–2–C below. The other end of the Teflon hose terminates in a three-way valve enabling the calorimeter vessel to be connected either to a vacuum pump or to a calibrated filling reservoir.

To fill the twin calorimeter vessels, the frame is first weighed with the calorimeter vessels empty and dry. A measured quantity of the solution to be studied is then transferred as liquid from the calibrated reservoir to the previously evacuated first calorimeter vessel. This vessel is then sealed and the frame is weighed again. The amount of pure solvent to be put into the second calorimeter vessel is then calculated from the weight of solution in the first vessel and known weight composition of the solution. Approximately the correct amount of solvent is then transferred from the second calibrated reservoir to the previously evacuated second calorimeter vessel. After the second vessel is sealed the frame is again weighed. If the two charges are not exactly matched, the amount of solvent to be added to the second vessel is calculated. This vessel is then re-connected to the filling apparatus and the additional solvent put in. Again the vessel is sealed and the frame is weighed once more. This procedure is repeated until the previously calculated mass of the filled system is confirmed by the final weighing. If too much solvent has been added, the filling nipple is unscrewed completely and removed and solvent is sucked out through a Teflon siphon.

To clean the calorimeter vessel which has contained solution, it is repeatedly filled with pure solvent and then emptied by suction through the siphon. The final residue of solvent is then removed by evacuation.

C. Sealing the Calorimeter Vessel

The method of sealing the filling inlet to the calorimeter vessel is subject to three main requirements:
- (*i*) It minimizes atmospheric contamination of the specimen after the vessel is detached from the filling apparatus and before the seal is completed.
- (*ii*) It is completely leak-proof when the surrounding insulation space is evacuated.
- (*iii*) The total heat capacity of the inlet and seal assembly is small in comparison with that of the vessel proper (this is necessary to ensure

uniformity of temperature over the whole external surface of the calorimeter vessel).

A straightforward way of meeting the first requirement is to design the filling valve so that it may be completely sealed *before* the calorimeter vessel is detached from the filling apparatus. In fact, although this is comparatively easy to do when the vessel proper is made of glass and used with a vapor space above the liquid, it is very difficult in the case of a metal vessel without the heat capacity of the filling valve being unacceptably large.

(1) *Glass Vessels*

A glass vessel which is to be incompletely filled with liquid is usually fitted with an ampoule seal[2, 48, 102]. A small-diameter glass tube having a constriction near its inner end is first blown on to the end of the calorimeter vessel. This filling tube is then attached to the filling apparatus by a ground-glass socket with the calorimeter vessel suspended filling end uppermost After filling, the constriction is sealed off with a small flame. The rest of the filling tube is then detached from the filling apparatus and put on the balance with the filled calorimeter vessel for final weighing.

Flame sealing cannot be used if the vessel is completely filled with liquid, but we know of no apparatus in which the combination of corrosive liquid and no vapor space has been studied.

(2) *Metal Vessels*

From the point of view of requirement (i) above, the simplest procedure is to fit a valve to the filling end of the calorimeter vessel. This procedure has several grave disadvantages, however. The external shape of the calorimeter vessel is greatly complicated, so the radiative heat transfer to the shield is increased, then the outermost parts of the valve are not in good thermal contact with the rest of the calorimeter vessel, and in addition the valve has, inevitably, a relatively large heat capacity. Consequently it is often better in practice to compromise on requirement (i) in order to meet requirement (iii). Depending on where this compromise is drawn in a particular case (this will be influenced by, among other things, the size of the calorimeter vessel) one or other of three schemes might form the basis of the seal design.

If the filling pressure is 1 atm and a small amount of atmospheric contamination may be tolerated, the simple pad seal shown schematically in *Figure 8(a)* should be satisfactory. After the vessel has been filled completely, the pressure on the liquid is raised to atmospheric. The calorimeter vessel A is then unscrewed from the filling apparatus E and the sealing pad B is quickly slipped into position in the recess in the top of the calorimeter vessel inlet. The sealing cap D is then screwed on and tightened down to make the final seal. *Figure 8(b)* shows a variation of this scheme, in which the stainless steel ball C is dropped into position on the Nylon seating before the vessel A is unscrewed from the filling apparatus E, thus effecting a partial seal. After the calorimeter vessel is detached from the filling apparatus the screwed stopper is tightened down, compressing the stainless steel ball C into the seating B to make the final seal. The main disadvantage of this scheme is that the filling orifice has to be very carefully machined to accommodate the

HEAT CAPACITY OF LIQUIDS AND SOLUTIONS

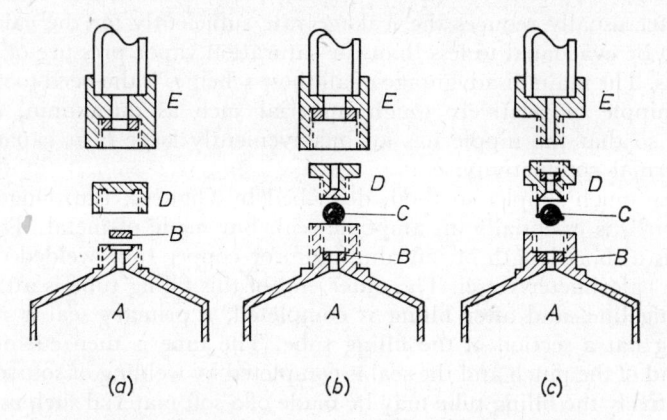

Figure 8. Seals for calorimeter vessel filling nipples.
A, Body of calorimeter vessel
B, Sealing pad or ring (Nylon, Teflon, Polythene, Neoprene or soft metal)
C, Stainless steel ball
D, Cap or stopper
E, Outlet-end of filling apparatus

Nylon seating, especially with copper vessels, when only a limited pressure can be transmitted by the sealing screw.

The scheme shown in *Figure 8(c)* is in fact a simple valve. When the filling nipple D is partially unscrewed the stainless steel ball C falls away from the central hole in the nipple, but the wire loop prevents it from falling far enough to block the hole through the Nylon seating B into the interior of the calorimeter vessel. The filling apparatus connection E screws into the top of the filling nipple. When filling is completed, the calorimeter vessel is rotated to tighten down the filling nipple, and the stainless steel ball C is compressed into the Nylon seating B to make the main seal, just as in the scheme of *Figure 8(b)*. A simpler version of the scheme of *Figure 8(c)* is used on the calorimeter vessel illustrated in *Figure 10*. The filling nipple screws into the boss V at the bottom of the calorimeter vessel. The outer part of the filling nipple is bored and tapped to accommodate the screwed nozzle of the Teflon hose connection to the filling apparatus. The inner part of the filling nipple has two skewed holes drilled as shown in *Figure 10*. When the nipple is tightened down the orifices of these holes are sealed against the Teflon gasket which also seals the nipple to the boss V. When the nipple is partially unscrewed the orifices of the two skewed holes are removed from the Teflon gasket so that the passage from the interior of the calorimeter vessel to the filling connection is opened. In principle these arrangements avoid atmospheric contamination completely. In practice the main difficulty is to avoid inward leakage of air around the thread of the filling nipple when it is partially unscrewed. This may be prevented by sealing the exposed part of the filling nipple thread with hard vacuum wax or low vapor-pressure plastic; this is then removed after the nipple has been tightened. Alternatively a soft silicone ring may be used instead of wax or plastic, or it may be sufficient to lubricate the thread of the filling nipple with graphite powder.

The latter usually reduces the leakage rate sufficiently for the calorimeter vessel to be evacuated to less than the saturation vapor pressure of aqueous solutions. The main disadvantage of all these schemes is the need to make the filling nipple of relatively tough material such as duralumin, brass or bronze, so that the nipple has an inconveniently large heat capacity and low thermal conductivity.

A very much simpler method, described by Giguère, Liu, Dugdale and Morrison[61], is essentially an ampoule seal, but made of metal. The filling nipple is a short length of soft aluminum or copper tube welded or sealed into the calorimeter vessel. The outer end of this filling tube is attached to the filling line, and after filling is completed, a primary seal is made by pinching flat a section of the filling tube. The tube is then cut off at the outer end of the pinch and the seal is completed by welding or soft-soldering. Alternatively the filling tube may be made of a soft material such as gold, to obviate the necessity of welding.

If it has not been especially important to avoid atmospheric contamination, a variety of simple seals has been used. One rather ingenious method, developed by Staveley, Tupman and Hart[125] is simply a screw plug fitting into a tapped hole in the calorimeter vessel. Before the calorimeter vessel is filled, the two threaded surfaces are coated with bismuth solder, and after filling the plug is screwed into the hole with a heated screwdriver. A plug seal which avoids the application of heat has been developed for heat-of-mixing calorimeter vessels by Brown and Fock[16]. In this case the filling hole in the calorimeter vessel is smooth-bored, and the plug is a cylinder of Teflon mounted on the shank of a small bolt. The ends of this cylinder are constrained by metal washers whose outer diameter is the same as that of the hole. The plug is inserted into the filling hole with the bolt head inside, and the seal is made by tightening the projecting nut; this compresses the Teflon against the bore of the hole and the shank of the bolt to make the seal.

D. Heating the Calorimeter Vessel and the Liquid Specimen

An essential requirement of the design of the heating element in any calorimeter vessel is that the heat liberated in the heating element must be distributed quickly through the specimen and the vessel proper. Two distinct considerations are important in this respect. First, the temperature difference between the actual resistance element of the heater and the calorimeter vessel proper must be minimized to ensure constancy of the heater resistance; a small temperature difference also simplifies the problems of tempering the external heater leads. Second the temperature gradients in the liquid sample must be kept small; small gradients help to reduce the thermal equilibration period after heating ceases (in the case of intermittent heating) and to reduce the change in the apparent total heat capacity due to the change with temperature of the thermal configuration inside the vessel proper (this is important in the case of continuous heating).

Since the configuration of the thermal gradients inside the calorimeter vessel depends strongly on the mode of heat transport through the liquid, it is convenient to consider the latter aspect first.

(1) *Heat Transport in the Liquid*

In a liquid-filled calorimeter vessel there are three main processes of heat transport, namely, conduction through the metal parts, conduction through the liquid, and convection; the convection may be either free, due to thermal expansion of the heated liquid, or forced, resulting from the action of a stirrer. Of these three processes, conduction through the liquid will almost always be the least effective, because the thermal conductivities of most liquids are much smaller than those of solid metals.

It is interesting to examine the hypothetical situation in which convection and metallic conduction are excluded. From the point of view of heat transport, the simplest situation is that in which the heating element is a resistance wire wound on to the outside of the calorimeter vessel proper—this arrangement has in fact been used recently by Staveley, Tupman and Hart[125]. Consider a metal calorimeter vessel about 4 cm in diameter and about 8 cm long, filled with water and provided with this type of external heating element. The total heat capacity of the filled vessel will be about 400 joule deg^{-1}. If the heating rate is set at 0·00025 deg sec^{-1}, the rate of heat transfer to the water will be about 10^{-3} watt cm^{-2}. Given that the thermal conductivity of water near room temperature is about 0·007 watt cm^{-1} deg^{-1}, the temperature difference between the vessel proper and the water at the axis of the vessel during steady heating will be about 0·15 deg, the temperature profile being parabolic. The time taken to attain temperature uniformity (within 10^{-4} deg, say) after heating ceases will be at least several hours. The temperature difference between the centers of the end-plates of the vessel proper and the circumference, on the other hand, will be only 0·0007 deg if the vessel proper is made of copper. If the vessel proper is made of glass 2 mm thick the temperature difference across the glass walls during steady heating at 0·00025 deg sec^{-1} will be only about 0·03 deg, but the temperature lag at the centers of the end-plates will be close to 0·15 deg.

If, in this example, we now allow free convection to occur, the net result is that the temperature profile is levelled out over a region in the center of the vessel, whereas the temperature gradient next to the outer wall is unaltered. The temperature boundary layer, which contains the effective temperature difference between the wall and the central core of liquid at more or less uniform temperature, is relatively wide, but the rising velocity in the liquid has its maximum value quite close to the wall. The detailed analysis of this situation is extremely difficult because the flow pattern is inevitably complicated. Some indications can nevertheless be got by using the approximations described in standard texts on Heat Transfer (see, for example, Rogers and Mayhew[113], Chapter 21). Estimation of the probable range of values of the Grashof Number (see also Chapter 4, Section III–3 above) in the present example indicates that the temperature difference across the boundary layer with free convection will be at least a few hundredths of a degree. The corresponding thermal equilibration time is unlikely to be less than 1 hour. It must be stressed at this point that the efficacy of free convection can very easily be much less than indicated by these figures. The problem of thermal lag-times in unstirred calorimeter vessels was investigated by Sturtevant[128], and for more detailed information the reader is referred to this paper.

The heat transfer will always be improved by an efficient stirring system, but it is important to note that the primary object of stirring is to provide forced convection relative to the hot surface (the walls of the vessel in the present discussion), rather than merely to agitate the liquid. Even so, heat transport theory indicates that to reduce the temperature across the boundary layer in this type of calorimeter vessel to less than 0·01 deg would require high liquid velocities close to the walls, certainly greater than 100 cm sec^{-1}. Such velocities would be very hard to achieve without at the same time setting up severe turbulence around the impeller. This consideration means that it is very difficult to improve much on the heat transfer given by ideally free convection without the heat of stirring being unacceptably large and probably neither constant nor reproducible. In practice a reasonable compromise is to use a cylindrical flow-guide about 0·5 cm inside the walls of the vessel proper, with a centrally sited, slowly revolving stirring impeller drawing the liquid downwards inside the flow-guide and forcing it upwards in the annular space between the flow-guide and the wall. This arrangement both stabilizes and enhances the convection without producing excessive heating. The heating effect of the stirrer will, however, be significant in high-precision work, and so it must be accounted for. Since it is quite difficult to evaluate, a special calibration procedure is necessary; but once the heating power of the stirrer is known, provided that it is sufficiently small, random variations in its value may be ignored.

A second way of reducing the temperature gradients inside a metal calorimeter vessel is to extend the internal surface in contact with the liquid, so that no part of the liquid is more than a few millimeters from a metal surface providing a conduction path to the heater (see Sturtevant[128]). This may be done in a calorimeter vessel of relatively large diameter by a set of metal vanes whose extremities are in good thermal contact with the vessel proper. This is similar to the technique often used with solid specimens (see Trowbridge and Westrum[139]) and applied to liquids at low temperatures by Ruehrwein and Huffman[115]. With stirred liquids the internal metal vanes must of course be so designed as not to impede the circulation. Alternatively, if the calorimeter vessel includes internal metal parts (thermometer units, etc.) it may suffice to make these parts to span the regions where the temperature gradients would otherwise be greatest. In fact these principles can to some extent be incorporated into the design of the heating element.

(2) *Heating Elements*

The figures on which the foregoing discussion is based all relate to a metal calorimeter vessel with external heater windings. This arrangement is wholly satisfactory only with metal vessels of capacity less than about 50 ml. Instead of discussing all the possible alternative arrangements, we have selected from the literature those which seem to us the most satisfactory from the point of view of heat dissemination, taking account also of practical considerations such as ease of manufacture and maintenance. This selection is inevitably subjective, but we think it will provide more useful guidance than would an exhaustive review.

HEAT CAPACITY OF LIQUIDS AND SOLUTIONS

(a) *Heating Elements for Metal Vessels.* The simplest design from the point of view of making the calorimeter vessel is undoubtedly that already discussed, namely the external heater in which the resistance wire is wound on to the outside of the vessel proper. The problem here is to combine good thermal contact with good electrical insulation. One basic principle is to use the greatest possible length of resistance wire, to minimize the heat transfer rate per unit length of wire. Although the more usual forms of electrical insulation, e.g. sheet mica, introduce large thermal lags and so are not really suitable for high-precision work, limiting the working temperature to about 120°C enables some simple but effective methods of insulation to be used. For example, Sturtevant[129] used a thin sheet of "Cellophane" instead of mica. The low insulation resistance, only 20 kΩ, introduced no significant error into the power measurement, whereas the heat transfer was much better than with mica because of the pliability of the material. If the vessel proper is made of duralumin, the external surface may be grooved and then anodically oxidized; the resistance wire may then be wound directly into the grooves without further insulation[99] (see Ackermann[2]). This arrangement gives very good thermal contact, but unless the range of working temperatures is quite small the thermal expansivity of the resistance wire has to be close to that of duralumin, otherwise some distortion of the wire will result from repeated heating and cooling, with consequent deterioration of the thermal contact. This effect may be partially counteracted by setting the resistance wire in place with a resilient adhesive; "Bakelite" or epoxide type resins are suitable. Another way of providing electrical insulation is simply to wind enamelled resistance wire on to a surface previously treated with "Bakelite" varnish. With all these methods, however, the effective radiative emissivity of the external surface of the vessel is higher than that of smooth metal, and it would seem always worthwhile to provide a close-fitting sheath of thin copper or silver; such a sheath also improves the temperature uniformity of the radiating surface. Staveley, Tupman and Hart[125] made the outer sheath of tinfoil.

For larger calorimeter vessels (capacity greater than 50 ml, say) internal heater units have important advantages. In particular, the effective heating area in contact with the liquid can be made relatively large and of such a shape that no part of the liquid is more than about 1 cm from the heating surface. In this way the heater assembly itself can be used to assist in the dissemination of heat through the liquid. Further, since the resistance wire of an internal heater need not continue to the exterior of the vessel proper, relatively large temperature differences between the resistance wire and the heater casing may be tolerated, especially if the total heat capacity of the heating element is small. This permits more robust electrical insulation to be used. The principle of using the longest possible resistance wire applies here also, but it has to be balanced against the need to keep the heat capacity of the wire and its former small, to minimize internal thermal lags which influence the thermal equilibration time of the whole calorimeter vessel.

For the investigation of liquids which are chemically unreactive (e.g. hydrocarbons) a satisfactory internal heater may be made by winding the resistance wire on to the outer surface of a thin-walled cylinder of anodized duralumin whose diameter is about two-thirds of that of the vessel proper.

The relative flexibility of the duralumin former permits the resistance wire to be tightly wound at the highest working temperature (140°C, say) without its being too highly stressed at lower temperatures, and so the problem of the slackening of the resistance wire is avoided. The complete unit has a low heat capacity while giving a large effective heating area, and it forms an efficient flow-guide for an axial-flow stirrer, ensuring that the greatest liquid velocities occur next to the heating surfaces. The main problem with this design is how to make the electrical connections between the resistance element and the exterior of the calorimeter vessel. Two reasonably satisfactory ways of doing this are suggested in *Figure 9*, which is self-explanatory. It should be noted, however, that the glass/metal tubular seals have to be joined to the vessel proper with soft solder, see Chapter 4, Section II-2.

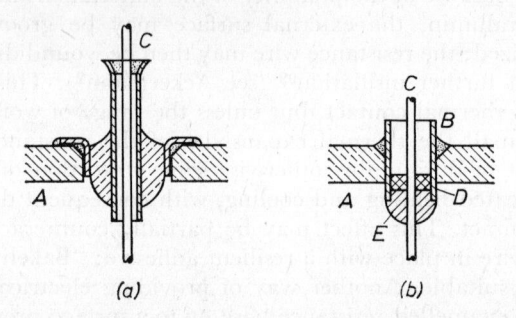

Figure 9. Vacuum-tight conductor seals: (*a*) metal-glass-metal tubular seal, soft-soldered to the vessel and the conductor; (*b*) fabricated seal—the tube is welded or brazed to the vessel, the conductor is positioned centrally by the plastic washer *D*, and the seal is completed by thermosetting epoxide resin *E*

An alternative design for an internal heater which avoids the difficulties of making the external electrical connections, and has the additional advantage of easy replacement of the resistance elements, has been described by Ackermann[5]. The main structure of the heater unit may be seen in *Figure 10*. Four thin-walled metal casings are joined to the cylindrical shell *T* by soft solder. These casings are open to the external surface of the vessel proper; they contain the heating elements, which are easily removable from the casings for replacement. The shell *T* serves to distribute the heat to the liquid and also as a flow-guide for the axial-flow stirrer which is actuated magnetically. The heating elements comprise constantan filaments, insulated by spun glass sleeving, wound on to silver rods. The resistance wire terminates well below the tops of the silver rods, so that the relatively large temperature difference which builds up between the resistance wire and the rod and casing does not result in any uncompensated heat loss to the exterior, while the low heat capacity of the resistance wire ensures that it comes rapidly to thermal equilibrium with the casing and the silver rod once heating ceases. Probably the most important design considerations with this scheme are to ensure adequate tempering of the low-resistance heater leads to the silver rods, and good thermal contact between these rods and the heater casings.

A third type of internal heater, suitable for use with mildly corrosive liquids, is made from a compound resistance element consisting of an inner resistance wire surrounded by electrical insulation, the whole being contained in a tubular outer sheath of metal. This type of heater appears to have been developed as long ago as 1925, when it was described by Richards and Gucker[110]. Their resistance element was constantan, insulated with asbestos, in long brass tubes of 1·6 mm outside diameter. A modern commercially available version uses compressed magnesia as the electrical insulation, giving relatively good thermal conductivity, while the outer sheath is of stainless steel or Inconel, having various outside diameters between 1·0 and 2·0 mm.† This element may conveniently be wound into a close coil on a diameter about two-thirds of that of the vessel proper, and used as wound; alternatively the outer sheath may be soft-soldered to a cylinder of thin stainless steel or copper. Hard solder may also be used for this, but some care is required. The ends of the outer sheath of the resistance element may either be soldered directly through the end walls of the calorimeter vessel and the electrical connections to the resistance wire made externally, or the ends of the outer sheath may be soldered into thin tubes of brass, bronze or stainless steel after insulated copper leads have been welded to the ends of the resistance wire. The second arrangement eliminates hot spots on the calorimeter vessel exterior, but manufacture is more difficult.

(*b*) *Heating Elements for Glass Vessels*. The use of a glass calorimeter vessel for high-precision work presupposes that corrosive liquids are being studied, so that only completely protected heating elements need be considered. External heaters present serious problems because of the low thermal conductivity of glass, and the internal heater in an annular glass well is undoubtedly the most satisfactory solution to the problem. The most recent form of this type of heater is that described by Ackermann[2]. This is in fact a simple adaptation of one of the designs described above. Constantan wire is wound on to the grooved outer surface of an anodized aluminum tube, which is designed to fit inside an annular glass casing whose median diameter is about two-thirds of that of the vessel proper. External connections are made by leads passing down the glass tubes which connect the annular casing to the exterior of the top of the calorimeter vessel (see *Figure 10a*). Thus the heating element is completely protected from the solution, while taking the leads out through the glass supports ensures adequate tempering despite the relatively large thermal resistance between the heating element and the glass of the heater casing. This resistance is minimized by the heat-spreading effect of the aluminum former, and the large area of heated glass gives reasonable thermal contact with the liquid.

E. Accommodating the Thermal Expansion of the Liquid Specimen

(1) *The Vapor Space Method*

In this method the calorimeter vessel is incompletely filled by the liquid specimen, so that there remains a vapor space of a few milliliters. In early work this vapor space usually contained a mixture of air and the vapor of the

† Made by Société Anonyme d'Etudes et Réalisations Nucléaires (SODERN), 10, Rue de la Passerelle, Suresnes (Seine), France, under the trade name "Thermocoax".

liquid, but in recent high-precision work air has been excluded by using filling techniques like those described in Section II–2–B above. In this case the liquid specimen is sealed into the calorimeter vessel under its own saturation vapor pressure, and the composition of the vapor depends only on the composition of the liquid and, of course, on the temperature. With a single-component system, as the temperature is increased the volume of the vapor space is diminished by the thermal expansion of the liquid phase. The saturation vapor pressure also increases with increasing temperature and, depending on the relative rates of these two processes, liquid will either evaporate or condense. Then the total energy input per unit temperature increase is given by

$$\frac{dQ}{dT} = n^l C_S^l + n^v C_S^v + \Delta H v \frac{dn^v}{dT} + C_{cal}, \qquad (30)$$

in which n^l and n^v are the numbers of moles in the liquid and vapor phases respectively, C_S indicates molar heat capacity at saturation vapor pressure and C_{cal} is the heat capacity of the calorimeter vessel by itself. The first term on the right-hand side of Equation (30) is then the heat capacity of the liquid phase under the experimental conditions. The second term gives the heat capacity of the vapor phase. The third term, giving the heat of phase equilibration per degree temperature rise, is usually much larger than the second, but it may easily become negative near the upper end of the working range of temperature. Both n^l and n^v are complicated functions of temperature, depending on the temperature derivative of the saturation vapor pressure and the thermal expansivity of the liquid, as well as on the physical dimensions of the calorimeter vessel and the mass of specimen, so that exact evaluation of Equation (30) would be extremely tedious. It might appear that the evaluation of C_S^l could be simplified by choosing the initial volume of vapor space such that its diminution due to the thermal expansion of the liquid increased the pressure of the vapor by an amount just sufficient to maintain liquid–vapor equilibrium without change in n^l and n^v, i.e. without evaporation or condensation. In fact this could be achieved only over a small temperature increment since the vapor space volume decreases nearly linearly with increasing temperature whereas the saturation vapor pressure increases proportionately to $\exp(-1/T)$.

In practice it is convenient to charge the calorimeter vessel with that mass of specimen liquid which will fill it completely at a temperature 5 or 10 deg above the upper limit of the measurements. In this case the third term on the right-hand side of Equation (30) is positive at the lower end of the temperature range, increases to a maximum in the upper part of the temperature range, and then decreases rapidly, becoming negative at the highest working temperatures. If the thermal expansivity of the liquid is small, adopting this procedure ensures that n^v is always much smaller than n^l, and negligible error in C_S^l results from taking the sum of the liquid and vapor heat capacities to be adequately approximated by $C_S^l(n^l + n^v)$, the quantity $(n^l + n^v)$ being calculated from the total mass of the charge. Although the third term has to be taken into account, at least in the temperature range where its maximum values occur, a rough approximation suffices. Water is a familiar example of a low-expansivity liquid, and this argument is well illustrated by the figures in *Table 2*, which refer to a calorimeter vessel whose total volume is 206 ml, charged with 200 g of water. The total heat capacity will be around 900 joule deg^{-1}, and if we assume a precision of about 0·08 joule deg^{-1}, heat effects equivalent to less than 0·03 joule per unit temperature rise may evidently be neglected. The vapor volumes in *Table 2* have been calculated approximately by equating n^l to the total amount of water; the consequent errors in the quoted values of n^v are of the order of 10^{-8} mole. The figures in the right-hand column represent the average (per 1 deg temperature rise) of the heat of phase equilibration over the preceding 20 deg.

Table 2. Vapor space data for a 206 ml vessel, charged with 200 g water (11·1012 moles)

Temp. (°C)	Vapor vol. (ml)	Moles vapor ($n^v \times 10^6$)	Vapor ht. capy. (J deg^{-1} × 10^{-4})	Amount evapd. ($\Delta n^v \times 10^6$)	$\Delta H v \dfrac{\Delta n^v}{\Delta T}$ (J deg^{-1})
20	5·64	5·4	2·0	—	—
40	4·42	12·4	4·5	+ 7·0	+0·015
60	2·58	18·5	6·7	+ 6·1	+0·013
80	0·20	3·2	1·2	−15·3	−0·032

Table 2 indicates that the greatest error consequent upon ignoring the difference between the specific heats of the liquid and vapor phases is less than $(C_S{}^l - C_S{}^v)/50000$, i.e. less than 10^{-3} joule deg^{-1}. The evaporation correction is also negligible below about 70°C, but should probably be taken into account above this temperature. In the case of comparison measurements (twin calorimeter), even with moderately concentrated electrolyte solutions, the *difference* between the heat of evaporation terms for the two calorimeter vessels is much smaller than the required precision, so that no correction need be made. This is especially important in method 3c of Section II–1–B, in which the vapor space method must be used in order to have the same mass of water in the specimen vessel as in the reference vessel. In this case the vapor space is relatively rather larger than indicated in *Table 2*.

For comparison, *Table 3* gives the corresponding data for a calorimeter vessel whose total volume is 246 ml, charged with 200 g of benzene. Benzene is chosen because of its high thermal expansivity.

Table 3. Vapor space data for a 248 ml vessel, charged with 200 g benzene (2·5607 moles)

Temp. (°C)	Vapor vol. (ml)	Moles vapor ($n^v \times 10^6$)	Vapor ht. capy. (J deg^{-1} × 10^{-4})	Amount evapd. ($\Delta n^v \times 10^6$)	$\Delta H v \dfrac{\Delta n^v}{\Delta T}$ (J deg^{-1})
20	20·42	79	77	—	—
40	14·67	137	134	+58	+0·093
60	8·72	162	159	+25	+0·040
80	2·32	80	78	−81	−0·128

In this case, if the desired precision is assumed to be about 0·03 joule deg^{-1} on the total heat capacity (about 420 joule deg^{-1}), errors exceeding 0·01 joule deg^{-1} should be taken into account. With 0·000162 mole of benzene in the vapor phase the error in the heat capacity of the benzene consequent upon ignoring the difference between the specific heats of the liquid and vapor phases will be close to 0·01 joule deg^{-1}. The heat of evaporation has obviously to be estimated fairly accurately, but the number of moles in the vapor phase may safely be estimated as in *Table 3*, the vapor volume being calculated from that occupied by the total sample as liquid.

The quantity obtained from measurements with vapor phase present is the specific heat of the liquid phase at saturation vapor pressure, c_S. This quantity differs from the specific heat under standard conditions (1 atm applied pressure) by two terms, one depending on the vapor pressure and the other on its temperature derivative. The first term may be calculated from the relation

$$\left(\frac{\partial c_P}{\partial P}\right)_T = -Tv\left[\alpha^2 + \left(\frac{\partial \alpha}{\partial T}\right)_P\right] \qquad (31)$$

in which v is the specific volume and α is the thermal expansivity defined by Equation (6). Negligible error results from putting v equal to its value under 1 atm. The correction indicated by Equation (31) is negligible only in the neighborhood of the normal boiling point. Thus for water at 50°C, while α^2 is very small, $(\partial \alpha/\partial T)$ approaches 10^{-5} deg^{-2}, and in fact $(\partial c_P/\partial P)_T$ is $-0·0020$ joule deg^{-1} g^{-1} atm^{-1}. Thus the specific heat at 1 atm is less than the *isobaric* specific heat at saturation vapor pressure, $c_P(0·1214$ atm$)$, by 0·0017 joule deg^{-1} g^{-1}, or 0·05 per cent. For benzene at 50°C, $(\partial c_P/\partial P)_T$ is $-0·0023$ joule deg^{-1} g^{-1} atm^{-1}, and $c_P(1$ atm$)$ is 0·0015 joule deg^{-1} g^{-1} less than $c_P(v.p.)$, or nearly 0·1 per cent. The second correction term arises because every increment of temperature increases the vapor pressure. The difference between c_S and $c_P(v.p.)$ is given by [77]

$$c_P = c_S + Tv\alpha(dP/dT)_S \qquad (32)$$

This correction increases continuously with temperature; for water at 100°C it is $+\,0·0011$ joule deg^{-1} g^{-1}.

(2) *The Variable Volume Method*

The basis of this method is to allow the volume available to the liquid specimen to vary with pressure, so that a relatively large volume change is accommodated with only a small increase in the applied pressure. The applied pressure must of course always exceed the vapor pressure of the liquid. The specific heat at 1 atm pressure is then calculated from the measured specific heat by means of Equation (31). With this method it is necessary to know the volume available to the liquid specimen accurately as a function of applied pressure, but provided that the applied pressure does not exceed 2 atm the compressibility of the liquid may be ignored.

Two different ways of making a variable-volume calorimeter vessel have been used in recent years. The simplest scheme is to replace the ends of the calorimeter vessel by flexible diaphragms. This scheme was used by Brown and Fock[16] to accommodate the volume of mixing in a calorimeter vessel for heats of mixing, but it is equally applicable to heat capacity work. Brown and Fock made the ends of their calorimeter vessel of beryllium–copper alloy sheet, 0·003 in. thick, with annular corrugations similar to those of an aneroid barometer capsule. As the internal pressure increases, the ends of the calorimeter vessel bulge outwards. Todd[137] has used a similar arrangement. The disadvantages of these schemes are that only small volume changes can be accommodated (about 0·5 ml per end on a 2·5 cm diameter vessel), whereas the pressure increase to effect this expansion is relatively large and not always reproducible.

The second method is to have a highly compressible metal capsule inside the calorimeter vessel. A suitable capsule is shown diagrammatically in *Figure 11*. It is made from a flexible metal bellows, one end of which is sealed by a re-entrant well. This arrangement reduces the internal volume of the capsule and also serves as a safety stop. A brass or phosphor bronze bellows of the dimensions indicated in *Figure 11* ($\frac{15}{16}$ in. o.d., 13 convolutions per in., 0·004 in. walls)† can reproducibly accommodate volume changes up to 3 ml. The capsules used by Cruickshank and Timimi[38] have a flexibility, after rhodium-plating, of 3·25 ml atm^{-1}. The thermal expansion of 80 g of water between 20°C and 80°C is 2·2 ml, so that the pressure increase over this temperature range with a capsule like that just described is about 0·68 atm and, correspondingly, the measured specific heat at 80°C is smaller than that under 1 atm by about 0·00017 joule deg^{-1} g^{-1} according to Equation (24). It is apparent that in this case this correction may be ignored *below* about 60°C, whereas with the vapor space method the correction to standard pressure can be ignored only at temperaures *above* about 70°C.

The incorporation of this type of capsule inside the calorimeter vessel makes it difficult to arrange mechanical stirring; but provided that the heating element surrounds the capsule over most of its length, and the annular space between the two is about as wide as that between the heating element and the outer wall of the vessel, thermal conduction through the fabric of the capsule will maintain an adequate rate of heat dissemination without stirring. The main limitation of the capsule method becomes apparent with liquids of high thermal expansivity, for which the volume of the capsule has to be a significant fraction of that of the liquid specimen. Thus 70 g of benzene expands from 79·6 ml at 20°C to 86·0 ml at 80°C, so that in this case a capsule of Timimi's design would have to have 28 convolutions, making its overall length about 2$\frac{1}{2}$ in. A capsule of this size could be fitted into a vessel of 95 ml total volume only by allowing the thermometer (see *Figure 11*) to project into the re-entrant well in the capsule, and this would certainly increase the effect on the thermometer of local temperature inhomogeneities. It would be better therefore to use a smaller capsule and to limit the working temperature range to 30 deg or so, even though this involves adjusting the amount of liquid specimen at about 50°C.

More elaborate realizations of the variable volume method including, for example, calorimeter vessels constructed on the piston-and-cylinder principle, all require fairly elaborate pressure-control equipment, but since these are

† Made by Drayton Controls Ltd., Drayton, Middlesex, England, under the trade name "Hydroflex".

generally necessary only for high-temperature work, they will not be discussed further here. Gucker's[72] high-pressure calorimeter, designed for measurements on aqueous solutions up to the critical temperature, exemplifies one approach to these problems.

F. Measuring the Temperature of the Calorimeter Vessel and Contents

(1) *Choice of Method*

In high-precision adiabatic calorimetry the two most important requirements for thermometric elements are: rapid thermal response, so that the element may be assumed always to indicate the same temperature (within 10^{-4} deg, say) as its immediate environment; and also that the element and its external connections do not significantly increase the rate of heat transfer between the calorimeter vessel and the adiabatic shield. For measuring accurately the mean temperature of the calorimeter vessel and its contents, only thermocouples and resistance thermometers have been widely used.

Although a multi-junction thermocouple is easier to make and to use than a resistance thermometer, it usually contributes more to the heat transfer coefficient because of the larger number of wires passing between the calorimeter vessel and the exterior of the main apparatus. In addition, the provision of accurate thermostatting for the reference junctions often presents practical difficulties. In recent years, only Staveley, Tupman and Hart[125] have used thermocouples to measure the absolute temperature in a single-vessel calorimeter for heat capacity work, and they claim an accuracy only of 0·002 deg. With a properly constructed four-lead platinum resistance thermometer, on the other hand, the accuracy should be better than 10^{-4} deg. The main disadvantage of the platinum thermometer is that its thermal response time is usually much longer than that for a thermocouple. On balance, it seems that when the temperature increment of the calorimeter vessel, or its heating rate, is a primary measurement, as in the direct method (single-vessel calorimeter), a resistance thermometer is to be preferred to a multi-junction thermocouple. This is true also of the comparison method when the heating rate has to be accurately controlled, as in method 3–*c* of Section II–1–B above.

For adiabatic control and for measuring accurately the temperature difference between twin calorimeter vessels, on the other hand, thermocouples are especially suitable because of the proximity of the surfaces whose temperatures have to be compared. In a twin-vessel calorimeter the condition that the external surfaces of the twin vessels are at the same temperature (so that the heat leak is the same for both vessels) should ensure, if the two vessels are properly matched, that the mean temperatures also are equal, the internal temperature gradients being the same in both vessels. Thus a multi-junction thermocouple between the outer surfaces of the twin vessels suffices to compare their mean temperatures, and this method is probably the most accurate. Externally wound resistance thermometers have, however, been used in twin-vessel calorimeters to compare the temperature of the two vessels, notably in the reaction calorimeter of Buzzell and Sturtevant[20]. They use a third resistance thermometer for adiabatic shield control, the resistance of this third thermometer being compared with the average resistance of the other two by a double bridge circuit. This scheme has the advantages (i) that since resistances are compared a low-frequency a.c. bridge may be used with a high-impedance a.c. amplifier and phase-sensitive d.c. converter whose output actuates the regulator servomechanism; the low noise-to-signal ratio attainable with a.c. amplifiers gives better temperature discrimination than is attainable with d.c. circuits, and (ii) either calorimeter vessel thermometer may be used also to measure the absolute temperature increase or heating rate. The main disadvantage is that the metering equipment is rather more elaborate than that required with thermocouple detectors.

A third method of measuring the mean temperature of a calorimeter vessel and its contents is to use a thermistor thermometer. These thermometers have two important advantages for work near room temperature. First, because their resistance is relatively high throughout this temperature range, two-lead systems can be used satisfactorily, and second, because of their compactness and the ease with which they can be electrically insulated, thermistors generally show a faster response (comparable with thermocouples) than resistance thermometers. An important limitation arises from the fact that although thermistors have large negative temperature coefficients of resistance, they are subject to secular changes of resistance unless the current is kept much less than the value corresponding to the recommended maximum power dissipation. Skinner, Sturtevant and Sunner[120] recommend that the current through a 2 kΩ thermistor rated at 10 mW should not exceed 0·1 mA. With such small currents the temperature discrimination obtainable with a bridge and galvanometer is inferior to that obtainable with a 50 Ω platinum thermometer and a Smith or Mueller bridge reading to 10^{-5} ohm. In fact, to obtain discrimination to better than 10^{-4} deg with a thermistor the bridge galvometer must be replaced by a low-noise linear amplifier. Recently, however, Argue, Mercer and Cobble[6], by using high-resistance thermistors (10^5 Ω nominal), have achieved a temperature discrimination close to 10^{-6} deg with a thermistor current of only 5 μA. In this case the power dissipated in the thermistor is only 2·5 μW, or nearly 100 times less than that in a 50 Ω platinum thermometer with a measuring current of 2 mA. At this power rating recently developed types of thermistor are nearly as stable as platinum thermometers, especially when a low-frequency a.c. bridge is used. There remains the difficulty that the functional dependence of the thermistor resistance upon temperature is far from linear, and although Beakley[10] and Cole[33] have described modified Wheatstone bridges for thermistor thermometers which give roughly linear calibrations over 30 deg ranges [see also Chapter 2, Section III–2–A–(2)–(a)] thermistor thermometers are not really suitable for automatic temperature-program control. Thus it seems that the most appropriate application for high-resistance, high-stability thermistor thermometers is likely to be the measurement of the temperature increments in an intermittently heated single-vessel calorimeter, and in this field their use might well lead to a significant gain in experimental precision.

(2) *Resistance Thermometers*

The relative advantages of various metals which might be used for the resistance element are discussed in Chapter 2. In fact only platinum has been widely used for high-precision measurements on liquids, although Brown and Fock[16] use an externally wound nickel resistance thermometer on their heat-of-mixing calorimeter vessel. Since the general principles of constructing platinum resistance thermometers are also discussed in Chapter 2, we confine ourselves here to the siting of the resistance thermometer in the calorimeter vessel, and the methods of measuring the resistance.

(*a*) *Siting the Resistance Thermometer.* There are, broadly speaking, two alternatives: either the resistance element is wound circumferentially on to the outside of the calorimeter vessel and protected by an outer sheath, or a self-contained resistance thermometer is sited within the region of the liquid specimen. A circumferentially wound element will generally have the faster response, but to eliminate mechanical strain and to keep the recommended slightly oxidizing atmosphere it is necessary to mount the element on a special former in a separate annular compartment which can be gas filled and sealed. This method is especially suited to calorimeter vessels of less than about 5 ml capacity, and a very elegant design has been described by Westrum, Hatcher and Osborne[142]. This design has an additional advantage that the temperature gradients in the outermost wall are independent of whether the vessel is full or empty, cf. Chapter 4, Section IV-3. From the point of view of manufacture, however, there is the disadvantage that the thermometer leads have to be brought to the outside through special seals, to avoid spurious e.m.f.

The siting of internal resistance thermometers in liquid-filled calorimeter vessels depends on consideration of the effects of small temperature gradients on the determination of the true mean temperature change during the heating period. The temperature gradients during heating are discussed in general terms in Section II-2-D above. It is immediately obvious that the thermometer sheath should have the largest possible area in contact with the liquid. Further, in view of the relatively slow thermal response of the thermometer unit, it should be sited in a region whose temperature is close to the mean of the vessel and the liquid. In an unstirred vessel heated from the outside, this region is in fact at about two-thirds of the way out from the center towards the wall. In a vessel with an internal annular heater and a stirrer, the optimum position for the thermometer would seem to be at about two-thirds of the radius of the heater annulus, but not so close to the latter that the two boundary layers overlap. It is often found in practice, however, that slight overheating of the thermometer occurs wherever it is placed, presumably due to radiation from the heater, and more recently described calorimeter vessels of this type have the resistance thermometer at the axis.

(*b*) *Mounting Internal Resistance Thermometers.* These divide into two groups: (*i*) small metal or glass encapsulated types (of which a variety are now commercially available) which may conveniently be mounted in re-entrant wells extending far into the liquid specimen and open to the exterior of the vessel at their outer ends, and (*ii*) integral thermometers of which the metal sheath or capsule is part of the calorimeter vessel itself, as in *Figure 11*. The first method is exemplified by the calorimeter of Giguère, Liu, Dugdale and Morrison[61]. In this apparatus the thermometer capsule is cast with Wood's metal into a brass cylinder, tapered to fit closely into the tapered well of the calorimeter vessel. Others have cast the thermometer capsule directly into the inner region of the thermometer well. This arrangement enables the thermometer leads to be tempered to the inside of the well[139, 146]. Alternatively, the end of the capsule may project from the open end of the well, being protected by a combined radiation shield and tempering ring in good thermal contact with the vessel proper[61, 141]. The main disadvantage of this method is the relatively large mass of the complete thermometer unit (in the calorimeter of Giguère *et al.* the brass cylinder containing the thermometer capsule is fitted into the well after filling). There remains the difficulty that the inevitably large heat capacity of the thermometer unit, together with the high thermal impedances of the various interfaces, results in relatively long response times.

These disadvantages may be minimized in the second method, but in this case the resistance element has to be made up and properly annealed, and the thermometer well has to be gas-filled and sealed. The well has to be made as part of a detachable lid, so that the thermometer can be calibrated [cf. Section II–2–A–(2)]. Also, the thermometer leads have to be tempered externally.

(c) Measuring the Resistance. The various d.c. bridge and potentiometer circuits suitable for resistance thermometery are discussed in detail in Chapter 2. When measuring temperatures not far removed from that of the resistance measuring instrument, the stray thermal e.m.f. in the leads usually remains quite constant over periods of hours, and so its effects are eliminated by lead reversal. Consequently bridge methods can safely be used under conditions of maximum sensitivity. With continuously heated calorimeters it is obviously advantageous to be able to make measurements quickly, and a Smith's Type III bridge offers advantages here, provided that the cooling of the thermometer element during current reversal switching can be made small and reproducible, e.g. by the use of snap switches. In intermittently heated calorimeters this problem can be eliminated by passing the bridge current for only a fraction of a second at a time to test each balance position. This procedure virtually eliminates drift due to heating of the calorimeter vessel by the thermometer, but it requires a very fast-response bridge detector. Several suitable versions of Preston's[107] taut-suspension liquid-damped galvanometer with photocell amplifier are commercially available. By using one of these galvanometers with a Smith's Type II bridge (broadly similar to the Mueller bridge with reversing commutator) reading to 10^{-5} Ω it is possible to discriminate to better than 5 $\mu\Omega$. To realize this degree of precision, however, careful maintenance of the bridge is essential, and the bridge coils and leaf switches must be thermostatted to within ± 0.01 deg.

The bridge used by Cruickshank and Timimi[38] is thermostatted in oil at 22·00°C; the leaf switches are cleaned and the mercury contacts are renewed every six months. A complete self-calibration of the bridge is carried out, and the steam point and triple point resistances of the calorimeter vessel thermometers are re-determined after each bridge calibration.

When a continuous thermometer current can be tolerated, as with twin calorimeters or continuously heated single vessel calorimeters, the use of a self-balancing low-frequency a.c. bridge has important advantages[58, 76].

G. Details of Representative Calorimeter Vessels

Space does not permit an extensive review of the many calorimeter vessels reported in the literature, so the reader is referred to the bibliography. It will be useful, however, to describe briefly the three calorimeter vessels shown in *Figures 10* and *11*. *Figure 10(a)* shows the construction of a vessel

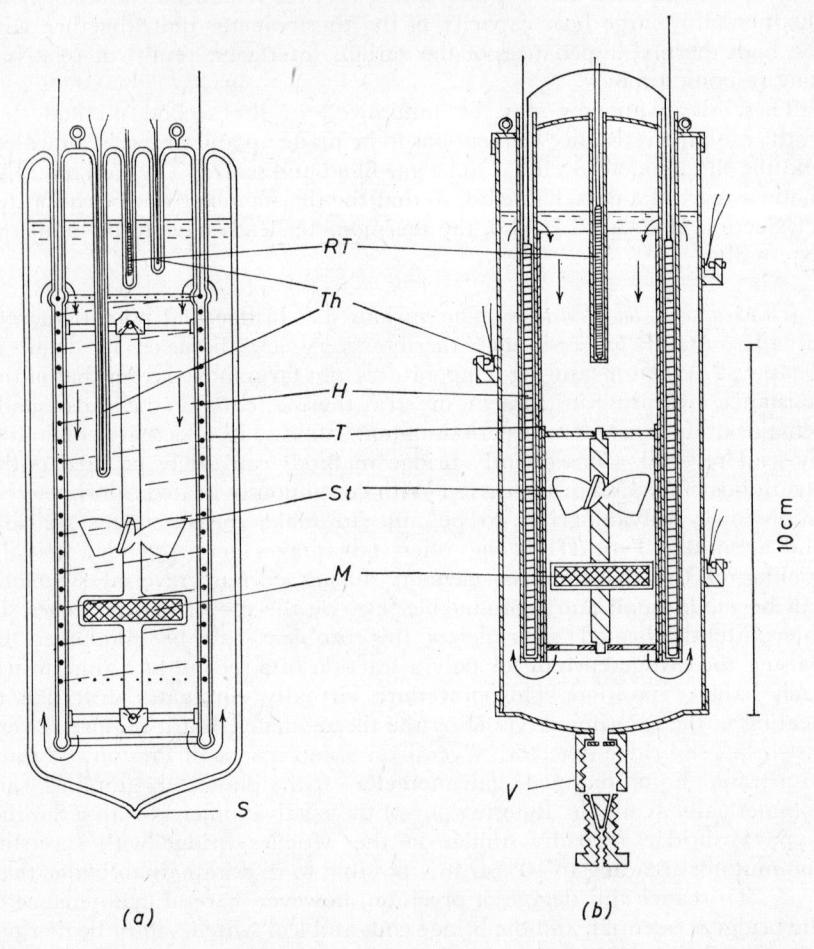

Figure 10. Vertical cross-sections of 300 ml calorimeter vessels[2]: (*a*) Pyrex vessel; (*b*) silver vessel. The heating elements *H* are isolated from the liquid by the casings *T*; the stirrers *St* are actuated by the glass sheathed magnets *M*. *Th*, thermocouple junctions; *RT*, glass encapsulated platinum thermometers. The seals *S* and *V* are discussed in Section II–2–C above.

made of hard borosilicate glass and *Figure 10(b)* illustrates a similar vessel made of silver. Both vessels have an internal capacity about 300 ml, suitable for liquid samples of volume between 275 and about 290 ml. The heating elements H are encased in the tubular casing T, surrounding the magnetic stirrer system. The stirring magnets M are hermetically sealed in glass or silver respectively and fastened to the central axle of the propeller St, so that the assembly of heater and stirrer forms an electrically heated liquid-circulation pump. The small arrows indicate the direction of circulation. Platinum resistance thermometers RT (glass encapsulated, 3 cm long by 3 mm wide, 100 Ω) are inserted in central wells in the upper parts of both vessels. The thermocouple junctions Th are screw-clamped on to the outer surface of the silver vessel; in the glass vessel the junctions are inserted in wells open to the exterior and in thermal contact with the liquid. The glass vessel has provision for an ampoule-type seal at its lower end, whereas the silver vessel is fitted with a valve V which seals against a polytetrafluoroethylene gasket. The heat capacity of the empty glass vessel is 170 joule deg^{-1}, and the total heat capacity when it is charged with 175 g of water is 1325 joule deg^{-1}. The corresponding figures for the silver vessel are 150 joule deg^{-1} and 1305 joule deg^{-1}. Thus the heat capacity of the empty vessel is in each case less than 15 per cent of the total when it is filled with water or dilute aqueous solution. The glass vessel was originally designed for use with a single-vessel calorimeter, but the silver vessel is suitable for either single- or twin-vessel calorimeters. In the former case heating elements of 5 Ω each are satisfactory, but for use with twin-vessel calorimeters high resistance values (1000 Ω) are to be preferred. The optimum magnitude of the heating power depends on the heat capacity of the liquid specimen as well as on the problem being studied. The reader is therefore referred to Section II–5 below.

Figure 11 shows a smaller calorimeter vessel designed for direct method measurements on hydrocarbon mixtures and adapted for dilute aqueous solutions of aminoalcohols and their hydrochlorides. The body A of the vessel proper [*Figure 11(a)*] was turned from a single billet of pure copper, and is rhodium-plated both inside and outside—it would in fact be better to gold-plate the outside, see Section II–4–B below. The lid B is made demountable to facilitate cleaning and occasional reorganization of the internal layout or repair to the internal plating. It is sealed to the body A by the sealing ring C, of silicone rubber or Teflon, which is compressed by tightening the twelve *10 BA* screws into the tapped holes D. The lid is so designed that when the screws are fully tightened the sealing ring is totally enclosed by metal, to prevent creeping of the Teflon ring. The sheath of the heating element E (Thermocoax 1 NcI 10) is soft-soldered into the lid B as shown in *Figure 11(b)*. After rhodium plating, the soft solder is additionally protected by polyurethane varnish, cured at 100°C. The joints between the exposed ends of the resistance wire of the Thermocoax and the 0·273 mm (32 s.w.g.) enamelled copper current leads are made as close as possible (about 1 mm) to the ends of the Inconel sheath, and are sited just inside the outer surface of the lid B. These joints, and the exposed insulation of the Thermocoax, are protected by the Araldite cement shown F. The thermometer well G is fabricated from thin-walled copper tubing, silver-soldered to the lid B. The

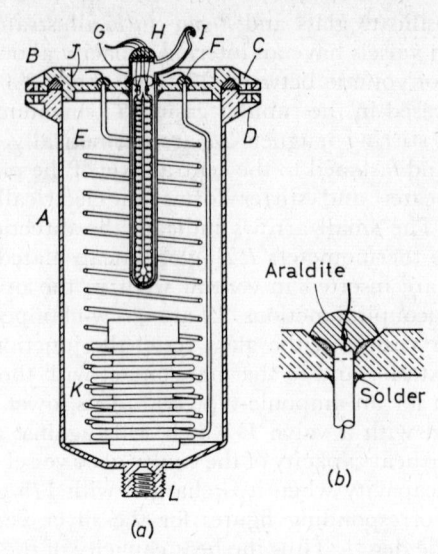

Figure 11. Copper calorimeter vessel (80 ml) with opening lid[38]: (*a*) vertical cross-section, *C*, sealing ring; *E*, heating element; *G*, platinum thermometer; *K*, expansion bellows; *L*, filling nipple; (*b*) detail of termination of heating element in lid.

platinum resistance element is made up according to Barber's[8] design, using Pyrex glass tubing of 2 mm bore and approximately 0·1 mm wall thickness; the triple point resistance is about 50 Ω. The four 0·122 mm (40 s.w.g.) enamelled copper leads pass through fine holes in the Tufnol (reinforced Bakelite-type resin) cap *H*, and join the internal platinum leads just below this cap, which is sealed to the well *G* by Araldite cement. After manufacture the thermometer well was pumped out to 10^{-6} torr at 100°C for several days and then filled with a mixture of helium and oxygen (15 per cent) at about 1·5 atm (as recommended in Chapter 2) through the copper side arm *I*, which was then pinched off and sealed with soft solder. The thermometer leads and the current leads from the heater are bound together with a few turns of cotton thread. The bundle is tempered by taking it nearly a full turn around the circumference of the recess *J*; it is held in good thermal contact with the lid metal by Araldite adhesive. For convenience the potential leads to the heater are joined to the current leads at the point where the lead bundle leaves the recess *J*. The free length of the lead bundle is about 4 cm (see Section II–4–C below) terminating in an 8-pin light weight miniature plug which fits into a 10-hole socket mounted in the entrance to the adiabatic shield vacuum line (the remaining 2 holes in the socket accommodate the adiabatic control thermocouple leads). The flexible capsule *K* is described in Section II–2–E above; the two parts of the capsule (bellows and well) are joined with hard solder. The capsule is located inside the heater coils by three lugs. The filling nipple *L* can be adapted to either of the sealing methods shown in *Figures 8(a)* and *8(b)*. Mounting a small magnetically operated stirrer on the inner end of the thermometer well slightly improves the rate of thermal equilibration, but its operation distorts the functioning of the adiabatic control and, of course, account has to be taken of the heat of stirring. Attaching the heater coils to a thin metal cylinder ought to improve the rate of thermal equilibration, but this has not so far been tried. With a heating power of rather less than 0·1 W (heating at about 0·0002 deg sec^{-1}) the time elapsing after heating ceases before the thermometer readings are uniform within 10^{-4} deg is about 40 or 50 min. The empty vessel weighs 209·2 g, and its heat capacity increases from 83·5 joule deg^{-1} at 20°C to 88·0 joule deg^{-1} at 60°C. When filled with 77·6 g of water at 20°C, 1 atm, it can be used up to 95°C, when the internal pressure is 1·9 atm. The total heat capacity is then 408·0 joule deg^{-1} at 20°C, so that the heat capacity of the vessel itself is less than 21 per cent of the total. The adiabatic control thermocouple junctions (not shown in *Figure 11*) are arranged to make touching contact with the outer surface of the calorimeter vessel; this point is discussed further in Section II–4–D below.

3. *Adiabatic shields and regulators*

A. Regulator Theory and Shield Control

In the room temperature range a liquid may be used to transport heat to and from the shield structure, and this is the principle of the "wet" shield calorimeter in which the shield is a vessel forming the outer boundary of the evacuated insulating space surrounding the calorimeter vessel; the shield vessel is immersed in a bath of liquid whose temperature is controlled by the shield regulator.

The alternative method—the "dry" shield, widely used at high temperatures—has also been used in work near room temperature. The dry shield may either be a separate structure inside the insulating jacket, or it may form part of the jacket vessel; in both cases it is heated by attached electrical windings in which the current is controlled by the shield regulator.

The choice between wet and dry shields for a particular application is governed by the shield-control requirements for the various adiabatic calorimeter regimes applicable to liquids and solutions, and the characteristics of the regulator systems and shield types which can be used.

(1) *Definitions*

The technical terms used in the following discussion are often used elsewhere with slightly different meanings, so that it is necessary first to define the most important of these.

  (*i*) The Datum Surface: The physical surface to which the controlling action of the regulator refers; it normally coincides with the outer surface of the calorimeter vessel.

  (*ii*) The Controlled Surface: The surface whose temperature it is the object of the regulator to control relative to that of the datum surface; it is normally the inner surface of the shield.

  (*iii*) The Regulator: We take this term to cover all the apparatus involved in setting the temperature of the controlled surface. All regulators applicable to adiabatic shield control comprise three main parts. These are (*iv*), (*v*) and (*vi*).

(*iv*) The Signal Source: This measures the temperature difference (the signal, denoted by ζ) between the controlled and datum surfaces.

(*v*) The Amplifier: This may be either an electronic circuit, an optical amplifier, or a combination of both.

(*vi*) The Servomechanism: This transmits energy to the shield in fixed relation to the signal.

(*vii*) The Regulator Response: Strictly, this is the instantaneous value of the power transmitted by the servomechanism.

(*viii*) The Conduction Link: All thermoregulators have the characteristic that the servomechanism necessarily acts at a finite distance from the controlled surface, so that the final link in the control sequence is always a thermal conduction process. This conduction link is thus a fourth part of the regulator. In adiabatic calorimetry it almost always suffices to describe the conduction link by a first-order differential equation, so that it may be characterized by a *response coefficient* $\psi(\sec^{-1})$; best regulator performance is obtained by maximizing ψ.

(*ix*) The Effectively Adiabatic Condition: The condition which ensures zero net heat leak during some finite period. When all heat transfer processes between the calorimeter vessel and the adiabatic shield depend only on the temperatures of the datum and controlled surfaces, i.e. when all conduction paths terminate at these surfaces and may be assumed always to be close to steady conduction with uniform temperature gradient (see Section II–4–A below) the effectively adiabatic condition corresponds to zero time integral of the signal over the relevant finite period.

(*x*) The Heat-Leak Coefficient: The theoretical rate of heat leak *into* the calorimeter vessel when the signal is $+1$ deg; it is measured in watt deg^{-1} and denoted by ξ.

(2) *Control Requirements*

These differ widely, especially between the comparison and direct methods. The simplest way to demonstrate how to estimate them is to discuss numerical examples.

(*a*) *Comparison Methods.* Great precision is unnecessary. In the first method of Section II–1–B, the proportional error in the temperature difference between the twin calorimeter vessels due to imperfectly adiabatic conditions will depend on the shield system used. If the reference vessel has its own shield and control system, independent of the specimen vessel, then provided that both control systems have the same characteristics, the proportional error due to heat leak on the temperature difference between the two vessels is similar to those on the total temperature changes and so is much smaller than the observational uncertainty. When only a single shield is used, controlled relative to the reference vessel, however, the error on the temperature difference is much larger than with the two-shield system. This point has in fact been overlooked in several designs published in the last ten years. In the second and third methods the heat leaks from both calorimeter vessels are, in principle, the same, so that there is no resultant error on the difference between the energies supplied to the twin calorimeter vessels.

In the case of the first method, assume that the temperatures of the two calorimeter vessels are raised in 1 deg steps with equilibration periods between each step, the uncertainty on each temperature measurement being 0·005 deg. If the duration of each experiment is 5000 sec and the heat-leak coefficient is 0·05 watt deg^{-1}, the total heat capacity of each vessel being about 800 joule deg^{-1}, then the rate of drift of the calorimeter vessel temperature with unit signal (1 deg) is $+62\cdot5$ microdeg sec^{-1}. The resultant error on the temperature increase should not exceed about 0·003 deg, so the average magnitude of the signal should not exceed about 0·01 deg.

When the temperature of the twin calorimeter vessels is raised continuously, as in the third method of Section II–1–B, heat leaks less than about 0·1 per cent of the power of the calorimeter vessel heaters may be ignored, since these produce errors only in $\Delta T/\Delta t$—see Equation (29); the required precision of adiabatic shield control then depends on the programmed heating rate, but it is unlikely to be better than $\pm 0\cdot005$ deg.

(*b*) *Direct Methods.* If the calorimeter vessel is heated continuously the heat leak (or its uncertainty if the heat leak itself is evaluated) must not exceed about 0·005 per cent of the power input to the calorimeter vessel when the heat capacity is required with a precision of 0·01 per cent. Assuming the heating rate of the calorimeter vessel is about 10^{-3} deg sec^{-1} (cf. Todd[137]) and its total heat capacity about 800 joule deg^{-1}, the uncertainty in the net power input has to be less than 40 microwatt, on average. If the heat-leak coefficient is taken to be 0·05 watt deg^{-1}, the average magnitude of the signal should not exceed 0·0008 deg. In the case where the signal is recorded and used to evaluate the heat leak, the uncertainty on the signal should be less than 0·0004 deg, allowing for integration errors.

For stepwise heating, with both the heating and equilibration periods assumed to be about 2000 sec and the heating steps assumed to be 1 deg, the initial and final temperatures of the calorimeter vessel have to be evaluated to within 50 microdeg. Taking the heat-leak coefficient again to be 0·05 watt deg^{-1} and the total heat capacity to be 800 joule deg^{-1}, gives the permissible average magnitude of the signal (or its uncertainty if evaluated) as 0·0002 deg.

Although these figures may differ by a factor as much as 5 from those in an actual apparatus, they show clearly that when direct measurements of total heat capacity are used to obtain partial molal heat capacities in dilute solution, the requirements for adiabatic shield control are much more stringent than the corresponding requirements in the case of high-temperature entropy evaluation, for example.

(3) *Regulator Systems*

It is convenient to discuss these n terms of the various practicable relations between signal and response.

(*a*) *The Two-Valued Response Regulator.* Either of two fixed values is open to the regulator response. This is the simplest of all adiabatic calorimeter regulators. It selects one or other of its two responses according to the sign of the signal; there is inevitably a threshold magnitude of the signal, called the backlash, below which the amplifier cannot detect that a change of sign has occurred. An approximate theory of the behavior of some practical versions of the simple two-valued response regulator has been developed by Cruickshank[37], assuming the conduction link to be adequately described by a first-order differential equation. It is concluded (*i*) that the effectively adiabatic condition can be maintained when the temperature of the datum surface remains constant only if the negative response removes heat from the shield at the same rate as the positive response adds heat to the shield; (*ii*) that for a particular apparatus the heat leak is uniquely related to the ratio of positive to negative half-cycle periods, provided only that the sum of the magnitudes of the positive and negative responses remains constant (the positive and negative half-cycle periods can of course be observed directly); and (*iii*) that to minimize the heat leak at a given heating rate the time-delay in the changeover from positive to negative response, the switching backlash and the response time $1/\psi$ of the conduction link must be kept as small as possible. When the datum temperature increases continuously, however, the condition for zero mean signal, that the positive and negative half-cycle periods are equal, is met when the shield heating rate during a positive half cycle is twice that at the datum surface and the net power input to the shield is zero during a negative half cycle. A modification of the two-valued response regulator without shield cooling which maintains this condition was developed by Gucker, Pickard and Planck[73]. This device does not, however, operate quickly enough to be adaptable to a high-precision, single-vessel calorimeter heated stepwise.

A typical two-valued response regulator uses a multi-junction thermocouple as the signal source and a d.c. galvanometer-photocell amplifier (with a gain about 10^8) operating an electronic or electro-magnetic relay which in turn actuates the positive and negative responses. If the response coefficient of the conduction link is greater than 0.1 sec^{-1} this arrangement is capable of keeping the magnitude of the mean signal (over a complete regulator cycle) always less than 0·005 deg, see Section II–4–E below, and this may be evaluated to better than 10 per cent, so that the time-average of the uncertainty on the mean signal should not exceed 0·0002 deg, which meets the most stringent requirements listed above. The performance of a particular two-valued response regulator may be improved by either of two modifications. The first of these uses the same principle as Gouy's[67, 121] modification of the mercury contact regulator. In the case of an electronic regulator the procedure is to superimpose upon the amplifier output an oscillating (saw-tooth) e.m.f. of fixed amplitude and period. The amplitude must exceed the amplification times the sum of the magnitudes of the switching backlash ϕ (deg) and the expected mean signal ζ (deg), and the period should not be less than twice the total time delay χ (sec), see Section II–4–E below. The effect of this modification is to give approximately linear response over the range of signal magnitudes corresponding to the amplitude of the oscillating e.m.f., with consequent improvement in flexibility and diminution in the magnitude of the mean signal corresponding to a particular heating rate of the calorimeter vessel. When a galvanometer-photocell amplifier is used, Gouy modulation may be achieved either by imparting a mechanical oscillation to the photocell mounting, or electronically, see Section II–3–D–(2). The second modification is the provision of an auxiliary signal source like that described by Dole and others[44]. This type of modification is often confused with Gouy modulation, but in fact it differs both in principle and in effect from the latter. According to Cruickshank[37] a well designed auxiliary signal source improves the performance of the regulator, but an incorrectly designed one increases the magnitude of the mean signal corresponding to a particular heating rate of the calorimeter vessel.

(b) *The Linear Response Regulator.* In this the magnitude of the response is proportional to the signal, but it has the opposite sign. It is the simplest of the regulators in which the response varies continuously rather than changing discontinuously from one to the other of two fixed values, but even so a relatively elaborate amplifier is required[41]. A rigorous theory, taking full account of the thermal transients in the conduction link, of the behavior of a thermostat regulator of this type, has been developed by Turner[140]. This treatment can be extended to shield control, but it is much easier to use Cruickshank's[37] approximate description of the conduction link (assuming quasi-steady conduction), which is quite adequate for thin shields. The qualitative conclusions of both treatments are as follows. While the effectively adiabatic condition is not maintained during an increase in the datum temperature, the total time integral of the signal (and hence the net heat leak) is directly proportional to the total increase in the datum temperature; the constant of proportionality is -1 times the reciprocal of the amplifier gain expressed in $\sec^{-1}$; this is simply the ratio of the first time derivative of the shield temperature to the signal under steady conditions. This proposition ignores the effects of random perturbations, but within the uncertainty consequent upon these, the net heat leak can be calculated very simply.

The use of a linear response regulator is subject to two important limitations. When the regulator gain is large enough to meet the shield-control requirements for a single-vessel calorimeter, the temperature at the controlled surface oscillates about that at the datum surface whenever the first time derivative of the latter alters; the amplitude of this oscillation decreases exponentially with increasing time. Thus, if this regulator is used with a discontinuously heated calorimeter, provision must be made whereby the servomechanism can *cool* the shield in response to a positive signal, just as with a two-valued response regulator. Also, when a linear response regulator is used with a single-vessel calorimeter, the shield must be so designed that the servomechanism temperature always remains constant when the response remains zero. Otherwise a residual heat loss from the shield has to be counterbalanced by the action of the regulator in response to a continuing negative signal. This behavior in turn gives rise to a residual cooling of the calorimeter vessel not included in the relation between total heat leak and total temperature rise, and difficult to detect. In this respect the linear response regulator is less satisfactory than the two-valued response regulator, in which a non-zero mean signal is immediately indicated by the ratio of positive to negative half-cycle periods departing from unity. The shield-control requirements indicate that the rate of heat removal from the shield must not change during a complete heating by more than the amount which corresponds to a drift in the shield temperature of 10^{-4} deg $\sec^{-1}$. The tolerance is about ten times larger with a twin-vessel calorimeter.

(c) *The Linear-Plus-Rate Response Regulator.* The incorporation of differential or rate response, i.e. the inclusion in the regulator response of a term proportional to the first time derivative of the signal can, in principle, give 'no overshoot' control with adequate gain, thereby obviating the need for shield cooling, but we do not know of such a regulator having been used. A different way of achieving a similar result is to make the response proportional to the square of the signal. This type of regulator has been used by Buzzell and Sturtevant[20] to control the temperature of both the adiabatic shield and the reference vessel in a twin-vessel reaction calorimeter.

Both of these regulators are, like the simple linear response regulator, subject to the requirement that the shield temperature remains constant at zero signal when they are used with single-vessel calorimeters.

(d) *The Regulator With Reset Response.* In this case the response includes a term proportional to the time integral of the signal. West and Ginnings[141] have shown that a linear term also is necessary for stable operation, and that rate response increases the flexibility. Irrespective of whether or not rate response is included, this regulator achieves the effectively adiabatic condition whenever the datum temperature is a linear function of time, or remains constant. The time integral of the signal over the initiation period of a heating is of course negative, but this is balanced by a positive time integral over the termination period. The relationships between the three terms in the response necessary to ensure stability have been derived by West and Ginnings[141]. In practice a regulator of this type requires controlled shield cooling only if the calorimeter vessel is heated in steps. The regulator incorporating reset response has the further important advantage that any drift of the shield temperature when the signal is zero is corrected by a small increment in the time integral of the signal, the signal itself rapidly returning to zero.

(4) *Discussion*

Of the possible combinations of the calorimetric methods and the regulator systems just described, the requirement of controlled shield cooling arises in six cases, namely the application of the two-valued response regulator, the simple linear response regulator and the reset response regulator to either of the methods in which the heating of the calorimeter vessel is carried out in steps. With either the linear or the linear-plus-rate response regulators applied to a discontinuously heated single-vessel calorimeter the shield temperature must not drift more than 10^{-4} deg sec^{-1} at zero signal.

With continuous heating of the calorimeter vessel or vessels, none of the four regulators requires controlled shield cooling, but in the case of a single-vessel calorimeter the same limitation on the shield drift as with discontinuous heating applies to both the linear and the linear-plus-rate response regulators.

B. Characteristics of Wet and Dry Shields

In choosing between the wet and dry types of shield for a particular application the most important considerations are: (*i*) whether or not the shield temperature has to be decreased in the course of normal operation of the regulator, (*ii*) whether or not it is necessary that the shield temperature always remain constant at zero signal, and (*iii*) the practical difficulties of applying the various regulators to each type of shield. It is convenient therefore to discuss the characteristics of wet and dry shields in relation to these three considerations.

(1) *Dry Shields*

The simplest type of dry shield is a cylindrical metal vessel which completely surrounds the calorimeter vessel or vessels; in high-precision calorimetry the insulating space between the two is always evacuated. The shield, in turn, is enclosed in a jacket vessel whose temperature is more or less accurately controlled. The space between the shield and the jacket is usually evacuated also, so that the shield itself does not need to be vacuum-tight. The shield is heated by electrical resistance wires; if cooling is provided, it operates continuously, by means of radiation to the jacket, which is then kept at a temperature lower than that of the shield.

The shape and structure of the shield are determined by the need to ensure uniformity of temperature over its inner surface, and high heating efficiency, i.e. small heat loss by direct radiation from the heating element. The second of these requirements is met most simply by locating the heating wires in grooves in the outer surface of the shield and covering the whole with a sheath having a low-emittance outer surface; the sheath has to be in good thermal contact with the main part of the shield. Irrespective of whether or not a sheath is used, the first requirement sets a minimum to the thickness of the shield body, at roughly three times the pitch of the heater windings.

The effective time-constant (see Chapter 4, Section III–3–B) of the transient term in thermal conduction across a 5 mm thick copper slab is less than 0·03 sec, so that in the type of dry shield just described the temperature distribution in the main part of the shield may be assumed always to be uniform. Also, the time-constant for conduction across an insulating lamina of mica or resin 0·1 mm thick is less than 0·002 sec. Thus the only significant thermal lag is that associated with the 'overheating' of the resistance wire necessary to drive the heat across the thermal insulation. Then, under quasi-steady conditions the temperature in the shield follows the equation

$$\frac{dT_s}{dt} = \psi(T_o - T_s), \qquad (33)$$

in which T_o is the temperature the shield would have attained if the thermal resistance of the insulating layer had been zero. The *response coefficient* $\psi(\text{sec}^{-1})$ is adequately approximated by

$$\psi = \lambda_l a / C_w x_l \qquad (34)$$

in which λ_l and x_l are the thermal conductivity and thickness of the insulation, a is the area of contact and C_w is the total heat capacity of the resistance wire, see Chapter 4, Section II–3. Since C_w increases with the cross-sectional area of the resistance wire, whereas a increases only with the wire diameter, it is evident that ψ can be made relatively large by using a great length of thin resistance wire. For example, if the shield body is a cylindrical copper shell 3 mm thick, wound with ten turns of 0·0152 mm diameter (38 s.w.g.) Manganin wire per centimeter over thin mica, the response coefficient ψ should be between 3 and 6 sec^{-1}.

When controlled shield cooling is not required, the jacket temperature need not be controlled precisely; instead, the heat-leak coefficient for the shield is reduced by interposing one or more radiation screens between the shield and the jacket. With this scheme the requirement (when linear or linear-plus-rate response regulators are used with single-vessel calorimeters) that the shield temperature drift be less than 10^{-4} deg sec^{-1} may be met simply by keeping the jacket temperature within 5 deg of that of the shield. This arrangement has the further advantage that the main part of the calorimeter (vessel, shield and jacket) is compact, so that it is quite easy to cool the jacket enough to permit measurements down to 0°C, while working above 100°C requires only a simple manually controlled jacket heater. In addition, the total heat capacity of this type of shield is small, so the heating power which has to be disposed of by the regulator is also small, and consequently continuous response regulators are especially suitable. This is true also of the dry shield in which cooling is effected by radiation to the jacket, but in this case there is a practical lower limit to the working temperature range. Thus if the shield is made of silver or copper 2 mm thick, and the emittance of its outer surface is as high as 0·2, the temperature difference between the shield and the jacket necessary to cool the shield at 0·005 deg sec^{-1} is about 25 deg. At working temperatures below 40°C, it is very difficult to keep the jacket temperature at a constant interval (within $\pm \frac{1}{2}$ deg, so that the shield temperature drift is less than 10^{-4} deg sec^{-1}) below that of the shield, because the jacket temperature is below ambient. This problem occurs only with discontinuously heated calorimeters. While the problem is by no means insoluble, in practice a discontinuously heated calorimeter working below 40°C might more conveniently be equipped with a wet shield, sacrificing control precision in order to gain accuracy and reliability.

An interesting method of providing controlled shield cooling at working temperatures below 40°C without recourse to a sub-ambient jacket thermostat has been described by Dole and others[44, 146]. This method consists in attaching to the outer surface of the shield a labyrinth of small-diameter piping, through which chilled air is blown continuously. The maximum power of the conventional electrical heater is adjusted to be twice that of the cooling, so that with a two-valued response regulator the heating and cooling rates are equal. This system could be used also with a continuous response regulator by making the heating power just to balance the cooling at zero signal. The main advantage is that the cooling power depends only slightly on the shield temperature, so that the drift-rate of the shield temperature at zero signal is always small. The disadvantages are that it is difficult to keep the shield temperature uniform because the chilled air is necessarily much colder than the shield, and that the temperature and flow-rate of the chilled air must be regulated.

Finally, the Peltier effect may, in principle, be used to extract heat from the shield at a precisely controlled rate. The development of semi-conductor thermocouples[65, 81] makes this method specially attractive, and its use in precision calorimetry cannot be long delayed.

(2) *Wet Shields*

In this system the shield is a thin-walled, vacuum-tight metal vessel enclosing the calorimeter vessel and the evacuated insulating space. The shield vessel is completely immersed in a liquid bath, and the shield temperature is varied solely as a consequence of changing the bath temperature. This implies that in general there are three thermal transport processes interposed between the servomechanism and the controlled surface. These are: (*i*) transfer of heat from the servomechanism to the bath liquid, (*ii*) transfer of heat from the bath liquid to the outer surface of the shield, and (*iii*) conduction of heat across the shield to its inner surface.

Provided that the thermal conductivity of the shield material is relatively high, the temperature differential across the shield may be taken as approximately zero. In fact, with a copper shield less than 5 mm thick, the temperature differential across the shield under steady conditions is at least 100 times smaller than the temperature difference between the bath liquid and the outer surface of the shield. The rate of heat transfer between the liquid and the shield is determined primarily by the thermal resistance of the liquid-to-metal boundary layer, and this rate in turn depends upon the thermal conductivity, viscosity, etc., of the liquid and the rate of stirring (see also Section II–2–D above). The time-constant of the boundary layer will not usually exceed 0·01 sec, so the variation of the shield temperature (assumed uniform) is adequately described by a steady conduction equation formally similar to Equation (33),

$$\frac{dT_s}{dt} = \psi(T_b - T_s), \qquad (35)$$

in which T_b is the temperature in the bulk of the liquid and T_s is the temperature of the shield body. In this case ψ is given by

$$\psi = \frac{\lambda_b}{x_b c_P^* w}, \qquad (36)$$

in which λ_b is the thermal conductivity of the liquid and x_b the thickness of the boundary layer; c_P^* is the volumetric specific heat of the shield material and w its thickness. Thus, given the physical properties of the bath liquid and the rate of stirring, the response coefficient ψ is inversely proportional to the heat capacity of the shield per unit area in contact with the bath liquid. For example, if the shield is made of copper, 2 mm thick, and the bath liquid is water, then

$$0 \cdot 1 < \frac{\psi}{\text{sec}^{-1}} < 0 \cdot 3,$$

depending on the rate of stirring, the surface smoothness of the copper, etc. Note that these values of ψ are very much smaller than the corresponding values for the type of dry shield described by Equation (34) (cf. Benson, Goddard and Hoevo[12]). Consequently, with the same regulator, the mean thermal signal corresponding to a given rate of heating the calorimeter vessel is an order of magnitude larger with the wet shield than with the dry shield.

The remaining factor is the rate of heat transfer between the servo-mechanism and the bath liquid. Provided that this is fast in comparison with $1/\psi$ it may be treated as introducing a small fixed time delay, χ, into the regulator system. Under these conditions the behavior of the various regulators applicable to this shield conforms well to the predictions of Cruickshank's analysis[37]. The value of χ is in fact the best criterion for assessing the possible methods of changing the bath temperature.

It is of historical interest to note that wet shields were used in some of the earliest versions of the adiabatic method, and the four most widely used methods of changing the bath temperature were all developed before 1930. The simplest of these methods is that used in a conventional water thermostat, namely that the regulator controls the power released by immersed electrical heaters, continuous cooling being provided by radiation and atmospheric convection from the bath. Although this method has been widely used[11, 73, 125] it has grave disadvantages. The working temperature range is restricted by the need to provide cooling, just as with the simple dry shield. Also, the large regulator output power requirement (at least 30 watts per liter of water) almost precludes the use of continuous response regulators. More important is the great difficulty of making the heat transfer between the heaters and the bath faster than that between the bath and the shield. This last disadvantage is almost completely overcome if the bath liquid is heated directly, by chemical reaction[109], for example, or by electrolysis. The latter method was first described by Carroll and Matthews[28]; the bath liquid is a dilute solution of ferric chloride in a mixture of glycerol and water, and the shield vessel itself forms one electrode. The time rate of change of the average bath temperature is always effectively proportional to the output power of the regulator, with negligible time-lag, but the method involves a number of practical difficulties. The fourth method of changing the bath temperature is by the controlled addition of hotter or colder liquid from thermostatted reservoirs. Originally[31, 95] the temperatures in the reservoir thermostats were fixed, so that the method could not be applied to continuously heated calorimeters. Recently however, Cruickshank and Timimi[38] have devised a method of controlling the reservoir thermostat temperatures relative to that in the shield bath, see Section II–3–E below. With this arrangement the temperatures of the hot and cold reservoirs need not differ by more than 1 deg from that in the shield bath. Flowrates up to 10 ml sec^{-1}, with efficient stirring in the shield bath, give a half-time for the thermal equilibration of the shield bath of less than 1 second, so that the behavior of the bath temperature can be adequately described by assuming a small fixed time-delay in the servomechanism. Additional time-delays arise in the liquid circulation system and in the operation of the valve gear, but the total can easily be kept less than 5 sec. Although this arrangement is not amenable to control by a continuous response regulator, the incorporation of a low-inertia electro-magnetically operated valve enables a Gouy[67, 121] modulation to be used to obtain an effectively linear response; this arrangement incidentally reduces the time-delay in the flow system.

C. Choice of Shield and Regulator

Since the shield type and the regulator system are inter-dependent, both must be selected in relation with the calorimetric method chosen. It is relevant first to make some general remarks about wet and dry shields.

The dry shield is particularly suited to continuous response regulators, but the application of a linear response regulator to a dry-shield, single-vessel calorimeter heated stepwise involves the difficulties that the rate of heat removal from the shield has to be relatively large and constant. Only the first applies with a reset response regulator, and only the second with a linear-plus-rate response regulator. The dry shield has advantages also with the two-valued response regulator in continuously heated carlorimeters (if shield cooling is not required), but again in the case of stepwise heating there is the difficulty of providing constant shield cooling at working temperatures below 40°C.

Although the use of the various versions of the wet shield in heat capacity calorimeters (and in reaction calorimeters also) for the room-temperature range is probably due more to tradition than to physical analysis, the liquid circulation type with hot and cold reservoirs is more easily applicable than the dry shield to calorimeters which are heated stepwise when the working temperature range extends below about 40°C. With a two-valued response regulator, the lowest temperature at which any part of the apparatus has to be controlled (that of the cold reservoir) needs to be only 1 or 2 deg below the working temperature. (With a dry shield, only the linear-plus-rate response regulator gives this advantage.) The wet shield is applicable to continuously heated calorimeters only if the hot and cold reservoirs are programmed to follow the shield temperature, but these are in fact the cases where the dry shield is most efficient.

The factors influencing the choices of shield type and regulator system in each of the four calorimetric methods discussed above may be summarized as follows.

(1) *The Twin-Vessel Calorimeter Heated Intermittently*

With a two-valued response regulator, controlled shield cooling is required, but changes of less than 20 per cent in the cooling rate between the initial and final equilibration periods can be tolerated. Thus a dry shield cooled by radiation to an insulated jacket should be satisfactory. If the range of working temperature for each filling of the calorimeter vessel is greater than 10 deg, however, variable jacket heating is necessary, and for working below 40°C the jacket must be cooled. Linear response regulators offer no important advantage here, but the inclusion of rate response should allow working down to about 20°C if the jacket is water-cooled. The reset response regulator has shield-cooling requirements similar to those with the two-valued response regulator, but the tolerance on the shield cooling rate is greater in the former case. Thus the optimum choice appears to be the simple dry shield, radiation cooled, with a reset response regulator, but for working below about 40°C the jacket thermostat has to be capable of operating below room temperature. The shield control requirements do not justify the use of an elaborate wet shield.

(2) *The Twin-Vessel Calorimeter Heated Continuously*

In this case controlled shield cooling is unnecessary, irrespective of which regulator is used. Thus a dry shield with radiation screens and a simple, insulated jacket might be used with a two-valued response regulator, in order to achieve the greatest simplicity in the electronic parts of the apparatus. Alternatively the radiation screens might be left out and the Gucker, Pickard and Planck modification incorporated in the regulator. In this case the regulator incorporating reset response has no significant advantage. When the heating rates of the twin vessels are matched and both held constant, as in Ackermann's[5] recording adiabatic twin calorimeter, a dry shield with a simple two-valued response regulator should be adequate.

(3) *The Single-Vessel Calorimeter Heated Intermittently*

Controlled shield cooling is essential, but it need not be large in the case of the linear-plus-rate response regulator. The combination of dry shield with radiation cooling and reset response regulator requires a jacket thermostat capable of being adjusted between each heating step, but for working below 40°C the jacket thermostat has to operate below room temperature. The same applies to both the linear and the two-valued response regulators, and in addition the jacket thermostat has to control to better than $\pm\frac{1}{2}$ deg to keep the shield cooling rate constant. Thus neither of these regulators is to be recommended. Although the linear-plus-rate response regulator requires only minimal shield cooling, so that the combination can work down to a few deg above service water temperature without difficulty, again the cooling rate of the shield has to remain constant within 10^{-4} deg sec^{-1}, so that an adjustable jacket thermostat is required. The most important of the alternative combinations is the liquid-circulation type of wet shield with a two-valued response regulator. This also can be used at working temperatures down to about 20°C. The incorporation of an auxiliary regulator which programs the temperatures of the hot and cold reservoirs to follow that of the shield bath greatly increases the effectiveness of this combination; it could in principle be applied to the jacket thermostat in the case of the dry shield with linear-plus-rate response regulator. Thus the practical choice seems to be between these two combinations, depending on whether the emphasis is put on compactness of the main apparatus, with relatively elaborate electronic gear (dry shield, linear-plus-rate response regulator) or on simplicity of electronic control gear with rather elaborate liquid circulation apparatus (wet shield with two-valued response regulator and special regulator for the hot and cold reservoirs). Finally, it is worth stressing that if the range of working temperature does not extend below about 40°C, the advantages of the dry shield with the reset response regulator become significant. This combination has the further advantage that there is no correction term to be added to the observed increment in the temperature of the calorimeter vessel.

(4) *The Single-Vessel Calorimeter Heated Continuously*

Since only minimal shield cooling is required, irrespective of which regulator system is used (in principle no cooling at all is required) the dry shield with radiation screens has important advantages. In this application the linear-plus-rate response regulator has no significant advantage over the simple linear response regulator; both, however, require that the shield cooling rate, small though it be, remain constant within about 10^{-4} deg sec^{-1}. Thus the jacket temperature has to be programmed to increase at about the same rate as that of the calorimeter vessel. The jacket temperature control, however, needs to be accurate only to about ± 5 deg when radiation screens are used. Similar considerations apply also with a two-valued response regulator, but the Gucker, Pickard and Planck[73] modification is certainly capable of offsetting the effect of the increase in the rate of shield cooling over a 15 or 20 deg heating. Probably the most satisfactory regulator in this application is that incorporating reset response. With this combination the use of a simple jacket cryostat (an ice bath, for example) should allow working down to about 5°C.

D. Dry Shields for Continuously Heated Adiabatic Calorimeters

(1) *Single-Vessel Calorimeter*

This shield, shown schematically in *Figure 12*, was developed for heat capacity measurements on aqueous electrolyte solutions over the temperature range from 300°K to 400°K. Designed for use in conjunction with the types of calorimeter vessel illustrated in *Figure 10*, to be controlled by a two-valued response regulator (either "on-off" or "high-low" responses may be used with continuously heated calorimeters, depending on the rate of heating of the calorimeter vessel), this shield has been shown to give a precision better than 0·1 per cent on the heat capacity results throughout the quoted temperature range[2].

The body of the shield shown in *Figure 12* is made from a cylindrical copper tube of 1 cm wall thickness. The massive copper lid of the shield is fixed to the mounting bar by two tubular supports which are made of german silver (agentan) to minimize the heat conduction between the shield and the mounting bar. The shield lid supports the entire inner part of the calorimeter, including the calorimeter vessel. The latter is suspended from the underside of the shield lid by thin steel wires (0·1 mm diameter). The stepped inner ring of the shield lid is threaded to screw into the top of the shield body. This arrangement facilitates the assembling and dismantling of the apparatus while ensuring good thermal contact between the lid and the body.

The radial location of the calorimeter vessel inside the shield is adjusted by means of the three glass locating pins shown LP in *Figure 12*, so that the calorimeter vessel is always aligned with the axis of the shield. The rate of heat leak from the calorimeter vessel into these locating pins is no greater than that through the electrical leads. To ensure the smooth motion of the internal stirrer St, the driving magnet M_{II} is placed inside the lower part of the shield body so that the gap between the driving and driven magnets is as small as possible. The shaft of the driving magnet is journalled in the

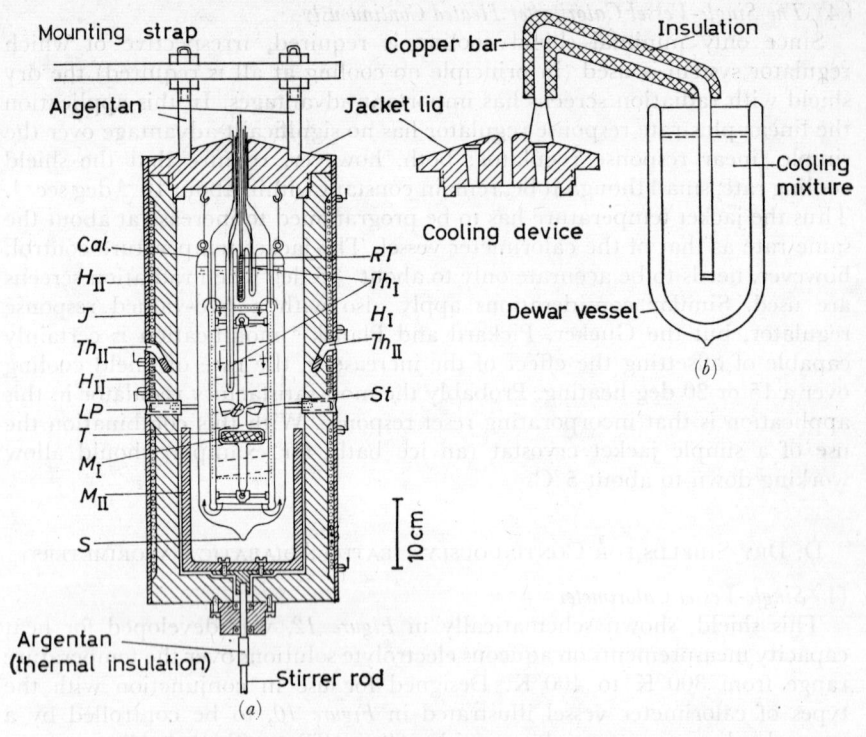

Figure 12. Dry shield for single calorimeter vessel[2]: (a) vertical cross-section with glass calorimeter vessel; (b) schematic diagram of cooling device.

H_{II}, Shield heater windings
J, Massive copper shield body
LP, Locating pins to center calorimeter vessel
Th_I, Inner thermocouple junctions
Th_{II}, Outer thermocouple junctions
M_{II}, Stirrer driving magnet
Other symbols as in Figure 10

ballraces and connected through an argentan insulating section to the rotor of a 100 rev/min synchronous motor. The shield heating element is a Teflon-insulated resistance wire wound in grooves on the outer surface of the heavy copper tube. The heater windings are covered by a closely fitting sheath of 1 mm thick copper sheet. Additional heating tapes (not shown in Figure 12) are attached to the base and to the lid of the shield. The three heating elements are normally connected in series, and so the resistances of the lid and bottom heaters are chosen so as to partially compensate the extra heat loss through the supporting tubes and the stirrer magnet shaft. Since the copper body of the shield may be assumed to be at effectively uniform temperature, the shield junctions of the control thermocouple are located in the mica-insulated sockets shown Th_{II}. A central hole through the lid provides passage for the leads to the calorimeter vessel heater and thermometer and to the calorimeter vessel thermocouple junctions. The entire shield assembly is surrounded by glass wool insulation in an outer container (not shown in Figure 12). The insulating space between the calorimeter vessel and the shield is not evacuated because of the practical difficulties of vacuum sealing the stirrer shaft without generating an unacceptably large amount of heat in the rotating seal. Despite its large ratio of total heat capacity to heated surface area the thermal lag of the shield does not exceed about 15 seconds. The degree of temperature uniformity over the inner surface of the shield is adequate for the precision quoted above provided that the shield heating rate does not exceed 0·01 deg sec^{-1}.

The two-valued response regulator is similar to that for the twin-vessel shield described below. A detailed description is given elsewhere[144, 145].

(2) Twin-Vessel Calorimeter

The adiabatic shield designed for the automatic recording twin calorimeter[5] discussed in Section II–1–E above is shown in simplified vertical cross-section in *Figure 13*. In this case the interior of the shield vessel is evacuated. The general design is similar to the single-vessel shield just described, so that the main differences arise from the provision of vacuum-tight seals and the need to accommodate two calorimeter vessels.

The shield is fixed to the shield body by clamping their flanges together, vacuum sealing being effected by the Teflon gasket shown T. Thermal contact between the lid and the body is improved by the clamps C, of which the inner part is 6 mm thick copper, contained in a steel former to give the requisite strength. The electrical leads pass to the outside of the shield lid through metal-to-glass seals similar to those illustrated in *Figure 9*. The shield vessel is connected to the vacuum line by the argentan pipe at the bottom. The stationary pressure in the insulating space inside the shield vessel does not exceed 10^{-4} torr at 400°K. The inner frame which supports the calorimeter vessels and the control thermocouples [see Section II–2–B–(2) above] is shown F in *Figure 13*. This frame fits into grooves in the body of the shield, as can be seen from the plan view. The rotating magnetic field which drives the stirrers in the two calorimeter vessels is generated in two sets of induction coils. These coils are wound on to copper-lined core sockets to prevent excessive local heating. The outer ends of the core sockets are located in the dovetail grooves in the body of the shield vessel, which may be seen in the plan view, *Figure 13*. The positions of the induction coils can be adjusted to give the maximum field at the level of the stirrer magnet inside each calorimeter vessel.

The arrangement of the shield heater is essentially the same as that in the single-vessel shield (see *Figure 12*). The principle of operation of the shield regulator is explained in Section II–1–E, and the main parts of the circuit are shown in *Figure 5*. The signal source is the multi-junction thermocouple mounted between the outer surface of the upper calorimeter vessel and the shield frame. This thermocouple is connected to a mirror galvanometer the reflected image from which actuates the photocell relay. The photocell relay is arranged as a quasi-linear response regulator; it is essentially a two-valued response regulator incorporating an electronic circuit which effects a Gouy type modulation[67] [see also Section II–3–A–(3)]. The "high" and "low" heating powers are defined by the series resistances, R in *Figure 5*, a constant d.c. voltage being applied to the whole heater circuit. The positive response of the regulator (corresponding to a negative signal, i.e. shield colder than upper calorimeter vessel) is to short-circuit one of the series

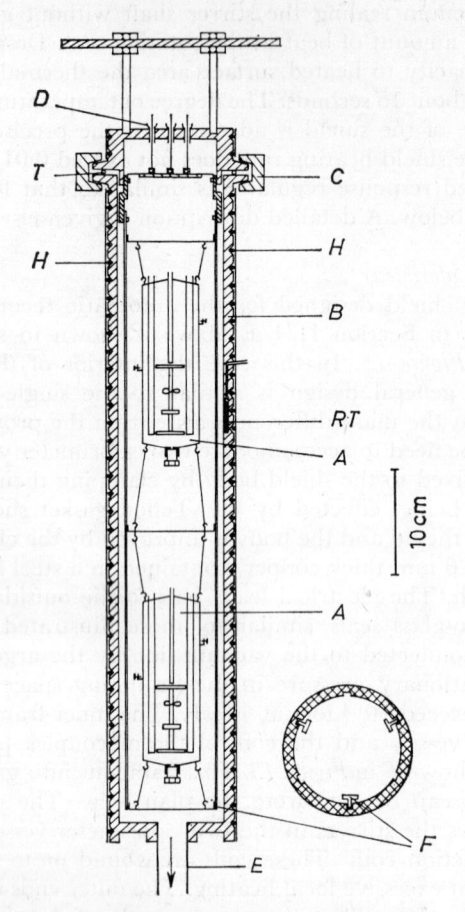

Figure 13. Evacuated dry shield for twin calorimeter vessels[5]: (left) vertical cross-section; (right) plan section, central.

A, A, Calorimeter vessels
B, Shield body
C, Lid securing clamps
D, Electrical lead exit seals
E, Vacuum line
F, Calorimeter vessel supporting frame (see *Figure 7*)
H, Heater windings
RT, Auxiliary resistance thermometer for calibration measurements
T, Teflon gasket

resistances, thereby increasing the shield heating power. A second quasi-linear response regulator (not shown in *Figure 5*) is used to control the lid of the adiabatic shield, because of the limited thermal contact between the lid and the body. This is not necessary in the single-vessel shield shown in *Figure 12* since the thermal contact is much better in this case.

The method of incorporating the Gouy modulation in the relay circuit in this apparatus is especially suited to calorimeter regulators by virtue of the low frequency of the modulating signal, so it is worthwhile to describe it in detail. The circuit is shown in *Figure 14*; it was developed by Wittig and his colleagues[144, 145]. Instead of the two photocells being connected in a simple bridge whose out-of-balance is amplified to operate the electromagnetic relay, each is connected to the grid of the controlling pentode of a

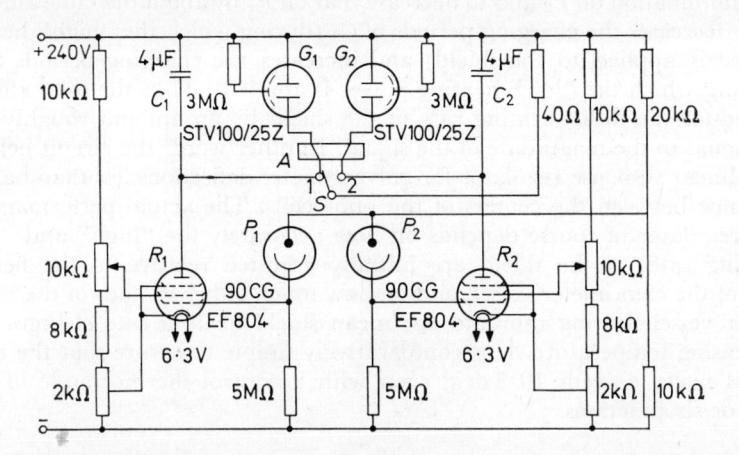

Figure 14. Photocell relay with quasi-linear response (Gouy modulation)[145] for shield regulator of recording twin-vessel adiabatic calorimeter. The electronic tubes are all German; replacement by equivalent British or American tubes necessitates adjustment of other components.

$\quad$ A, Polarized double-coil relay
$\quad$ P, 90 CG photocells
$\quad$ R, EF 804 pentodes
$\quad$ G, STV 100/25 discharge tubes
$\quad$ C, Matched capacitors

sweep circuit. The two sweep circuits are connected to a polarized double-coil relay.

In *Figure 14* the photocells are denoted by P_1 and P_2. When the two-way contact S of the polarized double-coil relay A is switched to position 1, as shown in *Figure 14*, the capacitor C_1 is shunted by the 40 Ω discharging resistor. The charging rate of the capacitor C_2 is then controlled by the suppressor-grid current in the pentode R_2, and the cathode current, i.e., by the intensity of the light on the photocell P_2. When the potential across the capacitor C_2 attains the ignition potential of the glow tube G_2, the capacitor is discharged and the discharge current switches the relay contact S over to position 2. The capacitor C_2 is now shunted by the 40 Ω resistor, while C_1 commences to charge at a rate depending upon the illumination of the photocell P_1. This charging continues until the potential across C_1 attains the ignition potential of G_1, when the contact S is switched back to position 1, and the whole sequence is repeated. The function of the photocell relay contacts shown in *Figure 5* is effected by a second two-way contact linked to S (not shown in *Figure 14*). In position 1, this second contact short-circuits the series resistance in the shield heater circuit, whereas in position 2 the heater series resistance remains in circuit.

The zero position of the galvanometer image is adjusted to give equal illumination of the photocells P_1 and P_2, and the 10 kΩ potentiometers are then set so as to make the ratio of the periods of the two sweep circuits effectively unity. Under these conditions, the mean heating rate of the shield at zero signal is midway between the "high" and "low" heating rates. The effect of a positive signal (shield hotter than calorimeter vessel) is to increase the illumination on P_2 and to decrease that on P_1 by about the same amount. This decreases the charging periods of C_2 (during which the "high" heating power is applied to the shield) and increases the charging periods of C_1 (during which the "low" heating power is applied). Thus the total effect is to reduce the mean heating rate of the shield by an amount roughly proportional to the magnitude of the signal. In other words the circuit behaves as a linear response regulator for galvanometer deflections less than half the distance between the centers of the photocells. The actual performance of this regulator of course depends on how accurately the "high" and "low" heating rates of the shield are initially adjusted relative to the heating rate of the calorimeter vessel, and on how much either or both of the calorimeter vessel heating rate and the mean shield heating rate changes with increasing temperature. It is comparatively simple to ensure that the mean signal never exceeds 10^{-3} deg, even with a control thermocouple of only four or six junctions.

E. A Wet Shield Controlled by Liquid Flow

This shield[38] exemplifies the type in which the temperature of the controlling bath is changed by the inflow of hotter or colder liquid, the corresponding outflow being fed back into a continuously circulating system. The design is suitable for a single-vessel calorimeter to be heated either in 1 deg steps or continuously at rates less than 10^{-3} deg sec^{-1}. In normal operation the mean value of the signal [see Section II–3–A–(1) above] never exceeds 0·005 deg. The heat-leak coefficient for the calorimeter vessel lies between 0·010 and 0·020 watt deg^{-1}, depending on the layout of the shield thermocouple and the radiation emissivity of the calorimeter vessel. Consequently the residual heat-leak rate never exceeds 10^{-4} watt. Over a 1 deg heating (including the subsequent equilibration period) the net heat leakage is thus of the order of 0·2 joule, and this may be estimated within about 10 per cent. When the calorimeter vessel is heated continuously, a simple manual adjustment of the temperature regulators in the liquid circulation system enables the mean signal to be kept close to zero ($\pm 0·0002$ deg). The corresponding uncertainty on the total heat capacity is then about $\pm 0·02$ joule deg^{-1}.

(1) Shield Vessel and Controlling Bath

The arrangement of the main part of the calorimeter is indicated by *Figure 15*, which gives a simplified sectional elevation. The calorimeter vessel A is mounted centrally inside the evacuated shield vessel B, which is completely surrounded by the well stirred water in the controlling bath C. This bath has to be thermally insulated from the exterior to ensure that the net heating and cooling rates depend only on the temperatures and flowrates of the incoming water. The external connections to C necessarily include two water inlets and one outlet, stirrer shafts, and the vacuum line, all of which constitute heat conduction pathways. Consequently, instead of heavily lagging C, it is better to surround it by a water thermostat whose temperature can be kept within 1 deg of that inside C. The inner vessel of this water thermostat is shown D in *Figure 15*, where the broken line indicates the normal water level.

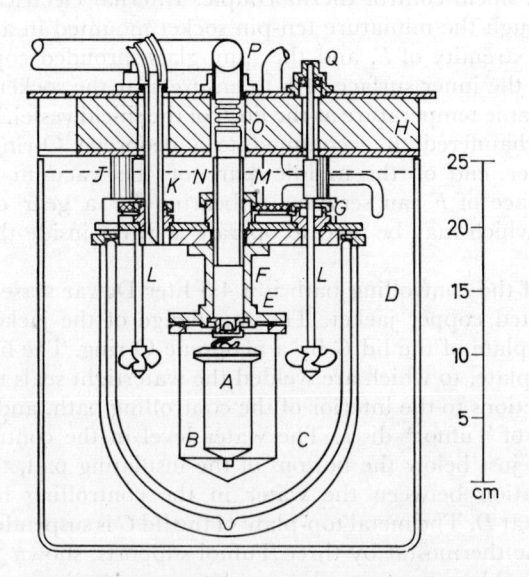

Figure 15. Liquid-flow wet shield with surrounding water thermostat[38].
    A, Calorimeter vessel
    B, Rhodium-plated copper shield vessel
    C, Shield bath (Dewar vessel in metal case)
    D, Thermostat bath
    G, Shield bath lid
    J, Insulating supports
    K, Water inlets
    M, Siphon overflow
    P, Vacuum line

Each of the three outer vessels, B, C and D, consists of a jar-shaped removable body which attaches to a fixed lid. The three lids form a rigid structure, and all interior service connections pass through seals on the upper surfaces of the lids. The apparatus is dismantled by detaching successively the bodies of the vessels D, C and B from their respective lids.

The body of the shield vessel B is made from 2 mm thick copper, welded to a 5 mm deep brass top ring which is grooved to locate a neoprene sealing ring and tapped to receive the bolts which attach it to the lid E. The body of the vessel is rhodium-plated overall. Its heat capacity is only 270 joule deg^{-1}, so that the response coefficient is about 0·2 sec^{-1} [cf. Equation (36)]. The time delay, χ, is about 1 sec, depending on the water flowrate, etc. The design of the ring seal requires that the lid E be made of 4 mm thick brass, so it has a smaller response coefficient than the body B, but it is found that this circumstance does not significantly affect the rate of heat leakage from the calorimeter vessel during normal operation of the adiabatic regulator.

The lid E is made integral with the boss F, which is attached by four bolts to the insulated underside of the lid G of the controlling bath. The boss F is almost completely immersed in the water of the controlling bath and is well insulated from the exterior. This boss forms the lower part of the vacuum line, and also carries the electrical leads from the calorimeter vessel and the shield-control thermocouple. Internal electrical connections are made through the miniature ten-pin socket mounted in a special fitting at the lower extremity of F, and the spun-glass-shrouded copper leads are coiled around the inner surface of F to ensure that the socket connection is always at the same temperature as the lid E of the shield vessel. The top, inner edge of F is chamfered to accommodate a neoprene O-ring which seals on to the lower end of the middle part of the vacuum line. Finally, the outer surface of F can serve as a bearing for a gear driven permanent magnet which may be used to operate a stirrer inside the calorimeter vessel.

The body of the controlling bath is a 4·5 liter Dewar vessel sealed into a chromium-plated copper jacket. The top flange of the jacket is sealed to the metal top-plate of the lid G by a neoprene O-ring. The lid G consists of the metal top-plate, to which are welded the watertight seals through which pass all connections to the interior of the controlling bath, and an insulating pad made up of Tufnol† discs. The water level in the controlling bath is adjusted to be just below the bottom of the insulating pad, to ensure good thermal insulation between the water in the controlling bath and that in the thermostat D. The metal top-plate of the lid G is suspended rigidly from the lid H of the thermostat by three Tufnol supports, shown J in *Figure 15*. The hot and cold water from the regulator circulating system enter the controlling bath through two glass-jacketed copper pipes, one of which is shown K in *Figure 15*, sited symmetrically on either side of the stirrer shaft L. This stirrer effects the mixing of the inflowing water with that in the bath. A second stirrer, sited diametrically opposite to the first, produces an upward flow to complete the internal circulation. Both stirrer shafts have nylon sections between the lids G and H to minimize thermal conduction to the exterior. The water outflow leaves the controlling bath through a constant-level siphon, of which the inner end is shown M. The side arm on M carries a sixteen-junction thermocouple which measures the difference in temperature between the controlling bath and the thermostat D.

The middle part N of the vacuum line is of precision-bore Pyrex glass. Its lower end is sealed into the boss F as described above, and a second O-ring seal above the top-plate of the lid G ensures that no water leaks into the controlling bath C from the thermostat. The glass tube N is joined by a stainless steel-to-Pyrex seal and expansion bellows O to the brass tube P forming the external part of the vacuum line. This three-part construction of the vacuum line facilitates assembly, and locates accurately the major temperature gradients. The electrical leads pass to the outside of P through tubular glass-to-metal seals similar to that shown in *Figure 9(b)*. In

---

† A laminated material bonded with synthetic phenolic resin, made and marketed by Tufnol Ltd., Perry Bar, Birmingham, England.

use the shield vessel B is continuously evacuated to less than 10^{-6} torr. This is attained initially in about 15 h pumping after the apparatus is assembled.

The stirrer shafts terminate in bearing units Q. Flexible drive cables from the separately mounted motors connect through these bearing units.

(2) *Water Circulation System*

The purpose of this system is to deliver to the controlling bath a constant flow of water from one or other of two auxiliary thermostats, respectively hotter and colder than the controlling bath. It is of prime importance to minimize the overall time-delay in the changeover from hotter to colder inflow and vice versa. This means that a fast acting selector valve must be used, and it must be sited as close as possible to the controlling bath. It is essential also that after each changeover, the temperature of the inflow to the controlling bath converges rapidly to that of the auxiliary thermostat from which it comes. Consequently the water entering the selector valve must be maintained at the temperature of the auxiliary thermostat. This is achieved by a high rate of outflow from each of the auxiliary thermostats to the selector valve unit. The greater part of this outflow is used to thermostat the valve unit, from whence it passes, together with the overflow from the controlling bath, to a sump, and finally it is pumped back to the cold auxiliary thermostat. The selector valve simply diverts a fraction of either the hot or the cold water flow through the controlling bath, while the whole of the other flow passes direct to the sump. The two delivery pipes from the selector valve to the controlling bath are then the only parts of the system through which the flow is intermittent. The delivery pipes cannot of course be kept at constant temperature when no water is flowing through them, but the cooling or heating during the no-flow periods is minimized by making the pipes of thin-walled copper tubing (except for short sections of flexible plastic tube); the upper sections are embedded in foamed polystyrene and the lower sections are glass-jacketed (see *Figure 15*).

The complete circulation system is shown schematically in *Figure 16*. A and B are, respectively, the cold and hot auxiliary thermostats. The Pyrex siphon tubes C have internally silvered vacuum-jackets, and terminate in electromagnetically operated cut-offs D in the selector valve unit. The insulated outer case of this unit is indicated by the broken line. The cut-offs D each deliver water at 22·5 ml sec^{-1} to the inner vessels of the constant-level reservoirs E which, in turn, supply the selector valve through the flow-control valves F. These valves can be set for flowrates between 0 and 8 ml sec^{-1}. The remainder of the flow from the cut-offs D serves first to thermostat the reservoirs E, and then passes into the selector valve jacket, G. The selector valve H diverts the flow from one or other of the flow-control valves F into the corresponding delivery pipe J. These pipes lead directly into the controlling bath via the inlets K (see also *Figure 15*). The flow from the other valve F goes directly into the jacket G. From G the mixed hot and cold water passes into the sump L, together with the outflow from the controlling bath through the constant-level siphon M. The water in the sump is cooled, when necessary (by tap-water passing inside the coils N), and returned by the pump O to the cold auxiliary thermostat. The siphon P

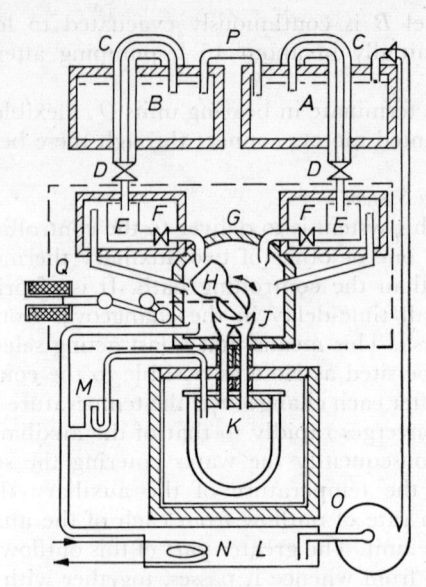

Figure 16. Water circulation system for liquid flow wet shield[38]—schematic, not to scale.

    A and B, "Cold" and "hot" auxiliary thermostats
    E and F, Flow control units
    H, Selector valve
    K, Inlets to shield bath
    M, Siphon overflow
    L, Sump
    O, Automatic re-circulating pump

completes the circulation. The speed of the pump O is controlled by a level-float in A; a second float in B stops the pump and closes the cut-offs D in the event of breakdown, indicated by the water in B falling below a pre-set level. The selector valve H is moved from one to the other of its two positions by the spring-loaded armature of the solenoid Q, which is energized by the adiabatic regulator.

(3) *Regulators*

The adiabatic regulator has as its signal source the multi-junction thermocouple between the outer surface of the calorimeter vessel and the inner surface of the shield vessel. A galvanometer-photocell amplifier similar to those referred to in Section II-3-D above actuates a relay which energizes the solenoid of the selector valve. The system may be protected against overshoot either by using lock-on photocells[148] or, more simply, by using half-parabolic reflectors (with the photocells at the foci) so that the light-sensitive area is still illuminated after the galvanometer image has passed beyond it. It is desirable also to provide the photocell unit with a fine control zero adjustment so that stray e.m.f. in the external part of the thermocouple circuit can be allowed for[38].

The two auxiliary thermostats are controlled independently of each other by two-valued response regulators. Basic heating is supplied by immersion elements, with an adjustable extra element in the hot thermostat which just balances the cooling due to the water coming in from the cold thermostat. Because of the relatively large flowrate, the controlling heater of the cold thermostat has to deliver at least 150 watts with very small time-delay on switching. This is achieved by using electric lamps mounted outside the Pyrex thermostat vessels, and surrounded by cylindrical metal reflectors. This arrangement is rather inefficient—only 45 per cent of the rated power of the lamps is absorbed by the water—but the switching time-lag is apparently less than 0·1 second. Each regulator comprises a pair of matched thermistors (one in the auxiliary thermostat and the other in the shield controlling bath) connected in an a.c. bridge circuit, an a.c. amplifier and a thyratron relay supplying the heating lamps around the thermostat tank. The thermistor bridge includes a variable balancing resistance which is used to adjust the thermostat temperature to a value differing from that in the controlling bath by a fixed interval. For normal operation with the temperature of the calorimeter vessel remaining constant, the hot auxiliary thermostat is run at 0·8 degree above the temperature of the controlling bath, and the cold thermostat is run at 1·0 degree below, the flow control valves (*F* in *Figure 16*) being set to deliver 4 ml sec^{-1}. The heating produced by the controlling bath stirrers then just suffices to keep the total heating and cooling rates equal, at just less than 0·0015 deg sec^{-1} (the effective heat capacity of the controlling bath is about 12000 joule deg^{-1}). The precise values of the heating and cooling rates are most easily determined by measuring the change in the ratio of heating to cooling half-cycle periods when the calorimeter vessel is heated at a known rate. To attain the effectively adiabatic condition when the calorimeter vessel is being heated continuously, the temperature difference between the hot auxiliary thermostat and the controlling bath is increased (and that between the cold thermostat and the bath decreased by the same amount) until the heating and cooling half-cycle periods of the main adiabatic regulator are again equal. Once the correct adjustment is found, the effectively adiabatic condition is maintained as long as the heating rate of the calorimeter vessel remains constant; a periodic small adjustment has to be made to offset the change in the temperature coefficient of resistance of the thermistors. For example, if the calorimeter vessel is heated at 0·0003 deg sec^{-1}, the hot auxiliary thermostat is set at 1·0 degree hotter than the controlling bath and the cold one at 0·8 degree colder than the controlling bath. The performance of this shield depends, of course, on the control thermocouple and galvanometer, but with a sensitive galvanometer (coil resistance less than 50 Ω, short period) and a properly designed thermocouple (see Section II–4–E below) the numerical value of the mean signal (in deg) should not exceed six times that of the calorimeter vessel heating rate (in deg sec^{-1}). If the operating parameters of the regulator are known with adequate precision, the proportional uncertainty on the calculated values of the mean signal ζ (arising from the approximate nature of the theory) should not exceed 10 per cent. Thus if the calorimeter vessel is heated at 0·0005 deg sec^{-1}, the magnitude of the mean signal will be less than 0·003 deg and its uncertainty will be less than 0·0003 deg. The latter figure represents the limits within which ζ can be approximated to zero by keeping the half-cycle periods equal.

4. Control thermocouples for adiabatic shields

A. Heat Transfer between Vessel and Shield

A basic factor in the design of the control thermocouple is the sensitivity (volt deg^{-1}) required. This depends to some extent upon the overall rate of heat transfer between the calorimeter vessel and the adiabatic shield; the greater the rate of heat transfer for a given temperature difference, the more precise must be the adiabatic control, and hence the higher the thermocouple sensitivity. In addition, if the residual heat leak is to be evaluated it is necessary to be able to estimate accurately the *effective* value of the heat-leak coefficient ξ, see Section II–3–A–(1). It is therefore of interest to examine the mechanism and probable magnitude of each of the major contributions to ξ.

The major contributions to ξ are (*i*) radiation across the insulating space, (*ii*) conduction along the electrical leads and the supports of the calorimeter vessel, and (*iii*) conduction along the wires of the control thermocouple.

In high precision calorimetry, we assume that the insulating space is evacuated, so that there is no gas-convection term. Other minor contributions include residual gas conduction and the two minor thermocouple effects, heat transport by the Peltier effect and Joule heating. In practice these last effects are very much smaller than the first three, and gas conduction may generally be neglected if the residual pressure is less than 10^{-5} torr in an insulating space more than 1 cm across.

It has been widely assumed by calorimetrists that provided all thermal conductors crossing the insulating space are in good thermal contact with the calorimeter vessel and, at the other end, with the shield, the rate of heat leak is determined by the thermal signal ζ [see Section II–3–A–(1)]. This is, in general, only approximately true in relation to thermal conduction processes, because these are propagated at a finite rate. Consequently the effective value of the heat-leak coefficient ξ when the signal ζ is small and changing rapidly is generally less than that determined experimentally, normally by measuring the "cooling rate" of the calorimeter vessel with a relatively large negative magnitude of ζ which is changing only slowly. The following discussion includes the precautions which must be taken to minimize this discrepancy.

The theory of the three main heat transfer processes is covered in Chapter 4, and only brief recapitulation is necessary here. The rate of heat transfer by radiation is completely determined by the absolute temperature of the outer surface of the calorimeter vessel and the signal ζ, and it is zero whenever ζ is zero. Detailed calculations may be based on equations given in Chapter 4, Section III–2. The rate of heat leak by conduction depends upon the temperature gradient in that part of the conductor nearest to the calorimeter vessel, rather than directly upon ζ, see Chapter 4, Section III–4. The only situation in which the conduction process is amenable to simple treatment is when the temperatures at the ends of the conduction path remain constant or change only slowly, i.e. steady-state conduction. Let T_1 be the temperature at the calorimeter vessel end of the conduction path

HEAT CAPACITY OF LIQUIDS AND SOLUTIONS

and T_2 be that at the shield end. Then the steady-state rate of heat transfer out of the calorimeter vessel is given by

$$\frac{dq}{dt} = \frac{\pi r^2 \lambda}{l}(T_1 - T_2). \tag{37}$$

In one-dimensional thermal conduction with temperature constraints at both ends, a departure from the steady state is diminished by 63 per cent in a period equal to the time-constant of the conduction path, given by the square of the path length divided by π^2 times the thermal diffusivity of the material (see Chapter 4, Section III–3–B). The thermal diffusivity is defined by

$$\kappa = \lambda/\rho c_P, \tag{38}$$

in which ρc_P is the volumetric specific heat, as in *Table 1*. If T_2 varies according to a periodic function of time, the steady state is never attained, and dq/dt is also periodic. In general, dq/dt is diminished in amplitude and lags in phase relative to the value given by Equation (37), both effects increasing with increasing time-constant.

Thermal conduction along the wires of the control thermocouple is complicated by the necessity that the ends of the wires be electrically insulated from the contiguous surfaces. This insulation is often achieved by thin laminae of mica, glass, or synthetic resin; usually the material has low thermal conductivity, but the thinness of the lamina makes the time-constant very small [see Section II–3–B–(1) above]. Consequently the thermal conduction across the thermocouple insulation is always effectively steady. The heat transfer across each insulating lamina then follows the equation

$$\frac{dq}{dt} = a_i \lambda_i \zeta_i / x_i, \tag{39}$$

in which ζ_i is the temperature difference across the lamina and a_i is the effective contact area. In glass calorimeter vessels, the thermocouple junctions are usually located in glass pockets, making contact with the glass through a liquid. In this case also, dq/dt may be assumed to depend only on ζ_i, but the time-constant is usually much larger than with thin solid insulation, although still much smaller than the time-constant for the thermocouple wires themselves. In either case, whenever the temperature gradient along the wire is uniform, the temperature difference between the inner and outer junctions (which determines the observed thermocouple e.m.f.) is a fixed fraction of $T_1 - T_2$, the fraction depending on the relative magnitudes of $a_1\lambda_1/x_1$, $a_2\lambda_2/x_2$ and $\pi r^2\lambda/l$.

It is evident that the three main heat transfer processes (between the calorimeter vessel and the shield) are determined by $\zeta = T_2 - T_1$ whenever steady conduction may be assumed, provided of course that the overall change in temperature is insufficient to significantly alter the coefficient of the radiation term. Thus whenever ζ is a periodic function of time, provided that the most important conduction paths between the calorimeter vessel and the adiabatic shield have time-constants that are small in comparison with the period of $\zeta = \zeta(t)$, ζ is a satisfactory measure of the departure from the adiabatic condition, $dq/dt = 0$, and so

$$\frac{dq}{dt} = \zeta \xi, \qquad (40)$$

in which, for a particular apparatus, the heat-leak coefficient, ξ, is a function of the operating temperature only, varying according to

$$\xi = A + BT^3. \qquad (41)$$

The departures from the steady state whenever T_1 or T_2 changes will generally be greatest when the value of $\pi r^2 \lambda/l$ for that conductor is least. The design of the control thermocouple and of the calorimeter vessel supports and electrical leads thus involves a compromise between minimizing the time-constant [to ensure conformity to Equation (40)] and minimizing the heat-leak coefficient [i.e. making A small in Equation (41)]. The best compromise evidently depends on what fractions of the total heat-leak coefficient are due to each of the three main heat transfer processes. Consequently it is useful, before discussing the design of the thermocouple, to establish the probable magnitude of the part of the heat-leak coefficient due to radiation and conduction along the calorimeter vessel supports.

B. Radiation and Conduction along Supports and Electrical Leads

The calorimeter vessels and shield used by Cruickshank and Timimi[38] are probably representative of the all-metal designs included in the scope of this chapter. The figures for these vessels indicate at least the range of magnitudes of the radiation and lead conduction coefficients at temperatures around 300°K.

The vessels were smooth-turned from pure copper stock, carefully polished, and then rhodium-plated. If the total emissivity of rhodium in this temperature range is 0·05, the radiation coefficient should be, ideally, 31 μW deg^{-1} cm^{-2} at 300°K. This value has to be adjusted by a factor which takes account of surface roughness. Cooling curve observations indicate that this factor is about 2, so that the emittance is 0·1. The corresponding value of the radiation coefficient is 62 μW deg^{-1} cm^{-2}. Taking the superficial area of the 80 ml capacity vessels as 160 cm^2 gives the value 0·010 watt deg^{-1} for $\xi_{\text{radiation}}$.

If the calorimeter vessel is supported inside the shield by pillars, the optimum length of the pillars is a compromise between long pillars (high time-constant, large uncertainty on small heat leak) and short pillars (large steady-state heat leak). The cross-sectional area of pillar supports has to be relatively large to avoid buckling under the weight of the calorimeter vessel. The rate of heat leak through the supports can, however, be reduced without increasing the time-constant, by one or both of two methods. The calorimeter vessel may be *suspended* by thin wires of high diffusivity material, such as copper; also, a large thermal resistance may be introduced at the point of contact between the calorimeter vessel and the support. For example, the rounded inner end of a pillar or strut support may be coated with a thin film of hard glass, enamel or resin. With wire suspensions it probably suffices to use enamelled wire. For an order-of-magnitude calculation the thermal transfer coefficient may be taken as 1 watt deg^{-1} cm^{-2}. The contact area depends on the loading and on the surface hardness of the materials. If the loading is 100 g per support, and the softer material is copper the contact area will be 0·0001 or 0·0002 cm^2. Allowing for the spreading of the heat flow through the insulating film, and taking account of the temperature gradient along the support indicate that the rate of heat transfer into a 34 s.w.g. copper wire support is unlikely to exceed 0·0003 watt deg^{-1} under steady conditions. In any case the rate of heat leak due to the supports is unlikely to exceed 10 per cent of that due to radiation, so it is best to use short supports whose outer ends are in good thermal contact, preferably metallic, with the inner surface of the shield, ensuring a small time-constant for the whole conduction path.

The electrical leads to the calorimeter heater and thermometer probably make a contribution to the total heat-leak coefficient much more important than that made by the calorimeter vessel supports. The special problem of the current leads to the calorimeter vessel heater is discussed in detail in Chapter 4, Section IV–4, so that we need deal only with the heat-leak aspect here. Consequently all the heater and thermometer leads can be considered together. A primary consideration is the need to ensure that the time-constant does not exceed 10 per cent of the expected period of $\zeta = \zeta(t)$. In practice this usually means that copper leads are used and that the free length between the calorimeter vessel and the shield is kept as small as is convenient; the balance between thermal and electrical resistance is then determined by the wire diameter. If the leads are not subjected to flexing whenever the apparatus is disassembled, then aluminum is superior to copper as a lead material; comparing wires of the same length, if the diameters are adjusted to equalize electrical resistance, thermal resistance is higher in the aluminum wire than in the copper. Consider lead wires of 5 cm free length, whose electrical resistance must not exceed 0·01 Ω: the aluminum wire is 0·399 mm in diameter and its heat transfer rate is 0·00059 watt deg^{-1}; the copper wire is 0·318 mm in diameter but its heat transfer rate is 0·00061 watt deg^{-1}. The time-constants are 2·5 sec for aluminum and 2·3 sec for copper.

In the calorimeter of Cruickshank and Timimi, enamelled copper leads are used, the free length being about 4 cm. The current leads are 30 s.w.g. (0·315 mm) and the other six leads are 38 s.w.g. (0·152 mm). The calculated heat leak along the lead bundle is then 0·0025 watt deg^{-1}. Thus the sum of the radiation and conduction heat leaks, excluding that due to the control thermocouple should be about 0·0135 watt deg^{-1}. By using wire suspension instead of resin-topped copper pillars, and using aluminum electrical leads this might be reduced to 0·0130 watt deg^{-1}, and electrolytic polishing of the calorimeter vessel before rhodium-plating would certainly reduce it still further.

C. Thermocouple Layout and Materials

(1) *Design Criteria*

The purpose of the control thermocouple is to indicate as accurately as possible the value of the signal, $\zeta = T_2 - T_1$, in which T_1 and T_2 are the temperatures at the outer surface of the calorimeter vessel and the inner surface of the shield respectively. It must conduct only a small amount of heat between the calorimeter vessel and the shield, and the conduction processes must be always close to steady conditions. The design criteria are then:

(*i*) The ratio of thermocouple e.m.f. to signal must be large.
(*ii*) The thermocouple e.m.f. must be always effectively proportional to the signal, even when the latter is changing rapidly.
(*iii*) The steady-state heat-leak coefficient for the complete thermocouple must be small in comparison with that due to radiation.
(*iv*) The total electrical resistance must not exceed the optimum value for the control circuit (see Section II–4–E below).

Of these criteria, (*iii*) is not especially restrictive, since the heat leak due to the control thermocouple is in any case unlikely to exceed 25 per cent of of that due to radiation (above 300°K), and (*iv*) is hardly ever important as the optimum resistance for the control circuit is always large. The first two criteria are interdependent, and meeting them also ensures that departures from steady-state thermal conduction do not cause significant errors.

(2) *Layout Characteristics*

In relation to the criteria listed above, the most important characteristics of a multi-junction thermocouple layout are the rate of response, the efficiency and the heat-leak coefficient.

(*a*) *The Rate of Response.* The rate at which the thermocouple e.m.f. approaches its steady-state value is governed by parameters related to the time-constant of the thermocouple wires, the effective response-time of the insulated junction assemblies and the heat capacities of the wires and the junction assemblies. The probable magnitude of the thermocouple response coefficient (assuming that the response is adequately approximated by a first-order differential equation) may be estimated by considering two extreme cases. If the junction assemblies have negligible heat capacity, then provided that conduction across the insulation is always effectively steady (time-constant less than 0·01 sec), the overall response coefficient is roughly proportional to the reciprocal of the time-constant of the wires, given by

$$\tau_W = l^2/\pi^2 \kappa \qquad (42)$$

in which l is the length of the wire and κ is the thermal diffusivity of its material. Departures from the steady state are directly proportional to the total heat capacity of the thermocouple, but decrease as the thermocouple efficiency (see below) increases. If, on the other hand, the junctions have heat capacities large in comparison with those of the wires, then the response coefficient may be estimated using Equations (33) and (34) (ignoring for this purpose the heat conducted along the wires). In a practical case the response coefficient will be less than the smaller of the two values estimated by assuming the extreme cases to apply.

(b) *The Efficiency.* We define the efficiency of a thermocouple as the ratio of the temperature difference indicated by the instantaneous e.m.f. to the actual temperature difference between the two surfaces, under steady-state conditions. In the case of a multi-junction thermocouple with both sets of junctions insulated, the efficiency is given by

$$\eta = \frac{l}{\lambda_\mathrm{W} \pi r^2} \bigg/ \left(\frac{l}{\lambda_\mathrm{W} \pi r^2} + \frac{2x_1}{a_1 \lambda_1} + \frac{2x_2}{a_2 \lambda_2} \right), \tag{43}$$

in which l and r are the length and radius of the thermocouple wires and λ_W is the conductivity of the material, and a_i, x_i and λ_i are, respectively, the contact area, thickness and conductivity of the insulation between the junctions and the contiguous surfaces; the factor 2 takes account of the fact that there are two wires per junction.

In an adiabatic calorimeter, while the temperature of the inner surface of the shield increases and decreases cyclically with a relatively short period, that of the outer surface of the calorimeter vessel changes slowly and monotonically. Consequently the inner and outer junctions are not equivalent from the point of view of conduction processes. Thus the performance of a low-efficiency thermocouple depends significantly on whether the inefficiency arises at the inner (calorimeter vessel) or the outer (shield) junctions.

(3) *Discussion*

The complicated nature of the relationship between response-time and efficiency makes analytical discussion difficult, but useful guidance for thermocouple design can be got from qualitative examination of practical cases. Consider first a high-efficiency thermocouple. Although the e.m.f. under steady conditions corresponds closely to the true signal, if the junction assemblies have large heat capacities the consequent slowness in attaining steady conditions after the shield temperature changes causes the thermocouple e.m.f. to lag in phase relative to the signal ζ, thereby distorting the control action. Although the relatively large heat leaks into and out of the calorimeter vessel nearly balance over a complete regulator cycle, the residual unbalanced heat leak along the thermocouple wires is related only indirectly to the instantaneous value of the e.m.f. Reducing the heat capacities of the junction assemblies increases the rate of thermal response up to the limit set by the wires themselves (the response coefficient cannot exceed the reciprocal of the time-constant τ_W of the wires). Further improvement requires that τ_W be decreased, by reducing the length and, if necessary, the cross-sectional area by the same proportion, to keep the efficiency the same. This indicates the most obvious way of achieving good thermocouple performance, namely designing for high efficiency together with fast response, i.e. low r_W and low total heat capacity; this means using short, thin wires and thin junction plates of high thermal conductivity.

Because for high efficiency the junctions have to be rigidly attached to the contiguous surface it is not always practicable to use a high-efficiency thermocouple. Consider a multi-junction thermocouple of which one set of junctions has $a_i\lambda_i/x_i$ small in comparison with $\lambda_W r_W^2/l_W$ [cf. Equation (43)]. To make the thermocouple response coefficient reasonably large the quantity $a_i\lambda_i/x_iC_i$ for the low-efficiency junctions must be made large, and hence C_i, the effective heat capacity of these junctions, must be made as small as possible. In fact the only way to make the effective heat capacity of these junctions small, since the pre-supposition of low efficiency implies that the wire radius is large in comparison with the junction contact area, is to have high-efficiency junctions at the other ends of the wires. This means that the temperature of the thermocouple wires is determined mainly by the surface at the high-efficiency junctions, only the small temperature gradient across the wires being related to the temperature of the surface at the low-efficiency junctions. The general behavior of such a thermocouple is then rather similar to that of the pillar supports discussed in Section II–4–B above, the effective thermal boundary being located between the low-efficiency junctions and the adjacent surface.

The disadvantages of having low-efficiency junctions at the shield surface are well illustrated by the special case of the fast-response dry shield [Section II–3–B–(1)]. If we assume the contact junctions described in Section II–4–D to be typical of low-efficiency junctions, the effective response coefficient for such junctions is unlikely to exceed 0.5 sec^{-1}. The response coefficient of this type of shield, on the other hand, may be as high as 5 sec^{-1}. Thus with this combination, a large part of the conduction link lag (which determines the behavior of the regulator) would be, in effect, inside the junction assembly. In this situation the mean value of the difference between the temperatures of the surfaces of the calorimeter vessel and the shield during a heating (which determines the radiation heat leak) may easily be of opposite sign to the mean signal calculated from regulator theory, which determines (approximately) the heat leak due to conduction along the thermocouple wires.

If the low-efficiency junctions are at the calorimeter vessel end, however, provided that both the time-constant of the thermocouple wires and the total heat capacity of the whole thermocouple assembly are small, the departures from steady-state conduction are small (even though the overall thermocouple response is relatively slow). This is because the time-derivative of the surface temperature of the calorimeter vessel changes only slowly. With unsymmetrical thermocouples the assumption of a first-order response breaks down completely; in the case under consideration the thermocouple e.m.f. actually leads in phase relative to the true signal when the latter changes periodically, but the heat leak from the calorimeter vessel always lags in phase.

Since the temperature of the inner (calorimeter vessel) junctions follows that of the outer junctions, and hence that of the shield, only a small part of the energy required to heat the thermocouple assembly at the same net rate as the calorimeter vessel comes from the vessel itself. Thus the lag signal associated with the change in the effective thermal boundary of the calorimeter vessel during heating is very small, and it is probably allowed for by the fact that the regulator theory slightly over-estimates the net heat leak during heating.

To summarize: the two arrangements which can, in principle, meet the design criteria have in common that the efficiency of the shield junctions is high, and that the time-constant, thermal conductance and heat capacity of the wires are all made as small as possible. The alternative requirements for the inner junctions are either (i) high efficiency (large contact area), when the greater part of the thermal resistance is located in the wires, or (ii) low efficiency (small contact area) but minimal junction heat capacity. In the latter case the greater part of the thermal resistance is between the surface of the calorimeter vessel and the inner junctions, just as with the calorimeter vessel supports.

(4) *Thermocouple Materials*

The relevant properties of some commonly used thermocouple metals are listed in *Table 4*, which includes also the properties of metals suitable for potential leads and junction plates.

Table 4. Electrical and thermal properties of commonly used thermocouple and lead materials

Material	E (μV deg^{-1})	κ (sec^{-1} cm^2)	ω (μ ohm cm)	λ (watt deg^{-1} cm^{-1})
Alumel	−13	0.09	30	0.29
Antimony	+48	0.012	40	0.18
Bismuth	−73	0.007	110	0.08
Chromel	+28	0.055	70	0.18
Constantan	−33	0.063	48	0.23
Copper	+ 7	1.102	1.6	3.84
Iron	+19	0.210	9	0.75
Nichrome	+11	0.038	103	0.13
Nickel	−15	0.226	7	0.90
Manganin	+ 6	0.066	42	0.23
Platinum	0	0.234	10	0.70
Silver	—	1.69	1.6	4.18
Aluminum	—	1.00	2.5	2.36

The figures tabulated under E give the thermal e.m.f. per degree temperature difference across a pair of junctions made between platinum and the materials listed. The difference between the E values for any two metals then gives the thermoelectric power per pair of junctions between these two metals. κ is the thermal diffusivity of the material listed, ω its electrical resistivity and λ its thermal conductivity. The effective time-constant of a pair of dissimilar wires may be taken as equal to that of the wire with the lower thermal diffusivity; it is given by Equation (42).

It is evident that the antimony–bismuth couple has the highest thermoelectric power, but both metals have very low diffusivity and poor mechanical properties. A more realistic choice is the chromel–constantan couple (61 μV deg^{-1}). This couple has the advantage from the point of view of construction that the optimum ratio of the wire diameters[79] to give the best balance between thermal and electrical resistance,

$$r_1/r_2 = (\lambda_2\omega_1/\lambda_1\omega_2)^{\frac{1}{4}}, \qquad (44)$$

is close to unity. Both metals are resilient and are easy to weld or braze, but they cannot be soft-soldered. The major disadvantage of this couple is the low thermal diffusivity of the metals.

Especially in the case of a dry shield (fast response) it is good practice to sacrifice thermoelectric power in order to get high thermal diffusivity, provided of course that the mechanical properties are satisfactory. On this count the iron–nickel couple (35 μV deg^{-1}) is to be preferred to the chromel–constantan couple, even though the first-mentioned metals are both less resilient than the second pair.

Table 5 compares the properties of matched multi-junction thermocouples made of these two pairs of metals.

Table 5. Comparison of matched chromel–constantan and nickel–iron multi-junction thermocouples

Materials	n	nE (mV deg^{-1})	τ_W (sec)	Diameters [mm (s.w.g.)]	ξ (wires) (mW deg^{-1})	R (wires) (ohm)
Chromel–constantan	36	1·098	7·4 6·4	0·3454[29] 0·2946[31]	2·927	5·22
Nickel–iron	64	1·088	1·8 1·9	0·1118[41] 0·1219[40]	2·816	9·50

The numbers of junctions are chosen to give the same overall thermoelectric power E, and the wire diameters, matched according to Equation (44), are chosen to give roughly the same total heat-leak coefficient ξ. The figures quoted under ξ are based on a free length of 2 cm for all wires, and 100 per cent efficiency is assumed; reducing the efficiency would reduce both E and ξ by the same proportion. The electrical resistance of the nickel–iron couple is 80 per cent higher than that of the chromel–constantan couple, but this is unlikely to be significant because even with a low-resistance galvanometer the optimum circuit resistance will probably exceed 50 Ω. The much thinner nickel and iron wires will certainly be less stiff than the chromel and constantan wires, but this disadvantage is far outweighed by the fact that the time-constant of the nickel–iron couple is smaller than that of the chromel–constantan couple by a factor close to 4.

With both these couples, the thermocouple heat-leak coefficient will be less than 25 per cent of the total heat-leak coefficient (see Section II–4–B above) so that some increase in the wire diameters may be accepted if mechnical considerations make it desirable. It must be stressed, however, that if the sensitivity is to be doubled while keeping the electrical resistance constant, the number of junctions and the wire cross-section have both to be doubled, so that the heat-leak coefficient is increased fourfold.

D. Mounting the Thermocouple Junctions

(1) *Fixed Junctions*

One well tried way of attaching thermocouple junctions semi-permanently to a metal surface is to make the junctions on gold tags which are then secured to the surface by mica-insulated clamping screws. The details of the method are described in Chapter II Section III. It has the disadvantages that, especially when the thermocouple wires are short, the overall thermocouple efficiency is low, and also the clamping screws inevitably increase the radiating area of the surface. Although the simplicity of the method and the temperature-stability of the insulation make it almost uniquely suitable for high-temperature work, the disadvantages outweigh the advantages for work in the room-temperature range, where other insulating materials can be used. In particular, resin adhesives can be used both to insulate the junctions and to attach them to the metal surface.

It is always advantageous to make the junctions on thin plates of high-conductivity material which does not tarnish, such as gold. In the first place this increases the ratio a_i/λ_i for the insulation, and thus increases both the overall efficiency and the response coefficient, and also partially offsets the effect of the thermal lag of the thermocouple junctions by ensuring that a part of the radiating surface (the free surface of the junction plates) is at the same temperature as the junctions themselves. Ideally, then, the junction plates should cover most of the radiating surface, as in Calvet's[26, 27] conduction microcalorimeter, for example. In fact this is the best way of coping with the problems posed by the fast-response dry shield [Section II–3–B–(1), Section II–4–C–(3)]. The response coefficient which is effective in determining the behavior of the regulator is then that which refers to the whole conduction path between the shield heater and the outer thermocouple junction plates.

Figure 17 indicates one method of making this type of junction. The bent-over and flattened ends of the thermocouple wires are first spot-welded or soldered to the junction plate. The underside of the plate is then insulated by spraying or painting with a polyurethane or similar varnish, and finally the plate is attached to the metal surface by a thin layer of resin adhesive (the thermosetting types are preferable because they have lower vapor pressures than the self-curing types). It is important that the junction plates be firmly clamped to the metal surface while the resin adhesive is setting to ensure a uniform thickness; the use of a rubber or cork clamping pad to distribute the pressure gives satisfactory results. Alternatively a piece of thin tissue paper impregnated with resin adhesive (Araldite, for example, flows freely when warmed to about 60°C) may be used to attach the junction plate to the metal surface with adequate electrical insulation.

Although it is difficult to determine the optimum thickness of this type of junction plate, some guidance can be got from studying the temperature gradients in circular plates of different radii and thicknesses. If the plates are too thin, the radial temperature gradient is relatively steep close to the terminations of the thermocouple wires, but falls off very rapidly with increasing distance. This reduces the efficiency of the thermocouple, and its response is significantly slower than would be indicated by the time-constant

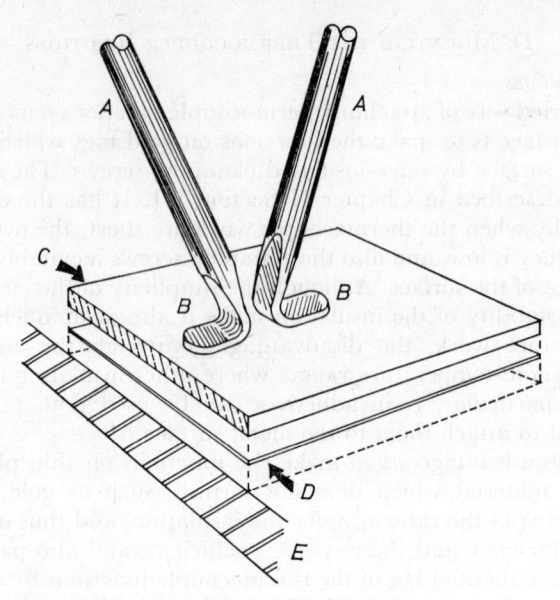

Figure 17. Fixed thermocouple junction—constructional diagram.
A, Thermocouple wires
B, Flattened ends, spot-welded to silver or gold plates (0·002 in. to 0·004 in. thick)
D, Adhesive/insulation (epoxide resin with tissue-paper or plastic-sheet spacer)
E, Inner wall of shield

of the wires. That is to say, only a small area of the plate, close to the wires, is effective in relation to the thermocouple performance. The temperature of the rest of the plate nevertheless lags slightly in phase relative to the underlying metal of the shield. If the plates are too thick, then although the thermocouple efficiency is close to 100 per cent and its response is as fast as the time-constant of the wires allows, the phase lag of the plate temperature is relatively large so that the effective shield response is slowed down; this effect will be serious only in the case of a fast-response dry shield, but any significant diminution in the shield response coefficient should be avoided. More important is the practical difficulty of attaching thick plates to the shield surface so as to keep uniform the thickness of the insulating and/or adhesive layer. It seems that the best compromise is to make the plate thickness about equal to the wire diameter times the ratio of the thermal conductivities of the wire and plate materials.

When the response of the shield itself is much slower than that of the junction plates, as is usually the case with a wet shield, for example, minimal advantage is gained by making the junction plates to cover the whole inner surface of the shield, and construction is facilitated by making the junction plates just large enough to optimize the thermocouple efficiency, i.e. of such a size that the radial temperature gradient under steady conditions nearly vanishes at the outer edge. For plates of optimum thickness as above, the minimum area is of the order of 100 times the sum of the cross-sectional areas of the thermocouple wires.

(2) Free-Contact Junctions

The discussion in the preceding Section indicates that having relatively inefficient junctions at the surface of the calorimeter vessel need not seriously affect the operation of the regulator, provided that the junctions at the shield surface are efficient, that the time-constant of the wires is small, and that the heat capacity of the inner (inefficient) junctions is made as small as possible. In this case the inner junctions need not be rigidly attached to the surface of the calorimeter vessel; a satisfactory alternative is that they be made to press against the surface with sufficient force to ensure that the contact area is at least comparable with the cross-sectional area of the thermocouple wires. With the latter arrangement the calorimeter vessel can be made completely removable from the shield. *Figure 18* shows a way of making this type of thermocouple.

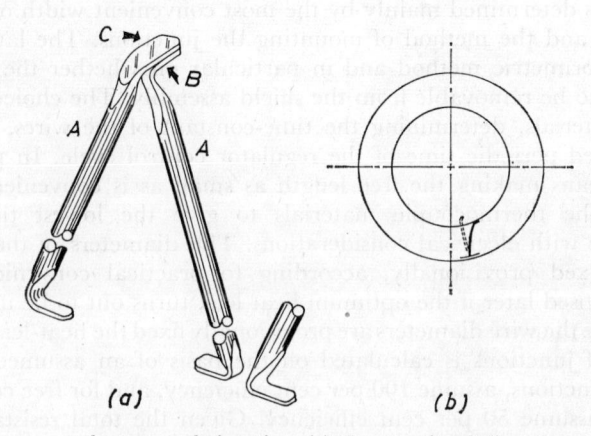

Figure 18. Free-contact thermocouple junction. (*a*) Constructional diagram.
    A, Thermocouple wires
    B, Flattened ends, spot-welded
    C, Insulation (polyurethane varnish)
(*b*) Plan section, showing orientation with calorimeter vessel in place (full line) and removed (broken line).

After making the outer (shield) junctions on plates as described above, the bent-over and flattened inner ends of the wires are spot-welded together and then filed flat to give a bearing surface. This surface is painted or sprayed with a thin film of polyurethane varnish. The configuration of the thermocouple when the calorimeter vessel is in its working position is indicated by the solid lines in *Figure 18(b)*. When the calorimeter vessel is removed, the inner thermocouple junctions spring back into their natural positions, indicated by the broken lines. The diameters of the thermocouple wires and the thickness of the shield junction plates are chosen so as to give the greatest mechanical resilience compatible with an acceptably low rate of heat leak.

The calorimeter vessel must be made with a rounded shoulder at its bottom end. Then inserting the calorimeter vessel with a turning motion makes the inner thermocouple junctions ride over this rounded shoulder, bending the thermocouple wires into the configuration indicated by the solid lines in *Figure 18(b)*. Note that the thermocouple junctions serve to locate the calorimeter vessel radially, so that only a vertical support is needed to take the weight.

A thermocouple of this type, made up of iron and nickel wires, 36 s.w.g. (0·193 mm diameter) 1·5 cm free length, might reasonably be expected to have an overall efficiency [Equation (43)] between 20 and 50 per cent. The thermocouple e.m.f. actually leads slightly in phase relative to the true signal when the latter varies sinusoidally at 0·02 cycle per second, but the heat-leak lags in phase by 3 or 4 seconds.

E. The Optimum Number of Junctions

In designing a high-performance multi-junction thermocouple, it is probably simplest to decide first the free length of the thermocouple wires, since this is determined mainly by the most convenient width of the insulating space and the method of mounting the junctions. The latter depends on the calorimetric method and in particular on whether the calorimeter vessel has to be removable from the shield assembly. The choice of thermocouple materials, determining the time-constant of the wires, depends on the expected periodic time of the regulator control cycle. In practice this usually means making the free length as small as is convenient and then choosing the thermocouple materials to give the lowest time-constant compatible with electrical considerations. The diameters of the wires may then be fixed provisionally, according to practical convenience. These may be revised later if the optimum heat leak turns out to be unacceptably large. Once the wire diameters are provisionally fixed the heat-leak coefficient per pair of junctions is calculated on the basis of an assumed efficiency; for fixed junctions, assume 100 per cent efficiency, and for free contact inner junctions assume 50 per cent efficiency. Given the total resistance for the thermocouple circuit, this leaves only the number of junctions to be decided. Once this is fixed the junction design may be finalized and the actual efficiency measured. The number of junctions is then adjusted accordingly (this does not alter the estimated heat-leak coefficient), and finally the electrical resistance of the thermocouple is checked against the total circuit resistance.

If the thermocouple circuit includes a galvanometer as primary amplifier it is usually necessary to consider the optimum number of junctions and the total circuit resistance together, since both these variables affect the sensitivity of the circuit, whereas the heat-leak coefficient depends only on the number of junctions. The choice of values for the circuit resistance R and the number of junctions n depends, of course, on the specifications of the whole control system, and will entail a trial and error procedure unless the theory of the control system has been worked out in detail and the other operating parameters evaluated.

The two-valued response regulators discussed in Section II–3–A–(3) above, in which the primary amplifier and detector are a sensitive galvanometer and a photocell bridge, have the further complication that changing the thermocouple circuit resistance R also changes the total time delay, χ, in the regulator, and hence its performance. Consequently finding the optimum values of R and n is especially difficult in these cases. Nevertheless it is instructive to carry out this exercise because it illustrates a methodology applicable to all regulators using thermocouple circuits as signal sources. Also it leads, in the particular case of the two-valued response regulator, to useful general conclusions.

The operating parameters of a two-valued response regulator are[37]: the shield response coefficient, ψ, defined by Equations (33) and (34) for a dry shield and by Equations (35) and (36) for a wet shield; the shield heating and cooling rates v' and v''; the regulator backlash ϕ; and the total time-delay χ. The first two are determined, respectively, by the shield design and by the expected rate of heating of the calorimeter vessel (and in a wet shield by the need to minimize the time-delay due to liquid mixing); changing v' and v'' does not in fact alter the optimum values of R and n. Both ϕ and χ also depend on n and R.

For a given rate of heating of the calorimeter vessel, the calorimetrically significant quantity is the total heat leak rate, determined by the mean value ζ of the thermal signal (temperature out-of-balance between calorimeter vessel and shield) and by the total heat leak coefficient ξ. ζ is a function of ψ, $v' + v''$, ϕ and χ; ξ is determined by the geometry of the shield and the thermocouple, and by the number of junctions n. Thus, by taking ψ and $v' + v''$ as fixed, together with the main parts of ξ and χ, we can establish, for a given rate of heating of the calorimeter vessel, the values of ξ and ζ for various values of n and R, and hence minimize the product $\xi\zeta$. The sequence of calculations is first to select a value for R and thence estimate the corresponding value of χ as the sum of the regulator time-delay and that due to the galvanometer when the circuit resistance is R, then to select a value for n and to calculate the corresponding values for ϕ and ξ. Cruickshank's[37] Equation 5.13 is then solved for the parameter μ by successive approximation, whence ζ is calculated according to Equation[37] 5.24. Finally the product $\xi\zeta$ is calculated. The procedure is repeated for various values of n, and the optimum value of n is found by graphical interpolation, corresponding to the minimum of $\xi\zeta$ as a function of n for that particular value of R. The entire procedure is then repeated for other values of R, and plotting the optimum values $\xi\zeta$ for each R against R indicates the optimum value of R.

Consider, for example, a wet shield like that described in Section II–3–C, for which ψ is 0·20 sec^{-1} and $v' + v''$ is 0·005 deg sec^{-1}. Suppose that the calorimeter vessel is being heated at 0·0005 deg sec^{-1}, so that the ratio of positive to negative half-cycle periods is 1·5; since ζ is a nearly linear function of the calorimeter vessel heating rate, a value other than 0·0005 deg sec^{-1} would do as well. We take the time-delay in the liquid-flow system to be 2·2 sec. With a Tinsley type 4500LS magnetically damped galvanometer used with minimum magnetic damping at 1·8 meters (sensitivity 75·0 cm μA^{-1}) the galvanometer time-delay varies with the circuit resistance R from 2·9 sec when R is 50 Ω to 0·1 sec when R is 600 Ω, so that the total time-delay χ varies with R as indicated in *Table 6*.

Taking the optical backlash in the photocell bridge as 0·15 cm, the control backlash ϕ (deg) is related to n and R by

$$\phi = 0 \cdot 3R/nGE, \qquad (45)$$

in which G is the galvanometer sensitivity in cm μA^{-1} and E is the thermocouple sensitivity in μV deg^{-1} per pair of junctions. With nickel–iron thermocouples E is about 35 μV deg^{-1}, so in this case Equation (45) becomes

$$\phi = (1\cdot143 \times 10^{-4})R/n. \qquad (46)$$

If the thermocouple wires are both 38 s.w.g. (0·152 mm) and 1·5 cm long, then taking the heat-leak coefficient due to radiation, etc., as 0·0140 watt deg^{-1} (see Section II–4–C above), the total heat-leak coefficient is given by

$$\xi = 0\cdot0140 + n(9\cdot99 \times 10^{-5})\eta. \qquad (47)$$

For simplicity we take η to be 50 per cent. Changing η in the following calculations alters no other variable except n, which is inversely proportional to η. The results of the program of calculations just described are summarized in *Table 6*.

Table 6. Calculated rate of heat leak for various values of thermocouple circuit resistance and number of junctions for a calorimeter heated at 0·0005 deg sec^{-1}

R (ohm)	χ (sec)	n	$10^3 \phi$ (deg)	$10^3 \zeta$ (deg)	$10^3 \xi$ (watt deg^{-1})	$10^5 \zeta\xi$ (watt)
150	3·0	16	2·14	2·87	14·8	4·25
150	3·0	24	1·43	2·73	15·2	4·15
150	**3·0**	**30**	**1·143**	**2·672**	**15·50**	**4·143**
150	3·0	32	1·07	2·66	15·6	4·15
150	3·0	40	0·86	2·62	16·0	4·19
300	2·5	24	2·86	2·67	15·2	4·06
300	2·5	36	1·90	2·51	15·8	3·96
300	**2·5**	**42**	**1·635**	**2·446**	**16·10**	**3·938**
300	2·5	48	1·43	2·40	16·4	3·95
300	2·5	60	1·15	2·34	17·0	3·98
400	2·4	30	3·05	2·64	15·5	4·09
400	2·4	40	2·28	2·51	16·0	4·02
400	**2·4**	**50**	**1·830**	**2·425**	**16·50**	**4·001**
400	2·4	60	1·52	2·36	17·0	4·02
400	2·4	70	1·31	2·31	17·5	4·05

The bold figures in *Table 6* refer to the optimum number of junctions for the value of R in the left-hand column. The trends evident in *Table 6* can be seen more clearly in *Figure 19*, on which the optimum number of junctions n^* and the corresponding heat-leak rate $\xi\zeta$ are plotted against the circuit resistance R.

The minimum value of $\xi\zeta$ evidently occurs near $R = 275$ Ω, and the corresponding optimum number of junctions n^* is about 40 (4 banks of 10, say). If the junctions were 80 per cent efficient, n^* would be 24. In either case the heat-leak rate under optimum conditions is just less than 40 μW for a calorimeter vessel heating rate of 0·0005 deg sec^{-1}. That is to say the net heat leak during a 1 degree heating will be $-0\cdot08$ joule.

In fact the optimum conditions (n^* and R^*) always occur with the thermocouple circuit resistance considerably larger than the critical damping resistance, and the optimum number of junctions always increases as the total time-delay is increased. The shallowness of the minimum of the curve of $\xi\zeta$ against R suggests that for most purposes it is sufficient to select R arbitrarily in the range of 2 or 3 times the critical damping resistance (if

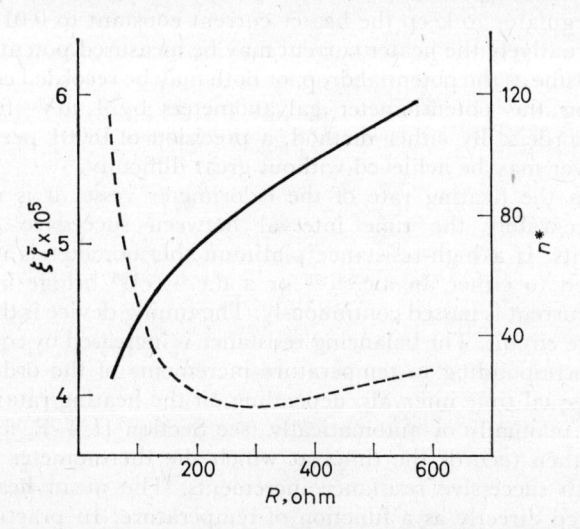

Figure 19. Optimum number of junctions and heat-leak rate as functions of galvanometer circuit resistance for a two-valued response regulator. Tinsley type 4500 LS galvanometer, nickel–iron thermocouples (38 s.w.g., 1·5 cm long), 80 ml calorimeter vessel heating at 5×10^{-4} deg sec^{-1}.

the galvanometer time-delay is small) and calculate the optimum number of junctions for the selected R. In the case considered above the thermocouple resistance is close to $0.75\, n\, \Omega$, so that electrical resistance is not a primary consideration in the design of the thermocouple.

It is interesting to note that the application of a correctly specified Gouy modulation to the regulator to which *Table 6* refers should reduce the magnitude of the mean signal to about 0·0005 deg when the calorimeter vessel heating rate is 0·0005 deg sec^{-1}, so that the heat leak over a 1 degree heating would be only about 0·016 joule. With a specimen of 80 ml of water, this heat loss represents an error in the specific heat of less than 0·005 per cent, while the error corresponding to the *uncertainty* on the heat loss is less than 0·002 per cent.

5. Experimental procedure

A. Continuously Heated Calorimeters

(1) *Measuring Procedures*

In an automatic adiabatic calorimeter the primary quantities to be measured are the rate of temperature increase and the heating power. The rate of heat leak is either computed from secondary measurements (single-vessel calorimeter) or neglected (twin-vessel calorimeter). The resistance element of the calorimeter vessel heater will usually have only a small temperature coefficient of resistance. Consequently for many purposes it may suffice to measure at intervals the potential drop across the heater resistance. For very precise work it will be necessary, in addition, to use an electronic regulator to keep the heater current constant to 0·01 per cent or better. Alternatively the heater current may be measured potentiometrically at the same time as the potential drop, or both may be recorded continuously, by replacing the potentiometer galvanometers by 1 mV (full-scale deflection) recorders. By either method, a precision of 0·001 per cent in the heating power may be achieved without great difficulty.

To obtain the heating rate of the calorimeter vessel it is necessary to measure accurately the time interval between successive temperature measurements. If a high-resistance platinum thermometer is used, it may be connected to either an a.c.[58, 76] or a d.c.[4, 7, 126] bridge in which the measuring current is passed continuously. The timing device is then included in the bridge circuit. The balancing resistance is increased by equal small increments corresponding to temperature increments of the order of 1 deg, at roughly equal time intervals, depending on the heating rate; this can be done either manually or automatically (see Section II–1–E, for example). The timer then records the times at which the thermometer bridge is in balance with successive resistance increments. The mean heating rate is then observed directly as a function of temperature. In practice it is convenient to use an automatic timer and print-out unit, and a typical apparatus is described below.

The principle is similar to that of the temperature-program regulator described in Section II–1–E, but instead of determining the bridge out-of-balance at fixed time intervals, the time intervals between successive bridge balance points are measured. The balancing arm of the Wheatstone bridge thermometer circuit is a decade-dial resistance box, operated by a telephone-type step relay. If the initial resistance of the decade arm is greater than that of the resistance thermometer, then as the temperature rises the light beam from the mirror galvanometer in the Wheatstone bridge circuit moves towards its zero position, where it illuminates a photocell. This triggers a series of operations whose sequence is controlled by a servomotor driven camshaft operating mechanical switches. First the count from a double electronic timer is transferred to a tape print-out unit, and the camshaft commences to turn. Next the photocell is locked to prevent retriggering on the return sweep of the galvanometer image, the bridge voltage is transferred to a dummy resistance, and the step relay is operated, increasing the bridge balancing resistance by one unit. The bridge power supply is then reconnected (swinging the galvanometer image to maximum deflection) and the trigger photocell is resensitized, ready for the next balance point of the thermometer bridge. The pulses which are fed continuously into the counter unit are supplied by a quartz crystal oscillator, so that the precision of the time measurements is potentially better than that of all the other measured properties.

Probably the most important source of error in this type of timing unit is due to drifting of the galvanometer zero. The image position corresponding to zero coil current in the galvanometer is always subject to changes due to variation in ambient temperature and humidity, and with most galvanometers a continued large deflection (such as occurs immediately after the bridge resistance is increased) shifts the zero towards the deflection. The situation for which zero bridge e.m.f. corresponds to the original image position at zero coil current can be regained by including a small e.m.f. in the galvanometer shunt circuit. A device for doing this automatically was developed by Todd[137]. The basic circuit is shown in *Figure 20*.

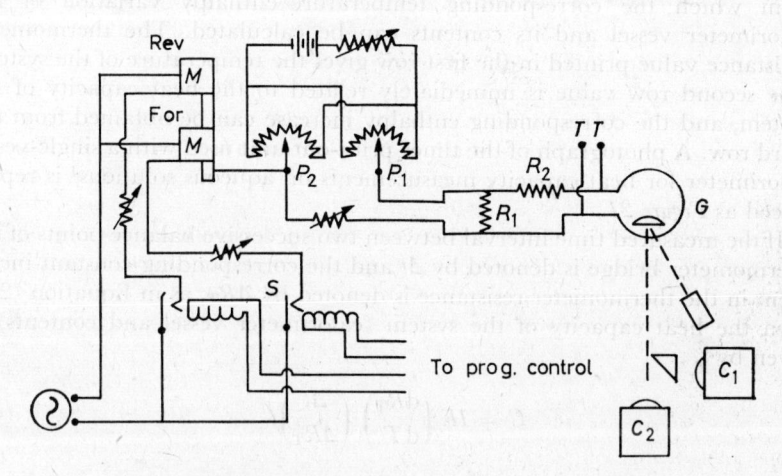

Figure 20. Automatic galvanometer-zero adjusting circuit[137]
C_1, Zero control photocell
C_2, Relay photocell
G, Galvanometer
M, Servomotor
P_1, Manually operated potentiometer (coarse control)
P_2, Servomotor-driven compensating potentiometer
S, Two-way relay actuated by C_1
T's, Connections to measuring potentiometer or bridge

The e.m.f. across the 1 Ω resistance R_1 is determined by the relative settings of the two helical potentiometers P_1 and P_2. P_1 is used for initial manual setting of the galvanometer zero point. P_2 is driven by a reversing servomotor which is controlled by a subsidiary program regulator. As the thermometer bridge approaches its balance point, the galvanometer image sweeps across the photocell C_1, initiating the zero-adjusting procedure. First the servomotor turns P_2 so as to increase the galvanometer deflection away from the main trigger photocell C_2, and at the same time the bridge is switched to a dummy resistance, leaving the galvanometer connected only to the shunt resistance R_2 and the zero-adjusting circuit. The servomotor is then reversed, and continues running until the galvanometer image activates the trigger photocell C_2. This completes the adjustment, and the bridge is reconnected to the galvanometer, C_2 reverts to its normal function (to trigger the timer print-out) and C_1 remains de-sensitized until the time recording sequence has been completed. This circuit may easily be adapted to an a.c. bridge with a phase-sensitive detector.

The print-out unit is designed to work as an adding machine, and prints three rows of digits for each equilibrium passage of the bridge circuit. The first row gives the programmed resistance value (i.e. the resistance of the platinum thermometer when the bridge passes through its balance point). The second row gives the number of time-pulses counted between successive balance points, and the third row gives the total time, equal to the sum of all the heating periods from the commencement of the experiment. Thus the printed tape represents a numerically recorded time *vs.* resistance function from which the corresponding temperature–enthalpy variation of the calorimeter vessel and its contents can be calculated. The thermometer resistance value printed in the first row gives the temperature of the system. The second row value is immediately related to the heat capacity of the system, and the corresponding enthalpy increase can be obtained from the third row. A photograph of the timer print-out unit used with a single-vessel calorimeter for heat capacity measurements on aqueous solutions[4] is reproduced as *Figure 21*.

If the measured time interval between two successive balance points of the thermometer bridge is denoted by Δt and the corresponding constant increment in the thermometer resistance is denoted by ΔR_T, as in Equation (29), then the heat capacity of the system (calorimeter vessel and contents) is given by

$$C = IE \left(\frac{dR_T}{dT}\right) \left(\frac{\Delta t}{\Delta R_T}\right), \tag{48}$$

in which (dR_T/dT) is the temperature coefficient of the thermometer resistance at that temperature, I is the current through the heater and E is the recorded potential drop across the heater. Equation (48) shows that for a particular heat capacity the time interval Δt varies inversely with the heating power EI. Making EI small increases the accuracy of the timing but decreases the accuracy with which EI can be measured and at the same time reduces the rate at which experiments can be completed. Making EI large, on the other hand, increases the difficulty of adiabatic control and may cause errors due to variations in the temperature distribution inside the calorimeter vessel as the temperature increases. For heat capacity measurements on aqueous solutions in calorimeter vessels of the type described in Section II–2–G (see *Figure 10*) a heating power of about 5 watts has proven satisfactory. With benzene and diphenylmethane, Todd[137] used various heating rates, the greatest being about 0·0015 deg sec^{-1}, and found that within this range the measured heat capacities were independent of heating rate.

The method of evaluating the heat leak depends on which types of adiabatic shield and regulator are used. With a dry shield and either a continuous reponse regulator incorporating reset response or a two-valued response regulator with heating rate adjustment [Section II–3–A–(3)] the net heat leak approximates to zero; with a simple two-valued response regulator the heat-leak rate is roughly proportional to the heating rate, provided that the shield is surrounded by a controlled jacket[37]; otherwise it will generally increase with increasing temperature, as the rate of heat

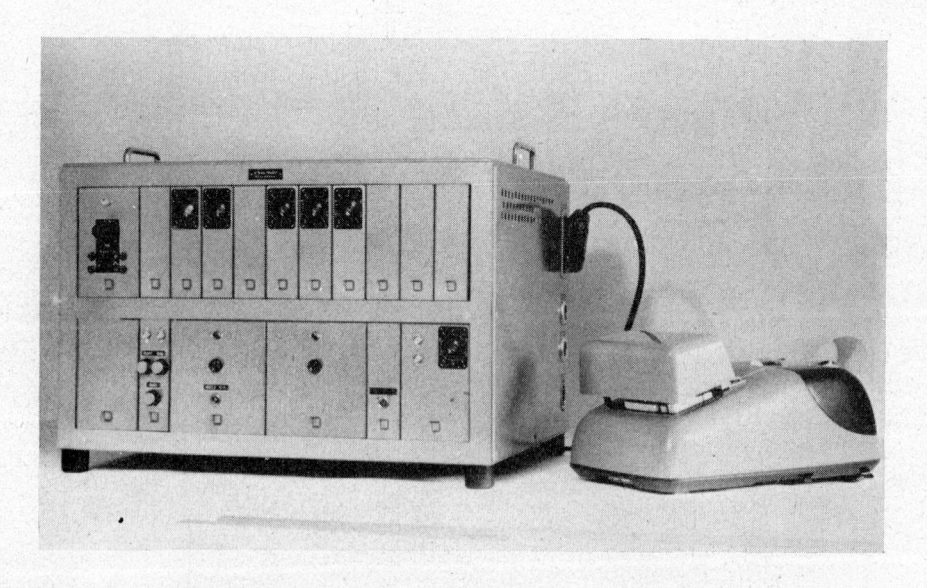

Figure 21. Automatic timer and print-out unit[4] for continuously heated calorimeters. Made by Dr. Brandt and Company, Bochum-Dahlhausen, Germany.

loss from the shield increases. In the latter case it is necessary to observe and record the average ratio of the "on" and the "off" periods of the regulator from time to time, so that the rate of heat leak may be calculated.

(2) *Calculations*

The complexity of the numerical calculations to be carried out on the primary data depends upon the nature of the problem being studied, and also upon the precision required. For example, if high precision is unnecessary, Equation (48) may be used to evaluate the heat capacity (vessel plus contents) over a single temperature increment directly from the measured value of Δt. If the total heat capacity changes only very slowly with increasing temperature, it may suffice to use the arithmetic mean of a series of measured Δt values (this can be done automatically by the adding machine of the print-out unit described above). In general, however, a series of values of $\Delta t/\Delta R_T$ will be fitted to a polynomial in T, using a "least squares", or other equivalent procedure. In this case a card or tape punching unit attached to the print-out unit may be used to transfer the primary data to a digital computer. Various arrangements of this kind have been developed in recent years[1, 53, 127, 137]. In the case of a simple two-valued response regulator the durations of "on" and "off" half cycles can be recorded and a subsidiary computer program used to evaluate the heat leak; this is then fed into the main program. When the heat leak is simply related to the heating rate the appropriate correction can be included in the main program.

Since sharp transitions in structure do not normally occur in liquids in the temperature range 250°K to 400°K, a relatively large number of successive Δt values may be used for one computation, thereby improving the precision of the heat capacity value at any one temperature, as interpolated from the best-fit curve of $\Delta t/\Delta R_T$ vs. T. In some special cases where relatively sharp transitions do occur, a small heating rate must be used, to minimize errors due to changes in the temperature configuration inside the calorimeter vessel, and it is better to fit small sections of the curve of $\Delta t/\Delta R_T$ vs. T by simple equations than to attempt to fit a region where the heat capacity changes rapidly by a many-term polynomial in T. This problem arises, for example, with the helix-to-random-coil transition of polypeptides in aqueous solution[4].

The experimental procedures and methods of calculation just described depend on the nature of the investigation rather than upon the use of the various automatic devices. In particular, the experimental procedure to be followed with a manually controlled calorimeter is implicit in the foregoing discussion, and the extent to which the data are processed automatically will be determined more by economic factors than by the methods to be employed.

B. Intermittently Heated Calorimeters

(1) *Measuring Procedures*

In this case the procedure for measuring or recording the *power* input to the calorimeter vessel is just the same as in the continuously heated calorimeter. The time interval which has to be measured, on the other hand, is that during which the heating power is applied, rather than that corresponding to a pre-selected temperature increment during a steady heating. This means that the timing circuit has to be linked to the energy circuit in an intermittently heated calorimeter. The basic parts of the heater timing unit are, as in the case of the temperature-rise timing unit, a quartz crystal oscillator or valve maintained tuning fork, and a pulse counter. The two events to be timed are, respectively, the transfer of the power supply from a ballast resistance to the calorimeter vessel heater, and the reverse process. For this, a fast acting mechanical switch is necessary to minimize the effects of switching transients, and so it is convenient to use a coupled switch to start and stop the pulse counter. This scheme has the advantage that only one counter reading has to be recorded for each temperature increment, and this reading may be transcribed manually, or fed into a printing or card punching unit with the counter stopped.

With intermittent heatings, separated by equilibration periods of duration comparable to the heating periods, each heat capacity measurement over a 1 deg temperature increment takes at least three times as long as in a continuously heated calorimeter. Thus if the total heating (over 30 or 40 deg) is to occupy about the same time in both cases, the number of measurements made with the intermittently heated calorimeter cannot be greater than one third of the number made with the continuously heated calorimeter (assuming that in the former case the heatings between the measured increments are relatively fast). Consequently, to achieve the same precision on the smoothed results, it is necessary that the error on the individual temperature measurements in the intermittently heated calorimeter does not exceed about half that in the continuously heated calorimeter. In the former case, however, measurements are made under steady conditions, so that several observations may be made at each temperature; also a bridge may be used which is fully compensated for changes in lead resistance, etc. Smith's type III, or type II, or Mueller bridges are well suited to this work, whereas only the first is suitable for stepping operation. With these three bridges it is advantageous to pass the bridge current (either a.c. or d.c., but preferably the former) only when the bridge is being balanced. Short pulses (0·1 or 0·2 sec duration) suffice to indicate the direction of the out-of-balance. This mode of operation eliminates the problem of zero drift in the indicating device, the thermometer current need not be stabilized, and its heating effect is negligibly small. Two practical inconveniences arise: to achieve the best precision the temperature from which a measured increment begins must be chosen such that the expected resistance increment can be accommodated without changing the settings of the higher decades of the bridge, since even auto-calibration cannot reduce the uncertainties on the 1 Ω and 10 Ω steps to less than 10^{-4} Ω; secondly the bridge has to be kept balanced with respect to capacitance (within a few tens of picofarads) or switching transients become obtrusive.

Unless the heat leak is kept negligibly small (by a modified two-valued response regulator or a regulator with reset response) it has to be evaluated, preferably as a correction to the input energy. With any two-valued response regulator it is worthwhile to observe the variation with time of the ratio of the "on" to the "off" half-cycle periods, so that tables of the mean signal ζ as a function of this ratio[37] may be used to estimate ζ as a function of time.

(2) Calculations

The first operation is necessarily the conversion of the means of the measured resistance values into temperatures. This may be done using a routine program in which the adjustable factors are determined by the characteristics of the resistance thermometer. The result of this program is a list of temperatures and temperature increments.

The computation of the total energy inputs is more complicated, unless the potential drop and heater current are continuously recorded. If the values of E and I are observed simultaneously at roughly equal time-intervals, the first step is to evaluate the products EI. If E and I are measured alternately, it is better first to interpolate values of whichever varies less with time, to correspond to the times at which the other was measured. Next, the smallest value of EI is found, and multiplied by the total time of heating to give the main part of the heating energy. Finally, the residuals of EI, being at roughly equal time intervals, are integrated by means of a program based on Simpson's rule. Only insignificant error is introduced by assuming that the first and last of these residuals extend, respectively, to the beginning and the end of the heating period.

After adding in the heat leak term, division of the total energy by the corresponding Δt gives the total heat capacity at that temperature. The collected values of the total heat capacity (at 4 or 5 deg intervals) are then fitted to a polynomial in T, using a standard program. It is in most cases convenient to limit the number of terms to the third or fourth power, depending on the calculated standard deviations.

III. Isothermal Drop Calorimetry

1. *Introduction*

It has been pointed out in Chapter 1 that drop calorimetry may be looked upon as a particular application—the most common one, in fact—of the method of mixtures. It consists in bringing the sample, usually in a container, to a uniform, and accurately known temperature T_1 inside a thermostat. The sample is then quickly transferred, usually by lowering on a windlass or by dropping, into a receiver. The receiver is initially at a temperature T_0 different from, and generally lower than T_1, and acts as a heat metering device. This method, one of the first to be used in calorimetry[112], is ideally suited to the measurement of heat capacities of one-component systems, as exemplified by the classical work of the National Bureau of Standards group on liquid metals (see Chapter 7). In most cases the isobaric heat capacity is measured, and the primary experimental data are the average molal isobaric heat capacities between the initial and final temperatures. To get the true isobaric heat capacity at a particular temperature, measurements are made at a series of initial temperatures, and the temperature–enthalpy relation so obtained is differentiated with respect to temperature.

Obviously the total heat effect must be corrected for the contribution of any possible phase transition of the sample in the temperature interval covered. However, since we are concerned here only with liquids and solutions, this complication arises only in unusual cases such as liquid sulfur[94] which undergoes a *lambda*-point transition in the neighborhood of 160°C. On the other hand, working within the narrow temperature range considered in this chapter imposes rather severe demands from the viewpoints of precision and accuracy. The heat quantities involved are small, and furthermore the heat capacities sought for are obtained from differences between much larger quantities. Because of the obvious limitations on the size of the sample, the other variables (temperature, mass, etc.) must be known with the best possible accuracy, and the heat metering device must be especially sensitive and precise.

In principle any type of calorimeter may be used as a receiver in the drop method, provided of course that it can be made with a top opening through which the sample can be introduced. This condition is not easily met in certain instruments, such as the Tian–Calvet microcalorimeter[26] which, in fact, has not been used so far in this fashion. Except for recent applications of the mercury-filled isoperibol calorimeter[106] and the copper block adiabatic calorimeter[101], the drop calorimeter receivers have generally been of the isothermal (phase transformation) type. An important advantage of these calorimeters is the great simplicity of construction and operation. Only one temperature, that of the "furnace", needs to be measured accurately, and the measurement of the heat input to the receiver in terms of the weight of mercury displaced, or the length of a mercury thread in a capillary tube, avoids the need for elaborate and costly electrical instruments. Operation of an isothermal calorimeter in the neighborhood of room temperature makes heat leak control easier, and better, than at extreme temperatures, and the attainable accuracy is, relatively speaking, greater in the former case.

On the negative side, the fixed temperature of the receiver is a limitation, although this limitation may be partly overcome by using several receivers with different working substances, as described in Section III–2 below. More serious, from the point of view of heat capacity measurements, is the rather modest precision (about 0·1 per cent) and accuracy (of the order of 1 per cent) attained so far with liquid/solid isothermal calorimeters. As suggested below the sensitivity of this type of calorimeter might be improved somewhat, possibly by a factor of 2 or 3, through the use of specially selected substances, but the prospects for considerable improvement are poor. The use of other phase changes, such as vaporization[30, 90, 138] or sublimation[100] can give much greater sensitivity, albeit at the cost of increased experimental complexity. Certainly for the investigation of fine effects, such as the variation of partial molal heat capacity with temperature and concentration in dilute solutions, drop calorimetry is not particularly suitable; for this work a comparison method using, for instance, the twin calorimeter would no doubt be a better choice. The drop method is much better adapted to the rapid determination of enthalpy increments, and the isothermal calorimeter by itself is well suited to the measurement of such quantities as heats of reaction or dilution and heats of fusion, vaporization etc.

2. Isothermal drop calorimeter receivers

The prototype of isothermal drop calorimeter receivers, and the one most frequently used in heat capacity measurements, is the Bunsen ice calorimeter. A modern version of this instrument[63] is fully described in Chapter 7. Now this calorimeter suffers from certain drawbacks. For instance, on the practical side, its operation requires the handling of large amounts of pure crushed ice. Possibly this tedium could be relieved by using an ice machine as container. Incidentally, the use of an ice bath for the latter, as has been customary so far, precludes any control of the heat leak from the calorimeter itself by adjusting the temperature of its immediate environment. A more serious disadvantage is the rather low sensitivity, namely 0·00370 g of mercury displaced per joule. The use of some organic compounds as working substance gives sensitivities three or four times as great as this. This peculiarity of the ice calorimeter is due mainly to the large heat of fusion of ice; the change in specific volume on melting is of similar magnitude (0·1 ml mole^{-1}) for most substances. This change is negative in the case of ice, so that the mercury "displaced" as a result of the absorption of heat is in fact sucked into the calorimeter from a weighed reservoir. There is also the limitation common to all isothermal calorimeters of a fixed operating temperature. Therefore, it is desirable to have a series of such calorimeters with different working temperatures.

So far about half a dozen organic compounds have been used in isothermal calorimeters. Their relevant physical properties are listed in Table 7. To be suitable for use in a fusion calorimeter a substance must meet

Table 7. Melting points and calibration factors of compounds used in isothermal calorimeters

Compound	References	m.p. (°C)	ΔHf (joule g^{-1})	k [g(Hg) joule^{-1}]
Water	45, 63, 64	0·00	333·5	0·00370
Acetic acid	42	16·58	187·0	0·01114
Anethole	68	22·5	—	0·00825
Diphenyl methane	17, 49, 66, 119, 123	26·3	105·4	—
Diphenyl ether	39, 62, 78, 85, 87, 116	26·90	101·15$_2$	0·01264
Benzalacetone	29	40·02	128·9	0·01087
Phenol	69, 118	40·35	121·3	0·00547
Naphthalene	13, 32, 98, 132	80·3	149·0	0·01398

several requirements. First, it must be readily available in sufficient quantity, and be stable at its melting point so that it can be prepared in a very high degree of purity by fractional distillation and/or crystallization. Light, oxygen, and moisture should not affect it appreciably, nor should mercury or other metals (copper, brass) sometimes used in the construction of the calorimeter vessel. Once in the calorimeter, it must remain unaltered indefinitely, as otherwise "premelting" will result from concentration of impurities on gradual solidification. Furthermore, it must exhibit sharp melting, at the desired temperature, and freeze in a single crystalline modification, preferably in large, clear crystals.

This combination of desirable features in the same substance is not very common. Only two of the organic compounds tried until now may be considered as satisfactory, namely naphthalene ($C_{10}H_8$) and diphenyl ether ($C_6H_5OC_6H_5$), also called phenoxybenzene; the latter particularly so, because its melting point (26·90°C) is quite near to room temperature and it is relatively inert to oxidation. Consequently, not only are the heat leaks minimized, but any extrapolation of the results to the standard temperature, 25°C, is also small. Three different versions of this isothermal calorimeter

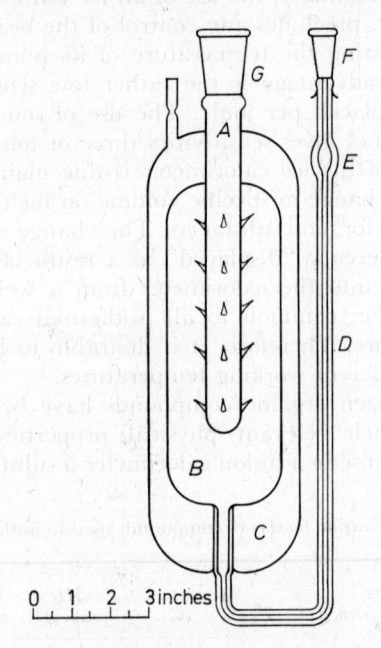

Figure 22. All-glass diphenyl ether isothermal calorimeter[62]—scaled section. The dimensions are optima for a thermal capacity about 2500 joules.

    B, Diphenyl ether compartment
    C, Insulating jacket
    F, Connection to mercury measuring device

have been described in the literature recently. The first one, by Giguère et al.[62], was intended for the measurement of various thermodynamic properties of hydrogen peroxide and its aqueous solutions. The all glass construction (see *Figure 22*) has the great advantage that the formation of the mantle of solid around the central well may be observed directly, and defects such as holes, and "bridges" of solid between the two walls, may be detected. The dimensions shown here are not critical, except possibly for the glass studs, ⅜ in. long, on the outside of the central tube A, which prevent the mantle of frozen ether from slipping down when the first layer of solid is melted from the inside. If made too long they would eventually puncture the mantle because the thermal conductivity of glass is appreciably higher than that of the ether; if too short, the total heat-absorption capacity of the calorimeter would be reduced.

The tube connecting the middle chamber B with the capillary D may be straight as shown, if the walls are thick enough, or shaped into a spiral. A small bulb E blown on the capillary tube has been found useful for trapping any air bubbles sneaking in accidentally through the mercury measuring device connected at F. The outer jacket is evacuated to 10^{-5} torr before sealing off for better insulation from the surrounding medium.

The diphenyl ether is introduced into the middle chamber of the calorimeter under low pressure directly from the purification apparatus, after thorough degassing. This latter precaution is extremely important on account of the great solubility of air in most organic liquids (from five to ten times that in water). The poor performance and low calibration factor of early instruments[66, 123] were no doubt due mostly to this effect. As a rule three successive freezings and meltings *in vacuo* are sufficient to achieve complete deaeration. In contrast, the solubility of air in mercury is so small that it causes no noticeable complication once the calorimeter is filled. The displaced mercury may be accounted for by simple weighing, to 0·1 mg, in small beakers, or by measuring the length of a column in a constant-bore capillary. The former method is sufficient when only the total heat effect is required, as in heat capacity measurements. In that case a useful adjunct consists of a short length of fine capillary tubing connected to the main mercury tube through a three-way valve[63] for evaluating the heat leak or ascertaining the end of the reaction. Incidentally, it is worth noting here that the fusion calorimeter is another instance of a physico-chemical instrument made possible by the unique properties of metallic mercury.

Since the diphenyl ether calorimeter operates in the neighborhood of the room temperature, a water thermostat is suitable as a container, provided its temperature is kept constant to better than 10^{-3} deg. This constancy is readily achieved in a large water bath with efficient stirring and a sensitive thermoregulator. The thermoregulator should be free from drifts and adjustable continuously. A commercial mercury regulator of the "Magnaset" type has been found satisfactory when combined with a sensitive electronic relay and an electric immersion heater of suitable capacity and fairly large surface. Long metal-tube heaters are preferable as they can be placed at the periphery of the water bath for a more even heat distribution. Very sensitive thermoregulators have been realized by combining mercury with a second liquid of very high thermal expansivity, such as dibutyl phthalate, the latter being contained in a coil of thin-walled copper tubing whereas the mercury section is of glass or stainless steel[36]. As for stirrers, they are usually of the impeller or the turbine types. Experience has shown that high torque is more profitable than high speed [cf. Section II–2–D–(1)].

A very effective, and at the same time extremely simple device, which has been generally ignored in the search for improved temperature regulation, is that of internal insulation by multiple enclosures discussed long ago by Tian[134]. To quote one particular instance[62], merely surrounding the calorimeter vessel with a Plexiglas box was sufficient to reduce temperature fluctuations by a factor of 50. Since this device may be multiplied at will, it seems like an ideal method of achieving a truly isothermal medium. For the latest developments in the art of liquid thermostats the reader is referred to the various monographs published from time to time.

For isothermal calorimeters operating at somewhat higher temperatures, Coffin[32] has devised an enclosure made of a glass jacket (see *Figures 23* and *24*) filled with the vapor of an appropriate liquid boiling in a reflux system under a constant pressure. This is achieved by the combination of a mercury manostat with a surge vessel and an electric heater. The following pairs of substances, one of which melts at about the normal boiling point of the other, have been used in this fashion: benzalacetone–methylal[29], at 40°C; naphthalene–benzene[32], at 80°C; anthracene–naphthalene, at 218°C.

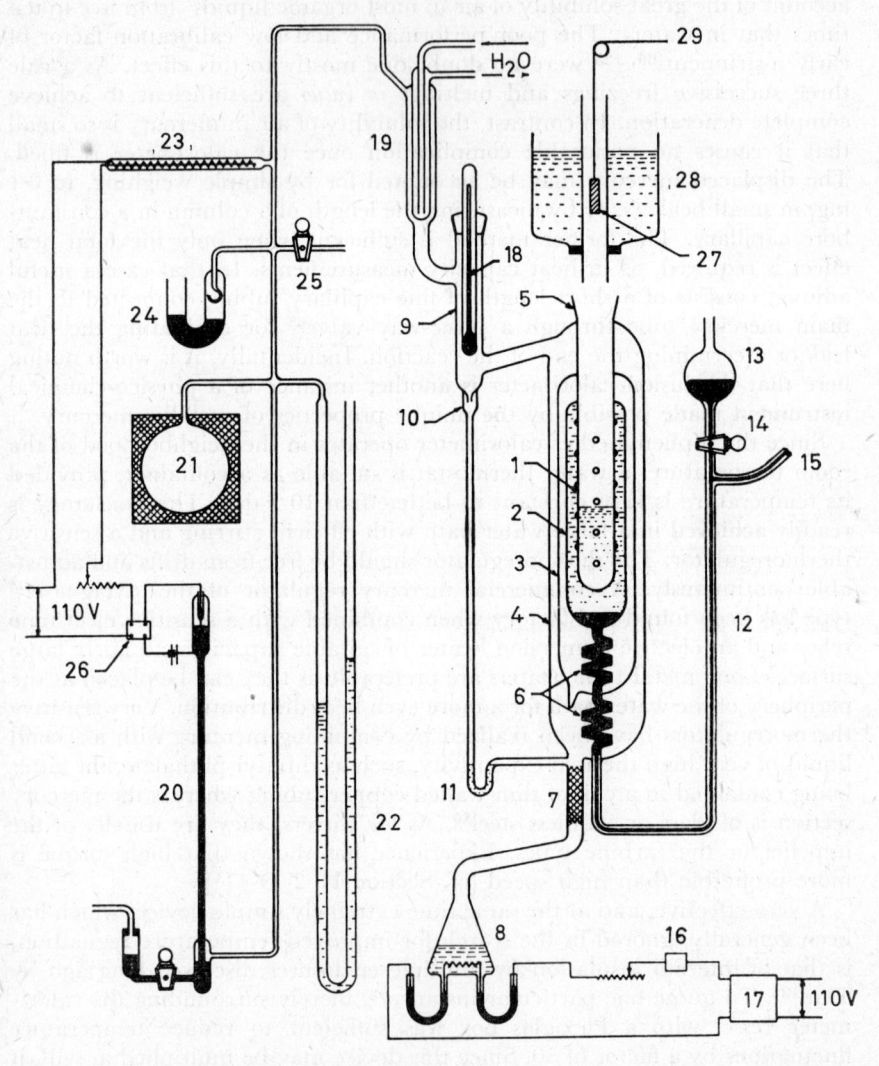

Figure 23. Naphthalene isothermal calorimeter set up for drop calorimetry[32]. The sample, 27, is in the tube of the furnace, 28. The oil in the inner tube, 1, facilitates heat transfer from the sample to the naphthalene. The horizontal measuring capillary is joined on at 15, and the manostat, 20–26, controls the benzene-vapor thermostatic jacket, 4.

Figure 24. Naphthalene isothermal calorimeter[32] with its outer lagging removed and all the naphthalene melted; the oil in the inner tube and the mercury in the lines leading to the measuring capillary are clearly visible.

It is obviously desirable to locate the calorimeter in a temperature-controlled room, a basement or at least a room with northern exposure only. A radiation shield made of aluminum foil, with a small window cut out for viewing the mantle of solid, may be wrapped around the calorimeter. It is easily removed, for instance, to melt the mantle with an infrared lamp. Formation of a mantle of solid by introduction of a cold object in the central well requires no special precaution, although it is advisable to minimize the effect of supercooling by remelting almost completely the first solid formed. If a mantle is frozen rapidly (one hour or less) it must be allowed a few hours (five or six) to come to equilibrium with the liquid. The slightest trace of impurity (e.g. dissolved air) will increase enormously the time necessary for obtaining a stable mantle.

An instrument quite similar to the above has been built by Dainton and his coworkers[39, 40] for measuring heats of polymerization. The container proper was narrower and somewhat taller than the previous one in order to accommodate a dilatometer type of reaction vessel. Because of this, and the fact that much smaller heat effects were measured in one run (some 125 joules instead of 1500 to 2000), the top part of the solid mantle probably did not melt, and there was no need for glass studs to hold it. It was even possible to melt a hole in the top part of the mantle for observing the meniscus in the dilatometer stem. The authors also found that evacuation of the central well had no significant effect on the heat loss. Since the instrument was used to follow the rate of reaction, the heat effect was measured by the horizontal capillary method. Depending on the sensitivity desired and the total heat absorption capacity of the calorimeter, both the size of the mercury thread and the method of evaluating its displacement will vary. For instance, with a very fine capillary visual observation against a graduated scale may be satisfactory; with a bigger bore capillary a travelling microscope or a horizontal cathetometer will be needed. Each method presents its own experimental problems. Thus, a fine capillary may be so long as to require folding, hence a loss of readability in the bends, whereas a very slow moving mercury meniscus is subject to "sticking". To prevent the "thermometer effect" the capillary should be adequately thermostatted.

The third instrument, by Jessup[85], is different in that it is made of metal (*Figure 25*) and is contained in a massive aluminum jacket. One serious inconvenience of this type of construction is that the mantle of solid ether cannot be seen, and one is reduced to guesses as to its condition and behavior. On the other hand, the metal central tube gives a much better thermal conductivity, still enhanced by the addition of a set of copper fins on the outside. This is an important consideration in the drop method, especially when the initial temperature of the sample is fairly high. In fact it has been observed that sudden addition of too large a heat quantity in an all-glass diphenyl ether calorimeter leads to erratic results[60]. The calibration and operation of the Jessup calorimeter seemed much more tedious than the other two, partly on account of a trace of air which had leaked accidentally into the ether compartment.

From these investigations the characteristics of the diphenyl ether calorimeter are now well known. The calibration factor k (weight of mercury displaced by one joule) may be taken as $0·01264 \pm 0·00002$ g (Hg) joule^{-1}

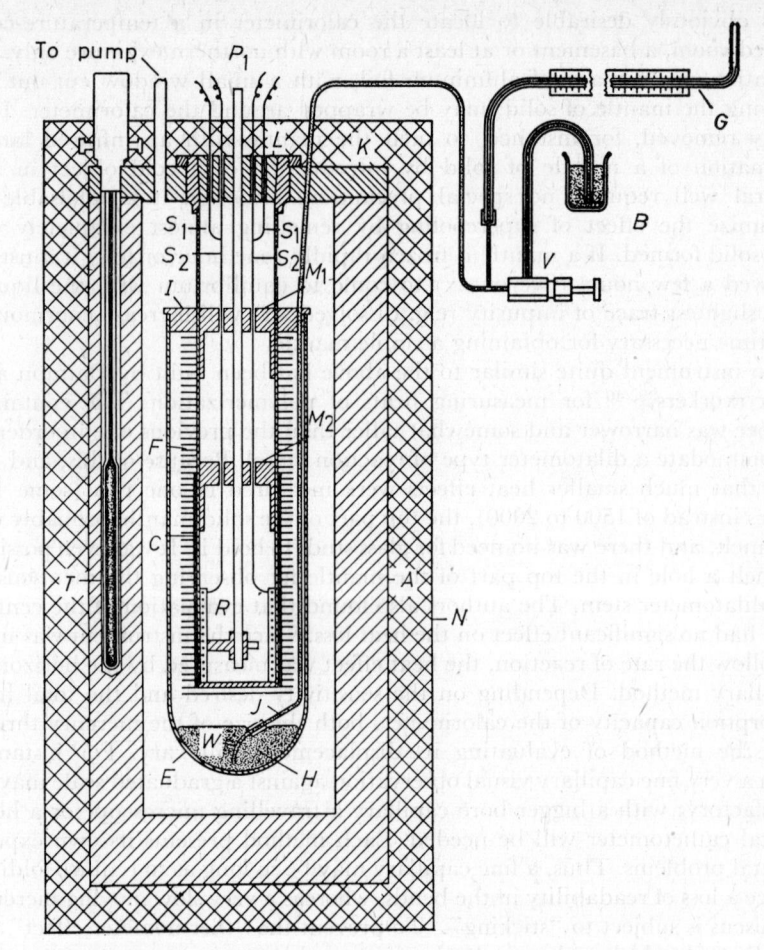

Figure 25. All-metal diphenyl ether isothermal calorimeter[85]—sectional diagram

A, Aluminum jacket
B, Beaker
C, Copper spacers
E, Outer wall of calorimeter
F, Copper fins
G, Graduated capillary
H, Mercury
I, Top of calorimeter
J, Stainless steel tube
K, Copper jacket cover
L, Insert in top of jacket
M_1, M_2, Copper cylinders
N, Thermal insulation
P, Copper–nickel tube
R, Electric heater
S_1, Potential leads
S_2, Current leads
T, Thermoregular bulb
V, Valve in mercury line
W, Platinum tube

from the average of the two direct calibrations by the electrical method[39, 85]. This value, combined with the heat of fusion ΔH[59] of the ether m, 101·152 joule g^{-1} and the density of mercury at 26·9°C, leads to a value of 0·0945 cm^3 g^{-1} for the change of specific volume on melting. Since the density of liquid ether is 1·0692 g cm^{-3} at the melting temperature, from extrapolation[62], that of the solid must be 1·1894 g cm^{-3}, and the coefficient dT/dP of the melting point[39], 0·0284 deg atm^{-1}. The instrument is capable of a precision of 0·1 per cent or slightly better[85], and the accuracy is of the order of 1 per cent, even for heat flows as small as 5 joules per hour.

Although the calibration factor of the diphenyl ether calorimeter is now well known, a new instrument must always be calibrated, at least by a comparative method, if only to check its performance.

In conclusion, the diphenyl ether isothermal calorimeter appears as a very valuable instrument, and one cannot help feeling that it should be more widely used. By its very simplicity of construction and operation it offers little opportunity for "gadgeteering". In fact, most of the isothermal calorimeters referred to in *Table 7* were built by non-specialists. Quite possibly other working substances could be found, still better adapted to that purpose. At least from the viewpoint of sensitivity one might consider compounds with very small enthalpy of fusion, such as the so-called "globular molecules"[136]. To quote examples, malononitrile ($CNCH_2CN$), m.p. 32·1°C, and succinonitrile ($CNCH_2CH_2CN$), m.p. 54·5°C; the latter, with a ΔHm of only[147] 46·19 joule g^{-1}, would give more than twice the sensitivity of the diphenyl ether calorimeter. Still, the plastic crystals formed by these substances on solidification may not be suitable in that connection.

Swietoslawski[130] has remarked: "The ice calorimeter is not a convenient apparatus for microcalorimetric measurements because a stable thermodynamic equilibrium cannot be established." Indeed, owing to the hydrostatic pressure of the liquid itself, there is a slight increase in pressure from the top to the bottom of the mantle. For instance, in a mantle some 13 cm long the melting point of ice at the bottom end is about 10^{-4} deg higher than at the top (because the coefficient dP/dT is negative). Thus an ice mantle tends to melt at the bottom and to grow at the top. Obviously, in a diphenyl ether calorimeter this effect is reversed, and the corresponding temperature difference is about four times larger than with ice. As far as the relative rates of ageing of the two mantles are concerned, the lower heat of fusion of the organic compound (see *Table 7*) is offset by its lower thermal conductivity (roughly one third). At any rate this phenomenon is of no great consequence unless the measurements extend over very long periods of time. This is certainly not the case for the heat capacity of liquids.

To be complete it must be mentioned here that other transformations besides fusion have been used in isothermal calorimetry: e.g. the boiling of a liquid, such as ammonia[90] at $-33°C$, or carbon tetrachloride[138] at 76·7°C, or the sublimation of a solid, carbon dioxide[100]. So far none of these has been applied to the determination of the heat capacity of liquids or solutions.

3. Drop calorimeter furnaces and vessels

The term "furnace" is used here by analogy with the high-temperature drop calorimetry (Chapter 8); it must be understood in its broadest sense to designate the enclosure wherein the specimen is brought to a constant, and accurately known temperature, before dropping into the calorimeter receiver. In fact, for the present purpose the "furnace" might very well be at room temperature or even below $0°C$[54]. Accordingly the construction is much simplified. In most instances a thermostatted jacket with circulation of an appropriate liquid from a large constant-temperature bath is entirely adequate. Also such accessories as a radiation gate[63] may be dispensed with, or at least greatly simplified. For the same reason there is little use for a swinging mount such as used with high-temperature furnaces to bring them in line with the calorimeter vessel just for the dropping of the specimen.

On the other hand, for measurements on liquids and solutions, the temperature of the "furnace" must be controlled and measured with greater accuracy than in high-temperature drop calorimetry. With temperature differentials of the magnitude considered here, the uncertainty should not exceed a few hundredths of a degree. *Figure 26(a)* shows a furnace made of glass tubes for measurements up to about 80°C with water as the heat transfer medium[54]. For higher temperature a less volatile liquid is used,

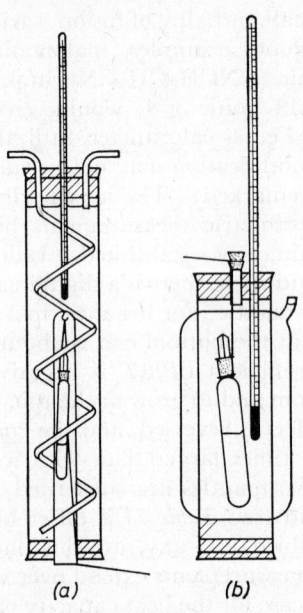

Figure 26. Constant-temperature enclosures ("furnaces") used in conjunction with an isothermal calorimeter for measuring the heat capacity of liquids and solutions near room temperature[54]. (*a*) Water from a large thermostat is circulated through the jacket at 100 ml per minute. (*b*) For temperatures below 0°C the whole enclosure is immersed in a cryostat.

and below 0°C, some anti-freeze solution. In both cases, it may be advisable to wrap the "furnace" with some lagging. The outer tube in *Figure 26* is about 3 in. in diameter and 10 in. or 12 in. long, which is enough to avoid end effects. As for the inner tube, it may be closed at the upper end as shown to improve the temperature uniformity in the middle section. A small glass hook is then provided at the top to carry the silk thread which holds the specimen. For easier alignment and also to minimize cooling of the specimen while dropping, the inner tube is made to extend down to the calorimeter well, from which it is separated by a plastic diaphragm.

The thermometer, for instance of the platinum resistance type, should preferably be mounted close to the inner tube but the thermoregulator may be located in the water bath, provided that the connecting tubes to the "furnace" are short and that the room is thermostatted. For a good temperature control the circulation of the liquid in the "furnace" should be brisk.

For temperatures below zero (Celsius) it may be simpler, albeit less accurate, to have the sample tube hanging inside a double walled glass container, with a glycerol solution in the jacket, and closed at both ends with rubber bungs, *Figure 26(b)*. The whole is immersed in a low-temperature bath until thermal equilibration is achieved; then it is quickly taken out, wiped dry, and the bottom rubber bung removed just before lowering the sample tube into the calorimeter proper.

The vessels used to contain the samples require special attention from the point of view of chemical inertness towards the liquids to be measured, as well as thermal stability and conductivity. Except on the latter count glass is the ideal material for holding most liquids near room temperature. Small ampoules made of thin walled glass tubing, with a drawn out neck for filling by means of a syringe or a thin pipette, are generally suitable. Volatile liquids require special precautions to minimize any errors due to vaporization. One way to achieve this is to seal the glass vessel, leaving the free space above the liquid as small as is feasible. The liquid is first degassed, then frozen under vacuum, and the neck of the ampoule sealed with a thin flame. This technique complicates somewhat the problem of finding the exact weight of both the vessel and the specimen. With solutions there is the added difficulty of possible concentration changes during this operation. For measurements at temperatures appreciably above the normal boiling point of the liquid specimen one might consider using thick-walled capillaries, with strong seals at both ends, or else a metal vessel with tight closure. However, in such cases it would be more difficult to keep the weight, as well as the heat capacity, ratio between container and sample within reasonable limits. Expansion devices, sometimes used in adiabatic calorimetry [see Section II–2–E–(2)] are not easily adaptable to the drop method.

Non-volatile liquids and solutions raise no such problems. A small glass vial, with ground-glass stopper (*Figure 26*) is then entirely satisfactory, as it is easily filled, weighed, and cleaned, and may be used repeatedly. Certain compounds require special treatment of the glass surface, for instance to reduce its catalytic activity. Thus with hydrogen peroxide[80] cleaning with a strong oxidizing agent was essential, as the slightest decomposition would vitiate entirely all measurements. In other cases, e.g. hydrogen disulfide[19], traces of alkali in the glass must first be neutralized (with gaseous hydrochloric acid) or else low alkali glass (Vycor) or quartz be substituted for Pyrex. Exceptionally, hydrofluoric acid should be handled in appropriate metal (copper, platinum, etc.) or plastic (Teflon) vessels. Calibration of the vessels for heat capacity may be achieved through a few measurements with the vessels either empty or filled with a reference liquid (e.g. water) as described in Section II–1 above.

4. *Procedure and calculations*

In carrying out systematic heat capacity determinations, say on a binary system of liquids, it is expedient (assuming that the components are stable enough) first to prepare a number of specimens covering the whole concentration range. Steps of 10 or even 20 per cent for the latter should be sufficient, save in the exceptional cases of pronounced maxima or minima in the C_P vs. concentration curves. Measurements on the various specimens are then made at a given temperature of the "furnace" since readjusting the latter to a new predetermined value is time consuming. As a rule temperature intervals of the order of 20° or 25°C are close enough because the heat capacity of most liquids and solutions varies rather slowly and regularly with temperature.

With the isothermal calorimeter in place, ready for operation (preferably with a fresh mantle of solid) the "furnace" containing the specimen is aligned with the calorimeter central tube. The ensemble is then "rated", i.e. the thermal leak under the operating conditions is determined. This should allow ample time for the specimen to come into equilibrium within the "furnace". Then the plastic diaphragm between "furnace" and calorimeter is briefly removed and the sample dropped, or rather quickly lowered into the latter (no sophisticated braking device needed here!). It may be necessary to use some heat transfer liquid in the calorimeter well to avoid heat losses due to poor thermal conductivity, especially with an all-glass calorimeter and at higher temperatures of the "furnace". This obviously complicates the operations because of the danger of contamination of the specimen, and the need to clean it after each determination. Furthermore, this liquid must be thermally and chemically stable, as well as non-volatile. Paraffin oil has been used successfully for that purpose[54]. One possible source of error in drop calorimetry near room temperature is that of atmospheric moisture condensation, either on the specimen or in the calorimeter well. This may require circulating in these enclosures a stream of dry gas at exactly the right temperature.

The usual precautions are *"de rigueur"* to protect the specimen from alteration by such agents as light, excessive heat, catalysts, etc. One must also guard, as in other methods, from complications due to changes in the specimen, such as precipitation of the solute, miscibility gaps (as in the water–phenol system), etc. As for the purity of the samples, generally speaking the current "C.P." or "Reagent Grade" chemicals may be used as such because the concentration of residual impurities is so low, and their heat capacity is so near that of the main constituent that no appreciable error ensues. Likewise, the concentration of solutions is almost always known with much higher accuracy than that of the calorimetric measurements, even when obtained from some physical property such as the refractive index.

The treatment of experimental data in the drop method is straightforward and similar to that in other methods. The measured enthalpy changes are converted into average values of C_P, the molal heat capacity of the solution over the temperature interval covered. These experimental points are then plotted on a large graph against the mole fraction x to get the smoothed values and the partial molal heat capacity of each component [see Equation (15)] by the usual tangent construction. In addition to these quantities it is customary to report the excess heat capacity of the solution [see Equation (20)] as well as its partial molar equivalent for both components 1 and 2, which are useful in calculating the heat of mixing, the partial molar heats of mixing and other functions of significance in the theory of solutions.

IV. References

1. Ackermann, Th., unpublished work.
2. Ackermann, Th., *Z. Elektrochem.*, **62**, 441 (1958).
3. Ackermann, Th. and H. Rüterjans, *Angew. Chem.*, **75**, 925 (1963).
4. Ackermann, Th. and H. Rüterjans, *Z. Elektrochem. Ber. Bunsenges. physik. Chem.*, **68**, 850 (1964).
5. Ackermann, Th., H. Rüterjans, U. Sage, O. F. Schmitz and F. Schreiner, (A.C.S. Meeting, Washington 1962), *J. Amer. Chem. Soc.*, to be published.
6. Argue, G. R., E. E. Mercer and J. W. Cobble, *J. Phys. Chem.*, **65**, 2041 (1961).
7. Armstrong, G. T., P. K. Wong and L. A. Krieger, *Rev. Sci. Instr.*, **30**, 339 (1959).
8. Barber, C. R., *J. Sci. Instr.*, **27**, 47 (1950).
9. Barieau, R. E. and W. F. Giauque. *J. Amer. Chem. Soc.*, **72**, 5676 (1950).
10. Beakley, W. R., *J. Sci. Instr.*, **28**, 176 (1951).
11. Belaga, M. W., D. M. Coddington and H. Marcus, *Rev. Sci. Instr.*, **27**, 948 (1956).
12. Benson, G. C., E. D. Goddard and C. A. J. Hoevo, *Rev. Sci. Instr.*, **27**, 725 (1956).
13. Beyon, J. H. and A. R. Humphries, *Trans. Faraday Soc.*, **51**, 1065 (1955).
14. Boissonnas, C. G., *Ind. Eng. Chem.*, **31**, 761 (1939).
15. Bridgman, P. W., *Rev. Mod. Phys.*, **18**, 1 (1946).
16. Brown, I. and W. Fock, *Australian J. Chem.*, **14**, 387 (1961).
17. Budnikoff, P. O. and L. Gulinowa, *Kolloid Z.*, **67**, 88 (1934).
18. Burlew, J. S., *J. Amer. Chem. Soc.*, **62**, 681, 690, 696 (1940).
19. Butler, K. H. and O. Maass, *J. Amer. Chem. Soc.*, **52**, 2184 (1930).
20. Buzzell, Anne and J. M. Sturtevant, *J. Amer. Chem. Soc.*, **73**, 2454 (1951).
21. Callendar, H. L. and H. T. Barnes, *Phil. Trans. Roy. Soc.* (London), **A199**, 55 (1902).
22. Calvet, E., in F. D. Rossini, ed., *Experimental Thermochemistry*, Vol. I, Interscience, New York, 1956, Chapter 12.
23. Calvet, E. and E. Brouty, *Compt. Rend.*, **238**, 1879 (1954).
24. Calvet, E. and E. Brouty, *Compt. Rend.*, **239**, 672 (1954).
25. Calvet, E. and C. Ditheil, *Compt. Rend.*, **253**, 1207 (1961).
26. Calvet, E. and H. Prat, *Microcalorimetrie*, Masson, Paris, 1956.
27. Calvet, E. and H. Prat, *Recents progres en microcalorimetrie*, Dunod, Paris, 1958.
28. Carroll, B. H. and J. H. Matthews, *J. Amer. Chem. Soc.*, **46**, 30 (1924).
29. Caule, E. J. and C. C. Coffin, *Can. J. Research*, **28B**, 639 (1950).
30. Chessick, J. J., G. S. Young and A. C. Zettlemoyer, *Trans. Faraday Soc.*, **50**, 587 (1954).
31. Cleland, W. W. and R. S. Harding, *Rev. Sci. Instr.*, **28**, 696 (1957).
32. Coffin, C. C., J. C. Devins, J. R. Dingle, J. H. Greenblatt, T. R. Ingraham, and S. Schrage, *Can. J. Research*, **28B**, 579 (1950).
33. Cole, K. S., *Rev. Sci. Instr.*, **28**, 326 (1957).
34. Coleman, D. S. and A. J. B. Cruickshank, unpublished work; see also Coleman, D. S., M.Sc. Thesis, Bristol 1958.
35. Criss, C. M. and J. W. Cobble, *J. Amer. Chem. Soc.*, **83**, 3223 (1961).
36. Cruickshank, A. J. B., unpublished work.
37. Cruickshank, A. J. B., *Phil. Trans. Roy. Soc.*, **A253**, 407 (1961).
38. Cruickshank, A. J. B. and B. A. Timimi, unpublished work; see also B. A. Timimi, Ph.D. Thesis, Bristol 1964.
39. Dainton, F. S., J. Diaper, K. J. Ivin and D. R. Sheard, *Trans. Faraday Soc.*, **53**, 1269 (1957).
40. Dainton, F. S. and K. J. Ivin, in H. A. Skinner, ed., *Experimental Thermochemistry*, Vol. II, Interscience, London, 1961, Chapter 12.
41. Dauphinee, T. M. and S. B. Woods, *Rev. Sci. Instr.*, **26**, 693 (1955).
42. de Visser, L. E. O., *Z. physik. Chem.*, **9**, 767 (1892).
43. Diaz-Peña, M. and M. L. McGlashan, *Trans. Faraday Soc.*, **57**, 1511 (1961).
44. Dole, M., W. P. Hettinger, N. Larson, J. A. Wethington and A. E. Worthington, *Rev. Sci. Instr.*, **22**, 812 (1951).
45. Douglas, T. B., A. F. Ball and D. C. Ginnings, *J. Res. Natl. Bur. Std.*, **46**, 334 (1951).
46. Duff, G. M. and D. H. Everett, *Trans. Faraday Soc.*, **52**, 753 (1956).
47. Egan, E. P. Jr., *J. Phys. Chem.*, **60**, 1344 (1956).
48. Eucken, A. and M. Eigen, *Z. Elektrochem.*, **55**, 343 (1951).
49. Evstrop'ev, K. S., *Zh. Fiz. Khim.*, **8**, 130 (1936).
50. Eyraud, C., *Compt. Rend.*, **238**, 1511 (1954); see also '*Differential Scanning Calorimetry*', Perkin-Elmer Inst. Div., Norwalk, Conn.
51. Eyraud, C. and R. Goton, *Compt. Rend.*, **240**, 423 (1955).
52. Ferguson, A. H. and J. T. Miller, *Proc. Phys. Soc.* (*London*), **45**, 194 (1933).
53. Flotow, H., D. W. Osborne and K. Otto (Chemistry Division, Argonne Natl. Lab.), unpublished work.

[54] Foley, W. T. and P. A. Giguère, *Can. J. Chem.*, **29**, 895 (1951).
[55] Forrest, H. D., E. W. Brugmann and L. W. T. Cummings, *Ind. Eng. Chem.*, **23**, 37 (1931).
[56] Franck, E. U. and W. Spalthoff, *Z. Elektrochem.*, **61**, 348 (1957).
[57] Freyer, E. B., J. C. Hubbard and D. H. Andrews, *J. Amer. Chem. Soc.*, **51**, 759 (1929).
[58] Frisch, M. A. and H. Mackle, *J. Sci. Instr.*, **42**, 186 (1965).
[59] Furukawa, G. T., D. C. Ginnings, R. E. McCoskey and R. A. Nelson, *J. Res. Natl. Bur. Std.*, **46**, 195 (1951).
[60] Giguère, P. A. and J. L. Carmichael, *J. Chem. Eng. Data*, **7**, 526 (1962).
[61] Giguère, P. A., I. D. Liu, J. S. Dugdale and J. A. Morrison, *Can. J. Chem.*, **32**, 117 (1954).
[62] Giguère, P. A., B. G. Morissette and A. W. Olmos, *Can. J. Chem.*, **33**, 657 (1955).
[63] Ginnings, D. C. and R. J. Corruccini, *J. Res. Natl. Bur. Std.*, **38**, 583 (1947).
[64] Ginnings, D. C., T. B. Douglas and A. F. Ball, *J. Res. Natl. Bur. Std.*, **45**, 23 (1950).
[65] Goldsmid, H. J., *Applications of Thermoelectricity*, Methuen, London, 1960, Chapter VIII.
[66] Gordon, M. M., *Zement*, **5**, 39 (1937); *Chem. Zentralblatt*, **1938**, II, 561.
[67] Gouy, M., *J. Physique* (3me ser.), **6**, 479 (1897).
[68] Grassi, U., *Atti. accad. Lincei.* **22**, I, 494 (1913); *Chem. Abstr.*, **7**, 3254 (1913).
[69] Gregg, S. J., *J. Chem. Soc. (London)*, **1927**, 1494.
[70] Gucker, F. T., *Ann. N.Y. Acad. Sci.*, **51**, 696 (1949); *Chem. Abstr.*, **43**, 7317^h (1949).
[71] Gucker, F. T., F. D. Ayres and T. R. Rubin, *J. Amer. Chem. Soc.*, **58**, 2118 (1936).
[72] Gucker, F. T. and J. M. Christens, *Proc. Indiana Acad. Sci.*, **64**, 97 (1955); *Chem. Abstr.* **50** 4562^g (1956).
[73] Gucker, F. T., H. B. Pickard and R. W. Planck, *J. Amer. Chem. Soc.*, **61**, 459 (1939).
[74] Guggenheim, E. A., *Mixtures*, Oxford Univ. Press, 1952, §3.01.
[75] Heydweiller, A., *Ann. Physik*, **46**, 253 (1915).
[76] Hládek, L., *J. Sci. Instr.*, **42**, 198 (1965).
[77] Hoge, H. J., *J. Res. Natl. Bur. Std.*, **36**, 111 (1946).
[78] Holmberg, T., *Soc. Sci. Fennica Commentationes Phys.-Math.*, **9**, No. 17 (1938); *Chem. Abstr.* **32**, 6107 (1938).
[79] Hornig, D. F. and B. J. O'Keefe, *Rev. Sci. Instr.*, **18**, 474 (1947).
[80] Huckaba, C. E. and F. G. Keyes, *J. Amer. Chem. Soc.*, **70**, 2578 (1948).
[81] Ioffe, A. F., *Semiconductor Thermoelements and Thermoelectric Cooling*, Infosearch, London, 1957.
[82] Jauch, K., *Z. Physik.*, **4**, 441 (1921).
[83] Jeener, J., *Rev. Sci. Instr.*, **28**, 263 (1957).
[84] Jepson, W. B. and J. S. Rowlinson, *J. Chem. Phys.*, **23**, 1599 (1955).
[85] Jessup, R. S., *J. Res. Natl. Bur. Std.*, **55**, 317 (1955).
[86] Joule, J. P., *Mem. Proc. Manchester Lit. Phil. Soc.*, **2**, 559 (1845).
[87] Klemm, W., W. Tilk and H. Jacobi, *Z. anorg. allgem. Chem.*, **207**, 187 (1932).
[88] Koch, W., *Forsch. Gebiete Ingenieurwes.*, **3**, 1 (1932); *ibid*, **5B**, 138 (1934); *Chem. Abstr.*, **28**, 7136^2 (1934).
[89] Kouvel, J. S., *J. Appl. Phys.*, **27**, 639 (1956).
[90] Kraus, C. A. and J. A. Ridderhof, *J. Amer. Chem. Soc.*, **56**, 79 (1934).
[91] Kuntz, R. R. and G. J. Mains, *J. Phys. Chem.*, **68**, 408 (1964).
[92] Leech, J. W., *Proc. Phys. Soc.*, **62B**, 390 (1949).
[93] Lewis, G. N. and M. Randall, *Thermodynamics*, 2nd ed. (revised K. S. Pitzer and L. Brewer), McGraw-Hill, 1961, p. 375 *et seq.*, equation 25–1.
[94] Lewis, G. N. and M. Randall, *J. Amer. Chem. Soc.*, **33**, 476 (1911).
[95] Lipsett, S. G., F. M. G. Johnson and O. Maass, *J. Amer. Chem. Soc.*, **49**, 925 (1927).
[96] Low, D. I. R. and E. A. Moelwyn-Hughes, *Proc. Roy. Soc. (London)*, **A267**, 384 (1962).
[97] McGlashan, M. L. and K. W. Morcom, *Trans. Faraday Soc.*, **57**, 581 (1961).
[98] Malcolm, G. N. and J. A. Rowlinson, *Trans. Faraday Soc.*, **53**, 921 (1957).
[99] Murphy, G. W., *Rev. Sci. Instr.*, **20**, 372 (1949).
[100] Oelsen, W., W. Tebbe and O. Oelsen, *Arch. Eisenhüttenw.*, **27**, 689 (1956); *Chem. Abstr.*, **51**, 4807^c (1957).
[101] Olette, M., *Compt. Rend.*, **244**, 891 (1957).
[102] O'Neal, H. E. and N. W. Gregory, *Rev. Sci. Instr.*, **30**, 434 (1959).
[103] Osborne, N. S., H. F. Stimson and D. C. Ginnings, *J. Res. Natl. Bur. Std.*, **23**, 197 (1939).
[104] Owen, B. B., J. R. White and J. S. Smith, *J. Amer. Chem. Soc.*, **78**, 3561 (1956).
[105] Pfaundler, L., *Sitzungsberichte Akad. Wiss. Wein. math.-naturwiss, kl.*, **59**, 145 (1869).
[106] Plester, D. W., S. E. Rogers and A. R. Ubbelohde, *J. Sci. Instr.*, **33**, 211 (1956).
[107] Preston, J. S., *J. Sci. Inst.*, **23**, 173, 301 (1946).
[108] Randall, M. and F. D. Rossini, *J. Amer. Chem. Soc.*, **51**, 323 (1929).
[109] Richards, T. W. and L. L. Burgess, *J. Amer. Chem. Soc.*, **32**, 431 (1910).
[110] Richards, T. W. and F. T. Gucker, *J. Amer. Chem. Soc.*, **47**, 1876 (1925).
[111] Richards, T. W. and F. T. Gucker, *J. Amer. Chem. Soc.*, **51**, 712 (1929).
[112] Richmann, G. W., *Novi Comment. Acad. Petropl.*, **1**, 152 (1750); quoted in Partington, J. R. *An Advanced Treatise on Physical Chemistry* Vol. III, Longmans Green, London, 1952, p. 266.

[113] Rogers, G. F. C. and Y. R. Mayhew, *Engineering Thermodynamics, Work and Heat Transfer*, Longmans Green, London, 1957, Chapter 21.
[114] Rossini, F. D., *J. Res. Natl. Bur. Std.*, **7**, 47 (1931).
[115] Ruehrwein, R. A. and H. M. Huffman, *J. Amer. Chem. Soc.*, **65**, 1620 (1943).
[116] Sachse, H., *Z. physik. Chem.*, **A143**, 94 (1929).
[117] Sage, B. H. and E. W. Hough, *Anal. Chem.*, **22**, 1304 (1950).
[118] Shchukarev, A. N., T. V. Ass and N. I. Putilin, *Med. exptl. (Ukraine)*, **6**, 114 (1936); *Chem. Abstr.*, **31**, 1835 (1937).
[119] Shchukarev, A. N., I. P. Krivobabko and L. A. Shchukareva, *Physik. Z. Sowjetunion*, **5**, 722 (1934).
[120] Skinner, H. A., J. M. Sturtevant and Stig Sunner, in H. A. Skinner, ed., *Experimental Thermochemistry*, Vol. II, Interscience, London, 1962, pp. 168–170.
[121] Sligh, T. S., Jr., *J. Amer. Chem. Soc.*, **42**, 60 (1920).
[122] Spear, N. H., *Anal. Chem.*, **24**, 938 (1952).
[123] Sreerangachar, H. B. and M. Sreenivasaya, *Biochem. J.*, **29**, 295 (1935).
[124] Staveley, L. A. K. and D. N. Parham, *Changements de Phases, Compt. rend. 2^e reunion ann. avec. comm. thermodynam.*, Union intern. phys. (Paris), 1952, p. 366; *Chem. Abstr.* **47**, 7278^c (1953).
[125] Staveley, L. A. K., W. I. Tupman and K. R. Hart, *Trans. Faraday Soc.*, **51**, 323 (1955).
[126] Stull, D. R., *Rev. Sci. Instr.*, **16**, 318 (1945); see also *Anal. Ed. Ind. Eng. Chem.*, **18**, 234 (1946).
[127] Stull, D. R., *Anal. Chim. Acta*, **17**, 133 (1957).
[128] Sturtevant, J. M., *Physics*, **7**, 232 (1936).
[129] Sturtevant, J. M., *J. Amer. Chem. Soc.*, **59**, 1528 (1937).
[130] Swietoslawski, W., *Microcalorimetry*, Reinhold, New York, 1946.
[131] Taylor, R. D., B. H. Johnson and J. E. Kilpatrick, *J. Chem. Phys.*, **23**, 1225 (1955).
[132] Thomas, A., *Trans. Faraday Soc.*, **47**, 569 (1951).
[133] Thomson, W. (Lord Kelvin), *Trans. Roy. Soc. (Edinburgh)*, **20**, 261 (1853); see also, Joule, J. P., *Phil. Mag.*, **17**, 364 (1859).
[134] Tian, A., *J. chim. phys.*, **20**, 132 (1922–23).
[135] Tian, A., *Compt. Rend.*, **178**, 705 (1924).
[136] Timmermans, J., *J. Phys. Chem. Solids*, **18**, 1 (1961).
[137] Todd, L. J., Ph.D. Thesis, Johns Hopkins University, 1955.
[138] Tong, L. K. J. and W. O. Kenyon, *J. Amer. Chem. Soc.*, **67**, 1278 (1945).
[139] Trowbridge, J. C. and E. F. Westrum, Jr., *J. Phys. Chem.*, **67**, 2381 (1963).
[140] Turner, L. B., *Proc. Cambridge Phil. Soc.*, **32**, 663 (1936); see also *J. Inst. Elec. Engrs.*, **81**(2), 399 (1937).
[141] West, E. D. and D. C. Ginnings, *Rev. Sci. Instr.*, **28**, 1070 (1957).
[142] Westrum, E. F., Jr., J. B. Hatcher and D. W. Osborne, *J. Chem. Phys.*, **21**, 419 (1953).
[143] Wilhoit, R. C. and A. Amador, IUPAC Symposium on Thermodynamics and Thermochemistry, Lund, July 1963; *Chem. Ing. Tech.*, **37**, 171 (1965).
[144] Wittig, F. E. and G. Kemeny, *Chem. Ing. Tech.*, **32**, 635 (1960).
[145] Wittig, F. E. and W. Schilling, *Chem. Ing. Tech.*, **33**, 554 (1961).
[146] Worthington, A. E., P. C. Marx and M. Dole, *Rev. Sci. Inst.*, **26**, 698 (1955).
[147] Wulff, C. A. and E. F. Westrum, Jr., *J. Phys. Chem.*, **67**, 2376 (1963).
[148] Zabetakis, M. G., R. S. Craig and K. F. Sterrett, *Rev. Sci. Inst.*, **28**, 497 (1957).

CHAPTER 13

Calorimetric Studies of Some Physical Phenomena at Low Temperatures

H. Chihara

Department of Chemistry, Osaka University, Toyonaka, Osaka, Japan

J. A. Morrison

Division of Pure Chemistry, National Research Council, Ottawa, Canada

Contents

I.	Introduction	537
II.	Thermodynamic Properties of Films Adsorbed on Solids	538
	1. Measurable Quantities	538
	2. Adsorption Calorimeters of Moderate Precision	538
	3. High Precision Adiabatic Calorimeters	540
	A. Thermodynamics	540
	B. Experimental Details	543
	4. Measurements at Liquid Helium Temperatures	545
III.	Particle Size Effects	545
	1. Heat Conduction between Solid Particles	545
	2. Calorimetric Measurements	546
IV.	Stored Energy Experiments	547
	1. General Remarks	547
	2. Radiation Damage at Low Temperatures	547
	3. Clustering in Alloys	548
V.	References	548

I. Introduction

Calorimetric techniques that are used to determine thermodynamic properties of substances in the region below room temperature are described in Chapters 5, 6, and 7. The justification for a fourth chapter on somewhat similar lines is that low-temperature calorimeters are sometimes used to study phenomena that place rather special requirements on the calorimetric technique. A few examples of such studies will be described in this chapter. Some measurements of the thermodynamic properties of adsorbed phases, for example, may involve no more than the refinement of existing methods; others, such as measurements of the energy associated with radiation damage in solids, may require the development of new methods. It should be stressed at the outset that only a limited number of examples will be given; the chapter is not intended to be an exhaustive treatment of "off-beat" low-temperature calorimetry.

II. Thermodynamic Properties of Films Adsorbed on Solids

1. *Measurable quantities*

Studies of adsorption on solids are usually concerned with the surface structure of the solid, the nature of the interaction between the solid adsorbent and the adsorbate, and with the state of aggregation or the mode of motion of the molecules in the adsorbate. Usually, the experimental measurements are non-calorimetric and consist of determinations of adsorption isotherms. If isotherms are obtained at different temperatures, then, in principle, heats of adsorption and other thermodynamic parameters may be computed. In practice, however, the accuracy obtainable (of the magnitude of ± 1 per cent in heats of adsorption at best) is not as high as that of direct calorimetric determinations (of the magnitude of $\pm 0 \cdot 1$ per cent). The greater accuracy of the direct measurements is of great value when attempts are made to interpret the thermodynamic properties statistically[41].

The experimental details to be discussed in this section pertain to the determination, below room temperature, of heats of adsorption and desorption, heat capacities of the adsorption system (adsorbent + adsorbed film + vapor), equilibrium pressures and the amount of vapor adsorbed. Obviously, these quantities cannot all be determined experimentally at all temperatures, but, since they can be related thermodynamically, this is not necessary. By a suitable choice of experiments, a complete thermodynamic description of an adsorbing system may be obtained, as will be outlined in sub-section II–3. Before discussing this more complex method, a somewhat simpler type of calorimeter assembly, designed to measure heats of adsorption only, will be described.

2. *Adsorption calorimeters of moderate precision*

A series of relatively simple adsorption calorimeters was developed by Beebe and colleagues[6-11]. The original model was a vacuum-jacketed metal calorimeter vessel into which adsorbate gas was admitted through a thin platinum tube. The calorimeter vessel was suspended inside a glass jacket that was immersed in a constant temperature bath. The measurement of the heat of adsorption required the recording of the temperature of the calorimeter vessel as a function of time before and after the admission of gas. The heat capacities of the adsorbent and the calorimeter vessel were not measured directly but were estimated from the known weights.

A later model, which is shown in *Figure 1*, contains a means for electrical calibration and some other useful features. The thermal shield E (2 cm in diameter and 7 cm long) is connected to the Pyrex glass inlet tube by a Pyrex–Kovar–platinum seal. This arrangement makes it easier to change the adsorbent. A second cylinder, B, contains the adsorbent and has six vertical vanes of perforated copper foil and a thin perforated metal tube, D, by means of which the incoming gas is distributed radially into the bulk of the adsorbent. The perforations serve to bring the incoming gas into simultaneous contact with a large part of the available surface, thus minimizing possible troublesome effects of non-uniform adsorption. The heater coil for the electrical calibration is imbedded in the adsorbent. A glass plug, not illustrated in *Figure 1*, is normally placed inside the inlet tube A to reduce the dead space. The amount of gas adsorbed is determined by standard methods. Smith and Beebe[38] give a description of the gas-handling apparatus.

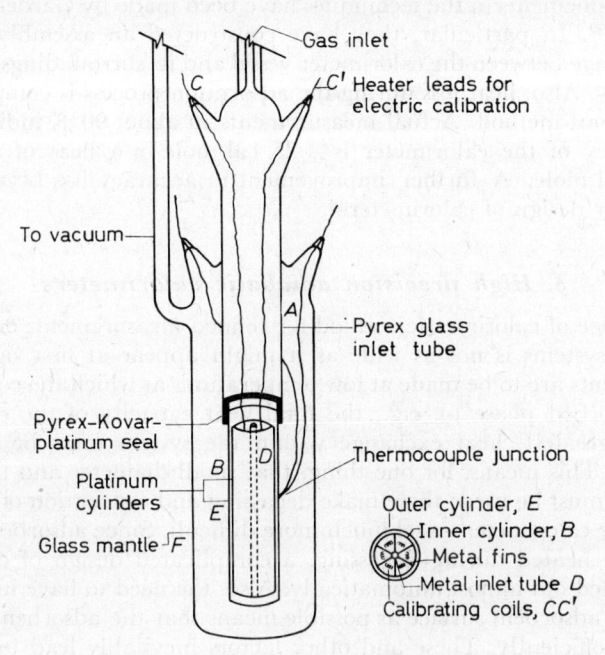

Figure 1. Adsorption calorimeter of moderate precision (after Beebe et al.[6]). Reproduced by courtesy of the American Chemical Society

The measurements of heats of adsorption are made by recording the temperature difference ΔT between the cylinder, E, and the outside bath by means of a thermocouple. The e.m.f. of the thermocouple is amplified with a stabilized d.c. chopper amplifier and recorded continuously. When the space around the calorimeter vessel is evacuated, the major heat exchange between the calorimeter vessel and its surroundings is by heat conduction along the inlet tube A and through the adsorbing gas, and by radiation between E and F. The reproducibility, particularly at higher temperatures, appears to be better if the space between E and F is filled with helium or a helium–nitrogen mixture to a pressure of 7 or 8 torr. The quantity,

$\Delta T - k \int_0^t \Delta T \, \mathrm{d}t$, is proportional to the heat evolved during adsorption or electrical calibration; k is the Newton's cooling law constant determined in the course of electrical calibration, and the value of the integral can be obtained rather easily from graphs of ΔT against t by using a planimeter.

Additional details about this kind of calorimetry as well as many results of measurements of heats of adsorption of simple gases on solids, such as carbon blacks or silicas, down to liquid nitrogen temperatures, are given in the papers of Beebe et al.[1, 6-11, 38]. The papers also contain discussions of possible errors which, of course, depend upon the particular method used, the amount of adsorbing surface in the calorimeter vessel, and the kind of adsorbate–adsorbent pair investigated. The highest accuracy attainable appears to be of the magnitude of a few tenths of 1 per cent.

Other refinements in the techniques have been made by Garden, Kington and Laing[22]. In particular, they have constructed an assembly in which heat exchange between the calorimeter vessel and its surroundings is reduced appreciably. Also, heat loss during the adsorption process is computed by a more rigorous method. Actual measurements at about 90°K indicate[22] that the accuracy of the calorimeter is ± 25 cal/mole in a heat of adsorption of 3800 cal/mole. A further improvement in accuracy has been achieved with a later design of calorimeter[26].

3. *High precision adiabatic calorimeters*

The choice of calorimetric method for refined measurements on gas–solid adsorbing systems is not as wide as it might appear at first sight. If the measurements are to be made at low temperatures at which the contribution of the adsorbed phase to, e.g., the total heat capacity of the calorimeter system is greatest, heat exchange within the system must be accurately controlled. This means, for one thing, that small diameter and thin-walled inlet tubes must be used; these make degassing and evacuation of the adsorbents in the calorimeter vessel much more difficult. Since adsorbents usually need to be heated during degassing, a complicated design of calorimeter vessel is ruled out almost automatically. Also, the need to have in the vessel as large an adsorbent surface as possible means that the adsorbent has to be packed in efficiently. These and other factors inevitably lead to a type of calorimeter vessel within which the distribution of heat or of adsorbing gas is not particularly rapid. This limitation in turn leads to a choice of the adiabatic style of assembly since its accuracy is not greatly affected by the lengths of equilibrium times. In fact, all of the comprehensive measurements of the thermodynamic properties of adsorbed films in the temperature region above 10°K have been made with low-temperature adiabatic calorimeters. Therefore, only these calorimeters, and the thermodynamic relations relevant to their operation, will be discussed.

A. Thermodynamics

If a calorimeter vessel contains known amounts of an adsorbent and an adsorbable gas, measurements can be made of the temperature of the vessel, the amounts of heat supplied or taken away from it, and the pressure of the gas. These are all measurements that, in effect, are made from the outside of the adsorbing system, and they do not indicate how the gas is divided between the gas phase and the adsorbed phase. Therefore, without assumptions or further information, the thermodynamic description of the system is a rather barren one and would hardly justify extensive precise measurements.

The simplest assumption that can be made is that the adsorbent is unperturbed by the gas adsorbed on it. Although this assumption cannot be strictly justified[41], it is a plausible one for the example of physical adsorption on a crystalline solid that has a reasonably large cohesive energy (e.g., ionic crystals). This is the same thing as saying that, if the heat of adsorption is only a few per cent of the cohesive energy of the adsorbent, it is a good approximation to treat the adsorbed phase as a one component system in equilibrium with the unadsorbed gas. The thermodynamic properties of the adsorbed phase obtained in this way can be interpreted statistically in a straightforward manner[41], and it is this possibility that has stimulated detailed calorimetric investigations.

Although the thermodynamics is that of a one component system, one needs to note that the molar properties of the adsorbed phase depend upon the amount of that phase. For example, the molar entropy of the adsorbed gas is given by $S_S = f(P, T, n_S)$, in which n_S is the number of moles adsorbed, rather than by $S = f(P, T)$ as is the case for an ordinary one component system. In addition to molar quantities, differential molar quantities can be defined, such as the differential molar entropy:

$$\bar{S}_S = \left(\frac{\partial(n_S S_S)}{\partial n_S}\right)_T = S_S + n_S \left(\frac{\partial S_S}{\partial n_S}\right)_T. \qquad (1)\dagger$$

The enthalpy of the adsorbed phase is usually referred to that of the gas phase in equilibrium with it. This quantity is then closely related to the heat evolved on adsorption. Several different heats of adsorption can be defined but the following two are adequate for our present purposes:

$$Q' = H_G - H_S$$
$$= \text{integral heat of adsorption,} \qquad (2)$$

and

$$q_{ST} = H_G - \bar{H}_S$$
$$= \text{isosteric (differential) heat of adsorption,} \qquad (3)$$

in which H_G is the molar enthalpy of the gas and H_S and $\bar{H}_S$ are respectively the molar and differential molar enthalpies of the adsorbed phase. The isosteric heat is so called because it occurs in the Clausius–Clapeyron type of relation for a constant amount adsorbed:

$$\left(\frac{\partial \ln P}{\partial T}\right)_{n_S} = \frac{q_{ST}}{RT^2}. \qquad (4)$$

It may therefore also be computed from isotherm data obtained at different temperatures.

In an actual experiment, the total enthalpy of the calorimeter system is given by:

$$H = H_{cal} + n_S H_S + n_G H_G, \qquad (5)$$

† Because of special requirements, a nomenclature and symbolism peculiar to adsorption thermodynamics has grown up. For the most part, it is preserved here so that this chapter can be easily connected to the original papers.

in which H_{cal} means the enthalpy of the calorimeter vessel plus the solid adsorbent, which can be determined in blank experiments. For a reversible change:

$$dH = q + V_G\, dP. \tag{6}$$

In this equation, q is the heat supplied to the calorimeter vessel and $V_G\, dP$ is the work done on the gas phase in the vessel. The apparent heat capacity of the system is given by:

$$\frac{q}{dT} = C_{cal} + \frac{d(n_G H_G)}{dT} + \frac{d(n_S H_S)}{dT} - V_G \frac{dP}{dT}$$

$$= C_{cal} + C_{P_G} + C_{N_S} - q_{ST}\left[\frac{dn_S}{dT}\right]_{expt} - V_G\left[\frac{dP}{dT}\right]_{expt} \tag{7}$$

in which C_{cal} = heat capacity (at constant pressure) of the calorimeter vessel plus the adsorbent; C_{P_G} = heat capacity at constant pressure of the unadsorbed gas in the calorimeter vessel; and C_{N_S} = heat capacity of the adsorbed phase at constant amount adsorbed. The subscript "expt" means differentiation under the actual conditions of the experiment. The penultimate term in Equation (7) takes into account desorption of gas during the experiment. Actually by a simple modification, Equation (7) can be adapted to describe the determination of the heat capacity at saturation pressure of a condensed substance.

For measurements of heats of adsorption under adiabatic conditions, $q = 0$ and therefore:

$$\int_{T_1}^{T_2} (C_{cal} + C_{P_G} + C_{N_S})\, dT = \int_{n_1}^{n_2} q_{ST}\, dn_S + V_G(P_2 - P_1). \tag{8}$$

Alternatively, the process can be described in terms of the integral heat of adsorption [Equation (2)] if $n_1 = 0$, $n_2 = n_S$, $P_1 = 0$, and $P_2 = P_f$. If $Q'(n_S)$ is the integral heat of adsorption of n_S moles of gas, then:

$$\int_0^{n_S} q_{ST}\, dn_S = Q'(n_S) \tag{9}$$

and

$$Q'(n_S) + P_f V_G = (C_{cal} + C_{P_G} + C_{N_S})\, \varDelta T, \tag{10}$$

in which P_f is the final pressure. The heat capacity of the adsorbed phase may also be obtained from the temperature coefficient of Q':

$$\left(\frac{\partial Q'}{\partial T}\right)_{N_S} = C_{P_G} - C_{N_S}. \tag{11}\dagger$$

† Since numerical results are not given, units are not specified and no distinction is made between molar and non-molar quantities.

CALORIMETRIC STUDY OF SOME PHYSICAL PHENOMENA

In summary, the calorimetric experiments have the object of determining C_{N_S}, q_{ST} (and Q') and P for as wide a range of the variables n_S and T as is possible. From the point of view of subsequent interpretation, results for values of n_S less than the equivalent of an adsorbed monolayer are particularly valuable[41]. In this region, C_{N_S} may be as little as 0·5 per cent of the total heat capacity[17, 31] and so the precision of the calorimetric experiments needs to be better than 0·1 per cent.

B. Experimental Details

Several different designs of adiabatic adsorption calorimeters suitable for measurements in the temperature region $T > 10°K$ have been described in some detail[24, 30, 32, 35]. The experimental details will be given primarily for one of these only, although a little will be mentioned of the others.

A sectional drawing of the calorimeter assembly described by Morrison and Los[30] is shown in *Figure 2*. Not all the parts of the cryostat will be discussed because the principles of its design are detailed adequately in Chapter 5. Particular attention will now be paid to the calorimeter vessel and its associated parts.

The calorimeter vessel, Q, and the perforated internal spacers between which the adsorbent is packed are made of pure aluminum. Although this manner of construction is advantageous from the point of view of ease of fabrication (mostly by machining) and of sealing by arc-welding, it is sometimes difficult to fasten aluminum to other metals by special hard solders. The platinum resistance thermometer, R, and the brass sub-assembly into which it is cast with Woods' metal are demountable (for constructional details, see Morrison and Los[30]) as is the adiabatic shield, O. With the thermometer assembly and shield set out of the way, the calorimeter vessel can be heated *in situ*, and so the adsorbent can be degassed at a moderately high temperature ($200° < T < 400°C$). This step is important for producing a reproducible (although not necessarily a clean) surface.

A further important feature of adsorption calorimeters is that they must contain a means for bringing the temperature of incoming gas to that of the calorimeter vessel. If the incoming gas is at a different temperature, an additional term is required in Equations (8) and (10). However, even if the term is added, it is probable that its numerical value cannot be obtained reliably. In the calorimeter assembly being discussed (*Figure 2*), a section of the filling tube, M, is used as a heat exchanger. This section is a length of about 15 cm that extends above the junction point, N, with the shield. The temperature of the section is accurately adjusted by balancing the output of a heater against a cold leak provided by a thermal shunt. More refined heat exchangers have been built into other calorimeters[24, 35], and the one described by Greyson and Aston[24] has been tested directly by measuring the temperature of gas before and after it passed through the exchanger. The efficacy of the heat exchange can also be determined indirectly by comparing heats of adsorption and desorption, since the latter are independent of the gas-filling arrangement. Such comparisons have been made for two different assemblies[31, 34]; for both, the heats of adsorption and desorption agreed within the probable experimental error.

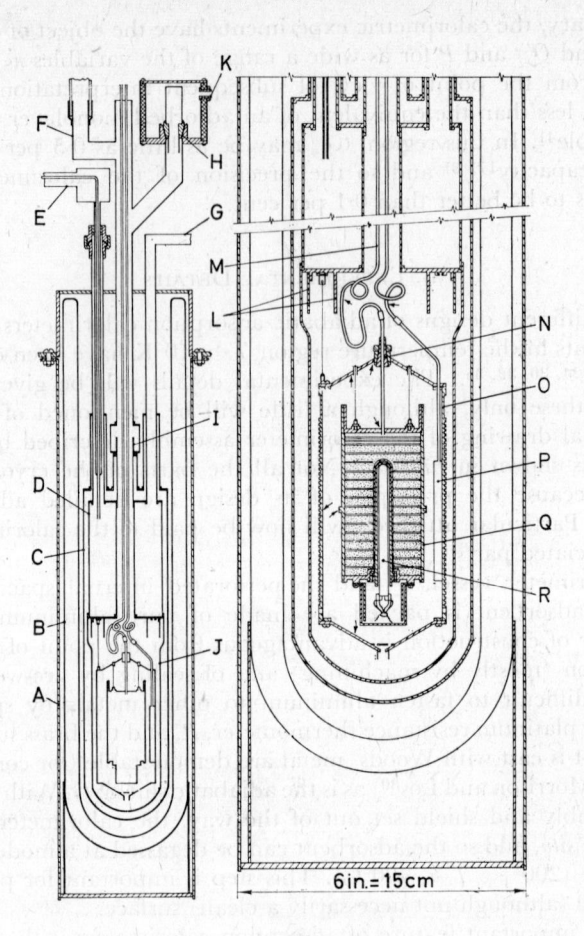

Figure 2. Low-temperature adiabatic calorimeter assembly (after Morrison and Los[30]). Reproduced by courtesy of the Faraday Society

A, Monel metal case
B, Silvered Pyrex Dewar flask, 4·25 in. int. diam. × 24 in. inside depth
C, Brass container ($\frac{3}{64}$ in. wall thickness)
D, Brass liquid hydrogen container, 680 cm^3 volume
E, Vacuum-jacketed metal syphon
F, Vacuum line connection
G, Hydrogen vent line
H, Vacuum seal for lead wires
I, Brass anchoring ring at liquid nitrogen temperature
J, Copper container ($\frac{1}{32}$ in. wall thickness)
K, Brass clamps for lead wires
L, Anchoring rings at temperature of hydrogen container–inner ring $\frac{1}{8}$ in. thick brass, outer ring $\frac{1}{32}$ in. thick copper
M, Copper nickel filling tube (2 mm ext. diam. and 1 mm int. diam.)
N, Brass tapered joint
O, Adiabatic shield, $\frac{1}{16}$ in. thick copper
P, Vapor pressure thermometer bulb, 7 cm^3 volume
Q, Calorimeter vessel, $\frac{1}{16}$ in. thick aluminum, 90 cm^3 volume
R, Platinum resistance thermometer.
    The arrows indicate the positions of thermocouple junctions

In nearly every other respect, low-temperature adsorption calorimeters are like the more precise of the adiabatic calorimeters described in Chapter 5. The accuracies of measurements of total heat capacities are probably similar, but it is difficult to translate this comparison into a general statement about the accuracy with which C_{N_S} can be determined. In the first place, C_{N_S} can vary from being the largest term to being the smallest term on the right-hand side of Equation (7). When it is very small, the precision of the calorimeter is the important thing because C_{N_S} is then determined as the difference between two much larger numbers. In the second place, the accuracy with which C_{N_S} is determined depends upon the accuracies to which the other quantities C_{P_G}, q_{ST}, V_G and P are known. No generalizations seem possible although it can be remarked that the heats of adsorption are usually obtained with an accuracy of ± 10 cal/mole or better[17, 34].

4. *Measurements at liquid helium temperatures*

In one respect the measurements of thermodynamic properties of adsorbed films in the region $T < 10°K$ become easier than those described in the preceding sub-section. The apparent characteristic temperatures of the films tend to be much smaller than those of the calorimeter vessel and the adsorbent, and so, as the temperature decreases, C_{N_S} becomes a larger and larger fraction of the total heat capacity. In other respects, however, the measurements are much more difficult. As is emphasized in Chapter 7, heat leaks interfere seriously in the region $T < 20°K$ and accurate working temperature scales are difficult to establish.

Heat capacities of helium adsorbed on jeweller's rouge[20] and on titanium dioxide[28] have been measured and a few experimental details are given in the papers cited. A fuller account of the work discussed in reference 20 is contained in a thesis[21]. The object of the experiments was in part to study the helium II–helium I transition in the adsorbed films int he vicinity of $2°K$. In bulk helium, the transition produces a λ-type anomaly in the heat capacity. The shape of the anomaly is modified and the maximum heat capacity is decreased in the adsorbed films. In the experiments on jeweller's rouge[20], the heat capacity of the helium was so large that the heat capacities of the calorimeter vessel and the adsorbent could be neglected. A non-adiabatic calorimeter assembly was used.

III. Particle Size Effects

1. *Heat conduction between solid particles*

In Chapter 7 (section V), it is pointed out that helium gas is often used to establish thermal contact between a calorimeter vessel and a solid contained inside it. The possible disadvantage of this procedure is also emphasized: at very low temperatures, helium begins to be adsorbed and, as a consequence, appreciable errors can be made in heat capacity determinations. In principle, if the amount of helium adsorbed (or desorbed) and the heat of adsorption are known, the experimental measurements can be corrected. Or more generally, in the language of section II, the last four terms of Equation (7) would be viewed as corrections in the determination of C_{cal}. However, the calorimeter vessel is not usually connected with the outside so that the pressure of helium gas cannot be monitored. Besides, the heat of adsorption of helium appears to vary strongly with the amount adsorbed and may be as much as ten or more times the heat of liquefaction[3]. As a result, reliable corrections are very difficult to make. The experimental alternative is to reduce the amount of helium to the point where there is just enough to conduct heat between the pieces of solid but not enough to produce significant effects on the energies. Under some circumstances such an amount can be found by trial and error[5].

The difficulties that have just been described would be expected to occur in an exaggerated form in experiments designed to measure effects of particle size on the heat capacity of solids because of the much larger adsorbing surfaces provided by the particles. Measurements of this kind are of value for comparisons with theoretical discussions of effects of boundaries on lattice vibrations[19, 29, 40] and of effects of gross motion of paricles on their heat capacities[25]. The average particle sizes required are, however, microscopic (in the range 0·01–1·0 μ). In addition, the heat capacity measurements should be made to as low temperatures as is possible because the theories only become tractable and realistic for the limiting condition $T \to 0°K$.

2. *Calorimetric measurements*

Possible ways of measuring the heat capacity of microscopic particles at very low temperatures without the use of helium gas for heat conduction have been fully discussed by Giauque[23]. The conclusion reached is that measurements might only be possible for certain paramagnetic substances, to which energy could be added by performance of magnetic work. With such a technique, each particle would be heated separately and by an amount proportional to its mass, but no example of the use of the technique has been reported. Instead, the heat capacity of assemblies of microscopic particles can be measured at low temperatures in a more or less conventional calorimeter without the use of helium exchange gas. In particular, Lien and Phillips[27] have obtained heat capacities of magnesium oxide particles (average size $\sim$0·01 μ) in the temperature region 1·5–4°K. The specimen was loosely packed in a sealed thin-walled copper calorimeter vessel with heater and thermometer on the outside. The only new feature in the behavior of the calorimeter system was that the equilibrium times after heating were longer (3–5 min instead of perhaps seconds as for an empty calorimeter vessel). Similar observations were made by Barkman *et al.*[4] in experiments with sodium chloride particles in the region $4° < T < 30°$ K.

Why the experiments work is not at all clear. Lien and Phillips[27] suggest that motion of the particles assists thermal conduction. Alternatively, the particles may be covered with a fairly tightly-bound adsorbed film that might "glue" the particles together. No doubt, magnesium oxide is a favorable example: since its characteristic temperature is high ($\sim$950°K), its heat capacity is very small at low temperatures (e.g., it is approximately

6×10^{-4} J/mole deg at 2°K). The amount of heat that has to be conducted through a specimen is therefore very small.

Measurements of particle size effects at somewhat higher temperatures ($T > 9°K$) have been made by using controlled amounts of helium exchange gas in sealed-off calorimeter vessels[18, 33, 36, 39]. The only additional experimental point which needs to be noted is that the vessels had to have relatively large volumes (as much as 300 cm³) because the bulk densities of the specimens (titanium dioxide and sodium chloride) were very low.

IV. Stored Energy Experiments

1. General remarks

When solids (including non-crystalline solids) are subjected to high energy irradiation or are deformed, atoms or molecules are displaced. If and when the atoms or molecules return to their normal positions, energy is released. Similarly, energy is released when impurity atoms precipitate out of a solid to form clusters or new phases, but the temperature region over which an energy change is measurable varies considerably. No general guides can be laid out other than to note that energies stored in point imperfections (e.g. in interstitial atoms) will often be released in the region below room temperature. Some calorimetric studies of energy released at low temperatures have been made, and a few examples are briefly discussed.

2. Radiation damage at low temperatures

Special techniques must obviously be used for cryogenic experiments in a nuclear reactor. There are many materials problems: e.g., the apparatus will usually have to be long and thin, and so the construction material needs to be strong. At the same time, many strong materials are ruled out because, under irradiation in the reactor, they are converted into inconveniently long-lived radioactive substances. These and other problems are discussed by Blewitt et al.[12, 13] who designed a cryostat assembly for installation in a graphite-moderated reactor at Oak Ridge. The space available to et al.[12, 13] Blewitt had a least cross-section of 4 in.² and the center of the reactor was about 20 ft from the outer face. The apparatus installed consisted of a long central tube connected to the sample chamber, on the outside of which was a heat exchanger through which passed helium gas brought in from the outside and returned in tubes connected to the heat exchanger. The entire assembly was surrounded by a vacuum jacket that provided the only insulation. The lower 17 ft of the cryostat was constructed of 2 S aluminum because, upon radiation, it forms only short-lived radioactive species. The circulating helium gas, which could remove energy at the rate of 500 W, was cooled by passing it through a helium refrigerator. With the reactor off, the temperature of the sample chamber could be reduced to 10°–11°K; at full power, the temperature rose by about 5°.

Blewitt et al.[12, 13] illustrate the operation of the cryostat by giving the results of measurements of energy stored in a single crystal of aluminum that had been irradiated for about 6 days at 19·2°K. The measurements comprised following the temperature of the sample chamber (as determined with a copper–constantan thermocouple) as a function of time as the sample was heated by γ-rays from the reactor operating at about 1/10 power. The stored energy was found from the difference between two series of measurements over the range 19·2–60°K, the first immediately after the irradiation and the second after the sample had been heated to 60°K.

Experiments with a similar principle have been made in connection with studies of trapped radicals produced by γ-irradiation of several condensed substances[2]. The apparatus consisted of an adiabatic calorimeter assembly mounted in a Dewar flask, that was sunk to the bottom of a 16 ft water-filled tank where reactor fuel elements were placed. The calorimeter could be operated from 4–300°K.

A warning about some hazards involved in the use of cryogenic apparatus in reactors is given by Blewitt et al.[12, 13]. Two explosions are described briefly, and it is suggested that possibly both were caused by ozone formed by the irradiation of small amounts of oxygen in the system.

To conclude this section, a different kind of study of radiation damage is briefly mentioned. In this example[37], radiation damage is self-induced in plutonium metal encased in a copper calorimeter vessel, and the release of stored energy as a function of temperature ($4° < T < 300°K$) is studied. The calorimeter assembly differs substantially from more conventional types only in the respect that the specimen is self-heating.

3. Clustering in alloys

De Sorbo et al.[14–16] have made several studies of the kinetics of precipitation reactions in metals and alloys in the temperature region below room temperature using an adiabatic microcalorimeter. Three different modifications of the calorimeter with full details are described by them. The specimens can be "homogenized" outside or inside the assembly and quickly brought to a desired temperature within an adiabatic enclosure. The amount and rate of energy release can then be determined.

The experiments mentioned in section IV are not intended to be highly accurate. Some of the processes being studied are not reproducible and therefore do not warrant highly accurate thermal measurements. Others, which appear to be reproducible, could be studied more carefully if required. In principle, there would appear to be no great difficulties about further refinement of the calorimetric techniques.

V. References

[1] Amberg, C. H., W. B. Spencer, and R. A. Beebe, *Can. J. Chem.*, **33**, 305 (1955).
[2] Arnett, R. L., Phillips Petroleum Co., Bartlesville, Oklahoma. Private communication.
[3] Aston, J. G., and J. Greyson, *J. Phys. Chem.*, **61**, 613 (1957).
[4] Barkman, J. H., R. L. Anderson, and T. E. Brackett, *J. Chem. Phys.*, **42**, 1112 (1965).
[5] Barron, T. H. K., W. T. Berg, and J. A. Morrison, *Proc. Roy. Soc. (London)*, **A250**, 70 (1959).
[6] Beebe, R. A., J. Biscoe, W. R. Smith, and C. B. Wendell, *J. Am. Chem. Soc.*, **69**, 95 (1947).

[7] Beebe, R. A., and D. A. Dowden, *J. Am. Chem. Soc.*, **60**, 2912 (1938).
[8] Beebe, R. A., B. Millard, and J. Cynarski, *J. Am. Chem. Soc.*, **75**, 839 (1953).
[9] Beebe, R. A., and H. M. Orfield, *J. Am. Chem. Soc.*, **59**, 1627 (1937).
[10] Beebe, R. A., M. H. Polley, W. R. Smith, and C. B. Wendell, *J. Am. Chem. Soc.*, **69**, 2294 (1947).
[11] Beebe, R. A., and N. P. Stevens, *J. Am. Chem. Soc.*, **62**, 2134 (1940).
[12] Blewitt, T. H., and R. R. Coltman, in *Experimental Cryophysics*, Butterworths, London, 1961, pp. 274–284.
[13] Coltman, R. R., T. H. Blewitt, and T. S. Noggle, *Rev. Sci. Instr.*, **28**, 375 (1957).
[14] De Sorbo, W., *Phys. Rev.*, **117**, 444 (1960).
[15] De Sorbo, W., H. N. Treaftis, and D. Turnbull, *Acta Met.*, **6**, 401 (1958).
[16] De Sorbo, W., and D. Turnbull, *Acta Met.*, **4**, 495 (1956).
[17] Drain, L. E., and J. A. Morrison, *Trans. Faraday Soc.*, **49**, 654 (1953).
[18] Dugdale, J. S., J. A. Morrison, and D. Patterson, *Proc. Roy. Soc. (London)*, **A224**, 228 (1954).
[19] Dupuis, M., R. Mazo, and L. Onsager, *J. Chem. Phys.*, **33**, 1452 (1960).
[20] Frederikse, H. P. R., *Physica*, **15**, 860 (1949).
[21] Frederikse, H. P. R., *Adsorption of Helium*, Thesis, Leiden, 1950.
[22] Garden, L. A., G. L. Kington, and W. Laing, *Trans. Faraday Soc.*, **51**, 1558 (1955).
[23] Giauque, W. F., *Phys. Rev.*, **103**, 1607 (1956).
[24] Greyson, J., and J. G. Aston, *J. Phys. Chem.*, **61**, 610 (1957).
[25] Jura, G., and K. S. Pitzer, *J. Am. Chem. Soc.*, **74**, 6030 (1952).
[26] Kington, G. L., and P. S. Smith, *J. Sci. Instr.*, **41**, 145 (1964).
[27] Lien, W. H., and N. E. Phillips, *J. Chem. Phys.*, **29**, 1415 (1958).
[28] Mastrangelo, S. V. R., and J. G. Aston, *J. Chem. Phys.*, **19**, 1370 (1951).
[29] Montroll, E., *J. Chem. Phys.*, **18**, 183 (1950).
[30] Morrison, J. A., and J. M. Los, *Discussions Faraday Soc.*, **8**, 321 (1950).
[31] Morrison, J. A., J. M. Los, and L. E. Drain, *Trans. Faraday Soc.*, **47**, 1023 (1951).
[32] Morrison, J. A., and G. J. Szasz, *J. Chem. Phys.*, **16**, 280 (1948).
[33] Morrison, J. A., and D. Patterson, *Trans. Faraday Soc.*, **52**, 764 (1956).
[34] Pace, E. L., E. L. Heric, and K. S. Dennis, *J. Chem. Phys.*, **21**, 1225 (1953).
[35] Pace, E. L., L. Pierce, and K. S. Dennis, *Rev. Sci. Instr.*, **26**, 20 (1955).
[36] Patterson, D., J. A. Morrison, and F. W. Thompson, *Can. J. Chem.*, **33**, 240 (1955).
[37] Sandenaw, T. A., *J. Phys. Chem. Solids*, **23**, 1241 (1962).
[38] Smith, W. R., and R. A. Beebe, *Ind. Eng. Chem.*, **41**, 1431 (1949).
[39] Sorai, M., A. Kosaki, H. Suga, and S. Seki, *Bull. Chem. Soc. Japan*, **41**, 536, (1968).
[40] Stratton, R., *Phil. Mag.*, **44**, 519 (1953).
[41] Young, D. M., and A. D. Crowell, *Physical Adsorption of Gases*, Chapter 3, Butterworths, London, 1962.

CHAPTER 14

High-speed Thermodynamic Measurements and Related Techniques

CHARLES W. BECKETT and ARED CEZAIRLIYAN

National Bureau of Standards, Washington, D.C., U.S.A.

Contents

I.	Introduction	552
II.	Generation of Heat	554
	1. General Considerations	554
	2. Heating by Electrical Pulses	555
	A. Repeating Pulse Generators	555
	B. Single Pulse Generators	555
	3. Heating by Pulsed Radiation	556
	4. Electrical Circuitry Associated with Single Pulse Generators	557
III.	Measurement of Heat	558
	1. General Considerations	558
	2. Current Measurements	559
	3. Voltage Measurements	559
	4. Conclusions Regarding Measurement of Heat	561
IV.	Measurement of Temperature	561
	1. General Considerations	561
	2. Photoelectric Temperature Measurement Techniques	564
	3. Photographic Temperature Measurement Techniques	565
	4. Other Temperature Measurement Techniques	567
V.	Other Measurements	568
	1. High-Speed Photographic Techniques	568
	2. Dynamic Pressure Measurement Techniques	569
	3. X-ray and Interferometric Techniques for Density and Velocity Measurements	571
VI.	Applications of High-Speed Measurements	572
	1. General Considerations	572
	2. Measurement of Properties	572
	A. Specific Heat	572
	B. Thermal Diffusivity	574
	(1) Utilizing Direct-Contact Heaters	575
	(2) Utilizing Radiation Heaters	575
	C. Thermodynamic Properties at Very High Pressures	575
	D. Electrical and Thermoelectrical Properties	577
	3. Other Applications	577
	A. Exploding Conductors	577
	B. High-Temperature Sources	578
VII.	Appendices	579
VIII.	References	581

'Time is the most undefinable yet paradoxical of things; the past is gone, the future is not come, and the present becomes the past, even while we attempt to define it, and like the flash of the lightning, at once exists and expires'.

Colton

I. Introduction

The interest in fast measurement techniques, not only in thermodynamics but in chemical kinetics and other scientific and engineering fields, arises from the need to measure properties of matter and to study transient phenomena under extreme conditions of temperature and pressure. The reasons for using high-speed measurement techniques are that at high temperatures, heat losses, chemical reaction rates, sublimation rates, etc. increase so rapidly that conventional methods may not yield accurate results. In high-speed studies the need for accurate measurement techniques goes far beyond the range of conventional methods; hence, high-speed measurement of thermodynamic and related quantities is in an exploratory state of development. It is expected that intensive investigations of time-resolved measurement techniques in the millisecond and microsecond range will lead to the measurement of a sufficient number of variables to determine changes in thermodynamic states of solids, liquids, and dense gases at temperatures above 2000°K with accuracies of perhaps 5–10 per cent.

Heat measurements are ordinarily carried out in times of the order of minutes. In specific heat calorimetry, heat is usually introduced electrically for periods in the range of 120–10 000 sec with approximately constant power, delivering energy to the sample in the range of 10–10 000 joules depending upon whether or not the sample is in a transition region. The temperature rise in the sample is usually determined by very accurate measurements of the temperature before and after the heating period. From these measurements, the quantity of sample, the evaluations of heat losses and the specific heat of the other substances in the system, the specific heat or heat of transition is obtained as described in other chapters of this book.

The characteristic times of the measurement, such as the heating period and the times required for temperature measurements, are limited by several processes. For example, time for the equilibration of the temperature depends largely on the length of heat flow path and the thermal diffusivity of the solids within the calorimeter. The thermal relaxation time (to come to 63 per cent of change) is proportional to L^2 where L is the linear dimension in cm. If the calorimeter heater is approximately 1 cm from the center of the sample, many seconds are required for the system to almost attain thermal equilibrium. However, if the dimensions could be reduced from centimeters to microns, the time would be of the order of microseconds. If the sample is an electrical conductor under the conditions of the experiment, the electrical energy can be introduced directly into the sample, thereby permitting the sample to be larger. In the ideal case, the temperature change in the sample will be uniform throughout the sample if the heating period is long as compared to the electrical characteristic time constant, which depends upon the specific resistance and dimensions. This electrical characteristic time for metallic conductors of ordinary dimensions and shape is less than one microsecond. Thus in such cases one could design systems that have heating periods from milliseconds to microseconds, provided that one has the means of measuring the electrical energy input with the required accuracy and also means of measuring the temperature rise in a time short compared to the cooling rate of the system, so that the effects of heat losses from the sample to its environment are small.

HIGH-SPEED THERMODYNAMIC MEASUREMENTS

With the pulse generators now available, one can deliver energies in the range of 10–10 000 joules and even higher in times ranging from milliseconds to microseconds, and with techniques now being investigated, the pulse energy can be measured with accuracies of about 1–2 per cent. These techniques might also be adapted to the study of the thermodynamic properties of metals at very high pressures in the ordinary temperature range. However, the chief difficulty would appear to be in the temperature measurement, since an electrical technique for measuring the temperature, utilizing either the temperature dependence of resistance or thermoelectrical properties would require rather high precision. The feasibility of carrying out such an experiment with the accuracy that one might desire has not been fully explored. The application of fast measurement techniques to heat measurements at temperatures where conventional methods become increasingly difficult is being explored quite actively at normal pressures. As is discussed later in this chapter, several groups have published work in which electrical or other energy has been delivered to the sample in times of less than 1 sec, and in which the temperature measurements have been made in times of the order of tenths of seconds. In this way thermodynamic data on solids have been obtained with accuracies of about 2 per cent at 1000°K and 5 per cent at 2500°K. The discussion in this chapter is concerned with the measurement of thermodynamic and related quantities having time resolution in the millisecond to microsecond domains at temperatures from 2000–6000°K. However, in order to preserve the continuity between conventional and high-speed measurements, in some cases, studies are also presented for temperatures lower than 2000°K and speeds of the order of seconds. Among the measurements of various thermodynamic quantities prime emphasis is placed on those closely related to high-speed calorimetry.

A review of experimental problems at temperatures above 2500°K, the approximate upper limit for conventional furnace techniques, indicates that the most serious difficulties arise from the exponential increase in the rate of evaporation of the sample and from corresponding increases in the rate of other physical and chemical processes. In typical cases the characteristic temperature in the exponential rate function is in the range of 20 000–100 000°K. For evaporation, this is merely the heat of vaporization divided by the gas constant. For the emission of electrons or ions from a hot surface, the work function also falls in this range. Furthermore, the rate of emission of photons in the visible region of the spectrum has characteristic temperatures in this range. For example, 22 000°K is the characteristic temperature corresponding to 6500 Å, the wavelength of light used in ordinary pyrometry, and 30 000°K is that for the blue region of the spectrum. Solid state diffusion, which may have serious effects due to contamination of the sample with materials from the walls of the container, also has an exponential temperature coefficient. Thus, many of the time-dependent phenomena that influence thermodynamic experiments change in the same manner; other properties such as thermal diffusivity and the electrical resistance of metals and alloys have a temperature dependence that is very nearly linear.

From such considerations of the influence of time-dependent phenomena upon thermodynamic measurements and particularly the difficulties encountered in obtaining thermodynamic data above 2000°K, it is concluded that the optimum time for an experimental measurement changes in an exponential manner. With the apertures and target areas now being used in photoelectric pyrometry, a sufficient number of photons is available in a short time (millisecond or microsecond) to permit accurate measurement of the temperature, provided that a measuring system can be designed and properly calibrated.

Generation and measurement of heat become increasingly difficult as the heating period is reduced from minutes to milliseconds and then to microseconds. In Sections II and III of this chapter, various methods of high-speed generation and measurement of heat are presented. Section IV is devoted to high-speed temperature measurement techniques. Factors which enter into the design of time-resolved temperature-measuring equipment are also discussed. High-speed photography, which is a useful research tool in high-speed measurement of thermodynamic quantities is reviewed briefly in Section V. Since high-speed investigations of thermodynamic, transport, and other properties of matter at high temperatures may be associated with high pressures, a review of high-pressure measurement techniques is included in Section V. Emphasis is placed on the various investigations related to the pressure dependence of electrical resistivity of substances of scientific and engineering interest. In Section VI, the application of high-speed measurements to the determination of properties, such as specific heat, thermal diffusivity, and electrical conductivity, is discussed. Measurement of thermodynamic properties by shock wave techniques is also reviewed. Other applications of high-speed measurement techniques in the areas of exploding conductors and high-temperature sources are also discussed in Section VI.

II. Generation of Heat

1. *General considerations*

In high-speed measurement of thermodynamic quantities, the heat required in the sample can be generated either internally in the sample or externally at the boundaries of the sample. Internal generation of heat is based upon the Joulean heat dissipation during the passage of an electric current through the sample. Therefore, this method applies only to samples that are electrical conductors. Since many of the materials of scientific and technological interest (metals, alloys, semiconductors) fall into this category, this method has fairly broad applications. For materials that are nonconductors (ceramics, etc.) heat generation must take place outside the sample. This can be accomplished most easily by external electric heaters. In high-speed measurements, the use of steady state heat generation is not feasible, and, therefore, utilization of transient techniques is essential.

There are several methods for generating transient heat. Some of the common ones fall into the following two categories: (1) heating by electrical pulses, (2) heating by pulsed radiation. The first method can be used for internal and external heat generation, whereas the second one can be used only in the case of generation of heat at the outer surface of the sample. In the following paragraphs these two methods and the electrical circuitry associated with single pulse generators are briefly reviewed.

2. *Heating by electrical pulses*

Pulsed electrical energy can be used to heat the sample either internally or externally through resistance heaters. Electrical pulse generators fall into two general classess: (1) repeating pulse generators, (2) single pulse generators. The distinction is somewhat arbitrary. In single pulse generators the repetition rate is of the order of one discharge per minute or longer time; whereas, with the repeating pulse generators, the repetition rate may be from 10–10 000 pulses/sec. Early and Walker[77] have discussed various means of storing great quantities of electrical energy. They have shown that capacitive storage may be used for discharges in the microsecond range, whereas inductive storage has advantages for discharges in the millisecond range. Pulsed electrical energy can also be obtained from storage batteries with proper switching attachments. However, this kind of energy source cannot be used for very fast measurement work. Its application can, at most, be extended to studies in the millisecond time domain.

A. Repeating Pulse Generators

In these generators, the output of electrical energy is in the form of repeated pulses. There are various types of repeated pulse generators. Some utilize transformers and high-speed relays, others utilize direct current sources with repeating on–off switching systems, etc. Such pulse generators have been discussed in detail, and pertinent references are given by Lewis and Wells[127] and by Glasoe and Lebacqz[89]. Typical pulse powers for repeating pulse generators are in the range of 1–20 MW. This means that the corresponding pulse energy is from 1–20 J/pulse if the period is 1 microsecond. However, because of the limitations imposed by the skin effect, very high-frequency pulse generators may not be used. In typical cases the energy appears to vary from one pulse to the next by about 1 per cent. If a fairly large number of pulses can be utilized during the heating period, this variation should not introduce serious errors into the heat measurement. Furthermore, the average pulse energy can be calibrated by the use of techniques similar to those that are used in primary electrical calibration where the precision is about 0·01 per cent. If the inductance, capacitance, and resistance in the calibration experiment are selected to closely simulate the values that occur in the thermodynamic experiment, then it would be expected that accuracies of about 0·1 per cent could be achieved in heat measurements. The difficulties arise from the fact that these properties (inductance, capacitance, and resistance) are functions of time, and furthermore, in general, variations occur not only in the sample but also in other parts of the measurement system, such as switches, so that the effects cannot be analysed in a precise way.

B. Single Pulse Generators

In single pulse generators, the energy is stored in a capacitor or a bank of capacitors and then discharged to the sample through a switch (spark gap) and associated electrical equipment. A single pulse may be oscillatory in nature, and depending upon the circuit parameters it may decay after 1–10 or more oscillations. Very high pulse energies are possible with single pulse generators, ranging up to a million joules for pulse periods in the range 10–100 μsec, tens of thousands of joules at 1–10 μsec, or 100–1000 J in 0·01 μsec. These figures correspond to pulse powers in the 1000 MW range. The achievement of very short pulse periods with relatively large amounts of energy requires increasingly high voltages, starting at about 10 kV and going up to about 1000 kV or more. Most of the present activities are concerned with the voltage range from about 10–50 kV. Some rather large single pulse generators are now being operated at about 100 kV, and a few are operating at 500 to 1000 kV. The use of very fast, high-voltage, single-pulse generators in investigations on exploding conductors was reported by O'Rourke[149].

Tucker[192] has described the construction and operation of a 100 kV coaxial cable square-wave generator producing a 2000 A, 3 μsec duration, 6 nanosecond rise–time current pulse. High-capacity, single-pulse generators are used in studies on plasmas, exploding conductors, etc., where a high energy per unit volume must be imparted in a very short time. Additional information on single pulse generators can be found in publications on exploding conductors discussed in Section VI–3–A of this chapter. The utility of these pulse generators in high-speed measurement of thermodynamic quantities depends upon the ability to measure the pulse power with an accuracy that is acceptable in such work, that is 1–2 per cent.

More conventional single pulse generators are also used to heat the boundary of the sample in connection with the measurement of thermal diffusivity of substances (electrical conductors or non-conductors). References to the above are given in Section VI–2–B of this chapter.

3. Heating by pulsed radiation

This method utilizes pulsed radiation energy from a radiation source to heat the surface of the sample. As mentioned earlier, heating by pulsed radiation is used to generate heat near the surface of an opaque sample. This has advantages over the external heating by electrical pulse generators since some of the difficulties resulting from contact resistance and heat transfer are partially eliminated. However, because of the finite size and appreciable magnitude of the thermal time constant of the sample, this method may be used only with very thin samples in the microsecond time range.

In some cases, a steady radiation source is used, and the pulsing is accomplished by means of a rotating shutter placed between the radiation source and the sample. The pulsing rate is determined by the shutter geometry and the speed of rotation. The radiation source may be an electrically controlled resistance. For high-temperature studies arc-image and plasma-image furnaces have proved to be very useful as radiation sources.

It is also possible to use single pulse radiation sources such as flash lamps. However, the energy per pulse from the above sources is usually relatively small. Kuebler and Nelson[118] have studied the measurement of radiant energy of capacitor discharge flash lamps. They have reported that energies of about 25 J/cm^2 are obtainable from some commercially available flash lamps. Other energy sources may be high-intensity electron beams and laser pulses. Studies along these lines were made by Cowan[57], Cerceo and

Childers[46], and Deem and Wood[67] in connection with thermal diffusivity measurements.

4. *Electrical circuitry associated with single pulse generators*

Since most of the high-speed work conducted to measure properties of matter at high temperatures requires relatively large powers, single-pulse generators are of great interest. Because sudden delivery of a great amount

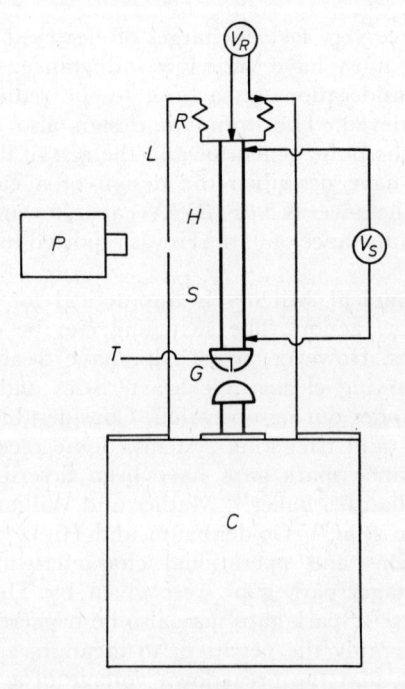

Figure 1. Schematic diagram of a high-speed thermodynamic measuring system

C, Capacitor
G, Spark gap
H, Hole
L, Tubular return conductor
P, Pyrometer
R, Tubular resistance
S, Tubular sample
T, Triggering system
V_R, Voltage measurement
V_S, Voltage measurement

of energy requires high voltage and low impedance, the design of the electrical circuitry associated with the pulse generator and the sample presents unusual problems.

A typical simple circuit consists of the sample in series with a switch and the pulse generator which, in this case, may be a single capacitor or a bank of capacitors. A schematic diagram of such a system is shown in *Figure 1*.

The sample, S, is in the form of a thin-walled tube. A small hole, H, at the middle of the tube provides a blackbody source for the pyrometer, P. The switching is accomplished by means of a spark gap, G, in conjunction with a triggering system, T. The energy is supplied by the capacitor, C.

The circuit and components for charging the capacitor are not shown in the figure. In order to calculate the energy dissipated in the sample the voltage, V_S, and the current, I, flowing through the sample must be known. The current, I, may be obtained by measuring the voltage, V_R, across a known resistance, R. The resistance, R, is of cylindrical shape and the voltage measuring lead is placed along the axis of the cylinder to avoid errors resulting from induced voltages. The recording of the electrical quantities is accomplished using oscilloscopes. The return conductor, L, is also of cylindrical shape.

In order to achieve very fast discharges of electrical energy through the sample, the circuit must have very low inductance. However, electrical and mechanical considerations set a limit to the reduction of inductance which may be achieved. The optimum design also requires impedance matching between the pulse generator and the rest of the circuit. Recently, Dike and Kemp[69] have described the design of a capacitor and switch assembly of low inductance. A 5 µF, 25 kV capacitor and the switch assembly having a total inductance of 0·04 µH was reported to deliver its electrical energy in 0·75 µsec.

The use of mechanical switches is impractical for microsecond pulses because of their slow action. The switching can be done by thyratron-triggered spark gaps. However, spark gaps have disadvantages, the chief ones being their varying electrical characteristics and the loss of energy between their electrodes during operation. Considerable research has been done on spark gaps and triggering systems. Some recent studies on high-voltage, low-inductance spark gaps have been described by Richeson[164], Hagerman and Williams[96], Baker[16], Mather and Williams[135], Cormack and Barnard[54], Goldman et al.[91], Goldenbaum and Hintz[90], Jahn et al.[104], and Smith[182]. Descriptions and operational characteristics of fast-triggering systems of high-voltage spark gaps were given by Theophanis[189], Levine et al.[126], and Hancox[97]. Spark gaps may also be triggered by heating one of the electrodes (preferably the negative) to incandescence. Broadbent and Wood[35] have shown that the breakdown voltage of a two-electrode spark gap is considerably reduced with the presence of a small source of heat at the surface of one electrode. Recently, Broadbent and Shlash[34] have performed experiments with a hot-wire triggered spark gap at voltages up to 1 MV and gap spacings up to 80 cm. Specifications and constructional and operational details of the electrical circuitry associated with single-pulse generators are given in the publications on exploding conductors cited in Section VI–3–A of this chapter.

III. Measurement of Heat

1. *General considerations*

In high-speed, high-temperature studies of properties of matter, heat that must be imparted to the sample may be generated most effectively by Joulean heat dissipation. This is particularly advantageous for samples that are electrical conductors, since the flow of electric current through the sample ideally provides uniform and sudden temperature rise. The amount of generated heat in the sample can be calculated from electrical quantities such as current, voltage and electrical resistance. In principle, this can be accomplished by integrating instantaneous power over the time of current flow, either by taking the product of the square of the current and the resistance of the sample, or by taking the product of the current and the voltage across the sample.

In high-speed measurements, a difficulty arises when one tries to use the former method to calculate the input energy. The reason is that the resistance of the sample is not constant and varies with time during the current flow through the sample. Therefore, it seems that one must employ the latter method which requires simultaneous measurement of the voltage across the sample and the current flowing through it.

In order to deliver large amounts of energy to the sample in a short time, the electrical energy source must be capable of supplying large currents at high voltages, which presents difficulties both in the design and in the measurements.

Summarized below are some of the methods used to measure pulse currents having peak values of about 10^4–10^5 A at frequencies of 0·01–1 Mc and voltages up to 10^5 V.

2. *Current measurements*

One of the early studies in measuring heavy surge currents of the order of 10^5 A and of effective duration greater than 100 μsec was described by Bellaschi[26]. However, recent direct measurement of surge currents, based upon the method developed by Park[150], is quite reliable.

In this method a tubular shunt with coaxial potential leads is utilized. The leads are connected to an oscilloscope and the measurements are recorded by photographing the screen. By this technique an accuracy of about 1 per cent can be achieved. Recently, the transient response of a coaxial current-measuring shunt was analysed by Bennett and Marvin[27].

In addition to the above, pulse currents may be measured by an indirect method which is based upon the measurement of an induced voltage in a coil placed in the vicinity of the circuit through which the current flows. Because of the nature of the current, a voltage is induced in the coil whose magnitude is proportional to the rate of change of the main current. The advantage of this is that the measuring circuit is not connected to the main circuit. A disadvantage is that it is not an absolute measuring method and requires calibration. Since the calibration is sensitive to the circuit parameters and the geometry of the system, the correct interpretation of the induced voltage in the coil presents some difficulties. Also, since the measurements are based upon the rate of change of the current, this method may be applied successfully only to cases in which there are no abrupt changes in the current giving large values for the time derivative of the current.

3. *Voltage measurements*

The direct measurement of the transient voltage across the sample may be accomplished with the use of a resistive or capacitive voltage divider in conjunction with a recording system, such as an oscilloscope.

One of the early studies in measuring high surge voltages using dividers was described by Bellaschi[25]. There has been considerable research on precision measurement of static high voltages and of high-frequency low voltages. However, high-frequency high-voltage or transient high-voltage measurement techniques are not developed to yield accurate (uncertainty of 1 per cent or less) results. Details of primarily static high-voltage measurement techniques and instrumentation were reviewed by Defandorf[68]. High-frequency (upper audio and radio) voltage measurement techniques were described by Selby[174]. Tsai and Park[191] have reported results on measurement of transient high voltages utilizing a resistive voltage divider of the type described by Park and Cones[152]. Brady and Dedrick[29] have performed measurements up to 350 kV with a capacitive voltage divider. Resistive voltage dividers are limited by the maximum allowable voltage gradient along the divider components, whereas capacitive voltage dividers are limited in the maximum voltage by the type of dielectric used.

One of the main sources of error in the measurement of transient high voltages with a voltage divider is the voltage divider itself. Park and Cones[151] have discussed various possible sources of divider errors, such as residual resistance, stray capacitance, impedance drop in the leads, etc. The voltage at the terminal of the voltage divider may be recorded by using oscillographic techniques. Problems in connection with oscillograph recording were discussed briefly by Park and Cones[151]. Recently, some difficulties in obtaining readings of high-frequency transient voltages and techniques for overcoming them were described by Lord[129].

Transient high-voltage measurements may not be reliable as a result of possible errors introduced by induced voltages in the voltage-measuring circuit due to inductive coupling with the current change in the main circuit. Moses and Korneff[145] have described a voltage-measuring system that may eliminate the error introduced by the induced voltage. Their method is based upon generating a voltage which is exactly equal in magnitude and in phase with the induced voltage and subtracting it from the output of the measuring system.

A rather new technique of measuring pulsed high voltages of the order of 10^5 V at frequencies up to 50 Mc is described by Ettinger and Venezia[83]. This method utilizes a Kerr cell placed in the voltage-measuring circuit The variation of the voltage across the cell causes changes in the intensity of a light beam passing through it. The response time of this measuring system is very small (of the order of 0·01 μsec); however, it requires a proper calibration of light intensity as a function of voltage. Recently, Wunsch and Erteza[213] have developed a system, based on the application of the Kerr effect, to measure high-voltage pulses with magnitudes 3×10^4–1×10^5 V at frequencies up to 100 Mc. They have reported that peak pulse magnitude could be measured with a resultion of better than 0·1 per cent, and after proper calibration, the accuracy of peak voltage measurement could be ±1 per cent. Beers and Strine[24] have discussed another possible method of measurement of direct voltages which is based upon the Stark effect.

Since the measurement of voltage across the sample is rather inaccurate at the present time, and since the resistance varies with time, it may be convenient to determine the energy imparted to the sample indirectly by subtracting the energy losses in the circuit (omitting the sample) from the total electrical energy supplied by the source. However, this method has the disadvantage that it requires a precise knowledge of the circuit characteristics. Baker and Warchal[15] have described experiments to measure the energy imparted to samples by the above method. Experiments based on a somewhat similar technique have been discussed by Bondarenko et al.[28] and Kvartskhava et al.[120].

4. *Conclusions regarding measurement of heat*

At the present time, transient energy measurement methods are still in a developmental stage. There are various factors that affect the accuracy and reliability of the results. Some of the most critical sources of error are given below.

The induced voltages in the electrical measuring circuits distort the actual values of the electrical quantities. At the high frequencies present in pulse systems, the skin effect may play an important role in changing the resistance ratios of the two resistances of the voltage divider, should the geometries or the materials of the two resistances be different from each other. The improper grounding of different sections of the measuring system, as well as the test section, may also cause errors in the results. The electrical instruments present another problem in measurements, but this can be minimized by proper calibration and compensation. The interpretation of the results from the oscilloscope on photographic films and the integration of the instantaneous voltage and current product present other possible sources of error.

For successful high-speed measurement of thermodynamic quantities, it is essential that heat measurements by electrical means be improved to yield results with accuracies better than 1 per cent. In the near future 2 per cent should be attainable, although all measurements of electrical energy dissipation by fast pulse generators to date probably are less accurate. Uncertainties reported in the literature vary from 2 to 10 per cent or more.

IV. Measurement of Temperature

1. *General considerations*

Precision measurement of thermodynamic quantities requires high precision in temperature measurements as well as heat and other related measurements. There are different techniques for measuring temperature, and the proper selection for a particular experimental case depends upon the temperature region of interest in addition to other physical and geometrical considerations, such as speed, accuracy, versatility, size, etc.

There are various well established conventional techniques for measuring temperature up to 2000°K based upon thermoelectric, resistance and thermodynamic phenomena. These methods are discussed in Chapter 2 of this book. In many experiments above 1500°K, pyrometry techniques are employed for the measurement of the temperature of opaque objects. Photographic temperature measurement techniques are also possibilities for high-speed, high-temperature measurements. Although not developed sufficiently to yield very accurate results at present, they are feasible with adoptions of high-speed photographic equipment now available.

Depending on their method of detection and operation, pyrometers may be grouped under the following classes: radiation, optical, and photoelectric pyrometers. It is also possible to classify the photographic temperature-measurement techniques under photographic pyrometry. In the literature, this is also referred to as photothermometry. Radiation and optical pyrometers are not suitable for high-speed temperature measurements; therefore, this section is confined primarily to the discussion of photoelectric and photographic pyrometry techniques.

Before delving into the details of each method, problems common to most pyrometry techniques are discussed.

The principal requirements of a radiation source are that its temperature be highly reproducible and constant throughout an area of at least 0·25 mm^2. Most common radiation sources used in pyrometry are blackbody sources and electrically controlled tungsten lamp reference sources. Furnace-type blackbody sources can be operated satisfactorily at temperatures up to 2700°K, and with modern temperature measurement and control equipment they should be reproducible to at least 0·1 per cent. The evaluation of the quality of a blackbody was discussed by De Vos[65]. Because of simple constructional and operational characteristics, tungsten lamps are more frequently used than blackbodies. The lamps can be calibrated to indicate "brightness" temperature as a function of current at temperatures up to about 2600–2700°K. However, the calibration becomes difficult at the high temperatures. Tungsten lamps operating at their maximum allowable temperature level do not provide sufficient intensity to permit accurate measurements in the microsecond region. For this purpose, the best available source is the crater of the low intensity carbon arc that is reproducible to approximately 0·1 per cent when the measurements are made with a target area of 0·12 mm^2 with a photoelectric pyrometer having a time constant of 1 sec. Null and Lozier[148] have defined operating conditions for a carbon arc that make it possible to maintain crater radiance equivalent to $\pm 10°$K. They have reported that the spectral radiance of the crater of the carbon arc is close to that of a blackbody at a temperature of 3800°K throughout the spectral range 3000–42 000 Å. However, the carbon arc source is not useful for temperature measurements in the spectral range 3500–4500 Å, owing to the occurrence of emission bands in the arc flame. A study of temperature variations in the millisecond to microsecond time range is needed in order to determine its usefulness as a radiation source for high-speed temperature measurements. Preliminary results available indicate that short term fluctuations are likely to be less than 15°C when measured over an area of about 0·3 mm^2. A 15°C temperature variation is equivalent to a variation of about 2·5–3 per cent in the intensity or 0·01–0·015 per cent disin optical density. At higher temperatures both steady state and transient charges in high density gases may be used as light sources for the visible and near ultraviolet region of the spectrum. However, they are difficult to control. In Appendix 1 a list of various radiation sources and associated errors is given.

HIGH-SPEED THERMODYNAMIC MEASUREMENTS

Visual methods have a precision which is equivalent to the measurement of an intensity ratio with an accuracy of about 2·3 per cent at 6500 Å, the wavelength ordinarily used in visual optical pyrometry. When expressed in optical density units it is 0·01, a number which is customarily used in the certification of photographic densitometry measurements. It should be pointed out that this is an estimate of the precision of brightness temperature measurements under ordinary conditions. Greater precision is possible with the expenditure of correspondingly greater effort with high precision visual instruments and also with some of the newer photoelectric instruments now being investigated. The accuracy achieved in thermodynamic measurements above 2000°K is usually less than this owing to the existence of gradients in the sample container or to inaccuracies that arise from errors in effective emittance. High-speed measurements will be subject to these difficulties and, in addition, other problems will arise owing to the greater speed of the measurement.

During the last two decades rather rapid advances in ultra high-speed optical equipment have taken place, utilizing a great variety of techniques that are usually either photographic, photoelectric, or a combination thereof, so that one may obtain optical equipment having a time resolution throughout the range from milliseconds to nanoseconds. Most of this equipment is designed primarily for the study of the motion of rapidly moving objects. Relatively little effort has been devoted to high precision intensity measurements or to temperature measurements.

In pyrometry, there are various parameters that may be adjusted in order to obtain near optimum design for specific experimental conditions. For example, the target area for both the reference and unknown temperature sources will usually be approximately 0·25 mm^2. The solid angle will be approximately 0·01 steradian or less, although larger values are desirable from the viewpoint of fast measurements of temperature. However, considerations such as the power requirements and the desire for obtaining uniform temperature throughout the surface area of a known spectral distribution tend to limit the values that one can use conveniently. The selected wavelength is determined by the temperature range under consideration and the spectral sensitivity of the detectors, together with other considerations such as the interference of the emission bands in the source. At the lowest temperature (2000°K) considered here, 6500 Å is satisfactory for photoelectric detectors having an S 20 type photo-cathode. A band in the red region may also be satisfactory with red-sensitive photographic film. Use of this wavelength has the advantage that the usual pyrometer radiation sources are calibrated at this wavelength, and hence do not require corrections for effective emittance. For temperatures in the range 3000–6000°K a wavelength in the blue region of 4500–5000 Å is probably near optimum for both photographic and photoelectric detectors. At still higher temperatures the optimum wavelength is probably in the ultraviolet region. However, when the carbon arc is used as a reference source the wavelength region of 3500–4500 Å is usually excluded, owing to the occurrence of the emission bands in the arc flame in this wavelength range. Furthermore, the decreased transmission coefficient of optical windows in the near ultraviolet region tends to decrease the overall efficiency of the detectors in this region. Perhaps the most serious problem is the low intensity of conventional radiation sources in the short wavelength region.

Spectral band-widths used in pyrometry usually vary from 2–8 per cent of the chosen wavelength. The tendency in precision pyrometry is to select the lower value in order to minimize effects of variations in the spectral sensitivity of the detector and in the emittance of the temperature sources. However, in fast measurements the tendency is to select wider band-widths in order to obtain a larger number of photons.

Lovejoy[130] has discussed the accuracy of optical pyrometers operating in the temperature range 800–4000°C. Recently, optical pyrometry methods and advances were reviewed by Kostkowski and Lee[116], and Lovejoy[131].

2. Photoelectric temperature measurement techniques

The need for more precise and accurate as well as fast measurement of temperatures above the gold freezing point was a major factor in the development of photoelectric pyrometry techniques. Since in photoelectric pyrometers visual matching of the brightness of the two radiation sources is replaced by a photoelectric–electronic system, human errors inherent in the conventional optical pyrometers are partially eliminated. Middlehurst and Jones[137] have pointed out that, while optical pyrometers can hardly be improved to yield reproducibility better than $\pm 1\cdot 5$°C at the gold point, the corresponding reproducibility for the photoelectric pyrometers may be $\pm 0\cdot 1$°C.

At present, the response time of photoelectric pyrometers is approximately in the range 0·1–1 sec. In the future it will be possible to develop such pyrometers capable of measuring temperatures in the millisecond and even microsecond time range. However, the success of a very fast, high-temperature measuring, photoelectric pyrometer lies in the careful analysis and design of every optical, electrical, electronic and mechanical component of the instrument. The most accurate photoelectric pyrometers also employ an internal radiation source and optical systems similar to those of optical pyrometers. In addition, they have a detector consisting of a photomultiplier tube and associated electronic system, to match the brightness of the internal source and the unknown outside object.

In recent years, photomultiplier tubes have found many applications in various fields, particularly those that require the measurement of weak light sources, or strong light sources in a short period of time. A comprehensive discussion on the characteristics of photomultiplier tubes was given by Engstrom[80]. It was shown that photoelectric emission from the cathode varies linearly (within his experimental error of 3 per cent) with light flux over a wide range, 10^{-10}–10^{-4} lumen. Studies on photomultiplier tubes may also be found in the articles by Engstrom et al.[81], Engstrom et al.[82], Morton et al.[144], Cathey[45], Marschall et al.[134], and Keene[111]. A major limitation of photomultiplier tubes is that they have poor stability and reproducibility.

In recent years a number of photoelectric pyrometers have been developed in various countries. Lee[125] has discussed a photoelectric pyrometer whose precision for a radiance temperature measurement at the gold point is

0·02°C. The photomultiplier tube used in the above pyrometer is a 14 stage type with an S 20 spectral response curve. The response time of this instrument is about 1 sec, and it operates at 6530 Å with a band-width of 110 Å. Middlehurst and Jones[137] have reported results on a photoelectric pyrometer which at the gold point has a sensitivity of ±0·02°C, stability of ±0·07°C, and a reproducibility of ±0·1°C. Rudkin et al.[170] have described the details and operation of a photoelectric pyrometer that was used in connection with the measurement of thermal properties of metals at high temperatures (up to about 3000°K). A photon-counting pyrometer utilizing a photomultiplier tube was described by Treiman[190] to be used in the temperature range 1000–4000°C. It was mentioned that it would be possible to measure temperatures to better than 0·1°C at 1063°C, 0·25°C at 2063°C, 0·5°C at 3063°C, and 0·8°C at 4063°C. A detailed discussion on the dynamic properties of photoelectric pyrometers for measuring unstable temperatures was given by Katys[110].

3. *Photographic temperature measurement techniques*

At present, photoelectric pyrometry is not fully developed to provide high-speed (millisecond or less) measurement of high temperatures. Measurement of temperatures in the millisecond or microsecond time range requires fast responding detectors and recorders. Photographic techniques have this advantage; however, there remains the problem of calibrating the exposed film. The correlation of the density of the exposure on the film with the temperature of the photographed object presents difficulties, and the results are not very accurate. However, it is believed that in the future photographic pyrometry will become one of the reliable techniques in measuring high temperatures at high speeds.

Some of the important conditions that are needed for optimum precision in photographic pyrometry were summarized in an early work on photographic photometry by Dobson et al.[70]. The quantum efficiency of the photographic film varies rapidly with exposure reaching a maximum of about 1 per cent in a region near the lower end of the linear portion of the film's characteristic curve at an optical density of approximately 0·5, and decreasing to 0·01 per cent or less near the threshold and at a density of about one. Thus, optimum results are obtained in a relatively small intensity region with an average quantum efficiency of about 0·1 per cent. This undoubtedly varies somewhat with different photographic materials, but the best results appear likely for optical densities in the range 0·4–1·0. For precision measurement of optical density and also for minimization of effect of local variation in the temperature of the source, one would choose a film area in the range 0·2–0·5 mm^2. This is approximately the same area that was selected for the radiation source. Hence, the magnification will be close to unity. A selection of other characteristics will depend somewhat on the specific application.

Under optimum conditions, one may expect to measure photographically, in the microsecond time range, a temperature difference of about 500°C in the vicinity of the carbon arc temperature, with an accuracy of about 1·5–2 per cent. Thus, it appears that photographic materials and techniques are likely to be sufficiently sensitive and reproducible to give results at high speeds that are comparable to the results that one can obtain by visual detection methods, for which much longer times are required to record the information, with about the same care being used to design and calibrate the measurement system.

Photographic film is essentially a photon integrator. Under the selected conditions approximately one hundred million photons per square millimeter of image strike the film. If the overall quantum efficiency, including losses in the optical system that are due to absorption, is approximately 0·01 per cent, one would expect to measure effectively 10 000 photons/mm^2, giving a precision of 1 per cent. Unfortunately, the quantum efficiency not only varies with the optical density and wavelength but also with the rate of exposure, giving rise to an effect which is known as reciprocity failure. This means, in general, that the efficiency decreases as the exposure time decreases, although by a small amount which varies considerably depending upon the nature of the film and its processing. This dependence on time is a very weak one corresponding to about 1/5 to 1/10 power of the time. Nevertheless, the variation in quantum efficiency, together with other variations arising from non-uniformity in the developing process, imposes rather severe restraints on the methods by which the intensity of the light from an unknown source is calibrated. In order to achieve the greatest accuracy very nearly the same number of photons must be measured in the same time on adjacent areas of the same film. Therefore, the reference source and any unknown source should be photographed simultaneously. Also, the intensity should be modulated in such a manner that the optical density of the two images or portions thereof are nearly identical. Thus optimum results can be obtained by using the photographic film as an equality detector. This will require the design of a general purpose instrument which might be described as being the optical analog of a potentiometer. For a coverage of a wide range of intensities corresponding to a temperature range of 2000–10 000°K or higher, this "optical potentiometer" should have a range of eight decades. However, in the near future, it would be sufficient to have an instrument suitable for a range of three decades with time resolution in either the millisecond or microsecond range. It would be desirable for the instrument to include both photographic and photoelectric recording techniques. In broad outline, such a temperature measuring system will consist of one or more standard radiation sources selected for high reproducibility, together with a detector and a means of adjusting its exposure time to radiation over a broad range of measurement periods. The system should also have a suitable means for adjusting the intensity of the reference and unknown sources, so that they will be approximately matched. Exploratory studies along these lines are being conducted by various groups, the initial emphasis being placed on the study of those factors which tend to limit the precision of the measurements at high speeds. Of particular concern are the limitations of the detector and associated recording techniques, rather than the improvement of radiation sources which, to a large extent, can be studied at ordinary speeds. However, short term fluctuations of radiation sources must be investigated before one can employ results from a slow measurement technique in the calibration of a fast measurement system. It is expected that some aspects of the standardization of the measurements can be carried out by utilizing static calibration techniques. In particular, the spectral transmission coefficients of the optics and filters should be very nearly independent of time. Hence, these coefficients can be measured with high precision equipment operating at speeds of the order of minutes.

Several investigations have been reported related to the application of photography to temperature measurements. Simmons and De Bell[179] have described a system employing a photographic technique to measure the temperature of rocket exhaust flames. Astheimer and Wormser[13] have discussed an instrument for thermal photography. Simmons and De Bell[180] have modified their previous photographic pyrometry system by using a color-separation camera. Photographic photometry of a flash tube using a high-speed streak camera is discussed by Milne and Miller[138].

In connection with photographic pyrometry, existing photographic techniques must be understood thoroughly. For information on photographic theory, materials, processes and devices, such as densitometers, etc., reference may be made to the books by Clerc[49], and James and Higgins[105].

4. *Other temperature measurement techniques*

In addition to the conventional temperature measuring techniques, such as gas thermometry, thermoelectric and resistance methods, and optical pyrometry, and rather newer techniques, such as photoelectric and photographic pyrometry, there are various other methods that may be employed to measure temperature over different ranges. In this section it is intended to mention only a few selected methods that may be used for the rapid measurement of high temperatures.

The fast measurement of temperature in hot gases and plasmas can be accomplished by spectroscopic techniques. A description of these methods was given by Lochte-Holtgreven[128] and recently a review discussion was presented by Penner[157]. In particular, the application of the line broadening theory to the measurement of temperature in dense plasmas was given by Shumaker and Wiese[176]. The spectral line reversal and other techniques for the measurement of flame temperatures were discussed by Bundy and Strong[39]. Similar spectroscopic techniques used for the measurement of gas temperatures in shock tubes was described by Gaydon and Hurle[87]. The design and application of a high-speed, time-resolving spectrograph and related spectroscopic instruments are given by Moore and Crosby[143] and by Harrington[98], respectively. Applications of spectroscopic techniques to the measurement of the temperature of plasmas may also be found in the various articles in the book edited by Maecker[132].

A combination of interferometry and photography may yield useful results in temperature measurements. Early work on interferometry in connection with heat transfer was done by Kennard[112]. A brief summary of this technique as well as photothermometry was given by Weber[203].

From the point of view of fast temperature measurement it is of interest to note the modified resistance thermometer developed by Chabai and Emrich[47]. The latter utilizes the temperature dependence of the electrical resistance of a gold foil only about 10^{-5} cm in thickness. The reported response time of this thermometer was less than 1 μsec. A resistance thermometer utilizing a platinum film about 10^{-6} cm thick sputtered over an insulator was described by Rabinowicz et al.[161]. They report response times that are less than 1 μsec. A similar thermometer with platinum film painted on to a ceramic base was constructed by Laderman et al.[123] to be used in detonation research. They have estimated that the response time was of the order of millimicroseconds. In the case of samples heated uniformly by electric current (direct, alternating, or transient), the electrical resistance of the sample may indicate its temperature. However, this requires a separate steady state determination of resistance as a function of temperature.

Special thermocouples have been developed for the rapid measurement of temperatures. A review on this and also the design of new thermocouples for the measurement of transient surface temperatures were given by Moeller[142]. The reported response time for these thermocouples was of the order of 10 μsec.

Bolometric temperature measurement techniques are also possibilities. Camac and Feinberg[41] have developed a high-speed infrared bolometer capable of measuring transient temperatures of plasmas in strong electric and magnetic fields. The temperature transferring element is a thin opaque surface, one side of which is in contact with the plasma, and the other side is viewed by the bolometer for measurements of infrared emission. It was reported that the opaque layer was thin enough to permit the determination of the temperature of the front surface in times of the order of 0·1 μsec.

Other specialized temperature measurement methods utilizing ultrasonic techniques, eddy currents, etc., may be found in the books edited by Herzfeld[100] and Wolfe[211]. However, it is unlikely that these methods will be applied to fast temperature measurements.

V. Other Measurements

In addition to the measurement of heat and temperature, there are other quantities, such as pressure, density, etc., that one needs to measure, depending upon the nature and goal of the experiment. In the following paragraphs some typical high-speed techniques used to measure various quantities are summarized.

1. *High-speed photographic techniques*

The dynamics of a sample can be studied by photographing the event. There are different kinds of high-speed cameras that one may use to meet the requirements of a particular experiment. Most of them fall into the following three general categories.

Streak cameras produce a continuous record of the object over time. These cameras are particularly useful in studying the kinematics of the sample. According to constructional and operational differences they are divided into two subgroups: (*i*) drum and spinning-mirror cameras and (*ii*) strip cameras. The time resolution of drum and spinning cameras can be of the order of 1 and 0·01 μsec, respectively.

Framing cameras record the event on successive frames with fixed time intervals between each frame. They are particularly useful in studying explosions, shock intensities, and combustion phenomena. They may have speeds of the order of 10 frames/μsec.

Image converter cameras produce the image of the event on the fluorescent screen of a special electronic tube. In order to have a permanent record the screen may be photographed. They are used to study very rapid events. Exposure times per frame may be as short as 0·001 μsec.

Recent advances in high-speed photography are given in the publications by Courtney-Pratt[55, 56], and Hyzer[103].

In high-speed photography, in addition to high-speed cameras, high-speed shutters are of great importance. In the microsecond region and below, mechanical shutters cannot be used. There are various techniques developed to achieve fast acting non-mechanical shutters. Most of these are based on magneto-optical and electro-optical effects. A summary of these is given by Hyzer[103]. Details on the operation of magneto-optical shutters may be found in a paper by Edgerton and Germeshausen[79] and discussions on electro-optical shutters (Kerr cells) in an article by Zarem *et al.*[215]. The response times of the magneto-optical and electro-optical shutters are of the order of 0·01 μsec; however, they have the disadvantage of absorbing most of the incident light. A high-transmittance, large-aperture opening shutter was developed by Cassidy and Tsai[44]. Its operation is based upon the crumpling of thin foils by electromagnetic forces. Its response time is of the order of 10 μsec. A closing shutter that operates on the blackening of a piece of glass at the camera aperture by the deposition of the fine lead particles resulting from the explosion of a loop of lead wire was developed by Edgerton[78]. However, the last two kinds of shutters have the disadvantage of requiring replacement of parts after each operation.

Techniques in high-speed photography and related fields are in continuous development to meet the demands of high-speed studies of mechanical, thermodynamical, electrical, and other physical phenomena.

2. Dynamic pressure measurement techniques

At about 500 kilobars and above, the generation and measurement of static pressures become difficult and inaccurate. For this reason, dynamic pressure generation methods, namely, shock techniques are used for very high pressure studies. Dynamic pressure measurement methods are of interest in this chapter primarily for the high-speed characteristics of shock techniques and also because certain high-speed thermodynamic measurement work may employ explosions and shocks. A recent review on the measurement of shock pressures in solids is given by Doran[73].

One method of determining shock pressures is based on the measurements of shock velocity and material velocity. The average shock velocity can be determined by measuring the time difference between the arrival of the shock at the front and back surfaces of the material of known thickness. The detection of arrival of the shock waves can be accomplished by optical methods, having time resolutions of the order of 0·01 μsec. For optical measurements, rotating mirror or image converter cameras may be used to detect the motion of the material. The illumination of the surface may be achieved by having a very small gap just above the surface filled with air or argon which becomes luminous during the shock owing to high temperatures resulting from the sudden high pressures. Reflected light techniques are also employed to illuminate the surface of the material. The surface velocity of the material can be measured by either optical or electrical methods. Doran[73] has reviewed three principal optical methods that are used at the present time to make continuous measurements of surface velocities. These methods are classified under inclined mirror, knife-edge or moving-image, and optical-lever techniques. The electrical methods can also be divided into three categories: pin, slanted resistance wire, and variable capacitor techniques. These were discussed by Doran[73]. The electrical methods mentioned above can also be used, with some modifications, to measure shock velocities as well as surface velocities. For specific references on shock pressure measurement techniques, the reader may refer to the references given in Section VI–2–C of this chapter in relation to the measurement of thermodynamic properties at very high pressures. Attenuation of shockwaves in solids, which is important for a thorough understanding of the behavior of waves, was discussed by Fowles[85].

Recently Cook *et al.*[53] have discussed the possibility of using water as a "pressure gauge" for measuring transient high pressures in shock and detonation waves, since the Rankine–Hugoniot curves for water are known with sufficient accuracy.

In addition to the measurement of high pressures by shock velocity techniques, one may consider using the piezoelectric method. However, its applicability is mainly in the normal and moderate pressure region (below one kilobar). Komel'kov and Sinitsyn[115] have described a piezoelectric transducer made of $BaTiO_3$ to record the shape and intensity of a normal pressure impulse arriving on the walls of a discharge chamber. Recently, Graham[93, 94] has discussed the possibility of using synthetic alpha-quartz as a transducer for transient pressure measurements in the range 5–70 kilobars.

Another possible method of pressure measurement utilizes the piezoresistive effect. It was noted[73] that sulfur wafers a few mils thick can respond to pressure changes in about 0·1 μsec. In this study, pressure dependence of electrical resistance of substances is of interest from the point of view of both pressure measurements and the design of high-speed, high-temperature apparatus for the measurement of thermodynamic quantities. The reason for the latter is that in certain experiments high pressures may result and this, in turn, may change the electrical characteristics of the various components of the system.

It was noted that insulators become relatively good conductors at very high pressures. Alder and Christian[2, 3] have reported that for some ionic and molecular crystals the electrical resistance around 250 kilobars is less than that under normal conditions by a factor of about 10^6. The most extensive investigation of the pressure dependence of electrical resistance of elements, alloys, and various compounds at pressures up to about 100 kilobars was made by Bridgman[31]. His pioneering work still remains one of the most accurate studies reported in the literature.

Pressure dependence of electrical resistance of various metallic elements was investigated by Balchan and Drickamer[18], Vereshchagin et al.[194, 195, 196, 197], and Stager and Drickamer[183]. Pressure dependence of electrical resistance of certain non-metallic elements, semiconductors, metallic oxides, salts and insulators was investigated by Joigneau and Thouvenin[108], David and Hamann[62], Kozyrev[117], Balchan and Drickamer[19], Minomura et al.[141], Young et al.[214], Minomura and Drickamer[140], Al'tshuler et al.[8], and Brish et al.[33]. The temperature coefficients of resistance of certain semiconductors and compounds were determined by Minomura et al.[141] at pressures up to about 350 kilobars and temperatures up to about 120°C. Similar studies on iodine and selenium at low temperatures were conducted by Riggleman and Drickamer[165] at pressures up to about 300 kilobars. In Appendix 2 of this chapter, a summary of experimental research on the pressure dependence of electrical resistance of various substances is given.

A review of the effect of static pressure on the electrical resistance of metals was given by Lawson[124]. Recently, Duvall[76] has discussed electrical effects in solids associated with shock waves. Techniques and equipment for static high-pressure (up to 500 kilobars) electrical resistance studies were described by Drickamer and Balchan[74]. A recent review of the various methods of measurement of pressure in shock tubes was given by Gaydon and Hurle[87]. For details of very high pressure research the reader may refer to the books edited by Bundy et al.[38], Wentorf[207], Paul and Warshauer[156], and Giardini and Lloyd[88].

3. X-ray and interferometric techniques for density and velocity measurements

Flash x-ray techniques can be used to measure various parameters and properties of materials. The x-ray method for measuring density of gases was discussed by Winkler[209]. Schall[172] and Dapoigny et al.[61] have utilized the x-ray technique to determine the velocity of the shock wave and the density of the material behind the shock wave. Grundhauser et al.[95] have discussed the operation of a 50 nanosecond flash x-ray system for hypervelocity research. Kistiakowsky and Kydd[113] have developed an x-ray absorption photometer capable of measuring gas densities with extreme rapidity. Knight and Venable[114] have described an apparatus based on the technique introduced by Kistiakowsky to measure densities behind gaseous shock and detonation waves with time resolution of 1 μsec. Schall[173] has developed an x-ray technique to detect shocks in opaque media. Al'tshuler and Petrunin[10] have discussed a method to determine the pressure and density of light solids and liquids behind the front of the reflected shock waves. Balchan[17] has extended the flash x-ray technique and studied the behavior of Armco iron under shock conditions. In addition to the determination of shock velocity and density, x-ray techniques can be used to obtain data on thermal expansion, lattice constants and crystal structure of materials. In connection with high-speed measurements, Marschall et al.[134] have developed a photomultiplier x-ray detector that can follow x-ray events in times as short as 10 μsec.

In addition to x-ray methods, the density of dense gases may be determined by interferometric techniques. By placing a drum camera at the place of the photographic plate a time-resolved recording of the interference patterns can be obtained and the density changes in the gas deduced.

Summaries of interferometric techniques and various kinds of interferometers used for density measurements were given by Ladenburg and Bershader[122], and Gaydon and Hurle[87].

VI. Applications of High-Speed Measurements

1. *General considerations*

With the increasing interest in high-temperature studies, high-speed measurement of thermodynamic and related quantities has become a necessary research technique. The applications of high-speed measurements may be classified under two general areas. (*i*) Direct applications related to the measurement of properties of matter at high temperatures. In this category, thermodynamic and transport properties—specifically, specific heat, compressibility, thermal diffusivity, thermal conductivity—also electrical and thermoelectrical properties, are typical examples. (*ii*) Somewhat indirect applications related to the study of various phenomena at high temperatures. In this category one may include phenomena such as exploding conductors, high-pressure studies, chemical kinetics, fusion, high-temperature sources, etc. Some of the important applications considered at the present time are summarized below.

2. *Measurement of properties*

A. Specific Heat

The conventional approach to specific heat measurements is the method of mixtures. However, at high temperatures this method fails to give correct results, partly because of the possible phase changes and the various types of imperfections that the sample material may experience in the wide temperature range of operation. To overcome the difficulties presented by the method of mixtures, pulse techniques have been employed to measure specific heats of substances at high temperatures. An early attempt along this line was made by Worthing[212]. The first successful pulse method was developed by Avramescu[14], and used after modifications and refinements by other investigators.

Two general procedures in obtaining the specific heat of electrical conductors by pulse heating are outlined below.

In the first method, the sample is heated in a furnace to the desired temperature level and then a capacitor is discharged through the sample to change its temperature by a small amount. The measurement of the temperature rise in conjunction with the energy supplied to the sample during the discharge enables one to calculate the specific heat of the sample. For such measurements, the sample may be connected to one of the arms of a Wheatstone or a similar bridge. The bridge may be balanced when the sample is at the desired temperature level. After the discharge the bridge will be unbalanced owing to the change of resistance resulting from the temperature change. The magnitude of this unbalance may be obtained from the deflection of a ballistic galvanometer. By having a proper calibration, the actual temperature rise may be determined from this deflection. The energy delivered to the sample may be calculated from the capacitor and circuit characteristics. This method of measuring specific heat is suitable particularly for small samples and for temperatures up to about 2000°K. The reason for this is that the initial steady state heating of samples above about 2000°K gives rise to certain thermal, electrical, chemical and mechanical complications. Also, large samples may not be used since the measurement system is not suitable for high energies.

In the second method, the sample is heated from room temperature to the desired temperature level by discharging a capacitor through the sample. In this case, the complete heating is done in a very short time. However, in comparison with the first method, a greater capacitor is required, which, in turn, complicates the electrical circuitry and the measurements. For this reason, the energy imparted to the sample cannot be determined accurately since the capacitor and the circuit characteristics may not be known. Instead, the energy may be determined from the measurements of the current and the voltage across the sample. These measurements may be recorded on an oscilloscope; thus, instantaneous values of electrical energy imparted to the sample may be calculated. In addition to this, if instantaneous values of the temperature are measured, the specific heat of the sample may be computed at any temperature. The measurement of temperature may be accomplished either by electrical resistance or optical methods. The electrical resistance method requires a separate determination of the electrical resistance of the sample as a function of temperature. In this case, from the instantaneous measurements of the current and voltage the resistance, and thus the instantaneous temperature of the sample, can be calculated. However, the temperature obtained by this method may not be very accurate since the temperature coefficient of resistance of metallic conductors is small.

Optical methods are not developed to the extent that they yield dependable fast temperature measurements. A possible practical method for the measurement of temperatures above about 2000°K may be the use of a photoelectric pyrometer tube output in conjunction with an oscilloscope. The results may be interpreted by having a proper calibration of the photoelectric tube output for various temperature settings of a standard radiation source under steady state conditions. The details of both heat and temperature measurements are given in the earlier part of this chapter.

Some of the recent work on the measurement of specific heat of substances by pulse heating is summarized below.

Kurrelmeyer *et al.*[119] have determined the specific heat of platinum near room temperature by placing the sample in one of the arms of a Wheatstone bridge and recording the deflection of a ballistic galvanometer after the discharge of a capacitor through the sample. Baxter[21] has heated the sample close to its melting point by an electric current passing through it in about 0·05 sec. He has shown that by recording the current and the voltage across the sample the resistance at each instant can be calculated, and, from the known relationship between electrical resistance and temperature and the energy imparted to the sample, specific heat can be determined. Nathan[146] has described a similar technique with the main difference that the temperature measurements were made with a thermocouple. The heating rate of the steel sample was 1000°C/sec. Pochapsky[158, 159] has performed specific heat and electrical conductivity measurements on aluminum and lead from room temperature to the melting point of each metal, using a Wheatstone bridge and a ballistic galvanometer to detect the change of the sample characteristics when a pulse of current flows through it. Wallace et al.[199] have discussed a dynamic pulse heating method that utilizes a Kelvin bridge to measure the resistance of the sample throughout each pulse. From a separately determined resistance–temperature relationship and energy input they have calculated the specific heat of pure iron in the temperature range 25–1050°C. Wallace[198] has also used this method to measure the specific heat of pure thorium in the temperature range 25–1000°C. Taylor and Finch[188] have performed measurements from very low temperatures to the melting points of molybdenum, tantalum and rhenium utilizing a pulse heating method. Pasternak et al.[154] have described a flash heating method to determine the specific heat of electrical conductors. They have checked their technique by making measurements on platinum from room temperature to about 800°C. Pasternak et al.[155] have performed preliminary experiments to develop a dynamic method for measuring the specific heat and total emissivity of electrical conductors. Their measurements of specific heat and emissivity of platinum were in the temperature ranges 400–1300°K and 1000–1250°K, respectively. Recently, Holland[101] has used a low-frequency (about 10 c/s) alternating current heating system for specific heat measurements of titanium over the temperature range 600–1345°K. Unlike some other pulse techniques, this method confined the temperature of the sample to a very small fluctuation about a steady value, giving a high resolution along the temperature axis.

Prophet and Stull[160] have conducted specific heat measurements on boron nitride and aluminum oxide in the temperature range 1300–2300°K employing the cooling-rate method. They have used an arc-image furnace as the heat source. Although their approach was not that of pulse heating, the arc-image furnace can be used with a proper rotating shutter arrangement to deliver heat pulses to the sample. This method can be applied to measurements on samples that are electrically non-conductors.

B. Thermal Diffusivity

Heat transfer in a substance under transient conditions is characterized by the property thermal diffusivity, α. Since α is related to thermal conductivity, k, by the relation $\alpha = k/c\rho$ (where c is the specific heat per unit mass and ρ the density of the sample material) it is convenient to determine thermal conductivity from measurements of thermal diffusivity, specific heat, and density at high temperatures.

Practically all the various techniques employed for the measurement of thermal diffusivity are, in principle, based upon the method developed by Angstrom[11]. This method consists of generating a time-dependent thermal energy at one end of the sample and measuring the time required for the heat wave to travel in the sample. Although the general principle is the same, different investigators have used different heat sources, different pulsing schemes, and have introduced refinements to compensate for the heat losses from the surface of the sample. Some of the recent investigations are summarized below.

(1) *Utilizing Direct-Contact Heaters.* Sidles and Danielson[177] have made measurements on copper, nickel, and thorium over the temperature range 0–500°C with a heater whose temperature varied sinusoidally. Sidles and Danielson[178] have made improvements in their technique and have extended the measurements on copper, iron, and nickel to 1000°C. Woisard[210] has used a single heat pulse for measurements on steel around room temperature. Shanks *et al.*[175] have made measurements on pure silicon at 300–1400°K with a heat source turned on instantaneously.

(2) *Utilizing Radiation Heaters.* For high-temperature studies it is advantageous to heat the end of the sample remotely by radiation. Various flash methods utilizing high-intensity, short-duration pulses to heat the end surface of the sample have been described in the literature. Parker *et al.*[153] have performed experiments on copper, silver, nickel, aluminum, tin, and zinc in the temperature range 22–135°C. Rudkin *et al.*[169] have determined thermal diffusivity of Armco iron, molybdenum, and titanium over the temperature range 300–1800°C. Recently, Rudkin[168], utilizing the same technique, has made measurements on Pyroceram, beryllium oxide, aluminum oxide, and magnesium oxide at temperatures up to 1780°C. Cape and Lehman[42] have presented a mathematical analysis of a flash thermal diffusivity measurement technique. Deem and Wood[67] have used a laser as a radiation source for measurements on reactor fuel materials. Cowan[57] has made a study of the utilization of high-energy electron beams as radiation sources. Cerceo and Childers[46] have performed experiments on Al_2O_3 and carbon based upon the above method. Recently, Cape *et al.*[43] have developed a technique to determine thermal diffusivity from the time interval separating two radial isotherms in a cylindrical specimen. They have obtained results on Ta, ZrC, and TiC in the temperature range 1300–1650°C. The specimen was heated by radiation by using an induction heater. Cutler and Cheney[60] have made a mathematical analysis for a case in which one end of the sample was heated by radiation or by direct-contact heaters. Recently, Cowan[58] has studied the temperature distribution in the sample for the case where energy losses at the surface owing to radiation and convection are not negligible. Becker[22] has suggested various techniques for measuring thermal diffusivity of semiconductors utilizing the bolometric, Nernst, and Seebeck effects. Transient surface heating, either by direct contact or by radiation heaters, may generate elastic waves in the specimen. A study on this subject was given by White[208].

C. THERMODYNAMIC PROPERTIES AT VERY HIGH PRESSURES

At about 500 kilobars and above, the measurement of thermodynamic properties by static pressure techniques developed by the pioneering work of Bridgman becomes difficult and yields inaccurate results. Therefore, in very high pressure studies, dynamic pressure methods, namely, shock techniques, are employed. Almost all high-pressure thermodynamic investigations undertaken at the present time are confined to the determination of the equation of state of substances. A review of the developments in the equation of state work at very high pressures and temperatures as of 1956 is given by Beckett et al.[23]. A summary of research on shock data on thermodynamic properties of various substances is given in Appendix 3 of this chapter.

Recently, a photoelectric technique was described by Taylor[187] for measuring the residual temperature of shocked metals at pressures up to about 1·5 megabars. Experiments on the hydrodynamic behavior of rapidly moving gas behind a shock in argon under a magnetic field were described by Dolder and Hide[71]. A thorough study on the compression characteristics of solids by strong shock waves was given by Rice et al.[163]. A review on the techniques of generation and measurement of dynamic high pressures was presented by Deal[66]. Recently, a review on physics experiments with strong pressure pulses was given by Alder[1].

There are other investigations related to the properties of matter at high pressures and high temperatures. Alder and Christian[4] have studied the behavior of strongly shocked carbon to pressures up to 800 kilobars and have found evidence of the conversion of graphite to carbon in less than a microsecond. De Carli and Jamieson[64] reported the formation of diamond by exposing pure graphite to explosive shock pressures of about 300 kilobars for 1 μsec.

There are also static high-pressure investigations of properties of matter in conjunction with transient high temperatures. Bundy[37] has used capacitor discharges through graphite samples at static pressures up to 200 kilobars to convert graphite to diamond directly. It was reported that the discharge was non-oscillatory, with 90 per cent of the energy delivered in about 3–6 msec, raising the temperature of the sample to about 5000°K. Bundy[36] has studied the melting characteristics of graphite at high static pressures using flash heating methods with discharge times less than 7 msec. He has found that the melting point of the sample increased from about 4100°K at 9 kilobars to about 4600°K at 125 kilobars. Bundy and Wentorf[40] have used capacitor discharges through a sample of hexagonal boron nitride under static pressures up to 130 kilobars to transform it directly to denser forms at temperatures up to 4200°K.

Successful interpretation of shock data requires a knowledge of the phase of the sample under shock conditions. In their experiments up to 600 kilobars, De Carli and Jamieson[63] have investigated the phase behavior of quartz and have found that single crystal quartz becomes amorphous at high shock pressures. The polymorphism of iron up to 200 kilobars was studied by Bancroft et al[20]. Polymorphic transition in bismuth at about 27 kilobars was observed by Duff and Minshall[75]. Recently, pressure-induced polymorphic transitions in some Group II–VI compounds were studied by Jayaraman et al.[106]. Johnson et al.[107] have investigated the shock-induced phase transformation in iron in the temperature range 78–1158°K. They have observed a discontinuity in the temperature–pressure relationship at 115 kilobars and

775°K. Static pressure-induced phase transitions in silicon, germanium and some Group III–V compounds were studied by Minomura and Drickamer[139] at pressures up to about 300 kilobars. Similar investigations were conducted by Samara and Drickamer[171] on some Group II–VI compounds. Effects of shock pressures up to about 500 kilobars on a gold–silver alloy were described by Appleton *et al.*[12]. They have noticed that hardness of the alloy increased after subjecting it to an explosive shock.

Electrical conductivity of the explosion products of certain explosives were measured by Brish *et al.*[32]. They have found that electrical conductivity of the explosion products increases with increasing density of the explosive and with increasing intensity of the detonation wave. They have suggested that the high values of electrical conductivity of the explosion products are due to the high densities and pressure that occur on the front of the detonation wave, in addition to thermal ionization. For the theories of detonation and the science of high explosives in general, the reader may refer to the studies by Evans[84] and Cook[52], respectively.

D. Electrical and Thermoelectrical Properties

By measuring the transient current and the voltage across a sample connected to an electrical discharge system, electrical resistance can be obtained as a function of imparted energy or discharge time. Bondarenko *et al.*[28] have made various measurements of this nature. If a record of the temperature variations in the sample as a function of time can also be obtained, the temperature dependence of electrical resistance can be calculated. Such a method permits the measurement of electrical resistance of samples in the liquid as well as in the solid phase. However, at the present time, the response of temperature measuring systems is not fast enough for the investigations mentioned above to be carried out.

High-temperature thermoelectric properties—specifically the Seebeck coefficient—can also be obtained by having records of temperature and potential differences across the sample. In this case, the measurements may start after the termination of the capacitor discharge and continue over the cooling period.

3. *Other applications*

A. Exploding Conductors

In recent years, theoretical and practical interest in exploding conductors (wires and strips) has opened a new avenue to the application of high-speed measurement techniques. Exploding conductors are used to investigate phenomena associated with the behavior of materials under very high energy densities applied in very short periods of time. Specifically, one may include the following: high-temperature studies, simulation of nuclear explosions, studies of shock waves, and as a tool, high-speed photography, etc. Several articles on exploding conductors can be found in the books edited by Chace and Moore[48].

In addition to the above, one may consider publications by Tucker[193] on exploding gold wires, by Cnare[50, 51] on the striation of exploding copper foils and exploded aluminum tubes, by Jones and Earnshaw[109] on a wire exploder for generating cylindrical shock waves, and recently, by Stevenson et al.[184] on the characteristics of exploded wires for optical maser excitation.

Since in exploding conductor experiments the resistance of the sample changes rapidly over the wide temperature range of operation, it is not convenient to determine the input energy by taking the integral of the product of the square of the current and the resistance over the time of current flow. Instead, the energy can be determined by taking the integral of the product of the current and the voltage across the sample over the time of the capacitor discharge. The details of current and voltage measurement techniques are discussed in Section III of this chapter. During the explosion, the geometry of the sample changes very rapidly which behavior, in turn, gives rise to theoretical and experimental difficulties in the investigations of exploding conductors.

In the above experiments electrical energy was imparted to the sample directly. There is also the possibility of transferring energy to the sample by radiation. Nelson and Kuebler[147] have described investigations on thin tungsten wires and strips undergoing thermally-induced explosion by means of a flash lamp. Their results show that the speed of explosion was in the millisecond time range.

A great proportion of the research in high-speed measurement techniques and exploding conductors is of a complementary nature. It is evident that studies in one field will have considerable contributions to the other.

B. High-Temperature Sources

High-speed measurement of various quantities may be required in connection with experiments with transient high-temperature sources. An excellent review on methods of producing high temperatures is presented by Lochte-Holtgreven[128]; therefore only a brief summary is given here.

High temperatures can be obtained by electric arcs, pulsed discharges, exploding conductors, plasma jets, shock waves, chemical reactions and nuclear explosions. The temperature attained by a continuous electric arc depends on the electrical characteristics as well as geometrical arrangement and the material of the system. Depending on the techniques used, arc temperatures may vary in the range 3000–30 000°K. However, continuous arcs are not practical for fast, transient experiments. Using pulsed arc techniques, one may establish transient conditions offering the advantage of delivering greater energies per unit time. However, the techniques for transient measurements of various quantities limit the extent to which pulsed arc techniques can be used. Exploding conductors offer potentialities for the production of high temperatures, but only for a short duration. Kvartskhava[121] has pointed out that conductors exploded in air or vacuum produce temperatures of the order of 30 000°K, whereas when exploded in a liquid or solid medium, temperatures of the order of 100 000°K may be attained. Plasma jets are of continuous nature and are capable of producing temperatures of 20 000°K or more. Shock-wave techniques may yield transient temperatures as high as 20 000°K. It is possible that in the future much higher temperatures will be attainable by shock waves. Chemical reactions are other possible sources of high temperatures, but in most cases 6000°K is the upper limit. Higher temperatures can be attained by detonating high explosives. Nuclear explosions, though at the present time not practical at the laboratory level, are capable of producing much higher temperatures than is possible by any of the above-mentioned methods. Brickwedde[30] has discussed the possibility of attaining 10–50 million °K (comparable to the temperature of the interiors of stars) by nuclear explosions. It seems that, in the future, for extremely high temperature research related to astrophysical problems, controlled nuclear reactions on the laboratory level may be utilized.

VII. Appendices

Appendix I

Various radiation sources and associated errors†

Element	Φ	θ_φ	$\delta\theta_\varphi$	Δ	δ_c	I_r
Au	f.p.	1336·15	±0·1	±0·01	±1	1
Pd	f.p.	1825	±1		±2	83
Pt	f.p.	2042	±1		±2	300
Rh	f.p.	2233	±2		±3	750
Ir	f.p.	2716	±3		±4	4350
C	crater of arc	3797	±10	±3	±15	44 000

Definition of Symbols

Φ = Point at which radiation source is used;
f.p. = Freezing point;
θ_φ = Temperature at Φ on International Practical Temperature Scale (°K);
$\delta\theta_\varphi$ = Uncertainty in temperature at Φ (°K);
Δ = Reproducibility of radiation source (°K);
δ_c = Uncertainty in best routine calibration (°K);
I_r = Intensity ratio at 6530 Å referred to the intensity at the gold point.

† For constructional and operational details, and classification and error analysis of various blackbody radiation sources and related subjects, such as melting points, reference may be made to the publications by Wensel et al.[205], Roeser et al.[166], Swanger and Caldwell[186], Henning and Wensel[99], Roeser and Wensel[167], Wensel et al.[204], Wensel et al.[206], and Stimson[185]. Additional references on these topics are given in the text.

CHARLES W. BECKETT and ARED CEZAIRLIYAN

Appendix II

Summary of experimental research on pressure dependence of electrical resistance of various substances (pressures exceeding 100 kilobars)

No.	Investigator	Reference	Year	Method	Substance	Approximate pressure range (kb)
1	Balchan and Drickamer	18	1961	Static	Ba, Ca, Fe, Pb, Rb	50–500
2	Balchan and Drickamer	19	1961	Static	I, Se	60–450
3	Vereshchagin et al.	194	1961	Static	Sb, As, Ca	20–250
4	Vereshchagin et al.	195	1961	Static	Cd, Pb, Nb, Re, Sn, Y, Zr	30–250
5	Vereshchagin et al.	196	1961	Static	Ce, La, Nd	20–250
6	Vereshchagin et al.	197	1962	Static	Dy, Er, Pr, Yb	40–250
7	Stager and Drickamer	183	1963	Static	Ba, Ca, Mg, Sr	20–650
8	Alder and Christian	2	1956	Shock	CsBr, CsI, I_2, $LiAlH_4$, P(red), NaCl, Teflon	250
9	Brish et al.	33	1960	Shock	Paraffin, Plexiglas	100–1000
10	Al'tshuler	8	1961	Shock	NaCl	50–800
11	Young et al.	214	1961	Static	NiO, CoO	10–200
12	Minomura et al.	141	1962	Static	CdSe	20–550
13	Minomura and Drickamer	140	1963	Static	CoO, Co_3O_4, CrO_3, CuO, CuS, FeS, FeS_2, MnO, MnO_2, MnO_3, MoS_2, NiO, Ni_2O_3, NiS, TiO, TiO_2, VO_2, V_2O_3, V_2O_5	20–500

Appendix III

Summary of research on shock data on thermodynamic properties of various substances

No.	Investigator	Reference	Year	Substance	Approximate pressure range (kb)
1	Mallory	133	1955	Al	30–400
2	Walsh and Christian	200	1955	Al, Cu, Zn	150–450
3	Walsh et al.	201	1957	Al, Be, Bi, Cd, Cr, Co, Cu, Au, Fe, In, Pb, Mg, Mo, Ni, Nb, Pd, Pt, Rh, Ag, Ta, Tl, Th, Sn, Ti, Zn, Zr, Brass	150–500
4	Al'tshuler et al.	7	1958	Fe	400–5000
5	Al'tshuler et al.	6	1958	Bi, Cd, Cu, Au, Pb, Ag, Sn, Zn	400–4000
6	McQueen and Marsh	136	1960	Sb, Bi, Cd, Cr, Co, Cu, Au, Fe, Mo, Ni, Pb, Ag, Sn, Th, Ti, Tl, V, W, Zn	200–2000
7	Al'tshuler et al.	5	1960	Al, Cu, Pb	50–4000
8	Fowles	86	1961	Al	25–50
9	Hughes et al.	102	1961	Bi, Fe	5–120
10	Skidmore and Morris	181	1962	U	340–6500
11	Doran	72	1963	Graphite (pyrolytic)	5–280
12	Goranson et al.	92	1955	Duralumin	100–300
13	Curran	59	1961	Invar	35–160
14	Al'tshuler et al.	8	1961	NaCl	50–800
15	Al'tshuler et al.	9	1963	CsI, LiF, KBr, KCl, NaI	20–1100
16	Walsh and Rice	202	1957	Water 14 liquids	30–400 50–150
17	Rice	162	1957	Water	25–250

VIII. References

[1] Alder, B. J., "Physics Experiments with Strong Pressure Pulses," in W. Paul and D. M. Warschauer, eds., *Solids Under Pressure*, McGraw-Hill, New York, 1963, p. 385.
[2] Alder, B. J., and R. H. Christian, *Phys. Rev.*, **104**, 550 (1956).
[3] Alder, B. J., and R. H. Christian, *Discussions Faraday Soc.*, No. 22, 44 (1956).
[4] Alder, B. J., and R. H. Christian, *Phys. Rev. Letters*, **7**, 367 (1961).
[5] Al'tshuler, L. V., S. B. Kormer, A. A. Bakanova, and R. F. Trunin, *Soviet Phys. JETP*, **11**, 573 (1960).
[6] Al'tshuler, L. V., K. K. Krupnikov, and M. I. Brazhnik, *Soviet Phys. JETP*, **7**, 614 (1958).
[7] Al'tshuler, L. V., K. K. Krupnikov, B. N. Ledenev, V. I. Zhuchikhin, and M. I. Brazhnik, *Soviet Phys. JETP*, **7**, 606 (1958).
[8] Al'tshuler, L. V., L. V. Kuleshova, and M. N. Pavlovskii, *Soviet Phys. JETP*, **12**, 10 (1961).
[9] Al'tshuler, L. V., M. N. Pavlovskii, L. V. Kuleshova, and G. V. Simakov, *Soviet Phys.-Solid State*, **5**, 203 (1963).
[10] Al'tshuler, L. V., and A. P. Petrunin, *Soviet Phys.-Tech. Phys.*, **6**, 516 (1961).
[11] Angstrom, A. J., *Phil. Mag.*, **25**, 130 (1863).
[12] Appleton, A. S., G. E. Dieter, and M. B. Bever, *Trans. Met. Soc. AIME*, **221**, 90 (1961).
[13] Astheimer, R. W., and E. M. Wormser, *J. Opt. Soc. Am.*, **49**, 184 (1959).
[14] Avramescu, A., *Z. Tech. Physik.*, **20**, 213 (1939).
[15] Baker, L., Jr., and R. L. Warchal, "Studies on Metal–Water Reactions by the Exploding Wire Technique," in W. G. Chace and H. K. Moore, eds. *Exploding Wires*, Vol. II, Plenum Press, New York, 1962, p. 207.
[16] Baker, W. R., *Rev. Sci. Instr.*, **30**, 700 (1959).
[17] Balchan, A. S., *J. Appl. Phys.*, **34**, 241 (1963).
[18] Balchan, A. S., and H. G. Drickamer, *Rev. Sci. Instr.*, **32**, 308 (1961).
[19] Balchan, A. S., and H. G. Drickamer, *J. Chem. Phys.*, **34**, 1948 (1961).
[20] Bancroft, D., E. L. Peterson, and S. Minshall, *J. Appl. Phys.*, **27**, 291 (1956).
[21] Baxter, H. W., *Nature*, **153**, 316 (1944).
[22] Becker, J. H., *J. Appl. Phys.*, **31**, 612 (1960).
[23] Beckett, C. W., M. S. Green, and H. W. Wooley, "Thermochemistry and Thermodynamics of Substances," in H. Eyring, C. J. Christian, and H. S. Johnston, eds., *Annual Review of Physical Chemistry*, Volume VII, Annual Reviews, Inc., Palo Alto, California, 1956, p. 287.
[24] Beers, Y., and G. L. Strine, *Inst. Radio Eng.*, *Trans. Instr.*, **11**, 171 (1962).
[25] Bellaschi, P. L., *Trans. AIEE*, **52**, 544 (1933).
[26] Bellaschi, P. L., *Elec. Eng.*, **53**, 86 (1934).
[27] Bennett, F. D., and J. W. Marvin, *Rev. Sci. Instr.*, **33**, 1218 (1962).
[28] Bondarenko, V. V., I. F. Kvartskhava, A. A. Pliutto, and A. A. Chernov, *Soviet Phys. JETP*, **1**, 221 (1955).
[29] Brady, M. M., and K. G. Dedrick, *Rev. Sci. Instr.*, **33**, 1421 (1962).
[30] Brickwedde, F. G., "Temperatures in Atomic Explosions," *Temperature, Its Measurement and Control in Science and Industry* (Ed. H. C. Wolfe), Vol. II, p. 395, Reinhold, New York (1955).
[31] Bridgman, P. W., *Proc. Am. Acad. Arts Sci.*, **81**, 165 (1952).
[32] Brish, A. A., M. S. Tarasov, and V. A. Tsukerman, *Soviet Phys. JETP*, **10**, 1095 (1960).
[33] Brish, A. A., M. S. Tarasov, and V. A. Tsukerman, *Soviet Phys. JETP*, **11**, 15 (1960).
[34] Broadbent, T. E., and A. H. A. Shlash, *Brit. J. Appl. Phys.*, **13**, 596 (1962).
[35] Broadbent, T. E., and J. K. Wood, *Brit. J. Appl. Phys.*, **6**, 368 (1955).
[36] Bundy, F. P., *J. Chem. Phys.*, **38**, 618 (1963).
[37] Bundy, F. P., *J. Chem. Phys.*, **38**, 631 (1963).
[38] Bundy, F. P., W. R. Hibbard, Jr., and H. M. Strong, eds., *Progress in Very High Pressure Research*, Wiley, New York, 1961.
[39] Bundy, F. P., and H. M. Strong, "Measurement of Flame Temperature, Pressure, and Velocity," in R. W. Ladenburg, B. Lewis, R. N. Pease, and H. S. Taylor, eds., *Physical Measurements in Gas Dynamics and Combustion*, Princeton University Press, New Jersey, 1954, p. 343.
[40] Bundy, F. P., and R. H. Wentorf, Jr., *J. Chem. Phys.*, **38**, 1144 (1963).
[41] Camac, M., and R. M. Feinberg, *Rev. Sci. Instr.*, **33**, 964 (1962).
[42] Cape, J. A., and G. W. Lehman, *J. Appl. Phys.*, **34**, 1909 (1963).
[43] Cape, J. A., G. W. Lehman, and M. M. Nakata. *J. Appl. Phys.*, **34**, 3550 (1963).
[44] Cassidy, E. C., and D. H. Tsai, *J. Res. Nat. Bur. Std.*, **67C**, 65 (1963).
[45] Cathey, L., *Inst. Radio Eng.*, *Trans. Nucl. Sci.*, **5**, 109 (1958).
[46] Cerceo, M., and H. M. Childers, *J. Appl. Phys.*, **34**, 1445 (1963).
[47] Chabai, A. J., and R. J. Emrich, *J. Appl. Phys.*, **26**, 779 (1955).

[48] Chace, W. G., and H. K. Moore, eds., *Exploding Wires*, Plenum Press, New York, Vol. I (1959), Vol. II (1962).
[49] Clerc, L. P., *Properties of Photographic Materials*, Fountain Press, London, 1950.
[50] Cnare, E. C., *J. Appl. Phys.*, **32**, 1043 (1961).
[51] Cnare, E. C., *J. Appl. Phys.*, **32**, 1275 (1961).
[52] Cook, M. A., *The Science of High Explosives*, Reinhold, New York, 1958.
[53] Cook, M. A., R. T. Keyes, and W. O. Ursenbach, *J. Appl. Phys.*, **33**, 3413 (1962).
[54] Cormack, G. D., and A. J. Barnard, *Rev. Sci. Instr.*, **33**, 606 (1962).
[55] Courtney-Pratt, J. S., "A Review of the Methods of High-Speed Photography," in *Reports on Progress in Physics*, Vol. XX, The Physical Society, London, 1957, p. 379.
[56] Courtney-Pratt, J. S., ed., *Proceedings of the Fifth International Congress on High-Speed Photography*, Society of Motion Picture and Television Engineers, New York, 1962.
[57] Cowan, R. D., *J. Appl. Phys.*, **32**, 1363 (1961).
[58] Cowan, R. D., *J. Appl. Phys.*, **34**, 926 (1963).
[59] Curran, D. R., *J. Appl. Phys.*, **32**, 1811 (1961).
[60] Cutler, M., and G. T. Cheney, *J. Appl. Phys.*, **34**, 1902 (1963).
[61] Dapoigny, J., J. Kieffer, and B. Vodar, *Compt. Rend.*, **245**, 1502 (1957).
[62] David, H. G., and S. D. Hamann, *J. Chem. Phys.*, **28**, 1006 (1958).
[63] De Carli, P. S., and J. C. Jamieson, *J. Chem. Phys.*, **31**, 1675 (1959).
[64] De Carli, P. S., and J. C. Jamieson, *Science*, **133**, 1821 (1961).
[65] De Vos, J. C., *Physica*, **20**, 669 (1954).
[66] Deal, W. E., Jr., "Dynamic High-Pressure Techniques," in R. H. Wentorf, ed., *Modern Very High Pressure Techniques*, Butterworths, London, 1962, p. 200.
[67] Deem, H. W., and W. D. Wood, *Rev. Sci. Instr.*, **33**, 1107 (1962).
[68] Defandorf, F. M., *J. Wash. Acad. Sci.*, **38**, 33 (1948).
[69] Dike, R. S., and E. L. Kemp, *The Design of a Capacitor and Switch Assembly for Low Inductance*, Los Alamos Scientific Laboratory Report LA-2957, 1963.
[70] Dobson, G. M. B., I. O. Griffith, and D. N. Harrison, *Photographic Photometry: A Study of Methods of Measuring Radiation by Photographic Means*, Clarendon Press, Oxford, 1926.
[71] Dolder, K., and R. Hide, *Rev. Mod. Phys.*, **32**, 770 (1960).
[72] Doran, D. G., *J. Appl. Phys.*, **34**, 844 (1963).
[73] Doran, D. G., "Measurement of Shock Pressures in Solids," in A. A. Giardini and E. C. Lloyd, eds., *High-Pressure Measurement*, Butterworths, London, 1963, p. 59.
[74] Drickamer, H. G., and A. S. Balchan, "High Pressure Optical and Electrical Measurements," in R. H. Wentorf, ed., *Modern Very High Pressure Techniques*, Butterworths, London, 1962, p. 25.
[75] Duff, R. E., and F. S. Minshall, *Phys. Rev.*, **108**, 1207 (1957).
[76] Duvall, G. E., *Appl. Mech. Rev.*, **15**, 849 (1962).
[77] Early, H. C., and R. C. Walker, "Inductive Energy Storage—A Tool for High Temperature Research," in H. Fischer and L. C. Mansur, eds., *Conference on Extremely High Temperatures*, Wiley, New York, 1958, p. 61.
Edgerton, H. E., *Rev. Sci. Instr.*, **27**, 162 (1956).
Edgerton, H. E., and K. J. Germeshausen, *J. Soc. Motion Picture Television Engrs.*, **61**, 286 (1953).
[80] Engstrom, R. W., *J. Opt. Soc. Am.*, **37**, 420 (1947).
[81] Engstrom, R. W., R. G. Stoudenheimer, and A. M. Glover, *Nucleonics*, **10**, 58 (1952).
[82] Engstrom, R. W., R. G. Stoudenheimer, H. L. Palmer, and D. A. Bly, *Inst. Radio Eng., Trans. Nuc. Sci.*, **5**, 120 (1958).
[83] Ettinger, S. Y., and A. C. Venezia, *Rev. Sci. Instr.*, **34**, 221 (1963).
[84] Evans, M. W., and C. M. Ablow, *Chem. Rev.*, **61** 129 (1961).
[85] Fowles, G. R., *J. Appl. Phys.*, **31**, 655 (1960).
[86] Fowles, G. R., *J. Appl. Phys.*, **32**, 1475 (1961).
[87] Gaydon, A. G., and I. R. Hurle, *The Shock Tube in High-Temperature Chemical Physics*, Reinhold, New York, 1963.
[88] Giardini, A. A., and E. C. Lloyd, *High-Pressure Measurement*, Butterworths, London, 1963.
[89] Glasoe, G. N., and J. V. Lebacqz, *Pulse Generators*, McGraw-Hill, New York, 1948.
[90] Goldenbaum, G., and E. Hintz, *Pressurized Trigatrons with a 10 kV–50 kV Low Jitter Operating Range*, University of Maryland, Tech. Rept. 314, 1963.
[91] Goldman, L. M., H. C. Pollock, J. A. Reynolds, and W. F. Westendorp, *Rev. Sci. Instr.*, **33**, 1041 (1962).
[92] Goranson, R. W., D. Bancroft, B. L. Burton, T. Blechar, E. E. Houston, E. F. Gittings, and S. A. Landeen, *J. Appl. Phys.*, **26**, 1472 (1955).
[93] Graham, R. A., *J. Appl. Phys.*, **32**, 555 (1961).
[94] Graham, R. A., *Rev. Sci. Instr.*, **32**, 1308 (1961).
[95] Grundhauser, F. J., W. P. Dyke, and S. D. Bennett, "A Fifty-Millimicrosecond Flash X-Ray System for Hypervelocity Research", in J. S. Courtney-Pratt, ed., *Proceedings of the*

Fifth International Congress on High-Speed Photography, Society of Motion Picture and Television Engineers, New York, 1962, p. 149.
[96] Hagerman, D. C., and A. H. Williams, *Rev. Sci. Instr.* **30**, 182 (1959).
[97] Hancox, R., *Rev. Sci. Instr.* **33**, 1239 (1962).
[98] Harrington, F. D. "High-Speed Time-Resolved Spectroscopic Instruments", in J. S. Courtney-Pratt, ed., *Proceedings of the Fifth International Congress on High-Speed Photography*, Society of Motion Picture and Television Engineers, New York 1962, p. 277.
[99] Henning, F., and H. T. Wensel, *J. Res. Natl. Bur. Std.* **10**, 809 (1933).
[100] Herzfeld, C. M., *Temperature, Its Measurement and Control in Science and Industry*, Vol. III, Parts 1 and 2, Reinhold, New York, 1962.
[101] Holland, L. R., *J. Appl. Phys.*, **34**, 2350 (1963).
[102] Hughes, D. S., L. E. Gourley, and M. F. Gourley, *J. Appl. Phys.*, **32**, 624 (1961).
[103] Hyzer, W. G., *Engineering and Scientific High-Speed Photography*, MacMillan, New York, 1962.
[104] Jahn, R. G., W. Jaskowsky, and A. J. Casini, *Rev. Sci. Instr.*, **34**, 1439 (1963).
[105] James, T. H., and G. C. Higgins, *Fundamentals of Photographic Theory*, Morgan and Morgan, New York, 1960.
[106] Jayaraman, A., W. Klement, Jr., and G. C. Kennedy, *Phys. Rev.*, **130**, 2277 (1963).
[107] Johnson, P. C., B. A. Stein, and R. S. Davis, *J. Appl. Phys.*, **33**, 557 (1962).
[108] Joigneau, S., and J. Thouvenin, *Compt. Rend*, **246**, 3422 (1958).
[109] Jones, D. L., and K. B. Earnshaw, "A Wire Exploder for Generating Cylindrical Shock Waves in a Controlled Atmosphere," *Technical Note* 148, *Natl. Bur. Std.* (1962).
[110] Katys, G. P. *Meas. Tech.* (*USSR*) No. 3, 230 (1959).
[111] Keene, J. P., *Rev. Sci. Instr.*, **34**, 1220 (1963).
[112] Kennard, R. B., *J. Res. Natl. Bur. Std.*, **8**, 787 (1932).
[113] Kistiakowsky, G. B., and P. H. Kydd, *J. Chem. Phys.*, **25**, 824 (1956).
[114] Knight, H. T., and D. Venable, *Rev. Sci. Instr.*, **29**, 92 (1958).
[115] Komel'kov, V. S., and V. I. Sinitsyn, "A Piezo-Electric Method for Investigation of a Powerful Gas Discharge," in M. A. Leontovich, ed., *Plasma Physics and the Problem of Controlled Thermonuclear Reactions*, Vol. I, p. 284, Pergamon Press, Oxford, 1961, p. 284.
[116] Kostkowski, H. J., and R. D. Lee, "Theory and Methods of Optical Pyrometry," in C. M. Herzfeld, ed., *Temperature, Its Measurement and Control in Science and Industry*, Vol. III, Reinold, New York 1962, Part 1, p. 449.
[117] Kozyrev, P. T., *Soviet Phys.-Solid State*, **1**, 94 (1959).
[118] Kuebler, N. A., and L. S. Nelson, *J. Opt. Soc. Am.*, **51**, 1411 (1961).
[119] Kurrelmeyer, B., W. H. Mais, and E. H. Green, *Rev. Sci. Instr.*, **14**, 349 (1943).
[120] Kvartskhava, I. F., V. V. Bondarenko, A. A. Pliutto, and A. A. Chernov, *Soviet Phys. JETP*, **4**, 623 (1957).
[121] Kvartskhava, I. F., A. A. Pliutto, A. A. Chernov, and V. V. Bondarenko, *Soviet Phys. JETP*, **3**, 40 (1956).
[122] Ladenburg, R. W., and D. Bershader, "Interferometry," in R. W. Ladenburg, B. Lewis, R. N. Pease, and H. S. Taylor, eds., *Physical Measurements in Gas Dynamics and Combustion*, Princeton University Press, New Jersey, 1954, p. 47.
[123] Laderman, A. J., G. J. Hecht, and A. K. Oppenheim, "Thin Film Thermometry in Detonation Research," in C. M. Herzfeld, ed., *Temperature, Its Measurement and Control in Science and Industry*, Vol. III, Reinhold, New York, 1962, Part 2, p. 943.
[124] Lawson, A. W., "The effect of Hydrostatic Pressure on the Electrical Resistivity of Metals." in B. Chalmers and R. King, eds., *Progress in Metal Physics*, Vol. 6, Pergamon, Oxford, 1956 p.1.
[125] Lee, R. D., "The NBS Photoelectric Pyrometer of 1961," in C. M. Herzfeld, ed., *Temperature, Its Measurement and Control in Science and Industry*, Vol. III, Reinhold, New York, 1962, Part 1, p. 507.
[126] Levine, M. A., L. S. Combes, and C. C. Gallagher, *Rev. Sci. Instr.* **32**, 1054 (1961).
[127] Lewis, I. A. D., and F. H. Wells, *Millimicrosecond Pulse Techniques*, 2nd ed., Pergamon Press, New York, 1959.
[128] Lochte-Holtgreven, W., "Production and Measurement of High Temperatures," *Reports on Progress in Physics*, Vol. XXI, The Physical Society, London, 1958.
[129] Lord, H. W., *Elec. Eng.*, **82**, 121 (1963).
[130] Lovejoy, D. R., *Can. J. Phys.*, **36**, 1397 (1958).
[131] Lovejoy, D. R., "Recent Advances in Optical Pyrometry", in C. M. Herzfeld, ed., *Temperature, Its Measurement and Control in Science and Industry*, Vol. III, Reinhold, New York, 1962, Part 1, p. 487.
[132] Maecker, H., ed., *Ionization Phenomena in Gases*, North-Holland, Amsterdam, 1962.
[133] Mallory, H. D., *J. Appl. Phys.*, **26**, 555 (1955).
[134] Marschall, F. H., J. W. Coltman and L. P. Hunter, *Rev. Sci. Instr.*, **18**, 504 (1947).
[135] Mather, J. W., and A. H. Williams, *Rev. Sci. Instr.*, **31**, 297 (1960).
[136] McQueen, R. G., and S. P. Marsh, *J. Appl. Phys.*, **31**, 1253 (1960).

[137] Middlehurst, J., and T. P. Jones, "A Precision Photoelectric Optical Pyrometer", in C. M. Herzfeld, ed., *Temperature, Its Measurement and Control in Science and Industry*, Vol. III, Reinhold, New York, 1962, Part 1, p. 517.
[138] Milne, G. G., and N. D. Miller, *J. Opt. Soc. Am.*, **49**, 1213 (1959).
[139] Minomura, S., and H. G. Drickamer, *J. Phys. Chem. Solids*, **23**, 451 (1962).
[140] Minomura, S., and H. G. Drickamer, *J. Appl. Phys.*, **34**, 3043 (1963).
[141] Minomura, S., G. A. Samara, and H. G. Drickamer, *J. Appl. Phys.*, **33**, 3196 (1962).
[142] Moeller, C. E., "Thermocouples for the Measurement of Transient Surface Temperatures", in C. M. Herzfeld, ed., *Temperature, Its Measurement and Control in Science and Industry*, Vol. III, Reinhold, New York, 1962, Part 2, p. 617.
[143] Moore, D. B., and J. K. Crosby, "Design and Application of a High-Speed Time-Resolving Spectrograph", in J. S. Courtney-Pratt, ed., *Proceedings of the Fifth International Congress on High-Speed Photography*, Society of Motion Picture and Television Engineers, New York, 1962, p. 273.
[144] Morton, G. A., R. M. Matheson, and M. H. Greenblatt, *Inst. Radio Eng., Trans. Nuc. Sci.*, **5**, 98 (1958).
[145] Moses, K. G., and T. Korneff, *Rev. Sci. Instr.*, **34**, 849 (1963).
[146] Nathan, A. M., *J. Appl. Phys.*, **22**, 234 (1951).
[147] Nelson, L. S., and N. A. Kuebler, *Rev. Sci. Instr.*, **34**, 806 (1963).
[148] Null, M. R., and W. W. Lozier, "The Carbon Arc as a Radiation Standard", in C. M. Herzfeld, ed., *Temperature, Its Measurement and Control in Science and Industry*, Vol. III, Reinhold, New York, 1962, Part 1, p. 551.
[149] O'Rourke, R. C., Edgerton, Germeshausen and Grier, Inc. Private Communication.
[150] Park, J. H., *J. Res. Natl. Bur. Std.*, **39**, 191 (1947).
[151] Park, J. H., and H. N. Cones, *Sphere–Gap Volt–Time Curves–Reference Standards For Steep Front Measurements*, AIEE Conference Paper, 57–215 (1957).
[152] Park, J. H., and H. N. Cones, *J. Res. Natl. Bur. Std.*, **66C**, 197 (1962).
[153] Parker, W. J., R. J. Jenkins, C. P. Butler, and G. L. Abbott, *J. Appl. Phys.*, **32**, 1679 (1961).
[154] Pasternak, R. A., E. C. Fraser, B. B. Hansen, and H. U. D. Wiesendanger, *Rev. Sci. Instr.* **33**, 1320 (1962).
[155] Pasternak, R. A., H. U. D. Wiesendanger, and B. B. Hansen, *J. Appl. Phys.*, **34**, 3416 (1963).
[156] Paul, W., and D. M. Warschauer, *Solids Under Pressure*, McGraw-Hill, New York, 1963.
[157] Penner, S. S., "Spectroscopic Methods of Temperature Measurements", in C. M. Herzfeld, ed., *Temperature, Its Measurement and Control in Science and Industry*, Vol. III, Reinhold, New York, 1962, Part 1, p. 561.
[158] Pochapsky, T. E., *Acta Met.* **1**, 747 (1953).
[159] Pochapsky, T. E., *Rev. Sci. Instr.*, **25**, 238 (1954).
[160] Prophet, H., and D. R. Stull, *J. Chem. Eng. Data*, **8**, 78 (1963).
[161] Rabinowicz, J., M. E. Jessey, and C. A. Bartsch, *J. Appl. Phys.*, **27**, 97 (1956).
[162] Rice, M. H., *J. Chem. Phys.*, **26**, 824 (1957).
[163] Rice, M. H., R. G. McQueen, and J. M. Walsh, "Compression of Solids by Strong Shock Waves", in *Solid State Physics*, Vol. VI, Academic Press, New York, 1958, p. 1.
[164] Richeson, W. E., *Rev. Sci. Instr.*, **29**, 99 (1958).
[165] Riggleman, B. M., and H. G. Drickamer, *J. Chem. Phys.*, **37**, 446 (1962).
[166] Roeser, W. F., F. R. Caldwell, and H. T. Wensel, *J. Res. Natl. Bur. Std.*, **6**, 1119 (1931).
[167] Roeser, W. F., and H. T. Wensel, *J. Res. Natl. Bur. Std.*, **12**, 519 (1934).
[168] Rudkin, R. L., "Thermal Diffusivity Measurements on Metals and Ceramics at High Temperatures", ASD–TDR–62–24 (1963).
[169] Rudkin, R. L., R. J. Jenkins, and W. J. Parker, *Rev. Sci. Instr.*, **33**, 21 (1962).
[170] Rudkin, R. L., W. J. Parker, and R. J. Jenkins, "Measurements of the Thermal Properties of Metals at Elevated Temperatures", in C. M. Herzfeld, ed., *Temperature, Its Measurement and Control in Science and Industry*, Vol. III, Reinhold, New York, 1962, Part 2, p. 523.
[171] Samara, G. A., and H. G. Drickamer, *J. Phys. Chem. Solids*, **23**, 457 (1962).
[172] Schall, R., *Z. Angew. Phys.*, **2**, 252 (1950).
[173] Schall, R., *Explosivstoffe*, **6**, 120 (1958).
[174] Selby, M. C., "High-Frequency Voltage Measurements", *Natl. Bur. Std. (U.S.), Circular* 481, (1949).
[175] Shanks, H. R., P. D. Maycock, P. H. Sidles, and G. C. Danielson, *Phys. Rev.*, **130**, 1743 (1963).
[176] Shumaker, J. B., Jr., and W. L. Wiese, "Measurement of Electron Density and Temperature in Dense Plasmas by Application of Line Broadening Theory", in C. M. Herzfeld, ed., *Temperature, Its Measurement and Control in Science and Industry*, Vol. III, Reinhold, New York, 1962, Part 1, p. 575.
[177] Sidles, P. H., and G. C. Danielson, *J. Appl. Phys.*, **25**, 58 (1954).

[178] Sidles, P. H., and G. C. Danielson, "Thermal Diffusivity Measurements at High Temperatures", in P. H. Egli, ed., *Thermoelectricity*, Wiley, New York, 1960, p. 270.
[179] Simmons, F. S., and A. G. De Bell, *J. Opt. Soc. Am.* **48**, 717 (1958).
[180] Simmons, F. S., and A. G. De Bell, *J. Opt. Soc. Am.*, **49**, 735 (1959).
[181] Skidmore, I. C., and E. Morris, "Experimental Equation of State Data for Uranium and Its Interpretation in the Critical Region", in *Thermodynamics of Nuclear Materials*, Proceedings of the Symposium on Thermodynamics of Nuclear Materials Held by the International Atomic Energy Agency in Vienna, International Atomic Energy Agency, Vienna, 1962, p. 173.
[182] Smith, O. E., *Rev. Sci. Instr.*, **35**, 134 (1964).
[183] Stager, R. A., and H. G. Drickamer, *Phys. Rev.*, **131**, 2524 (1963).
[184] Stevenson, M. J., W. Reuter, N. Braslau, P. P. Sorokin, and A. J. Landon, *J. Appl. Phys.*, **34**, 500 (1963).
[185] Stimson, H. F., *J. Res. Natl. Bur. Std.*, **65A**, 139 (1961).
[186] Swanger, W. H., and F. R. Caldwell, *J. Res. Natl. Bur. Std.*, **6**, 1131 (1931).
[187] Taylor, J. W. *J. Appl. Phys.*, **34**, 2727 (1963).
[188] Taylor, R. E., and R. A. Finch, *The Specific Heats and Resistivities of Molybdenum, Tantalum, and Rhenium from Low to Very High Temperatures*, Atomics International, California, Report NAA–SR–6034, 1961.
[189] Theophanis, G. A., *Rev. Sci. Instr.*, **31**, 427 (1960).
[190] Treiman, L. H., "A Precision Photon Counting Pyrometer", in C. M. Herzfeld, ed., *Temperature, Its Measurement and Control in Science and Industry*, Vol. III, Reinhold, New York, 1962, Part 1, p. 523.
[191] Tsai, D. H., and J. H. Park, "Calorimetric Calibration of the Electrical Energy Measurement in an Exploding Wire Experiment", in W. G. Chace and H. K. Moore, eds., *Exploding Wires*, Vol. II, Plenum Press, New York, 1962, p. 97.
[192] Tucker, T. J., *Rev. Sci. Instr.*, **31**, 165 (1960).
[193] Tucker, T. J., *J. Appl. Phys.*, **32**, 1894 (1961).
[194] Vereshchagin, L. F., A. A. Semerchan, N. N. Kuzin, and S. V. Popova, *Soviet Phys. "Doklady"*, **6**, 41 (1961).
[195] Vereshchagin, L. F., A. A. Semerchan, N. N. Kuzin, and S. V. Popova, *Soviet Phys. "Doklady"*, **6**, 391 (1961).
[196] Vereshchagin, L. F., A. A. Semerchan, and S. V. Popova, *Soviet Phys. "Doklady"*, **6**, 488 (1961).
[197] Vereshchagin, L. F., A. A. Semerchan, and S. V. Popova, *Soviet Phys. "Doklady"*, **6**, 609 (1962).
[198] Wallace, D. C., *Phys. Rev.*, **120**, 84 (1960).
[199] Wallace, D. C., P. H. Sidles, and G. C. Danielson, *J. Appl. Phys.*, **31**, 168 (1960).
[200] Walsh, J. M., and R. H. Christian, *Phys. Rev.*, **97**, 1544 (1955).
[201] Walsh, J. M., M. H. Rice, R. G. McQueen, and F. L. Yarger, *Phys. Rev.*, **108**, 196 (1957).
[202] Walsh, J. M., and M. H. Rice, *J. Chem. Phys.*, **26**, 815 (1957).
[203] Weber, R. L., *Heat and Temperature Measurement*, Prentice-Hall, New York, 1950.
[204] Wensel, H. T., D. B. Judd, and W. F. Roeser, *J. Res. Natl. Bur. Std.*, **12**, 527 (1934).
[205] Wensel, H. T., W. F. Roeser, L. E. Barbrow, and F. R. Caldwell, *J. Res. Natl. Bur. Std.*, **6**, 1103 (1931).
[206] Wensel, H. T., W. F. Roeser, L. E. Barbrow, and F. R. Caldwell, *J. Res. Natl. Bur. Std.*, **13**, 161 (1934).
[207] Wentorf, R. H., ed., *Modern Very High Pressure Techniques*, Butterworths, London, 1962.
[208] White, R. M., *J. Appl. Phys.*, **34**, 3559 (1963).
[209] Winkler, E. M., "X-ray Technique", in R. W. Landenburg, B. Lewis, R. N. Pease, and H. S. Taylor, eds., *Physical Measurements in Gas Dynamics and Combustion*, Princeton University Press, New Jersey, 1954, p. 97.
[210] Woisard, E. L., *J. Appl. Phys.*, **32**, 40 (1961).
[211] Wolfe, H. C., ed., *Temperature, Its Measurement and Control in Science and Industry*, Vol. II, Reinhold, New York, 1955.
[212] Worthing, A. G., *Phys. Rev.*, **12**, 199 (1918).
[213] Wunsch, D. C., and A. Erteza, Sandia Corporation and University of New Mexico, private communication.
[214] Young, A. P., W. B. Wilson, and C. M. Schwartz, *Phys. Rev.*, **121**, 77 (1961).
[215] Zarem, A. M., F. R. Marshall, and F. L. Poole, *Elec. Eng.* **68**, 282 (1949).

Author Index

Abbott, G. L., 575, 584
Abel, W. R., 272, 290
Ablow, C. M., 582
Abraham, B. M., 291
Ackermann, Th., 463, 464, 465, 488, 533
Adams, L. H., 56
Alder, B. J., 570, 576, 580, 581
Alieva, F. Z., 56
Altman, H., 213
Al'tshuler, L. V., 571, 580, 581
Amador, A., 535
Amberg, C. H., 548
Ambler, E., 270, 290
Anderson, A. C., 272, 288, 290
Anderson, R. L., 546, 548
Andon, R. J. L., 393
Andrews, D. H., 137, 213, 214, 365, 366, 428, 534
Angstrom, A. J., 574, 581
Appleton, A. S., 577, 581
Archibald, R. C., 260
Argue, G. R., 471, 533
Armstrong, G. T., 533
Armstrong, L. D., 361, 365
Arnett, R. L., 213, 548
Ass, T. V., 535
Astheimer, R. W., 567, 581
Aston, J. G., 137, 211, 217, 260, 543, 548, 549
Astrov, D. N., 56
Aven, M. H., 211
Avramescu, A., 572, 581
Awbery, J. H., 361, 365
Ayres, F. D., 439, 534

Backhurst, I., 361, 365
Bagrov, N. N., 366
Bakanova, A. A., 580, 581
Baker, L., Jr., 558, 561, 581
Baker, W. R., 558, 581
Balchan, A. S., 571, 580, 581, 582
Ball, A. F., 13, 140, 211, 313, 324, 330, 533, 534
Bancroft, D., 576, 580, 581, 582
Barber, C. R., 28, 56, 476, 533
Barbrow, L. E., 579, 585
Barger, J. W., 361, 365

Barieau, R. E., 431, 533
Barkman, J. H., 546, 548
Barnard, A. J., 558, 582
Barnes, C. B., 269, 291
Barnes, H. T., 370, 393, 434, 533
Barron, T. H. K., 548
Barry, F., 104, 130
Bartsch, C. A., 568, 584
Baxter, H. W., 573, 581
Beakley, W. R., 471, 533
Beattie, J. A., 56, 211
Becker, F., 13
Becker, J. H., 575, 581
Beckett, C. W., 576, 581
Beebe, R. A., 538, 539, 548, 549
Beehler, R. E., 211
Beers, Y., 560, 581
Behrens, W. U., 365
Bekkedahl, N., 213
Belaga, M. W., 361, 365, 533
Bellaschi, P. L., 559, 560, 581
Benedict, M., 56
Bennett, F. D., 559, 581
Bennett, S. D., 571, 582
Bennewitz, K., 371, 393
Benson, G. C., 486, 533
Berg, W. T., 393, 548
Berge, P., 361, 365
Berman, R., 130, 290
Bernstein, S., 290
Bershader, D., 572, 582, 583
Bestul, A. B., 140, 213
Bever, M. B., 577, 581
Beyon, J. H., 533
Bijl, D., 260
Birky, M. M., 365
Biscoe, J., 538, 539, 548
Black, W. C., 272, 290
Blackburn, D. H., 140, 213
Blackburn, G. F., 56
Bláha, O., 361, 365
Blaisdell, B. E., 56
Blake, C., 284, 290
Blanc, G., 361, 365
Blanpied, W. A., 260
Blechar, T., 580, 582
Blewitt, T. H., 547, 548, 549
Blue, R. W., 224, 260
Bly, D. A., 564, 582

AUTHOR INDEX

Boissonnas, C. G., 533
Bolze, C. C., 365
Bondarenko, V. V., 561, 577, 581, 583
Booth, S. F., 330
Borisov, N., 292
Born, M., 211
Borovick-Romanov, A. C., 56
Brackett, T. E., 546, 548
Brady, M. M., 560, 581
Braslau, N., 578, 585
Braun, M., 362, 365
Braun, R. M., 213
Brazhnik, M. I., 580, 581
Bredig, M. A., 296, 330
Brewer, L., 260, 534
Brickwedde, F. G., 56, 57, 130, 137, 211, 212, 213, 256, 260, 291, 579, 581
Bridgman, P. W., 336, 365, 428, 533, 570, 576, 581
Brinkworth, J. H., 370, 393
Brish, A. A., 571, 577, 580, 581
Broadbent, T. E., 558, 581
Brock, J. C. F., 290
Brooks, C. R., 364, 365
Bros, J. P., 335, 365
Brouty, E., 533
Brown, I., 460, 468, 469, 472, 533
Brugmann, E. W., 435, 534
Brunot, A. S., 130
Bruynooghe, W. M., 364, 367
Buckingham, E., 56
Buckingham, R. A., 211
Buckland, F. F., 130
Budnikoff, P. O., 533
Bunch, M. D., 57, 291
Bundy, F. P., 567, 571, 576, 581
Burgess, G. K., 56
Burgess, L. L., 534
Burk, D. L., 211
Burlew, J. S., 426, 533
Burns, G. W., 57
Burns, J. H., 211, 290
Burton, B. L., 580, 582
Busey, R. H., 137, 211, 217, 259, 260
Busse, J., 56
Butler, C. P., 575, 584
Butler, K. H., 533
Buzzell, A., 471, 482, 533

Caldwell, A. G., 130
Caldwell, F., 130

Caldwell, F. R., 56, 579, 584, 585
Callendar, H. L., 13, 370, 393, 434, 533
Calvet, E., 2, 13, 335, 365, 432, 509, 533
Camac, M., 568, 581
Campbell, A. N., 212
Campbell, I. E., 130
Cape, J. A., 575, 581
Carlson, H. G., 355, 365
Carmichael, J. L., 534
Carroll, B. H., 486, 533
Carslaw, H. S., 117, 122, 130
Casini, A. J., 558, 583
Cassidy, E. C., 569, 581
Cataland, G., 56, 292
Catalano, E., 227, 245, 260
Cathey, L., 564, 581
Caule, E. J., 533
Caywood, L. P., Jr., 57
Cerceo, M., 556, 575, 581
Chabai, A. J., 567, 581
Chalmers, B., 583
Chang, S. S., 140, 213
Chase, C. E., 284, 290
Chase, W. G., 577, 581, 582, 585
Chekhovskoi, V. Y., 296, 331
Cheney, G. T., 575, 582
Cheney, R. K., 393
Chernov, A. A., 561, 577, 581, 582, 583
Chessick, J. J., 533
Childers, H. M., 557, 575, 581
Chipman, J., 365
Chisholm, R. C., 248, 249, 261
Christens, J. M., 439, 534
Christian, C. J., 581
Christian, R. H., 570, 576, 580, 581, 585
Churchill, R. V., 130
Cines, M. R., 211
Clarke, J. T., 137, 211, 212, 217, 270
Cleland, W. W., 533
Clement, J. R., 15, 36, 56, 211
Clerc, L. P., 567, 582
Clever, H. L., 211
Clusius, K., 211, 217, 260
Cnare, E. C., 578, 582
Cobble, J. W., 471, 533
Cockett, A. H., 211
Cochran, J. F., 213, 292
Coddington, D. M., 361, 365, 533
Code, M. A., 570, 577, 582
Coffin, C. C., 533

AUTHOR INDEX

Cole, A. G., 211, 217, 219, 227, 228, 239, 245, 259, 260
Cole, K. S., 471, 533
Coleman, D. S., 428, 429, 533
Coltman, J. W., 564, 571, 583
Coltman, R. R., 547, 548, 549
Combes, L. S., 558, 583
Cones, H. N., 560, 584
Cook, M. A., 582
Cooke, A. H., 290
Coon, E. D., 419
Corak, W. S., 211, 290
Cordes, A. W., 211
Cormack, G. D., 558, 582
Corruccini, R. J., 57, 291, 296, 299, 309, 313, 329, 330, 534
Coulter, L. V., 217, 260
Courtney-Pratt, J. S., 569, 582, 583, 584
Cowan, R. D., 556, 575, 582
Cox, J. D., 378, 393
Cragoe, C. S., 57, 213, 325, 331
Craig, R. S., 137, 211, 214, 279, 290, 362, 366, 535
Criss, C. M., 533
Crosby, J. K., 567, 584
Crowell, A. D., 549
Croxon, A. A. M., 290
Cruickshank, A. J. B., 211, 428, 429, 469, 473, 480, 481, 486, 502, 503, 513, 533
Cummings, L. W. T., 435, 534
Curran, D. R., 580, 582
Cutler, M., 575, 582
Cynarski, J., 538, 539, 549

Dabbs, J. W. T., 285, 290, 291
Dahl, A. I., 56, 57, 130
Dainton, F. S., 527, 533
Daniels, F., 365, 367, 419
Danielson, G. C., 13, 574, 575, 584, 585
Dapoigny, J., 571, 582
Daunt, J. G., 269, 278, 286, 291
Dauphinee, T. M., 48, 57, 211, 284, 290, 533
David, H. G., 571, 582
Davis, R. S., 576, 583
Dawson, J. P., 393
Day, A. L., 23, 57
Deal, W. E., Jr., 576, 582
De Bell, A. G., 567, 585
Debye, P., 259, 260

De Carli, P. S., 576, 582
Dedrick, K. G., 560, 581
Deem, H. W., 13, 296, 299, 302, 331, 557, 575, 582
Defandorf, F. M., 560, 582
de Klerk, D., 57, 270, 284, 285, 290
de Launay, J., 212
Dench, W. A., 362, 365
Dennis, K. S., 549
de Nevers, N. H., 363, 366
De Nobel, J., 211
De Sorbo, W., 211, 217, 260, 279, 290, 548, 549
Dettre, R. H., 137, 214
Devins, J. C., 533
de Visser, L. E. O., 533
De Vos, J. C., 562, 582
Diaper, J., 527, 533
Diaz-Peña, M., 428, 533
Dickinson, H. C., 336, 365
Dieter, G. E., 577, 581
Dike, R. S., 558, 582
Dillinger, J. R., 211
Din, F., 211
Dingle, J. R., 533
Ditheil, C., 533
Ditmars, D. A., 130
Dixon, H. D., 212
Dobson, G. M. B., 565, 582
Doescher, R. N., 393
Dolder, K., 576, 582
Dole, M., 137, 211, 362, 365, 367, 480, 484, 533, 535
Dolecek, R. L., 56
Domen, S. R., 13
Donjon, A., 211
Donovan, P., 364, 367
Doran, D. G., 569, 570, 580, 582
Dornte, R. W., 212
Douglas, T. B., 13, 140, 211, 212, 296, 309, 313, 324, 329, 330, 533, 534
Douslin, D. R., 393
Dowden, D. A., 538, 539, 549
Drain, L. E., 549
Drickamer, H. G., 571, 577, 580, 581, 582, 584, 585
Droms, C. R., 57
Drucker, C., 365
du Chatenier, F. J., 211
Duff, G. M., 426, 533, 582
Duff, R. E., 582
Dugdale, J. S., 290, 460, 473, 534, 549

AUTHOR INDEX

Duke, W. M., 214
Dunfee, B. L., 211
Dupuis, M., 549
Durieux, M., 56, 57, 211
Duvall, G. E., 571, 582
Dworkin, A. S., 296, 330
Dyke, W. P., 571, 582

Early, H. C., 555, 582
Earnshaw, K. B., 578, 583
Eastman, E. D., 216, 260
Edgerton, H. E., 569, 582
Edlow, M. H., 56, 57, 291, 292
Egan, C. J., 212, 221, 223, 224, 226, 243, 245, 255, 260
Egan, E. P., Jr., 431, 533
Egli, P. H., 585
Eidinoff, M. L., 137, 211
Eigen, M., 533
Elder, G. E., 366
Elliott, J. H., 364, 366
Emrich, R. J., 567, 581
Enagonio, D., 212
Engstrom, R. W., 564, 582
Erickson, R. A., 285, 291
Erteza, A., 560, 585
Essen, L., 212
Estle, T. L., 292
Ettinger, S. Y., 560, 582
Eucken, A., 135, 212, 216, 260, 370, 393, 533
Evans, J. P., 57
Evans, M. W., 577, 582
Everett, D. H., 426, 533
Evstrop'ev, K. S., 533
Eyraud, C., 533
Eyring, H., 581

Feinberg, R. M., 568, 581
Ferguson, A. H., 431, 533
Ferrier, A., 330
Filimonov, A. I., 291
Finch, R. A., 574, 585
Findlay, A., 212
Finke, H. L., 211, 212, 393
Fiock, E. F., 393, 399, 401, 402, 419
Fischer, H., 582
Fisher, L. H., 261
Flotow, H. E., 213, 264, 291, 264, 291, 533
Flügge, S., 57, 212, 290
Fock, W., 460, 469, 472, 533

Foley, W. T., 534
Fomichev, E. N., 296, 300, 302, 330
Ford, P. J.,
Forrest, H. D., 435, 534
Förster, F., 365
Forsythe, G. E., 212
Fowles, G. R., 570, 580, 582
Fox, D., 214
Franck, E. U., 453, 534
Franck, J. P., 291
Fraser, E. C., 574, 584
Frederikse, H. P. R., 549
Freyer, E. B., 428, 534
Friedberg, S. A., 57, 211
Frisch, M. A., 534
Frow, F. R., 393
Furukawa, G. T., 130, 133, 137, 140, 211, 212, 259, 260, 324, 329, 330, 355, 365, 534

Gaede, W., 135, 212
Gaines, J. M., Jr., 56, 211
Gal'chenko, G. L., 363, 366
Gallagher, C. C., 558, 583
Garber, M., 213
Garden, L. A., 540, 549
Garfunkel, M. P., 211, 290, 291
Garner, C. S., 137, 214
Garwin, R. L., 291
Gautier, M., 45, 57
Gaydon, A. G., 567, 571, 572, 582
Geballe, T. H., 57, 212, 291
Gerkin, R. E., 260
Germeshausen, K. J., 569, 582
Giardini, A. A., 571, 582
Giauque, W. F., 212, 213, 216, 217, 221, 223, 224, 225, 226, 232, 242, 243, 245, 255, 259, 260, 393, 431, 533, 546, 549
Gibson, G. E., 212, 216, 260
Giguère, P. A., 460, 473, 524, 534
Gilbert, R. A., 296, 330
Ginnings, D. C., 13, 85, 130, 131, 137, 140, 212, 213, 214, 260, 296, 299, 307, 309, 313, 329, 330, 331, 359, 363, 365, 366, 367, 393, 399, 401, 402, 419, 420, 437, 482, 533, 534, 535
Gittings, E. F., 580, 582
Glasgow, A. R., Jr., 212
Glasoe, G. N., 555, 582
Glasstone, S., 212
Glover, A. M., 564, 582

AUTHOR INDEX

Gniewek, J. J., 291
Goddard, E. D., 486, 533
Goldenbaum, G., 558, 582
Goldman, L. M., 558, 582
Goldsmid, H. J., 534
Good, W. D., 393
Goranson, R. W., 580, 582
Gordon, M. M., 534
Gorter, C. J., 260, 278, 291
Goton, R., 533
Gourlay, L. E., 580, 583
Gourley, M. F., 580, 583
Gouy, M., 480, 487, 534
Graham, R. A., 570, 582
Grant, D. A., 57
Grant, N. J., 365
Grassi, U., 534
Gratch, S., 214
Graus, C. A., 13
Green, E. H., 573, 583
Green, M. S., 516, 581
Greenblatt, M. H., 533, 564, 584
Greensfelder, B. S., 217, 260
Gregg, S. J., 534
Gregory, N. W., 363, 365, 366, 367, 534
Greyson, J., 543, 548, 549
Griffel, M., 291
Griffing, D. F., 292
Griffith, I. O., 565, 582
Griffiths, E., 361, 365
Grønvold, F., 360, 365
Gross, F. J., 213
Gross, M. E., 212, 393
Grundhauser, F. J., 571, 582
Gucker, F. T., 438, 439, 465, 470, 480, 488, 489, 534
Guenther, R. A., 291
Guggenheim, E. A., 534
Gulinowa, L., 533
Gupta, A. K., 213
Guthrie, G. B., 213, 393

Haas, E. G., 217, 260
Hagerman, D. C., 558, 583
Hales, J. L., 378, 393
Hall, H. E., 292
Hall, J. A., 57
Hall, R. G., 212
Hamann, S. D., 571, 582
Hancox, R., 558, 583
Hansen, B. B., 574, 584
Harding, R. S., 533

Harrington, F. D., 567, 583
Harris, F. K., 212
Harrison, D. N., 565, 582
Harrison, R. H., 394
Hart, K. R., 460, 461, 463, 470, 535
Hartshorn, L., 59, 285, 291
Haseda, T., 289, 290, 291
Hatcher, J. B., 214, 292, 472, 535
Haworth, E., 366
Hecht, G. J., 568, 583
Heer, C. V., 269, 278, 291
Hellwege, K. H., 362, 366
Henning, F., 579, 583
Henning, J. M., 212
Heric, E. L., 549
Herzfeld, C. M., 56, 57, 130, 211, 291, 568, 583, 584, 585
Hettinger, W. P., Jr., 137, 211, 362, 365, 480, 484, 533
Heuse, W., 371, 393
Heydweiller, A., 534
Hibbard, W. R., Jr., 571, 581
Hickes, W. F., 57
Hicks, J. F. G., Jr., 217, 260
Hide, R., 576, 582
Higgins, G. C., 567, 583
Hildebrand, F. B., 212
Hildebrand, J. H., 212
Hildenbrand, D. L., 260
Hill, R. W., 137, 212, 291
Hintz, E., 558, 582
Hirabayashi, M., 363, 366
Hirschfelder, J. O., 393
Hládek, L., 534
Hoare, F. E., 212, 265, 291
Hoch, M., 296, 302, 316, 329, 330
Hoevo, C. A. J., 486, 533
Hoge, H. J., 130, 212, 256, 260, 366, 419, 534
Holland, L. R., 574, 583
Holmberg, T., 534
Holton, W. B., 393
Horman, J., 140, 213
Hornig, D. F., 534
Horowitz, K. H., 364, 366
Horton, A. T., 212
Hossenlopp, I. A., 393
Hough, E. W., 535
Householder, A. S., 212
Houston, E. E., 580, 582
Howling, D. H., 288, 291
Hubbard, J. C., 428, 534
Hubbard, W. N., 393
Huckaba, C. E., 534

AUTHOR INDEX

Hudson, R. P., 270, 290, 292
Huffman, H. M., 137, 212, 213, 291, 394, 462, 534, 535
Hughes, D. S., 580, 583
Hughes, R. L., 361, 365
Hull, G. W., Jr., 57, 212, 291
Hull, R. A., 290
Hulscher, W. S., 286, 292
Hultgren, R., 3, 330
Humphries, A. R., 533
Hunter, L. P., 564, 571, 583
Huntley, D. J., 285, 290
Hurle, I. R., 567, 571, 572, 582
Hutchens, J. O., 211, 217, 219, 227, 228, 239, 245, 259, 260
Hwang, Yu-Tang, 363, 366
Hyzer, W. G., 569, 583

Ingraham, T. R., 533
Ioffe, A. F., 534
Ivin, K. J., 527, 533

Jackson, L. C., 212, 265, 291
Jacob, M., 103, 104, 107
Jacobi, H., 534
Jacobs, R. B., 130
Jacobus, D. D., 56, 211
Jaeger, J. C., 117, 122, 130
Jahn, R. G., 558, 583
Jakob, M., 130
James, T. H., 567, 583
Jamieson, J. C., 576, 582
Janz, G. J., 211
Jaskowsky, W., 558, 583
Jastram, P. S., 286, 291
Jauch, K., 534
Jayaraman, A., 576, 583
Jeener, J., 534
Jenkins, R. J., 565, 575, 584
Jepson, W. B., 456, 534
Jessey, M. E., 568, 584
Jessup, R. S., 527, 534
Jewell, R. C., 330
Johnson, B. H., 217, 230, 261, 535
Johnson, C. L., 212
Johnson, F. M. G., 534
Johnson, P. C., 576, 583
Johnston, H. L., 137, 211, 212, 213, 217, 224, 230, 232, 260, 296, 302, 316, 329, 330, 581
Johnston, H. S., 581
Johnston, W. V., 362, 366
Joigneau, S., 571, 583
Jones, D. L., 578, 583

Jones, F. W., 364, 366
Jones, T. P., 564, 565, 584
Jones, W. M., 393
Jost, W., 362, 366
Joule, J. P., 437, 534, 535
Judd, D. B., 579, 585
Jura, G., 549
Justice, B. H., 212

Kaeser, R. S., 292
Kandyba, V. V., 296, 300, 302, 330
Kangro, W., 362, 366
Kantor, P. B., 296, 300, 302, 330
Katys, G. P., 565, 583
Katz, C., 393
Kay, A. E., 362, 366
Keene, J. P., 564, 583
Keesom, P. H., 213, 276, 291
Keesom, W. H., 212, 260, 274, 291
Kelley, K. K., 296, 316, 320, 322, 323, 327, 329, 330, 331
Kelvin, Lord, 16, 535
Kemeny, G., 493, 535
Kemp, E. L., 558, 582
Kemp, J. D., 212, 224, 260
Kemp, W. R. G., 275, 291
Kennard, R. B., 567, 583
Kennedy, G. C., 576, 583
Kenyon, W. O., 13, 535
Kerr, E. C., 137, 212, 217, 260
Keyes, F. G., 534
Keyes, R. T., 570, 582
Khomyakov, K. G., 364, 367
Khotkevich, V. I., 366
Kieffer, J., 571, 582
Kilpatrick, J. E., 217, 261, 535
Kincheloe, T. C., 393
King, E. G., 320, 322, 330
King, G. J., 212, 259, 260
King, R., 583
Kingery, W. D., 360, 361, 366
Kington, G. L., 540, 549
Kirillin, V. A., 296, 331
Kisel, A. N., 330
Kistiakowsky, G. B., 370, 393, 571, 583
Kistler, S. S., 130
Klement, W., Jr., 576, 583
Klemm, W., 534
Kleppa, O. J., 335, 366
Knappe, W., 362, 366
Knight, H. T., 571, 583
Knowles, E. G., 330
Kobe, K. A., 393

AUTHOR INDEX

Koch, W., 435, 534
Kohlhaas, R., 362, 365
Kok, J. A., 212, 274, 291
Komel'kov, V. S., 570, 583
Kono, H., 366
Kormer, S. B., 580, 581
Korn, G. A., 119, 130
Korn, T. M., 119, 130
Korneff, T., 560, 584
Kosaki, A., 549
Kostkowski, H. J., 53, 57, 564, 583
Kostryukova, M. O., 212, 291
Kouvel, J. S., 432, 534
Kozyrev, P. T., 571, 583
Kramers, H. C., 286, 292
Krasovitskaya, R. M., 330
Kraus, C. A., 13, 534
Krauss, F., 362, 366
Krieger, L. A., 533
Krivobabko, I. P., 535
Krupnikov, K. K., 580, 581
Kubaschewski, O., 3, 362, 366
Kuebler, N. A., 556, 578, 583, 584
Kuleshova, L. V., 571, 580, 581
Kuntz, R. R., 456, 534
Kunzler, J. E., 57, 212, 291
Kurrelmeyer, B., 573, 583
Kurti, N., 57, 212, 265, 291
Kuzin, N. N., 571, 580, 585
Kvartskhava, I. F., 561, 577, 578, 581
Kydd, P. H., 571, 583

Labes, M. M., 214
Lachman, J. C., 57
Ladenburg, R. W., 572, 581, 583, 585
Laderman, A. J., 568, 583
Laing, W., 540, 549
Land, T., 330
Landeen, S. A., 580, 582
Landon, A. J., 578, 585
Lange, F., 137, 212
Larson, N., 137, 211, 362, 365, 480, 484, 533
Latimer, W. M., 216, 217, 260
Lauritzen, J. I., 57
Lawson, A. W., 571, 583
Leake, L. E., 311, 313, 331
Leaver, V. M., 57
Lebacqz, J. V., 555, 582
Ledenev, B. N., 580, 581
Lee, R. D., 57, 564, 583

Leech, J. W., 431, 534
Lees, E. B., 378, 393
Lehman, G. W., 575, 581
Leidenfrost, W., 364, 366
Leontovich, M. A., 583
Levine, M. A., 558, 583
Levinson, L. S., 296, 331
Lewis, B., 581, 583, 585
Lewis, G. N., 260, 534
Lewis, I. A. D., 555, 583
Liburg, M., 292
Lien, W. H., 291, 546, 549
Lifkin, E. B., 217, 260
Lifshits, I. M., 212
Lindenfeld, P., 57, 291
Lipsett, S. G., 534
Liu, I. D., 460, 473, 534
Lloyd, E. C., 571, 582
Loasby, R. G., 362, 366
Lochte-Holtgreven, W., 567, 578, 583
Logan, J. K., 56, 211
Lohr, H. R., 213, 259, 260
Lonberger, S. T., 57
Lonsdale, H. K., 214
Lord, H. W., 560, 583
Los, J. M., 543, 544, 549
Lounasmaa, O. V., 277, 291
Lovejoy, D. R., 15, 53, 57, 564, 583
Low, D. I. R., 428, 534
Lozier, W. W., 562, 584
Lucina, J. L., 393
Lucks, C. F., 13, 296, 299, 302, 331
Lyman, T., 130
Lyusternik, V. E., 362, 366

Maass, O., 533, 534
MacDonald, D. K. C., 211, 212, 290
Mackle, H., 534
Maecker, H., 567, 583
Maier, C. G., 319, 327, 331
Mains, G. J., 456, 534
Mais, W. H., 573, 583
Malcolm, G. N., 534
Mallory, H. D., 580, 583
Mallya, R. W., 279, 290
Manchester, F. D., 274, 291
Mann, W. B., 13
Mansur, L. C., 582
Marcus, H., 361, 365, 533
Markowitz, W., 212
Marschall, F. H., 564, 571, 583
Marsh, S. P., 580, 583

AUTHOR INDEX

Marshall, F. R., 569, 585
Martin, D. L., 212, 280, 281, 291
Martin, J. J., 363, 366
Martin, W. H., 363, 366
Marvin, J. W., 559, 581
Marx, P. C., 362, 365, 367, 484, 535
Masi, J. F., 366, 370, 393
Massena, C. W., 279, 290
Mastrangelo, S. V. R., 212, 549
Mate, C. F., 130
Mather, J. W., 558, 583
Matheson, R. M., 564, 584
Matthews, J. H., 486, 533
Maxwell, E., 284, 286, 290, 291
Maycock, P. D., 575, 584
Mayhew, Y. R., 461, 535
Mazo, R., 549
McClure, F. T., 393
McCoskey, R. E., 140, 211, 212, 259, 260, 324, 329, 330, 534
McCullough, J. P., 133, 213, 214, 393
McElroy, D. L., 364, 366
McGlashan, M. L., 428, 533, 534
McGurty, J. A., 57
McKim, F. R., 285, 291
McNish, A. G., 59
McQueen, R. G., 576, 580, 583, 584, 585
Meads, P. F., 260
Mebs, R. W., 131
Medved, T. M., 361, 365
Mellors, J. W., 371, 393
Mendelssohn, K., 212, 278, 291
Mendoza, E., 288, 291
Mercer, E. E., 471, 533
Messer, C. E., 367
Messerly, G. H., 211, 213, 217, 260
Messerly, J. F., 213, 393
Meyers, C. H., 57, 213, 227, 260, 366
Michels, A., 370, 393
Middlehurst, J., 564, 565, 584
Miedema, A. R., 289, 290, 291
Millar, R. W., 217, 260
Millard, B., 538, 539, 549
Miller, J. T., 431, 533
Miller, N. D., 567, 584
Milne, G. G., 567, 584
Milne, W. E., 213
Milne-Thomson, L. M., 213
Milner, R. T., 213
Minomura, S., 571, 577, 580, 584
Minshall, F. S., 582
Minshall, S., 576, 581

Mockler, R. C., 211
Moeller, C. E., 568, 584
Moelwyn-Hughes, E. A., 428, 534
Montroll, E., 549
Moore, D. B., 567, 584
Moore, H. K., 577, 581, 582, 585
Moore, R. T., 393
Mooser, E., 284, 290
Morcom, K. W., 534
Morissette, B. G., 524, 534
Morris, E., 580, 585
Morrison, C. F., 213
Morrison, J. A., 460, 473, 534, 543, 544, 548, 549
Morton, G. A., 564, 584
Moser, H., 57, 363, 366
Moses, K. G., 560, 584
Mueller, E. F., 13, 57, 213, 419
Mullins, J. C., 214
Murch, L. E., 225, 255, 260
Murphy, G. W., 534
Murphy, W. K., 296, 312, 314, 331
Murrell, T. A., 57

Nagasaki, S., 363, 366
Nakata, M. M., 575, 581
Nathan, A. M., 366, 574, 584
Natrella, M. G., 213
Naylor, B. F., 296, 331
Neganov, B., 292
Neighbor, J. E., 292
Nelson, L. S., 556, 578, 583, 584
Nelson, R. A., 534
Nernst, W., 135, 213, 260
Newcombe, P., 330
Noggle, T. S., 547, 548, 549
Nölting, J., 363, 366
Normann, W., 364, 366
North, J. M., 363, 366
Null, M. R., 562, 584

Oel, H. J., 362, 366
Oelsen, O., 534
Oelson, W., 534
Oetting, F. L., 363, 366
Ohlmeyer, P., 366
O'Keefe, B. J., 534
Olette, M., 296, 331, 534
Olmos, A. W., 524, 533, 534
Olsen, J. L., 278, 291
O'Neal, H. E., 363, 365, 366, 534
O'Neal, H. R., 291
Onsager, L., 549
Oppenheim, A. K., 568, 583

AUTHOR INDEX

Orfield, H. M., 538, 539, 549
Oriani, R. A., 296, 312, 314, 331
Orlova, M. P., 56
O'Rourke, R. C., 556, 584
Orr, R. L., 330
Osborn, A., 393
Osborne, D. W., 137, 211, 213, 214, 259, 260, 290, 291, 292, 393, 472, 533, 535
Osborne, N. S., 13, 213, 324, 325, 331, 363, 366, 371, 393, 395, 396, 397, 399, 401, 402, 407, 419, 420, 437, 534
Ott, J. B., 260
Otto, J., 57
Otto, K., 533
Owen, B. B., 426, 534

Pace, E. L., 549
Pallister, P. R., 366
Palmer, H. L., 564, 582
Parham, D. N., 428, 535
Park, J. H., 559, 560, 584, 585
Parker, W. J., 565, 575, 584
Parkinson, D. H., 213, 291
Parks, G. S., 216, 217, 260, 366
Parry, J. V. L., 212
Partington, J. R., 57, 370, 393, 534
Pasternak, R. A., 574, 584
Patterson, D., 549
Paul, W., 571, 581, 584
Pavlovskii, M. N., 571, 580, 581
Pawel, R. E., 364, 366
Payne, W. H., 296, 309, 330
Pearlman, N., 213, 291
Pearsall, G. W., 213
Pease, R. N., 581, 583, 585
Penner, S. S., 567, 584
Pennington, R. E., 393
Person, W. B., 393
Peshkov, V. P., 272, 291
Peterson, E. L., 576, 581
Petkof, B., 393
Petree, B., 131
Petree, M. C., 364, 367
Petrunin, A. P., 571, 581
Pfaundler, L., 534
Phillips, N. E., 270, 271, 275, 291, 546, 549
Phipps, L. W., 363, 366
Piccirelli, J. H., 212
Pickard, H. B., 480, 488, 489, 534
Picklesimer, M. L., 364, 366
Pierce, L., 549

Pilcher, G., 213
Pillinger, W. L., 286, 291
Pimentel, G. C., 213
Pitzer, K. S., 213, 217, 260, 371, 372, 393, 534, 549
Planck, R. W., 480, 488, 489, 534
Plester, D. W., 534
Pliutto, A. A., 561, 577, 581, 583
Plumb, H., 290, 292
Plumb, H. H., 56, 57, 291, 292
Pochapsky, T. E., 574, 584
Polley, M. H., 538, 539, 549
Pollock, H. C., 558, 582
Poole, F. L., 569, 585
Popov, M. M., 363, 366
Popova, S. V., 571, 580, 585
Popp, L., 217, 260
Postma, H., 291
Powell, R. L., 57, 260, 291
Prat, H., 13, 365, 533
Preston, J. S., 473, 534
Preston-Thomas, H., 57, 211
Prophet, H., 574, 584
Pruitt, J. S., 13
Putilin, N. I., 535

Quarrington, J. E., 213, 291
Quinnell, E. H., 36, 56

Rabinowicz, J., 568, 584
Raine, H. C., 363, 366
Randall, M., 534
Rands, R. D., Jr., 213
Ramanathan, K. G., 213, 291
Randall, M., 260
Rayne, J., 291
Rayne, J. A., 213, 275, 291
Redheffer, R. M., 261
Reich, H. A., 291
Reilly, M. L., 212
Reuter, W., 578, 585
Reynolds, J. A., 558, 582
Rice, M. H., 576, 580, 584, 585
Rice, W. W., 370, 393
Richards, R. B., 363, 366
Richards, T. W., 438, 465, 534
Richeson, W. E., 558, 584
Richmann, G. W., 534
Ridderhof, J. A., 13, 534
Rider, E. A., 131
Rifkin, E. B., 137, 212
Riggleman, B. M., 571, 584
Riggs, F. B., Jr., 366
Rixon, J., 363, 366

Roach, P. R., 291
Roach, W. R., 290
Robbrecht, G. G., 364, 367
Roberts, H. S., 366
Roberts, L. D., 285, 290, 291
Roberts, T. R., 57
Robie, R. A., 211, 217, 219, 227, 228, 239, 245, 259, 260
Robinson, F. N. H., 57
Robinson, G., 214
Rodebush, W. H., 216, 260
Roebuck, J. R., 57
Roeser, W. F., 131, 304, 331, 579, 584, 585
Rogers, G. F. C., 461, 535
Rogers, S. E., 534
Rose-Innes, A. C., 213, 265, 291
Rosengren, K., 213
Ross, G. S., 212
Rossini, F. D., 13, 213, 533, 534, 535
Rossner, W., 371, 393
Roth, W. A., 13
Rowlinson, J. S., 534, 370, 393, 456, 534
Rubin, T., 213
Rubin, T. R., 214, 439, 534
Rudin, B. D., 213
Rudkin, R. L., 565, 575, 584
Ruehrwein, R. A., 137, 213, 291, 462, 535
Russell, H., Jr., 214
Russell, K. E., 211
Rüterjans, H., 533
Ryder, H., 363, 366

Saba, W. G., 212
Sachse, H., 535
Sage, B. H., 535
Sage, U., 533
Salinger, G. L., 288, 290
Samara, G. A., 571, 577, 580, 584
Samoilov, B. N., 288, 291
Sandenaw, T. A., 549
Satterthwaite, C. B., 211, 290
Saylor, C. P., 212
Scarborough, J. B., 213
Schall, R., 571, 584
Scheele, K., 371, 393
Scheil, E., 364, 366
Schilling, W., 13, 493, 535
Schmidt, E. O., 364, 366
Schmitz, O. F., 533
Schniedermann, G., 362, 366
Schrage, S., 533

Schreiner, F., 291, 533
Schroeder, C. M., 291
Schulze, O., 371, 393
Schumann, S. C., 211
Schwartz, C. M., 571, 580, 585
Schwers, F., 216, 260
Scott, D. W., 393, 394
Scott, R. B., 57, 131, 213, 260, 371, 393
Scott, R. L., 212
Seekamp, H., 366
Seidel, G., 276, 291
Seki, S., 549
Selby, M. C., 560, 584
Semerchan, A. A., 571, 580, 585
Senin, M. D., 363, 366
Shanks, H. R., 575, 584
Shchukarev, A. N., 535
Shchukareva, L. A., 535
Sheard, D. R., 527, 533
Sheindlin, A. E., 296, 331
Shenker, H., 57
Sherman, R. H., 57
Shiffman, C. A., 213, 292
Shilling, W. G., 370, 393
Shlash, A. H. A., 558, 581
Shmidt, N. E., 350, 352, 364, 366
Shomate, C. H., 296, 327, 328, 331
Shumaker, J. B., Jr., 567, 584
Sidles, P. H., 13, 574, 575, 584, 585
Silsbee, F. B., 213
Simakov, G. V., 580, 581
Simmons, F. S., 567, 585
Simon, F., 57, 216, 260
Sinel'nikov, N. N., 364, 366
Sinitsyn, V. I., 570, 583
Skidmore, I. C., 580, 585
Skinner, H. A., 13, 471, 527, 533, 535
Skochdopole, R. E., 291
Sligh, T. S., Jr., 213, 325, 331, 371, 393, 535
Smith, F. D., 43, 44, 57
Smith, J. C., 393, 394
Smith, J. S., 427, 534
Smith, N. O., 212
Smith, O. E., 558, 585
Smith, P. L., 212, 291
Smith, P. S., 549
Smith, R. H., 366
Smith, W. R., 538, 539, 548, 549
Snider, C. S., 211
Sokolnikoff, I. S., 261
Sokolov, V. A., 350, 352, 364, 366

AUTHOR INDEX

Sommerfeld, A., 213
Sorai, M., 549
Sorokin, P. P., 578, 585
Sosman, R. B., 23, 57
Southard, J. C., 137, 213, 296, 308, 309, 318, 331
Spalthoff, W., 453, 534
Spear, N. H., 433, 535
Spencer, W. B., 548
Spitzer, R., 393
Spohr, D. A., 57
Sreenivasaya, M., 535
Sreerangachar, H. B., 535
Srinivasan, T. M., 213, 291
Stager, R. A., 571, 580, 585
Stansbury, E. E., 364, 365, 366
Starr, C., 130
Staveley, L. A. K., 213, 428, 460, 461, 462, 470, 535
Steenland, M. J., 291
Stegeman, G., 217, 260
Stein, B. A., 576, 583
Sterrett, K. F., 137, 140, 213, 214, 362, 366, 535
Stevens, N. P., 538, 539, 549
Stevenson, M. J., 578, 585
Steyert, W. A., 288, 290
Stijland, J. C., 370, 393
Stimson, H. F., 13, 15, 57, 213, 255, 261, 325, 331, 363, 366, 371, 393, 399, 401, 402, 419, 420, 437, 534, 579, 585
Stone, H., 214
Stoudenheimer, R. G., 564, 582
Stout, J. W., 211, 217, 219, 227, 228, 239, 245, 248, 249, 259, 260, 261
Stow, F. S., Jr., 364, 366
Stratton, R., 549
Strelkov, P. G., 56, 212, 291
Strine, G. L., 560, 581
Strong, H. M., 567, 571, 581
Stull, D. R., 137, 213, 366, 535, 574, 584
Sturtevant, J. M., 451, 462, 463, 471, 482, 533, 535
Suga, H., 549
Suits, E., 214
Sunner, S., 471, 535
Swanger, W. H., 579, 585
Swann, W. F. G., 371, 393
Swietoslawski, W., 13, 529, 535
Sydoriak, S. G., 57
Sykes, C., 364, 366
Szasz, G. J., 549

Takagi, Y., 366
Takahashi, Y., 213, 363, 366
Tarasov, M. S., 571, 577, 580, 581
Tarasov, V. V., 213
Taylor, H. S., 581, 583, 585
Taylor, J. W., 576, 585
Taylor, R. D., 217, 230, 261, 535
Taylor, R. E., 574, 585
Tebbe, W., 534
Theophanis, G. A., 558, 585
Thomas, A., 535
Thomas, J. L., 213
Thomas, S. B., 366
Thomas, W., 57
Thompson, F. W., 549
Thompson, K., 292
Thompson, R. D., 57
Thomson, W., 16, 426, 535
Thouvenin, J., 571, 583
Tian, A., 335, 367, 432, 525, 535
Tilk, W., 534
Timimi, B. A., 469, 473, 486, 502, 503, 533
Timmerhaus, K. D., 56, 214
Timmermans, J., 535
Todd, L. J., 137, 214, 469, 479, 517, 518, 535
Todd, S. S., 213, 393, 394
Tong, L. K. J., 13, 535
Tooke, J. W., 211
Treaftis, H. N., 548, 549
Treiman, L. H., 565, 585
Tret'yakov, Yu. D., 364, 367
Troshkina, V. A., 364, 367
Trowbridge, J. C., 214, 353, 359, 364, 367, 462, 535
Trunin, R. F., 580, 581
Tsai, D. H., 366, 560, 569, 581, 585
Tschentke, G., 365
Tsukerman, V. A., 571, 577, 580, 581
Tucker, T. J., 556, 577, 585
Tunnicliff, D. D., 214
Tupman, W. I., 460, 461, 463, 470, 535
Turkdogan, E. T., 311, 313, 331
Turnbull, D., 548, 549
Turner, L. B., 481, 535

Ubbelohde, A. R., 534
Ursenbach, W. O., 570, 582

Vance, R. W., 214
Van den Ende, J. N., 212, 291

AUTHOR INDEX

van Dijk, H., 56, 211
Vaughen, J. V., 211
Venable, D., 571, 583
Venezia, A. C., 560, 582
Vereshchagin, L. F., 571, 580, 585
Verhaeghe, J. L., 364, 367
Vilches, O. E., 292
Villa, H., 366
Vlugt, N. J. v.d., 291
Vodar, B., 571, 582
von Kármán, T., 211
von Lude, K., 370, 393

Wacker, P. F., 393
Waddington, G., 212, 213, 393, 394
Wadsö, I., 419, 420
Walker, R. C., 555, 582
Wallace, D. C., 13, 574, 585
Wallace, W. E., 211, 362, 366
Walsh, J. M., 576, 580, 584, 585
Walter, A., 366
Warchal, R. L., 561, 581
Ward, G., 131
Warfield, R. W., 364, 367
Warncke, H., 362, 366
Warner, L., 330
Warschauer, D. M., 571, 581, 584
Webb, F. J., 214, 274, 277, 279, 291
Weber, R. L., 567, 585
Weber, S., 57
Week, I. F., 393
Weills, N. D., 131
Weinstock, B., 291
Weissberger, A., 13, 214
Wells, F. H., 555, 583
Wendell, C. B., 538, 539, 548, 549
Wenner, F., 214, 229
Wensel, H. T., 304, 331, 579, 583, 584, 585
Wentorf, R. H., Jr., 571, 576, 581, 582, 585
West, E. D., 57, 85, 117, 130, 131, 137, 212, 214, 331, 337, 359, 365, 367, 482, 535
Westendorp, W. F., 558, 582
Westrum, E. F., Jr., 133, 211, 213, 214, 259, 260, 261, 290, 292, 353, 359, 364, 367, 462, 472, 535
Wethington, J. A., 137, 211, 362, 365, 480, 484, 533
Wetzel, W., 362, 366
Wexler, A., 211, 290, 291
Wheatley, J. C., 272, 288, 290, 292
White, G. K., 57, 214, 265, 272, 292

White, J. R., 426, 534
White, R. M., 575, 585
White, W. P., 2, 13, 50, 57, 99, 104, 131, 229, 260, 261, 296, 331, 360, 367
Whittaker, E. T., 214
Wiebe, R., 212, 232, 242, 260
Wiebes, J., 286, 292
Wien-Harms, 260
Wiese, W. L., 567, 584
Wiesendanger, H. U. D., 574, 584
Wilhoit, R. C., 535
Wilks, J., 214, 274, 277, 279, 291
Williams, A. H., 558, 583
Williams, J. W., 365, 367
Williamson, K. D., 394
Willihnganz, E., 211
Wilson, W. B., 571, 580, 585
Winckler, J. R., 360, 367
Winkler, E. M., 571, 585
Wisely, H. R., 331
Wittig, F. E., 367, 493, 535
Wittig, V. F. E., 13
Woisard, E. L., 575, 585
Wolf, W. P., 285, 291
Wolfe, H. C., 56, 57, 568, 581, 585
Wong, P. K., 533
Wood, J. K., 558, 581
Wood, W. D., 557, 575, 582
Woods, S. B., 214, 533
Wooley, H. W., 57, 576, 581
Wormser, E. M., 567, 581
Worthing, A. C., 367
Worthing, A. G., 572, 585
Worthington, A. E., 137, 211, 362, 365, 367, 480, 484, 533, 535
Wulff, C. A., 535
Wunsch, D. C., 560, 585
Wydeven, T. J., 363, 367

Yarger, F. L., 580, 585
Yost, D. M., 137, 214, 393
Young, A. P., 571, 580, 585
Young, D. M., 549
Young, G. S., 533

Zabetakis, M. G., 137, 214, 535
Zarem, A. M., 569, 585
Zettlemoyer, A. C., 533
Zhuchikhin, V. I., 580, 581
Ziegler, W. T., 214, 367
Zimmerman, J. E., 288, 291
Zinov'eva, K. N., 291

Subject Index

Absolute determination of ampere, 70–72
 ohm, 72–75
Absolute electrical standards, 66–68
Acceleration of gravity, 64
Accuracy, 6
 of drop calorimetry, 329
 of heat-capacity measurements, 418, 445
 of heat-of-vaporization measurements, 390, 417
 of high-temperature adiabatic calorimetry, 359
 of low-temperature calorimetry, 208, 259
 of second-virial-coefficient values, 391
 of vapor-heat-capacity measurements, 391
Adiabatic calorimeter, 2, 11
Adiabatic calorimetry, 209, 210, 133–214, 333
 comparison with other methods, 209, 210
 from 300 to 800°K, 333
 low-temperature, 133–214
Adiabatic demagnetization apparatus, 271
Adiabatic shield, 146–148, 342, 347, 410, 477–515
 choice of, 487
 design of, 146, 342, 355, 482, 489
 dry shields, 483
 environmental control, 347
 heaters, 147
 saturated steam, 410
 "wet" shield, 477, 485
Adiabatic shield control, 168, 169
 automatic, 168, 345, 355
 control thermocouples, 500
 definitions, 477
 regulator systems, 479, 498
 theory, 477
 "wet" shield, 494
 with saturated steam, 410
Alloys, clustering in, 548
Ampere, absolute determination, 70–72
 definition, 65
Amplifiers, electronic, 97
Aneroid calorimeter, 4

Born and von Kármán theory, 204
Bridges, for semiconductors, 46
 for temperature measurement, 41
Bunsen ice calorimeter, 3

Cadmium standard cell, 78
Calibration factor, in block calorimeter, 322
 in isothermal drop calorimeters, 313
Callendar–Van Dusen equation, 201, 202
Calorie, 7
 International Table, 8
 thermochemical, 8
Calorimeter (cf. also Calorimeters, types of)
 calibration, 359
 classification, 2
 definition, 1
 filling, 173
 for condensed gases, 231
 for hydrocarbons, 413–418
 for measuring properties of water, 403–413
 for solids, 230
 methods of cooling, 273
 thermal isolation of, 272
 with external thermometer-heater, 242
 with internal thermometer-heater, 226, 245
 with spring-loaded contacts, 281
 with thermometer-heater on outer surface, 224
Calorimeter heater, 94–97
 circuit, 165
 lead problem, 127
Calorimeter resistance thermometer, 319
Calorimeters, types of
 adiabatic, 2, 11, 136, 440
  cross-sectional diagram, 354
 adsorption, 538, 543

SUBJECT INDEX

Calorimeters, types of—*contd*
  aneroid, 4
  block, 4
  Bunsen ice, 3, 310
  Calvet, 3, 432
  'clamp' type, 280
  cryogenic, 136
  Dewar, 3
  differential, 3
  drop, 11, 294
    design of, 523, 526
    diphenyl ether receiver, 523
    electrical calibration, 319
    for liquids and solutions, 521 (cf. Calorimeter, liquids and solutions)
    furnace, 296, 529 (cf. Furnace, for drop calorimeters)
    isothermal calorimeter, 310, 318
    naphthalene receiver, 526
    precision and accuracy, 329
    sample container, 306–308, 531
    suspension and dropping mechanisms, 4, 308
    temperature measurement in, 304
    vessels, 529
  flow, 2, 4, 13 (cf. Calorimeter, liquid-flow; Calorimeter, vapor-flow)
  gold, 225
  high-temperature adiabatic,
    for operation in the range 250–600°K, 353
      300–800°K, 338
    methods of calorimetry above 300°K, 333
    survey of types employed, 361
  isoperibol, 3
  isothermal, 2, 10, 12, 217
    all-glass diphenyl ether, 524
    all-metal diphenyl ether, 528
    naphthalene, 526
  liquid flow, 434
  liquids and solutions
    adiabatic heating methods, 435
    comparison methods, 437
    direct methods, 440
    dynamic methods, 431
    flow calorimeter, 434
    indirect methods, 425
  micro, 3
  Nernst, 3
  non-adiabatic, 372

radiation, 4
saturated fluid
  design of, 399
  for hydrocarbons, 413
  for water, 403
  theory of method, 396
silver, 356
single-vessel, 455
twin, 3, 12, 437
vapor-flow
  adiabatic, 371
  design of calorimeter vessel, 375
  design of vaporizer vessel, 374, 375
  non-adiabatic, 371, 372
  thermostat, 378
Calorimeter vessel, 2, 150–156
  calibration curves, 179
  design of, 150, 224, 339, 356, 449, 474
  filling nipples, 459
  filling of, 173, 454
  for adsorption calorimetry, 547
  for liquids, 156, 224, 449, 474
  for solids, 152, 225, 226
  for sulfur, 339
  heat capacity of, 151, 179, 187, 255, 440
  materials of construction, 451
  schematic, 476
  sealing, 457
  strengths of, 87
  temperature gradients in wall, 122
  variable volume, 468
  with external thermometer-heater, 224
  with integral radiation shields, 339
  with internal thermometer-heater, 226
  with shells, 125
Calorimetric apparatus, typical, 217
Calorimetric design
  applications to, 122
  chemical considerations, 86, 87
  electrical circuits, 91, 96
    considerations, 91–98
  heat flow considerations, 99–122
  intermediate temperature (250–600°K) thermostat, 353
  materials of construction, 86
  mechanical considerations, 87–91
  principles, 85–131
  soldered joints, 89
  supports, 89

600

SUBJECT INDEX

Calorimetric measurements of microscopic particles, 546
Calorimetric methods
　choice of, 9, 287
　comparison of, 10, 209, 259, 287, 295, 335, 361, 423, 443
　for above room temperature, 333
　for saturated fluids, 396
Calorimetric studies of physical phenomena at low temperatures, 537–548
Calorimetry
　adiabatic, 133–214
　　automatic, 143
　　from 300 to 800°K, 333
　　low-temperature, 133–214
　below 20°K, 263–290
　drop, furnace for, 296
　　high-temperature, 293–330
　　silver-core furnace, 301
　isothermal versus adiabatic, 287
　low-temperature, 215–260
　vapor-flow, 369–393
Calorimetry of saturated fluids, 395–419
Campbell mutual inductor, 72
Carbon thermometers, 35
Cells, standard, 77, 78
Celsius (Centigrade) scale, 17
Clustering in alloys, 548
Computer methods (cf. Data reduction)
　for an enthalpy determination, 199
　for temperature calculations, 201
Constants, 7
Control heaters, design of, 97
Convective heat transfer, 103
Corrections, 21
　curvature, 187, 257
　heat leak, 185, 232, 338, 386, 442
　heater lead, 127, 194
　heating by thermometer current, 254
　helium gas, 194, 255
　impurities, 322
　in gas thermometry, 21
　in heat-of-vaporization measurements, 384, 413
　in vapor-heat-capacity measurements, 387
　lead wire, 253, 409
　mass, 194, 255
　non-constant heating, 249
　premelting, 189
　pump energy, 412
　thermal gradients, 242
　transient heating, 252
　vaporization, 191, 323, 465
C_P, calculation of, 256
Cryogenic thermometry, 157–162
Cryostats
　adiabatic demagnetization, 269
　aneroid-type adiabatic, 137
　combination immersion-aneroid-type, 140
　filling-tube-type, 142, 221
　for calorimetry below 20°K, 264–272
　for calorimetry of condensed gases, 221
　for heat capacity measurement, 217–221
　for measurements below 1°K, 267–272
　for use between 4° and 20°K, 265
　for use in the liquid helium range, 266
　He^3, 267, 268
　He^3 dilution refrigerators, 269
　immersion-type, 140
　incorporating a simple mechanical heat switch, 275
　isothermal shield type, 217
　magnetic refrigerator, 268
　preparation of, 173
　typical, 217
Current measurements, 559
Curvature corrections, 257

Data reduction, 184
　application of high speed digital computers, 198
　in drop calorimetry, 322
　in isothermal-shield calorimetry, 232
　in vapor-flow calorimetry, 383
　reduction of observed data to molal basis, 186, 256
　treatment of experimental data, 184, 238
Debye theory, 204
Demagnetization, adiabatic, for magnetic cooling, 269
Density measurements, x-ray and interferometric techniques, 571
Derived thermodynamic quantities and experimental precision, 435

SUBJECT INDEX

Deviation, average, 7
  standard, 7
Dewar calorimeter, 3
Differential calorimeter, 3
Digital computer smoothing of the experimental data, 203

Electrical circuits
  for adiabatic shield control, 346, 492
  for galvanometer zero adjustment, 517
  for germanium or carbon resistance thermometers, 283
  for magnetic thermometry, 285
  for power measurements, 163, 229
  for pulsed generators, 557
  for temperature measurement, 229
  for temperature-program regulation, 446
  for time measurement, 229
Electrical, energy measurements, 163–167
  and calculations, 251–255
  measurements in thermometry, 40–50
  power, measurement of, 348
  properties, high-speed measurement of, 577
  pulses, heating by, 554–556
  standards, absolute, 66–68
  thermometry, 32
  units, 65, 66
Electronic amplifiers, 97, 98
Emissivity, definition, 101
Emittance, definition, 101
Enantiotropic substances, 181–183
Energy
  free, 6
  Gibbs, 6
  Helmholtz, 6
  input and measurement, 162–168
  internal, 4
  measurement and standardization, 59–84
  measurement, basis for, 60–63
  stored, by radiation damage, 547
Enthalpy, 5
  determination, energy calculation for, 199
  high-temperature, 294
  of adsorbed phases, 538
  of adsorption, 541
  of adsorption, integral, 541

of adsorption, isoteric, 541
of fusion, 6, 328
of phase changes, 180
of sublimation, 6, 232
of transition, 328
of vaporization, 6, 184, 232, 380, 383, 395–419
  transition method for measuring, 418
Equation of state, 389
Equilibration rates, 178, 247
Error, probable, 7
  standard, 7
Exploding conductors, 577

Films adsorbed on solids, 538
Flow calorimeter, 2, 4, 13
Fluids, saturated, 395
Free energy, 6
Furnace, for drop calorimeters, 296, 529–531
  automatic temperature control, 303
  design and operation, 296
  for operation above 1500°C, 302
  heaters, inductive, 302
    resistive, 298
  materials of construction, 298
  sample in the, 306
  silver core, 302

Galvanometers, 97, 98
  automatic zero adjustment, 517
Gas thermometer, 17
Germanium thermometer, 35, 282
Gibbs energy, 6
Gold calorimeter, 225

Hartshorn bridge, 285, 286
Haseda and Miedema's apparatus, 289
He^3 cryostat with a simple mechanical heat switch, 276
He^3 scale of temperature, 30
He^4 scale of temperature, 29
Heat (cf. enthalpy)
  conduction, one-dimensional, 121
  generation of, 554–558
  measurement of, 558
    conclusions regarding, 561
  of vaporization, 6, 390

SUBJECT INDEX

Heat capacity, 5
  apparent, 386
  apparent molal, of solute, 435
  below about 10°K, 203–206
  data, calculation of, 232–259
  determination, energy calculation for, 199
  extrapolation of, 203–206, 259
  ideal gas, 370, 389
  integral, of mixing, 436
  isobaric, 435
  measurements, accuracy of, 391
  of adsorbed phases, 538
  of calorimeter, 179
  of empty calorimeter, solder and helium, 255
  of liquids and solutions near room temperature, 421–532
  of microscopic particles, 546
  of saturated fluids, 396
  partial molal, of solute, 435
  real gas, 370, 385
  representation by equations, 326
  tare, 187
Heat content, see Enthalpy
Heat exchange in adiabatic calorimetry, 335
Heat flow, radial, 112
  steady-state, 111
  unsteady-state, 117
Heat leak
  below 20°K, 272
  corrections for, 232–251
  due to temperature gradients, 124
  evaluated, 215–260
  evaluation of, 99, 232, 400, 412, 518, 520
  in adiabatic calorimetry, 144, 502
  in vapor-flow calorimetry, 386
  reduction of, 124, 399
  values for multi-junction thermocouples, 514
Heat leak coefficient, 442
Heat of, see Enthalpy of
Heat switches, mechanical, 275, 276
  comparison of, 279
Heat transfer
  between solid particles, 545
  between vessel and shield, 500
  by mechanical vibration, 108, 264, 273
  by thermal acoustical oscillation, 108
  conductive, 108, 233
  convective, 103, 233
  effect of shields on, 105
  from radio-frequency sources, 264, 273
  in adiabatic calorimetry, 335, 343, 441, 500
  in furnaces, 300
  in liquids, 461
  principle of superposition, 119, 337
  radiative, 100, 233, 453
  time lags, 105, 121
Heaters
  auxiliary, 439
  for temperature control, 97
  independently controlled, 439
  variable-resistance, 439
Heating, by electrical pulses, 554–556
  by pulsed radiation, 556
  correction, non-constant, 249
  elements, 462–465
    for glass vessels, 465
    for metal vessels, 463
  rates, comparison of, 438
Helmholtz energy, 6
High-speed measurements, applications of, 572
High-temperature, drop calorimetry, 293
  sources, 578
Hysteresis effects, 183

Ice calorimeter, 301
  heat-leak, 316
Inductors, standard, 68–70
International temperature scales, 23–28
Isoperibol calorimeter, 3
Isothermal calorimeter, 2, 10, 12

Joule twin-calorimeter system, 3

Kelvin temperature scale, 16
Kilogram, definition of, 62

Lead wires, corrections for, 253

Magnetic and radio frequency fields, effects of, 37
Magnetic thermometry, 55, 284

SUBJECT INDEX

Measurement, 78, 79
  electrical, 78, 559
  of heat, 79, 558
  of power, 79, 163, 251, 348, 379, 411, 515, 519
  of shock pressures, 569
  of temperature (see Temperature measurement)
  of time, 167, 515, 519
  of voltage, 559–561
Measurements, high speed
  of specific heat, 572
  of thermal diffusivity, 574
Mechanical equivalent of heat, 8
  units, 63, 64
Melting point and purity determinations, 180, 350
Meter, definition of, 61, 62
Microcalorimeter, 3
Monotropic substances, 183, 184
Mueller bridge, 42, 161, 201, 520

Nernst calorimeter, 3
Newton–Raphson method, 202
Newton's law of cooling, 104, 232

Ohm, absolute determination of, 72–75
  definition of, 65
Optical pyrometry, 54

Particle size effects, 545
Particles, heat conduction between, 545
Photoelectric technique for measuring thermodynamic properties, 576
Photographic techniques, high speed, 568
Piezometer, 429
Piezo-thermometric method, 425–428
Planck quantum, 60
Platinum resistance thermometers, 32
Poiseuille's equation, 107
Polymeric substances, studies of, 181
Power, heater, 251
Precision, 6, 259
Procedures
  crystallization, 175, 176
  for continuously heated calorimeters, 515
  for intermittently heated calorimeters, 519
  for melting point and purity determinations, 180
  in calorimetry for liquids and solutions, 515
  in high-temperature adiabatic calorimetry, 349
  in isothermal drop calorimetry, 313, 531
  in low-temperature calorimetry, 173, 230
  in vapor-flow calorimetry, 379
Procedures, observational, 176, 180
  enthalpy measurement, 314, 320
  enthalpy of phase changes, 180, 350
  heat capacity measurements, 176, 349
  heat of vaporization, 381
  melting point, 180, 350
  solid–solid phase changes, 181, 352
  vapor heat capacity, 382
Provisional temperature scale, 10°K to 90°K, 28, 29
Pulse generators, electrical circuitry associated with, 557
  repeating, 555
  single, 555
Pulsed radiation, heating by, 556

Radiation calorimeter, 4
Radiation damage, at low temperatures, 547
  energy of, 547
Radiation shields, floating, 101, 105
  multiple, 101
Radiative heat transfer, 101
Reference block for calorimetric thermometers, 402
Regulators, adiabatic, 477
  choice of, 487
  linear-plus-rate response, 482
  reset response, 482
  two-valued response, 479
Regulatory circuits for the recording twin-vessel calorimeter, 446–448
Resistance thermometers, 472–474
  thermometry, 32
Resistors, standard, 76, 77

Sample, suspending and dropping of, 308

SUBJECT INDEX

Sample containers, 306–308
Saturated fluids, calorimetry of, 395–419
Second, definition of, 62
Second virial coefficient, 389
  values, accuracy of, 391
Semiconductors, 34–36
Shield, isothermal, 215–260
Shield control, adiabatic, 169
  idealized circuit, 170
Shields, cf. Adiabatic shield, Thermal shield, Radiation shields
  adiabatic, 477
  choice of, 487
  dry, 483, 489
  wet, 485
Smith bridge, 44
Smoothing experimental data, 195–203
  enthalpy increments, 198, 325
  heat capacity data, 195–197
Specific heat (cf. Heat capacity)
  definition, 5
  measurement by pulse techniques, 572–574
Standard, definition of, 61
Standard cells, 77, 78
Standard inductors, 68–70
Standard resistors, 76, 77
Standards
  aluminum oxide, 9
  benzoic acid, 9
  calorimetry conference, 8, 208, 264, 359
  copper, 264
  electrical, 66
  for dynamics, 64
  n-heptane, 9
  reference standards, 8, 392
  time, 167
  working, 75–78
Superconducting leads, 264
Superposition, principle of, 119

Tare heat capacity, 187
Temperature determination by potentiometric method, 201
Temperature gradients
  in calorimeter, 122, 247, 400
  in furnaces, 299
  in high-temperature adiabatic calorimeter, 336
Temperature measurement, 30–56, 79, 160, 283, 379, 515, 519, 561

bridge methods, 41, 284
by optical pyrometry, 50
by photoelectric pyrometry, 564
by photographic pyrometry, 565
by thermocouples, 304
choice of method, 470
high speed, 562
potentiometric methods, 47, 284
Temperature scales, 16–30, 157, 255, 256
  Celsius (centigrade), 17
  fixed points, 24
  international, 23, 25
  interpolation formulas, 24
  Kelvin (speed of sound), 22
  Kelvin (thermodynamic), 16
  practical, 23–30
  provisional temperature scale 10°K to 90°K, 28
  1958 He^4 scale, 29
  1962 He^3 scale, 30
Tempering
  continuous, 113
  electrical leads, 113, 140
  of fluid, 116
  step, 115
  wires carrying current, 115
Tempering ring, 148
Terminology, 1–4
Thermal, contact with solid samples, 279
  diffusivity, 574
  gradients in the calorimetric system, 242
  isolation of calorimeter, 272
  shield, 105, 216
  switching, 149, 150
    mechanical heat switch, 270, 274
    superconducting heat switch, 270, 278
  tiedowns, 115, 345, 401
Thermistors, 35
Thermocouples, 37, 91–94
  calibration, 92, 304
  Chromel–Constantan, 508
  efficiency, 505
  for adiabatic shields, 147, 504
  junctions, fixed, 509
    free-contact, 511
  materials, 92, 507
  mounting junctions, 509
  nickel–iron, 508
  optimum number of junctions, 512

SUBJECT INDEX

Thermocouples—*contd.*
  properties of common metals, 507
  response, 504
  sources of error, 305
  tempering, 93, 305
Thermodynamic measurements, application of electrical measurements to, 78–84
  high-speed, 551–580
Thermodynamic properties, 258 (cf. specific property)
  at very high pressures, 575
  calculation of, 206, 207
  definitions and symbols, 4
  from relative enthalpy, 328
  high-speed measurement of, 577
  of adsorbed phases, 538, 540
Thermoelectric properties, high-speed measurement of, 577
Thermoelectric thermometry, 37–40
Thermometer current, correction for heating by, 254
Thermometer resistance, measurement of, 160
Thermometers, 17, 22, 30, 32, 35, 157
  carbon, 35
  encapsulated, 158, 227
  gas, 17
  germanium-resistance, 35, 158, 282
  gold-resistance, 224
  helium-gas, 158
  liquid-in-glass, 30
  magnetic, 284
  platinum resistance, 32, 158, 376
  speed of sound, 22
  thermistors, 35
Thermometry, 25
  cryogenic, 157–162
  electrical, 32–50
  electrical measurements in, 40–50
  gas, methods of measurement, 19
  liquid-in-glass, 30
  magnetic, 55
  optical pyrometry, 50
  recommended procedures, 25
  resistance, 32–37
  thermoelectric, 37–40
  vapor pressure, 54

Thermostats, intermediate-temperature (250–600°K), 353
Thomas type standard resistor, 76
Tian–Calvet microcalorimeter, 432, 522
Transitions, detection of, 182, 352
Transition temperature, determination of, 352
Transpiration method for measuring enthalpy of vaporization, 418
Twin-calorimeter, 3, 12
Twin-vessel calorimeter, heated continuously, 488
  dry shields for, 491
  heated intermittently, 487

Units
  absolute and international, 67
  definition of, 61
  electrical, 65, 66, 70, 72
  of energy, 59, 60
  of heat, 7
  of length, 61
  of mass, 62
  of mechanical quantities, 63
  of time, 62
  S.I., 64

Vapour-pressure-thermometry, 54
Vapor pressure, measurement of, 184, 406
Vaporizer heater, 375
Vaporizing cycling, 372
Velocity measurements, x-ray and interferometric techniques, 571
Voltage measurements, 559–561
  high-frequency, 560
  high-surge, 560
  pulsed, 560
  transient, 559, 560

Wenner method, 73, 74
Weston cell, 77
Wet shield controlled by liquid flow, 494
Working standards, 75–78